# Atlas of Benthic Foraminifera

# Atlas of Benthic Foraminifera

Ann Holbourn, Andrew S. Henderson and Norman MacLeod

WILEY-BLACKWELL
A John Wiley & Sons, Ltd., Publication

NATURAL HISTORY MUSEUM

This edition first published 2013 © 2013 by Natural History Museum

*Registered office:* John Wiley & Sons, Ltd, The Atrium, Southern Gate, Chichester, West Sussex, PO19 8SQ, UK

*Editorial offices:* 9600 Garsington Road, Oxford, OX4 2DQ, UK
The Atrium, Southern Gate, Chichester, West Sussex, PO19 8SQ, UK
111 River Street, Hoboken, NJ 07030-5774, USA

For details of our global editorial offices, for customer services and for information about how to apply for permission to reuse the copyright material in this book please see our website at www.wiley.com/wiley-blackwell.



*Library of Congress Cataloging-in-Publication Data*
Holbourn, Ann E. L.
Atlas of benthic foraminifera / Ann Holbourn, Andrew S. Henderson and Norman MacLeod.
pages cm
Includes bibliographical references and index.
ISBN 978-1-118-38980-5 (cloth : alk. paper) – ISBN 978-1-118-45249-3 – ISBN 978-1-118-45250-9 (epdf) – ISBN 978-1-118-45251-6 (emobi) – ISBN 978-1-118-45252-3 (epub) 1. Foraminifera. 2. Benthos. 3. Marine organisms. I. Henderson, Andrew S. II. Natural History Museum (London, England) III. Title.
QL368.F6H66 2013
578.77'7–dc23

2013007034

A catalogue record for this book is available from the British Library.

Wiley also publishes its books in a variety of electronic formats. Some content that appears in print may not be available in electronic books.

Cover image: *Planulina wuellerstorfi*. "Meteor" Cruise, off West Africa; M25. Deposited in the Lutze Collection (West Africa), no. 200. Department of Geosciences, Kiel University, Germany. Spiral view, BM(NH) PM DI 0523, scale bar = 186 μm.
Cover design by Design Deluxe

Set in 9/11.5 pt Minion by Toppan Best-set Premedia Limited

1 2013

# Contents

# Acknowledgements

This work would not have been possible without the help of many people who generously provided taxonomic advice, access to collections, insightful reviews and technical support.

We would like to thank most sincerely Haydon Bailey, Steve Culver, Adam Gasinski, David Haig, Bruce Hayward, Brian Huber, Bob Jones, Clive Jones, Michael Kaminski, Miriam Katz, Wolfgang Kuhnt, Ewa Malata, Giles Miller, Michel Moullade, Tim Patterson, Stefan Revets, Joachim Schönfeld, Jane Swallow, Ellen Thomas and John Whittaker.

We are also especially grateful to Miriam Katz, Wolfgang Kuhnt and Ellen Thomas for their unfailing support and enthusiastic encouragement throughout this project.

This work was partially supported by a generous grant from the UK Natural Environmental Research Council (NERC) to Steve Culver and Norman MacLeod.

Grateful thanks go to the following institutions for the cooperation in loaning us important specimens and type material, and for affording us access to their collections:

- Institute of Geosciences, Kiel University, Germany
- Institute of Geological Sciences, Jagiellonian University, Kraków, Poland
- The Natural History Museum, London, UK
- Sedgwick Museum, Cambridge, UK
- United States National Museum, Washington, DC.

# Introduction

Benthic foraminifera inhabit all marine environments, living either above, at, or below the sediment water interface in water depths ranging from the intertidal zone to the deep ocean (Corliss, 1980; Jorissen, 1999; Pawlowski and Holzmann, 2008). In the deep sea, they represent the most important contributors to the meiofaunal biomass and their distribution is intricately linked to the flux of particulate organic matter from the upper ocean to the sea floor (Gooday et al., 1992; Altenbach and Struck, 2001; Gooday, 2003). Benthic foraminifera comprise the oldest group of deep-sea organisms, tracking back to 1150–690 million years according to molecular data and possibly 716–635 million years according to the fossil record (Culver, 1991; Pawlowski et al., 2003; Bosak et al., 2011), although they exhibit relatively slow evolution rates in comparison to planktonic foraminifera (Pawlowski et al., 1997; Kucera and Schönfeld, 2007). Thus, on account of their geographic ubiquity, their abundance in Mesozoic and Cenozoic deep-sea sediments and in Phanerozoic sediments overall, and their utility as indicators of past environmental conditions, benthic foraminifera continue to play a pivotal role in paleoceanographic and paleoclimatic research.

## 1. Test Morphology and Classification

### Taxonomy

Benthic foraminifera belong to a large group of marine "amoeboid" protists, recently classified as a stand-alone phylum within the eukaryote Supergroup Rhizaria in the Protist Kingdom (Adl et al., 2005). Foraminifera have granulo-reticular anastomosing pseudopodia and a single- or multi-chambered test, although naked or soft-shelled forms have been included at the higher taxonomic level (Lee, 1990; Pawlowski et al., 2003). These forms are traditionally poorly known, but recently have been shown to be both taxonomically diverse and abundant in many oceanic environments (Gooday et al., 2008). The classical definition of foraminifera is based on possession of three main diagnostic characters: (1) a test composed of calcite secreted by the cell or consisting of mineral grains embedded in the organic test, (2) a unique reproductive cycle with alternation of a uninucleate haploid generation and a diploid generation that can be multinucleate, and (3) the presence of streaming granulo-reticular (granular, branching, network-like) pseudopodia, generally called rhizopodia (Lee, 1990; Tendal, 1990; Bowser and Travis, 2002).

The suprageneric morphological classification of benthic foraminifera remains a matter of intense debate. Three main morphological characters are given priority in classification schemes: (1) the wall structure of the test (essentially agglutinated, porcelaneous, or hyaline), (2) the number of chambers (monothalamous versus polythalamous), and (3) the test morphology including mode of coiling, chamber arrangement, and apertural system. A comprehensive description of morphological features, which are used in foraminiferal classification, is provided in Hottinger (2006). While most of the recent classification schemes give priority to the composition and structure of the test wall (Wood, 1949; Loeblich and Tappan, 1984, 1987, 1989, 1992; Sen Gupta, 1999), alternative classification schemes view the test architecture as an equally or higher ranking criterion for the suprageneric classification of foraminifera (Mikhalevich, 2004; Tyszka, 2006; Kaminski et al., 2011). Molecular-genetic analyses support the notion that some species with different wall structures are phylogenetically closely related but indicate that monothalamous and polythalamous forms are distinct and may have separated in an early stage of the evolution of foraminiferids (Pawlowski et al., 2003).

*Atlas of Benthic Foraminifera*, First Edition. Ann Holbourn, Andrew S. Henderson, and Norman MacLeod.

### Wall structure

Benthic foraminifera (except for soft-bodied allogromids) possess three main types of wall structure: hyaline perforate calcitic or aragonitic, imperforate porcelaneous, and agglutinated with either calcareous or organic cement (Wood, 1949). The different wall structures of hyaline and porcelaneous foraminifera arise from different biomineralization processes ranging from extracellular crystal nucleation on a preformed organic template (hyaline tests) to intracellularly formed crystals released to the cell periphery (high Mg calcite of porcelaneous tests) (Hemleben et al., 1986; Debenay et al., 2000; Erez, 2003; de Nooijer et al., 2009). These different modes of calcification have major implications for the uptake of trace elements and the resulting geochemical signature of foraminiferal calcite in relation to ambient seawater. In general, hyaline benthic foraminifera have lower Mg/Ca than porcelaneous foraminifera, but their Mg/Ca ranges from lower to much higher values than in calcite precipitated inorganically from seawater (Bentov and Erez, 2006). Controlled calcification within a space of biologically regulated ion concentration has the advantage that the concentration of ions in the calcifying solution (including pH) is mainly controlled by the organism, although the energy cost to concentrate the essential ions is high (Erez, 2003; Bentov and Erez, 2005; de Nooijer et al., 2009). Ongoing research gradually provides new insights into the complex species-dependent and environment-dependent process of calcification and incorporation of various trace elements in living foraminifera (e.g., de Nooijer et al., 2007; Dissard et al., 2010; Raitzsch et al., 2010).

## 2. Biology

The living cell consists of two types of cytoplasm: cell-body cytoplasm and reticulopodia (or rhizopodia), which constantly exchange smaller organelles through cytoplasmic streaming (Lee and Anderson, 1991; Bowser and Travis, 2002). The most distinctive character of the living foraminifer is the reticulate network of branched pseudopodia or reticulopodia, which extends from a single aperture or multiple apertures in the test and plays a key role in most life processes, including food acquisition, digestion, test construction, signal transmission, locomotion, and anchoring to hard substrates (Travis and Bowser, 1991; Bowser and Travis, 2002; Murray, 2006). The network of pseudopodia constantly changes its form, producing dynamically branching and joining strands that exhibit bidirectional flow and transport granules (commonly mitochondria) both toward and away from the aperture. Detailed biological descriptions of living foraminifera are provided by Lee and Anderson (1991), Goldstein and Bernhard (1997), Sen Gupta (1999), Lee and Hallock (2000), Cedhagen et al. (2002) and Gooday et al. (2008).

There is still only limited understanding about the nutrition, lifespan, reproduction cycles, and test biomineralization processes of most deep-water benthic foraminifera. The scant information that is available points to a huge heterogeneity of life histories, growth patterns, and reproduction strategies ranging from extremely fast reproduction and short lifespans for opportunist taxa exploiting phytodetritus pulses (Gooday, 1988) to lifespans of several years with slow growth and extremely low reproduction rates for deep-water agglutinated foraminifera (Ohga and Kitazato, 1997; Hess et al., 2000; Kuhnt et al., 2005).

The feeding strategies of bathyal and abyssal foraminifera appear similarly diverse (Lee, 1980; Lipps, 1983; Gooday, 1990, 1993, 2003), including suspension, detritus feeding (particularly on diatoms, Suhr et al., 2003), and bacterial scavenging (Langezaal et al., 2005). In organic-rich sediments, detritus feeding takes place both at the surface of the sea floor and infaunally within the top few decimeters of the sediment, which become increasingly depleted in oxygen with depth. In oxygen-poor environments foraminifera may live symbiotically with prokaryotes (Buck and Bernhard, 2001). However, oxygen is also brought down to levels that would otherwise remain anoxic by the burrowing macrofauna, locally expanding the depth habitat of infaunal taxa (Murray, 1991; Loubere et al., 1995). In oligotrophic areas, where the organic export flux is low, surface dwellers consist mainly of epifaunal suspension feeders, while infaunal detritivores are scarce. Dissolved organic matter may additionally provide an important source of food for some species of benthic foraminifera (DeLaca et al., 1981; Nomaki et al., 2011), whereas carnivory (e.g., Dupuy et al., 2010) and parasitism (e.g., Sigwart, 2009) constitute less common modes of feeding (Murray, 1991).

## 3. Ecology

### Microhabitat

The life position and depth distribution of deep-water benthic foraminifera at the sediment–water interface has been a major field of ecological research since the importance of epifaunal and infaunal habitats and species-specific microhabitat preferences for isotope and trace element studies and paleoenvironmental reconstructions were recognized (Corliss, 1985; Jones and Charnock, 1985; Altenbach and Sarnthein, 1989). The initial concept related benthic foraminiferal morphology and pore distribution on the surface of the test to depth habitat at the sea floor ("morphotype"-concept, Corliss and Chen, 1988; Corliss, 1991). This was later modified to account for the capability of motile benthic deep-water foraminifera to rapidly adapt their microhabitat to changes in food availability and oxygenation (Linke and Lutze, 1993; Loubere, 1996; Ohga and Kitazato, 1997; Jorissen, 1999). This dynamic adaptation is also reflected in the so-called TROX model (TROX = TRophic OXygen model, Jorissen et al. 1995, 2007), which explains that the depth of the foraminiferal microhabitat is controlled by food availability in oligotrophic ecosystems and by oxygen concentration in eutrophic ecosystems. In oligotrophic environments, the microhabitat depth is limited by the low amount of food available within the sediment, whereas in eutrophic systems, the penetration depth of most taxa depends on the level of oxygen present in the sediment (Jorissen et al., 1995).

### Biogeography

Deep-water benthic foraminifera are generally cosmopolitan and their distribution within ocean basins is primarily controlled by environmental parameters such as flux rate of particulate organic matter from the upper ocean to the sea floor and carbonate dissolution, rather than by physical barriers or the physico-chemical properties of deep-water masses (Altenbach et al., 1999; Jorissen et al., 2007; Pawlowski and Holzmann, 2008; Gooday and Jorissen, 2012). One reason for the cosmopolitan distribution pattern of deep-water foraminifera is their capability for fast dispersal and recolonization of new substrates at the sea floor, which is unusual for other benthic organisms (Hess and Kuhnt, 1996; Alve, 1999; Murray, 2006). This rapid dispersal has been shown to be due to the occurrence of propagules (Alve and Goldstein, 2003; Goldstein and Alve, 2011). Living deep-sea benthic foraminifera, in contrast to shallow-water forms, show high genetic as well as morphological similarity over large distances in the world's oceans (e.g., from Arctic to Antarctic, Pawlowski et al., 2007). Deep-water benthic foraminiferal populations of the geological past were similarly cosmopolitan. Late Cretaceous abyssal agglutinated assemblages living below the calcite compensation depth (CCD) were first documented from Deep Sea Drilling Project (DSDP) sites in the Indian and West Pacific Oceans (Krasheninnikov, 1973, 1974) and later from Ocean Drilling Program (ODP) sites in the Central West and East Atlantic Ocean (Moullade et al., 1988; Kuhnt and Moullade, 1991; Kuhnt et al., 1996a).

### Population dynamics

Observational and experimental studies provided evidence for a close benthic–pelagic coupling linking deep-sea benthic foraminiferal population dynamics and changes in surface ocean productivity on various temporal scales (Gooday, 1988, 2002; Smart et al., 1994; Heinz et al., 2001). While eutrophic surface ocean conditions favor low diversity with blooms of opportunistic species in the deep sea (e.g., Ohga and Kitazato, 1997), oligotrophic environments sustain low population density, promoting high diversity with complex trophic relationships, morphological innovations, and full exploitation of ecological niches (Gooday, 1999). On longer timescales, changes in surface productivity during major climatic and paleoceanographic events influence the population structure of deep-sea communities, but probably also lead to speciation and long-term evolutionary turnover within small and geographically isolated communities (e.g., Hallock et al., 1991; Groves and Yue, 2009). It has, for instance, been argued that the latitudinal gradient in diversity in deep-sea benthic foraminifera is linked to seasonality of productivity at high latitudes (e.g., Corliss et al., 2009) and evolved with the establishment of continental ice-sheets on the Antarctic continent (Thomas and Gooday, 1996).

## 4. Evolution

From the Late Mesozoic to Cenozoic, three large-scale steps can be recognized in the evolution of deep-water benthic foraminifera.

(1) A stepwise middle Cretaceous evolutionary turnover from the early Aptian to early Turonian appears related to the development of oceanic anoxic events. Increases in organic export flux from the upper ocean triggering deep-sea anoxia may have enabled allopatric speciation within isolated small populations and led to the occupation of vacant niches. The middle Cretaceous turnover resulted in a major radiation with the emergence of several modern lineages including buliminids, cassidulinids, and rotalids (Brasier, 1980; Tappan and Loeblich, 1988) as well as the appearance of most "modern" genera of agglutinated deep-water foraminifera (Kuhnt et al., 1996b). Surprisingly, the Cretaceous–Paleogene boundary event, which had major repercussions on the marine plankton, had little evolutionary impact on deep-water benthic foraminifera (Culver, 2003; Thomas, 2007; Alegret et al., 2012).

(2) The most rapid and severe extinction in deep-water benthic foraminifera (30%–50% of species) occurred at the end of the Paleocene (Thomas and Shackleton, 1996; Thomas, 1998). Interestingly, this event did not concur with a major extinction event in the marine plankton, nor with global deep-water anoxia, nor a substantial change in organic export flux or carbonate dissolution (Thomas, 2003, 2007). However, the extinction appears related to a period of rapid deep ocean warming at the Paleocene–Eocene boundary affecting all major oceans. This deep ocean warming was possibly associated with ocean acidification and increased ocean stratification resulting in decreased upwelling and open-ocean productivity together with extension of oxygen minimum zones (Winguth et al., 2012).

(3) Following the early Eocene recovery, three smaller-scale episodes of accelerated turnover coincided with intervals of pronounced global cooling and cryosphere expansion (Thomas et al., 2000; Zachos et al., 2001), which occurred across the Eocene–Oligocene boundary, in the middle Miocene (Berggren, 1972; van Morkhoven et al., 1986; Berggren and Miller, 1989; Miller et al., 1992; Thomas et al., 2000; Zachos et al., 2001; Thomas, 2007) and in the middle Pleistocene ("*Stilostomella* extinction", Weinholz and Lutze, 1989; Schönfeld, 1996; Hayward, 2001; Hayward et al., 2010). Many long-lived genera of benthic foraminifera, which had decreased in abundance during earlier periods of cooling, finally became extinct during episodes of more intense global cooling.

## 5. Applications

### Biostratigraphy

The utility of benthic deep-water foraminifera as biostratigraphic index fossils is limited by their slow evolution rates, strong environmental dependency, and inconsistent taxonomy (Boltovskoy, 1980; Kucera and Schönfeld, 2007; Hayward et al., 2010). However, the availability of more complete DSDP and ODP sedimentary successions containing well-preserved benthic assemblages allowed development of bathyal and abyssal benthic foraminiferal zonations for the Cenozoic (Berggren and Miller, 1989) and Cretaceous (Geroch and Nowak, 1984; Moullade, 1984; Kuhnt et al., 1992). The latter zonations are mainly applied in high latitudes and sub-CCD abyssal environments, where planktonic foraminiferal zonations cannot be used.

### Paleobathymetry and sea-level reconstructions

Benthic foraminifera were recognized as paleobathymetric indicators after distribution studies along slope transects off California and in the Gulf of Mexico indicated meaningful changes in assemblage composition with water depth (Bandy, 1953, 1961; Bandy and Arnal, 1957; Pflum and Frerichs, 1976). Van Morkhoven et al. (1986) compiled globally applicable charts of paleobathymetric distributions for 125 Cenozoic cosmopolitan deep water benthic foraminifera, which were subsequently widely used in subsidence and sea-level history reconstructions (e.g., Kaiho, 1992; Katz and Miller, 1993, 1996). However, modern ecological investigations revealed that oxygenation and food availability (which generally diminishes with increasing water depth) rather than water depth per se were primary controls on benthic foraminiferal distribution (Jorissen et al., 1995, 2007; Van der Zwaan et al., 1999; Murray, 2001). It has also been found that some deep-dwelling species are only able to reproduce under high hydraulic pressure (Tiedemann et al., 2012).

## Paleoceanography and paleoclimate research

The reconstruction of past climates and ocean conditions relies heavily on geochemical analyses of the carbonate tests produced by benthic foraminifera. In particular, stable isotope and trace element analyses provide unique information on past variations in the carbon cycle, ocean pH, ventilation and circulation, productivity, temperature, salinity, ice volume, and seawater chemistry. Some challenging issues with the use of deep-sea benthic foraminifera in geochemical analysis are their relative scarcity, as compared to planktonic foraminifera, and the need to use monospecific specimens for analysis in order to avoid "vital effects" problems. Some recent studies, however, have perfected techniques, which only use very small quantities of calcite, (e.g., laser ablation). The development of geochemical proxies has progressed enormously in recent decades and an overview of potential applications for paleoclimatic and paleoceanographic reconstructions through time is presented below. A comprehensive review of traditional and emerging geochemical proxies in foraminifera by Katz et al. (2010) provides many details beyond the brief overview presented here.

### *Stable oxygen and carbon isotopes*

Benthic foraminiferal $\delta^{18}O$ generally reflects a mixed signal, which varies as a function of global ice volume, the source of deep-water formation, and bottom water temperature (Shackleton and Opdyke, 1973). Thus benthic $\delta^{18}O$ can be used to reconstruct past variations in global ice volume and water mass temperature and to identify the source regions of water masses. Benthic foraminiferal $\delta^{18}O$ has also proven to be a powerful stratigraphic tool, allowing the development of high-resolution timescales for the Pleistocene and Pliocene (Imbrie et al., 1984; Lisiecki and Raymo, 2005) and for older intervals of the Cenozoic (Holbourn et al., 2005, 2007; Pälike et al., 2006) through the correlation of glacial cycles to computed changes in the geometry of Earth's orbit (Laskar et al., 2004, 2011).

The $\delta^{13}C$ of epibenthic foraminiferal species is primarily a function of the dissolved inorganic carbon $\delta^{13}C$ in bottom water (Duplessy et al., 1984; Curry et al., 1988) and as such co-varies with the ventilation and nutrient content of deep-water masses. As water masses move away from their source areas, they progressively become enriched in nutrients and $CO_2$ with low $\delta^{13}C$ values as a result of organic matter oxidation through the water column. Epibenthic species of the genera *Cibicidoides*, *Nuttalides*, and *Planulina* and the infaunal sediment-dwelling genus *Uvigerina* ($\delta^{18}O$ only) are most commonly used for stable isotope analysis. The $\delta^{13}C$ of infaunal benthic foraminifera is mainly related to the $\delta^{13}C$ of pore water, thus carrying a signal influenced by decomposing organic matter at the ocean floor (Mackensen et al., 1993, 2000; Mackensen and Licari, 2004). On a global scale, carbon isotope values of benthic foraminifera may indicate changes in the carbon cycle such as release of isotopically light carbon into the ocean–atmosphere system, as documented for the Paleocene–Eocene maximum (e.g., Thomas and Shackleton, 1996; Zachos et al., 2001). In such instances, benthic foraminiferal $\delta^{13}C$ can additionally be used as a stratigraphic tool for detailed correlation of sedimentary successions (Saltzman and Thomas, 2012).

### *Minor and trace elements*

The variable uptake of minor and trace elements in the crystal lattice of calcareous foraminiferal tests reflects ambient environmental conditions at the sea-floor depth, either in bottom waters or pore waters, where the tests were formed. Thus, analysis of various minor and trace elements in foraminiferal tests can provide powerful tools to reconstruct past hydrographic and circulation changes.

Analysis of Mg/Ca in benthic foraminiferal calcite tests makes it possible in principle to independently reconstruct bottom water temperatures and in conjunction with $\delta^{18}O$ to separate the temperature and global $\delta^{18}O$ seawater (related to ice volume) signals in the $\delta^{18}O$ measured in tests. Studies have focused on a few taxa (*Cibicidoides pachyderma*, *Planulina wuellerstorfi*, *Oridorsalis umbonatus*, *Hoeglundina elegans*, and *Melonis* spp.), which showed relatively high temperature sensitivity (Lear et al., 2002, 2010; Martin et al., 2002; Rosenthal et al., 2006). However, uncertainty concerning the accuracy of calibration equations (Bryan and Marchitto, 2008) and the carbonate saturation effects altering the Mg/Ca relationship (Erez, 2003; Elderfield et al., 2006; Rosenthal et al., 2006; Yu and Elderfield, 2008) suggests that benthic foraminiferal paleothermometry especially at depth, where saturation may be of major importance, is not as straightforward as originally envisaged.

The distribution of the trace metals cadmium (Cd), barium (Ba), and zinc (Zn) follows the cycling of organic matter and varies as a function of water depth and deep-water circulation. Thus, the elemental ratios of Cd/Ca, Ba/Ca, and Zn/Ca in foraminiferal calcite can be used as tracers of water masses and circulation change (Boyle, 1988, 1992; Lea and Boyle, 1989, 1990; Marchitto et al., 2000). Benthic foraminiferal Cd/Ca varies proportionally to seawater Cd (which parallels the nutrient element phosphorus) and has proven useful for reconstructing glacial–interglacial variability in deep ocean circulation (Boyle and Keigwin, 1985; Rickaby et al., 2000; Zahn and Stüber, 2002; Marchitto and Broecker, 2006). Zn/Ca is a sensitive tracer of past ocean carbonate chemistry, as the partition coefficient in benthic foraminifera closely depends on the carbonate saturation state of the deep ocean (Marchitto et al., 2000, 2005). Benthic foraminiferal Ba/Ca has also been used to reconstruct glacial–interglacial changes in water mass distribution. However, its most promising application appears to be in planktonic foraminifera for the reconstruction of past variations in fluvial inputs to the ocean (Weldeab et al., 2007).

*Boron isotope ($\delta^{11}B$) and B/Ca*

The use of boron isotopes in benthic foraminiferal tests to reconstruct bottom water pH is still in an early stage, as initial measurements of large numbers of mixed benthic species produced unrealistic and highly scattered data. However, recent attempts to measure $\delta^{11}B$ in the epibenthic species *Planulina wuellerstorfi*, which has a high boron concentration, appeared more promising (Yu and Elderfield, 2007; Hönisch et al., 2008). Recent measurements of $\delta^{11}B$ in individual *Cibicidoides* species to reconstruct past changes in the ocean carbonate system were also encouraging (Rae et al., 2011). Analysis of B/Ca in calcitic benthic foraminiferal tests has been used as an alternative approach to reconstruct paleo-pH (Yu and Elderfield, 2007). Initial studies indicated that the uptake of boron in benthic foraminifera in the deep ocean was closely related to carbonate saturation in the deep ocean (Yu et al., 2007; Foster, 2008; Brown et al., 2011). However, some recent results indicated considerable scatter in the B/Ca of at least some co-occurring morphotype species, suggesting potential complications arising with the application of this proxy (Rae et al., 2011).

## 6. Impact of Deep Sea Drilling

During the first phase of ocean drilling, 72 papers on deep-water benthic foraminifera were published in the Initial Reports of the Deep Sea Drilling Project between 1969 and 1987 (Table 1). A large proportion of these papers were dedicated to Jurassic and Cretaceous records, since deep-sea drilling provided the first opportunity to study "deep-time" deep-water assemblages in a much better preserved state than in most land outcrops. Entirely new assemblages of Late Jurassic to Cretaceous abyssal agglutinated foraminifera from sub-CCD environments were described in particular, shedding new light on the ecology and evolution of this hitherto poorly known group (Krasheninnikov, 1973, 1974). Cenozoic calcareous deep-sea species were previously known from some land sections (e.g., Velasco Formation, Mexico; Oceanic Formation, Barbados (van Morkhoven et al., 1986), but the realization that many species are cosmopolitan and have long stratigraphic ranges came with the analysis of samples recovered by the DSDP. Benthic assemblages also were commonly much better preserved in the deep-sea material than in land outcrops.

The expanded and more complete sedimentary successions obtained during the follow-on phases of ocean drilling with the ODP and Integrated Ocean Drilling Program (IODP) offered, in particular, the opportunity to develop high-resolution, astronomically calibrated time series over extended intervals of the Cenozoic. These continuous records provided fresh insights into the evolution of deep-water benthic foraminifera, as speciation and extinction events could be more precisely constrained and correlated across different oceans. Furthermore, these records allowed development and testing of new hypotheses concerning global radiation and mass extinction events in the deep sea and their relations to ocean chemistry and climate change. A total of 51 papers on deep-water benthic foraminifera were published in ODP volumes between 1988 and 2004 (Table 2). In the later stages of ODP and IODP, the papers directly linked to specific drilling legs were commonly no longer published in the Scientific Results volumes, but in the general literature, thus being more difficult to trace.

## 7. Approach of this Atlas

A voluminous literature concerns the taxonomy of foraminifera, as more than 60,000 species have been described, the majority of which are benthic species (Culver, 1993). The standard of published illustrations and descriptions of type specimens, which form the basis of the taxonomy, varies enormously. This mainly reflects the lack of taxonomic standardization and the dispersion of descriptions and illustrations in a wide range of publications that have strong historical or regional associations. Thus, benthic foraminiferal taxonomy remains highly unwieldy, which severely compromises the application of benthic foraminifera for paleoceanographic and paleoclimatic research. It is especially difficult for geochemical paleoceanographers, who cannot consult benthic foraminiferal specialists, to exert quality control on specimens picked for analysis, as these are subsequently dissolved, prohibiting any further taxonomic check.

We have compiled a database of 300 deepwater benthic foraminifera, which comprises sets of fully focused, composite images and standardized taxonomic descriptions that consolidate previous taxonomic efforts. We illustrate whenever possible type specimens, mainly deposited in collections of the Smithsonian Institution, The Natural History Museum, London, and the Jagiellonian University, Kraków, for which only drawings or black and white illustrations were previously available. We additionally illustrate well-preserved material, mainly topotypic, from deep-sea cores, commercial wells, and land sections. For selected species, SEM micrographs are given for comparison and documentation of smaller specimens. The digital imaging techniques that we used involve taking a number of images per specimen. This image set constitutes a series of 'slices' taken sequentially at different focal plane depths such that all aspects of the specimens' morphology are captured within an in-focus slice. A composite image is then built up by combining the areas in each image slice that are in focus. The resulting composite image provides a true-color image of the specimen viewed with full focus throughout the field of view. These images, which are the most realistic view of what the micropaleontologist actually 'sees' through the microscope tube, are particularly useful for illustrating type specimens that were previously poorly illustrated. The technique additionally has the benefit of not altering the original specimens, and thus is particularly suitable for type specimens. We used a Kontron Electronic ProgRes 3012 camera scanner, attached to a Leica Diaplan Microscope or an Allen Compact Video Microscope, or a Zeiss Axiocam camera, attached to a Leica MZ16 stereoscopic microscope, to produce digital images. A fully in-focus composite image was generated with the help of an image manipulation application (Adobe Photoshop) or with the help of automated imaging software (Syncroscopy Automontage). By

**Table 1**
Monographic studies of deep-water benthic foraminifera published in the Initial Reports of the Deep Sea Drilling Project. Time slices are coded as: ju = Late Jurassic, ku = early Cretaceous, ko = late Cretaceous, pal = Paleogene, neo = Neogene, quart = Quarternary. Access to listed publications is available through the Integrated Ocean Drilling Program website (http://www.iodp.org).

| Author | Year | Leg | Chapter | Age |
|---|---|---|---|---|
| Luterbacher | 1972 | 11 | dsdp11_18 | ju–ku |
| Berggren | 1972 | 12 | dsdp12_14 | pal–neo |
| Douglas | 1973 | 17 | dsdp17_21 | ko–neo |
| Krasheninnikov | 1973 | 20 | dsdp20_11 | ku–ko |
| Maync | 1973 | 13 | dsdp13pt2_41_1 | ku |
| Bartenstein | 1974 | 27 | dsdp27_35 | ju–ku |
| Krasheninnikov | 1974 | 27 | dsdp27_32 | ko |
| Kuznetsova | 1974 | 27 | dsdp27_34 | ju–ku |
| Scheibnerova | 1974 | 27 | dsdp27_36 | ku |
| Vincent et al. | 1974 | 24 | dsdp24_21 | pal |
| Luterbacher | 1975 | 32 | dsdp32_25 | ku |
| Webb | 1975 | 29 | dsdp29_22 | pal |
| McNulty | 1976 | 33 | dsdp33_08 | ku–ko |
| Resig | 1976 | 34 | dsdp34_63 | pal–neo |
| Rögl | 1976 | 35 | dsdp35_33 | ko–neo |
| Boersma | 1977 | 39 | dsdp39_29 | pal–neo |
| Sliter | 1977 | 36 | dsdp36_10 | ku–ko |
| Sliter | 1977 | 39 | dsdp39_30 | ku–ko |
| Beckmann | 1978 | 40 | dsdp40_18 | ko |
| Cameron | 1978 | 40 | dsdp40_20 | neo |
| Gradstein | 1978 | 44 | dsdp44_32 | ku |
| Krasheninnikov & Pflaumann | 1978 | 41 | dsdp41_10 | ku–ko |
| Kuznetsova & Seibold | 1978 | 41 | dsdp41_08 | ju–ku |
| Lutze | 1978 | 41 | dsdp41_13 | neo |
| Proto Decima & Bolli | 1978 | 40 | dsdp40_19 | pal |
| Scheibnerova | 1978 | 40 | dsdp40_17 | ku |
| Wright | 1978 | 41 | dsdp41pt1_31 | neo |
| Bock | 1979 | 48 | dsdp48_14 | ku |
| Butt | 1979 | 47 | dsdp47pt1_04 | ku |
| Dupeuple | 1979 | 48 | dsdp48_18 | ku–ko |
| Lutze | 1979 | 47 | dsdp47pt1_11 | quart |
| Murray | 1979 | 48 | dsdp48_16 | pal–neo |
| Schnitker | 1979 | 48 | dsdp48_15 | pal–neo |
| Sigal | 1979 | 47 | dsdp27pt2_05 | ku–ko |
| Butt | 1980 | 55 | dsdp55_08 | pal–neo |
| Keller | 1980 | 56 | dsdp56_57p2_24 | pal–neo |
| Sliter | 1980 | 50 | dsdp50_09 | ju–ko |
| Thompson | 1980 | 56 | dsdp56_57pt2_22 | pal–neo |
| Matoba & Yamaguchi | 1982 | 64 | dsdp64pt2_45 | neo–quart |
| Basov & Krasheninnikov | 1983 | 71 | dsdp71pt2_28 | ku–quart |
| Blanc-Vernet | 1983 | 76 | dsdp76_18 | neo–quart |
| Dailey | 1983 | 72 | dsdp72_34 | ku–pal |
| Gradstein | 1983 | 76 | dsdp76_20 | ju |

Table 1
(*Continued*)

| Author | Year | Leg | Chapter | Age |
|---|---|---|---|---|
| Tjalsma | 1983 | 72 | dsdp72_33 | pal–neo |
| Boersma | 1984 | 75 | dsdp75_13 | neo |
| Clark & Wright | 1984 | 73 | dsdp73_13 | pal |
| Hemleben & Tröster | 1984 | 78 | dsdp78a_26 | ko |
| McNulty | 1984 | 75 | dsdp75_10 | ku–ko |
| Murray | 1984 | 81 | dsdp81_09 | pal–neo |
| Riegraf et al. | 1984 | 79 | dsdp79_26 | ju |
| Schnitker | 1984 | 81 | dsdp81_17 | neo |
| Boersma | 1985 | 82 | dsdp82_37 | pal |
| Caralp | 1985 | 80 | dsdp80pt2_26 | quart |
| McDougalll | 1985 | 84 | dsdp84_09 | neo–quart |
| Magniez & Sigal | 1985 | 80 | dsdp80pt1_18 | ku |
| Miller et al. | 1985 | 80 | dsdp80pt1_13 | pal |
| Poag & Low | 1985 | 80 | dsdp80pt1_12 | neo |
| Thomas | 1985 | 85 | dsdp85_17 | pal–quart |
| Boersma | 1986 | 90 | dsdp90pt2_20 | pal–neo |
| Kurihara & Kennett | 1986 | 90 | dsdp90pt2_21 | neo |
| Lagoe | 1986 | 87 | dsdp87_10 | neo–quart |
| Leckie & Webb | 1986 | 90 | dsdp90pt2_24 | pal–neo |
| Schröder | 1986 | 96 | dsdp96_33 | quart |
| Sliter | 1986 | 89 | dsdp89_09 | ku–ko |
| Blanc-Vernet & Moullade | 1987 | 93 | dsdp93_06 | neo–quart |
| Hart | 1987 | 95 | dsdp95_05 | ko |
| Hulsbos | 1987 | 93 | dsdp93_10 | pal |
| Katz & Miller | 1987 | 95 | dsdp95_08 | pal |
| Miller & Katz | 1987 | 95 | dsdp95_07 | ku–ko |
| Saint-Marc | 1987 | 93 | dsdp93_11 | pal |
| Scott | 1987 | 95 | dsdp95_09 | quart |
| Thomas | 1987 | 94 | dsdp94pt2_33 | pal–quart |

applying further standard digital processing techniques (e.g., sharpening, color and level balance, brightness/contrast, and the addition of a uniform background color), the final image was completed.

Our selection of deep-water benthic foraminifera focused on calcareous groups that are of stratigraphic or paleoecological significance for IODP studies and/or are increasingly being used for geochemical analyses. Many of these groups have a particularly confused taxonomic history, including the genera *Bolivina*, *Bulimina*, *Cibicidoides*, *Globobulimina*, *Planulina*, *Stilostomella*, and *Uvigerina*. We have additionally targeted agglutinated taxa to highlight morphologic features such as wall texture, grain and cement composition, test color and opacity, chamber arrangement, shape of sutures, and inner structure including early ontogenic stages, as these represent important taxonomic criteria that are not detectable in SEM illustrations. We present systematic data (descriptions, synonymies, biogeographic, paleoecologic, and stratigraphic ranges) that were assembled from primary research, from the literature and from consultation with specialists. The generic taxonomic classification follows Loeblich and Tappan (1987) with a few exceptions, which mainly reflect taxonomic updates published subsequently. We used the revised timescale of Gradstein et al. (2004) and biostratigraphic zonations therein. Our hope is that these new illustrations and revised descriptions of selected deep-water benthic foraminifera will contribute to the clarification of their taxonomy and will support efforts to develop reliable proxies for paleontological and geochemical research.

Table 2
Monographic studies of deep-water benthic foraminifera published in the Scientific Results of the Ocean Drilling Program. Time slices are coded as: ju = Late Jurassic, ku = early Cretaceous, ko = late Cretaceous, pal = Paleogene, neo = Neogene, quart = Quarternary. Access to listed publications is available through the Integrated Ocean Drilling Program website (http://www.iodp.org).

| Author | Year | Leg | Chapter | Age |
|---|---|---|---|---|
| Moullade et al. | 1988 | 103 | sr103_21 | ko |
| Osterman & Qvale | 1989 | 104 | sr104_37 | pal–neo |
| Cotalongo et al. | 1990 | 107 | sr107_30 | neo–quart |
| Hasegawa et al. | 1990 | 107 | sr107_29 | neo–quart |
| Sprovieri & Hasegawa | 1990 | 107 | sr107_28 | neo–quart |
| Weinholz & Lutze | 1990 | 108 | sr108_07 | quart |
| Clark | 1990 | 110 | sr110_10 | neo–quart |
| Resig | 1990 | 112 | sr112_16 | neo–quart |
| Thomas | 1990 | 113 | sr113_35 | ko–neo |
| Boersma | 1990 | 115 | sr115_20 | pal–neo |
| Iaccarino & Proto Decima | 1990 | 116 | sr116_18 | neo |
| Scott & Leger | 1990 | 116 | sr116_16 | neo–quart |
| Hermelin | 1991 | 117 | sr117_03 | quart |
| Schröder–Adams | 1991 | 119 | sr119_31 | pal–quart |
| Nomura | 1991 | 121 | sr121_01 | ko–pal |
| Nomura | 1991 | 121 | sr121_02 | pal–quart |
| Kaminski & Huang | 1991 | 124 | sr124_12 | pal |
| Mackensen & Berggren | 1992 | 120 | sr120_34 | pal |
| Mackensen | 1992 | 120 | sr120_36 | neo |
| Quilty | 1992 | 120 | sr120_23 | ko |
| Jones & Wonders | 1992 | 122 | sr122_33 | ku |
| Zaninetti et al. | 1992 | 122 | sr122_24 | trias |
| Haig | 1992 | 123 | sr123_14 | ku |
| Kaminski et al. | 1992 | 123 | sr123_13 | jur–ku |
| Milner | 1992 | 125 | sr125_04 | pal |
| Kaiho & Nishimura | 1992 | 126 | sr126_20 | quart |
| Kaiho | 1992 | 126 | sr126_19 | pal–quart |
| Brunner | 1992 | 127 | sr127–128_12 | neo–quart |
| Kato | 1992 | 127 | sr127–128_22 | neo–quart |
| Nomura | 1992 | 127 | sr127–128_29 | neo |
| Wightman & Kuhnt | 1992 | 129 | sr129_13 | ko |
| Katz & Miller | 1993 | 133 | sr133_06 | neo |
| Akimoto | 1994 | 134 | sr134_12 | pal–neo |
| Schönfeld & Spiegler | 1995 | 141 | sr141_15 | neo–quart |
| Dowsett & Ishman | 1995 | 145 | sr145_08 | neo |
| Zellers | 1995 | 146 | sr146pt1_05 | neo–quart |
| Collins et al. | 1996 | 149 | sr149_07 | ju |
| Collins et al. | 1996 | 149 | sr149_09 | neo–quart |
| Kuhnt & Collins | 1996 | 149 | sr149_08 | ko–pal |
| Katz & Miller | 1996 | 150 | sr150_05 | pal–neo |
| Osterman & Spiegler | 1996 | 151 | sr151_09 | pal–neo |
| Osterman | 1996 | 151 | sr151_10 | neo–quart |
| Bignot | 1998 | 159 | sr159_33 | pal |
| Kuhnt et al. | 1998 | 159 | sr159_31 | ko–pal |
| Holbourn & Kuhnt | 1998 | 159 | sr159_30 | ko |

**Table 2**
*(Continued)*

| Author | Year | Leg | Chapter | Age |
|---|---|---|---|---|
| Holbourn & Moullade | 1998 | 159 | sr159_28 | ku |
| Kaminski & Austin | 1999 | 162 | sr162_11 | pal |
| Eidvin & Nagy | 1999 | 162 | sr162_01 | neo |
| Katz | 2000 | 166 | sr166_12 | neo |
| Quilty | 2002 | 183 | sr183_03 | ko |
| Nomura & Takata | 2004 | 199 | sr199_223 | pal |

## References


Adl, S.M., Simpson, A.G.B., Farmer, M.A., et al., 2005, The new higher level classification of eukaryotes with emphasis on the taxonomy of protists. Journal of Eukaryotic Microbiology, vol. 52, pp. 399–451.

Alegret, L., Thomas, E., and Lohmann, K.C., 2012, End-Cretaceous marine mass extinction not caused by productivity collapse. Proceedings of the National Academy of Sciences, vol. 109, pp. 728–732.

Altenbach, A.V. and Sarnthein, M., 1989, Productivity record in benthic foraminifera. In Berger W.H., Smetacek V.S., and Wefer, G., ed., Productivity of the Ocean: Present and Past. New York: Wiley, pp. 255–269.

Altenbach, A.V. and Struck, U., 2001, On the coherence of organic carbon flux and benthic foraminiferal biomass. Journal of Foraminiferal Research, vol. 31, no. 2, pp. 79–85.

Altenbach, A.V., Pflaumann, U., Schiebel, R., Thies, A., Timm, S., and Trauth, M., 1999, Scaling percentages and distributional patterns of benthic foraminifera with flux rates of organic carbon. Journal of Foraminiferal Research, vol. 29, no. 3, pp. 173–185.

Alve, E., 1999, Colonization of new habitats by benthic foraminifera: a review. Earth Science Reviews, vol. 46, pp. 167–185.

Alve, E. and Goldstein, S.T., 2003, Propagule transport as a key method of dispersal in benthic foraminifera. Limnology and Oceanography, vol. 48, pp. 2163–2170.

Bandy, O.L., 1953, Ecology and paleoecology of some California Foraminifera, Part II. Foraminiferal evidence of subsidence rates in the Ventura Basin. Journal of Paleontology, vol. 27, pp. 200–203.

Bandy, O.L., 1961, Distribution of foraminifera, radiolaria and diatoms in sediments of the Gulf of California. Micropaleontology, vol. 7, pp. 1–26.

Bandy, O.L. and Arnal, R., 1957, Distribution of Recent foraminifera off the west coast of Central America. American Association of Petroleum Geologists Bulletin, vol. 41, pp. 2037–2053.

Bentov, S. and Erez, J., 2005, Novel cellular observations on the calcification mechanisms of perforate foraminifera. Geology, vol. 33, pp. 841–844.

Bentov, S. and Erez, J., 2006, Impact of biomineralization processes on the Mg content of foraminiferal shells: A biological perspective. Geochemistry, Geophysics, Geosystems, vol. 7, pp. 1–11.

Berggren, W.A., 1972, Cenozoic biostratigraphy and paleobiogeography of the North Atlantic. Initial Reports of the Deep Sea Drilling Project, vol. 12, pp. 965–1001.

Berggren, W.A. and Miller, K.G., 1989, Cenozoic bathyal and abyssal benthic foraminiferal zonations. Micropaleontology, vol. 35, pp. 308–320.

Boltovskoy, E., 1980, On the benthonic bathyal-zone foraminifera as stratigraphic guide fossils. Journal of Foraminiferal Research, vol. 10, pp. 163–172.

Bosak, T., Lahr, D.J.G., Pruss, S.B., MacDonald, F.A., Gooday, A.J., Dalton, L., and Matys, E.D., 2011, Possible early foraminiferans in post-Sturtian (716–635 Ma) cap carbonates. Geology, vol. 40, pp. 67–70.

Bowser, S.S. and Travis, J.L., 2002, Reticulopodia: structural and behavioral basis for the suprageneric placement of granuloreticulosan protists. Journal of Foraminiferal Research, vol. 32, pp. 440–447.

Boyle, E.A., 1988, Cadmium: chemical tracer of deep-water paleoceanography. Paleoceanography, vol. 3, pp. 471–490.

Boyle E.A., 1992, Cadmium and $\delta^{13}C$ paleochemical ocean distributions during the stage 2 glacial maximum. Annual Review of Earth and Planetary Sciences, vol. 20, pp. 245–287.

Boyle E.A. and Keigwin L.D., 1985, Comparison of Atlantic and Pacific paleochemical records for the last 215,000 years: changes in deep ocean circulation and chemical inventories. Earth and Planetary Science Letters, vol. 76, pp. 135–150.

Brasier, M.D., 1980, Microfossils. London: George Allen & Unwin, 193 pp.

Brown, R.E., Anderson, L.D., Thomas, E., and Zachos, J.C., 2011, A core-top calibration of B/Ca in the benthic foraminifera Nuttallides umbonifera and Oridorsalis umbonatus: reconstructing bottom water carbonate saturation. Earth and Planetary Science Letters, vol. 310, pp. 360–368.

Bryan, S.P. and Marchitto, T.M., 2008, Mg/Ca–temperature proxy in benthic foraminifera: New calibrations from the Florida Straits and a hypothesis regarding Mg/Li. Paleoceanography, vol. 23, pp. 1–17.

Buck, K.R. and Bernhard, J.M., 2001, Protistan–prokaryotic symbioses in deep-sea sulfidic sediments. In Seckbach J., ed., Symbiosis: Mechanisms and Model Systems (Cellular origin, life in extreme habitats and astrobiology), 4. Dordrecht, The Netherlands: Kluwer Academic Publishers, pp. 507–517.

Cedhagen, T., Goldstein, S.T., and Gooday, A.J., 2002, Biology and diversity of allogromiid foraminifera. Journal of Foraminiferal Research, vol. 32, pp. 331–452.

Corliss, B.H., 1980, Vertical distribution of foraminifera off the coast of Northumberland, England. Journal of Foraminiferal Research, vol. 10, pp. 75–78.

Corliss, B.H., 1985, Microhabitats of benthic foraminifera within deep-sea sediments. Nature, vol. 314, pp. 435–438.

Corliss, B.H., 1991, Morphology and microhabitat preferences of benthic foraminifera from the northwest Atlantic Ocean. Marine Micropaleontology, vol. 17, pp. 195–236.

Corliss, B.H. and Chen, C., 1988, Morphotype patterns of Norwegian Sea deep-sea benthic foraminifera and ecological implications. Geology, vol. 16, pp. 716–719.

Corliss, B.H., Brown, C.W., Sun, X., and Showers, W.J., 2009, Deep-sea benthic diversity linked to seasonality of pelagic productivity. Deep-Sea Research Part 1, vol. 56, pp. 835–841.

Culver, S.J., 1991, Early Cambrian foraminifera from West Africa. Science, vol. 254, pp. 689–691.

Culver, S.J., 1993, Foraminifera. In: Lipps, J.H., ed., Fossil Prokaryotes and Protists. Cambridge, MA: Blackwell Scientific Publications, pp. 203–247.

Culver, S.J., 2003, Benthic foraminifera across the Cretaceous–Tertiary (K–T) boundary: a review. Marine Micropaleontology, vol. 47, pp. 177–226.

Curry, W.B., Duplessy, J-C., Labeyrie, L.D., and Shackleton, N.J., 1988, Changes in the distribution of $\delta^{13}C$ of deep water $\Sigma CO_2$ between the last glaciation and the Holocene. Paleoceanography, vol. 3, no. 3, pp. 317–341.

Debenay, J.-P., Guillou, J.J., Geslin, E., and Lesourd, M., 2000, Crystallization of calcite in foraminiferal tests. Micropaleontology, vol. 46, no. 1, pp. 87–94.

DeLaca, T.E., Karl, D.M., and Lipps, J.H., 1981, Direct use of dissolved organic carbon by agglutinated benthic foraminifera. Nature, vol. 289, pp. 287–289.

de Nooijer, L.J.,. Reichart, G.J., Duenas-Bohorquez, A., et al., 2007, Copper incorporation in foraminiferal calcite: results from culturing experiments. Biogeosciences, vol. 4, pp. 493–504.

de Nooijer, L.J., Toyofuku, T., and Kitazato, H., 2009, Foraminifera promote calcification by elevating their intracellular pH. Proceedings of the National Academy of Sciences, vol. 106, pp. 15374–15378.

Dissard, D., Nehrke, G., Reichart, G.J., and Bijma, J., 2010, Impact of seawater $pCO_2$ on calcification and Mg/Ca and Sr/Ca ratios in benthic foramifera calcite: results from culturing experiments with Ammonia tepida. Biogeosciences, vol. 7, pp. 810–893.

Duplessy, J.C., Shackleton, N.J., Matthews, R.K., et al., 1984, 13C record benthic foraminifera in the last interglacial ocean: implications for the carbon cycle and the global deep water circulation. Quaternary Research, vol. 21, pp. 225–243.

Dupuy, C., Rossignol, L., Geslin, E., and Pascal, P.-Y., 2010, Predation on mudflat meio-macrofaunal metazoans by a calcareous foraminifer, Ammonia tepida (Cushman 1926). Journal of Foraminiferal Research, vol. 40, pp. 305–312.

Elderfield, H., Yu, J.M., Anand, P., Kiefer, T., and Nyland, B., 2006, Calibrations for benthic foraminiferal Mg/Ca palaeothermometry and the carbonate ion hypothesis. Earth and Planetary Science Letters, vol. 250, pp. 633–649.

Erez, J., 2003, The source of ions for biomineralization in Foraminifera and their implications for paleoceanographic proxies. Reviews in Mineralogy and Geochemistry, vol. 54, pp. 115–149.

Foster, G.L., 2008, Seawater pH, pCO2 and [CO2–3] variations in the Caribbean Sea over the last 130 kyr: A boron isotope and B/Ca study of planktic foraminifera. Earth and Planetary Science Letters, vol. 271(1–4), pp. 254–266.

Geroch, S. and Nowak, W., 1984, Proposal of zonation for the late Tithonian–late Eocene, based upon arenaceous foraminifera from the outer Carpathians, Poland. In Oertli, H., ed., Benthos '83, 2nd International Symposium on Benthic Foraminifera, Pau (France), April 11–15, 1983. Elf Aquitaine, ESSO REP and TOTAL CFP, Pau and Bordeaux, pp. 225–239.

Goldstein, S.T. and Alve, E., 2011, Experimental assembly of foraminiferal communities from coastal propagule banks. Marine Ecology Progress Series, vol. 437, pp. 1–11.

Goldstein, S.T. and Bernhard, J.M. (ed.), 1997, Theme Issue: Biology of Foraminiferida: Applications in Paleoceanography, Paleobiology, and Environmental Sciences. Journal of Foraminiferal Research, vol. 27, no. 4, pp. 253–328.

Gooday, A.J., 1988, A response by benthic foraminifera to the deposition of phytodetritus in the deep sea. Nature, vol. 332, pp. 70–73.

Gooday, A.J., 1990, Recent deep-sea agglutinated foraminifera: a brief review. In Hemleben, C., Kaminski, M.A., Kuhnt, W., and Scott, D.B., ed., Paleoecology, Biotratigraphy, Paleoceanography and Taxonomy of Agglutinated Foraminifera. Proceedings of the 3rd International Workshop on Agglutinated Foraminifera (IWAF III). NATO-ASI Series. Dordrecht: Kluwer, pp. 271–304.

Gooday, A.J., 1993, Deep-sea benthic foraminiferal species which exploit phytodetritus: Characteristic features and controls on distribution. Marine Micropaleontology, vol. 22, pp. 187–205.

Gooday, A.J., 1999, Biodiversity of foraminifera and other protists in the deep sea: scales and patterns. Belgian Journal of Zoology, vol. 129, pp. 61–80.

Gooday, A.J., 2002, Biological responses to seasonally varying fluxes of organic matter to the ocean floor: a review. Journal of Oceanography, vol. 58, no. 2, pp. 305–332.

Gooday, A.J., 2003, Benthic Foraminifera (Protista) as tools in deep-water paleoceanography: environmental influences of faunal characteristics. Advances in Marine Biology, vol. 46, pp. 1–90.

Gooday, A.J. and Jorissen, F.J., 2012, Benthic foraminiferal biogeography: controls on global distribution patterns in deep-water settings. Annual Review of Marine Science, vol. 4, pp. 237–262.

Gooday, A.J., Levin, L.A., Linke, P., and Heeger, T., 1992, The role of benthic foraminifera in deep-sea food webs and carbon cycling. In Rowe, G.T. and Patiente, V., ed., Deep-sea Food Chains and the Global Carbon Cycle. Dordrecht: Kluwer Academic Publishers, pp. 63–91.

Gooday, A.J., Nomaki, H., and Kitazato, H., 2008, Modern deep-sea benthic foraminifera: a brief review of their morphology-based biodiversity and trophic diversity. Geological Society of London, Special Publication, vol. 303, pp. 97–119.

Gradstein, F.M., Ogg, J.G., Smith, A.G., et al., 2004, A Geologic Time Scale. Cambridge: Cambridge Universty Press, 589 pp.

Groves, J.R. and Yue, W., 2009, Foraminiferal diversification during the late Paleozoic ice age. Paleobiology, vol. 35, pp. 367–392.

Hallock, P., Premoli Silva, I., and Boersma, A., 1991, Similarities between planktonic and larger foraminiferal evolutionary trends through Paleogene paleoceanographic changes. Palaeogeography, Palaeoclimatology, Palaeoecology, vol. 83, pp. 49–64.

Hayward, B.W., 2001, Global deep-sea extinction during the Pleistocene ice ages. Geology, vol. 29, pp. 599–602.

Hayward, B., Johnson, K., Sabaa, A.T., Kawagata, S., and Thomas, E., 2010, Cenozoic record of elongate, cylindrical deep-sea benthic Foraminifera in the North Atlantic and equatorial Pacific Oceans. Marine Micropaleontology, vol. 74, pp. 75–95.

Heinz, P., Kitazato, H., Schmiedl, G., and Hemleben, C., 2001, Response of deep-sea benthic foraminifera from the Mediterranean Sea to simulated phytoplankton pulses under laboratory conditions. Journal of Foraminiferal Research, vol. 31, no. 3, pp. 210–227.

Hemleben, C., Anderson, O.R., Berthold, W.U., and Spindler, M., 1986, Calcification and chamber formation in foraminifera – a brief overview. In Leadbeater, B.S.C. and Riding, R., ed., Biomineralization in Lower Plants and Animals. The Systematics Association, Special Volume, vol. 30, pp. 237–249.

Hess, S. and Kuhnt, W. 1996, Deep-sea benthic foraminiferal recolonization of the 1991 Mt. Pinatubo ash layer in the South China Sea. Marine Micropaleontology, 28, 171–197.

Hess, S., Kuhnt, W., Spivey, B., Kaminski, M.A., and Whittaker, J.E., 2000, Ash grains of the 1991 Mt. Pinatubo eruption as a tracer in Rose Bengal stained deep sea agglutinated foraminifera: How old is Freddy? In: Hart, M., Kaminski, M., and Smart, C.W., ed., Proceedings of the 5th International Workshop on Agglutinated Foraminifera (IWAF V), Plymouth, 1997, p. 126.

Holbourn, A., Kuhnt, W., Schulz, M., and Erlenkeuser, H., 2005, Impacts of orbital forcing and atmospheric carbon dioxide on Miocene ice-sheet expansion. Nature, vol. 438, pp. 483–487.

Holbourn, A., Kuhnt, W., Schulz, M., Flores, J., and Andersen, N., 2007, Orbitally paced climate evolution during the middle Miocene "Monterey" carbon-isotope excursion. Earth and Planetary Science Letters, vol. 261, pp. 534–550.

Hönisch, B., Bickert, T., and Hemming, N.G., 2008, Modern and Pleistocene boron isotope composition of the benthic foraminifer Cibicidoides wuellerstorfi. Earth and Planetary Science Letters, vol. 272, no. 1–2, pp. 309–318.

Hottinger, L., 2006, Illustrated Glossary of terms used in foraminiferal research. Carnets de Geologie – Memoir 2. http://paleopolis.rediris.es/cg/CG2006_M02/index.html.

Imbrie, J., Hays, J.D., Martinson, D.G., et al., 1984, The orbital theory of Pleistocene climate: support from a revised chronology of the marine $\delta^{18}O$ record. In Berger, A.L., Imbrie, J., Hays, J., Kukla, G., and Saltzman, B., ed., Milankovitch and Climate, Part 1. Hingham, MA: D. Riedel, pp. 269–305.

Jones, R.W. and Charnock, M.A., 1985, Morphogroups of agglutinating foraminifera, their life positions and feeding habits and potential applicability in (paleo)ecological studies. Revue de Paléobiologie, vol. 4, no. 2, pp. 311–320.

Jorissen, F.J., 1999, Benthic foraminiferal microhabitats below the sediment-water interface. In: Sen Gupta, B.K., ed., Modern Foraminifera. Dordrecht: Kluwer, pp. 161–179.

Jorissen, F.J., De Stigter, H.C., and Widmark, J.G.V., 1995, A conceptual model explaining benthic foraminiferal microhabitats. Marine Micropaleontology, vol. 26, pp. 3–15.

Jorissen, F.J., Fontanier, C., and Thomas, E., 2007, Paleoceanographical proxies based on deep-sea benthic foraminiferal assemblage characteristics. In Hillaire-Marcel, C. and De Vernal, A., ed., Proxies in Late Cenozoic Paleoceanography, Volume 1. Developments in Marine Geology. New York: Elsevier, pp. 263–325.

Kaiho, K., 1992, Eocene to Quaternary benthic foraminifers and paleobathymetry of the Izu-Bonin Arc, Legs 125 and 126. In Taylor, B., Fujioka, K., et al., ed., Proceedings of the Ocean Drilling Program, Scientific Results, vol. 126, pp. 285–310.

Kaminski, M.A., Cetean, C.G., and Tyszka, J., 2011, Nomenclature to describe the transition from multiserial to uniserial chamber arrangement in benthic foraminifera. Journal of Micropalaeontology, vol. 30, pp. 7–10.

Katz, M.E. and Miller, K.G., 1993. Neogene subsidence along the Northeast Australian margin: benthic foraminiferal evidence. In Davies, P.J., McKenzie, J.A., Julson, A., et al., ed., Scientific Results of the Proceedings of the Ocean Drilling Program, vol. 133, pp. 75–92.

Katz, M.E. and Miller, K.G., 1996, Eocene to Miocene oceanographic and provenance changes in a sequence stratigraphic framework: benthic foraminifers of the New Jersey Margin. In Mountain, G.S., Miller, K.G., Blum, P., Poag, C.W., and Twichell, D.C., ed., Scientific Results of the Proceedings of the Ocean Drilling Program, vol. 150, pp. 65–95.

Katz, M.E., Cramer, B.S., Honisch, B., Miller, K.G., Rosenthal, Y., and Wright, J.D., 2010, Traditional and emerging geochemical proxies in foraminifera. Journal of Foraminiferal Research, vol. 40, no. 2, pp. 165–192.

Krasheninnikov, V.A., 1973, Cretaceous benthonic foraminifera, Leg 20, Deep Sea Drilling Project. In Heezen, B.C,, MacGregor, I.D., et al., ed., Initial Reports of the Deep Sea Drilling Project, Washington (U.S. Government Printing Office), vol. 20, pp. 205–221.

Krasheninnikov, V.A., 1974, Upper Cretaceous benthonic agglutinated foraminifera, Leg 27 of the Deep Sea Drilling Project. In Veevers, J.J., Heirtzler, J.R., et al., ed., Initial Reports of the Deep Sea Drilling Project, Washington (U.S. Government Printing Office), vol. 27, pp. 531–662.

Kucera, M. and Schönfeld, J., 2007, The origin of modern oceanic foraminiferal faunas and Neogene climate change. In Williams, M., Haywood, A.M., Gregory, F.J., and Schmidt, D.N., ed., Deep-Time Perspectives on Climate Change: Marrying the Signal from Computer Models and Biological Proxies. The Micropalaeontological Society, Spec. Publ. vol. 2, pp. 409–426.

Kuhnt, W. and Moullade, M., 1991, Quantitative analysis of Upper Cretaceous abyssal agglutinated foraminiferal distribution in the North Atlantic – paleoceanographic implications. Revue de Micropaléontologie, vol. 34, no. 4, pp. 313–349.

Kuhnt, W., Geroch, S., Kaminski, M.A., Moullade, M., and Neagu, T., 1992, Upper Cretaceous abyssal claystones in the North Atlantic and the western Tethys – biostratigraphical correlation using agglutinated foraminifers and paleoceanographic events. Cretaceous Research, 13, 467–478.

Kuhnt, W., Hess, S., Holbourn, A., Paulsen, H., and Salomon, B., 2005, The impact of the 1991 Mt. Pinatubo eruption on deep-sea foraminiferal communities: a model for the Cretaceous–Tertiary (K/T) boundary? Palaeogeography, Palaeoclimatology, Palaeoecology, vol. 224, pp. 83–107.

Kuhnt, W., Moullade, M., and Kaminski, M.A., 1996a, Cretaceous paleoceanographic events and abyssal agglutinated foraminifera. In Moguilevsky, A.E. and Whatley, R.C. et al., ed., Microfossils and Oceanic Environments. Proceedings of the "ODP and the Marine Biosphere" International Conference, Aberystwyth, 19th–21st April 1994, Aberystwyth Press, University of Wales, pp. 63–76.

Kuhnt, W., Moullade, M., and Kaminski, M.A., 1996b, Ecological structuring and evolution of deep sea agglutinated foraminifera – a review. Revue de Micropaléontologie, vol. 39, pp. 271–281.

Langezaal, A.M., Jannink, N.T., Person, E.S., and van der Zwaan, G.J., 2005, Foraminiferal selectivity towards bacteria: An experimental approach using a cell-permeant stain. Journal of Sea Research, vol. 54, pp. 256–275.

Laskar, J., Fienga, A., Gastineau, M., and Manche, H., 2011, La2010: A new orbital solution for the long-term motion of the Earth. Astronomy and Astrophysics, vol. 532, pp. A89.

Laskar, J., Robutel, P., Joutel, F., Gastineau, M., Correia, A., and Levrard, B.A., 2004, Long term numerical solution for the insolation quantities of the Earth. Astronomy and Astrophysics, vol. 428, pp. 261–285.

Lea, D.W. and Boyle, E.A., 1989, Barium content of benthic foraminifera controlled by bottom water composition. Nature, vol. 338, pp. 751–753.

Lea, D.W. and Boyle, E.A., 1990, 200,000 year record of Barium in the deep western North Atlantic. Nature, vol. 347, pp. 269–272.

Lear, C.H., Mawbey, E.M., and Rosenthal, Y., 2010, Cenozoic benthic foraminiferal Mg/Ca and Li/Ca records: Toward unlocking temperatures and saturation states. Paleoceanography, vol. 25, pp. 1–11.

Lear, C.H., Rosenthal, Y., and Slowey, N., 2002, Benthic foraminiferal Mg/Ca paleothermometry: A revised core-top calibration. Geochimica et Cosmochimica Acta, vol. 66, no. 19, pp. 3375–3387.

Lee, J.J., 1980, Nutrition and physiology of the Foraminifera. In: Lewandowsky, M. and Hutner, S.H., (ed.), Biochemistry and Physiology of Protoza, Vol. 3. Academic Press, London, 43–66.

Lee, J.J., 1990, Phylum Granuloreticulosa (Foraminifera). In: Margulis, L., Corliss, J.O., Melkonian, M., Chapman, D.J., and H. McKhann, H., ed., Handbook of Protochtista, the Structure, Cultivation, Habits, and Life Histories of the Eukaryotic Microorganisms and Their Descendants Exclusive of Animals, Plants, and Fungi. Boston: Jones and Bartlett Publishers, pp. 524–548.

Lee, J.J. and Anderson, O.R., 1991, Biology of Foraminifera. Oxford: Academic Press, 368 pp.

Lee, J.J. and Hallock, P., 2000, Advances in the biology of foraminifera. Micropaleontology, vol. 46, supplement 1, pp. 1–198.

Linke, P. and Lutze, G., 1993, Microhabitat preferences of benthic foraminifera – a static concept or a dynamic adaptation to optimize food acquisition? Marine Micropaleontology, vol. 20, pp. 215–234.

Lipps, J.H., 1983, Biotic interactions in benthic foraminifera. In Tevesz, M.J.S. and McCall, P.L. ed., Biotic Interactions in Recent and Fossil Benthic Communities. New York: Plenum Publishing Co., pp. 331–376.

Lisiecki, L.E. and Raymo, M.E., 2005, A Pliocene–Pleistocene stack of 57 globally distributed benthic $\delta^{18}O$ records. Paleoceanography, vol. 20, pp. 1–16.

Loeblich, A.R. and Tappan, H., 1984, Suprageneric classification of the Foraminiferida (Protozoa). Micropaleontology, vol. 30, pp. 1–70.

Loeblich, A.R. Jr and Tappan, H., 1987, Foraminiferal Genera and Their Classification. New York: Van Nostrand Reinhold, 970 pp. + 847 pl.

Loeblich, A.R. and Tappan, H., 1989, Implications of wall composition and structure in agglutinated foraminifers. Journal of Paleontology, vol. 63, pp. 769–777.

Loeblich, A.R. and Tappan, H., 1992, Present status of foraminiferal classification. In Takayanagi, Y. and Saito, T., ed., Studies in Benthic Foraminifera, Proceedings of the Fourth International Symposium on Benthic Foraminifera, Sendai, 1990. Tokyo: Tokai University Press, pp. 93–101.

Loubere, P., 1996, The surface ocean productivity and bottom water oxygen signals in deep water benthic foraminiferal assemblages. Marine Micropaleontology, vol. 28, pp. 247–261.

Loubere, P., Meyers, P., and Gary, A., 1995, Benthic foraminiferal microhabitat selection, carbon isotope values, and association with larger animals; a test with Uvigerina peregrina. Journal of Foraminiferal Research, vol. 25, pp. 83–95.

Mackensen, A. and Licari, L. 2004, Carbon isotopes of live benthic foraminifera from the eastern South Atlantic Ocean: Sensitivity to organic matter rain rates and bottom water carbonate saturation state. In Wefer, G., Mulitza, S., and Rathmeyer, V. ed., The South Atlantic in the Late Quaternary: Reconstruction of Material Budget and Current Systems. Berlin: Springer-Verlag, pp. 623–644.

Mackensen, A., Hubberten, H.W., Bickert, F.G., and Fütterer D.K., 1993, The $?^{13}C$ in benthic foraminiferal tests of Fontbotia wuellerstorfi (Schwager) relative to the $?^{13}C$ of dissolved inorganic carbon in southern ocean deep water: Implications for glacial ocean circulation models. Paleoceanography, vol. 8, pp. 87–610.

Mackensen, A., Schumacher, S., Radke, J., and Schmidt, D.N., 2000, Microhabitat preferences and stable carbon iostopes of endobenthic foraminifera: clue to quantitative reconstruction of oceanic new production? Marine Micropaleontology, vol. 40, pp. 233–258.

Marchitto, T.M. and Broecker, W.S., 2006, Deep water mass geometry in the glacial Atlantic Ocean: A review of constraints from the paleonutrient proxy Cd/Ca. Geochemistry, Geophysics, Geosystems, vol. 7, no. 12, pp. 1–16.

Marchitto, T.M., Curry, W.B., and Oppo, D.W., 2000, Zinc concentrations in benthic foraminifera reflect seawater chemistry. Paleoceanography, vol. 15, pp. 299–306.

Marchitto, T.M., Lynch-Stieglitz, J., and Hemming, S.R., 2005, Deep Pacific $CaCO_3$ compensation and glacial–interglacial atmospheric $CO_2$. Earth and Planetary Science Letters, vol. 231, pp. 317–336.

Martin, P.A., Lea, D.W., Rosenthal, Y., Papenfuss, T.P., and Sarnthein, M., 2002, Late Quaternary deep-sea temperatures inferred from benthic foraminiferal magnesium. Earth and Planetary Science Letters, vol. 198, pp. 193–209.

Mikhalevich, V., 2004, On the heterogeneity of the former Textulariina (Foraminifera). In Bubik, M. and Kaminski, M.A., ed., Proceedings of the Sixth International Workshop on Agglutinated Foraminifera. Grzybowski Foundation Special Publication, vol. 8, pp. 317–349.

Miller, K., Katz, M., and Berggren, W., 1992, Cenozoic deep-sea benthic foraminifera: A tale of three turnovers. In Takyanagi, Y. and Saito, T., ed., Studies in Benthic Foraminifera, Benthos '90: Proceedings of the Fourth International Symposium on Benthic Foraminifera. Tokyo: Tokai University Press, pp. 67–75.

Moullade, M., 1984, The value of "deep" smaller benthic Foraminifera in the biostratigraphy and analysis of Mesozoic oceanic paleoenvironments. In: Benthos '83; 2nd International Symposium on Benthic Foraminifera, Pau, 1983, pp. 429–464.

Moullade, M., Kuhnt, W., and Thurow, J., 1988, Agglutinated benthic foraminifers from Upper Cretaceous variegated clays of the North Atlantic Ocean (DSDP Leg 93 and ODP Leg 103). In Boillot, G., Winterer, E.L., et al., ed., Scientific Results of the Proceedings of the Ocean Drilling Program, vol. 103, pp. 349–377.

Murray, J.W., 1991, Ecology and Palaeoecology of Benthic Foraminifera. New York: John Wiley & Sons, 397 pp.

Murray, J.W., 2001, The niche of benthic foraminifera, critical thresholds and proxies. Marine Micropaleontology, vol. 41, pp. 1–7.

Murray, J.W., 2006, Ecology and Applications of Benthic Foraminifera. Cambridge: Cambridge University Press, 426 pp.

Nomaki, H., Ogawa, N.O., Takano, Y., Suga, H., Ohkouchi, N., and Kitazato, H. 2011, Differing utilization of glucose and algal-particulate organic matter by deep-sea benthic organisms of Sagami Bay, Japan. Marine Environmental Progress Series, vol. 431, pp. 11–24.

Ohga, T. and Kitazato, H., 1997, Seasonal changes in bathyal foraminiferal populations in response to the flux of organic matter. Terra Nova, vol. 9, pp. 33–37.

Pälike, H., Norris, R.D., Herrle, J.O., et al., 2006, The heartbeat of the Oligocene climate system. Science, vol. 314, no. 5807, pp. 1894–1898.

Pawlowski, J. and Holzmann, M., 2008, Diversity and geographic distribution of benthic foraminifera: a molecular perspective. Biodviersity and Conservation, vol. 17, pp. 317–328.

Pawlowski, J., Bolivar, I., Fahrni, J., de Vargas, C., Gouy, M., and Zaninetti, L., 1997, Extreme differences in rates of molecular evolution of foraminifera revealed by comparison of ribosomal DNA sequences and the fossil record. Molecular Biology and Evolution, vol. 14, pp. 498–505.

Pawlowski, J., Fahrni, J., Lecroq, B., Longet, D., Cornelius, N., Excoffier, L., Cedhagen, T., and Gooday, A.J., 2007, Bipolar gene flow in deep-sea benthic foraminifera. Molecular Ecology, vol. 16, pp. 4089–4096.

Pawlowski, J., Holzmann, M., Berney, C., et al., 2003, The evolution of early Foraminifera. Proceedings of the National Academy of Sciences, vol. 100, no. 20, pp. 11494–11498.

Pflum, C.E. and Frerichs, W.E., 1976, Gulf of Mexico deep-water foraminifers. Special Publication of the Cushman Foundation, vol. 14, pp. 1–125.

Rae, J.W.B., Foster, G.L., Schmidt, D.N., and Elliott, T., 2011, Boron isotopes and B/Ca in benthic foraminifera: Proxies for the deep ocean carbonate system. Earth and Planetary Science Letters, vol. 302, pp. 403–413.

Raitzsch, M., Duenas-Bohorquez, A., Recihart, G.J., de Nooijer, L.J., and Bickert, T., 2010, Incorporation of Mg and Sr in calcite of cultured benthic foraminifera: impact of calcium concentration and associated calcite saturation state. Biogeosciences, vol. 7, pp. 869–881.

Rickaby, R.E.M., Greaves, M.J., and Elderfield, H., 2000, Cd in planktonic and benthic foraminiferal shells determined by thermal ionisation mass spectrometry. Geochimica et Cosmochimica Acta, vol. 64, pp. 1229–1236.

Rosenthal, Y., Lear, C.H., Oppo, D.W., and Linsley, B.K., 2006, Temperature and carbonation effect on Mg/Ca and Sr/Ca ratios in benthic foraminifera: aragonitic species Hoeglundina elegans. Paleoceanography, vol. 21, pp. 1–14.

Saltzman, M. and Thomas, E., 2012, Chapter 11: Carbon isotope stratigraphy. In Gradstein, F.M., Ogg, J.G., Schmitz, M., and Ogg, G., ed., A Geologic Timescale 2012. Elsevier, pp. 221–246.

Schönfeld, J., 1996, The 'Stilostomella Extinction'. Structure and dynamics of the last turnover in deep-sea benthic foraminiferal assemblages. In Moguilevsky, E.A. and Whatley, R., ed., Microfossils and Oceanic Environments. Aberystwyth Press, University of Wales, pp. 27–37.

Sen Gupta, B.K., 1999, Systematics of modern foraminifera. In Sen Gupta, B.K., ed., Modern Foraminifera, Chapter 2. Dordrecht: Kluwer Academic Publishers, pp. 7–35.

Shackleton, N.J. and Opdyke, N.D., 1973, Oxygen isotope and paleomagnetic stratigraphy of equatorial Pacific Core V28–238: oxygen isotope temperatures and ice volume on a 105 year and 106 year scale. Quaternary Research, vol. 3, pp. 39–55.

Sigwart, J.D., 2009, Parasitic foraminifers on a deep-sea chiton (Mollusca, Polyplacophora, Leptochitonidae) from Iceland. Marine Biology Research, vol. 5, pp. 193–199.

Smart, C.W., King, S.C., Gooday, A.J., Murray, J.W., and Thomas, E., 1994, A benthic foraminiferal proxy of pulsed organic matter paleofluxes. Marine Micropaleontology, vol. 23, pp. 89–99.

Suhr, S.B., Pond, D.W., Gooday, A.J., and Smith, C.R., 2003, Selective feeding by benthic foraminifera on phytodetritus on the western Antarctic Peninsula shelf: evidence from fatty acid biomarker analysis. Marine Ecology Progress Series, vol. 262, pp. 153–162.

Tappan, H. and Loeblich, A.R., 1988, Foraminiferal evolution, diversification, and extinction. Journal of Paleontology, vol. 62, pp. 695–714.

Tendal, O.S., 1990, Why are foraminiferida foraminifers? In Hemleben, C., Kaminski, M.A., Kuhnt, W., and Scott, D.B., ed., Paleoecology, Biostratigraphy, Paleoceanography and Taxonomy of Agglutinated Foraminifera. Proceedings of the 3rd International Workshop on Agglutinated Foraminifera (IWAF III). NATO-ASI Series. Dordrecht: Kluwer, pp. 13–18.

Thomas, E., 1998, The biogeography of the late Paleocene benthic foraminiferal extinction. In Aubry, M.-P., Lucas, S., and Berggren, W.A., ed., Late Paleocene–early Eocene Biotic and Climatic Events in the Marine and Terrestrial Records. Columbia University Press, pp. 214–243.

Thomas, E., 2003, Extinction and food at the sea floor: a high-resolution benthic foraminiferal record across the Initial Eocene Thermal Maximum, Southern Ocean Site 690. In Wing, S., Gingerich, P., Schmitz, B., and Thomas, E., ed., Causes and Consequences of Globally Warm Climates of the Paleogene, GSA Special Paper, vol. 369, pp. 319–332.

Thomas, E., 2007, Cenozoic mass extinctions in the deep sea: what perturbs the largest habitat on earth? In Monechi, S., Coccioni, R., and Rampino, M., ed., Large Ecosystem Perturbations: Causes and Consequences, Geological Society of America Special Paper, vol. 424, pp. 1–23.

Thomas, E. and Gooday, A.J., 1996, Deep-sea benthic foraminifera: tracers for Cenozoic changes in oceanic productivity? Geology, vol. 24, pp. 355–358.

Thomas, E. and Shackleton, N.J., 1996, The Palaeocene–Eocene benthic foraminiferal extinction and stable isotope anomalies. Geological Society London Special Publication, vol. 101, pp. 401–441.

Thomas, E., Zachos, J.C., and Bralower, T.J., 2000, Deep-sea environments on a warm Earth: latest Paleocene – early Eocene. In Huber, B., MacLeod, K., and Wing, S., ed., Warm Climates in Earth History. Cambridge: Cambridge University Press, pp. 132–160.

Tiedemann, R., Wollenburg, J. and Groeneveld, J., 2012, Culture experiments on the ?$^{13}$C values recorded in tests of benthic foraminifera from methane sources at high latitudes. http://www.spp-antarktisforschung.de/projects/Ralf-Tiedemann/Ralf-Tiedemann-TI-240-19.html.

Travis, J.J. and Bowser, S.S., 1991, The motility of foraminifera. In Lee, J.J. and Anderson, O.R., ed., The Biology of Foraminifera. London: Academic Press, pp. 91–155.

Tyszka, J., 2006, Morphospace of foraminiferal shells: results from the moving reference model. Lethaia, vol. 39, no. 1, pp. 1–12.

Van der Zwaan, G.J., Duijnstee, I.A.P., Den Dulk, M., Ernst, S.R., Jannink, N.T., and Kouwenhoven, T.J., 1999, Benthic Foraminifera: proxies or problems? A review of paleoecological concepts. Earth Science Reviews, vol. 46, pp. 213–236.

van Morkhoven, F.P.C.M., Berggren, W.A., Edwards, A.S., et al., 1986, Cenozoic cosmopolitan deep-water benthic foraminifera. Bulletin des Centres de Recherches Exploration-Production Elf-Aquitaine, Mémoire, vol. 11, pp. 1–421.

Weinholz, P. and Lutze, G.F., 1989, The Stilostomella Extinction. In Ruddiman, W., Sarnthein, M., et al., ed., Scientific Results of the Proceedings of the Ocean Drilling Program, vol. 108, pp. 113–117.

Weldeab, S., Lea, D.W., Schneider, R.R., and Andersen, N., 2007, 155,000 years of West African Monsoon and Ocean Thermal Evolution. Science, vol. 316, pp. 1303–1307.

Winguth, A., Thomas, E., and Winguth, C., 2012, Global decline in ocean ventilation, oxygenation and productivity during the Paleocene–Eocene Thermal Maximum – Implications for the benthic extinction. Geology, vol. 40, pp. 263–266.

Wood, A., 1949, The structure of the wall of the test in the foraminifera; its value in classification. Quarterly Journal of the Geological Society, vol. 104, pp. 229–252.

Yu, J.M. and Elderfield, H., 2007, Benthic foraminiferal B/Ca ratios reflect deep water carbonate saturation state. Earth and Planetary Science Letters, vol. 258, no. 1–2, pp. 73–86.

Yu, J.M. and Elderfield, H., 2008, Mg/Ca in the benthic foraminifera Cibicidoides wuellerstorfi and Cibicidoides mundulus: temperature versus carbonate ion saturation. Earth and Planetary Science Letters, vol. 276, no. 1–2, pp. 129–139.

Yu, J.M., Elderfield, H., and Hönisch, B., 2007, B/Ca in planktonic foraminifera as a proxy for surface water pH. Paleoceanography, vol. 22, pp. 1–17.

Zachos, J., Pagani, M., Sloan, L., Thomas, E., and Billups, K., 2001, Trends, Rhythms, and aberrations in global climate change 65 Ma to present. Science, vol. 292, pp. 686–693.

Zahn, R. and Stüber, A., 2002, Suborbital intermediate water variability inferred from paired benthic foraminiferal Cd/Ca and 13C in the tropical west Atlantic and linking with North Atlantic climates. Earth and Planetary Science Letters, vol. 200, pp. 191–205.

# Atlas of Benthic Foraminifera

# *Abyssamina poagi*

## Schnitker and Tjalsma, 1980

1. *Abyssamina poagi.* North Biscay, North Atlantic; DSDP Leg 48, Hole 400A, Core 47, Section 7, 75–79 cm. Spiral view, NHMUK PM DI 0001, scale bar = 45 μm.

2. *Abyssamina poagi.* North Biscay, North Atlantic; DSDP Leg 48, Hole 400A, Core 47, Section 7, 75–79 cm. Apertural view, NHMUK PM DI 0002, scale bar = 48 μm.

3. *Abyssamina poagi.* North Biscay, North Atlantic; DSDP Leg 48, Hole 400A, Core 47, Section 7, 75–79 cm. Umbilical view, NHMUK PM DI 0003, scale bar = 56 μm.

## Original Designation

*Abyssamina poagi* Schnitker and Tjalsma, 1980, pp. 237, 239, pl. 1, figs. 7–9.

## Type Specimen

Holotype (USNM 305101) deposited in the collections of the United States National Museum, Washington, DC.

## Type Level

Lower Eocene.

## Type Locality

DSDP Leg 48, Hole 400A, Core 55, core catcher, latitude 47° 22.9′ N, longitude 09° 11.9′ W, water depth 4399 meters, Bay of Biscay, Northeast Atlantic. Also found in cores from the Shatsky Rise (DSDP Site 47) and East Pacific (DSDP Site 321) by synonymy.

## Description

Test forms a small, low trochospire; ovoid in plan view and quadrate in edge view. Umbilical side involute with shallow, narrow umbilicus. Spiral side involute to slightly evolute. The four inflated chambers in the last whorl increase gradually in size, and are separated by moderately depressed sutures, radial on the spiral side and strongly curved on the umbilical side. Chamber walls are calcareous, finely perforate, and smooth. The primary aperture is a low, umbilical arch covered by a short, lobate extension of the last chamber.

## Synonyms and Citations

*Cystammina?* sp. Resig, 1976, p. 752, pl. 1, figs. 4–7.
*?Quadrimorphina* sp. 1 Schnitker, 1979, p. 400, pl. 6, figs. 8–14.
*Abyssamina poagi* Schnitker and Tjalsma. –van Morkhoven et al., 1986, p. 285, pl. 95.

## Remarks

*Abyssamina poagi* Schnitker and Tjalsma differs from *Abyssamina quadrata* Schnitker and Tjalsma by its longer, narrower chambers that tend to overlap in the umbilical region, by its more strongly curved umbilical sutures, and by its shorter, wider aperture. The test of *Abyssamina poagi* is also narrower in edge view and more distinctly trochospiral (see van Morkhoven et al., 1986).

## Biogeography

Worldwide. Recorded from the Pacific, the North and South Atlantic, the Caribbean, the Indian Ocean, and the Gulf of Mexico (see van Morkhoven et al., 1986).

## Bathymetry

Lower bathyal to abyssal. Wide bathymetric distribution during the Early Eocene; became restricted to greater depths during the Middle Eocene (Schnitker and Tjalsma, 1980).

## Chronostratigraphy

Middle Paleocene to Latest Eocene.

## References


Resig, J.M., 1976, Benthic foraminiferal stratigraphy, eastern margin, Nazca Plate. In Yeats, R.S., Hart, S.R., et al., ed. Initial Reports of the Deep Sea Drilling Project, Washington (U.S. Government Printing Office), vol. 34, pp. 743–759.

Schnitker, D., 1979, Cenozoic deep water foraminifers, Bay of Biscay. In Montadert, L., Roberts, D.G., et al., ed. Initial Reports of the Deep Sea Drilling Project, Washington (U.S. Government Printing Office), vol. 48, pp. 377–413.

Schnitker, D. and Tjalsma, L.R.C., 1980, New genera and species of benthic foraminifera from Paleocene and Eocene deep-water deposits. Journal of Foraminiferal Research, vol. 10, no. 3, pp. 235–241.

van Morkhoven, F.P.C.M., Berggren, W.A., Edwards, A.S., et al., 1986, Cenozoic cosmopolitan deep-water benthic foraminifera. Bulletin des Centres de Recherches Exploration-Production Elf-Aquitaine, Mémoire, vol. 11, pp. 1–421.

# *Abyssamina quadrata*

## Schnitker and Tjalsma, 1980

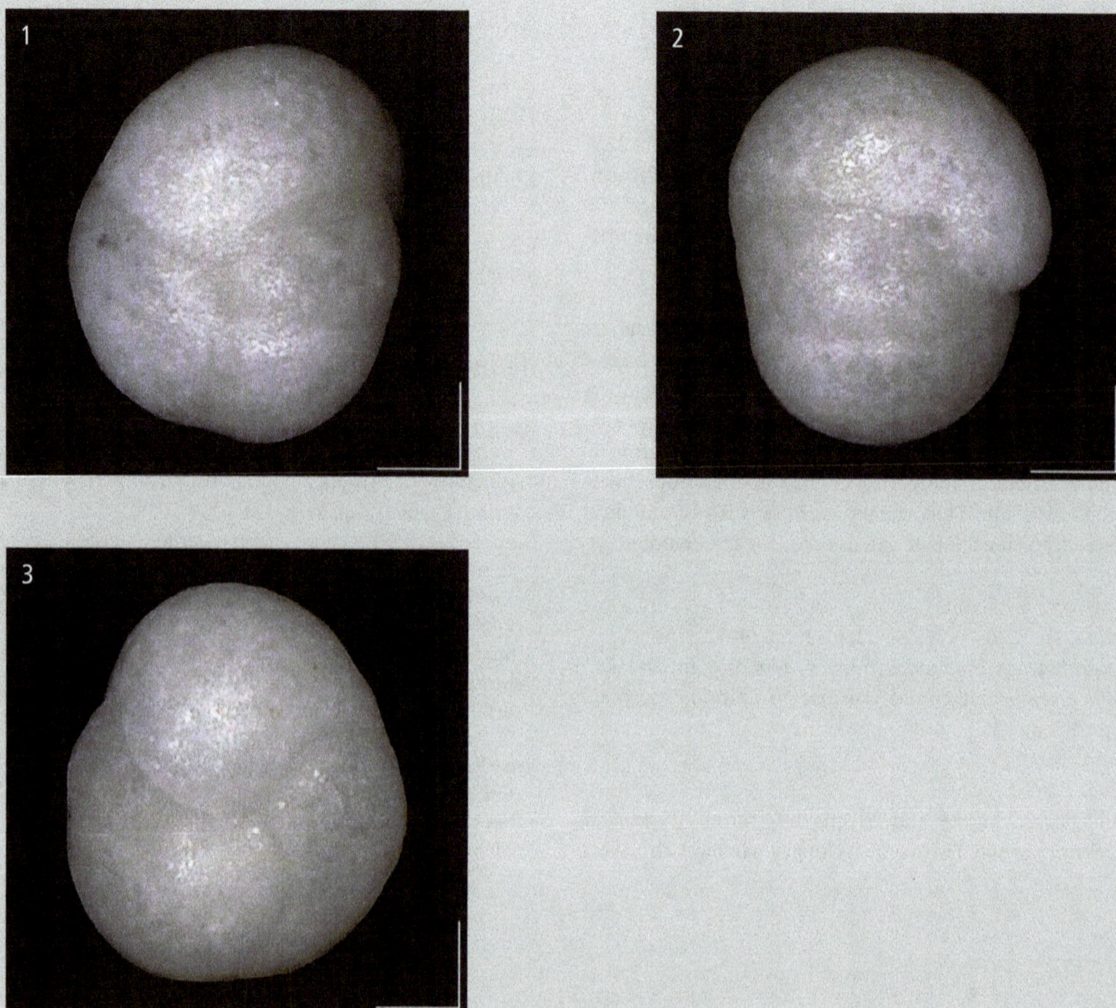


1. *Abyssamina quadrata*. Northeast Georgia Rise, South Atlantic; ODP Leg 114, Hole 700B, Core 6, Section 6, 72–76 cm. Spiral view, NHMUK PM DI 0004, scale bar = 50 μm.

2. *Abyssamina quadrata*. Northeast Georgia Rise, South Atlantic; ODP Leg 114, Site 700B, Core 6, Section 6, 72–76 cm. Apertural view, NHMUK PM DI 0005, scale bar = 47 μm.

3. *Abyssamina quadrata*. Northeast Georgia Rise, South Atlantic; ODP Leg 114, Hole 700B, Core 6, Section 6, 72–76 cm. Umbilical view, NHMUK PM DI 0006, scale bar = 47 μm.

## Original Designation

*Abyssamina quadrata* Schnitker and Tjalsma, 1980, p. 237, pl. 1, figs. 1–6.

## Type Specimen

Holotype (USNM 305103) deposited in the collections of the United States National Museum, Washington, DC.

## Type Level

Upper Paleocene.

## Type Locality

DSDP Leg 48, Hole 400A, Core 57, Section 2, 4–7 cm, latitude 47° 22.9′ N, longitude 09° 11.9′ W, water depth 4399 meters, Bay of Biscay, Northeast Atlantic. Found at same site in Core Sections 59-1 to 47-6.

## Description

Test forms a small, low, unequally involute trochospire; globular in outline and quadrate in cross-section. The umbilical side is involute with a shallow, narrow umbilicus; the spiral side is involute to slightly evolute. The four inflated, globular chambers present in the last whorl increase gradually in size, and are separated by radial, straight, or very slightly curved, slightly depressed sutures. Chamber walls are calcareous, finely perforate, and smooth. The primary aperture is a narrow, interiomarginal slit extending from near the periphery to the umbilicus.

## Synonyms and Citations

*Pullenia jarvisi* Cushman. –Hillebrandt, 1962, p. 94, pl. 7, figs. 1–2.

*Anomalina* sp. 1 Proto Decima and Bolli, 1978, p. 789, pl. 5, figs. 14–15.

?*Quadrimorphina* sp. 2. Schnitker, 1979, pl. 6, figs. 15–17.

*Abyssamina quadrata* Schnitker and Tjalsma. –Tjalsma and Lohmann, 1983, p. 22, pl. 4, fig. 5; pl. 19, figs. 1–2. –Loeblich and Tappan, 1987, p. 626, pl. 703, figs. 1–3.

## Remarks

Type species of the genus *Abyssamina* Schnitker and Tjalsma, 1980.

*Abyssamina poagi* Schnitker and Tjalsma differs from *Abyssamina quadrata* Schnitker and Tjalsma by its longer, narrower chambers that tend to overlap in the umbilical region, its more strongly curved umbilical sutures, and shorter, wider aperture. The test of *Abyssamina poagi* is also narrower in edge view and more distinctly trochospiral (see van Morkhoven et al., 1986).

## Biogeography

Worldwide. Recorded from the North and South Atlantic and the Indian Ocean (Schnitker and Tjalsma, 1980; Tjalsma and Lohmann, 1983).

## Bathymetry

Lower bathyal to abyssal.

According to Schnitker and Tjalsma (1980), this species became progressively more restricted to greater depths so that by the Middle Eocene it only occurred at paleodepths below 3000 meters.

## Chronostratigraphy

Middle Paleocene to Middle Eocene.

# References


Hillebrandt, A. von, 1962, Das Paleozaen und seine Foraminiferen-fauna im Becken von Reichenhall und Salzburg. Abhandlungen, Bayerische Akademie der Wissenschaften, Mathematisch-Naturwissenschaftliche Klasse, N.F., no. 108, pp. 9–182.

Loeblich, A.R. Jr and Tappan, H., 1987, Foraminiferal Genera and Their Classification. New York: Van Nostrand Reinhold, 970 pp.

Proto Decima, F. and Bolli, H.M., 1978, Southeast Atlantic Leg 40 Paleogene benthic foraminifers. In Bolli, H.M., Ryan, W.B.F., et al., ed., Initial Reports of the Deep Sea Drilling Project, Washington (U.S. Government Printing Office), vol. 40, pp. 783–809.

Schnitker, D., 1979, Cenozoic deep water foraminifers, Bay of Biscay. In Montadert, L., Roberts, D.G., et al., ed., Initial Reports of the Deep Sea Drilling Project, vol. 48, pp. 377–413.

Schnitker, D. and Tjalsma, L.R.C., 1980, New genera and species of benthic foraminifera from Paleocene and Eocene deep-water deposits. Journal of Foraminiferal Research, vol. 10, no. 3, pp. 235–241.

Tjalsma, R.C. and Lohmann, G.P., 1983, Paleocene–Eocene bathyal and abyssal benthic foraminifera from the Atlantic Ocean. Micropaleontology, Special Publication, no. 4, pp. 1–90.

van Morkhoven, F.P.C.M., Berggren, W.A., Edwards, A.S., et al., 1986, Cenozoic cosmopolitan deep-water benthic foraminifera. Bulletin des Centres de Recherches Exploration-Production Elf-Aquitaine, Mémoire, vol. 11, pp. 1–421.

# *Alabamina creta*

## (Finlay), 1940

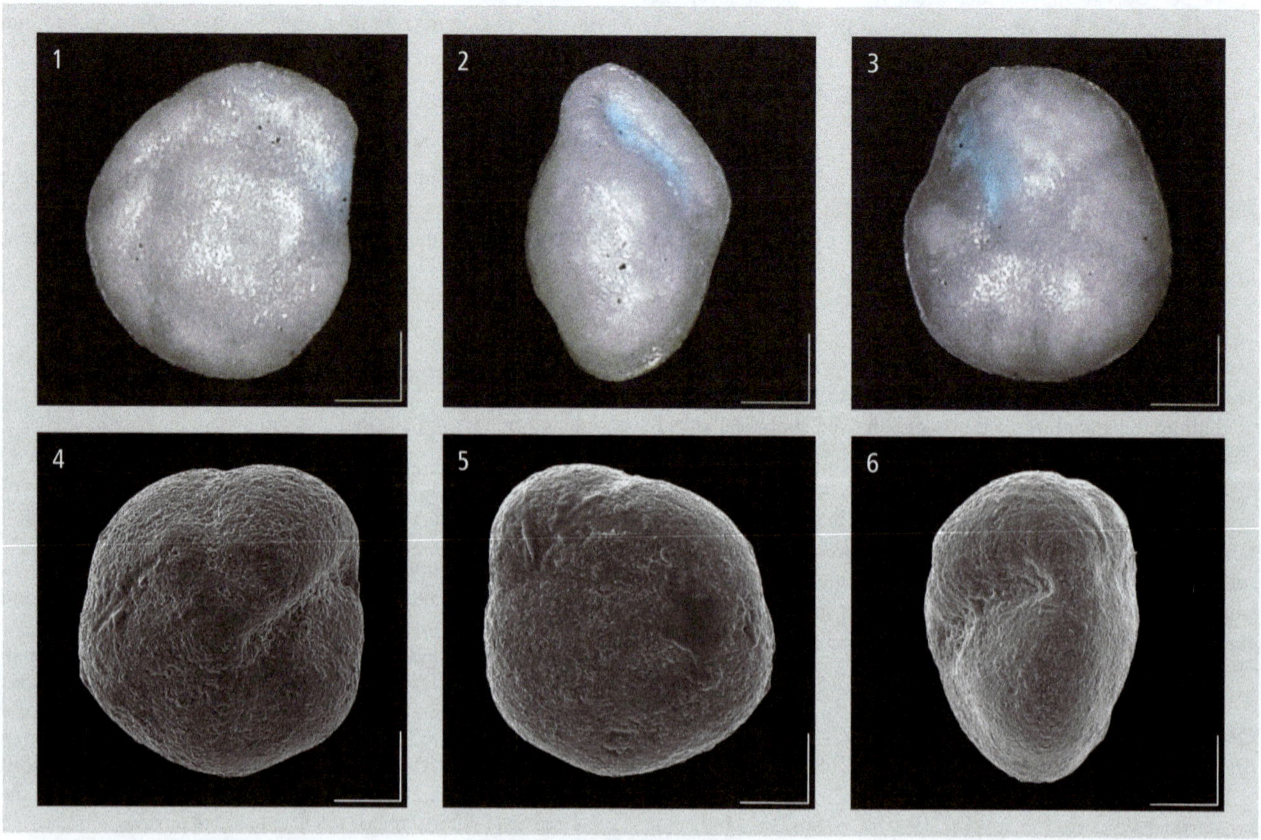

1. *Alabamina creta*. Northeast Georgia Rise, South Atlantic; ODP Leg 114, Hole 700B, Core 32, Section 4, 55–59 cm. Spiral view, NHMUK PM DI 0007, scale bar = 50 μm.

2. *Alabamina creta*. Northeast Georgia Rise, South Atlantic; ODP Leg 114, Hole 700B, Core 32, Section 4, 55–59 cm. Apertural view, NHMUK PM DI 0008, scale bar = 51 μm.

3. *Alabamina creta*. Northeast Georgia Rise, South Atlantic; ODP Leg 114, Hole 700B, Core 32, Section 4, 55–59 cm. Umbilical view, NHMUK PM DI 0009, scale bar = 52 μm.

4. *Alabamina creta*. Northeast Georgia Rise, South Atlantic; ODP Leg 114, Hole 700B, Core 32, Section 4, 55–59 cm. Umbilical view, NHMUK PM DI 0010, scale bar = 39 μm.

5. *Alabamina creta*. Northeast Georgia Rise, South Atlantic; ODP Leg 114, Hole 700B, Core 32, Section 4, 55–59 cm. Spiral view, NHMUK PM DI 0011, scale bar = 42 μm.

6. *Alabamina creta*. Northeast Georgia Rise, South Atlantic; ODP Leg 114, Hole 700B, Core 32, Section 4, 55–59 cm. Apertural view, NHMUK PM DI 0012, scale bar = 42 μm.

### Original Designation

*Pulvinulinella creta* Finlay, 1940, p. 463, pl. 66, figs. 187–192.

### Type Specimen

Deposited in the New Zealand Geological Survey Collection, Wellington, New Zealand. Cotypes deposited in the collections of the United States National Museum, Washington, DC, and in the Parr Collection at Melbourne, Australia.

### Type Level

Campanian, Whangai beds, marls in siliceous shale. Piripauan index.

### Type Locality

Locality 3250, mouth of Tuparoa Stream, Waipiro Survey District, New Zealand. Original figs. 188, 189 (holotype), 191, 192 from Locality 3250A. Figs. 187, 190 from Locality 5374, 148 chains, at 60° from Trig U, Mangaotoro Survey District. Other locality given: 5664 Mid-Wapara, North Canterbury, New Zealand.

### Description

Test forms a small, lenticular trochospire; pentagonal in outline and unequally biconvex in cross-section with an involute, convex, umbilical side, an evolute, flattened or slightly convex spiral side, and a rounded, imperforate periphery. The five inflated chambers present in the last whorl increase gradually in size, and are separated by curved and radial, flush sutures, on the umbilical side, and oblique sutures on the spiral side. Chamber walls are calcareous, finely perforate, and smooth. The primary aperture is an extraumbilical, interiomarginal slit bordered by a lip with the uppermost end extending into an inframarginal sulcus.

### Synonyms and Citations

*Eponides dorsoplana* Brotzen, 1940, p. 31, pl. 8, figs. 2.

*Alabamina solnasensis* Brotzen, 1948, p. 102, pl. 16, fig. 4a–c.

*Alabamina creta* (Finlay). –Tjalsma and Lohmann, 1983, p. 4, pl. 7, fig. 13. –Thomas, 1990, p. 589.

### Biogeography

Worldwide. Originally described from New Zealand. Recorded from the Caribbean and Falkland Plateau in the Atlantic by Tjalsma and Lohmann (1983) and from the Weddell Sea, Antarctica, by Thomas (1990).

### Bathymetry

Lower bathyal (Thomas, 1990).

### Chronostratigraphy

Late Cretaceous (Santonian) to Early Eocene. Late Cretaceous (Santonian to Campanian) (Finlay, 1940); Paleocene (Tjalsma and Lohmann, 1983); Late Cretaceous (Maastrichtian) to Paleocene (Thomas, 1990).

## References


Brotzen, F., 1940, Flintraennans och Trindelraennans geologi (Oeresund). (Die Geologie der Flint- und Trindelrinne (Oeresund)). Årsbok Sveriges Geologiska Undersökning, vol. 34, no. 5, pp. 1–33.

Brotzen, F., 1948, The Swedish Paleocene and its Foraminiferal Fauna. Årsbok Sveriges Geologiska Undersökning, vol. 42, no. 2, pp. 1–140.

Finlay, H.J., 1940, New Zealand Foraminifera; key species in stratigraphy – No. 4. Transactions of the Royal Society of New Zealand, vol. 69, no. 4, pp. 448–472.

Thomas, E., 1990, Late Cretaceous through Neogene benthic foraminifers. In Barker, P.F., Kennett, J.P., et al., ed., Scientific Results of the Proceedings of the Ocean Drilling Program, vol. 113, pp. 571–594.

Tjalsma, R.C. and Lohmann, G.P., 1983, Paleocene–Eocene bathyal and abyssal benthic foraminifera from the Atlantic Ocean. Micropaleontology, Special Publication, no. 4, pp. 1–90.

# *Alabamina dissonata*

## (Cushman and Renz), 1948

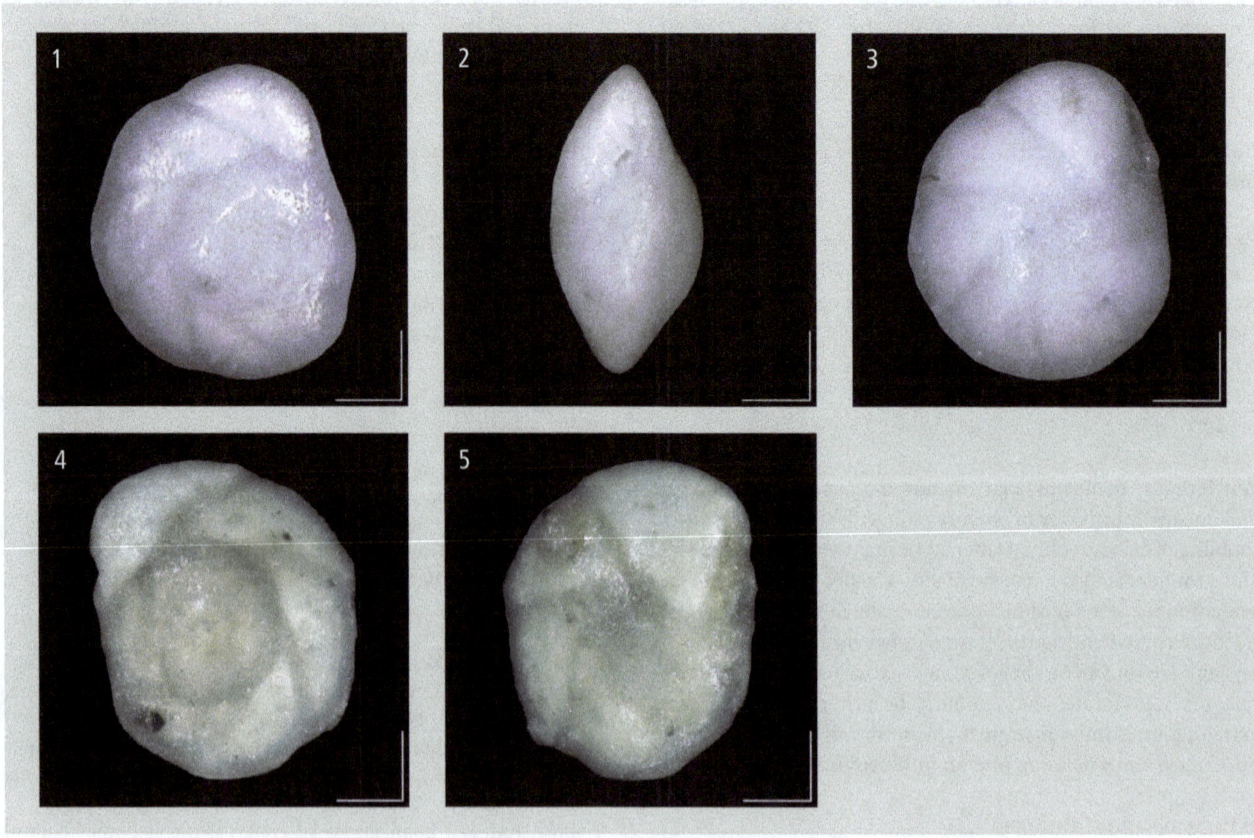


1. *Alabamina dissonata.* Islas Orcadas Rise, South Atlantic; ODP Leg 114, Hole 702B, Core 22, Section 2, 62–66 cm. Spiral view, NHMUK PM DI 0013, scale bar = 191 μm.

2. *Alabamina dissonata.* Islas Orcadas Rise, South Atlantic; ODP Leg 114, Hole 702B, Core 22, Section 2, 62–66 cm. Apertural view, NHMUK PM DI 0014, scale bar = 224 μm.

3. *Alabamina dissonata.* Islas Orcadas Rise, South Atlantic; ODP Leg 114, Hole 702B, Core 22, Section 2, 62–66 cm. Umbilical view, NHMUK PM DI 0015, scale bar = 231 μm.

4. *Alabamina dissonata.* Trinidad, West Indies; Navet Formation, Nariva River Marl, Nariva River, Eastern Central Range. Deposited in the collections of the United States National Museum, Washington, DC; CC 57353. Holotype. Spiral view, NHMUK PM DI 0016, scale bar = 224 μm.

5. *Alabamina dissonata.* Trinidad, West Indies; Navet Formation, Nariva River Marl, Nariva River, Eastern Central Range. Deposited in the collections of the United States National Museum, Washington, DC; CC 57353. Holotype. Oblique umbilical view, NHMUK PM DI 0017, scale bar = 224 μm.

## Original Designation

*Pulvinulinella atlantisae* Cushman var. *dissonata* Cushman and Renz, 1948, p. 35, pl. 7, figs. 11–12.

## Type Specimen

Holotype (CC 57353) deposited in the collections of the United States National Museum, Washington, DC.

## Type Level

Eocene, Navet Formation, Nariva River Marl Member; light gray, yellowish to khaki-weathering soft marl.

## Type Locality

Nariva River, Eastern Central Range, Trinidad; Trinidad Government Cadastral Coordinates: North 314,460 links, East 486,945 links.

## Description

Test forms a small, lenticular trochospire; nearly biconvex in cross-section with a subacute keeled periphery. The spiral side is evolute; the umbilical side involute with a closed umbilicus. The six crescent-shaped chambers in the last whorl increase gradually in size. Sutures are flush or slightly depressed, radial, and slightly curved on the umbilical side; tangential, limbate, and elevated on the spiral side. Chamber walls are calcareous and finely perforate except at the peripheral margin. The primary aperture is an interiomarginal–extraumbilical narrow slit with a thin lip extending into a deeply sunken inframarginal sulcus.

## Synonyms and Citations

*Alabamina dissonata* (Cushman and Renz). –Tjalsma and Lohmann, 1983, p. 22, pl. 17, fig. 3; pl. 20, fig. 5. –van Morkhoven et al., 1986, p. 283, pl. 94. –Bolli et al., 1994, p. 154, fig. 42.30.

## Remarks

Shallow-water ecophenotypes are smaller, have smaller keel, and lack raised sutures (van Morkhoven et al., 1986).

*Alabamina dissonata* (Cushman and Renz) differs from *Pulvinulinella atlantisae* Cushman by the keeled rather than just subacute periphery.

## Biogeography

Worldwide. Recorded in Barbados, Trinidad, the Caribbean, the North and South Atlantic, the Pacific, and Europe (van Morkhoven, 1981; Tjalsma and Lohmann, 1983; van Morkhoven et al., 1986; Bolli et al., 1994).

## Bathymetry

Primarily bathyal.

## Chronostratigraphy

Early Eocene (Ypresian) to Latest Eocene (Priabonian).

# References


Bolli, H.M., Beckmann, J.P., and Saunders, J.B., 1994, Benthic Foraminiferal Biostratigraphy of the South Caribbean Region. Cambridge: Cambridge University Press, 408 pp.

Cushman, J.A. and Renz, H.H., 1948, Eocene foraminifera of the Navet and Hospital Hill formations of Trinidad, BWI. Special Publications of the Cushman Laboratory for Foraminiferal Research, vol. 24, pp. 1–42.

Tjalsma, R.C. and Lohmann, G.P., 1983, Paleocene–Eocene bathyal and abyssal benthic foraminifera from the Atlantic Ocean. Micropaleontology, Special Publication, no. 4, pp. 1–90.

van Morkhoven, F.P.C.M., 1981, Cosmopolitan Tertiary bathyal benthic foraminifera. Transactions of the Gulf Coast Association of Geological Societies, Supplement 31, 445 abstracts with accompanying range chart.

van Morkhoven, F.P.C.M., Berggren, W.A., Edwards, A.S., et al., 1986, Cenozoic cosmopolitan deep-water benthic foraminifera. Bulletin des Centres de Recherches Exploration-Production Elf-Aquitaine, Mémoire, vol. 11, pp. 1–421.

# *Ammoanita ingerlisae*

## Gradstein and Kaminski, 1997

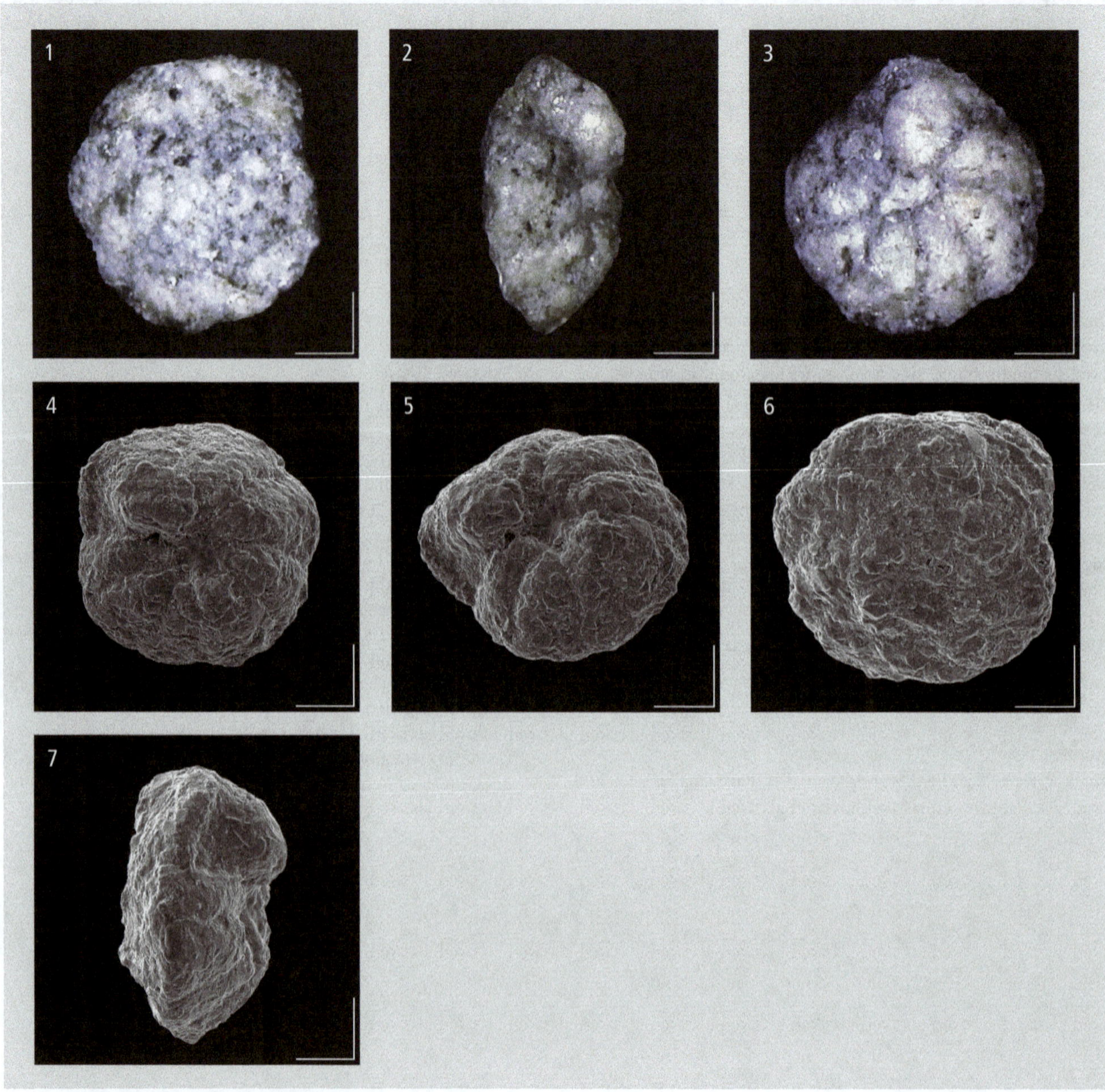

*Please see opposite for captions*

## Original Designation

*Ammoanita ingerlisae* Gradstein and Kaminski, 1997, p. 220, fig. 6, fig. 7 (1–3).

## Type Specimen

Holotype (NHMUK PF62944 Conoco 211/19-1, 5850 feet) housed in the micropaleontology collections of The Natural History Museum, London, UK.

## Type Level

Upper Paleocene green clays.

## Type Locality

Conoco Well 211/19-1, between 5450 and 5850 feet, Viking Graben.

## Description

Test forms a trochospire comprised of two to four whorls, with a circular to slightly lobulate outline, a flattened to slightly convex spiral side, a low, convex umbilical side with a narrow umbilical depression, and an acute periphery. The numerous chambers (six to nine, eight on average) in the last whorl increase slowly in size, and are separated by slightly depressed or flush, limbate sutures. Sutures curve slightly backwards on the umbilical side and are tangential on the spiral side. The test wall is finely agglutinated with scattered coarser grains, often well silicified. The primary aperture is an extraumbilical slit at the base of the final chamber, rarely visible.

## Synonyms (following Kaminski and Gradstein, 2005) and Citations

*Trochammina* aff. *albertensis* Wickenden. –Gradstein and Berggren, 1981, pl. 9, figs. 3–8 (not figs. 1–2).

*Trochammina (Ammoanita) ruthvenmurrayi* Cushman and Renz. –Charnock and Jones, 1990, pl. 22, figs. 2d–f.

*Trochammina (Insculptarenula) subvesicularis* Hanzlikova. –Charnock and Jones, 1990, pl. 10, figs. 10–12.

*Trochammina* cf. *subvesicularis* Hanzlikova. –Gradstein et al., 1994, pl. 1, fig. 4 (not fig. 5).

*Ammoanita ingerlisae* Gradstein and Kaminski. –Kaminski and Gradstein, 2005, pp. 452–455, fig. 110, pl. 110.

## Remarks

The species *Ammoanita rosea* Seiglie and Baker, *Ammoanita ingerlisae* Gradstein and Kaminski, and *Ammoanita ruthvenmurrayi* Cushman and Renz are morphologically close, and have overlapping stratigraphic ranges (see Remarks in Gradstein and Kaminski, 1997). Some authors consider *Ammoanita ruthvenmurrayi* to be the microspheric stage and *Ammoanita ingerlisae* to be the megalospheric stage of the same species.

## Biogeography

Reported from the North Sea, offshore Norway, and Trinidad (Lizard Springs Formation) by Gradstein and Kaminski (1997) and as *Trochammina (Ammoanita) ruthvenmurrayi* from the central North Sea by Charnock and Jones (1990).

## Bathymetry

Bathyal and abyssal.

## Chronostratigraphy

Middle to Late Paleocene. Typical in the Late Paleocene (Biozone NSR2A) of the North Sea and offshore Norway (Gradstein and Kaminski, 1997).

## References


Charnock, M.A. and Jones, R.W., 1990, Agglutinated foraminifera from the Palaeogene of the North Sea. In Hemleben, C., Kaminski, M.A., Kuhnt, W., and Scott, D.B., ed., Paleoecology, Biostratigraphy, Paleoceanography and Taxonomy of Agglutinated Foraminifera, NATO ASI Series, C: Mathematical and Physical Sciences. Dordrecht: Kluwer Academic, pp. 139–244.

Gradstein, F.M. and Berggren, W.A., 1981, Flysch-type agglutinated foraminifera and the Maestrichtian to Paleogene history of the Labrador and North Seas. Marine Micropaleontology, vol. 6, pp. 448–472.

Gradstein, F.M. and Kaminski, M.A., 1997, New species of Paleogene deep-water agglutinated foraminifera from the North Sea and Norwegian Sea. Annales Societatis Geologorum Poloniae, vol. 67, pp. 217–229.

Gradstein, F.M., Kaminski, M.A., Berggren, W.A., and D'Iorio, M.A., 1994, Cenozoic biostratigraphy of the Central North Sea and Labrador Shelf. Micropaleontology, vol. 40 (Supplement), pp. 1–152.

Kaminski, M.A. and Gradstein, F.M., 2005, Atlas of Palaeogene Cosmopolitan Deep-Water Agglutinated Foraminifera. Grzybowski Foundation Special Publication, no. 10, 547 + vii pp.


1. *Ammoanita ingerlisae*. North Sea; Conoco Well 211/19-1, depth: 5850 feet. Deposited in the micropaleontological collections of The Natural History Museum, London. NHMUK PF62944. Holotype. Spiral view, NHMUK PM DI 0018, scale bar = 61 μm.

2. *Ammoanita ingerlisae*. North Sea; Conoco Well 211/19-1, depth: 5850 feet. Deposited in the micropaleontological collections of The Natural History Museum, London. NHMUK PF62944. Holotype. Apertural view, NHMUK PM DI 0019, scale bar = 64 μm.

3. *Ammoanita ingerlisae*. North Sea; Conoco Well 211/19-1, depth: 5850 feet. Deposited in the micropaleontological collections of The Natural History Museum, London. NHMUK PF62944. Holotype. Umbilical view, NHMUK PM DI 0020, scale bar = 67 μm.

4. *Ammoanita ingerlisae*. Conoco Well 211/19-1, depth: 5850 feet, North Sea. Deposited in the micropaleontological collections of The Natural History Museum, London; NHMUK PF62944. Holotype. Umbilical view, NHMUK PM DI 0021, scale bar = 62 μm.

5. *Ammoanita ingerlisae*. Conoco Well 211/19-1, depth: 5850 feet, North Sea. Deposited in the micropaleontological collections of The Natural History Museum, London; NHMUK PF62944. Holotype. Oblique umbilical view, NHMUK PM DI 0022, scale bar = 61 μm.

6. *Ammoanita ingerlisae*. Conoco Well 211/19-1, depth: 5850 feet, North Sea. Deposited in the micropaleontological collections of The Natural History Museum, London; NHMUK PF62944. Holotype. Spiral view, NHMUK PM DI 0023, scale bar = 60 μm.

7. *Ammoanita ingerlisae*. Conoco Well 211/19-1, depth: 5850 feet, North Sea. Deposited in the micropaleontological collections of The Natural History Museum, London; NHMUK PF62944. Holotype. Apertural view, NHMUK PM DI 0024, scale bar = 57 μm.

# *Ammobaculites agglutinans*

## (d'Orbigny), 1846

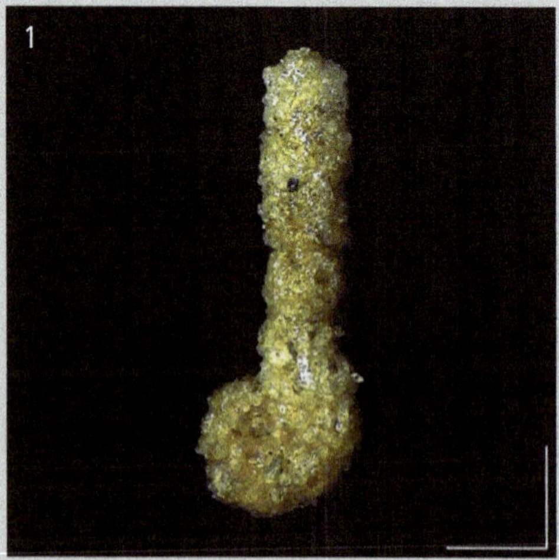

1. *Ammobaculites agglutinans.* "Albatross" Station North Atlantic; D 2043, latitude 39° 49′ 0″ N, longitude 68° 28′ 20″ W, 1467 fathoms. Deposited in the collections of the United States National Museum, Washington, DC; USNM 10499. Lateral view, NHMUK PM DI 0025, scale bar = 391 μm.

### Original Designation

*Spirolina agglutinans* d'Orbigny, 1846, p. 137, pl. 7, figs. 10–12.

### Type Specimen

Lectotype designated by Loeblich and Tappan (1964) and deposited in the collections of the Laboratoire de Paléontologie, Muséum d'Histoire Naturelle, Paris, France. The origin of the specimen is unclear. Papp and Schmid (1985) uncovered and re-illustrated d'Orbigny's original specimens in the micropaleontological collections of the Geologische Bundesanstalt in Vienna, Austria, and designated a lectotype, which is regarded as the true lectotype by Kaminski and Gradstein (2005). Metatype specimens from the d'Orbigny Collection are housed in the micropaleontological collections of the US Natural History Museum, Washington, DC.

### Type Level

Miocene (Badenian) blue marl.

### Type Locality

Clay pits in the town of Baden, near Vienna, Vienna Basin, Austria.

### Description

Test forms an initial planispire, later becoming a large, elongate, rectilinear or arched uniserial series usually consisting of three or more chambers. A small, depressed umbilicus is visible in the center of the planispire. Chambers are moderately inflated and separated by slightly depressed sutures. The test wall is non-alveolar and coarsely agglutinated with non-calcareous cement. The primary aperture is a terminal, round opening.

### Synonyms and Citations

*Haplophragmium agglutinans* (d'Orbigny). –Brady, 1884, pl. 32, figs. 19–20; 24–26.

*Ammobaculites agglutinans* (d'Orbigny). –Bartenstein, 1952, p. 318, pl. 1, fig. 1; pl. 2, figs. 10–16. –Thomas, 1985, p. 78, pl. 3, fig. 3. –Schröder, 1986, p. 50, pl. 21, figs. 1–4. –Loeblich and Tappan, 1987, p. 74, pl. 58, figs. 3–4. –Kuhnt and Kaminski, 1990, p. 461, pl. 3, figs. e–f. –Kaminski and Kuhnt, 1991, pp. 221–230, figs. 2–3. –Jones, 1994, p. 39, pl. 32, figs. 19–20; 24–26.

*Ammobaculites agglutinans* (d'Orbigny), 1846 emend. Bartenstein, 1952. –Kaminski and Gradstein, 2005, pp. 324–328, fig. 70, pl. 70.

### Remarks

Type species of the genus *Ammobaculites* Cushman, 1910.

Originally described from the Middle Miocene of the Vienna Basin, *Ammobaculites agglutinans* (d'Orbigny) has subsequently been reported from rocks ranging in age from Jurassic to Recent. Bartenstein (1952) considered that Jurassic and Cretaceous specimens referred to this species differed in their chamber arrangement, wall composition, and planispire formation. Bartenstein (1952) also noted that the species is dimorphic: the planispire may be flattened with six distinct chambers or have fewer (four or five), less distinct chambers. See also discussion of morphologic variability in Kaminski and Kuhnt (1991) and Kaminski and Gradstein (2005).

### Biogeography

Worldwide. See also Observed Occurrences in Kaminski and Gradstein (2005).

### Bathymetry

Middle neritic to abyssal (Kaminski and Kuhnt, 1991). Outer neritic to abyssal (Kaminski and Gradstein, 2005). Recorded by Brady (1884) from nine "Challenger" stations in the North Atlantic (from 530 to 2750 fathoms.), from two stations in the South Atlantic (from 2200 and 2900 fathoms), and from 12 stations in the Pacific Atlantic (from 2 to 3125 fathoms.). Reported by Cushman (1920) from 30 "Albatross" stations in the North Atlantic (from 27 to 2369 fathoms).

### Chronostratigraphy

Late Cretaceous (Campanian) to Recent. Kaminski and Gradstein (2005) reported that the distribution of *Ammobaculites agglutinans* in Cretaceous–Paleogene deep-water sediments is more restricted than in the modern ocean.

## References


Bartenstein, H., 1952, Taxonomische Bemerkungen zu den Ammobaculites, Haplophragmium, Lituola und verwandten Gattungen. (For.) Senckenbergiana, vol. 33, pp. 313–342.

Brady, H.B., 1884, Report on the foraminifera dredged by H.M.S. Challenger during the years 1873–1876. In Brady, H.B., ed., Report on the Scientific Results of the Voyage of the H.M.S. Challenger during the years 1873–1876. Zoology, vol. 9, pp. 1–814.

Cushman, J.A., 1920, The foraminifera of the Atlantic Ocean. Part 2. Lituolidae. Bulletin United States National Museum, vol. 104, no. 2, pp. 1–111.

d'Orbigny, A.D., 1846, Foraminifères fossiles du Bassin Tertiaire de Vienne (Autriche). Paris: Gide et Compe, xxxvii + 312 pp.

Jones, R.W., 1994, The Challenger Foraminifera. Oxford: Oxford University Press, 149 pp.

Kaminski, M.A. and Gradstein, F.M., 2005, Atlas of Palaeogene Cosmopolitan Deep-Water Agglutinated Foraminifera. Grzybowski Foundation Special Publication, no. 10, 547 + vii pp.

Kaminski, M.A. and Kuhnt, W., 1991, Depth-related shape variation in Ammobaculites agglutinans (d'Orbigny). Annales Societatis Geologorum Poloniae, vol. 61, pp. 221–230.

Kuhnt, W. and Kaminski, M.A., 1990, Paleoecology of Late Cretaceous to Paleocene deep-water agglutinated foraminifera from the North Atlantic and western Tethys. In Hemleben, C., Kaminski, M.A., Kuhnt, W., and Scott, D.B., ed., Paleoecology, Biostratigraphy, Paleoceanography and Taxonomy of Agglutinated Foraminifera. NATO ASI Series, Series C: Mathematical and Physical Sciences. Dordrecht: Kluwer Academic, vol. 327, pp. 433–505.

Loeblich, A.R. Jr and Tappan, H., 1964, Sarcodina, chiefly "Thecamoebians" and Foraminiferida. In Moore, R.C., ed., Treatise on Invertebrate Paleontology, Part C, Protista 2 (2 vols). Lawrence, KS: Geological Society of America and University of Kansas Press, pp. 1–900.

Loeblich, A.R. Jr and Tappan, H., 1987, Foraminiferal Genera and Their Classification. New York: Van Nostrand Reinhold, 970 pp.

Papp, A. and Schmid, M.E., 1985, Die fossilen Foraminiferen des tertiaeren Beckens von Wien: Revision der Monographie Alcide d'Orbigny (1846). Abhandlungen der Geologischen Bundesanstalt, vol. 37, pp. 1–311.

Schröder, C.J., 1986, Deep-Water Arenaceous Foraminifera in the Northwest Atlantic Ocean. Canadian Technical Report of Hydrography and Ocean Sciences No. 71. Atlantic Geoscience Centre, Bedford Institute of Oceanography, 191 pp.

Thomas, E., 1985, Late Eocene to Recent deep-sea benthic foraminifers from the central equatorial Pacific Ocean. In Mayer, L., Theyer, F., et al., ed., Initial Reports of the Deep Sea Drilling Project, Washington (U.S. Government Printing Office), vol. 85, pp. 655–694.

# *Ammobaculites jarvisi*

## Cushman and Renz, 1946

1. *Ammobaculites jarvisi*. Trinidad, Lizards Springs Formation in Ravine Ampelu. Deposited in the collections of the United States National Museum, Washington, DC; CC 46514. Holotype, Lateral view, NHMUK PM DI 0026, scale bar = 345 μm.

### Original Designation

*Ammobaculites jarvisi* Cushman and Renz, 1946, p. 19, pl. 2, figs. 8–9.

### Type Specimen

Holotype (CC 46514) and paratypes (CC 46515 and 46516) deposited in the collections of the United States National Museum, Washington, DC.

### Type Level

Upper Paleocene, Lizards Springs Formation, Trinidad.

### Type Locality

Ravine of the Ampelu River, near Lizards Springs, Trinidad.

### Description

Test forms a relatively small initial planispire followed by a large, elongate, broad, rectilinear or slightly arched uniserial series, usually with four chambers. Chambers are moderately inflated and separated by depressed sutures. Chamber walls are non-alveolar, well cemented, finely to coarsely agglutinated, and occasionally include carbonate grains. The primary aperture is a terminal round opening.

### Synonyms and Citations

*Ammobaculites jarvisi* Cushman and Renz. –Kaminski et al., 1988, p.188, pl. 4, fig. 4. –Kuhnt and Kaminski, 1990, p. 462, pl. 3, fig. d. –Kaminski and Gradstein, 2005, pp. 329–331, fig. 71, pl. 71.

### Remarks

*Ammobaculites jarvisi* Cushman and Renz may agglutinate carbonate material, and is characteristic of marly slope environments (Kuhnt and Kaminski, 1990). The holotype has an attached test of *Ammolagena clavata*.

### Biogeography

Worldwide. North and South Atlantic, Caribbean, Pacific. See also Observed Occurrences in Kaminski and Gradstein (2005).

### Bathymetry

Common in upper to middle bathyal deposits in Trinidad, the North African paleomargin, and the outer neritic updip wells on the Labrador Margin (Kuhnt and Kaminski, 1990).

### Chronostratigraphy

Late Cretaceous (Maastrichtian) to Early Eocene.

## References


Cushman, J.A. and Renz, H.H., 1946, The foraminiferal fauna of the Lizard Springs Formation of Trinidad, British West Indies. Special Publications of the Cushman Laboratory for Foraminiferal Research, vol. 18, pp. 1–48.

Kaminski, M.A. and Gradstein, F.M., 2005, Atlas of Palaeogene Cosmopolitan Deep-Water Agglutinated Foraminifera. Grzybowski Foundation Special Publication, no. 10, 547 + vii pp.

Kaminski, M.A., Gradstein, F.M., Berggren, W.A., Geroch, S., and Beckmann, J.P., 1988, Agglutinated foraminiferal assemblages from Trinidad: taxonomy, stratigraphy and paleobathymetry. In Gradstein, F.M. and Rögl, F., ed., Second International Workshop on Agglutinated Foraminifera, Vienna 1986, Abhandlungen der Geologischen Bundesanstalt, vol. 41, pp. 155–228.

Kuhnt, W. and Kaminski, M.A., 1990, Paleoecology of Late Cretaceous to Paleocene deep-water agglutinated foraminifera from the North Atlantic and western Tethys. In Hemleben, C., Kaminski, M.A., Kuhnt, W., and Scott, D.B., ed., Paleoecology, Biostratigraphy, Paleoceanography and Taxonomy of Agglutinated Foraminifera. NATO ASI Series, Series C: Mathematical and Physical Sciences. Dordrecht: Kluwer Academic, vol. 327, pp. 433–505.

# *Ammobaculoides carpathicus*

## Geroch, 1966

1. *Ammobaculoides carpathicus.* Poland, Subsilesian Unit, Stepina near Strzyzow, Sample 115/57 (black shales). Deposited in the collections of the United States National Museum, Washington, DC; USNM 460972. Paratype. Lateral view. NHMUK PM DI 0027, scale bar = 131 μm.

### Original Designation

*Ammobaculoides carpathicus* Geroch, 1966, p. 479, tf. 13, figs. 13–22 (fig. 14: holotype).

### Type Specimen

Housed in the Geroch Collection, Institute of Geological Sciences, Jagiellonian University, Kraków, Poland. Paratype (USNM 460972) deposited in the collections of the United States National Museum, Washington, DC.

### Type Level

Lower Verovice Shales (?Barremian), in black shale. Also occurs in the upper Cieszyn Shale (Valanginian), and the Grodziszcze Shale (Hauterivian).

### Type Locality

Stepina near Strzyzow, Silesia, Poland. Also found at Sulkowice-Ubionka; Lipnik, near Bielsko; and Leszna Gorna, near Goleszow.

### Description

Test forms an initial planispire followed by a short biserial series and (finally) elongate uniserial series with numerous chambers. The moderately inflated chambers are separated by depressed sutures in the uniserial part. This uniserial portion usually forms half to three-quarters of the total length of the test. Uniserial chambers are low and rounded in cross-section, increasing gradually in size initially and then becoming almost constant in size in the (presumptive) adult stage. The test wall is medium to coarsely agglutinated with organic cement. The primary aperture is a terminal opening in the uniserial part.

### Synonyms and Citations

*Ammobaculoides carpathicus* Geroch. –Geroch and Nowak, 1984, pl. 1, figs. 15–16; pl. 6, figs. 8–11.

### Remarks

One of the few stratigraphic indicators for Lower Cretaceous sediments from below the calcite compensation depth (CCD). Geroch (1966) noted that immersed specimens reveal interiors of chambers, cup-like in the uniserial portion and joined by tubes.

### Biogeography

First described from the Polish Carpathians by Geroch (1966).

Kaminski et al. (1992) recorded specimens from ODP Site 765 in the Argo Abyssal Plain off northwestern Australia that were similar in size, outline, and shape of chambers. As the wall structure of the Site 765 specimens was not determined, these authors adopted the designation *Ammobaculoides* cf. *carpathicus*.

### Bathymetry

Bathyal to abyssal. Characteristic of Early Cretaceous sub-CCD environments.

### Chronostratigraphy

Early Cretaceous (Berriasian to Aptian) in the Polish Carpathians (Geroch and Nowak, 1984).

## References

Geroch, S., 1966, Lower Cretaceous small foraminifera of the Silesian series, Polish Carpathians. Rocznik Polskiego Towarzystwa Geologicznego, vol. 36, pp. 413–480.

Geroch, S. and Nowak, W., 1984, Proposal of zonation for the late Tithonian–late Eocene, based upon arenaceous foraminifera from the outer Carpathians, Poland. In Oertli, H., ed., Benthos '83, 2nd International Symposium on Benthic Foraminifera, Pau (France), April 11–15, 1983. Elf Aquitaine, ESSO REP and TOTAL CFP, Pau and Bordeaux, pp. 225–239.

Kaminski, M.A., Gradstein, F. and Geroch, S., 1992, Uppermost Jurassic to Lower Cretaceous deep-water benthic foraminiferal assemblages from Site 765 on the Argo Abyssal Plain. In Gradstein, F., Ludden, J.N., et al., ed., Scientific Results of the Proceedings of the Ocean Drilling Program, vol. 123, pp. 239–269.

# *Ammodiscus infimus*

## Franke, 1936

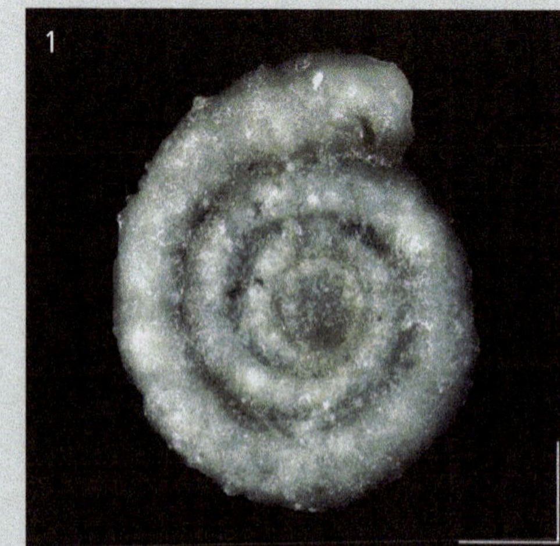


1. *Ammodiscus infimus*. Verovice Shales, Silesian Unit, Wiltowisko, East Myślenice, Poland, MK 260. Umbilical view, NHMUK PM DI 0028, scale bar = 149 μm.

### Original Designation

*Ammodiscus infimus* (Strickland) (non Strickland) Franke, 1936, p. 15, pl. 1, fig. 14.

### Type Specimen

Not designated.

### Type Level

Jurassic, Lower to Middle Lias.

### Type Locality

Not designated. Localities given by Franke (1936): Castle by Metz, France, and Wennigstedt by Mattierzoll, Germany.

### Description

Test forms a large, robust, tubular form with globular proloculus and planispirally or irregularly coiled second chamber. Chamber walls are thick and coarsely agglutinated. The primary aperture is a terminal and round opening at the end of the tube.

### Synonyms and Citations

*Ammodiscus giganteus* Myatliuk, 1939, p. 39, pl. 2, fig. 21.

*Ammodiscus infimus* Franke. –Geroch, 1966, p. 437, figs. 8.13–14. –Geroch and Nowak, 1984, pl. 1, fig. 11; pl. 5, fig. 13. –Weidich, 1990, p. 82, pl. 34, figs. 1–3. –Reicherter et al., 1994, pl. 7A, figs. G–H. –Holbourn and Kaminski, 1997, p. 35, pl. 4, figs. 9–11.

### Remarks

The second chamber may show a tendency to uncoil.

### Biogeography

Worldwide. Commonly reported from both the Tethyan and Boreal Lower Cretaceous.

### Bathymetry

Bathyal to abyssal.

### Chronostratigraphy

Early Jurassic to Late Cretaceous. Cretaceous (Barremian to Maastrichtian) in the Polish Carpathians (Geroch and Nowak, 1984). Jurassic to Late Cretaceous (Weidich, 1990).

## References

Franke, A., 1936, Die Foraminiferen des deutschen Lias Preuss. Abhandlungen der Preussischen Geologischen Landesanstalt, Neue Folge, vol. 169, pp. 1–138.

Geroch, S., 1966, Lower Cretaceous small foraminifera of the Silesian series, Polish Carpathians. Rocznik Polskiego Towarzystwa Geologicznego, vol. 36, pp. 413–480.

Geroch, S. and Nowak, W., 1984, Proposal of zonation for the late Tithonian–late Eocene, based upon arenaceous foraminifera from the outer Carpathians, Poland. In Oertli, H., ed., Benthos '83, 2nd International Symposium on Benthic Foraminifera, Pau (France), April 11–15, 1983. Elf Aquitaine, ESSO REP and TOTAL CFP, Pau and Bordeaux, pp. 225–239.

Holbourn, A.E.L. and Kaminski, M.A., 1997, Cretaceous deep-water benthic foraminifera of the Indian Ocean. Grzybowski Foundation Special Publication, no. 4, pp. 1–175.

Mjatliuk, E.V., 1939, Foraminifera of the Upper Jurassic and Lower Cretaceous of the region of the Middle Volga River and the Great Syrte (in Russian). Trudy Vsesoyuznogo Nauchno-Issledovatel'skogo Geologorazvedochnogo Instituta VNIGRI, vol. 120, pp. 1–76.

Reicherter, K., Pletsch, T., Kuhnt, W., Manthey, J., Homeier, G., Wiedmann, J., and Thurow, J., 1994, Mid-Cretaceous paleogeography and paleoceanography of the Betic Seaway (Betic Cordillera, Spain). Palaeogeography, Palaeoclimatology, Palaeoecology, vol. 107, pp. 1–33.

Weidich, K., 1990, Die kalkalpine Unterkreide und ihre Foraminiferenfauna. Zitteliana, vol. 17, pp. 1–312.

# *Ammodiscus latus*

## Grzybowski, 1898

1. *Ammodiscus latus*. Silesian Unit, Green shales, Karaś Stream 48. Biecz near Gorlice, Poland. Umbilical view, NHMUK PM DI 0029, scale bar = 189 μm.

### Original Designation

*Ammodiscus latus* Grzybowski, 1898, p. 282, pl. 10, figs. 27–28.

### Type Specimen

Depository not given. Lectotype from red clay outcropping in the right bank of the Wislok River at Kroscienko Nizne, designated and illustrated by Kaminski et al. (1993). Lectotype and syntypic material (No. UJ-132-P, 1/38) deposited in the Grzybowski Collection at the Jagiellonian University, Kraków, Poland.

### Type Level

Uppermost Eocene; red and gray clay.

### Type Locality

Not designated. Localities given: Hanover-Galicia Company wells nos 33 and 34, at Potok; Duniecki well no. 1 at Toroszówka; gray clay (rare) and red clay (common) in the bed of the Wislok River at Kroscienko Nizne; all in the vicinity of Krosno, Poland.

### Description

Test forms a robust, thick, evolute, planispiral coil increasing in size very slowly, with a distinct coil suture. In megalospheric individuals, the test consists of two to three whorls with an umbilicus that is depressed on one or both sides. In microspheric individuals, the test comprises up to five whorls and the umbilicus is usually not depressed. Some specimens display irregular coiling in the last whorl or exhibit a tendency to uncoil. When present, the uncoiled part is short, and bends away from the previous direction of coiling (Kaminski and Gradstein, 2005). Chamber walls are medium to coarsely agglutinated with organic cement. The primary aperture is simple, at the end of the open tube.

### Synonyms and Citations

*Ammodiscus* cf. *tenuis* Brady. –Rzehak, A., 1895, p. 214, pl. 7, fig. 5.

*Ammodiscus umbonatus* Grzybowski, 1898, p. 283, pl. 10, figs. 29–30.

*Lituotuba eocenica* Cushman and Renz, 1948, p. 7, pl. 1, figs. 20–21.

*Ammodiscus latus* Grzybowski. –Morgiel and Olszewska, 1981, p. 8, pl. 1, figs. 8–9. –Geroch and Nowak, 1984, pl. 1, figs. 12–13; pl. 5, figs. 9–1. –Kaminski and Geroch, 1993, p. 254, pl. 5, figs. 4a–6. –Bolli et al., 1994, p. 68, fig. 18.25. –Gradstein et al., 1994, pl. 9, figs. 23–25. –Kaminski and Gradstein, 2005, pp. 150–154, fig. 16, pl. 16.

### Remarks

Extremely variable in size (see Remarks in Kaminski and Gradstein, 2005).

### Biogeography

Worldwide. See also Observed Occurrences in Kaminski and Gradstein (2005).

### Bathymetry

Bathyal to abyssal. Common in flysch-type and sub-CCD assemblages. The size of tests is probably related to the available export organic flux. For instance, specimens from green organic-rich clays of Labrador may exceed 2 mm in diameter, whereas specimens from abyssal ODP Site 647 are smaller, between 600 and 800 μm in diameter (Kaminski et al., 1989).

### Chronostratigraphy

Middle Eocene to Middle Miocene. The first common occurrence of *Ammodiscus latus* Grzybowski marks the base of the *Ammodiscus latus* Biozone of Geroch and Nowak (1984). Kaminski and Gradstein (2005) reported its range as Middle Eocene to mid-Oligocene in the Tethys and Atlantic and as Oligocene to Middle Miocene in the Arctic.

## References


Bolli, H.M., Beckmann, J.P., and Saunders, J.B., 1994, Benthic Foraminiferal Biostratigraphy of the South Caribbean Region. Cambridge: Cambridge University Press, 408 pp.

Cushman, J.A. and Renz, H.H., 1948, Eocene foraminifera of the Navet and Hospital Hill formations of Trinidad, BWI. Special Publications of the Cushman Laboratory for Foraminiferal Research, vol. 24, pp. 1–42.

Geroch, S. and Nowak, W., 1984, Proposal of zonation for the late Tithonian–late Eocene, based upon arenaceous foraminifera from the outer Carpathians, Poland. In Oertli, H., ed., Benthos '83, 2nd International Symposium on Benthic Foraminifera, Pau (France), April 11–15, 1983. Elf Aquitaine, ESSO REP and TOTAL CFP, Pau et Bordeaux, pp. 225–239.

Gradstein, F.M., Kaminski, M.A., Berggren, W.A., and D'Iorio, M.A., 1994, Cenozoic biostratigraphy of the Central North Sea and Labrador Shelf. Micropaleontology, vol. 40 (Supplement), pp. 1–152.

Grzybowski, J., 1898, Otwornice pokładów naftonosnych okolicy Krosna. Rozprawy Wydziału Matematyczno-Przyrodniczego, Akademia Umiejętności w Krakowie, serya 2, vol. 33, pp. 257–305.

Kaminski, M.A. and Geroch, S., 1993, A revision of foraminiferal species in the Grzybowski Collection. In Kaminski, M.A., Geroch, S., and Kaminski, D.G., ed., The Origins of Applied Micropalaeontology: The School of Jósef Grzybowski (with a taxonomic revision of the Grzyboswski Collection). Grzybowski Foundation Special Publication, no. 1, pp. 239–330.

Kaminski, M.A., Geroch, S., and Kaminski, D.G., 1993, The Origins of Applied Micropalaeontology: The School of Jósef Grzybowski (with a taxonomic revision of the Grzyboswski Collection). Grzybowski Foundation Special Publication, no. 1, pp. 1–336.

Kaminski, M.A. and Gradstein, F.M., 2005, Atlas of Palaeogene Cosmopolitan Deep-Water Agglutinated Foraminifera. Grzybowski Foundation Special Publication, no. 10, 547 + vii pp.

Kaminski, M.A., Gradstein, F.M., and Berggren, W.A., 1989, Paleogene benthic foraminifer biostratigraphy and paleoecology at site 647, southern Labrador Sea. In Srivastava, S.P., Arthur, M.A., Clement, B., et al., ed., Scientific Results of the Proceedings of the Ocean Drilling Program, vol. 105, pp. 705–730.

Morgiel, J. and Olszewska, B., 1981, Biostratigraphy of the Polish External Carpathians based on agglutinated foraminifera. Micropaleontology, vol. 27, no. 1, pp. 1–30.

Rzehak, A., 1895, Ueber einige merkwürdige Foraminiferen aus österreichischen Tertiär. Annalen Naturhistorisches Hofmuseum, Wien, vol. 10, pp. 213–230.

# *Ammodiscus pennyi*

## Cushman and Jarvis, 1928

1. *Ammodiscus pennyi*. Lizard Springs well, near Guayaguayare, southeastern Trinidad, British West Indies, depth 720 feet. Collected by W. Jarvis. Deposited in the collections of the United States National Museum, Washington, DC; CC 9670. Holotype. Umbilical view, NHMUK PM DI 0030, scale bar = 230 μm.

### Original Designation

*Ammodiscus pennyi* Cushman and Jarvis, 1928, p. 87, pl. 12, figs. 4–5.

### Type Specimen

Holotype (CC 9670) deposited in the collections of the United States National Museum, Washington, DC.

### Type Level

Upper Paleocene, Lizard Springs Formation.

### Type Locality

Well at Lizard Springs at a depth of 720 feet, Trinidad.

### Description

Test is large and robust, composed of a few planispiral whorls surrounding a globular proloculus. The prominent planispirally coiled, tubular second chamber is wide. The diameter of the second chamber increases slowly in size. Some specimens display slightly irregular coiling. Chamber walls are thick, non-calcareous, and made up of coarse agglutinated particles joined together with much cement. The primary aperture is a simple opening at the end of the tube.

### Synonyms and Citations

*Ammodiscus pennyi* Cushman and Jarvis. –Kaminski et al., 1988, p. 184, pl. 3, figs. 9–10. –Kuhnt and Kaminski, 1990, p. 463. –Bolli et al., 1994, p. 68, figs. 18.28, 36. –Kaminski and Gradstein, 2005, pp. 155–157, fig. 17, pl. 17.

### Remarks

*Ammodiscus pennyi* Cushman and Jarvis differs from *Ammodiscus latus* Grzybowski by its larger size, thicker and coarser wall, and more numerous coils (Kaminski et al., 1988).

### Biogeography

Worldwide. See also Observed Occurrences in Kaminski and Gradstein (2005).

### Bathymetry

Bathyal to abyssal.

### Chronostratigraphy

Late Cretaceous (Campanian) to Paleocene.

## References


Bolli, H.M., Beckmann, J.P., and Saunders, J.B., 1994, Benthic Foraminiferal Biostratigraphy of the South Caribbean Region. Cambridge: Cambridge University Press, 408 pp.

Cushman, J.A. and Jarvis, P.W., 1928, Cretaceous foraminifera from Trinidad. Contributions from the Cushman Laboratory for Foraminiferal Research, vol. 44, pp. 85–103.

Kaminski, M.A. and Gradstein, F.M., 2005, Atlas of Palaeogene Cosmopolitan Deep-Water Agglutinated Foraminifera. Grzybowski Foundation Special Publication, no. 10, 547 + vii pp.

Kaminski, M.A., Gradstein, F.M., Berggren, W.A., Geroch, S., and Beckmann, J.P., 1988, Agglutinated foraminiferal assemblages from Trinidad: taxonomy, stratigraphy and paleobathymetry. In Gradstein, F.M. and Rögl, F., ed., Second International Workshop on Agglutinated Foraminifera, Vienna 1986, Abhandlungen der Geologischen Bundesanstalt, vol. 41, pp. 155–228.

Kuhnt, W. and Kaminski, M.A., 1990, Paleoecology of Late Cretaceous to Paleocene deep-water agglutinated foraminifera from the North Atlantic and western Tethys. In Hemleben, C., Kaminski, M.A., Kuhnt, W., and Scott, D.B., ed., Paleoecology, Biostratigraphy, Paleoceanography and Taxonomy of Agglutinated Foraminifera. NATO ASI Series, Series C: Mathematical and Physical Sciences. Dordrecht: Kluwer Academic, vol. 327, pp. 433–505.

# Ammodiscus tenuis

## (Brady), 1881

1. *Ammodiscus tenuis.* South China Sea, Station 17961-1B, latitude 8° 30.4′ N, longitude 112° 19.9′ E, water depth: 1795 meters, surface sample. Deposited in the South China Sea Collection, Institute of Geosciences, Kiel University, Germany. Umbilical view, NHMUK PM DI 0031, scale bar = 98 μm.

### Original Designation

*Trochammnina (Ammodiscus) tenuis* Brady, 1881, p. 51 (type description), illustrated as *Ammodiscus tenuis* in Brady, 1884, pl. 38, figs. 4–6.

### Type Specimen

Figured specimens housed in the micropaleontological collections of The Natural History Museum, London, UK.

### Type Level

Surface sediments.

### Type Locality

Not designated. Localities given: "Challenger" Station 46, North Atlantic (latitude 40° 17′ N, longitude 66° 48′ W, water depth 1350 fathoms), "Challenger" Station 168, off northeast coast of New Zealand (latitude 40° 28′ S, longitude 177° 43′ E, water depth 1100 fathoms).

### Description

Test is large and robust, composed of a globular proloculus surrounded by a thick, evolute overlapping, second chamber arranged in a planispiral coil. The diameter of the coiled chamber increases gradually. Chamber walls are thick, well cemented, and non-calcareous; composed of medium to coarse agglutinated particles. The primary aperture is a round opening at the end of the tube.

### Synonyms and Citations

*Ammodiscus tenuis* (Brady). –Brady, 1884, p. 332, pl. 38, figs. 4–6. –Charnock and Jones, 1990, p. 155, pl. 2, fig. 8; pl. 14, fig. 6. –Jones, 1994, p. 43, pl. 38, figs. 4–6.

### Biogeography

Worldwide.

### Bathymetry

Bathyal. North Atlantic bathymetric range: 110–2070 meters (Charnock and Jones, 1990).

### Chronostratigraphy

Eocene? to Recent (Jones, 1994).

### References

Brady, H.B., 1881, Notes on some of the reticularian Rhizopoda of the Challenger Expedition, Part III. Quarterly Journal of Microscopical Science (New Series), vol. 21, pp. 31–71.

Brady, H.B., 1884, Report on the foraminifera dredged by H.M.S. Challenger during the years 1873–1876. In Brady, H.B., ed., Report on the Scientific Results of the Voyage of the H.M.S. Challenger during the Years 1873–1876, Zoology, vol. 9, pp. 1–814.

Charnock, M.A. and Jones, R.W., 1990, Agglutinated foraminifera from the Palaeogene of the North Sea. In Hemleben, C., Kaminski, M.A., Kuhnt, W., and Scott, D.B., ed., Paleoecology, Biostratigraphy, Paleoceanography and Taxonomy of Agglutinated Foraminifera, NATO ASI Series, C: Mathematical and Physical Sciences. Dordrecht: Kluwer Academic, pp. 139–244.

Jones, R.W., 1994, The Challenger Foraminifera. Oxford: Oxford University Press, 149 pp.

# Ammolagena clavata

## (Jones and Parker), 1860

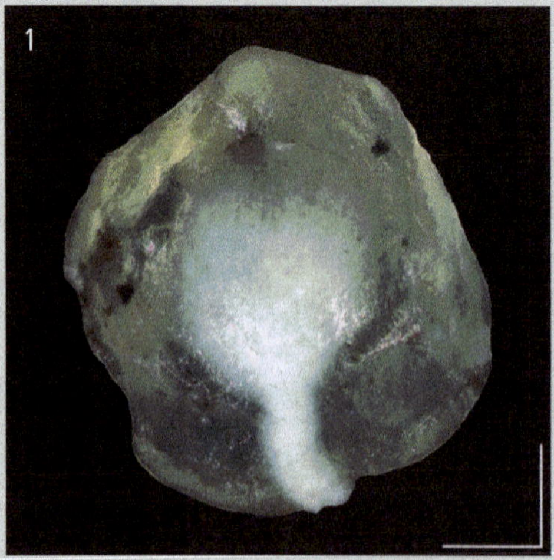

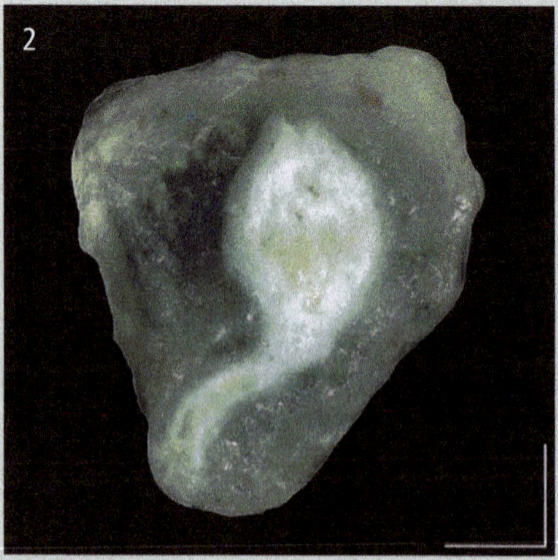

1. *Ammolagena clavata.* Iceland Plateau, Greenland Sea; ODP Leg 162, Site 985, Core 59, Section 4, 30–37 cm. Lateral view, NHMUK PM DI 0032, scale bar = 243 µm.

2. *Ammolagena clavata.* Iceland Plateau, Greenland Sea; ODP Leg 162, Site 985, Core 59, Section 4, 30–37 cm. Lateral view, NHMUK PM DI 0033, scale bar = 211 µm.

### Original Designation

*Trochammina irregularis* (d'Orbigny) var. *clavata* Jones and Parker, 1860, p. 304, illustrated in Carpenter et al. (1862, pl. 11, fig. 6).

### Type Specimen

Housed in the micropaleontological collections of The Natural History Museum, London, UK. Twelve slides containing syntypes, forming part of the W.K. Parker Collection were cataloged by T.R. Jones in 1892 (Slides 1894.4.3.507, 1894.4.3.608, 1894.4.3.615, 1894.4.3.650–651, 1894.4.3.705–708, 1894.4.3.748–751). Lectotype, designated by Kaminski and Gradstein (2005), is registered in slide ZF4873.

### Type Level

Recent, surface sediments.

### Type Locality

Not designated. Localities given: 90 fathoms off Syra; 170 fathoms off Serpho; 500 fathoms off Ipsara; 1100 fathoms between Crete and Santoria; Grecian Archipelago; 250 fathoms northwest of Crete; 360 fathoms off Crete, 1620 fathoms and 1650 fathoms between Malta and Crete; 1700 fathoms between Malta and Greek Archipelago, Mediterranean Sea.

### Description

Test forms a large, ovoid proloculus followed by a narrow, tubular chamber. Chamber walls are non-calcareous, finely agglutinated, and smooth. The aperture is a terminal and round opening at the end of the tube. A secondary aperture may also be present at the base of the proloculus.

### Synonyms and Citations

*Trochammina irregularis* (d'Orbigny) var. *clavata* Jones and Parker. –Carpenter, Parker, and Jones, 1862, p. 142, pl. 11, fig. 6.

*Ammolagena clavata* (Jones and Parker). –Loeblich and Tappan, 1987, p. 49, pl. 36, fig. 16. –Zheng, 1988, p. 39, pl. 6, fig. 6. –Kaminski et al., 1990, p. 365, pl. 3, fig. 1. –Charnock and Jones, 1990, p. 155, pl. 2, fig. 8; pl. 14, fig. 6. –Jones, 1994, p. 46, pl. 41, figs. 12–16. –Loeblich and Tappan, 1994, p. 14, pl. 4, figs. 1–4. –Holbourn and Kaminski, 1997, p. 36, pl. 5, fig. 6. –Kaminski and Gradstein, 2005, pp. 165–168, fig. 21, pl. 21. –Milker and Schmiedl, 2012, p. 28, fig. 9.5.

### Remarks

Type species of the genus *Ammolagena* Eimer and Fickert, 1899.

Test is often found attached to shell fragments or to other foraminiferal tests. Common in Upper Cretaceous to Paleogene flysch-type assemblages.

### Biogeography

Worldwide. See also Observed Occurrences in Kaminski and Gradstein (2005).

### Bathymetry

Bathyal and abyssal.

### Chronostratigraphy

Early Cretaceous to Recent.

## References


Carpenter, W.B., Parker, W.K., and Jones, J.P., 1862, Introduction to the Study of the Foraminifera. London: Ray Society, 319 pp.

Charnock, M.A. and Jones, R.W., 1990, Agglutinated foraminifera from the Paleogene of the North Sea. In Hemleben, C., Kaminski, M.A., Kuhnt, W., and Scott, D.B., ed., Paleoecology, Biostratigraphy, Paleoceanography and Taxonomy of Agglutinated Foraminifera, NATO ASI Series, C: Mathematical and Physical Sciences. Dordrecht: Kluwer Academic, pp. 139–244.

Holbourn, A.E.L. and Kaminski, M.A., 1997, Cretaceous deep-water benthic foraminifera of the Indian Ocean. Grzybowski Foundation Special Publication, no. 4, pp. 1–175.

Jones, R.W., 1994, The Challenger Foraminifera. Oxford: Oxford University Press, 149 pp.

Jones, J.P. and Parker, W.K., 1860, On the Rhizopodal fauna of the Mediterranean compared with that of the Italian and some other Tertiary deposits. Quarterly Journal of the Geological Society of London, vol. 16, pp. 292–307.

Kaminski, M.A. and Gradstein, F.M., 2005, Atlas of Palaeogene Cosmopolitan Deep-Water Agglutinated Foraminifera. Grzybowski Foundation Special Publication, no. 10, 547 + vii pp.

Kaminski, M.A., Gradstein, F.M., Goll, R.M., and Greig, D., 1990, Biostratigraphy and Paleoecology of deep-water aggutinated foraminifera at ODP Site 643, Norwegian–Greenland Sea. In Hemleben, C., Kaminski, M.A., Kuhnt, W., and Scott, D.B., ed., Paleoecology, Biostratigraphy, Paleoceanography and Taxonomy of Agglutinated Foraminifera. NATO ASI Series, Series C: Mathematical and Physical Sciences, vol. 327. Dordrecht: Kluwer Academic, pp. 345–386.

Loeblich, A.R. Jr and Tappan, H., 1987, Foraminiferal Genera and Their Classification. New York: Van Nostrand Reinhold, 970 pp.

Loeblich, A.R. Jr and Tappan, H., 1994, Foraminifera of the Sahul Shelf and Timor Sea. Special Publications of the Cushman Laboratory for Foraminiferal Research, vol. 31, pp. 1–661.

Milker, Y. and Schmiedl, G., 2012, A taxonomic guide to modern benthic shelf foraminifera of the western Mediterranean Sea. Palaeontologia Electronica, vol. 15, Issue 2;16A, 134 pp.

Zheng, S.Y., 1988, The Agglutinated and Porcelaneous Foraminifera of the East China Sea. Beijing: China Ocean Press, 337 pp.

# Amphicoryna scalaris

## (Batsch), 1791

1. *Amphicoryna scalaris.* West Africa, Lutze Collection (West Africa), no. 100. Deposited in the Institute of Geosciences, Kiel University, Germany. Lateral view, NHMUK PM DI 0034, scale bar = 195 μm.

### Original Designation

*Nautilus (Orthoceras) scalaris* Batsch, 1791, pp. 1, 4, pl. 2, fig. 4a–b.

### Type Specimen

No repository given.

### Type Level

Not given. Probably surface sediments.

### Type Locality

Shores of Italy (exact locality not given). Probably Rimini, on the Adriatic coast.

### Description

Test forms an apiculate base, and an initially coiled (microsphere) or globular (megalosphere) proloculus that later uncoils into a rectilinear, uniserial series. Chambers are inflated, slowly increasing in size, circular in cross-section, and separated by straight, constricted sutures in the rectilinear portion. Chamber walls are calcareous, optically radial, finely perforate, and ornamented by fine striations. The primary aperture is a terminal, radiate opening at the end of a distinct neck with ring-like concentric ridges.

### Synonyms and Citations

*Nodosaria longicauda* d'Orbigny, 1826, p. 254.
*Marginulina falx* Jones and Parker, 1860, p. 302.
*Nodosaria scalaris* (Batsch). –Brady, 1884, p. 510, pl. 63, figs. 28–31. –Cushman, 1913, p. 58, pl. 24, fig. 7. –Cushman, 1921, p. 199, pl. 35, fig. 6.
*Amphicoryna scalaris* (Batsch). –Barker, 1960, p. 134, pl. 63, figs. 28–31. –Jones, 1994, p. 75, pl. 63, figs. 28–31. –Milker and Schmiedl, 2012, p. 73, figs. 18.22–25.

### Biogeography

Worldwide.

### Bathymetry

Neritic to bathyal.

### Chronostratigraphy

Miocene to Recent (Jones, 1994).

## References

Barker, R.W., 1960, Taxonomic notes on the species figured by H.B. Brady in his report on the foraminifera dredged by H.M.S.

Challenger during the years 1873–1876. Society of Economic Paleontologists and Mineralogists Special Publication, vol. 9, pp. 1–238.
Batsch, A.I.G.C., 1791, Sechs Kupfertafeln mit Conchylien des Seesandes, gezeichnet und gestochen. Jena: von A.J.G.K. Batsch, 6 pls.
Brady, H.B., 1884, Report on the foraminifera dredged by H.M.S. Challenger during the years 1873–1876. In Brady, H.B., ed., Report on the Scientific Results of the Voyage of the H.M.S. Challenger during the years 1873–1876. Zoology, vol. 9, pp. 1–814.
Cushman, J.A., 1913, A monograph of the Foraminifera of the North Pacific Ocean. Part 3. Lagenidae. Bulletin of the United States National Museum, vol. 71, no. 3, pp. 1–125.
Cushman, J.A., 1921, Foraminifera of the Phillipine and adjacent seas. Bulletin of the United States National Museum, vol. 100, pp. 1–608.
d'Orbigny, A., 1826, Tableau méthodique de la classe des Céphalopodes. Annales des Sciences Naturelles, vol. 7, pp. 245–314.
Jones, J.P. and Parker, W.K., 1860, On the Rhizopodal fauna of the Mediterranean compared with that of the Italian and some other Tertiary deposits. Quarterly Journal of the Geological Society of London, vol. 16, pp. 292–307.
Jones, R.W., 1994, The Challenger Foraminifera. Oxford: Oxford University Press, 149 pp.
Milker, Y. and Schmiedl, G., 2012, A taxonomic guide to modern benthic shelf foraminifera of the western Mediterranean Sea. Palaeontologia Electronica, vol. 15, Issue 2;16A, 134 pp.

# Annectina biedai

## Gradstein and Kaminski, 1997

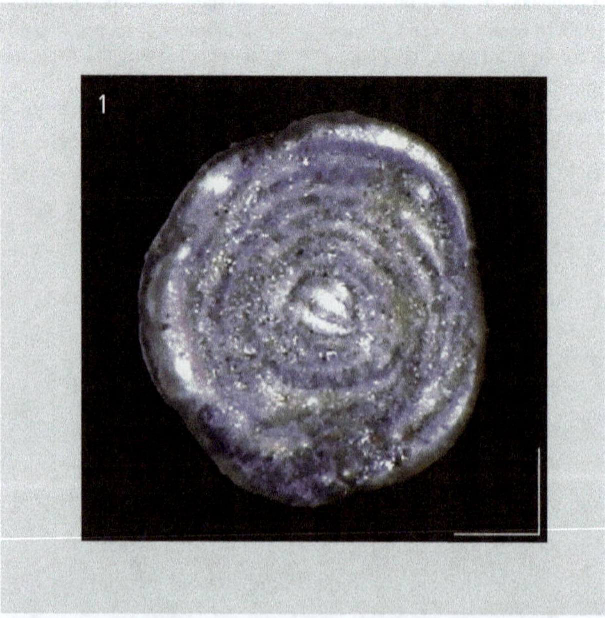


1. *Annectina biedai.* North Sea; British Petroleum Well 21/10-4, depth: 1830 meters. M.A. Kaminski Collection. Umbilical view, NHMUK PM DI 0035, scale bar = 121 μm.

### Original Designation

*Annectina biedai* Gradstein and Kaminski, 1997, p. 218, fig. 2; fig. 3 (1–2).

### Type Specimen

Housed in the micropaleontological collections of The Natural History Museum, London, UK. The holotype is registered in slide PF62942.

### Type Level

Oligocene.

### Type Locality

Norwegian well Saga 2/2-4, at a depth of 1900 meters.

### Description

Test slightly elliptical to circular in outline, composed of a proloculus followed by an undivided, tubular, second chamber. The tubular chamber remains thin, initially forming a milioline coil, then becoming planispiral with numerous whorls (ten or more). The planispiral portion is almost flat with a flush, coiled suture, which may exhibit small, wavy irregularities (Kaminski and Gradstein, 2005). The test wall is solid, very finely agglutinated, and silicified. The primary aperture is a round, terminal opening at the end of the planispiral tube.

### Synonyms and Citations

*Ammodiscus* sp. B, –King, 1989, pl. 9.1, fig. 5.

*Glomospira biedai* (Samuel). –Charnock and Jones, 1990, p. 156, pl. 14, fig. 8; pl. 2, fig. 10.

*Annectina biedai* Gradstein and Kaminski. –Kaminski and Gradstein, 2005, pp. 195–198, fig. 29, pl. 29.

### Remarks

*Annectina biedai* Gradstein and Kaminski differs from *Annectina grzybowskii* (Jurkiewicz) by its thinner, more compressed test, larger number of whorls in the planispiral part, and smaller initial triloculine portion. *Annectina grzybowskii* probably became extinct in the Early Eocene, whereas *Annectina biedai* mainly occurs in younger strata, becoming extinct in the Oligocene (Gradstein and Kaminski, 1997).

### Biogeography

North Sea and Norwegian Sea (Kaminski and Gradstein, 2005).

### Bathymetry

Bathyal (Kaminski and Gradstein, 2005).

### Chronostratigraphy

?Middle Eocene to Late Oligocene. Common in the Early Oligocene (Biozone NSR7A of Gradstein and Bäckström, 1996).

Total stratigraphic range may extend down to the Late Paleocene (Kaminski and Gradstein, 2005).

## References

Charnock, M.A. and Jones, R.W., 1990, Agglutinated foraminifera from the Paleogene of the North Sea. In Hemleben, C., Kaminski, M.A., Kuhnt, W., and Scott, D.B., ed., Paleoecology, Biostratigraphy, Paleoceanography and Taxonomy of Agglutinated Foraminifera, NATO ASI Series, C: Mathematical and Physical Sciences. Dordrecht: Kluwer Academic, pp. 139–244.

Gradstein, F.M. and Bäckström, S.A., 1996, Cenozoic biostratigraphy and paleobathymetry, northern North Sea and Haltenbanken. Norsk Geologisk Tidsskrift, vol. 76, pp. 3–32.

Gradstein, F.M. and Kaminski, M.A., 1997, New species of Paleogene deep-water agglutinated foraminifera from the North Sea and Norwegian Sea. Annales Societatis Geologorum Poloniae, vol. 67, pp. 217–229.

Kaminski, M.A. and Gradstein, F.M., 2005, Atlas of Palaeogene Cosmopolitan Deep-Water Agglutinated Foraminifera. Grzybowski Foundation Special Publication, no. 10, 547 + vii pp.

King, C., 1989, Cenozoic of the North Sea. In Jenkins, D.G. and Murray, J.W., ed., Stratigraphical Atlas of Fossil Foraminifera, 2nd edn. Chichester: Ellis Horwood, pp. 418–489.

# *Annectina grzybowskii*

## (Jurkiewicz), 1960

1. *Annectina grzybowskii.* Radziechowy (near Żywiec), Poland, Subsilesian Unit, RG 17; gray-green shales. Umbilical view, NHMUK PM DI 0036, scale bar = 172 μm.

### Original Designation

*Glomospira grzybowskii* Jurkiewicz, 1960, pp. 339, 342, pl. 38, figs. 7, 10, 11.

### Type Specimen

Depository not given. The Jurkiewicz Collection was originally deposited at the Polish Petroleum Company in Jaslo but the status of the holotype is unknown (Kaminski and Gradstein, 2005).

### Type Level

Upper Paleocene Czarnorzeki shales (=Istebna), Silesian Unit of the Polish Carpathians.

### Type Locality

Not designated. Found in boreholes at Roztoki, Jaszczew, and Osobnica, Carpathian Mountains, Poland.

### Description

Test large, elliptical to subcircular in outline, composed of a proloculus followed by an undivided, tubular, second chamber, initially forming a milioline coil. Final whorls (up to five) become progressively planispiral. Chamber walls are thick, medium to finely agglutinated, and well cemented. The primary aperture is a round opening that occupies the end of the planispiral tube.

### Synonyms and Citations

*Glomospirella biedai* Samuel, 1977, p. 29, pl. 3, fig. 16; pl. 21, fig. 3.
*Glomospirella grzybowskii* (Jurkiewicz). –Hemleben and Tröster, 1984, p. 519, pl. 1, fig. 24. –Charnock and Jones, 1990, p. 157, pl. 2, fig. 13; pl. 14, fig. 10. –Bubik, 1995, p. 81, pl. 9, fig.12.
*Annectina grzybowskii* (Jurkiewicz). –Kaminski and Gradstein, 2005, pp. 192–195, fig. 28, pl. 28.

### Remarks

*Annectina biedai* Gradstein and Kaminski differs from *Annectina grzybowskii* (Jurkiewicz) by its thinner, more compressed test, larger number of whorls in the planispiral part, and smaller initial triloculine portion. *Annectina biedai* mainly occurs in younger strata, becoming extinct in the Oligocene (Gradstein and Kaminski, 1997).

### Biogeography

Tethys and eastern Atlantic. Common in flysch-type assemblages. See also Observed Occurrences in Kaminski and Gradstein (2005).

### Bathymetry

Bathyal to abyssal.

### Chronostratigraphy

Late Cretaceous (Campanian) to Late Paleocene.

## References

Bubik, M., 1995, Cretaceous to Paleogene agglutinated foraminifera from the Bile Karpaty unit. In Kaminski, M.A., Geroch, S., and Gasinski, M.A., ed., Proceedings of the Fourth International Workshop on Agglutinated Foraminifera, Grzybowski Foundation Special Publication, no. 3, pp. 71–116.

Charnock, M.A. and Jones, R.W., 1990, Agglutinated foraminifera from the Paleogene of the North Sea. In Hemleben, C., Kaminski, M.A., Kuhnt, W., and Scott, D.B., ed., Paleoecology, Biostratigraphy, Paleoceanography and Taxonomy of Agglutinated Foraminifera, NATO ASI Series, C: Mathematical and Physical Sciences. Dordrecht: Kluwer Academic, pp. 139–244.

Gradstein, F.M. and Kaminski, M.A., 1997, New species of Paleogene deep-water agglutinated foraminifera from the North Sea and Norwegian Sea. Annales Societatis Geologorum Poloniae, vol. 67, pp. 217–229.

Hemleben, C. and Tröster, J., 1984, Campanian–Maestrichtian deep-water foraminifers from Hole 543A, Deep Sea Drilling Project. In Bij-Duval, B., Moore, J.C., et al., ed., Initial Reports of the Deep Sea Drilling Project, Washington (U.S. Government Printing Office), vol. 78A, pp. 509–532.

Jurkiewicz, H., 1960, Foraminifera in Czarnorzeki-Shales in the eastern part of the Silesian Unit (Carpathians). Rocznik Polskiego Towarzystwa Geologicznego, vol. 30, no. 3, pp. 333–343.

Kaminski, M.A. and Gradstein, F.M., 2005, Atlas of Palaeogene Cosmopolitan Deep-Water Agglutinated Foraminifera. Grzybowski Foundation Special Publication, no. 10, 547 + vii pp.

Samuel, O., 1977, Agglutinated foraminifers from Paleogene flysch formations in West Carpathians of Slovakia. Zapadne Karpaty, Serie Paleontologie, vol. 2–3, pp. 7–70.

# *Anomalinella rostrata*

## (Brady), 1881

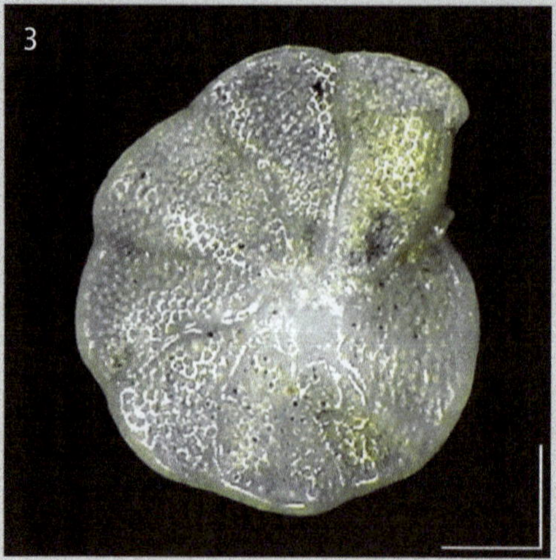



1. *Anomalinella rostrata.* Humboldt Bay, off Papua New Guinea, Pacific Ocean; "Challenger" Station 217A; depth: 2740 fathoms. Deposited in the micropaleontological collections of The Natural History Museum, London; NHMUK ZF2549. Lectotype. Spiral view, NHMUK PM DI 0037, scale bar = 174 μm.

2. *Anomalinella rostrata.* Humboldt Bay, off Papua New Guinea, Pacific Ocean; "Challenger" Station 217A; depth: 2740 fathoms. Deposited in the micropaleontological collections of The Natural History Museum, London; NHMUK ZF2549. Lectotype. Apertural view, NHMUK PM DI 0038, scale bar = 168 μm.

3. *Anomalinella rostrata.* Humboldt Bay, off Papua New Guinea, Pacific Ocean; "Challenger" Station 217A; depth: 2740 fathoms. Deposited in the micropaleontological collections of The Natural History Museum, London;Museum, London; NHMUK ZF2549. Lectotype. Umbilical view, NHMUK PM DI 0039, scale bar = 183 μm.

## Original Designation

*Truncatulina rostrata* Brady, 1881, p. 65 (type description), illustrated as *Truncatulina rostrata* in Brady, 1884, pl. 94, fig. 6.

## Type Specimen

Lectotype (ZF2549) designated by Loeblich and Tappan (1987) is housed in the micropaleontological collections of The Natural History Museum, London, UK.

## Type Level

Not designated.

## Type Locality

Lectotype is from "Challenger" Station 217A, off Papua New Guinea, Pacific.

## Description

Test lenticular, biumbonate, involute, initially trochospiral, then planispiral, with acute periphery bearing a double keel. Approximately nine to ten moderately inflated chambers in the last whorl, increasing gradually in size, are separated by curved, limbate sutures. Wall is calcareous, coarsely perforate, smooth. Aperture is a low interiomarginal, equatorial arch, bordered by a protruding lip. Secondary aperture forms an elongate slit between the two keels on the periphery.

## Synonyms and Citations

*Truncatulina rostrata* Brady. –Brady, 1884, p. 668, pl. 94, fig. 6.

*Anomalinella rostrata* (Brady). –Loeblich and Tappan, 1987, p. 623, pl. 700, figs. 8–12. –van Marle, 1991, p. 185, pl. 20, figs. 9–10. –Jones, 1994, p. 98, pl. 94, fig. 6.

## Remarks

Type species of the genus *Anomalinella* Cushman, 1927.

## Biogeography

Worldwide. Cosmopolitan in low to moderate latitudes.

## Bathymetry

Neritic to abyssal (2740 fathoms at "Challenger" Station 217A, off Papua New Guinea, Pacific, Jones, 1994).

## Chronostratigraphy

Early Miocene to Recent.

## References


Brady, H.B., 1881, Notes on some of the reticularian Rhizopoda of the Challenger Expedition, Part III. Quarterly Journal of Microscopical Science (New Series), vol. 21, pp. 31–71.

Brady, H.B., 1884, Report on the foraminifera dredged by H.M.S. Challenger during the years 1873–1876. In Brady, H.B., ed., Report on the Scientific Results of the Voyage of the H.M.S. Challenger during the years 1873–1876. Zoology, vol. 9, pp. 1–814.

Cushman, J.A., 1927, An outline of a re-classification of the foraminifera. Contributions from the Cushman Laboratory for Foraminiferal Research, vol. 31, pp. 1–105.

Jones, R.W., 1994, The Challenger Foraminifera. Oxford: Oxford University Press, 149 pp.

Loeblich, A.R. Jr and Tappan, H., 1987, Foraminiferal Genera and Their Classification. New York: Van Nostrand Reinhold, 970 pp.

van Marle, L.J., 1991, Eastern Indonesian Late Cenozoic smaller benthic foraminifera. Verhandelingen der Koninklijke Nederlanse Akademie van Wetenschappen, Afdeeling Natuurkunde, Eerste Reeks, vol. 34, pp. 1–328.

# *Anomalinoides capitatus*

## (Gümbel), 1868

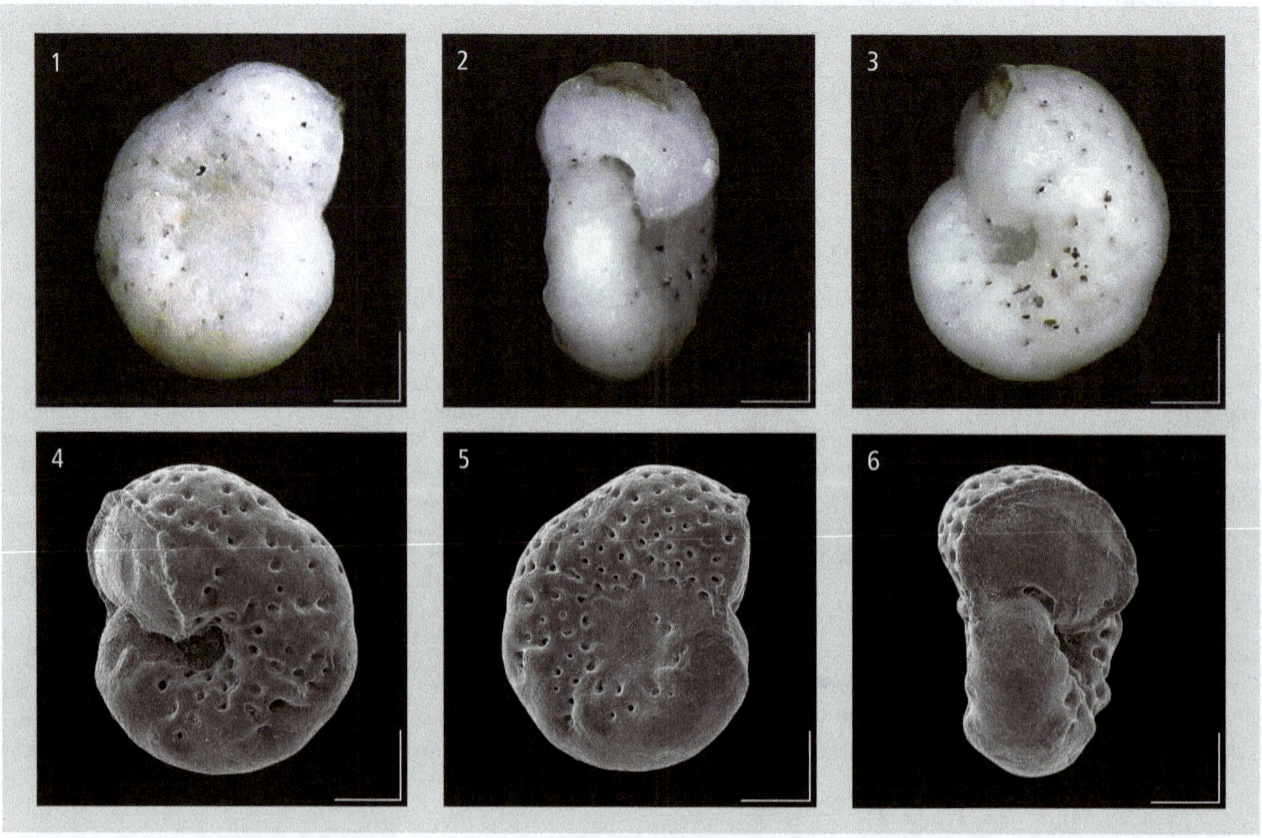

1. *Anomalinoides capitatus.* Northeast Georgia Rise, South Atlantic; ODP Leg 114, Hole 698A, Core 3, Section 1, 60–64 cm. Note specimen missing final chamber. Spiral view, NHMUK PM DI 0040, scale bar = 95 μm.

2. *Anomalinoides capitatus.* Northeast Georgia Rise, South Atlantic; ODP Leg 114, Hole 698A, Core 3, Section 1, 60–64 cm. Note specimen missing final chamber. Apertural view, NHMUK PM DI 0041, scale bar = 98 μm.

3. *Anomalinoides capitatus.* Northeast Georgia Rise, South Atlantic; ODP Leg 114, Hole 698A, Core 3, Section 1, 60–64 cm. Note specimen missing final chamber. Umbilical view, NHMUK PM DI 0042, scale bar = 97 μm.

4. *Anomalinoides capitatus.* Northeast Georgia Rise, South Atlantic; ODP Leg 114, Hole 698A, Core 3, Section 1, 60–64 cm. Note specimen missing final chamber. Umbilical view, NHMUK PM DI 0043, scale bar = 103 μm.

5. *Anomalinoides capitatus.* Northeast Georgia Rise, South Atlantic; ODP Leg 114, Hole 698A, Core 3, Section 1, 60–64 cm. Note specimen missing final chamber. Spiral view, NHMUK PM DI 0044, scale bar = 101 μm.

6. *Anomalinoides capitatus.* Northeast Georgia Rise, South Atlantic; ODP Leg 114, Hole 698A, Core 3, Section 1, 60–64 cm. Note specimen missing final chamber. Apertural view, NHMUK PM DI 0045, scale bar = 105 μm.

## Original Designation
*Rotalia capitata* Gümbel, 1868, p. 653, pl. 2, fig. 92.

## Type Specimen
Original material deposited in the collections of the Institute for Historical Geology and Paleontology of the University of Munich, Germany, was destroyed during the Second World War.

## Type Level
Upper Eocene, nummulitic marl.

## Type Locality
Not designated. Localities given: Hammer in Traunthal and Götzreuther Graben near Siegsdorf, Bavarian Alps, Germany.

## Description
Test forms a very low trochospire, slightly lobulate in outline and unequally biconvex in cross-section, with a broadly rounded periphery that is imperforate only in the early chambers of the final whorl. The spiral side is partially evolute and slightly convex with a central plug. The umbilical side is involute and more convex with an open, depressed umbilicus. In the ultimate whorl, five to six inflated chambers increase gradually in size, and are separated by thin, depressed, straight or slightly curved sutures in the final chambers and by thick, prominent, raised sutures in the earlier chambers. Chamber walls are calcareous and coarsely perforate on both sides of the test. The primary aperture is a low, equatorial arch, bordered by a thickened lip, extending on to the spiral side beneath the protruding apertural lip. Previous supplementary apertures on the spiral side are gradually closed off.

## Synonyms (following van Morkhoven et al., 1986) and Citations
*?Rosalina calymene* Gümbel, 1868, p. 658, pl. 2, fig.100.
*Anomalina dorri* Cole var. *aragonensis* Nuttall, 1930, p. 291, pl. 24, fig. 18; pl. 25, fig. 1.
*Anomalina visenda* Finlay, 1940, p. 458, pl. 65, figs. 116–119.
*Anomalina keijzeri* Montagne, 1941, p. 55, pl. 6, figs. 38–40.
*Anomalina packardi* Bandy, 1944, p. 373, pl. 61, fig. 7.
*Anomalinoides capitatus* (Gümbel). –Hagn, 1956, p. 176, pl.16, figs. 15–16. –van Morkhoven et al., 1986, p. 276, pl. 92, figs. 1–2. –Katz and Miller, 1991, pl. 3, figs. 6–8.
*Gavelinella capitata* (Gümbel). –Tjalsma and Lohmann, 1983, p. 31, pl. 16, figs. 4–5. –Miller and Katz, 1987, p. 286, pl. 4.

## Remarks
Van Morkhoven et al. (1986) considered *Anomalinoides rubiginosus* (Cushman), *Anomalinoides capitatus* (Gümbel), *Anomalinoides semicribratus* (Beckmann), and *Anomalinoides globulosus* (Chapman and Parr) to represent a chronocline ranging from the Campanian to the Recent.

## Biogeography
Worldwide. Recorded from Europe, Crimea, Turkmenia, Caribbean, the Gulf of Mexico, the North and South Atlantic, and Barbados. As *Anomalinoides aragonensis* recorded from Cuba, Mexico, Trinidad, Barbados, North Pacific, Indian Ocean, Labrador Sea, California, and Libya (see van Morkhoven et al., 1986).

## Bathymetry
Primarily bathyal, but also ranges to abyssal depths (van Morkhoven et al., 1986).

## Chronostratigraphy
Early Eocene (Ypresian) to Early Oligocene (earliest Rupelian). Doubtful in Early Oligocene "*Turborotalia*" *ampliapertura* Biozone (P19) and *Globigerina sellii* Biozone (P20) (van Morkhoven et al., 1986).

## References


Bandy, O.L., 1944, Eocene foraminifera from Cape Blanco, Oregon. Journal of Paleontology, vol. 18, no. 4, pp. 366–377.

Finlay, H.J., 1940, New Zealand Foraminifera; key species in stratigraphy – No. 4. Transactions of the Royal Society of New Zealand, vol. 69, no. 4, pp. 448–472.

Gümbel, C.W., 1868, Beitrage zur Foraminiferenfauna der nordalpinen Eocängebilde. Abhandlungen der K. Bayerischen Akademie der Wissenschaften, cl. 2 (1870), vol. 10, no. 2, pp. 581–730.

Hagn, H., 1956, Geologische und palaontologische Untersuchungen im Tertiar des Monte Brione und seiner Umgebung. Palaeontographica, vol. 107(A), pp. 67–210.

Katz, M.E. and Miller, K.G., 1991, Early Paleogene benthic foraminiferal assemblages and stable isotopes in the Southern Ocean. In Ciesielski, P.F., Kristoffersen, Y., et al., ed., Scientific Results of the Proceedings of the Ocean Drilling Program, vol. 114, pp. 481–512.

Miller, K.G. and Katz, M.E., 1987, Eocene benthic foraminiferal biofacies of the New Jersey transect. In Poag, C.W., Watts, A.B., et al., ed., Initial Reports of the Deep Sea Drilling Project, Washington (U.S. Government Printing Office), vol. 95, pp. 267–298.

Montagne, D.G., 1941, Geologie und Palaeontologie der Umgebung von Sestanovac, Dalmatien. Utrecht University, Geographische en Geologische Mededelingen, Physiographisch-Geologische Reeks, ser. 2, no. 1, pp. 9–93.

Nuttall, W.L.F., 1930, Eocene foraminifera from Mexico. Journal of Paleontology, vol. 4, pp. 271–293.

Tjalsma, R.C. and Lohmann, G.P., 1983, Paleocene–Eocene bathyal and abyssal benthic foraminifera from the Atlantic Ocean. Micropaleontology, Special Publication, no. 4, pp. 1–90.

van Morkhoven, F.P.C.M., Berggren, W.A., Edwards, A.S., et al., 1986, Cenozoic cosmopolitan deep-water benthic foraminifera. Bulletin des Centres de Recherches Exploration-Production Elf-Aquitaine, Mémoire, vol. 11, pp. 1–421.

# *Anomalinoides globulosus*

## (Chapman and Parr), 1937

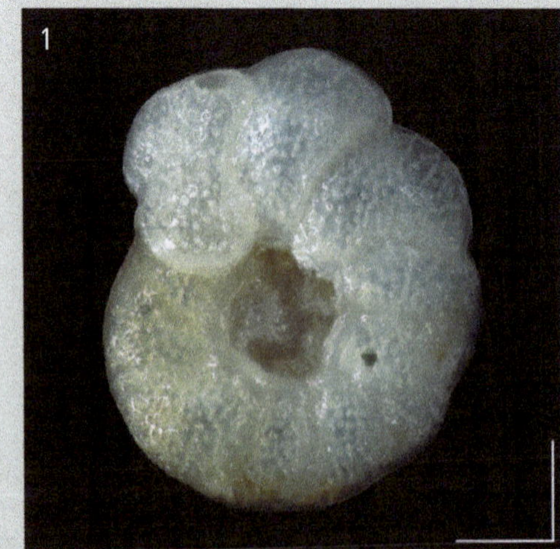

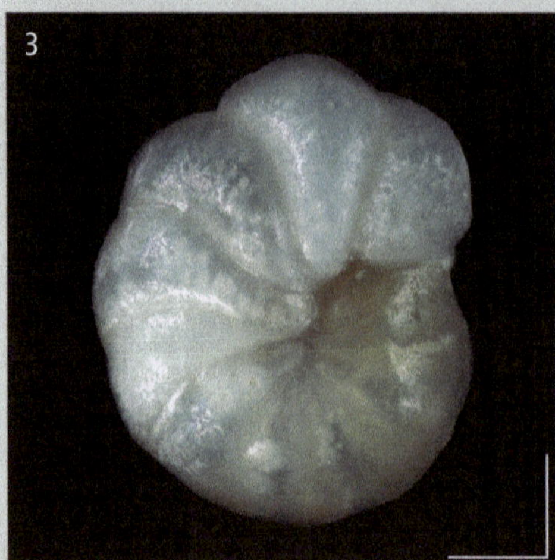

1. *Anomalinoides globulosus*. West Africa, Lutze Collection (West Africa), no. 195. Deposited in the Institute of Geosciences, Kiel University, Germany. Umbilical view, NHMUK PM DI 0046, scale bar = 202 μm.

2. *Anomalinoides globulosus*. West Africa, Lutze Collection (West Africa), no. 195. Deposited in the Institute of Geosciences, Kiel University, Germany. Apertural view, NHMUK PM DI 0047, scale bar = 194 μm.

3. *Anomalinoides globulosus*. West Africa, Lutze Collection (West Africa), no. 195. Deposited in the Institute of Geosciences, Kiel University, Germany. Spiral view, NHMUK PM DI 0048, scale bar = 199 μm.

## Original Designation

*Anomalina globulosa* Chapman and Parr, 1937, p. 117, pl. 9, fig. 27.

## Type Specimen

Deposited in the Australian Museum, Sydney, Australia.

## Type Level

Recent, brown terrigenous sand.

## Type Locality

Latitude 42° 53′ S, longitude 148° 25.25′ E, 675 fathoms, between Antarctica and Australia.

## Description

Test forms a very low trochospire; lobulate in outline and unequally biconvex in cross-section with a broadly rounded periphery. The spiral side is partially evolute and slightly convex, while the umbilical side is involute and more convex with a small, open umbilicus. The last whorl is composed of approximately seven inflated chambers that increase gradually in size and are separated by curved, depressed sutures. Chamber walls are calcareous and coarsely perforate on the spiral side, but relatively smooth and more finely perforate on the umbilical side. The primary aperture is a low, equatorial arch, bordered by a thickened lip, extending on to the spiral side beneath the protruding apertural lip. Previous supplementary apertures on the spiral side are gradually closed off.

## Synonyms and Citations

*Anomalina grosserugosa* (Gümbel). –Brady, 1884, p. 673, pl. 94, figs. 4–5.

?*Anomalina* sp. Phleger, Parker and Pierson, 1953, p. 48, pl. 10, figs. 26–28.

*Anomalinoides globulosus* (Chapman and Parr). –van Morkhoven et al., 1986, p. 36, pl. 9. –Hermelin, 1989, p. 84, pl. 17, figs. 1, 5. –Jones, 1994, p. 98, pl. 94, figs. 4–5. –Robertson, 1998, p. 240, pl. 97, fig. 2.

## Remarks

According to van Morkhoven et al. (1986), *Anomalinoides globulosus* (Chapman and Parr) is the youngest form of a chronocline ranging from the Campanian to Recent and including *Anomalinoides rubiginosus* (Cushman), *Anomalinoides capitatus* (Gümbel), *Anomalinoides semicribratus* (Beckmann), and *Anomalinoides globulosus*.

## Biogeography

Worldwide.

## Bathymetry

Middle bathyal to abyssal (van Morkhoven et al., 1986). Predominantly lower bathyal (no abyssal data) according to Katz and Miller (1993).

## Chronostratigraphy

Middle Miocene to Recent. Miller and Katz (1987) reported the first occurrence of *Anomalinoides globulosus* in Biozone M8 in Jamaica and at other locations, while van Morkhoven et al. (1986) reported its first occurrence in Biozone M12.

## References


Brady, H.B., 1884, Report on the foraminifera dredged by H.M.S. Challenger during the years 1873–1876. In Brady, H.B., ed., Report on the Scientific Results of the Voyage of the H.M.S. Challenger during the years 1873–1876, Zoology, vol. 9, pp. 1–814.

Chapman, F. and Parr, W.J., 1937, Foraminifera. In Johnston, T.H., ed., Australasian Antarctic Expedition 1911–1914, Scientific Reports, Series C (Zoology and Botany). Sydney: David Harold Paisley, vol. 1, pp. 1–190.

Hermelin, J.O.R., 1989, Pliocene benthic foraminifera from the Ontong-Java Plateau (Western Equatorial Pacific Ocean): faunal response to changing paleoenvironment. Special Publication of the Cushman Foundation for Foraminiferal Research, vol. 26, pp. 1–143.

Jones, R.W., 1994, The Challenger Foraminifera. Oxford: Oxford University Press, 149 pp.

Katz, M.E. and Miller, K.G., 1993, Latest Oligocene to Earliest Pliocene benthic foraminiferal biofacies of the northeastern Gulf of Mexico. Micropaleontology, vol. 39, no. 4, pp. 367–403.

Miller, K.G. and Katz, M.E., 1987, Oligocene to Miocene benthic foraminiferal and abyssal circulation changes in the north Atlantic. Micropaleontology, vol. 33, no. 2, pp. 97–149.

Phleger, F.B., Parker, F.L., and Pierson, J.F., 1953, North Atlantic foraminifera. Reports of the Swedish Deep-Sea Expedition, 1947–1948, vol. 7, pp. 3–122.

Robertson, B.E., 1998, Systematics and paleoecology of the benthic foraminiferida from the Buff Bay section, Miocene of Jamaica. Micropaleontology, vol. 44, supplement 2, pp. 1–266.

van Morkhoven, F.P.C.M., Berggren, W.A., Edwards, A.S., et al., 1986, Cenozoic cosmopolitan deep-water benthic foraminifera. Bulletin des Centres de Recherches Exploration-Production Elf-Aquitaine, Mémoire, vol. 11, pp. 1–421.

# *Anomalinoides rubiginosus*

## (Cushman), 1926

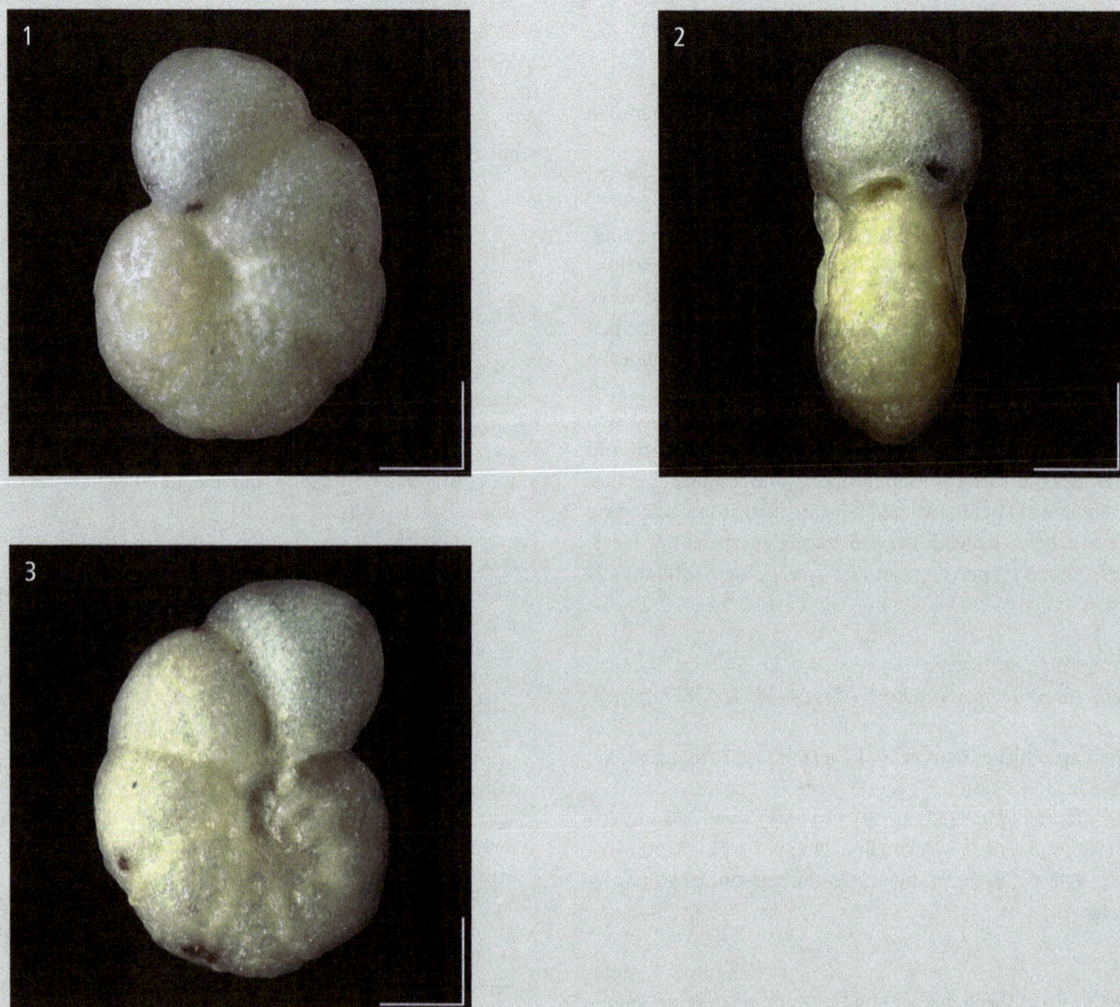

1. *Anomalinoides rubiginosus.* Upper Nile, Egypt, Outcrop sample 102. Deposited in the collections at the United States National Museum, Washington, DC. Early Paleocene (Zone P1/P2) specimen illustrated in van Morkhoven et al. 1986, Ref. no. 119. Umbilical view, NHMUK PM DI 0049, scale bar = 64 μm.

2. *Anomalinoides rubiginosus.* Upper Nile, Egypt, Outcrop sample 102. Deposited in the collections at the United States National Museum, Washington, DC. Early Paleocene (Zone P1/P2) specimen illustrated in van Morkhoven et al. 1986, Ref. no. 119. Apertural view, NHMUK PM DI 0050, scale bar = 61 μm.

3. *Anomalinoides rubiginosus.* Upper Nile, Egypt, Outcrop sample 102. Deposited in the collections at the United States National Museum, Washington, DC. Specimen illustrated in van Morkhoven et al. 1986, Ref. no. 119. Spiral view, NHMUK PM DI 0051, scale bar = 61 μm.

## Original Designation

*Anomalina rubiginosa* Cushman, 1926, p. 607, pl. 21, fig. 6.

## Type Specimen

Holotype (No. 5226) deposited in the collections of the United States National Museum, Washington, DC.

## Type Level

Paleocene, Velasco Shale. Originally given as Upper Cretaceous by Cushman (1926).

## Type Locality

Eastern part of the Hacienda El Limón, State of San Luis Potosi, Mexico.

## Description

Test forms a low trochospire; slightly lobulate in outline, and unequally biconvex in cross-section with a rounded periphery. The spiral side is slightly convex, partially evolute with a central plug; the umbilical side is convex and involute with a depressed umbilicus. Approximately nine to ten moderately inflated chambers form the last whorl, each increasing gradually in size. Chambers are separated by curved, flush or raised sutures that become slightly depressed in the last chambers. Chamber walls are calcareous, coarsely perforate on both sides, and may be irregularly pitted in the early portion. The primary aperture is a low equatorial arch, bordered by a thickened lip, and extending on to the spiral side beneath the protruding apertural lip. Previous supplementary apertures on the spiral side are gradually closed off.

## Synonyms (following van Morkhoven et al., 1986) and Citations

*Anomalina grosserugosa* (Gümbel). –Franke, 1927, p. 37, pl. 4, fig. 3.
*Anomalina westraliensis* Parr, 1938, p. 85, pl. 3, fig. 4.
*Cibicides danica* Brotzen, 1940, p. 31, pl. 7:2.
*Anomalinoides rubiginosus* (Cushman). –van Morkhoven et al., 1986, p. 366, pl. 119. –Bolli et al., 1994, p. 158, figs. 44.6, 13, 18–19.

## Remarks

In neritic and upper bathyal deposits this species is often identified as *Anomalinoides danicus* (Brotzen). Van Morkhoven et al. (1986) regarded these forms as depth-related ecophenotypes and used the name *Anomalinoides rubiginosus* (Cushman) for them.

Van Morkhoven et al. (1986) considered *Anomalinoides rubiginosus*, *Anomalinoides capitatus* (Gümbel), *Anomalinoides semicribratus* (Beckmann), and *Anomalinoides globulosus* (Chapman and Parr) to represent a chronocline ranging from the Campanian to the Recent.

## Biogeography

Worldwide. Recorded from Arkansas, California, Europe, Trinidad, and the South Atlantic. As *Cibicides* or *Gavelinella danica* (Brotzen), this species has been recorded from New Jersey, Europe, the Soviet Union, Tunisia, Egypt, the Gulf of Mexico, Caribbean, South Atlantic, and Indian Ocean. As *Anomalinoides westraliensis* (Parr), it has been recorded from Australia (McGowran, 1965).

## Bathymetry

Middle neritic to abyssal.

## Chronostratigraphy

Late Cretaceous (Campanian) to Early Eocene (Ypresian).

## References


Bolli, H.M., Beckmann, J.P., and Saunders, J.B., 1994, Benthic Foraminiferal Biostratigraphy of the South Caribbean Region. Cambridge: Cambridge University Press, 408 pp.

Brotzen, F., 1940, Flintraennans och Trindelraennans geologi (Oeresund). (Die Geologie der Flint- und Trindelrinne (Oeresund). Årsbok Sveriges Geologiska Undersökning, vol. 34, no. 5, pp. 1–33.

Cushman, J.A., 1926, The foraminifera of the Velasco Shale of the Tampico Embayment. Bulletin of the American Association of Petroleum Geologists, vol. 10, no. 6, pp. 581–612.

Franke, A., 1927, Die Foraminiferen und Ostracoden des Palaozans von Rugaard in Jiitland und Sundkrogen bei Copenhagen. Danmarks geologiske Undersøgelse, række 2, no. 46, 49 pp.

McGowran, B., 1965, Two Paleocene foraminiferal faunas from the Wangerrip Group, Pebble Point Coastal Section, Western Victoria. Proceedings of the Royal Society of Victoria, vol. 79, no. 1, pp. 9–74.

Parr, W.J., 1938, Upper Eocene foraminifera from deep borings in King's Park, Perth, Western Australia. Journal of the Royal Society of Western Australia, vol. 24, no. 8, pp. 69–101.

van Morkhoven, F.P.C.M., Berggren, W.A., Edwards, A.S., et al., 1986, Cenozoic cosmopolitan deep-water benthic foraminifera. Bulletin des Centres de Recherches Exploration-Production Elf-Aquitaine, Mémoire, vol. 11, pp. 1–421.

# *Anomalinoides semicribratus*

## (Beckmann), 1954

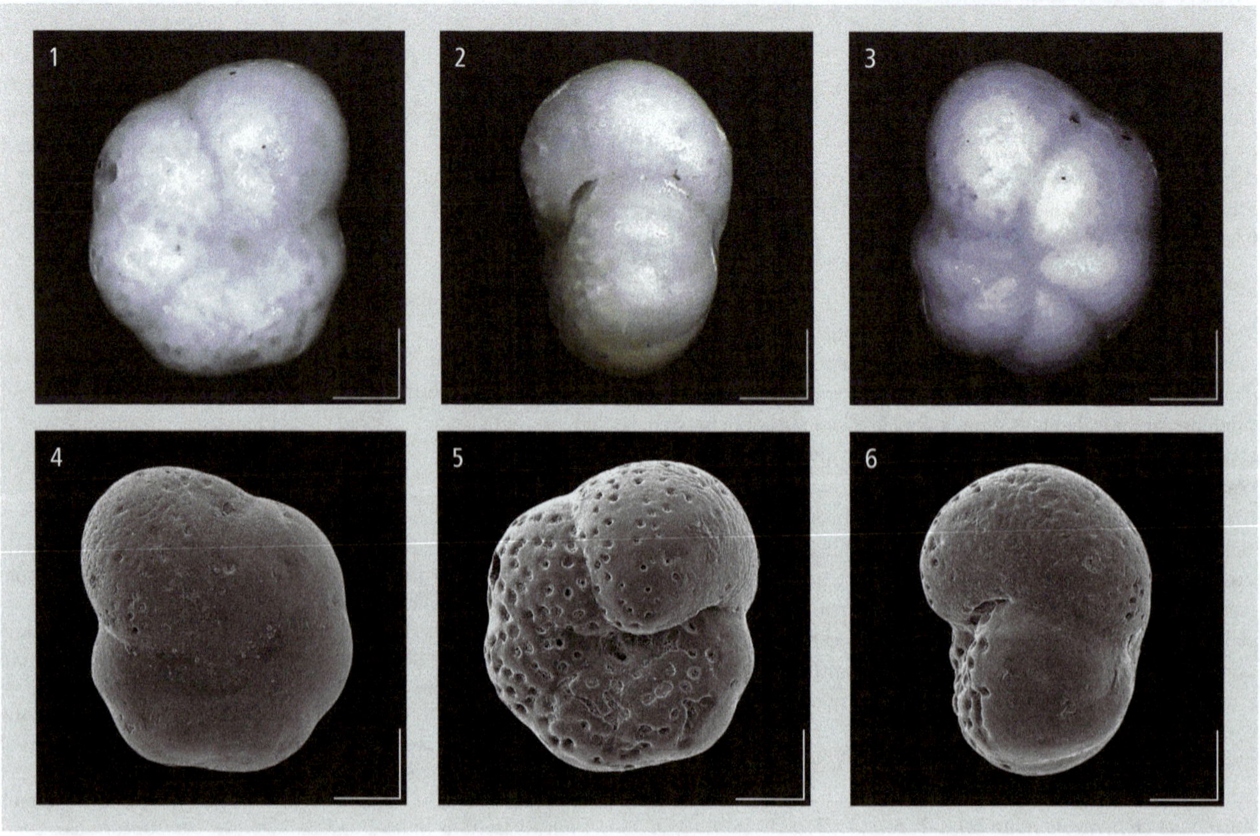


1. *Anomalinoides semicribratus.* Meteor Rise, South Atlantic; ODP Leg 114, Hole 703A, Core 18, Section 4, 60–64 cm. Umbilical view, NHMUK PM DI 0052, scale bar = 104 μm.

2. *Anomalinoides semicribratus.* Meteor Rise, South Atlantic; ODP Leg 114, Hole 703A, Core 18, Section 4, 60–64 cm. Apertural view, NHMUK PM DI 0053, scale bar = 109 μm.

3. *Anomalinoides semicribratus.* Meteor Rise, South Atlantic; ODP Leg 114, Hole 703A, Core 18, Section 4, 60–64 cm. Spiral view, NHMUK PM DI 0054, scale bar = 111 μm.

4. Anomalinoides semicribratus. Meteor Rise, South Atlantic; ODP Leg 114, Hole 703A, Core 18, Section 4, 60–64 cm. Spiral view, NHMUK PM DI 0055, scale bar = 117 μm.

5. *Anomalinoides semicribratus.* Meteor Rise, South Atlantic; ODP Leg 114, Hole 703A, Core 18, Section 4, 60–64 cm. Umbilical view, NHMUK PM DI 0056, scale bar = 117 μm.

6. *Anomalinoides semicribratus.* Meteor Rise, South Atlantic; ODP Leg 114, Hole 703A, Core 18, Section 4, 60–64 cm. Apertural view, NHMUK PM DI 0057, scale bar = 112 μm.

## Original Designation

*Anomalina pompilioides* Galloway and Heminway var. *semicribrata* Beckmann, 1954, p. 400, pl. 27, fig. 3, textfigs. 24–25.

## Type Specimen

Holotype (No. 305103) deposited in the paleontological collections of The Natural History Museum, Basel, Switzerland.

## Type Level

Lower Oligocene. Upper part of the Oceanic Formation, Codrington College Beds in 120 meters of yellowish, argillaceous, *Globigerina* marl. Occurs also throughout all the other beds of the Oceanic Formation (Middle Eocene to Lower Oligocene).

## Type Locality

Senn Locality S. 830 about 220 meters northeast of Codrington College, above cane track, southwest of Consets Bay and south of Bath railway station, in the southeastern part of the island of Barbados. Found also elsewhere in eastern Barbados, near Mount Hillaby and Bath.

## Description

The test forms a very low trochospire, lobulate in outline and unequally biconvex in cross-section, with a broadly rounded periphery. The spiral side is partially evolute and slightly convex, while the umbilical side is involute and more convex with a small, open umbilicus. Five to seven inflated chambers combine to form the last whorl, each increasing gradually in size. These chambers are separated by depressed, slightly curved sutures. Chamber walls are calcareous and coarsely perforate on the spiral side; relatively smooth and more finely perforate on the umbilical side. The primary aperture is a low, equatorial arch, bordered by a thickened lip, extending on to the spiral side beneath the protruding apertural lip. Previous supplementary apertures on the spiral side are gradually closed off.

## Synonyms and Citations

*Anomalina grosserugosa* (Gümbel). –Nuttall, 1928, p. 99, pl. 7, figs. 18–19.

*Gavelinella semicribrata* (Beckmann). –Miller et al., 1985, pl. 6, figs. 1–2.

*Anomalinoides semicribratus* (Beckmann). –van Morkhoven et al., 1986, p. 147, pl. 48. –Katz and Miller, 1991, pl. 3, figs. 9–11. –Robertson, 1998, p. 242, pl. 97, fig. 3. –Kuhnt et al., 2002, p. 138, pl. 6. figs. 3–8.

## Remarks

Van Morkhoven et al. (1986) considered *Anomalinoides rubiginosus* (Cushman), *Anomalinoides capitatus* (Gümbel), *Anomalinoides semicribratus* (Beckmann), and *Anomalinoides globulosus* (Chapman and Parr) to represent a chronocline ranging from the Campanian to the Recent. *Anomalinoides globulosus* has a distinct apertural lip, and is less perforate than *Anomalinoides semicribratus*. Transitional forms between *Anomalinoides capitatus* and *Anomalinoides semicribratus* and between *Anomalinoides globulosus* and *Anomalinoides semicribratus* occur at the beginning and end of the range.

## Biogeography

Worldwide. Recorded from the North and South Atlantic, Pacific, Caribbean, Gulf of Mexico, and Europe (van Morkhoven et al., 1986).

## Bathymetry

Middle bathyal to abyssal (van Morkhoven et al. 1986).

## Chronostratigraphy

Middle Eocene (Lutetian) to Middle Miocene (Serravallian).

# References


Beckmann, J.P., 1954, Die Foraminiferen der Oceanic Formation (Eocaen–Oligocaen) von Barbados, Kl. Antillen. Eclogae Geologicae Helvetiae, vol. 46, pp. 301–412.

Katz, M.E. and Miller, K.G., 1991, Early Paleogene benthic foraminiferal assemblages and stable isotopes in the Southern Ocean. In Ciesielski, P.F., Kristoffersen, Y., et al., ed., Scientific Results of the Proceedings of the Ocean Drilling Program, vol. 114, pp. 481–512.

Kuhnt, W., Holbourn, A.E., and Zhao, Q., 2002, The early history of the South China Sea: Evolution of Oligocene–Miocene deep water environments. Revue de Micropaléontologie, vol. 45, pp. 99–159.

Miller, K.G., Curry, W.B., and Ostermann, D.R., 1985, Late Paleogene (Eocene to Oligocene) benthic foraminiferal oceanography of the Goban Spur region, Deep Sea Drilling Project Leg 80. In de Graciansky, P.C., Poag, C.W., et al., ed., Initial Reports of the Deep Sea Drilling Project, Washington (U.S. Government Printing Office), vol. 80, pp. 505–538.

Nuttall, W.L.F., 1928, Tertiary foraminifera from the Naparima region of Trinidad (British West Indies). Quarterly Journal of the Geological Society of London, vol. 84, pp. 57–117.

Robertson, B.E., 1998, Systematics and paleoecology of the benthic foraminiferida from the Buff Bay section, Miocene of Jamaica. Micropaleontology, vol. 44, supplement 2, pp. 1–266.

van Morkhoven, F.P.C.M., Berggren, W.A., Edwards, A.S., et al., 1986, Cenozoic cosmopolitan deep-water benthic foraminifera. Bulletin des Centres de Recherches Exploration-Production Elf-Aquitaine, Mémoire, vol. 11, pp. 1–421.

# *Aragonia aragonensis*

## (Nuttall), 1930

1. *Aragonia aragonensis.* Allison Guyot, Mid-Pacific Mountains, Northwest Pacific; ODP Leg 142, Hole 865B, Core 10, Section 5, 111–116 cm. Lateral view, NHMUK PM DI 0058, scale bar = 46 μm.

2. *Aragonia aragonensis.* Allison Guyot, Mid-Pacific Mountains, Northwest Pacific; ODP Leg 142, Hole 865B, Core 10, Section 5, 111–116 cm. Lateral view, NHMUK PM DI 0059, scale bar = 47 μm.

3. *Aragonia aragonensis.* Allison Guyot, Mid-Pacific Mountains, Northwest Pacific; ODP Leg 142, Hole 865B, Core 10, Section 5, 111–116 cm. Lateral view, NHMUK PM DI 0060, scale bar = 47 μm.

### Original Designation

*Textularia aragonensis* Nuttall, 1930, p. 280, pl. 23, fig. 6.

### Type Specimen

Holotype (CC 25671) deposited in the collections of the United States National Museum, Washington, DC.

### Type Level

Lower Eocene, Aragon Formation.

### Type Locality

1850 meters north and 85° west of El Tule, Tampico region, Mexico.

### Description

Test forms an elongate, compressed, biserial series; suboval or fan-shaped in outline, elliptical in cross-section with a tapered initial portion, and broadly rounded apertural end. The widest portion of the test is approximately at the center. Chambers are broad and low (approximately three times as wide as high), increasing rapidly in breadth, and separated by oblique, thin, elevated, limbate sutures with short, straight projections extending on to the surface of the test. Chamber walls are calcareous, without pores, and ornamented by the pattern of elevated sutures with raised projections. The primary aperture is a small slit at the base of the last chamber without a toothplate.

### Synonyms and Citations

*Bolivina capdevilensis* Cushman and Bermúdez, 1937, p. 14, pl. 1, figs. 49–50.

*Bolivina semireticula* LeRoy, 1953, p. 20, pl. 8, fig. 26.

*Bolivinoides compressa* Olsson, 1960, p. 30, pl. 4, figs. 20–21.

*Aragonia aragonensis* (Nuttall). –Tjalsma and Lohmann, 1983, p. 23, pl. 11, fig. 2. –van Morkhoven et al., 1986, p. 308, pls. 101A–C. –Thomas, 1990, pl. 1, fig. 10. –Bolli et al., 1994, p. 130, fig. 35.5. –Robertson, 1998, p. 128, pl. 49, fig. 1. –Ortiz and Thomas, 2006, p. 112, pl. 4, figs. 1–3.

### Biogeography

Worldwide. This species has been widely recorded from many locations in Europe, the USA, Mexico, the Caribbean, the Polish Carpathians, Israel, the North and South Atlantic, the Gulf of Mexico, the Weddell Sea, and the Indian Ocean (see van Morkhoven et al., 1986).

### Bathymetry

Bathyal and abyssal.

### Chronostratigraphy

Late Paleocene (Thanetian) to Middle Eocene (Bartonian).

## References


Bolli, H.M., Beckmann, J.P., and Saunders, J.B., 1994, Benthic Foraminiferal Biostratigraphy of the South Caribbean Region. Cambridge, Cambridge University Press, 408 pp.

Cushman, J.A. and Bermúdez, P.J., 1937, Further new species of foraminifera from the Eocene of Cuba. Contributions from the Cushman Laboratory for Foraminiferal Research, vol. 13, pp. 1–29.

LeRoy, L.W., 1953, Biostratigraphy of the Maqfi section, Egypt. Memoirs of the Geological Society of America, vol. 34, pp. 1–73.

Nuttall, W.L.F., 1930, Eocene foraminifera from Mexico. Journal of Paleontology, vol. 4, pp. 271–293.

Olsson, R.K., 1960, Foraminifera of latest Cretaceous and earliest Tertiary age in the New Jersey Coastal Plain. Journal of Paleontology, vol. 34, pp. 1–58.

Ortiz, S. and Thomas, E., 2006, Lower–middle Eocene benthic foraminifera from the Fortuna Section (Betic Cordillera, southeastern Spain). Micropaleontology, vol. 52, pp. 97–150.

Robertson, B.E., 1998, Systematics and paleoecology of the benthic foraminiferida from the Buff Bay section, Miocene of Jamaica. Micropaleontology, vol. 44, supplement 2, pp. 1–266.

Thomas, E., 1990, Late Cretaceous through Neogene benthic foraminifers. In Barker, P.F., Kennett, J.P., et al., ed, Scientific Results of the Proceedings of the Ocean Drilling Program, vol. 113, pp. 571–594.

Tjalsma, R.C. and Lohmann, G.P., 1983, Paleocene–Eocene bathyal and abyssal benthic foraminifera from the Atlantic Ocean. Micropaleontology, Special Publication, no. 4, pp. 1–90.

van Morkhoven, F.P.C.M., Berggren, W.A., Edwards, A.S., et al., 1986, Cenozoic cosmopolitan deep-water benthic foraminifera. Bulletin des Centres de Recherches Exploration-Production Elf-Aquitaine, Mémoire, vol. 11, pp. 1–421.

# *Aragonia velascoensis*

## (Cushman), 1925

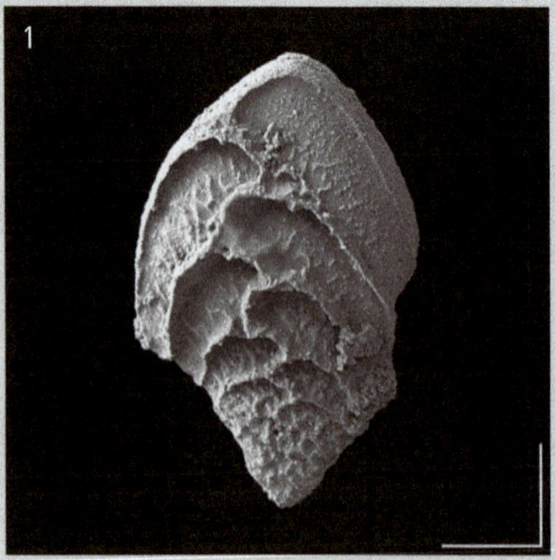

1. *Aragonia velascoensis.* Shatsky Rise, West Central Pacific; DSDP Leg 86, Site 577, Core 10, Section 5, 53–60 cm. Lateral view, NHMUK PM DI 0061, scale bar = 69 μm.

2. *Aragonia velascoensis.* North East Pacific Rise, North Pacific; ODP Leg 142, Hole 865B, Core 15, Section 3, 52–57 cm. Lateral view, NHMUK PM DI 0062, scale bar = 109 μm.

### Original Designation

*Textularia velascoensis* Cushman, 1925, p. 18, pl. 3, fig. 1.

### Type Specimen

Holotype (No. 4343) deposited in the collections of the United States National Museum, Washington, DC.

### Type Level

Paleocene, Velasco Shale. Originally given as Upper Cretaceous by Cushman (1925).

### Type Locality

Tamalte Arroyo, Hacienda El Limón, State of San Luis Potosi, Mexico.

### Description

Test is elongate, biserial, broadly flaring, suboval in outline, and elliptical in cross-section, with a tapered initial portion, and broadly rounded apertural end. The widest portion of the test is approximately at the center. Chambers are broad and low (approximately three to four times as wide as high), increasing rapidly in breadth, and separated by oblique, elevated, limbate sutures with raised projections extending on to the lateral chamber surfaces. The test wall is calcareous, non-porous, and ornamented by the pattern of elevated sutures with raised projections. The primary aperture is an elongate slit at the base of the last chamber without a toothplate.

### Synonyms and Citations

*Bolivinoides trinitatensis* Cushman and Jarvis, 1928, p. 99, pl. 14, fig. 10.

*Bolivinoides ouezzanensis* Rey, 1955, p. 210, pl. 12, fig. 2.

*Bolivinoides stewarti* Mallory, 1959, p. 203, pl. 16, fig. 27.

*Bolivinoides* sp. A Beckmann, 1960, p. 62, fig. 11.

*Aragonia velascoensis* (Cushman). –Tjalsma and Lohmann, 1983, p. 5, pl. 4, fig. 6. –van Morkhoven et al., 1986, p. 340, pl. 111A–B. –Bolli et al., 1994, p. 130, figs. 35.7, 15.

### Remarks

Degree of ornamentation varies significantly.

### Biogeography

Worldwide. Recorded in Mexico, the Gulf of Mexico, the Caribbean, Trinidad, Western and Eastern Europe, the Polish Carpathians, the North and South Atlantic, the Pacific, and the western Indian Ocean (van Morkhoven et al., 1986).

### Bathymetry

Upper bathyal to abyssal.

### Chronostratigraphy

Late Cretaceous (Campanian) to Early Eocene (Ypresian).

## References


Beckmann, J.P., 1960, Distribution of benthonic foraminifera at the Cretaceous–Tertiary boundary of Trinidad (West Indies). Report of the International Geological Congress, 21st Session, Norden, Copenhagen, 1960, part 5, The Cretaceous–Tertiary Boundary, pp. 57–69.

Bolli, H.M., Beckmann, J.P., and Saunders, J.B., 1994, Benthic Foraminiferal Biostratigraphy of the South Caribbean Region. Cambridge: Cambridge University Press, 408 pp.

Cushman, J.A., 1925, Some new foraminifera from the Velasco Shale of Mexico. Contributions from the Cushman Laboratory for Foraminiferal Research, vol. 1, no. 1, pp. 18–23.

Cushman, J.A. and Jarvis, P.W., 1928, Cretaceous foraminifera from Trinidad. Contributions from the Cushman Laboratory for Foraminiferal Research, vol. 4, pp. 85–103.

Mallory, V.S., 1959, Lower Tertiary Biostratigraphy of the California Coast Ranges. Tulsa, OK: American Association of Petroleum Geologists, 416 pp.

Rey, M., 1955, Description de quelques espèces nouvelles de foraminifères dans le Nummulitique nord-marocain. Bulletin de la Société géologique de France, series 6, vol. 4, no. 4–6, pp. 209–211.

Tjalsma, R.C. and Lohmann, G.P., 1983, Paleocene–Eocene bathyal and abyssal benthic foraminifera from the Atlantic Ocean. Micropaleontology, Special Publication, no. 4, pp. 1–90.

van Morkhoven, F.P.C.M., Berggren, W.A., Edwards, A.S., et al., 1986, Cenozoic cosmopolitan deep-water benthic foraminifera. Bulletin des Centres de Recherches Exploration-Production Elf-Aquitaine, Mémoire, vol. 11, pp. 1–421.

# *Astrononion stelligerum*

## (d'Orbigny), 1839

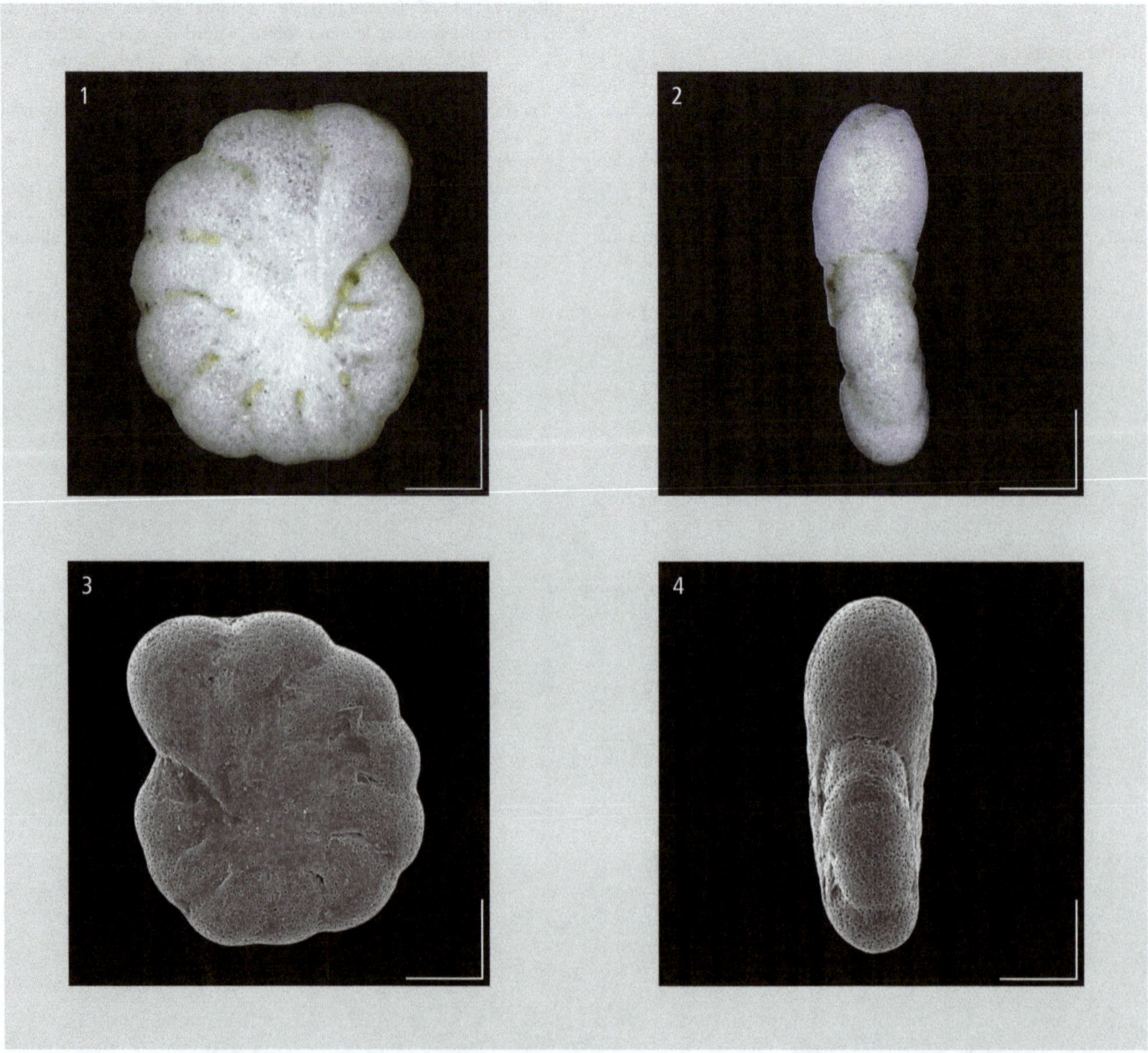


1. *Astrononion stelligerum*. East of the Shetland Islands, North Atlantic, "Porcupine" Station 67-68, depth: 64–75 fathoms. Deposited in the micropaleontological collections of The Natural History Museum, London; NHMUK ZF1998. Umbilical view, NHMUK PM DI 0063, scale bar = 81 μm.

2. *Astrononion stelligerum*. East of the Shetland Islands, North Atlantic, "Porcupine" Station 67-68, depth: 64–75 fathoms. Deposited in the micropaleontological collections of The Natural History Museum, London; NHMUK ZF1998. Apertural view, NHMUK PM DI 0064, scale bar = 83 μm.

3. *Astrononion stelligerum*. East of the Shetland Islands, North Atlantic, "Porcupine" Station 67-68, depth: 64–75 fathoms. Deposited in the micropaleontological collections of The Natural History Museum, London; NHMUK ZF1998. Umbilical view, NHMUK PM DI 0066, scale bar = 69 μm.

4. *Astrononion stelligerum*. East of the Shetland Islands, North Atlantic, "Porcupine" Station 67-68, depth: 64–75 fathoms. Deposited in the micropaleontological collections of The Natural History Museum, London; NHMUK ZF1998. Umbilical view, NHMUK PM DI 0068, scale bar = 91 μm.

### Original Designation
*Nonionina stelligera* d'Orbigny, 1839, p. 128, pl. 3, figs. 1–2.

### Type Specimen
Neotype designated by Le Calvez (1974) and deposited in the collections of the Laboratoire de Paléontologie, Muséum d'Histoire Naturelle, Paris, France.

### Type Level
Recent. Marine sands.

### Type Locality
Tenerife, Canary Islands.

### Description
Test forms an involute, compressed planispire; lobulate in outline, with a rounded periphery. Composed of approximately eleven moderately inflated chambers that increase gradually in size, each chamber is separated by a slightly curved, depressed suture. Chambers are also characterized by triangular flaps extending from the umbilicus. These flaps partially cover the sutures and form a stellate pattern. The test wall is thin, calcareous, finely perforate, and smooth. The primary aperture is a low interiomarginal, equatorial slit, bordered by a lip that extends to the umbilicus.

### Synonyms and Citations
*Nonionina stelligera* d'Orbigny. –Brady, 1884, p. 728, pl. 109, figs. 3–4.

*Astrononion stelligerum* (d'Orbigny). –Barker, 1960, pl. 109, figs. 3–4. –Jones, 1994, p. 107, pl. 109, figs. 3–4. –Milker and Schmiedl, 2012, p. 113, figs. 26.7–8.

### Remarks
Type species of the genus *Astrononion* Cushman and Edwards, 1937.

There is some doubt as to the conspecificity of Brady's specimen (northern cool-temperate) with d'Orbigny's original specimens (warm-temperate).

### Biogeography
Worldwide.

### Bathymetry
Middle neritic. Reported from 64 and 75 fathoms at "Porcupine" stations east of Shetland, North Atlantic (Jones, 1994).

### Chronostratigraphy
Recent.

## References


Barker, R.W., 1960, Taxonomic notes on the species figured by H.B. Brady in his report on the foraminifera dredged by H.M.S. Challenger during the years 1873–1876. Society of Economic Paleontologists and Mineralogists Special Publication, vol. 9, pp. 1–238.

Brady, H.B., 1884, Report on the foraminifera dredged by H.M.S. Challenger during the years 1873–1876. In Brady, H.B., ed., Report on the Scientific Results of the Voyage of the H.M.S. Challenger during the years 1873–1876. Zoology, vol. 9, pp. 1–814.

d'Orbigny, A.D., 1839, Foraminifères des Îles Canaries. In Barker-Webb, P. and Berthelot, S., ed., Historie Naturelle des Îles Canaries, vol. 2, part 2: Paris (Béthune), pp. 119–146.

Jones, R.W., 1994, The Challenger Foraminifera. Oxford: Oxford University Press, 149 pp.

Le Calvez, Y., 1974, Révision des foraminifères de la collection d'Orbigny. I. Foraminifères des Îles Canaries. Cahiers de Micropaléontologie, vol. 2, pp. 1–108.

Milker, Y. and Schmiedl, G., 2012, A taxonomic guide to modern benthic shelf foraminifera of the western Mediterranean Sea. Palaeontologia Electronica, vol. 15, Issue 2;16A, 134 pp.

# *Bigenerina nodosaria*

## d'Orbigny, 1826

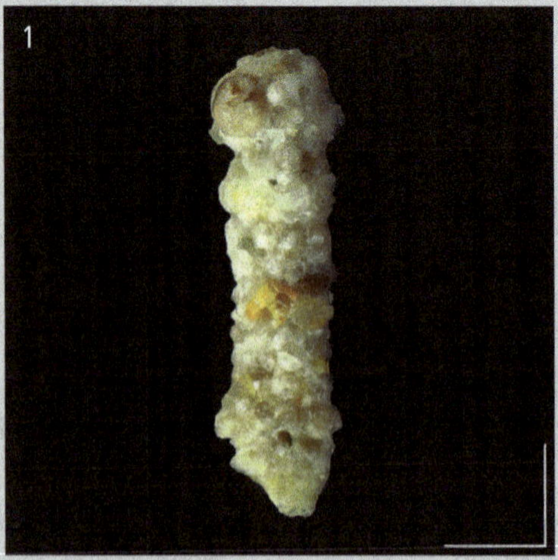


1. *Bigenerina nodosaria.* West Africa, Lutze Collection (West Africa), no. 59. Institute of Geosciences, Kiel University, Germany. Lateral view, NHMUK PM DI 0069, scale bar = 322 μm.

### Original Designation

*Bigenerina nodosaria* d'Orbigny, 1826, p. 261, pl. 11, figs. 9–12.

### Type Specimen

Deposited in the collections of the Laboratoire de Paléontologie, Muséum d'Histoire Naturelle, Paris, France.

### Type Level

Recent, probably marine sands.

### Type Locality

Adriatic Sea.

### Description

Test forms an initially elongate and rectilinear biserial series, becoming uniserial with circular cross-section in the later portion. Chambers are inflated, increase gradually in size, and are separated by depressed sutures that are straight and at right angles to the test axis in the uniserial part. Chamber walls are very coarsely agglutinated and canaliculate. The primary aperture is round and areal in the uniserial stage.

### Synonyms and Citations

*Bigenerina nodosaria* d'Orbigny. –Brady, 1884, p. 369, pl. 44, figs. 14–18. –Loeblich and Tappan, 1987, p. 172, pl. 191, figs. 1–2. –Jones, 1994, p. 49, pl. 44, figs. 14–18. –Loeblich and Tappan, 1994, p. 27, pl. 31, figs. 8–12; pl. 32, figs. 11–12. –Milker and Schmiedl, 2012, p. 38, figs. 10.10–12.

### Remarks

Type species of the genus *Bigenerina* d'Orbigny, 1826.

### Biogeography

Worldwide.

### Bathymetry

Upper bathyal.

### Chronostratigraphy

Late Pliocene (Gelasian) to Recent.

## References


Brady, H.B., 1884, Report on the foraminifera dredged by H.M.S. Challenger during the years 1873–1876. In Brady, H.B., ed., Report on the Scientific Results of the Voyage of the H.M.S. Challenger during the years 1873–1876. Zoology, vol. 9, pp. 1–814.

d'Orbigny, A., 1826, Tableau méthodique de la classe des Céphalopodes. Annales des Sciences Naturelles, vol. 7, pp. 245–314.

Jones, R.W., 1994, The Challenger Foraminifera. Oxford: Oxford University Press, 149 pp.

Loeblich, A.R. Jr and Tappan, H., 1987, Foraminiferal Genera and Their Classification. New York: Van Nostrand Reinhold, 970 pp.

Loeblich, A.R. Jr and Tappan, H., 1994, Foraminifera of the Sahul Shelf and Timor Sea. Special Publications of the Cushman Laboratory for Foraminiferal Research, vol. 31, pp. 1–661.

Milker, Y. and Schmiedl, G., 2012, A taxonomic guide to modern benthic shelf foraminifera of the western Mediterranean Sea. Palaeontologia Electronica, vol. 15, Issue 2;16A, 134 pp.

# *Bolivina decussata*

## Brady, 1881

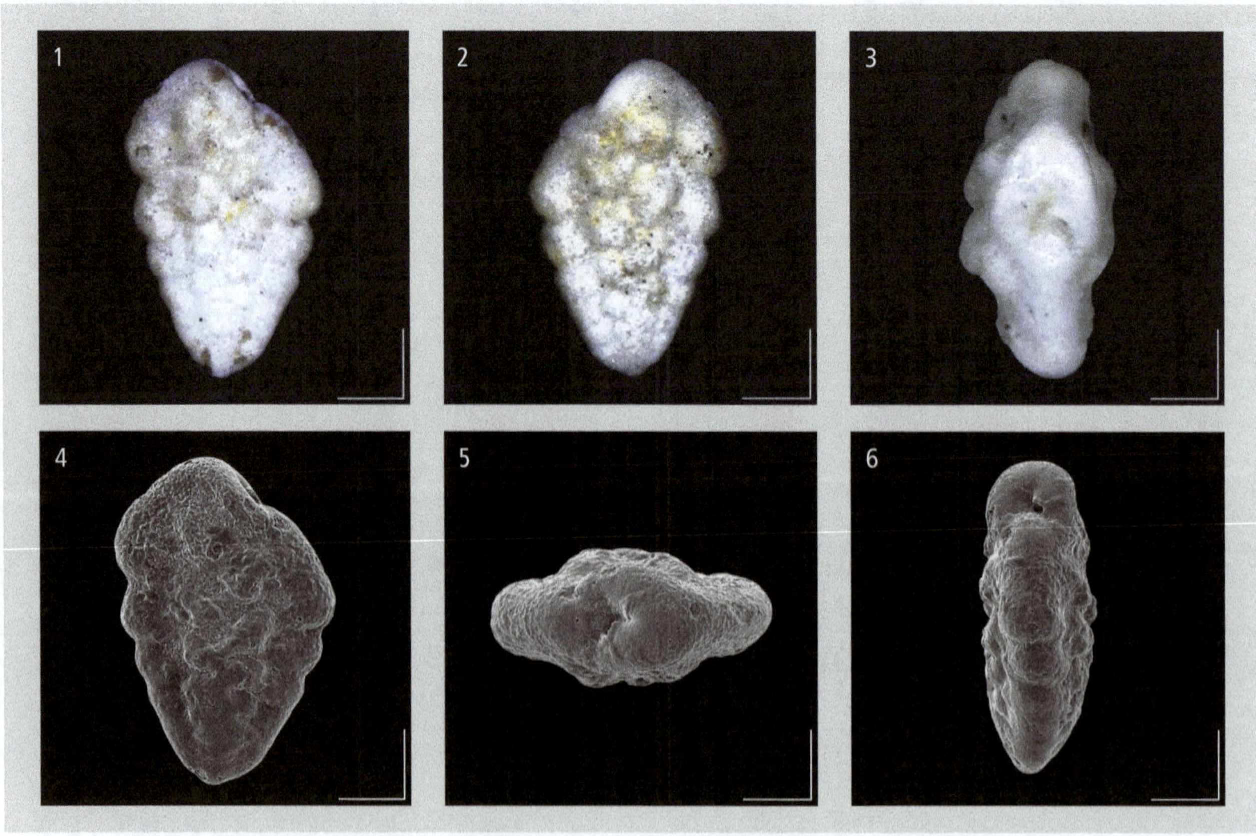


1. *Bolivina decussata*. North of Juan Fernandez, East Pacific; "Challenger" Station 300, depth: 1375 fathoms. Deposited in the micropaleontological collections of The Natural History Museum, London; NHMUK ZF1176. Syntype. Lateral view, NHMUK PM DI 0070, scale bar = 83 μm.

2. *Bolivina decussata*. North of Juan Fernandez, East Pacific; "Challenger" Station 300, depth: 1375 fathoms. Deposited in the micropaleontological collections of The Natural History Museum, London; NHMUK ZF1176. Syntype. Lateral view, NHMUK PM DI 0071, scale bar = 88 μm.

3. *Bolivina decussata*. North of Juan Fernandez, East Pacific; "Challenger" Station 300, depth: 1375 fathoms. Deposited in the micropaleontological collections of The Natural History Museum, London; NHMUK ZF1176. Syntype. Apertural view, NHMUK PM DI 0072, scale bar = 86 μm.

4. *Bolivina decussata*. North of Juan Fernandez, East Pacific; "Challenger" Station 300, depth: 1375 fathoms. Deposited in the micropaleontological collections of The Natural History Museum, London; NHMUK ZF1176. Syntype. Lateral view, NHMUK PM DI 0073, scale bar = 88 μm.

5. *Bolivina decussata*. North of Juan Fernandez, East Pacific; "Challenger" Station 300, depth: 1375 fathoms. Deposited in the micropaleontological collections of The Natural History Museum, London; NHMUK ZF1176. Syntype. Apertural view, NHMUK PM DI 0074, scale bar = 72 μm.

6. *Bolivina decussata*. North of Juan Fernandez, East Pacific; "Challenger" Station 300, depth: 1375 fathoms. Deposited in the micropaleontological collections of The Natural History Museum, London; NHMUK ZF1176. Syntype. Edge view, NHMUK PM DI 0075, scale bar = 92 μm.

## Original Designation

*Bulimina (Bolivina) decussata* Brady, 1881, p. 58 (type description), illustrated as *Bolivina decussata* in Brady, 1884, pl. 53, figs. 12–13.

## Type Specimen

Figured specimens are housed in the micropaleontological collections of The Natural History Museum, London, UK, and are registered in slide ZF1176.

## Type Level

Recent, surface sediments.

## Type Locality

Not designated. Syntypes are from "Challenger" Station 300, latitude 33° 42′ S, longitude 78° 18′ W, north of Juan Fernandez Island in the South Pacific, 1375 fathoms.

## Description

Test forms an elongate, flaring, slightly compressed, lobulate, triserial series; triangular in outline with a subrounded periphery. The globular chambers increase gradually in size and are separated by depressed sutures. Chamber walls are calcareous and finely perforate, with prominent round protuberances. The primary aperture is a loop-shaped opening and extends from the base of the last chamber, with a rounded lip and internal toothplate.

## Synonyms and Citations

*Bolivina decussata* Brady. –Brady, 1884, p. 423, pl. 53, figs. 12–13. –Barker, 1960, pl. 53, figs. 12–13. –Jones, 1994, p. 58, pl. 53, figs. 12–13.

## Biogeography

Worldwide.

## Bathymetry

Lower bathyal. Reported between 1375 and 1450 fathoms at "Challenger" stations in the South Pacific (Jones, 1994).

## Chronostratigraphy

Late Pliocene (Gelasian) – Recent.

## References


Barker, R.W., 1960, Taxonomic notes on the species figured by H.B. Brady in his report on the foraminifera dredged by H.M.S. Challenger during the years 1873–1876. Society of Economic Paleontologists and Mineralogists Special Publication, vol. 9, pp. 1–238.

Brady, H.B., 1881, Notes on some of the reticularian Rhizopoda of the Challenger Expedition, Part III. Quarterly Journal of Microscopical Science (New Series), vol. 21, pp. 31–71.

Brady, H.B., 1884, Report on the foraminifera dredged by H.M.S. Challenger during the years 1873–1876. In Brady, H.B., ed., Report on the Scientific Results of the Voyage of the H.M.S. Challenger during the years 1873–1876. Zoology, vol. 9, pp. 1–814.

Jones, R.W., 1994, The Challenger Foraminifera. Oxford: Oxford University Press, 149 pp.

# *Bolivina huneri*

## Howe, 1939

1. *Bolivina huneri.* South of Hayes Fracture Zone, west flank of the Mid-Atlantic Ridge, North Atlantic; DSDP Leg 82, Site 563, Core 22, Section 3, 110–116 cm. Lateral view, NHMUK PM DI 0076, scale bar = 50 μm.

2. *Bolivina huneri.* Allison Guyot, Mid-Pacific Mountains, Northwest Pacific; DSDP Leg 82, Site 563, Core 22, Section 3, 110–116 cm. Lateral view, NHMUK PM DI 0077, scale bar = 50 μm.

3. *Bolivina huneri.* Allison Guyot, Mid-Pacific Mountains, Northwest Pacific.; DSDP Leg 82, Site 563, Core 22, Section 3, 110–116 cm. Lateral view, NHMUK PM DI 0078, scale bar = 52 μm.

## Original Designation

*Bolivina huneri* Howe, 1939, p. 66, pl. 9, figs. 3–4.

## Type Specimen

Holotype (No.125) deposited in the author's collection at Louisiana State University, USA.

## Type Level

Eocene, Clairborne, Cook Mountain Formation. Very fossiliferous, slightly glauconitic marl, 12 feet thick.

## Type Locality

Winn Parish, Louisiana, USA. Type locality directions given by Howe (1939): five feet below top of slope leading down to creek branch, and in upper two feet of marl, in ditch on south side of improved road in the NE 1/4 SW 1/4 SW 1/4 sec. 28, T. 13 N, R. 3 W, Winn Parish, Louisiana, USA.

## Description

Test forms a small, elongate, compressed, biserial series; slightly tapered and oval in cross-section with a subrounded initial portion, and a rounded apertural end. The slightly inflated chambers are separated by slightly depressed sutures that are mostly obscured by surface ornamentation. Chamber walls are calcareous, finely perforate, and ornamented by a pattern of anastomosing costae, extending to all but the last chambers of the test. The aperture is a loop-shaped opening, extending from the base of the last chamber, with a rounded lip and internal toothplate.

## Synonyms and Citations

*Bolivina huneri* Howe. –Mallory, 1959, p. 201, pl. 16, fig. 21. –Miller, 1983, p. 431, pl. 4, fig. 6. –Tjalsma and Lohmann, 1983, p. 23, pl. 11, fig. 5. –Thomas, 1985, p. 675, pl. 3, fig. 1. –Miller and Katz, 1987, p. 124, pl. 1, figs. 7a–b.

## Biogeography

Worldwide. Recorded from Louisiana (Howe, 1939), California (Mallory, 1959), the North and South Atlantic (Miller, 1983; Tjalsma and Lohmann, 1983; Miller and Katz, 1987), and the Pacific (Thomas, 1985).

## Bathymetry

Bathyal to abyssal. Exclusively abyssal in Oligocene samples from the Atlantic (M. Katz, pers. comm. 1998).

## Chronostratigraphy

Middle Eocene (Lutetian) to Late Miocene (Tortonian).

# References

Howe, H.V., 1939, Louisiana Cook Mountain Eocene foraminifera. Bulletin of the Geological Survey of Louisiana, vol. 14, pp. 1–122.

Mallory, V.S., 1959, Lower Tertiary biostratigraphy of the California Coast Ranges. Tulsa, OK: American Association of Petroleum Geologists, 416 pp.

Miller, K.G., 1983, Eocene–Oligocene paleoceanography of the deep Bay of Biscay: benthic foraminiferal evidence. Marine Micropaleontology, vol. 7, pp. 403–440.

Miller, K.G. and Katz, M.E., 1987, Oligocene to Miocene benthic foraminiferal and abyssal circulation changes in the North Atlantic. Micropaleontology, vol. 33, no. 2, pp. 97–149.

Thomas, E., 1985, Late Eocene to Recent deep-sea benthic foraminifers from the central equatorial Pacific Ocean. In Mayer, L., Theyer, F., et al., ed., Initial Reports of the Deep Sea Drilling Project, Washington (U.S. Government Printing Office), vol. 85, pp. 655–694.

Tjalsma, R.C. and Lohmann, G.P., 1983, Paleocene–Eocene bathyal and abyssal benthic foraminifera from the Atlantic Ocean. Micropaleontology, Special Publication, no. 4, pp. 1–90.

# *Bolivinita quadrilatera*

## (Schwager), 1866

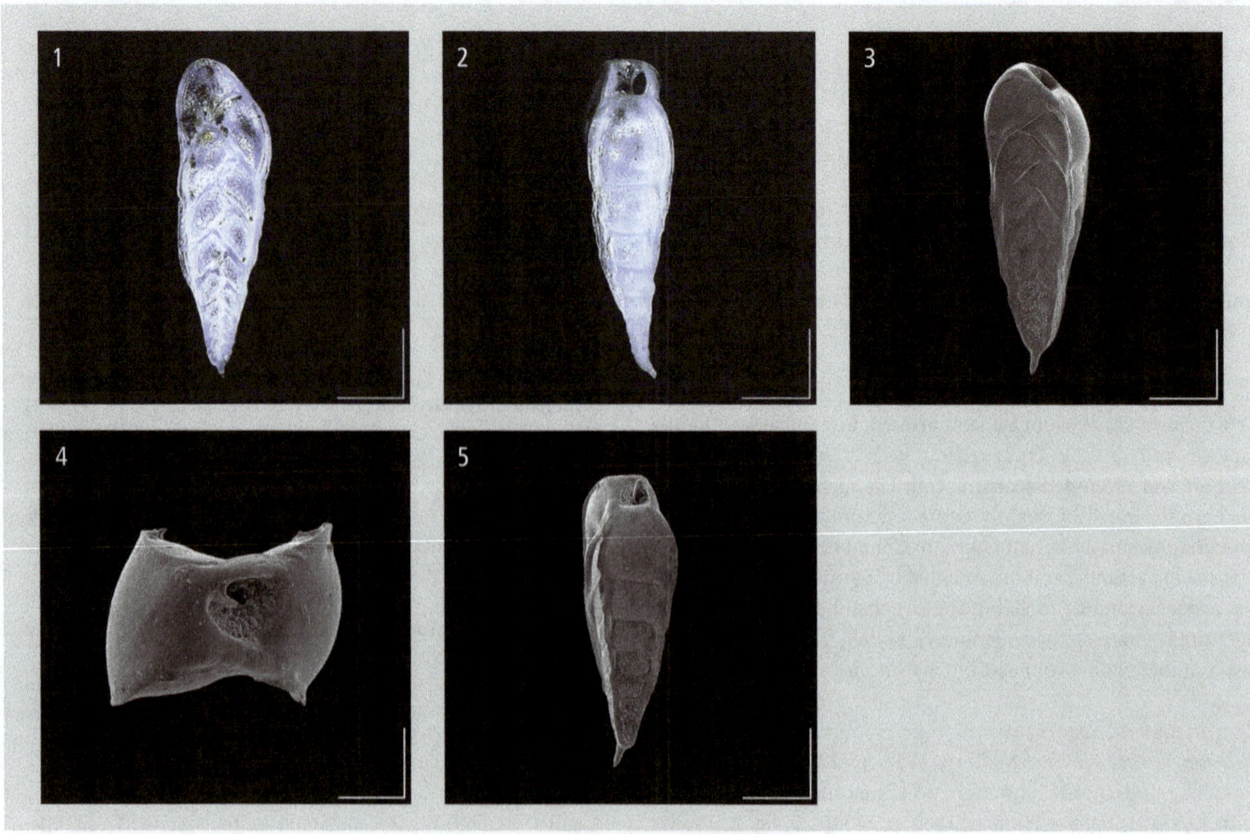


1. *Bolivinita quadrilatera*. Off Tahiti, Pacific Ocean; "Challenger" Station 279C, depth: 620 fathoms. Deposited in the micropaleontological collections of The Natural History Museum, London; NHMUK ZF2458. Lateral view, NHMUK PM DI 0079, scale bar = 231 μm.

2. *Bolivinita quadrilatera*. Off Tahiti, Pacific Ocean; "Challenger" Station 279C, depth: 620 fathoms. Deposited in the micropaleontological collections of The Natural History Museum, London; NHMUK ZF2458. Edge view, NHMUK PM DI 0080, scale bar = 236 μm.

3. *Bolivinita quadrilatera*. Off Tahiti, Pacific Ocean; "Challenger" Station 279C, depth: 620 fathoms. Deposited in the micropaleontological collections of The Natural History Museum, London; NHMUK ZF2458. Oblique lateral view, NHMUK PM DI 0081, scale bar = 238 μm.

4. *Bolivinita quadrilatera*. Off Tahiti, Pacific Ocean; "Challenger" Station 279C, depth: 620 fathoms. Deposited in the micropaleontological collections of The Natural History Museum, London; NHMUK ZF2458. Apertural view, NHMUK PM DI 0082, scale bar = 107 μm.

5. *Bolivinita quadrilatera*. Off Tahiti, Pacific Ocean; "Challenger" Station 279C, depth: 620 fathoms. Deposited in the micropaleontological collections of The Natural History Museum, London; NHMUK ZF2458. Edge view, NHMUK PM DI 0082, scale bar = 262 μm.

## Original Designation
*Textilaria quadrilatera* Schwager, 1866, p. 253, pl. 7, fig. 103.

## Type Specimen
Depository not given. Neotype designated by Srinivasan and Sharma (1980) deposited in the Department of Geology, Banaras Hindu University, India.

## Type Level
Jüngere Tertiärschichten. Obere Thone.

## Type Locality
Kar Nikobar, Indian Ocean.

## Description
Test forms an elongate, tapered, compressed, biserial series; quadrangular in cross-section with concave sides, and a keeled periphery. The low, broad chambers increase gradually in size and are separated by distinct, oblique, raised sutures. Frontal overlap of the chambers occupies the middle third of the test width. Chamber walls are calcareous, finely perforate, and smooth or with a few longitudinal striae in the early stage, and ornamented by lateral keels on the outer edges. A spine is often present on the proloculus. The oval primary aperture starts slightly above the basal suture, extends up the apertural face, and is surrounded by a curving lip that merges into the upper edge of the toothplate, thus delimiting a distinct sulcus.

## Synonyms and Citations
*Textularia quadrilatera* Schwager. –Brady, 1884, p. 358, pl. 42, figs. 8–12.

*Bolivinita quadrilatera* (Schwager). –Barker, 1960, pl. 42, figs. 8–12. –Srinivasan and Sharma, 1980, p. 42, pl. 6, figs. 22–23. –Loeblich and Tappan, 1987, p. 503, pl. 554, figs. 6–10. –Jones, 1994, p. 47, pl. 42, figs. 8–12. –Revets, 1996, p. 6, pl. 1, figs. 1–4.

## Remarks
Type species of the genus *Bolivinita* Cushman, 1927. See Revets (1996) for a detailed description of toothplate in topotype material.

## Biogeography
Worldwide.

## Bathymetry
Middle Bathyal. Reported from 410 and 620 fathoms at "Challenger" stations off Australia and Patagonia (Jones, 1994).

## Chronostratigraphy
Miocene to Recent (Jones, 1994). First occurrence reported from the Middle Miocene *Globorotalia fohsi robusta* Biozone by Srinivasan and Sharma (1980) in the Andaman Islands.

## References

Barker, R.W., 1960, Taxonomic notes on the species figured by H.B. Brady in his report on the foraminifera dredged by H.M.S. Challenger during the years 1873–1876. Society of Economic Paleontologists and Mineralogists Special Publication, vol. 9, pp. 1–238.

Brady, H.B., 1884, Report on the foraminifera dredged by H.M.S. Challenger during the years 1873–1876. In Brady, H.B., ed., Report on the Scientific Results of the Voyage of the H.M.S. Challenger during the years 1873–1876. Zoology, vol. 9, pp. 1–814.

Jones, R.W., 1994, The Challenger Foraminifera. Oxford: Oxford University Press, 149 pp.

Loeblich, A.R. Jr and Tappan, H., 1987, Foraminiferal Genera and Their Classification. New York: Van Nostrand Reinhold, 970 pp.

Revets, S.A., 1996, The generic revision of five families of Rotaliine Foraminifera. Special Publications of the Cushman Foundation for Foraminiferal Research, vol. 34, pp. 1–108.

Schwager, C., 1866, Fossile Foraminiferen von Kar Nikobar, Reise der Österreichischen Fregatte Novara um Erde in den Jahren 1857, 1858, 1859 unten den Befehlen des Commodore B. Von Wüllerstorf-Urbair. Geologischer Theil, vol. 2, no. 1, Geologische Beobachtungen, no. 2, Paläontologische Mittheilungen, pp. 187–268.

Srinivasan, M.S. and Sharma, V., 1980, Schwager's Car Nicobar Foraminifera in the Reports of the Novara Expedition – A Revision. New Delhi: Today & Tomorrow's Printers and Publishers, 83 pp.

# Bolivinoides delicatulus

## Cushman, 1927

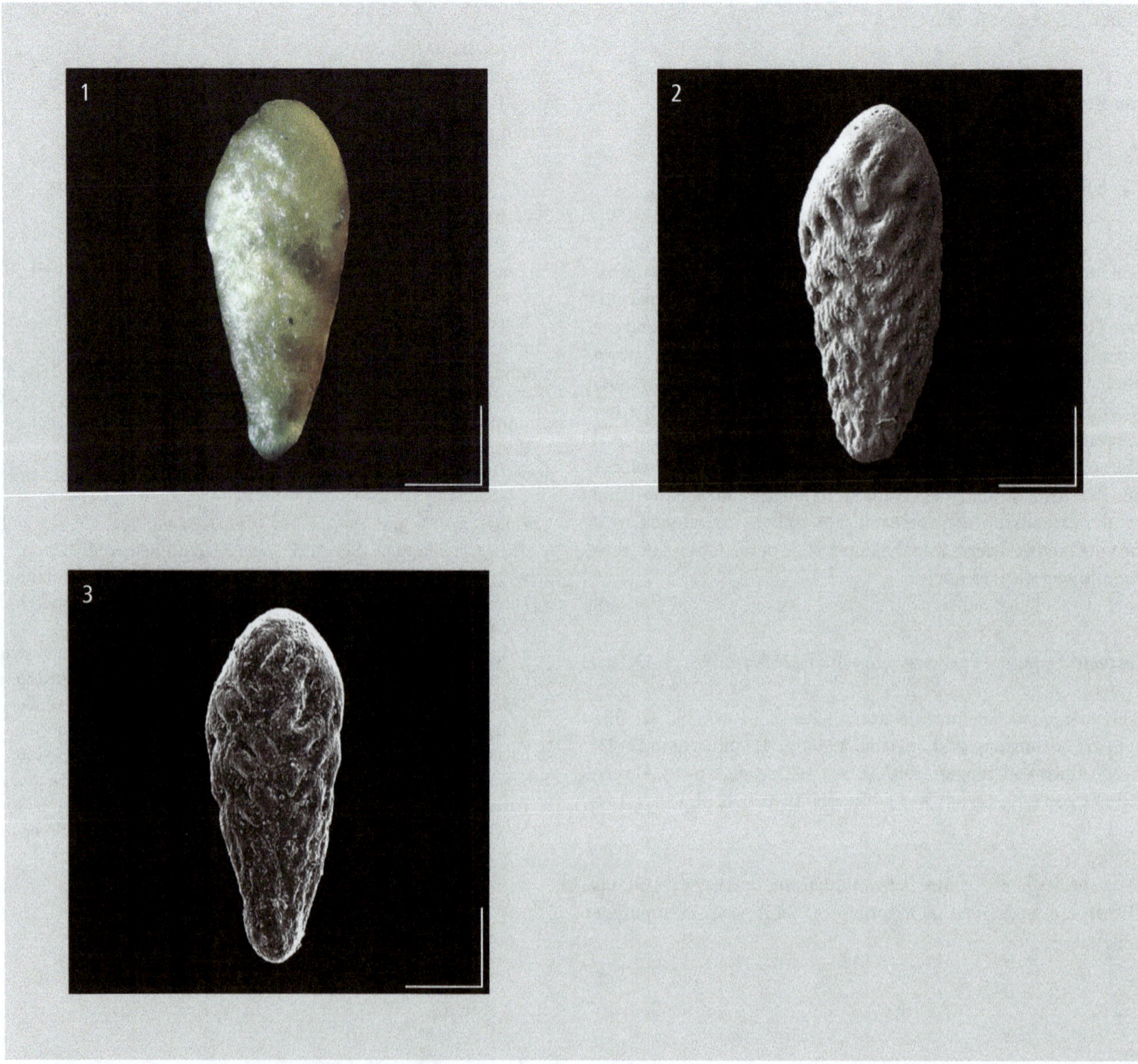

1. *Bolivinoides delicatulus.* Velasco Shale, Hacienda El Limon, Vera Cruz, Mexico, Franco Espanola no. 5, 510 feet, Marland Oil Co. Deposited in the collections of the United States National Museum, Washington, DC; No. 5139. Holotype. Lateral view, NHMUK PM DI 0084, scale bar = 81 μm.

2. *Bolivinoides delicatulus.* Maud Rise, Weddell Sea, Antarctica; ODP Leg 113, Hole 690B, Core 19, Section 3, 72–74 cm. Lateral view, NHMUK PM DI 0084, scale bar = 83 μm.

3. *Bolivinoides delicatulus.* Northeast Georgia Rise, Subantarctic South Atlantic; ODP Leg 114, Hole 699A, Core 54, Section 2, 22–24 cm. Lateral view, NHMUK PM DI 0086, scale bar = 97 μm.

## Original Designation

*Bolivinoides decorata* (Jones) var. *delicatula* Cushman, 1927, p. 90, pl. 12, fig. 8.

## Type Specimen

Holotype (No. 5139) deposited in the collections of the United States National Museum, Washington, DC.

## Type Level

Paleocene, Velasco Shale. Originally given as Upper Cretaceous by Cushman (1927).

## Type Locality

Hacienda El Limón, Vera Cruz, Mexico.

## Description

Test forms an elongate, compressed, biserial series; gently tapered in outline and oval in cross-section. The length/breadth ratio of the test varies from 2.0 to 2.5. This species' numerous, broad, low chambers are separated by sutures that are typically obscured by surface ornamentation. Chamber walls are calcareous, finely perforate, and ornamented by a distinctive pattern of lobes. The primary aperture is an areal slit with an internal toothplate.

## Synonyms and Citations

*Bolivina decorata* Jones. –Cushman, 1926, p. 586, pl. 15, fig. 11.

*Bolivina sidestrandensis* (Barr). –Boltovskoy and Watanabe, 1982, p. 32, fig. 2:15.

*Bolivinoides delicatulus* Cushman. –van Morkhoven et al., 1986, p. 337, pl. 110. –Mackensen and Berggren, 1992, p. 620, pl. 2, fig. 4.

*Bolivinoides delicatulus delicatulus* Cushman. –Bolli et al., 1994, p. 129, fig. 34.37.

## Remarks

*Bolivinoides delicatulus* Cushman has a complex taxonomic history (see Remarks in van Morkhoven et al., 1986). Swiecicki (1980) suggested two major evolutionary lineages for the genus *Bolivinoides*, originating in the Early Campanian from an ancestral species. One trend evolved increasingly compressed rhomboid tests with acute margins, and an increasing amount of surface ornamentation (*Bolivinoides draco* (Marsson)); the other lineage evolved more elongated, slender, less ornamented tests (*Bolivinoides delicatulus*).

## Biogeography

Worldwide. Recorded from Trinidad, Arkansas, Texas, Tennessee, the Gulf of Mexico, Europe, Israel, India, the western Indian Ocean, the Pacific, the Caribbean, and the Gulf of Mexico (van Morkhoven et al., 1986).

## Bathymetry

Mainly bathyal, occasionally neritic or abyssal.

## Chronostratigraphy

Late Cretaceous (Campanian) to Early Eocene (Ypresian).

# References

Bolli, H.M., Beckmann, J.P., and Saunders, J.B., 1994, Benthic Foraminiferal Biostratigraphy of the South Caribbean Region. Cambridge: Cambridge University Press, 408 pp.

Boltovskoy, E. and Watanabe, S., 1982, Foraminiferos bentónicos batiales del Cenozoico del Atlántico sudoriental. Physis (Buenos Aires), Secc. A, vol. 41, no. 100, pp. 29–41.

Cushman, J.A., 1926, The foraminifera of the Velasco Shale of the Tampico Embayment. Bulletin of the American Association of Petroleum Geologists, vol. 10, no. 6, pp. 581–612.

Cushman, J.A., 1927, American Upper Cretaceous species of Bolivina and related species. Contributions from the Cushman Laboratory for Foraminiferal Research, vol. 2, no. 4, pp. 85–91.

Mackensen, A. and Berggren, W.A., 1992, Paleogene benthic foraminifers from the southern Indian Ocean (Kerguelen Plateau): biostratigraphy and paleoecology. In Wise, S.W. Jr, Schlich, R., et al., ed., Scientific Results of the Proceedings of the Ocean Drilling Program, vol. 120, pp. 603–630.

Swiecicki, A., 1980, A foraminiferal biostratigraphy of the Campanian and Maastrichtian chalks of the United Kingdom. Unpublished PhD thesis, vol. 1, 358 pp.

van Morkhoven, F.P.C.M., Berggren, W.A., Edwards, A.S., et al., 1986, Cenozoic cosmopolitan deep-water benthic foraminifera. Bulletin des Centres de Recherches Exploration-Production Elf-Aquitaine, Mémoire, vol. 11, pp. 1–421.

# Bolivinoides draco

## (Marsson), 1878

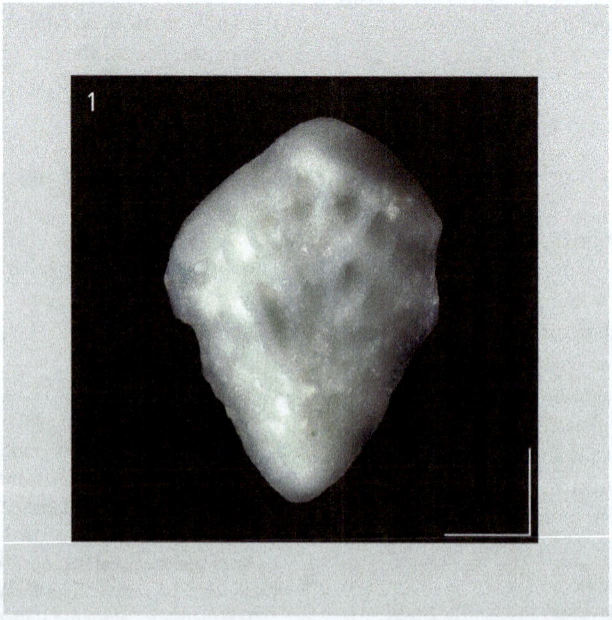

1. *Bolivinoides draco*. Gulf of Mexico; Eureka Core E68-139. Deposited in the collections at the United States National Museum, Washington, DC. Specimen illustrated in van Morkhoven et al. 1986, Ref. no. 124. Lateral view, NHMUK PM DI 0087, scale bar = 197 μm.

### Original Designation
*Bolivina draco* Marsson, 1878, p. 157, pl. 3, fig. 25.

### Type Specimen
Depository not given.

### Type Level
Cretaceous, Kreideformation, Weisse Schreibkreide.

### Type Locality
Island of Rügen, Germany.

### Description
Test forms an elongate, compressed, subrhombohedral, biserial series; tapered in outline, and oval in cross-section with a subacute initial portion, a rounded apertural end, and a subacute, keeled periphery. Up to sixteen low, slightly inflated chambers increase rapidly in width and are separated by oblique sutures, mostly obscured by surface ornamentation. Chamber walls are calcareous, finely perforate, and ornamented by small, rounded ponticuli spanning retral processes perpendicularly to the sutures. Ponticuli in earlier chambers join to form costae and a median canal, giving the test a ribbed appearance. The primary aperture is loop-shaped, surrounded by a thin lip, with an internal tooth-plate protruding slightly into the apertural space.

### Synonyms and Citations
*Bolivinoides draco* (Marsson). –van Morkhoven et al., 1986, p. 378, pl. 124. –Revets, 1996, p. 9, pl. 4, figs. 1–4.

### Biogeography
Worldwide. Widely recorded through the Campanian and Maastrichtian.

### Bathymetry
Lower neritic and bathyal.

### Chronostratigraphy
Late Cretaceous (Campanian to Maastrichtian).

## References

Marsson, T., 1878, Die Foraminiferen der weissen Schreibkreide der Insel Ruegen. Mitteilungen des Naturwissenschaftlichen Vereins für Neu-Vorpommern und Rugen in Greifswald, vol. 10, pp. 115–196.

Revets, S.A., 1996, The generic revision of five families of Rotaliine Foraminifera. Special Publications of the Cushman Foundation for Foraminiferal Research, vol. 34, pp. 1–108.

van Morkhoven, F.P.C.M., Berggren, W.A., Edwards, A.S., et al., 1986, Cenozoic cosmopolitan deep-water benthic foraminifera. Bulletin des Centres de Recherches Exploration-Production Elf-Aquitaine, Mémoire, vol. 11, pp. 1–421.

# *Brizalina alata*

## (Seguenza), 1862

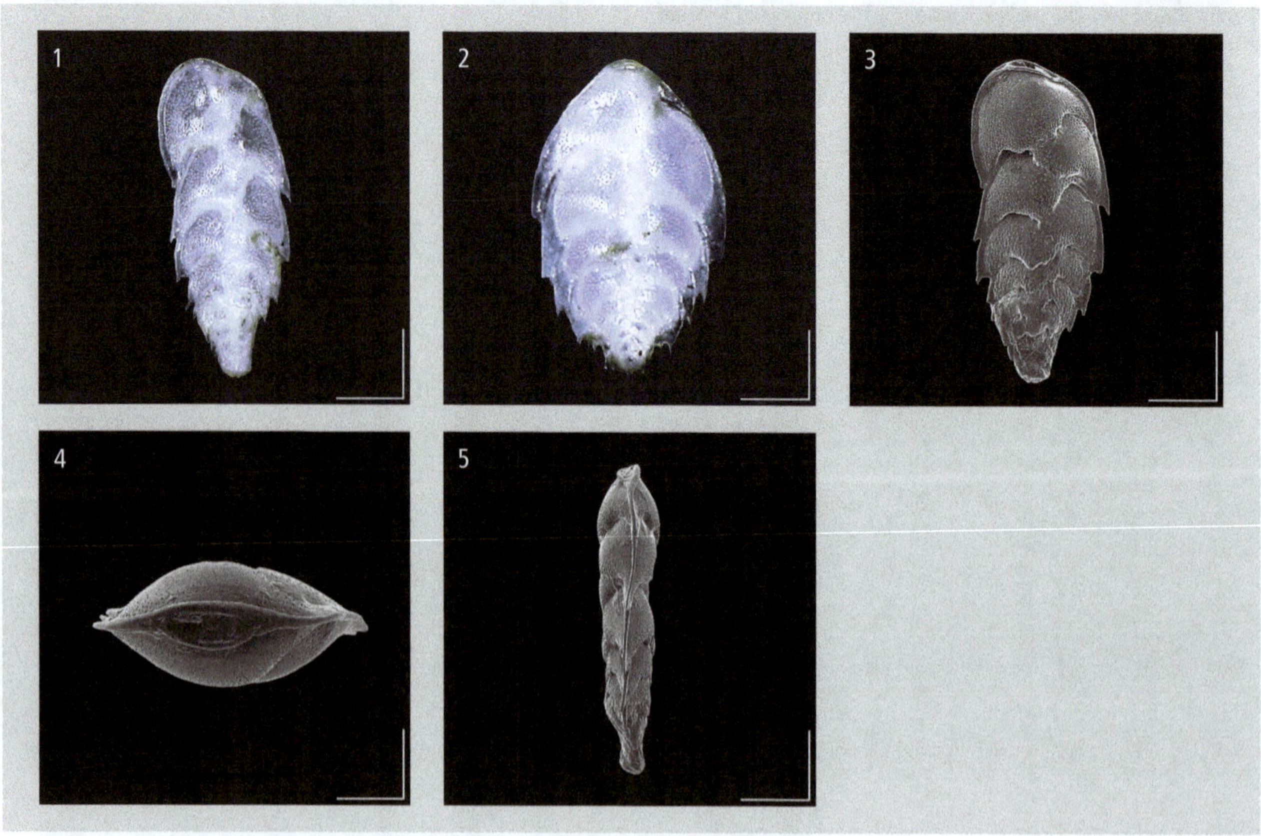


1. *Brizalina alata*. Arrou Islands, North Pacific Ocean; "Challenger" Station 191, depth: 800 fathoms. Deposited in the micropaleontological collections of The Natural History Museum, London; NHMUK ZF1173. Lateral view, NHMUK PM DI 0088, scale bar = 160 μm.

2. *Brizalina alata*. Philippines, Pacific Ocean; "Challenger" Station 209, depth: 95–100 fathoms. Deposited in the micropaleontological collections of The Natural History Museum, London; NHMUK ZF1174. Lateral view, NHMUK PM DI 0089, scale bar = 108 μm.

3. *Brizalina alata*. Arrou Islands, North Pacific Ocean; "Challenger" Station 191, depth: 800 fathoms. Deposited in the micropaleontological collections of The Natural History Museum, London; NHMUK ZF1173. Lateral view, NHMUK PM DI 0090, scale bar = 165 μm.

4. *Brizalina alata*. Arrou Islands, North Pacific Ocean; "Challenger" Station 191, depth: 800 fathoms. Deposited in the micropaleontological collections of The Natural History Museum, London; NHMUK ZF1173. Apertural view, NHMUK PM DI 0091 scale bar = 84 μm.

5. *Brizalina alata*. Arrou Islands, North Pacific Ocean; "Challenger" Station 191, depth: 800 fathoms. Deposited in the micropaleontological collections of The Natural History Museum, London; NHMUK ZF1173. Edge view, NHMUK PM DI 0092, scale bar = 173 μm.

### Original Designation
*Vulvulina alata* Seguenza, 1862, p. 115, pl. 2, fig. 5.

### Type Specimen
Depository not given.

### Type Level
Not designated. Level given: Pleistocene.

### Type Locality
Sicily.

### Description
Test forms an elongate, strongly compressed, biserial series; lobulate in outline and elliptical in cross-section with a sharply keeled periphery. The slightly inflated chambers increase rapidly in height and are separated by strongly curved, slightly depressed sutures. Chambers are prolonged by sharp, downward-pointing spines that merge into the peripheral keel. Chamber walls are calcareous and finely perforate. The primary aperture is loop-shaped opening with a lip and an internal toothplate.

### Synonyms and Citations
*Bolivina alata* (Seguenza). –Cushman, 1937, p. 106, pl. 13, figs. 3–11. –Renz, 1948, p. 116, pl. 6, fig. 26; pl. 12, fig. 12. –Bolli et al., 1994, p. 339, figs. 78.4–5; 73.

*Brizalina alata* (Seguenza). –van Marle, 1991, p. 166, pl. 17, figs. 1–2. –Jones, 1994, p. 58, pl. 53, figs. 2–4.

### Remarks
*Brizalina alata* (Seguenza) differs from *Bolivina pisciformis* Galloway and Morrey and *Bolivina alazanensis* Cushman in possessing fewer chambers, which are higher, slightly more inflated, and prolonged by sharp, downward-pointing spines.

*Brizalina alata* is a deep infaunal taxon, characteristic in oxygen-depleted sediments from the oxygen minimum zone (den Dulk et al., 2000).

### Biogeography
Worldwide.

### Bathymetry
Middle neritic.

### Chronostratigraphy
Miocene to Recent (Jones, 1994). Early Miocene to Recent in Indonesia (van Marle, 1991).

## References


Bolli, H.M., Beckmann, J.P., and Saunders, J.B., 1994, Benthic Foraminiferal Biostratigraphy of the South Caribbean Region. Cambridge: Cambridge University Press, 408 pp.

Cushman, J.A., 1937, A monograph of the subfamily Virgulininae of the foraminiferal family Buliminidae. Special Publications of the Cushman Laboratory for Foraminiferal Research, vol. 9, pp. 1–228.

den Dulk, M., Reichart, G.J., van Heyst, S., Zachariasse, W.J., and van der Zwaan, 2000, Benthic foraminifera as proxies of organic matter flux and bottom water oxygenation? A case history from the northern Arabian Sea. Palaeogeography, Palaeoclimatology, Palaeoecology, vol. 161, pp. 337–359.

Jones, R.W., 1994, The Challenger Foraminifera. Oxford: Oxford University Press, 149 pp.

Renz, H.H., 1948, Stratigraphy and fauna of the Agua Salada group, State of Falcón, Venezuela. Memoirs of the Geological Society of America, vol. 32, pp. 1–219.

Seguenza, G., 1862, Prime recerche intorno ai rizopodi fossili delle argille Pleistoceniche dei dintorni di Catania. Accademia Gioenia di Scienze Naturali di Catania, Atti, ser. 2, vol. 18, pp. 84–126.

van Marle, L.J., 1991, Eastern Indonesian Late Cenozoic smaller benthic foraminifera. Verhandelingen der Koninklijke Nederlanse Akademie van Wetenschappen, Afdeeling Natuurkunde, Eerste Reeks, vol. 34, pp. 1–328.

# *Brizalina aliformis*

## (Cushman), 1926

1. *Brizalina aliformis.* Rio Buena Vista, at bluff south and 25° east, 0.5 kilometer from Tumbadero Hacienda House, Vera Cruz, Mexico. Deposited in the collections of the United States National Museum, Washington, DC; CC 353838. Lateral view. Lectotype. NHMUK PM DI 0093, scale bar = 170 μm.

### Original Designation

*Bolivina mexicana* Cushman var. *aliformis* Cushman, 1926, p. 82, pl. 12, figs. 3–4.

### Type Specimen

Cotypes (CC 353838) deposited in the collections at the United States National Museum, Washington, DC. Van Morkhoven et al. (1986) designated as lectotype the microspheric specimen illustrated in pl. 12, fig. 4a, b by Cushman (1926).

### Type Level

Lower Oligocene, Alazan Shale.

### Type Locality

Left bank of Tuxpan River, at bluff south and 25° east, about 0.5 kilometer from Tumbadero Hacienda, Vera Cruz, Mexico.

### Description

Test forms an elongate, strongly compressed, biserial series; lobulate in outline and elliptical in cross-section, surrounded by a broad, transparent keel that may appear serrate in poorly preserved specimens. The broad, low chambers are slightly imbricate in the later portion of the test and separated by strongly curved, limbate, slightly depressed sutures, becoming thicker toward the central portion of the test. Chamber walls are calcareous and finely perforate. The primary aperture is loop-shaped with a lip and an internal toothplate.

### Synonyms and Citations

*Bolivina aliformis* Cushman. –van Morkhoven et al., 1986, p. 217, pl. 74. –Katz and Miller, 1993, pl. 1, fig. 4.

### Biogeography

Worldwide. Recorded from Cuba, Puerto Rico, Venezuela, the Dominican Republic, and the Gulf of Mexico (van Morkhoven et al., 1986). Recorded as *Bolivina* sp. 1 from Europe, the North and South Atlantic, and the Pacific (van Morkhoven, 1981).

### Bathymetry

Lower neritic to upper bathyal.

### Chronostratigraphy

Early Oligocene (Rupelian) to Early Miocene (Burdigalian).

## References

Cushman, J.A., 1926, Some fossil Bolivinas from Mexico. Contributions from the Cushman Laboratory for Foraminiferal Research, vol. 1, pp. 81–85.

Katz, M.E. and Miller, K.G., 1993, Latest Oligocene to earliest Pliocene benthic foraminiferal biofacies of the northeastern Gulf of Mexico. Micropaleontology, vol. 39, no. 4, pp. 367–403.

van Morkhoven, F.P.C.M., 1981, Cosmopolitan Tertiary bathyal benthic foraminifera. Transactions of the Gulf Coast Association of Geological Societies, Supplement 31, 445 abstracts with accompanying range chart.

van Morkhoven, F.P.C.M., Berggren, W.A., Edwards, A.S., et al., 1986, Cenozoic cosmopolitan deep-water benthic foraminifera. Bulletin des Centres de Recherches Exploration-Production Elf-Aquitaine, Mémoire, vol. 11, pp. 1–421.

# *Brizalina antegressa*

## (Subbotina), 1953

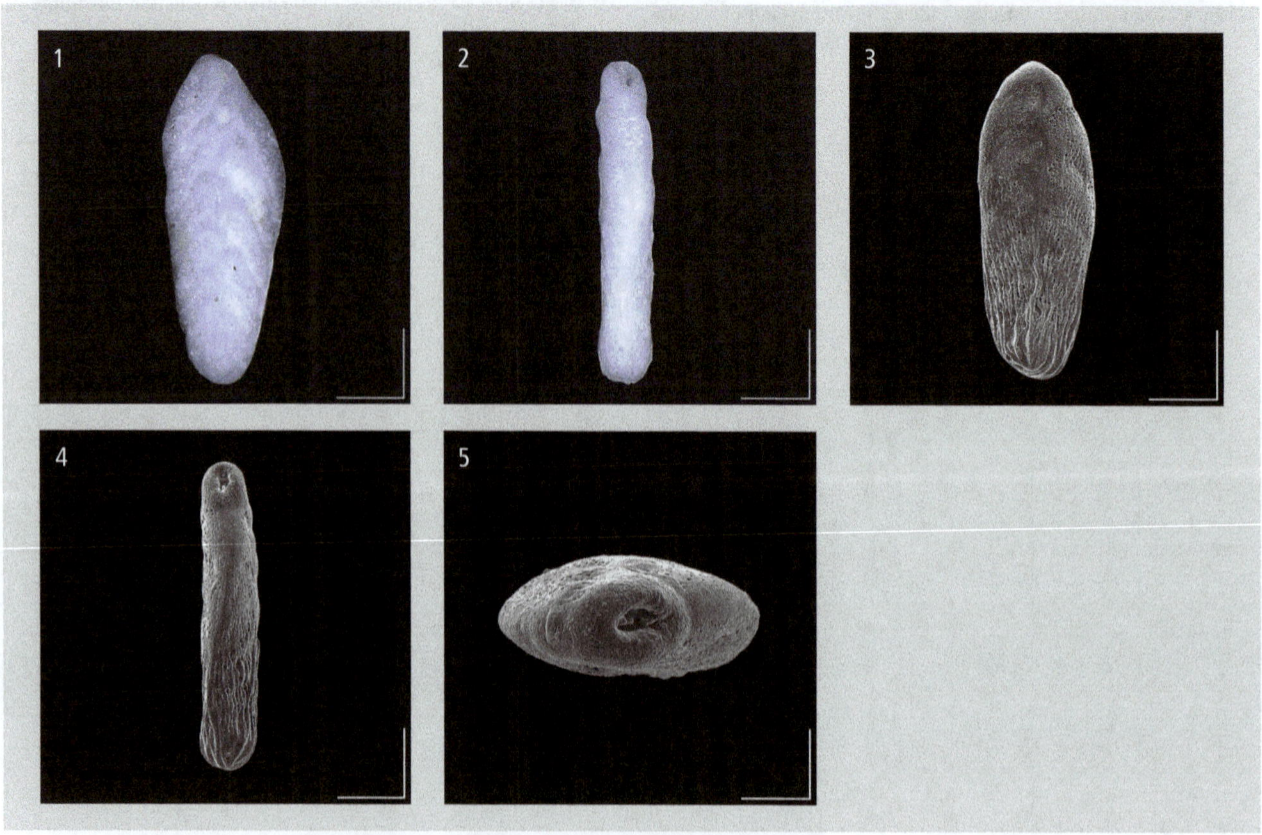


1. *Brizalina antegressa*. South of Pico Fracture Zone, west flank of the Mid-Atlantic Ridge, North Atlantic; DSDP Leg 82, Site 558, Core 24, Section 1, 34–38 cm. Lateral view, NHMUK PM DI 0096, scale bar = 94 μm.

2. *Brizalina antegressa*. South of Pico Fracture Zone, west flank of the Mid-Atlantic Ridge, North Atlantic; DSDP Leg 82, Site 558, Core 24, Section 1, 34–38 cm. Edge view, NHMUK PM DI 0096, scale bar = 105 μm.

3. *Brizalina antegressa*. South of Pico Fracture Zone, west flank of the Mid-Atlantic Ridge, North Atlantic; DSDP Leg 82, Site 558, Core 24, Section 1, 34–38 cm. Lateral view, NHMUK PM DI 0096, scale bar = 102 μm.

4. *Brizalina antegressa*. South of Pico Fracture Zone, west flank of the Mid-Atlantic Ridge, North Atlantic; DSDP Leg 82, Site 558, Core 24, Section 1, 34–38 cm. Edge view, NHMUK PM DI 0097, scale bar = 102 μm.

5. *Brizalina antegressa*. South of Pico Fracture Zone, west flank of the Mid-Atlantic Ridge, North Atlantic; DSDP Leg 82, Site 558, Core 24, Section 1, 34–38 cm. Apertural view, NHMUK PM DI 0098, scale bar = 47 μm.

## Original Designation

*Bolivina antegressa* Subbotina, 1953, p. 226, pl. 10, figs. 11–16.

## Type Specimen

Holotype (No. 3257) deposited in the collections of the All-Union Petroleum (Scientific Research) Geological Prospecting Institute (VNIGRI), St Petersburg.

## Type Level

Given as Upper Eocene, *Globigerinoides conglobatus* Zone.

## Type Locality

Simferopol, Crimea, Ukraine.

## Description

Test forms an elongate, strongly compressed, lanceolate, biserial series; tapered in outline and elliptical in cross-section with a sharp, keeled periphery. The broad, low chambers increase regularly in size and are separated by thick, curved sutures. The median suture pattern is almost straight or forms a slight zigzag. Chamber walls are calcareous, finely perforate, and ornamented by a thin, median ridge and by fine striae, extending at least halfway up the test or over the whole test. The aperture is loop-shaped, extending from the base of the last chamber, with a rounded lip and internal toothplate.

## Synonyms and Citations

*Bolivina tectiformis* Cushman. –Bermúdez, 1949, p. 195, pl. 12, fig. 47. –Tjalsma, 1983, p. 739, pl. 1, fig. 3.

*Bolivina antegressa* Subbotina. –Miller, et al., 1985, pl. 4, fig. 11. –Miller and Katz, 1987, p. 121, pl. 1, fig. 4.

## Remarks

Microspheric forms are more tapered with a more pointed initial portion than megalospheric forms. This species also exhibits some variation in the degree of ornamentation and the development of a keel.

## Biogeography

Worldwide. Recorded in the Caucasus, Emba, and Central Asia (Subbotina, 1953), and the North Atlantic (Miller et al., 1985; Miller and Katz, 1987). Recorded as *Bolivina tectiformis* Cushman, in the Dominican Republic (Bermúdez, 1949) and in the South Atlantic (Tjalsma, 1983).

## Bathymetry

Bathyal.

## Chronostratigraphy

Late Eocene (Bartonian) to Middle Miocene (Serravallian).

# References


Bermúdez, P.J., 1949, Tertiary Smaller Foraminifera of the Dominican Republic. Special Publications of the Cushman Laboratory for Foraminiferal Research, no. 25, pp. 1–322.

Miller, K.G. and Katz, M.E., 1987, Oligocene to Miocene benthic foraminiferal and abyssal circulation changes in the north Atlantic. Micropaleontology, vol. 33, no. 2, pp. 97–149.

Miller, K.G., Curry, W.B., and Ostermann, D.R., 1985, Late Paleogene (Eocene to Oligocene) benthic foraminiferal oceanography of the Goban Spur region, Deep Sea Drilling Project Leg 80. In de Graciansky, P.C., Poag, C.W., et al., ed., Initial Reports of the Deep Sea Drilling Project, Washington (U.S. Government Printing Office), vol. 80, pp. 505–538.

Subbotina, N.N., 1953, Microfauna of the USSR, Volume 6. Upper Eocene Lagenidae and Buliminidae of the southern USSR. Trudy Vsesoyuznogo Neftyanogo Nauchno-Issledovatel'skogo, Geologorazvedochnogo Instituta (VNIGRI) (new series), vol. 69, pp. 115–282.

Tjalsma, R.C., 1983, Eocene to Miocene benthic foraminifera from DSDP site 516, Rio Grande Rise, South Atlantic. In Barker, P.F., Carlson, R.L., Johnson, D. A., et al., ed., Initial Reports of the Deep Sea Drilling Project, Washington (U.S. Government Printing Office), vol. 72, pp. 731–755.

# *Brizalina subaenariensis* var. *mexicana*

## (Cushman), 1922

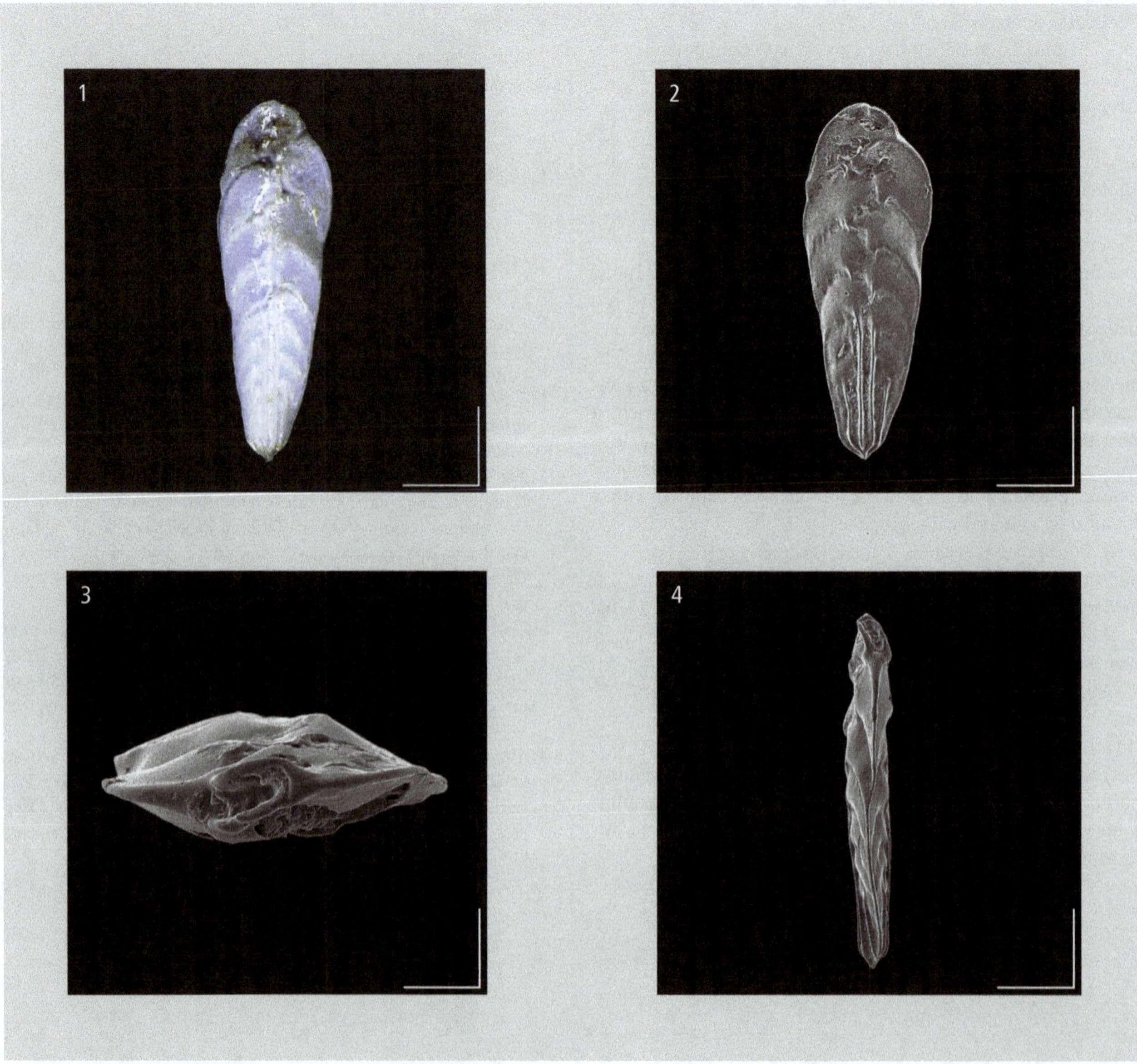


1. *Brizalina subaenariensis* var. *mexicana*. West of Ireland, North Atlantic Ocean; "Porcupine" Station 11, depth: 1630 fathoms. Deposited in the micropaleontological collections of The Natural History Museum, London; NHMUK ZF1170. Lateral view, NHMUK PM DI 0099, scale bar = 214 μm.

2. *Brizalina subaenariensis* var. *mexicana*. West of Ireland, North Atlantic Ocean; "Porcupine" Station 11, depth: 1630 fathoms. Deposited in the micropaleontological collections of The Natural History Museum, London; NHMUK ZF1170. Lateral view, NHMUK PM DI 0100, scale bar = 184 μm.

3. *Brizalina subaenariensis* var. *mexicana*. West of Ireland, North Atlantic Ocean; "Porcupine" Station 11, depth: 1630 fathoms. Deposited in the micropaleontological collections of The Natural History Museum, London; NHMUK ZF1170. Apertural view, NHMUK PM DI 0101, scale bar = 71 μm.

4. *Brizalina subaenariensis* var. *mexicana*. West of Ireland, North Atlantic Ocean; "Porcupine" Station 11, depth: 1630 fathoms. Deposited in the micropaleontological collections of The Natural History Museum, London; NHMUK ZF1170. Edge view, NHMUK PM DI 0102, scale bar = 190 μm.

## Original Designation

*Bolivina subaenariensis* var. *mexicana* Cushman, 1922, p. 47, pl. 8, fig. 1.

## Type Specimen

Deposited in United States National Museum, Washington, DC, USA, catalog numbers 17128, 17129 (type), 17130–17134.

## Type Level

Recent.

## Type Locality

Not designated. "Albatross" Station D2377, latitude 29° 07′ 30″ N, longitude 88° 08′ 00″ W, Gulf of Mexico; 210 fathoms.

## Description

Test is elongate, biserial, strongly compressed, tapered in outline, and elliptical in cross-section. Slightly inflated, broad and low, overlapping chambers, increasing slowly in height, are separated by curved, slightly depressed sutures. Wall is calcareous, finely perforate, ornamented by a few longitudinal costae, and more prominent in the initial part of the test. Aperture is loop-shaped, bordered by a lip, with an internal toothplate.

## Synonyms and Citations

*Bolivina aenariensis* Costa. –Brady, 1884, p. 423, pl. 53, figs. 10–11.

*Bolivina subaenariensis* Cushman var. –Barker, 1960, pl. 53, figs. 10–11.

*Brizalina subaenariensis* var. *mexicana* (Cushman). –Jones, 1994, p. 58, pl. 53, figs. 10–11.

## Biogeography

Worldwide.

## Bathymetry

Neritic to bathyal. Reported between 13 and 1630 fathoms at "Challenger" stations (Jones, 1994).

## Chronostratigraphy

Miocene to Recent.

# References


Barker, R.W., 1960, Taxonomic notes on the species figured by H.B. Brady in his report on the foraminifera dredged by H.M.S. Challenger during the years 1873–1876. Society of Economic Paleontologists and Mineralogists Special Publication, vol. 9, pp. 1–238.

Brady, H.B., 1884, Report on the foraminifera dredged by H.M.S. Challenger during the years 1873–1876. In Brady, H.B., ed., Report on the Scientific Results of the Voyage of the H.M.S. Challenger during the years 1873–1876. Zoology, vol. 9, pp. 1–814.

Cushman, J.A., 1922, The foraminifera of the Atlantic Ocean. Part III: Textulariidae. Bulletin of the United States National Museum, vol. 104, pp. 1–149.

Jones, R.W., 1994, The Challenger Foraminifera. Oxford: Oxford University Press, 149 pp.

# *Brizalina subspinescens*

## (Cushman), 1922

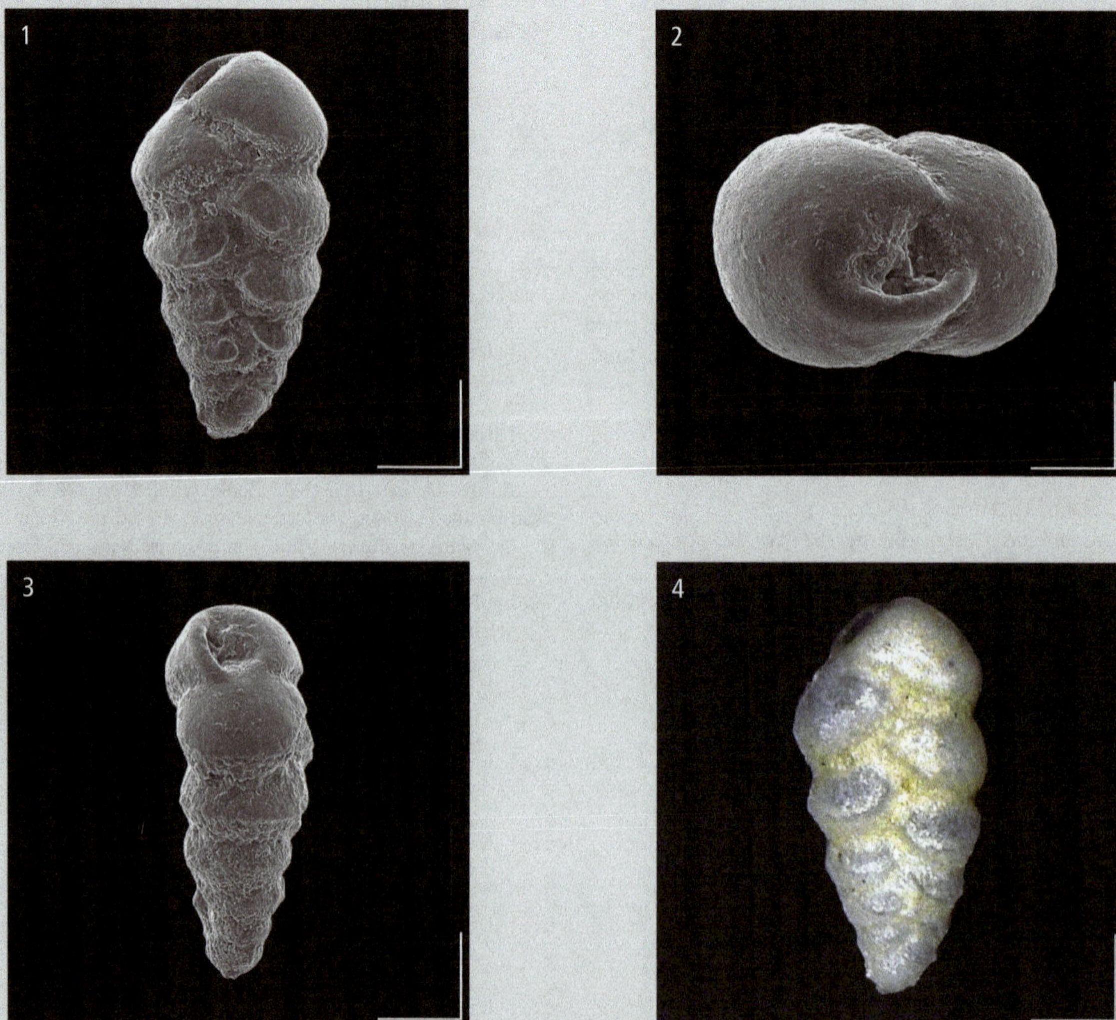


1. *Brizalina subspinescens.* West of Ireland, North Atlantic Ocean; "Porcupine" Station 18, depth: 183 fathoms. Deposited in the micropaleontological collections of The Natural History Museum, London; NHMUK ZF1200. Lateral view, NHMUK PM DI 0104, scale bar = 80 μm.

2. *Brizalina subspinescens.* West of Ireland, North Atlantic Ocean; "Porcupine" Station 18, depth: 183 fathoms. Deposited in the micropaleontological collections of The Natural History Museum, London; NHMUK ZF1200. Lateral view, NHMUK PM DI 0105, scale bar = 48 μm.

3. *Brizalina subspinescens.* West of Ireland, North Atlantic Ocean; "Porcupine" Station 18, depth: 183 fathoms. Deposited in the micropaleontological collections of The Natural History Museum, London; NHMUK ZF1200. Edge view, NHMUK PM DI 0106, scale bar = 83 μm.

4. *Brizalina subspinescens.* West of Ireland, North Atlantic Ocean; "Porcupine" Station 18, depth: 183 fathoms. Deposited in the micropaleontological collections of The Natural History Museum, London; NHMUK ZF1200. Lateral view, NHMUK PM DI 0103, scale bar = 74 μm.

## Original Designation

*Bolivina subspinescens* Cushman, 1922, p. 48, pl. 7, fig. 5.

## Type Specimen

Deposited in United States National Museum, Washington, DC, Catalog numbers 17080 (type), 17081–17083.

## Type Level

Recent.

## Type Locality

"Albatross" Station D2192 (39° 46′ 30″ N; 70° 14′ 45″ W) in 1060 fathoms, off the northeast coast of the USA.

## Description

Test forms an elongate, strongly compressed, biserial series; slightly lobulate, tapered in outline. The slightly inflated, broad and low, angular chambers increase slowly in height, and are separated by curved, slightly depressed sutures. Chamber walls are calcareous and finely perforate with short spines at the base of chambers. The aperture is a loop-shaped opening bordered by a lip, with an internal toothplate.

## Synonyms and Citations

*Bolivina textilarioides* Reuss. –Brady, 1884, p. 419, pl. 52, figs. 24–25.

*Bolivina subspinescens* Cushman. –Barker, 1960, pl. 52, figs. 24–25. –van Marle, 1991, p. 164, pl. 16, figs. 12–14.

*Brizalina subspinescens* (Cushman). –Jones, 1994, p. 57, pl. 52, figs. 24–25.

## Remarks

*Brizalina subspinescens* (Cushman) resembles *Bolivina spinescens* Cushman; however, it has a more tapering test and distinctly angular chambers.

## Biogeography

Worldwide.

## Bathymetry

Primarily bathyal. Reported from 220 fathoms, 345 fathoms, and 128 fathoms in the Atlantic (Cushman, 1937), from 130 to 2379 meters off New Zealand (Lewis, 1979), from 183 fathoms at "Porcupine" Station 18, west of Ireland (Jones, 1994), and at depths below 78 meters in eastern Indonesia (van Marle, 1991).

## Chronostratigraphy

Middle Miocene to Recent.

# References


Barker, R.W., 1960, Taxonomic notes on the species figured by H.B. Brady in his report on the foraminifera dredged by H.M.S. Challenger during the years 1873–1876. Society of Economic Paleontologists and Mineralogists Special Publication, vol. 9, pp. 1–238.

Brady, H.B., 1884, Report on the foraminifera dredged by H.M.S. Challenger during the years 1873–1876. In Brady, H.B., ed., Report on the Scientific Results of the Voyage of the H.M.S. Challenger during the years 1873–1876. Zoology, vol. 9, pp. 1–814.

Cushman, J.A., 1922, The foraminifera of the Atlantic Ocean. Part III: Textulariidae. Bulletin of the United States National Museum, vol. 104, pp. 1–149.

Cushman, J.A., 1937, A monograph of the subfamily Virgulininae of the foraminiferal family Buliminidae. Special Publications of the Cushman Laboratory for Foraminiferal Research, vol. 9, pp. 1–228.

Jones, R.W., 1994, The Challenger Foraminifera. Oxford: Oxford University Press, 149 pp.

Lewis, K.B., 1979, Foraminifera on the continental shelf and slope off Southern Hawke's Bay. Memoirs of the New Zealand Oceanographic Institute, vol. 84, pp. 1–45.

van Marle, L.J., 1991, Eastern Indonesian Late Cenozoic smaller benthic foraminifera. Verhandelingen der Koninklijke Nederlanse Akademie van Wetenschappen, Afdeeling Natuurkunde, Eerste Reeks, vol. 34, pp. 1–328.

# *Bulbobaculites problematicus*

## (Neagu), 1962

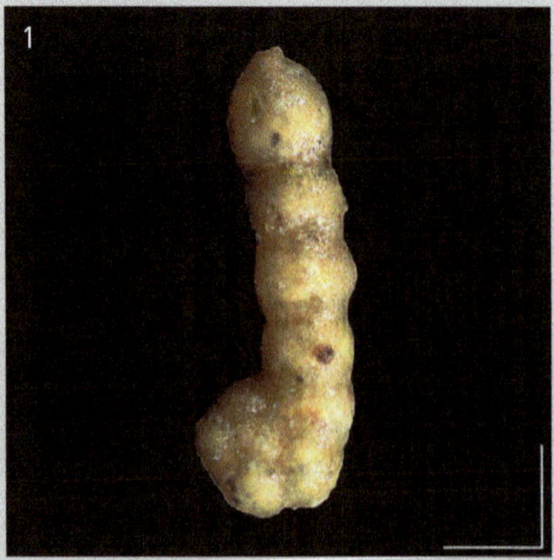


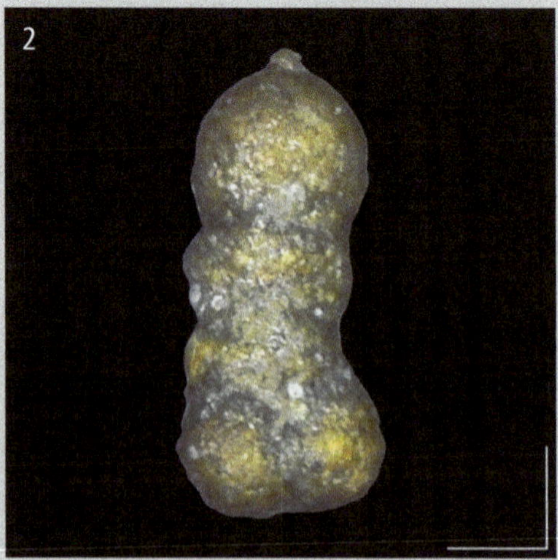


1. *Bulbobaculites problematicus.* Subsilesian Unit, Red shale, Węglówka (near Krosno), Poland; Borehole 130, 230 meters. Deposited in the Geroch Collection, Jagiellonian University, Kraków, Poland. Lateral view, NHMUK PM DI 0107, scale bar = 319 µm.

2. *Bulbobaculites problematicus.* Colon Formation, Northeastern Colombia. Deposited in the collections of the United States National Museum, Washington, DC; CC 37905. Holotype of *Ammobaculites lueckei* Cushman and Hedberg (1941), a suspected synonym of *Bulbobaculites problematicus.* Lateral view, NHMUK PM DI 0108, scale bar = 89 µm.

### Original Designation

*Ammobaculites agglutinans* (d'Orbigny) subspecies *problematicus* Neagu, 1962, p. 61, pl. 2, figs. 22–24.

### Type Specimen

Depository not given.

### Type Level

Upper Cenomanian to Lower Turonian, red clay.

### Type Locality

Pîrîul Fetii (valea superioara a Buzaului), Carpatii de curbura, red clay.

### Description

Test forms a small, initial, streptospiral stage, followed by a uniserial, rectilinear portion of variable length test, cylindrical in cross-section. The inflated chambers are separated by markedly depressed sutures. Chamber walls are non-alveolar, finely or coarsely agglutinated with organic cement, and occasionally with some carbonate grains. The last chamber tapers toward a rounded, terminal aperture, and generally has a smooth collar.

### Synonyms and Citations

*Haplophragmium aequicameratum* Huss, 1966, pp. 68–69, pl. 9, figs. 10–25.

*Ammobaculites problematicus* Neagu. –Geroch and Nowak, 1984, pl. 1, figs. 17–18; pl. 6, fig. 23.

*Haplophragmium lueckei* (Cushman and Hedberg). –Krasheninnikov, 1974, p. 639, pl. 4, figs. 3a, 3b, 4a, 5a. –Moullade et al., 1988, p. 363, pl. 3, figs. 1–6.

*Haplophragmium problematicum* (Neagu). –Kuhnt, 1990, p. 312, pl. 4, figs. 3–9.

*Bulbobaculites problematicus* (Neagu). –Kuhnt and Kaminski, 1990, p. 465, figs. 5, 5A.

### Remarks

*Bulbobaculites problematicus* (Neagu) (including the type specimens in the Neagu Collection) displays a wide range of morphologic variability, suggested to be substrate-dependent (see Kuhnt and Kaminski, 1990, figs. 5, 5A). Specimens from abyssal settings are usually fine-grained, whereas specimens found in turbidites are coarser-grained.

*Ammobaculites lueckei* Cushman and Hedberg, 1941, selected as the type species of the genus *Bulbobaculites* Maync, 1952, is most likely a synonym of *Bulbobaculites problematicus* (Neagu). However, the only specimen available for taxonomic comparison in the collections of the Smithsonian Institution is the holotype. Although Cushman and Hedberg (1941) reported in the initial description and illustration that the holotype possessed an elliptical aperture at the end of a short neck, a re-examination of the holotype showed that it has a rounded aperture, with a smooth collar. The holotype of *Ammobaculites lueckei* has a red color confirming its fossilization in deep-sea red clay. *Ammobaculites lueckei* probably represents one growth stage of *Bulbobaculites problematicus*.

*Bulbobaculites problematicus* is a very useful maker in the Atlantic, indicating a post-Cenomanian–Turonian boundary age.

### Biogeography

Worldwide.

### Bathymetry

Bathyal to abyssal. Common in flysch-type assemblages.

### Chronostratigraphy

Cretaceous (Cenomanian to Santonian) in the Polish Carpathians (Geroch and Nowak, 1984). However, the first occurrence of *Bulbobaculites problematicus* (Neagu) in the North Atlantic occurs after the Cenomanian–Turonian boundary (see Kuhnt and Kaminski, 1990, fig. 8). In the marginal basins of Zumaya, Gubbio, and the Penibetic the stratigraphic range of *Bulbobaculites problematicus* (Neagu) extends into the Santonian (see Kuhnt and Kaminski, 1990, fig. 8).

## References


Cushman, J.A. and Hedberg, H.D., 1941, Upper Cretaceous foraminifera from Santander del Norte, Colombia. Contributions from the Cushman Laboratory for Foraminiferal Research, vol. 17, no. 4, pp. 79–102.

Geroch, S. and Nowak, W., 1984, Proposal of zonation for the late Tithonian–late Eocene, based upon arenaceous foraminifera from the outer Carpathians, Poland. In Oertli, H., ed., Benthos '83, 2nd International Symposium on Benthic Foraminifera, Pau (France), April 11–15, 1983. Elf Aquitaine, ESSO REP and TOTAL CFP, Pau and Bordeaux, pp. 225–239.

Huss, F., 1966, Otwornice aglutynajace serii podslakiej jednostki roponosnej Weglowki (Polskie Karpaty Flisowe). Prace Geologiczne, Polska Akademia Nauk, no. 34, pp. 7–76.

Krasheninnikov, V.A., 1974, Upper Cretaceous benthonic agglutinated foraminifera, Leg 27 of the Deep Sea Drilling Project. In Veevers, J.J., Heirtzler, J.R., et al., ed., Initial Reports of the Deep Sea Drilling Project, Washington (U.S. Government Printing Office), vol. 27, pp. 531–662.

Kuhnt, W., 1990, Agglutinated foraminifera of western Mediterranean Upper Cretaceous pelagic limestones (Umbrian Apennines, Italy, and Betic Cordillera, Southern Spain). Micropaleontology, vol. 36, no. 4, pp. 297–330.

Kuhnt, W. and Kaminski, M.A., 1990, Paleoecology of Late Cretaceous to Paleocene deep-water agglutinated foraminifera from the North Atlantic and western Tethys. In Hemleben, C., Kaminski, M.A., Kuhnt, W., and Scott, D.B., ed., Paleoecology, Biostratigraphy, Paleoceanography and Taxonomy of Agglutinated Foraminifera. NATO ASI Series, Series C: Mathematical and Physical Sciences. Dordrecht: Kluwer Academic, vol. 327, pp. 433–505.

Moullade, M., Kuhnt, W., and Thurow, J., 1988, Agglutinated benthic foraminifers from Upper Cretaceous variegated clays of the North Atlantic Ocean (DSDP Leg 93 and ODP Leg 103). In Boillot, G., Winterer, E.L., et al., ed., Scientific Results of the Proceedings of the Ocean Drilling Program, vol. 103, pp. 349–377.

Neagu, T., 1962, Studiul foraminiferelor aglutinante din argilele cretacic-superioare de pe Valea Sadova (Cîmpulung Moldovenesc) si basinul superior al Vaii Buzaului. Studii Cercetari Geologie, Academia Republicii Populare Romine, Sectia de Geologie si Geografie si Istitutul de Geologie si Geografie, vol. 7, no. 1, pp. 45–81.

# *Bulimina aculeata*

## d'Orbigny, 1826

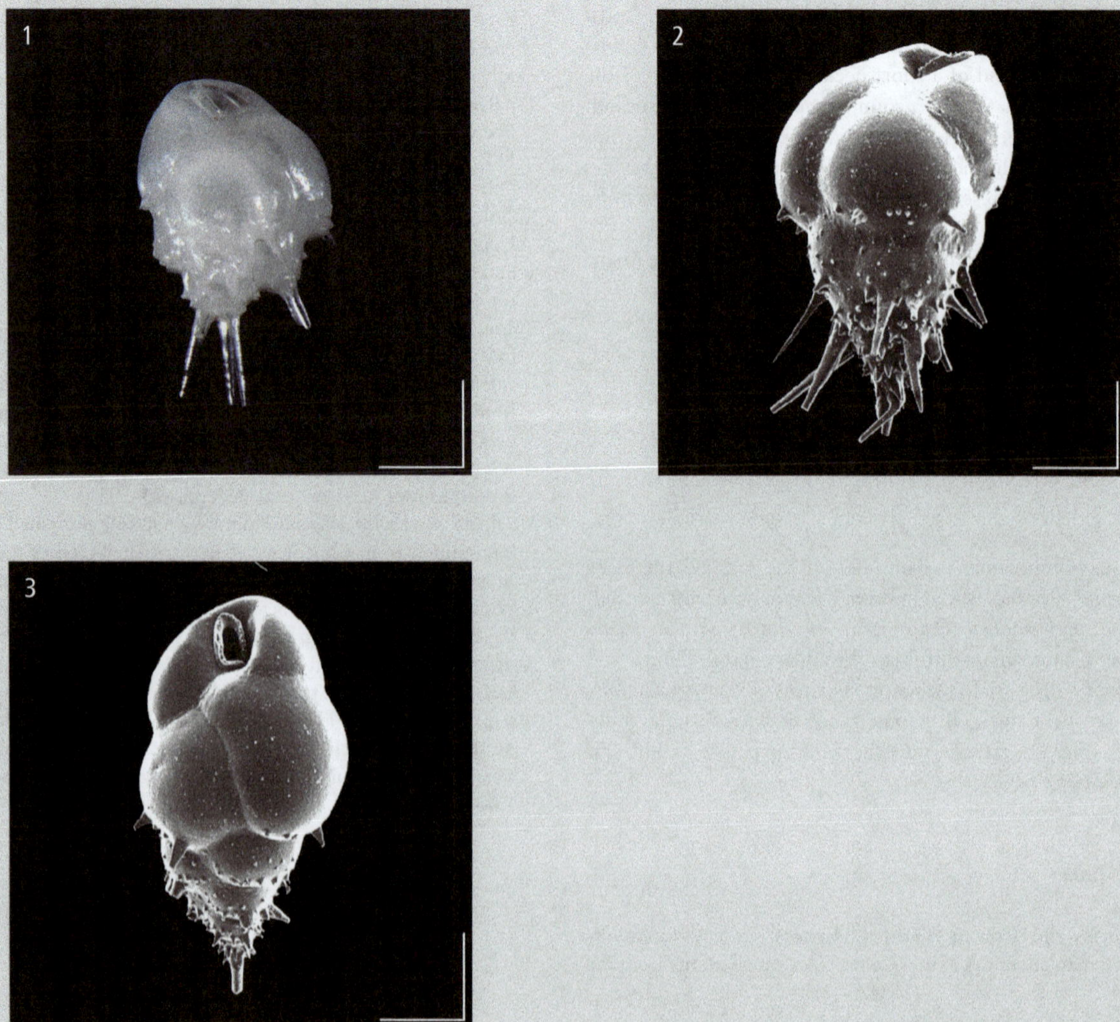

1. *Bulimina aculeata*. South China Sea; Station 17964, latitude 06° 09.50′ N, longitude 112° 12.80′ E, water depth: 1556 meters, surface sample. Deposited in the South China Sea Collection, Institute of Geosciences, Kiel University, Germany. Lateral view, NHMUK PM DI 0109, scale bar = 152 μm.

2. *Bulimina aculeata*. South China Sea; Station 17961, latitude 08° 30.40′ N, longitude 112° 19.90′ E, water depth: 1968 meters, surface sample. Deposited in the South China Sea Collection, Institute of Geosciences, Kiel University, Germany. Lateral view, NHMUK PM DI 0110, scale bar = 117 μm.

3. *Bulimina aculeata*. South China Sea; Station 17964, latitude 06° 09.50 N, longitude 112° 12.80 E, water depth: 1556 meters, surface sample. Deposited in the South China Sea Collection, Institute of Geosciences, Kiel University, Germany. Lateral view, NHMUK PM DI 0111, scale bar = 121 μm.

## Original Designation

*Bulimina aculeata* d'Orbigny, 1826, p. 269, type figure not given, illustrated in Parker et al., 1871, pl. 11, fig. 128.

## Type Specimen

Deposited in the collections of the Laboratoire de Paléontologie, Muséum d'Histoire Naturelle, Paris, France.

## Type Level

Recent.

## Type Locality

Adriatic Sea, near Rimini, Italy.

## Description

Test forms an elongate, triserial series; tapered in outline and subcircular in cross-section, with an acute initial portion, and a rounded apertural end; widest in the last whorl. The inflated chambers increase rapidly in height and are separated by distinct, depressed sutures. Chamber walls are calcareous, finely perforate, and often partially translucent and smooth, except at the outer margins of the basal chambers, which are fringed by sturdy spines. These spines may extend halfway up the test. Well-preserved specimens may also have a prominent basal spine. The primary aperture is a loop-shaped opening bordered by a lip and extending up from the base of the final chamber; with an internal toothplate.

## Synonyms (following van Morkhoven et al., 1986) and Citations

*Polymorpha pineïformis* Soldani, 1791, p. 119, pl. 269, fig. 1; pl. 130, fig. vv.

*Bulimina trilobata* d'Orbigny, 1826, p. 269.

*Bulimina patagonica* d'Orbigny, 1839, p. 50, pl. 1, figs. 8–9.

*Bulimina pupoides* var. *spinulosa* Williamson, 1858, p. 62, pl. 5, fig. 128.

*Bulimina spinosa* Seguenza, 1862, p. 107, 124, pl. 1, fig. 8.

*Bulimina aculeata* d'Orbigny. –Parker et al., 1871, pl. 11, fig. 128. –Brady, 1884, p. 406, pl. 51, figs. 7–9. –van Morkhoven et al., 1986, p. 31, pl. 7. –van Marle, 1991, p. 84, pl. 5, figs. 3–5. –Jones, 1994, p. 56, pl. 51, figs. 7–9. –Mendes et al., 2012, p. 38, fig. 3.8. –Milker and Schmiedl, 2012, p. 87, fig. 20.19.

## Remarks

Tapering of test and development of spines vary considerably.

## Biogeography

Worldwide.

## Bathymetry

Bathyal to abyssal. More common at lower bathyal depths. Reported between 1000 and 2740 fathoms at "Challenger" stations (Jones, 1994).

## Chronostratigraphy

Early Miocene (Aquitanian) to Recent. Early Miocene to Recent in the Indo-Pacific region (van Marle, 1991).

## References


Brady, H.B., 1884, Report on the foraminifera dredged by H.M.S. Challenger during the years 1873–1876. In Brady, H.B., ed., Report on the Scientific Results of the Voyage of the H.M.S. Challenger during the years 1873–1876, Zoology, vol. 9, pp. 1–814.

d'Orbigny, A., 1826, Tableau méthodique de la classe des Céphalopodes. Annales des Sciences Naturelles, vol. 7, pp. 245–314.

d'Orbigny, A.D., 1839, Voyage dans l'Amérique méridionale; foraminifères, vol. 5, pt 5. Paris and Strasbourg: P. Bertrand, 86 pp.

Jones, R.W., 1994, The Challenger Foraminifera. Oxford: Oxford University Press, 149 pp.

Mendes, I., Dias, J.A., Schönfeld, J., and Ferreira, O., 2012, Distribution of living foraminifera on the northern Gulf of Cadiz continental shelf. Journal of Foraminiferal Research, vol. 42, pp. 18–38.

Milker, Y. and Schmiedl, G., 2012, A taxonomic guide to modern benthic shelf foraminifera of the western Mediterranean Sea. Palaeontologia Electronica, vol. 15, Issue 2;16A, 134 pp.

Parker, W.K., Jones, T.R., and Brady, H.B., 1871, On the nomenclature of the foraminifera. Part XIV – The species enumerated by d'Orbigny in the "Annales des Sciences Naturelles" 1826. Annals and Magazine of Natural History, Series 4, vol. 7 (continued), – The species founded upon the figures in Soldani's "Testaceographia ac Zoophytographia", Annals and Magazine of Natural History, Series 4, vol. 8, pp. 145–179, 238–266.

Seguenza, G., 1862, Prime recerche intorno ai rizopodi fossili delle argille Pleistoceniche dei dintorni di Catania. Accademia Gioenia di Scienze Naturali di Catania, Atti, ser. 2, vol. 18, pp. 84–126.

Soldani, A., 1791, Testaceographiae ac Zoophytographiae parvae et microscopicae. Siena: F. Rossi, pp. 81–200.

van Marle, L.J., 1991, Eastern Indonesian Late Cenozoic smaller benthic foraminifera. Verhandelingen der Koninklijke Nederlanse Akademie van Wetenschappen, Afdeeling Natuurkunde, Eerste Reeks, vol. 34, pp. 1–328.

van Morkhoven, F.P.C.M., Berggren, W.A., Edwards, A.S., et al., 1986, Cenozoic cosmopolitan deep-water benthic foraminifera. Bulletin des Centres de Recherches Exploration-Production Elf-Aquitaine, Mémoire, vol. 11, pp. 1–421.

Williamson, W.C., 1858, On the Recent Foraminifera of Great Britain. London: Ray Society, 107 pp.

# Bulimina alazanensis

## Cushman, 1927

1. *Bulimina alazanensis*. Goban Spur, North Atlantic; DSDP Leg 80, Hole 548A, Core 15, Section 1, 79–81 cm. Lateral view, NHMUK PM DI 0112, scale bar = 46 μm.

2. *Bulimina alazanensis*. Goban Spur, North Atlantic; DSDP Leg 80, Hole 548A, Core 15, Section 1, 79–81 cm. Lateral view, NHMUK PM DI 0113, scale bar = 56 μm.

### Original Designation

*Bulimina alazanensis* Cushman, 1927, p. 161, pl. 25, fig. 4.

### Type Specimen

Holotype (No. 369307) deposited in the collections of the United States National Museum, Washington, DC. Ortiz and Thomas (2006) noted that the holotype of *Bulimina alazanensis* is now lost.

### Type Level

Upper Eocene, Alazan Clay.

### Type Locality

Rio Buena Vista, just south of crossing of Alazan to Moyutla Road, Vera Cruz, Mexico.

### Description

Test forms a small, elongate, triserial series; tapered in outline and subtriangular in cross-section, with a pointed apical end, a rounded apertural end, and with greatest breadth of test at the apertural end. The moderately inflated chambers are separated by sutures whose character is typically obscured by surface ornamentation. Chamber walls are calcareous, finely perforate, and ornamented by prominent longitudinal costae forming spinose projections at the initial portion. The chamber surfaces may be corrugated between costae. The primary aperture is a loop-shaped opening bordered by a lip and extending up from the base of the final chamber; with an internal toothplate.

### Synonyms and Citations

*Bulimina truncanella* Finlay, 1940, p. 455, pl. 64, figs. 89–91.

*Bulimina alazanensis* Cushman. –Cushman and Parker, 1947, p. 103, pl. 24, figs. 14–16. –Bermúdez, 1949, p. 180, pl. 12, fig. 1. –Tjalsma, 1983, p. 739, pl. 1, figs. 4–5. –Tjalsma and Lohmann, 1983, p. 24, pl. 14, fig. 4. –Miller, Curry, et al., 1985, pl. 4, fig. 6. –Miller and Katz, 1987, p. 124, pl. 2, figs. 7a–b. –van Marle, 1991, p. 85, pl. 5, figs. 1–2. –Bolli et al., 1994, p. 347, figs. 53.27, 78.30, 81.13. –Robertson, 1998, p. 148, pl. 56, fig. 5. –Kuhnt et al., 2002, p. 138, pl. 6, fig. 18. –Ortiz and Thomas, 2006, p. 114, pl. 4, figs. 12–13.

### Remarks

Miller and Katz (1987) noted considerable variability in the nature of the ornamentation (from subparallel costae to anastomosing striae), particularly in Oligocene specimens. Miller (1983) observed that abyssal forms lacked a distinctive triangular test, and Tjalsma and Lohmann (1983) found that the length to width ratio varied. This species appears to be morphologically close to *Bulimina rostrata* Brady, 1884.

### Biogeography

Worldwide. Recorded from the Caribbean and Mexico (Bolli et al., 1994), the North Atlantic (Miller and Katz, 1987), off Morocco (Lutze, 1980), and eastern Indonesia (van Marle, 1991).

### Bathymetry

Bathyal.

### Chronostratigraphy

Early Eocene (Ypresian) to Recent.

## References


Bermúdez, P.J., 1949, Tertiary smaller foraminifera of the Dominican Republic. Special Publications of the Cushman Laboratory for Foraminiferal Research, no. 25, pp. 1–322.

Bolli, H.M., Beckmann, J.P., and Saunders, J.B., 1994, Benthic Foraminiferal Biostratigraphy of the South Caribbean Region. Cambridge: Cambridge University Press, 408 pp.

Cushman, J.A., 1927, Some characteristic Mexican fossil foraminifera. Journal of Paleontology, vol. 1, pp. 147–172.

Cushman, J.A. and Parker, F.L., 1947, Bulimina and related foraminiferal genera. Professional Paper of the United States Geological Survey, vol. 210-D, pp. 55–176.

Finlay, H.J., 1940, New Zealand Foraminifera; key species in stratigraphy – No. 4. Transactions of the Royal Society of New Zealand, vol. 69, no. 4, pp. 448–472.

Kuhnt, W., Holbourn, A.E., and Zhao, Q., 2002, The early history of the South China Sea: Evolution of Oligocene–Miocene deep water environments. Revue de Micropaléontologie, vol. 45, pp. 99–159.

Lutze, G.F., 1980, Depth distribution of benthic foraminifera on the continental margin of NW-Africa. "Meteor" Forsch. Ergebnisse, Reihe C, vol. 32, pp. 31–80.

Miller, K.G., 1983, Eocene–Oligocene paleoceanography of the deep Bay of Biscay: benthic foraminiferal evidence. Marine Micropaleontology, vol. 7, pp. 403–440.

Miller, K.G. and Katz, M.E., 1987, Oligocene to Miocene benthic foraminiferal and abyssal circulation changes in the north Atlantic. Micropaleontology, vol. 33, no. 2, pp. 97–149.

Miller, K.G., Curry, W.B., and Ostermann, D.R., 1985, Late Paleogene (Eocene to Oligocene) benthic foraminiferal oceanography of the Goban Spur region, Deep Sea Drilling Project Leg 80. In de Graciansky, P.C., Poag, C.W., et al., ed., Initial Reports of the Deep Sea Drilling Project, Washington (U.S. Government Printing Office), vol. 80, pp. 505–538.

Ortiz, S. and Thomas, E., 2006, Lower–middle Eocene benthic foraminifera from the Fortuna Section (Betic Cordillera, southeastern Spain). Micropaleontology, vol. 52, pp. 97–150.

Robertson, B.E., 1998, Systematics and paleoecology of the benthic foraminiferida from the Buff Bay section, Miocene of Jamaica. Micropaleontology, vol. 44, supplement 2, pp. 1–266.

Tjalsma, R.C., 1983, Eocene to Miocene benthic foraminifera from DSDP site 516, Rio Grande Rise, South Atlantic. In Barker, P.F., Carlson, R.L., Johnson, D.A., et al., ed., Initial Reports of the Deep Sea Drilling Project, Washington (U.S. Government Printing Office), vol. 72, pp. 731–755.

Tjalsma, R.C. and Lohmann, G.P., 1983, Paleocene–Eocene bathyal and abyssal benthic foraminifera from the Atlantic Ocean. Micropaleontology, Special Publication, no. 4, pp. 1–90.

van Marle, L.J., 1991, Eastern Indonesian Late Cenozoic smaller benthic foraminifera. Verhandelingen der Koninklijke Nederlanse Akademie van Wetenschappen, Afdeeling Natuurkunde, Eerste Reeks, vol. 34, pp. 1–328.

# Bulimina callahani

## Galloway and Morrey, 1931

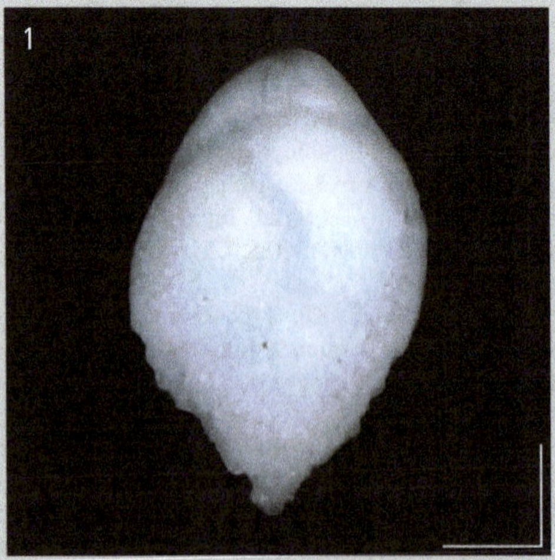



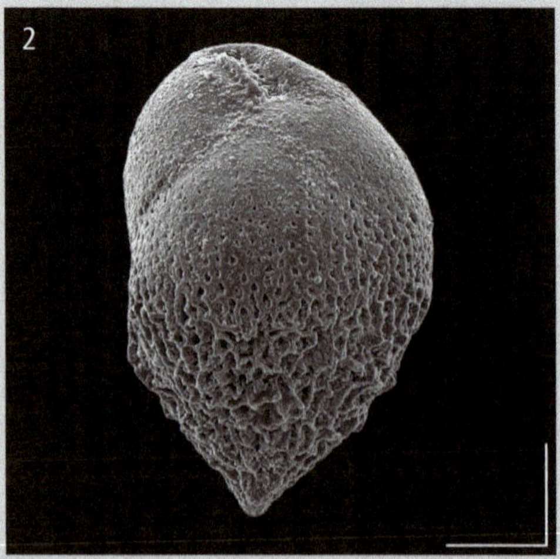



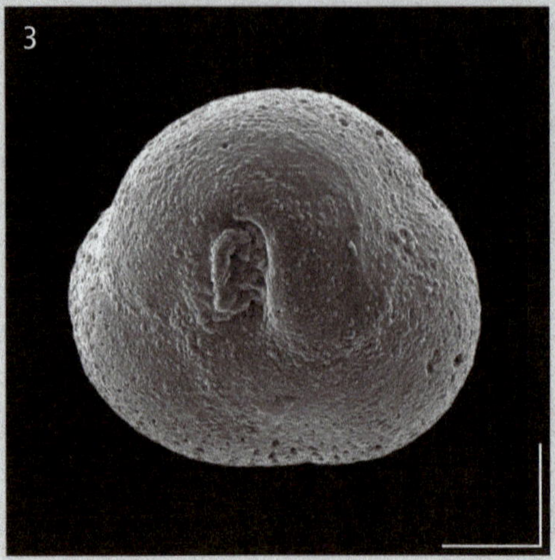



1. *Bulimina callahani.* Islas Orcadas Rise, South Atlantic; ODP Leg 114, Hole 702B, Core 25, Section 1, 40–44 cm. Lateral view, NHMUK PM DI 0114, scale bar = 50 μm.

2. *Bulimina callahani.* Islas Orcadas Rise, South Atlantic; ODP Leg 114, Hole 702B, Core 25, Section 1, 40–44 cm. Lateral view, NHMUK PM DI 0115, scale bar = 53 μm.

3. *Bulimina callahani.* Islas Orcadas Rise, South Atlantic; ODP Leg 114, Hole 702B, Core 25, Section 1, 40–44 cm. Apertural view, NHMUK PM DI 0116, scale bar = 40 μm.

## Original Designation

*Bulimina callahani* Galloway and Morrey, 1931, p. 350, pl. 40, fig. 6.

## Type Specimen

Holotype now deposited in the collections of the Department of Invertebrate Paleontology at the American Museum of Natural History, New York City.

## Type Level

Lower Eocene (given as Upper Cretaceous by Galloway and Morrey, 1931), surface shales.

## Type Locality

Near Puenta Piedra, on the Rio Puscatan, 19 kilometers south of Macuspana, Tabasco, Mexico (Lo. 5-61-3).

## Description

Test forms a small, elongate, flaring, triserial series; ovate in outline and subcircular in cross-section, with a subacute apical end and a rounded apertural end. The slightly inflated chambers increase rapidly in size and are separated by depressed sutures that are distinct only in the final chambers. Chamber walls are calcareous, perforate, and ornamented by a dense, coarse honeycomb reticulation. The primary aperture is a loop-shaped opening extending up from the base of the final chamber and bordered by a lip, with an internal toothplate.

## Synonyms and Citations

*Bulimina orphanensis* Berggren and Aubert, 1976, p. 337, pl. 2, figs. 1–11.

*Bulimina callahani* Galloway and Morrey. –Tjalsma and Lohmann, 1983, p. 24, pl. 11, figs. 6–7. –van Morkhoven et al., 1986, p. 322, pl. 105A–B.

## Biogeography

Worldwide. Recorded from California, the Labrador Sea, the Atlantic, Libya, the Gulf of Mexico, the East Pacific, and Europe (van Morkhoven et al., 1986).

## Bathymetry

Middle and lower bathyal. The upper depth limit of *Bulimina callahani* was approximately 600 meters according to Berggren and Aubert (1983).

## Chronostratigraphy

Early Eocene (Ypresian) to Middle Eocene (Lutetian).

# References


Berggren, W.A. and Aubert, J., 1976, Eocene benthonic foraminiferal biostratigraphy and paleobathymetry of Orphan Knoll (Labrador Sea). Micropaleontology, vol. 22, no. 3, pp. 327–346.

Berggren, W.A. and Aubert, J., 1983, Paleogene benthonic foraminiferal biostratigraphy and bathymetry of the Central Coast Ranges of California. In Brabb, E.E., ed., Studies in Tertiary Stratigraphy of the California Coast Ranges. Professional Paper of the United States Geological Survey, vol. 1213, pp. 4–21.

Galloway, J.J. and Morrey, M., 1931, Late Cretaceous foraminifera from Tabasco, Mexico. Journal of Paleontology, vol. 5, no. 4, pp. 329–354.

Tjalsma, R.C. and Lohmann, G.P., 1983, Paleocene–Eocene bathyal and abyssal benthic foraminifera from the Atlantic Ocean. Micropaleontology, Special Publication, no. 4, pp. 1–90.

van Morkhoven, F.P.C.M., Berggren, W.A., Edwards, A.S., et al., 1986, Cenozoic cosmopolitan deep-water benthic foraminifera. Bulletin des Centres de Recherches Exploration-Production Elf-Aquitaine, Mémoire, vol. 11, pp. 1–421.

# *Bulimina elongata*

## d'Orbigny, 1846

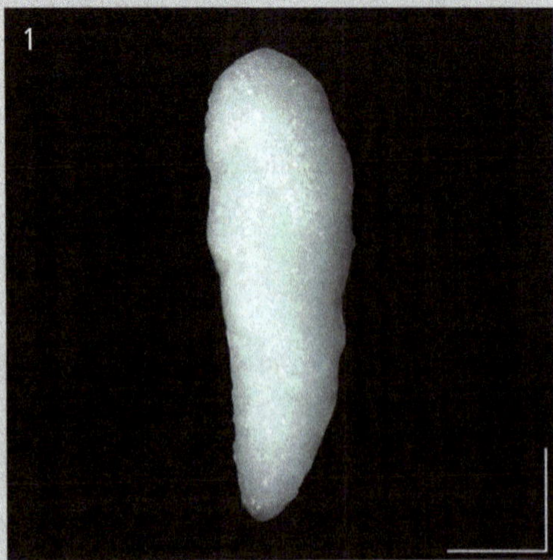


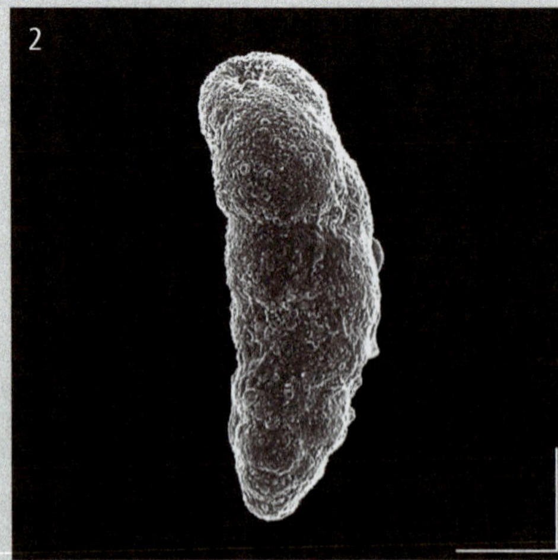


1. *Bulimina elongata.* Maud Rise, Weddell Sea, Antarctica; ODP Leg 113, Hole 690B, Core 12, Section 2, 40–43 cm. Lateral view, NHMUK PM DI 0117, scale bar = 57 μm.

2. *Bulimina elongata.* Maud Rise, Weddell Sea, Antarctica; ODP Leg 113, Hole 690B, Core 12, Section 2, 40–43 cm. Lateral view, NHMUK PM DI 0118, scale bar = 56 μm.

### Original Designation

*Bulimina elongata* d'Orbigny, 1846, p. 187, pl. 11, figs. 19–20.

### Type Specimen

Depository not given. Lectotype (1981/03/319) designated by Papp and Schmid (1985, pl. 63, figs. 5–9) and deposited in the Hauer d'Orbigny Collection at the Geologische Bundesanstalt, Vienna, Austria.

### Type Level

Tertiary, Tegel.

### Type Locality

Near the village of Nussdorf, north of Vienna, Austria.

### Description

Test forms an elongate, triserial series, slightly tapered in outline (greatest width in the last whorl or becoming parallel-sided in the last two whorls), and circular in cross-section, with a subrounded initial portion and a rounded apertural end. The moderately inflated chambers increase rapidly in height (with the last whorl approximately one-quarter of the length of the test), and are separated from one another by distinct, slightly depressed, curved sutures. Chamber walls are calcareous, finely perforate, and smooth. The primary aperture is a loop-shaped opening extending up from the base of the final chamber, and bordered by a lip that merges with an internal toothplate.

### Synonyms and Citations

*Bulimina elegans* d'Orbigny. –Brady, 1884, p. 398, pl. 50, figs. 3–4 (not figs. 1–2).

*Bulimina gibba* Fornasini. –Barker, 1960, pl. 50, figs. 3–4 (not figs. 1–2).

*Bulimina elongata* d'Orbigny var. *subulata* Cushman and Parker. –Barker, 1960, pl. 51, figs. 1–2.

*Bulimina elongata* d'Orbigny. Brady, 1884, p. 401, pl. 51, figs. 1, 2?. –Papp and Schmid, 1985, p. 73, pl. 63, figs. 5–9. –Thomas, 1990, pl. 1, fig. 2. –Jones, 1994, p. 54, pl. 50, figs. 3–4; p. 55, pl. 51, figs. 1–2.

### Remarks

Some authors have assigned this taxon to the genus *Caucasina* Khalikov 1951, based on the initial trochospire, but this has been shown by Nørvang (1968) to be present in the microspheric generation only (see Jones, 1994).

### Biogeography

Worldwide.

### Bathymetry

Primarily lower bathyal (Thomas, 1990). Reported between 630 and 1425 fathoms at "Challenger" stations in the Atlantic (Jones, 1994).

### Chronostratigraphy

Miocene to Recent (Jones, 1994), Middle Eocene to Pleistocene (Thomas, 1990).

## References


Barker, R.W., 1960, Taxonomic notes on the species figured by H.B. Brady in his report on the foraminifera dredged by H.M.S. Challenger during the years 1873–1876. Society of Economic Paleontologists and Mineralogists Special Publication, vol. 9, pp. 1–238.

Brady, H.B., 1884, Report on the foraminifera dredged by H.M.S. Challenger during the years 1873–1876. In Brady, H.B., ed., Report on the Scientific Results of the Voyage of the H.M.S. Challenger during the years 1873–1876. Zoology, vol. 9, pp. 1–814.

d'Orbigny, A.D., 1846, Foraminifères fossiles du Bassin Tertiaire de Vienne (Autriche). Paris: Gide et Compe, xxxvii + 312 pp.

Jones, R.W., 1994, The Challenger Foraminifera. Oxford: Oxford University Press, 149 pp.

Nørvang, A., 1968, Interior characteristics of Bulimina (Foraminifera). Proceedings of the 23rd International Geological Congress, pp. 415–422.

Papp, A. and Schmid, M.E., 1985, Die fossilen Foraminiferen des tertiaeren Beckens von Wien: Revision der Monographie Alcide d'Orbigny (1846). Abhandlungen der Geologischen Bundesanstalt, vol. 37, pp. 1–311.

Thomas, E., 1990, Late Cretaceous through Neogene benthic foraminifers. In Barker, P.F., Kennett, J.P., et al., ed., Scientific Results of the Proceedings of the Ocean Drilling Program, vol. 113, pp. 571–594.

# *Bulimina gibba*

## Fornasini, 1902

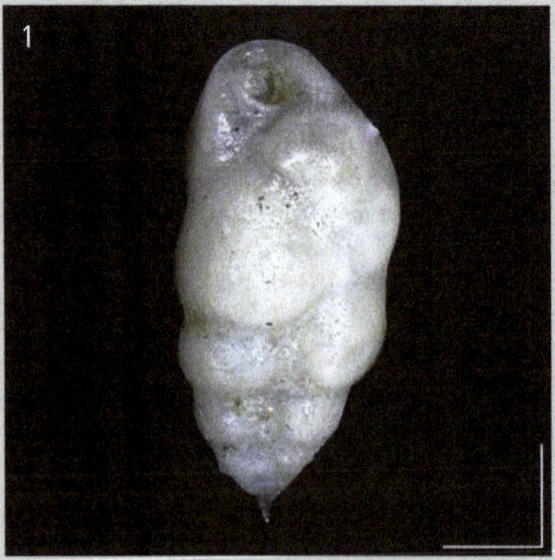

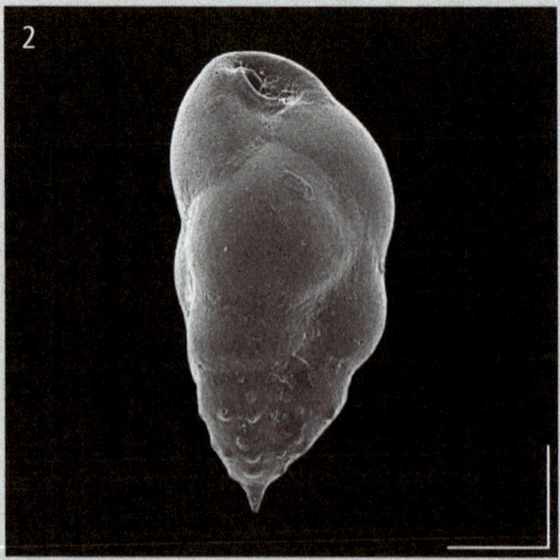

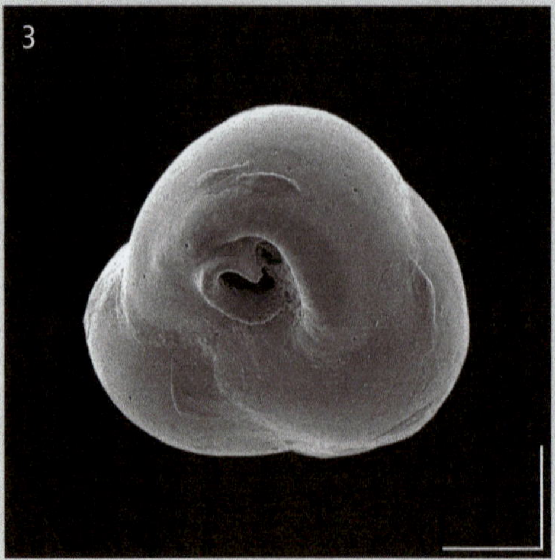

1. *Bulimina gibba.* West of Ireland, North Atlantic; "Porcupine" Station 11, depth: 1630 fathoms. Deposited in the micropaleontological collections of The Natural History Museum, London; NHMUK ZF1210. Lateral view, NHMUK PM DI 0119, scale bar = 104 μm.

2. *Bulimina gibba.* West of Ireland, North Atlantic; "Porcupine" Station 11, depth: 1630 fathoms. Deposited in the micropaleontological collections of The Natural History Museum, London; NHMUK ZF1210. Lateral view, NHMUK PM DI 0120, scale bar = 108 μm.

3. *Bulimina gibba.* West of Ireland, North Atlantic; "Porcupine" Station 11, depth: 1630 fathoms. Deposited in the micropaleontological collections of The Natural History Museum, London; NHMUK ZF1210. Apertural view, NHMUK PM DI 0121, scale bar = 66 μm.

### Original Designation

*Bulimina gibba* Fornasini, 1902, p. 378, pl. O, figs. 32, 34.

### Type Specimen

Depository not given.

### Type Level

Recent, surface sediments.

### Type Locality

Porto Corsini su la spiaggia di Ravenna, Italy.

### Description

Test forms an elongate, fusiform, triserial series; subcircular in cross-section with a finely pointed apical end and a rounded apertural end. The greatest breadth is approximately two-thirds of the way from the initial portion. The strongly inflated chambers increase gradually in size and are separated by distinct, depressed sutures. Chamber walls are calcareous, finely perforate, and smooth. The primary aperture is a loop-shaped opening that extends up from the base of the final chamber and is bordered by a lip that merges with an internal toothplate.

### Synonyms and Citations

*Bulimina elegans* d'Orbigny. –Brady, 1884, p. 398, pl. 50, figs. 1–2 (not figs. 3–4).

*Bulimina gibba* Fornasini. –Barker, 1960, pl. 50, figs. 1–2 (not figs. 3–4). –Jones, 1994, p. 54, pl. 50, figs. 1–2. –Milker and Schmiedl, 2012, p. 88, fig. 20.22.

### Biogeography

Atlantic, perhaps wider distribution.

### Bathymetry

Neritic to bathyal. Reported from 1630 fathoms at "Porcupine" Station 11, west of Ireland (Jones, 1994), and from shelf locations in the western Mediterranean Sea (Milker and Schmiedl, 2012).

### Chronostratigraphy

Miocene to Recent.

## References


Barker, R.W., 1960, Taxonomic notes on the species figured by H.B. Brady in his report on the foraminifera dredged by H.M.S. Challenger during the years 1873–1876. Society of Economic Paleontologists and Mineralogists Special Publication, vol. 9, pp. 1–238.

Brady, H.B., 1884, Report on the foraminifera dredged by H.M.S. Challenger during the years 1873–1876. In Brady, H.B., ed., Report on the Scientific Results of the Voyage of the H.M.S. Challenger during the years 1873–1876. Zoology, vol. 9, pp. 1–814.

Fornasini, C., 1902, Contributo a la conoscenza de le Bulimine adriatiche. Reale Academia delle Scienze dell'Instituto di Bologna, Memorie delle Scienze Naturali, ser. 5, vol. 9, pp. 371–381.

Jones, R.W., 1994, The Challenger Foraminifera. Oxford: Oxford University Press, 149 pp.

Milker, Y. and Schmiedl, G., 2012, A taxonomic guide to modern benthic shelf foraminifera of the western Mediterranean Sea. Palaeontologia Electronica, vol. 15, Issue 2;16A, 134 pp.

# *Bulimina glomarchallengeri*

## Tjalsma and Lohmann, 1983

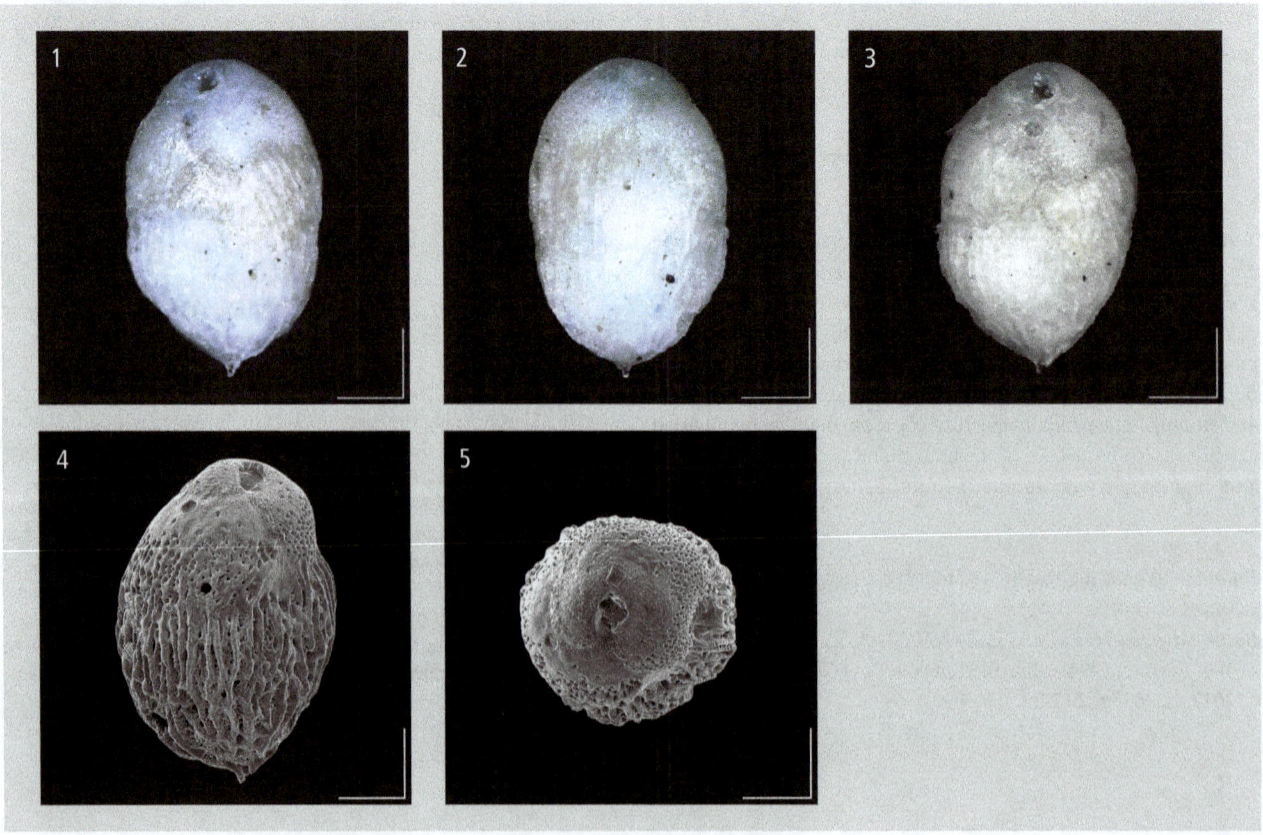

1. *Bulimina glomarchallengeri.* Shatsky Rise, West Central Pacific; DSDP Leg 86, Site 577, Core 8, Section 2, 80–87 cm. Lateral view, NHMUK PM DI 0122, scale bar = 97 μm.

2. *Bulimina glomarchallengeri.* Shatsky Rise, West Central Pacific; DSDP Leg 86, Site 577, Core 8, Section 2, 80–87 cm. Lateral view, NHMUK PM DI 0123, scale bar = 98 μm.

3. *Bulimina glomarchallengeri.* Shatsky Rise, West Central Pacific; DSDP Leg 86, Site 577, Core 8, Section 2, 80–87 cm. Lateral view, NHMUK PM DI 0124, scale bar = 97 μm.

4. *Bulimina glomarchallengeri.* Shatsky Rise, West Central Pacific; DSDP Leg 86, Site 577, Core 8, Section 2, 80–87 cm. Lateral view, NHMUK PM DI 0125, scale bar = 100 μm.

5. *Bulimina glomarchallengeri.* Shatsky Rise, West Central Pacific; DSDP Leg 86, Site 577, Core 8, Section 2, 80–87 cm. Apertural view, NHMUK PM DI 0126, scale bar = 100 μm.

## Original Designation

*Bulimina glomarchallengeri* Tjalsma and Lohmann, 1983, p. 25, pl. 13, figs. 8–12.

## Type Specimen

Deposited in the collections of the United States National Museum, Washington, DC.

## Type Level

Upper Eocene (Biozone P15).

## Type Locality

DSDP Leg 40, Site 363, Core 10, Section 2, 50–52 cm interval; latitude 19° 38.75′ S, longitude 09° 02.80′ E, Walvis Ridge, South Atlantic, 2109 meters water depth.

## Description

Test forms an elongate, triserial series; ovate in outline and subcircular in cross-section. The inflated subspherical chambers increase in size rapidly and are separated by sutures whose character is indistinct in the early whorls, but distinctly depressed in the final whorls. Chamber walls are calcareous and perforate except for the upper part of the final chambers. The earlier portion of the test is ornamented by fine, irregular longitudinal costae, ending in a short, basal spine. The primary aperture is a loop-shaped opening bordered by a raised lip that merges with an internal toothplate.

## Synonyms and Citations

*Bulimina glomarchallengeri* Tjalsma and Lohmann. –Miller et al., 1985, pl. 4, fig. 1. –van Morkhoven et al., 1986, p. 243, pl. 82A–B. –Miller and Katz, 1987, p. 125, pl. 2, fig. 8.

## Biogeography

Recorded in the North and South Atlantic (Tjalsma and Lohmann, 1983; van Morkhoven et al., 1986).

## Bathymetry

Bathyal to abyssal.

## Chronostratigraphy

Middle Eocene (Bartonian) to Early Miocene (Aquitanian). May occur in younger strata (see van Morkhoven et al., 1986).

## References


Miller, K.G. and Katz, M.E., 1987, Oligocene to Miocene benthic foraminiferal and abyssal circulation changes in the north Atlantic. Micropaleontology, vol. 33, no. 2, pp. 97–149.

Miller, K.G., Curry, W.B. and Ostermann, D.R., 1985, Late Paleogene (Eocene to Oligocene) benthic foraminiferal oceanography of the Goban Spur region, Deep Sea Drilling Project Leg 80. In de Graciansky, P.C., Poag, C.W., et al., ed., Initial Reports of the Deep Sea Drilling Project, Washington (U.S. Government Printing Office), vol. 80, pp. 505–538.

Tjalsma, R.C. and Lohmann, G.P., 1983, Paleocene–Eocene bathyal and abyssal benthic foraminifera from the Atlantic Ocean. Micropaleontology, Special Publication, no. 4, pp. 1–90.

van Morkhoven, F.P.C.M., Berggren, W.A., Edwards, A.S., et al., 1986, Cenozoic cosmopolitan deep-water benthic foraminifera. Bulletin des Centres de Recherches Exploration-Production Elf-Aquitaine, Mémoire, vol. 11, pp. 1–421.

# Bulimina impendens

## Parker and Bermúdez, 1937

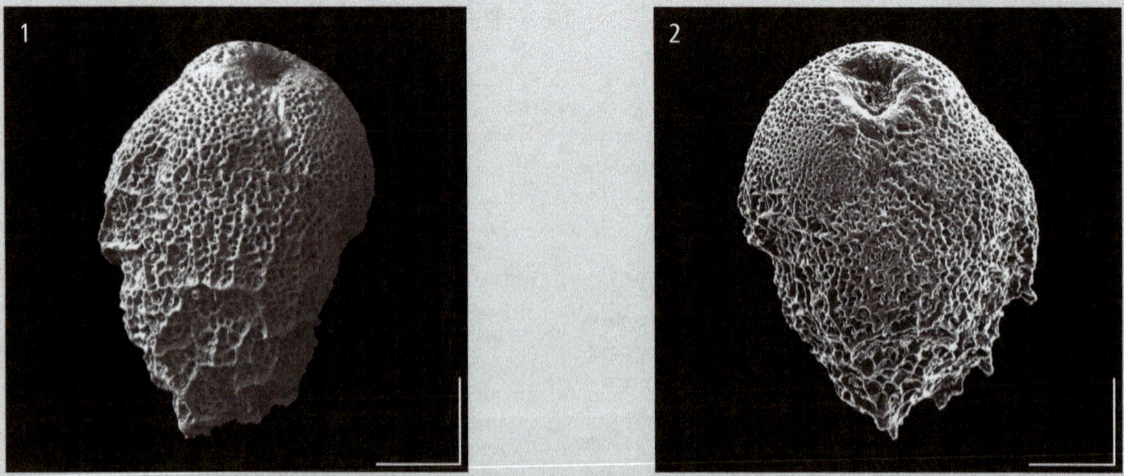


1. *Bulimina impendens*. North Biscay, North Atlantic; DSDP Leg 48, Site 401, Core 6, Section 3, 76–78 cm. Lateral view, NHMUK PM DI 0127, scale bar = 75 μm.

2. *Bulimina impendens*. North Biscay, North Atlantic; DSDP Leg 48, Site 401, Core 6, Section 3, 76–78 cm. Lateral view, NHMUK PM DI 0128, scale bar = 75 μm.

### Original Designation

*Bulimina impendens* Parker and Bermúdez, 1937, p. 514, pl. 58, figs. 7–8.

### Type Specimen

Holotype (No. 23335) deposited in the collections of the United States National Museum, Washington, DC.

### Type Level

Eocene.

### Type Locality

Alturas de Almendares quarry, Havana, Cuba.

### Description

Test forms a small, stout, elongate, triserial series slightly tapered in outline and subcircular in cross-section. The subspherical chambers are distinct in the last whorl only, increase rapidly in size, and are separated by curved, depressed sutures. Chambers of each whorl overhang previous chambers, thus producing a collared effect. Chamber walls are calcareous, coarsely perforate, and heavily ornamented by short, irregular spines in the early part. Later chambers are spinose or serrate in the basal part. The primary aperture is a loop-shaped opening, extending up from the base of the final chamber, and bordered by a lip that merges with an internal toothplate.

### Synonyms and Citations

*Bulimina impendens* Parker and Bermúdez. –Tjalsma and Lohmann, 1983, p. 25, pl. 14, fig. 2. –van Morkhoven et al., 1986, p. 236, pl. 79. –Miller and Katz, 1987, p. 125, pl. 2, fig. 3.

### Remarks

*Bulimina impendens* Parker and Bermúdez differs from *Bulimina trinitatensis* Cushman and Jarvis by its shorter, stouter test, the absence of ribs, and the presence of irregular spines on the initial part of the test.

### Biogeography

Worldwide.

### Bathymetry

Middle to lower bathyal with upper depth limit of about 500 meters (Berggren and Aubert, 1983).

### Chronostratigraphy

Middle Eocene (Lutetian) to Early Miocene (Burdigalian).

## References

Berggren, W.A. and Aubert, J., 1983, Paleogene benthonic foraminiferal biostratigraphy and bathymetry of the Central Coast Ranges of California. In Brabb, E.E., ed., Studies in Tertiary stratigraphy of the California Coast Ranges. Professional Paper of the United States Geological Survey, vol. 1213, pp. 4–21.

Miller, K.G. and Katz, M.E., 1987, Oligocene to Miocene benthic foraminiferal and abyssal circulation changes in the north Atlantic. Micropaleontology, vol. 33, no. 2, pp. 97–149.

Parker, F.L. and Bermúdez, P.J., 1937, Eocene species of the genera Bulimina and Buliminella from Cuba. Journal of Paleontology, vol. 11, pp. 513–516.

Tjalsma, R.C. and Lohmann, G.P., 1983, Paleocene–Eocene bathyal and abyssal benthic foraminifera from the Atlantic Ocean. Micropaleontology, Special Publication, no. 4, pp. 1–90.

van Morkhoven, F.P.C.M., Berggren, W.A., Edwards, A.S., et al., 1986, Cenozoic cosmopolitan deep-water benthic foraminifera. Bulletin des Centres de Recherches Exploration-Production Elf-Aquitaine, Mémoire, vol. 11, pp. 1–421.

# *Bulimina jacksonensis*

## Cushman, 1925

1

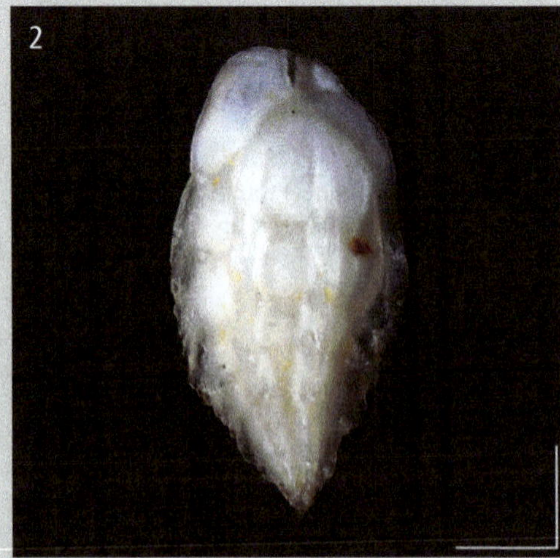

2

1. *Bulimina jacksonensis*. Gahannam Beds, Fayoum, Egypt; Sample H489. Deposited in the micropaleontological collections of The Natural History Museum, London; NHMUK P40903. Lateral view, NHMUK PM DI 0129, scale bar = 154 µm.

2. *Bulimina jacksonensis*. Cocoa Sand, Cocoa Post Office, Alabama. Deposited in the collections of the United States National Museum, Washington, DC. Specimen illustrated in van Morkhoven et al. 1986, Ref. no. 90. Lateral view, NHMUK PM DI 0130, scale bar = 206 µm.

### Original Designation

*Bulimina jacksonensis* Cushman, 1925, p. 6, pl. 1, figs. 6–7.

### Type Specimen

Holotype (CC 4333) deposited in the collections of the United States National Museum, Washington, DC.

### Type Level

Upper Eocene, yellow clay, Tantoyuca Formation.

### Type Locality

Palacho Hacienda, south of Panuco-Tampico R.R., State of Vera Cruz, Mexico.

### Description

Test forms an elongate, triserial series tapered in outline and subcircular in cross-section, with an acute initial portion and a broadly rounded apertural end. The numerous subspherical chambers are moderately inflated, increase slowly in size, and are separated by distinct, slightly depressed sutures. Chamber walls are calcareous, finely perforate, and ornamented by six to eight prominent, raised, longitudinal, serrate costae. The primary aperture is a loop-shaped opening that extends up from the base of the final chamber and is bordered by a lip that merges with an internal toothplate.

### Synonyms and Citations

*Bulimina jacksonensis* Cushman var. *cuneata* Cushman, 1926, p. 35, pl. 13, figs. 10–11.

*Bulimina forticosta* Finlay, 1940, p. 455, pl. 64, figs. 77–81.

*Bulimina jacksonensis* Cushman. –van Morkhoven et al., 1986, p. 271, pl. 90. –Bolli et al., 1994, p. 348, figs. 53.31–33.

### Biogeography

Worldwide.

### Bathymetry

Lower neritic to upper bathyal.

### Chronostratigraphy

Early Eocene (Ypresian) to Early Oligocene (Rupelian).

## References


Bolli, H.M., Beckmann, J.P., and Saunders, J.B., 1994, Benthic Foraminiferal Biostratigraphy of the South Caribbean Region. Cambridge: Cambridge University Press, 408 pp.

Cushman, J.A., 1925, New foraminifera from the upper Eocene of Mexico. Contributions from the Cushman Laboratory for Foraminiferal Research, vol. 1, no. 1, pp. 4–9.

Cushman, J.A., 1926, Some new foraminifera from the Upper Eocene of the southeastern Coastal Plain of the United States. Contributions from the Cushman Laboratory for Foraminiferal Research, vol. 2, no. 2, pp. 29–38.

Finlay, H.J., 1940, New Zealand Foraminifera; key species in stratigraphy – No. 4. Transactions of the Royal Society of New Zealand, vol. 69, no. 4, pp. 448–472.

van Morkhoven, F.P.C.M., Berggren, W.A., Edwards, A.S., et al., 1986, Cenozoic cosmopolitan deep-water benthic foraminifera. Bulletin des Centres de Recherches Exploration-Production Elf-Aquitaine, Mémoire, vol. 11, pp. 1–421.

# *Bulimina jarvisi*

## Cushman and Parker, 1936

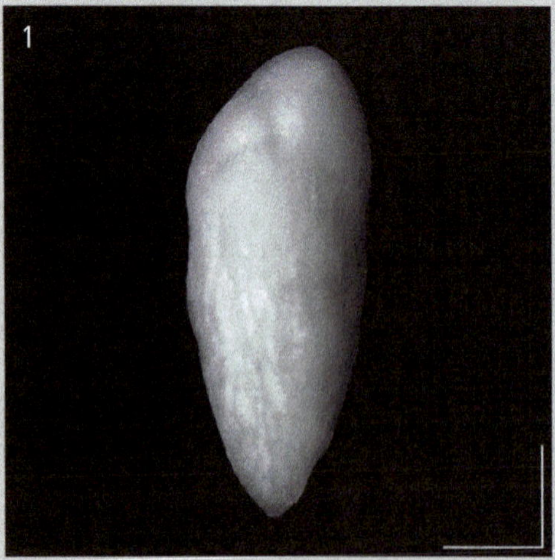

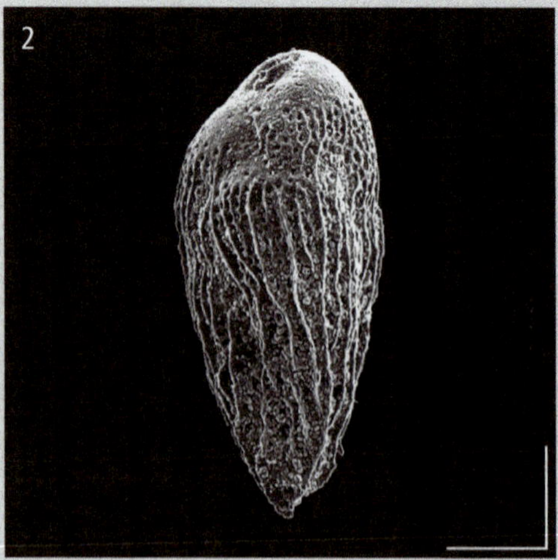

1. *Bulimina jarvisi*. Northeast Georgia Rise, South Atlantic; ODP Leg 114, Hole 700B, Core 10, Section 3, 60–64 cm. Lateral view, NHMUK PM DI 0131, scale bar = 52 μm.

2. *Bulimina jarvisi*. Northeast Georgia Rise, South Atlantic; ODP Leg 114, Hole 700B, Core 10, Section 3, 60–64 cm. Lateral view, NHMUK PM DI 0132, scale bar = 66 μm.

### Original Designation
*Bulimina jarvisi* Cushman and Parker, 1936, p. 39, pl. 7, fig. 1.

### Type Specimen
Holotype (CC 23128) deposited in the collections of the United States National Museum, Washington, DC.

### Type Level
Eocene, Lower Marl.

### Type Locality
Cipero Section. Station No. 10, Trinidad.

### Description
Test forms a large, slender, triserial series; tapered in outline and triangular with rounded corners in cross-section with an acute initial portion, and a subrounded apertural end. The numerous chambers are inflated, subspherical, increase gradually in size, and are separated by distinct, slightly curved, depressed sutures. The test is usually arranged in six or seven whorls. Chamber walls are calcareous, coarsely perforate, and ornamented by fine, irregular costae. Some specimens possess a prominent adapical spine. The primary aperture is a loop-shaped opening extending up from the base of the final chamber, and bordered by a lip that merges with an internal toothplate.

### Synonyms and Citations
*?Bulimina scobinata* Finlay, 1940, p. 455, pl. 64, figs. 82–83.

*Bulimina jarvisi* Cushman and Parker. –Tjalsma and Lohmann, 1983, p. 25, pl. 13, figs. 4–5. –van Morkhoven et al., 1986, p. 184, pl. 62. –Miller and Katz, 1987, p. 125, pl. 2, fig. 6. –Bolli et al., 1994, p. 349, figs. 53.30a–b, 50, 87.19, 86.

### Remarks
*Bulimina jarvisi* Cushman and Parker is larger and more slender, and has more inflated chambers, finer ornamentation, and a less triangular initial part than *Bulimina semicostata* Nuttall. Transitional forms between the two species are reported in Biozone P14 (Tjalsma and Lohmann, 1983).

### Biogeography
Worldwide.

### Bathymetry

Lower bathyal to abyssal.

### Chronostratigraphy

Middle Eocene (Bartonian) to Middle Miocene (Serravallian).

## References


Bolli, H.M., Beckmann, J.P., and Saunders, J.B., 1994, Benthic Foraminiferal Biostratigraphy of the South Caribbean Region. Cambridge: Cambridge University Press, 408 pp.

Cushman, J.A. and Parker, F.L., 1936, Some American Eocene Buliminas. Contributions from the Cushman Laboratory for Foraminiferal Research, vol. 12, no. 2, pp. 39–45.

Finlay, H.J., 1940, New Zealand Foraminifera; key species in stratigraphy – No. 4. Transactions of the Royal Society of New Zealand, vol. 69, no. 4, pp. 448–472.

Miller, K.G. and Katz, M.E., 1987, Oligocene to Miocene benthic foraminiferal and abyssal circulation changes in the north Atlantic. Micropaleontology, vol. 33, no. 2, pp. 97–149.

Tjalsma, R.C. and Lohmann, G.P., 1983, Paleocene–Eocene bathyal and abyssal benthic foraminifera from the Atlantic Ocean. Micropaleontology, Special Publication, no. 4, pp. 1–90.

van Morkhoven, F.P.C.M., Berggren, W.A., Edwards, A.S., et al., 1986, Cenozoic cosmopolitan deep-water benthic foraminifera. Bulletin des Centres de Recherches Exploration-Production Elf-Aquitaine, Mémoire, vol. 11, pp. 1–421.

# Bulimina macilenta

## Cushman and Parker, 1939

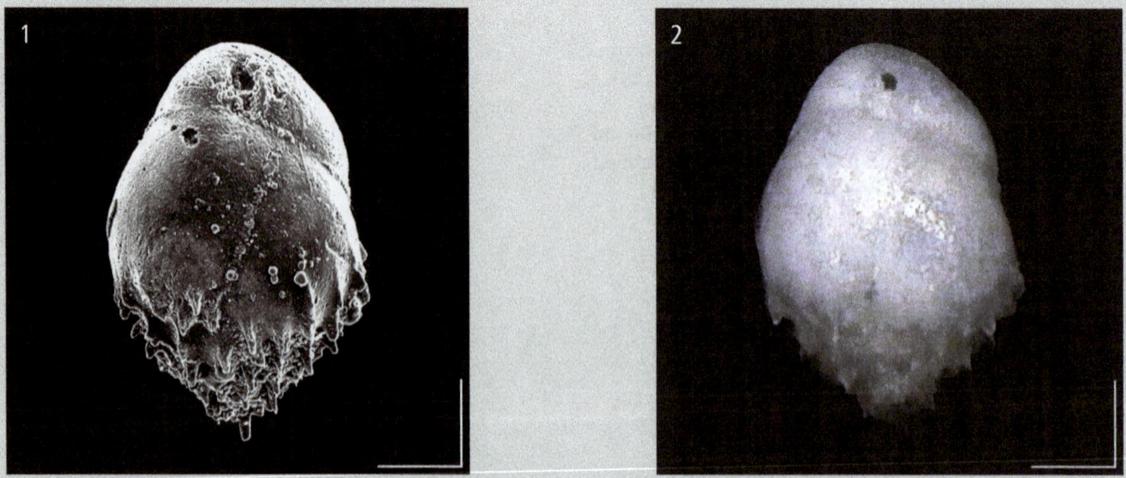


1. *Bulimina macilenta*. Meteor Rise, South Atlantic; ODP Leg 114, Hole 703A, Core 15, Section 6, 10–14 cm. Lateral view, NHMUK PM DI 0133, scale bar = 87 μm.

2. *Bulimina macilenta*. Meteor Rise, South Atlantic; ODP Leg 114, Hole 703A, Core 15, Section 6, 10–14 cm. Lateral view, NHMUK PM DI 0134, scale bar = 87 μm.

### Original Designation

*Bulimina macilenta* Cushman and Parker, 1939, p. 93, illustrated as *Bulimina denticulata* in Cushman and Parker, 1936, pl. 7, figs. 7–8.

### Type Specimen

Holotype (No. 23134) deposited in the collections of the United States National Museum, Washington, DC.

### Type Level

Martinez, brown shale, 2710 feet stratigraphically above the Cretaceous contact.

### Type Locality

Approximately 3 miles north 45° east of the town of Santa Susana, Poison Oak Canyon, north of Simi Valley, Ventura County, California, USA.

### Description

Test forms an elongate, triserial series; approximately one and a half times as long as broad, and strongly tapered in outline, with an acute apical end, a rounded apertural end, and a subcircular cross-section. The strongly inflated chambers increase very rapidly in size and are separated by distinct, slightly depressed sutures, except in the early portion where the sutures are obscured by surface ornamentation. Chambers overhang those of the previous whorl and have distinctive costate or fluted basal margins. Chamber walls are calcareous and finely perforate. The primary aperture is loop-shaped, extends up from the base of the final chamber, and is bordered by a lip that merges with an internal toothplate.

### Synonyms and Citations

*Bulimina denticulata* Cushman and Parker, 1936, p. 42, pl. 7, figs. 7–8.

*Bulimina macilenta* Cushman and Parker, 1939, p. 93. New name for *Bulimina denticulata* Cushman and Parker.

*Bulimina macilenta* Cushman and Parker. –Beckmann, 1954, p. 366, pl. 21, fig. 11. –Tjalsma, 1983, p. 739, pl. 1, fig. 6. –Tjalsma and Lohmann, 1983, p. 25, pl. 14, fig. 3. –Bolli et al., 1994, p. 349, figs. 53.24, 50, 87.20, 86.

### Remarks

Cushman and Parker (1939) gave the new name *Bulimina macilenta* to *Bulimina denticulata* Cushman and Parker, as this species name was already occupied by *Bulimina truncana* Gümbel var. *denticulata* Protescu.

*Bulimina macilenta* differs from *Bulimina midwayensis* Cushman and Parker by its less slender test, more costate and less spinose ornamentation at the base of its chambers, lack of basal spine, and more distinct overhang over previous whorls.

### Biogeography

Worldwide. Recorded from California (Cushman and Parker, 1936), the North and South Atlantic (Tjalsma and Lohmann, 1983), and the Caribbean (Beckmann, 1954; Bolli et al., 1994).

### Bathymetry

Bathyal.

### Chronostratigraphy

Early Eocene (Ypresian) to Early Miocene (Aquitanian). Early Eocene (P7) to Late Eocene (P15) in the Atlantic according to Tjalsma and Lohmann (1983). Early Eocene to Early Miocene in Trinidad; Late Eocene to Oligocene in Barbados according to Bolli et al. (1994).

## References


Beckmann, J.P., 1954, Die Foraminiferen der Oceanic Formation (Eocaen–Oligocaen) von Barbados, Kl. Antillen. Eclogae Geologicae Helvetiae, vol. 46, pp. 301–412.

Bolli, H.M., Beckmann, J.P., and Saunders, J.B., 1994, Benthic Foraminiferal Biostratigraphy of the South Caribbean Region. Cambridge: Cambridge University Press, 408 pp.

Cushman, J.A. and Parker, F.L., 1936, Some American Eocene Buliminas. Contributions from the Cushman Laboratory for Foraminiferal Research, vol. 12, no. 2, pp. 39–45.

Cushman, J.A. and Parker, F.L., 1939, Bulimina macilenta Cushman and Parker a new name. Contributions from the Cushman Laboratory for Foraminiferal Research, vol. 15, no. 4, p. 93.

Tjalsma, R.C., 1983, Eocene to Miocene benthic foraminifera from DSDP site 516, Rio Grande Rise, South Atlantic. In Barker, P.F., Carlson, R.L., Johnson, D.A. et al., ed., Initial Reports of the Deep Sea Drilling Project, Washington (U.S. Government Printing Office), vol. 72, pp. 731–755.

Tjalsma, R.C. and Lohmann, G.P., 1983, Paleocene–Eocene bathyal and abyssal benthic foraminifera from the Atlantic Ocean. Micropaleontology, Special Publication, no. 4, pp. 1–90.

# Bulimina marginata

## d'Orbigny, 1826

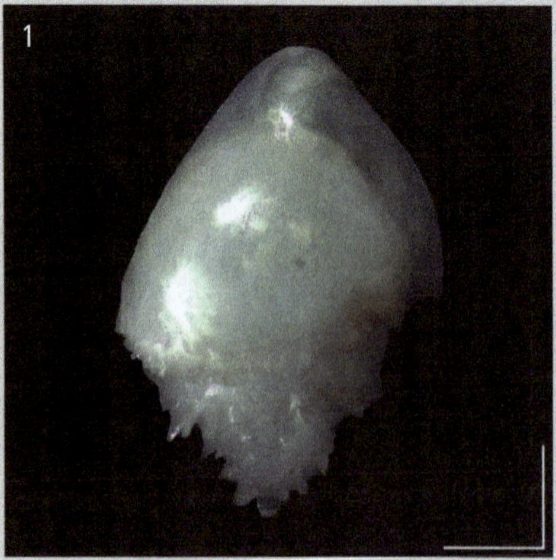

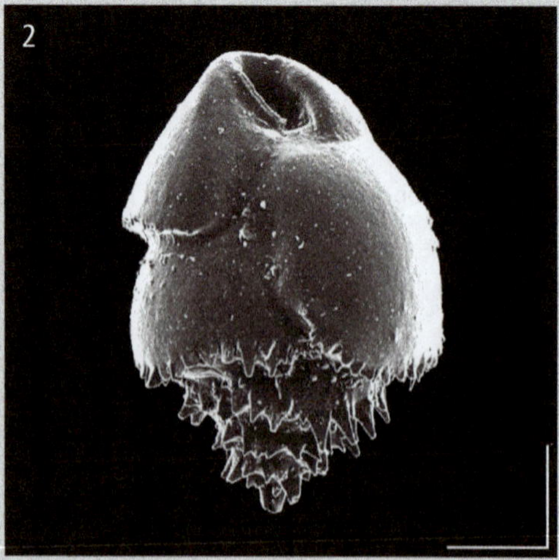

1. *Bulimina marginata.* Central Great Barrier Reef, Grafton Passage, Northeast Australian Margin, tropical Southwest Pacific; ODP Leg 133, Hole 820B, Core 1, Section 1, 0–5 cm. Lateral view, NHMUK PM DI 0135, scale bar = 157 μm.

2. *Bulimina marginata.* Central Great Barrier Reef, Grafton Passage, Northeast Australian Margin, tropical Southwest Pacific; ODP Leg 133, Hole 820B, Core 1, Section 1, 0–5 cm. Lateral view, NHMUK PM DI 0136, scale bar = 168 μm.

### Original Designation

*Bulimina marginata* d'Orbigny, 1826, p. 269, pl. 12, figs. 10–12.

### Type Specimen

Deposited in the collections of the Laboratoire de Paléontologie, Muséum d'Histoire Naturelle, Paris, France (Terquem, 1878, p. 2).

### Type Level

Recent, surface sediments.

### Type Locality

Adriatic Sea near Rimini, Italy.

### Description

Test forms an elongate, medium-sized, triserial series; tapered in outline, and subspherical in cross-section. The moderately inflated chambers are angled and undercut at the margin, increase rapidly in size, and are separated by distinct, strongly depressed sutures. Chamber walls are calcareous, smooth, finely perforate, and often partially translucent. The marginal surfaces of these chambers are ornamented by crenulations, often extending into short, sharp spines. The primary aperture is a loop-shaped opening that extends up from the base of the final chamber and is bordered by a lip that merges with an internal toothplate.

### Synonyms and Citations

*Buliminella pulchella* d'Orbigny, 1839, p. 50, pl. 1, figs. 6–7.

*Bulimina serrata* Bailey, 1851, p. 12, figs. 32–34.

*Bulimina tenuis*, Phleger and Parker, 1951, p. 16, pl. 7, figs. 33–34.

*Bulimina marginata* d'Orbigny, 1826. –Bermúdez, 1949, p. 182, pl. 12, fig. 11. –van Morkhoven et al., 1986, p. 18, pl. 2, fig. 1. –Loeblich and Tappan, 1987, p. 521, pl. 571, figs. 1–3. –van Marle, 1991, p. 87, pl. 5, figs. 9–10. –Jones, 1994, p. 55, pl. 51, figs. 3–5. –Mendes et al., 2012, p. 38, fig. 3.10. –Milker and Schmiedl, 2012, p. 88, fig. 20.23.

### Remarks

Type species of the genus *Bulimina* d'Orbigny 1826.

### Biogeography

Worldwide. Recorded from the North and South Atlantic, Europe, the Pacific, the Indian Ocean, and the Gulf of Mexico (van Morkhoven et al., 1986). Also recorded in the South China Sea

(Szarek et al., 2009), and the Mediterranean (Milker and Schmiedl, 2012).

### Bathymetry

Lower neritic to upper bathyal. Most abundant in upper bathyal depths on the Sunda Shelf in the South China Sea (Szarek et al., 2009). Recorded in water depths between 344 and 1097 meters in eastern Indonesia (van Marle, 1991).

### Chronostratigraphy

Late Miocene to Recent (Jones, 1994). Late Miocene (Subzone M13a) to Recent in Indonesia (van Marle, 1991). Late Pliocene to Recent (van Morkhoven et al., 1986).

## References


Bailey, J.W., 1851, Microscopical examination of soundings made by the United States Coast Survey, off the Atlantic Coast of the United States. Smithsonian Contributions to Knowledge, vol. 2, pp. 1–15.

Bermúdez, P.J., 1949, Tertiary smaller foraminifera of the Dominican Republic. Special Publications of the Cushman Laboratory for Foraminiferal Research, no. 25, pp. 1–322.

d'Orbigny, A., 1826, Tableau méthodique de la classe des Céphalopodes. Annales des Sciences Naturelles, vol. 7, pp. 245–314.

d'Orbigny, A.D., 1839, Voyage dans l'Amérique méridionale; foraminifères, vol. 5, pt 5. Paris and Strasbourg: P. Bertrand, 86 pp.

Jones, R.W., 1994, The Challenger Foraminifera. Oxford: Oxford University Press, 149 pp.

Loeblich, A.R. Jr and Tappan, H., 1987, Foraminiferal Genera and their Classification. New York: Van Nostrand Reinhold, 970 pp.

Mendes, I., Dias, J.A., Schönfeld, J., and Ferreira, O., 2012, Distribution of living foraminifera on the northern Gulf of Cadiz continental shelf. Journal of Foraminiferal Research, vol. 42, pp. 18–38.

Milker, Y. and Schmiedl, G., 2012, A taxonomic guide to modern benthic shelf foraminifera of the western Mediterranean Sea. Palaeontologia Electronica, vol. 15, Issue 2;16A, 134 pp.

Phleger, F.B. and Parker, F.L., 1951, Ecology of foraminifera, northwest Gulf of Mexico. Part II – Foraminifera species. Memoirs of the Geological Society of America, vol. 46, pp. 1–64.

Szarek, R., Kuhnt, W., Kawamura, H., and Nishi, H., 2009, Distribution of recent benthic foraminifera along the continental slope of the Sunda Shelf (South China Sea). Marine Micropaleontology, vol. 71, pp. 41–59.

Terquem, O., 1878, Les Foraminifères et les Entomostracés-Ostracodes du Pliocène Supérieur de l'île de Rhodes. Mémoires de la Société géologique de France, ser. 3, vol. 1, pp. 1–135.

van Marle, L.J., 1991, Eastern Indonesian Late Cenozoic smaller benthic foraminifera. Verhandelingen der Koninklijke Nederlanse Akademie van Wetenschappen, Afdeeling Natuurkunde, Eerste Reeks, vol. 34, pp. 1–328.

van Morkhoven, F.P.C.M., Berggren, W.A., Edwards, A.S., et al., 1986, Cenozoic cosmopolitan deep-water benthic foraminifera. Bulletin des Centres de Recherches Exploration-Production Elf-Aquitaine, Mémoire, vol. 11, pp. 1–421.

# Bulimina mexicana

## Cushman, 1922

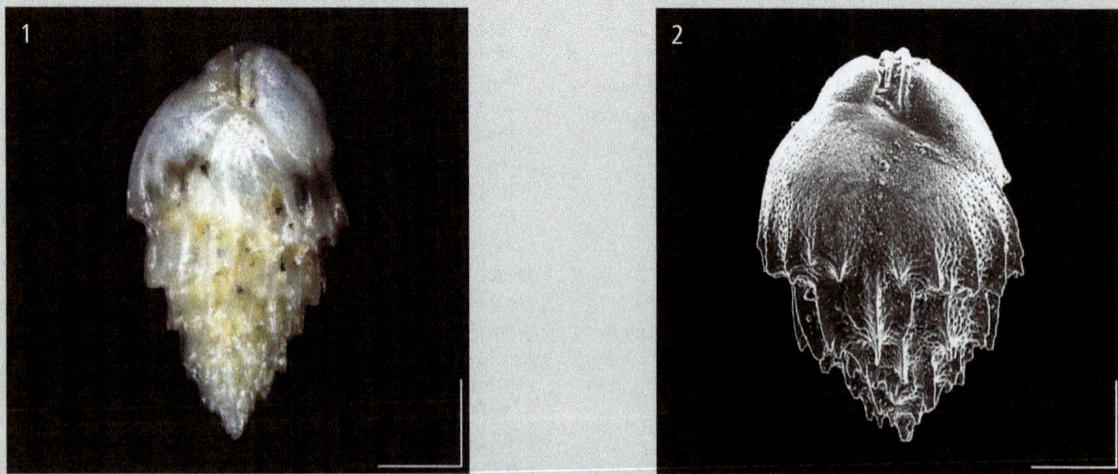


1. *Bulimina mexicana*. Gulf of Mexico; "Albatross" Station D2377, depth: 210 fathoms. Deposited in the collections of the United States National Museum, Washington, DC; CC 16402. Holotype. Lateral view, NHMUK PM DI 0137, scale bar = 148 µm.

2. *Bulimina mexicana*. Maud Rise, Weddell Sea, Antarctica; ODP Leg 113, Hole 690B, Core 19, Section 2, 118–124 cm. Lateral view, NHMUK PM DI 0138, scale bar = 182 µm.

### Original Designation

*Bulimina inflata* Seguenza var. *mexicana* Cushman, 1922, p. 95, pl. 21, fig. 2.

### Type Specimen

Holotype (No. 16402) and catalog numbers 16396–16401, 16403–16413 deposited in the collections of the United States National Museum, Washington, DC.

### Type Level

Recent, surface sediments.

### Type Locality

"Albatross" Station D2377, latitude 29° 07′ 30″ N, longitude 88° 08′ 00″ W, Gulf of Mexico; 210 fathoms.

### Description

Test forms an elongate, conical, triserial series; tapered in outline, subcircular in cross-section with an acute initial portion, and a rounded apertural end. The widest portion of this species' test is across the last whorl. Chambers are inflated, increasing in size rapidly, and slightly overhanging previous chambers. These are separated by sutures that are distinct and depressed in the last whorl, but obscured in previous whorls. Chamber walls are calcareous, finely perforate, and ornamented by five to seven longitudinal costae on the lower half of each chamber, ending in narrow spines. The primary aperture is a loop-shaped opening, extending up from the base of the final chamber, and bordered by a lip that merges with an internal toothplate.

### Synonyms (following van Morkhoven et al., 1986) and Citations

?*Bulimina striata* d'Orbigny, 1826, in Guérin-Méneville, 1843, p. 9, pl. 2, fig. 16.

?*Bulimina costata* d'Orbigny, 1826, in Fornasini, 1901, p. 174, pl. 1.

*Bulimina inflata* Seguenza –Cushman and Jarvis, 1930, p. 362, pl. 33, fig. 5.

*Bulimina subacuminata* Cushman et al., 1930, p. 65, pl. 5, figs. 2–3.

*Bulimina bleeckeri* Hedberg, 1937, p. 675, pl. 91, figs. 12–13.

*Bulimina striata* d'Orbigny var. *mexicana* Cushman and Parker, 1940, p. 16, pl. 3, fig. 9.

*Bulimina mexicana* Cushman. –van Morkhoven et al., 1986, p. 59, pl. 19. –Hermelin, 1989, p. 63, pl. 11, fig. 16. –Jones, 1994,

p. 56, pl. 51, figs. 10–13. –Ortiz and Thomas, 2006, p. 114, pl. 4, figs. 15–16.

### Remarks

Revised stratigraphy at the Eureka boreholes in the Gulf of Mexico (Katz and Miller, 1993) has extended the first occurrence of this species to Biozone M1, in contrast to the Biozone M4 first occurrence previously reported by van Morkhoven et al. (1986).

### Biogeography

Worldwide. Recorded from the Atlantic, the Caribbean, the Pacific, the Gulf of Mexico, and Venezuela (see Observed Occurrences in van Morkoven et al., 1986). Also recorded in the Mediterranean (as *Bulimina costata* d'Orbigny) by Jonkers (1984) and in the South China Sea (Szarek et al., 2009).

### Bathymetry

Primarily upper to middle bathyal (van Morkhoven et al., 1986). Common in upper and middle bathyal zones in the Gulf of California (Bandy, 1961), Gulf of Mexico (Bandy and Chierici, 1966; Poag, 1981; Hermelin, 1989), and the western Pacific (Hermelin, 1989), between 500 and 1500 meters in the Peru–Chile Trench (Ingle et al., 1980). Reported between 95 and 2435 fathoms at "Challenger" stations (Jones, 1994). Most abundant in upper bathyal depths on the Sunda Shelf in the South China Sea (Szarek et al., 2009).

### Chronostratigraphy

Early Miocene (Aquitanian) to Recent.

## References


Bandy, O.L., 1961, Distribution of foraminifera, radiolaria and diatoms in sediments from the Gulf of California. Micropaleontology, vol. 7, pp. 1–21.

Bandy, O.L. and Chierici, M.A., 1966, Depth–temperature evaluation of selected Californian and Mediterranean bathyal foraminifera. Marine Geology, vol. 4, pp. 259–271.

Cushman, J.A., 1922, The foraminifera of the Atlantic Ocean. Part III – Textulariidae. Bulletin of the United States National Museum, vol. 104, pp. 1–149.

Cushman, J.A. and Jarvis, P.W., 1930, Miocene foraminifera from Buff Bay, Jamaica. Journal of Paleontology, vol. 4, no. 4, pp. 353–368.

Cushman, J.A. and Parker, F.L., 1940, The species of the genus Bulimina having Recent types. Contributions from the Cushman Laboratory for Foraminiferal Research, vol. 16, no. 1, pp. 7–23.

Cushman, J.A., Stewart, R.E., and Stewart, K.C., 1930, Tertiary Foraminifera from Humboldt County, California. Transactions of the San Diego Society for Natural History, vol. 6, pp. 41–94.

d'Orbigny, A., 1826, Tableau méthodique de la classe des Céphalopodes. Annales des Sciences Naturelles, vol. 7, pp. 245–314.

Fornasini, C., 1901, Le Bulimine e le Cassiduline fossili d'Italia. Società Geologica Italiana, Bollettino, Rome, vol. 20, pp. 159–214.

Guérin-Meneville, F.E., 1843, Iconographie du Règne animal de G. Cuvier Mollusques. Paris: J.B. Baillière, 64 pp.

Hedberg, H.D., 1937, Foraminifera of the middle Tertiary Carapitana formation of northeastern Venezuela. Journal of Paleontology, vol. 11, pp. 661–697.

Hermelin, J.O.R., 1989, Pliocene benthic foraminifera from the Ontong-Java Plateau (Western Equatorial Pacific Ocean): faunal response to changing paleoenvironment. Special Publication of the Cushman Foundation for Foraminiferal Research, vol. 26, pp. 1–143.

Ingle, J.C. Jr, Keller, G. and Kolpack, R.L., 1980, Benthic foraminiferal biofacies, sediments and water masses of the southern Peru–Chile trench area, southeastern Pacific Ocean. Micropaleontology, vol. 26, pp. 113–150.

Jones, R.W., 1994, The Challenger Foraminifera. Oxford: Oxford University Press, 149 pp.

Jonkers, H.A., 1984, Pliocene benthonic foraminifera from homogeneous and laminated marls on Crete. Utrecht Micropaleontological Bulletins, vol. 31, pp. 1–179.

Katz, M.E. and Miller, K.G., 1993, Latest Oligocene to Earliest Pliocene benthic foraminiferal biofacies of the northeastern Gulf of Mexico. Micropaleontology, vol. 39, no. 4, pp. 367–403.

Ortiz, S. and Thomas, E., 2006, Lower-middle Eocene benthic foraminifera from the Fortuna Section (Betic Cordillera, southeastern Spain). Micropaleontology, vol. 52, pp. 97–150.

Poag, C.W., 1981, Ecologic Atlas of Benthic Foraminifera of the Gulf of Mexico. Stroudsburg, PA: Hutchinson Ross, 174 pp.

Szarek, R., Kuhnt, W., Kawamura, H., and Nishi, H., 2009, Distribution of recent benthic foraminifera along the continental slope of the Sunda Shelf (South China Sea). Marine Micropaleontology, vol. 71, pp. 41–59.

van Morkhoven, F.P.C.M., Berggren, W.A., Edwards, A.S., et al., 1986, Cenozoic cosmopolitan deep-water benthic foraminifera. Bulletin des Centres de Recherches Exploration-Production Elf-Aquitaine, Mémoire, vol. 11, pp. 1–421.

# *Bulimina midwayensis*

## Cushman and Parker, 1936

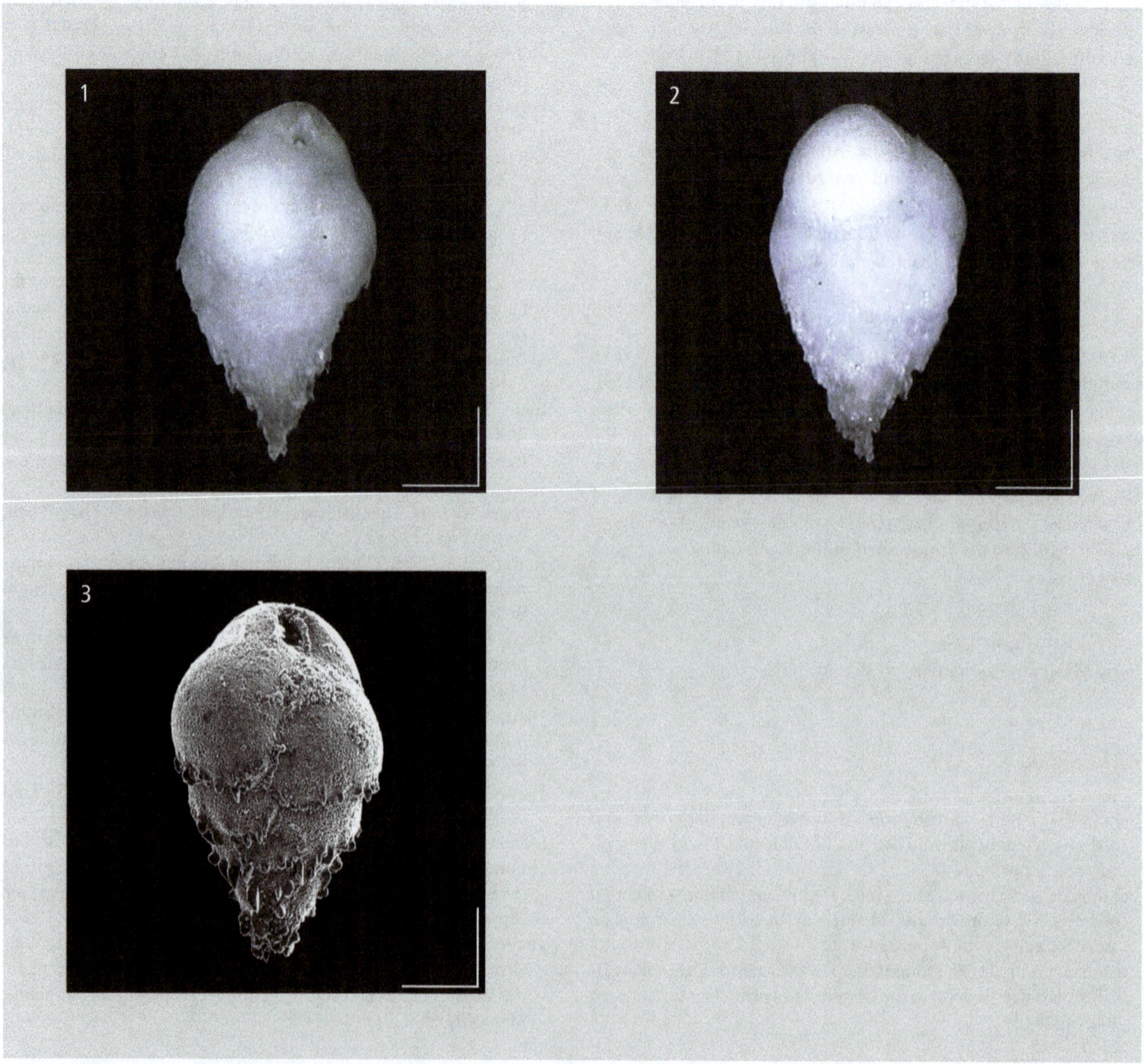


1. *Bulimina midwayensis*. Shatsky Rise, West Central Pacific; DSDP Leg 86, Site 577, Core 11, Section 3, 43–50 cm. Lateral view, NHMUK PM DI 0139, scale bar = 64 μm.

2. *Bulimina midwayensis*. Shatsky Rise, West Central Pacific; DSDP Leg 86, Site 577, Core 11, Section 3, 43–50 cm. Lateral view, NHMUK PM DI 0140, scale bar = 65 μm.

3. *Bulimina midwayensis*. Allison Guyot, Mid-Pacific Mountains, Northwest Pacific. ODP Leg 143, Hole 865B, Core 15, Section 3, 52–57 cm. Lateral view, NHMUK PM DI 0141, scale bar = 62 μm.

## Original Designation

*Bulimina arkadelphiana* var. *midwayensis* Cushman and Parker, 1936, p. 42, pl. 7, figs. 9–10.

## Type Specimen

Holotype (No. 23136) deposited in the collections of the United States National Museum, Washington, DC.

## Type Level

Originally given as Eocene, Midway.

## Type Locality

Road-cut south of reservoir 3.5 miles southeast of Corsicana, Texas, USA.

## Description

Test forms an elongate, triserial series; subcircular cross-section, about one to one and a half times as long as broad, strongly tapered in outline with a sharp apical end, and a rounded apertural end. The strongly inflated chambers increase very rapidly in size and are separated by distinct, slightly depressed sutures, except in the early chambers where the sutures are obscured by surface ornamentation. Chambers overhang those of the previous whorl and have distinctively spinose basal margins. Chamber walls are calcareous and finely perforate. The primary aperture is a loop-shaped opening that extends up from the base of the final chamber, and is bordered by a lip that merges with an internal toothplate.

## Synonyms and Citations

*Bulimina petroleana* Cushman and Hedberg var. *spinea* Cushman and Renz, 1946, p. 37, pl. 6, fig. 13.

*Bulimina stokesi* Cushman and Renz, 1946, p. 37, pl. 6, fig. 14.

*Bulimina midwayensis* Cushman and Parker. –Tjalsma and Lohmann, 1983, p. 6, pl. 3, fig. 1. –Bolli et al., 1994, p. 136, figs. 36.19–21.

## Remarks

*Bulimina macilenta* Cushman and Parker differs from *Bulimina midwayensis* Cushman and Parker by its less slender test, more costate and less spinose ornamentation at the base of its chambers, lack of basal spine, and more distinct overhang over previous whorls. Tjalsma and Lohmann (1983) included in the synonymy of *Bulimina midwayensis* the two species *Bulimina petroleana* Cushman and Hedberg var. *spinea* Cushman and Renz and *Bulimina stokesi* Cushman and Renz, which were reported by their original authors to have similar stratigraphic ranges. Bolli et al. (1994) distinguished *Bulimina midwayensis* and *Bulimina stokesi* as different species, reporting the occurrence of *Bulimina stokesi* in younger Eocene deposits in Trinidad.

## Biogeography

Mainly Atlantic. Recorded from Texas (Cushman and Parker, 1936), Poland (Szczechura and Pozaryska, 1974), the North and South Atlantic (Tjalsma and Lohmann, 1983), and Trinidad (Bolli et al., 1994).

## Bathymetry

Bathyal.

## Chronostratigraphy

Late Cretaceous (Campanian) to Early Eocene (Ypresian). Early Paleocene to Early Eocene according to Tjalsma and Lohmann (1983). Bolli et al. (1994) reported a range from the Campanian to Late Paleocene.

# References


Bolli, H.M., Beckmann, J.P., and Saunders, J.B., 1994, Benthic Foraminiferal Biostratigraphy of the South Caribbean Region. Cambridge: Cambridge University Press, 408 pp.

Cushman, J.A. and Parker, F.L., 1936, Some American Eocene Buliminas. Contributions from the Cushman Laboratory for Foraminiferal Research, vol. 12, no. 2, pp. 39–45.

Cushman, J.A. and Renz, H.H., 1946, The foraminiferal fauna of the Lizard Springs Formation of Trinidad, British West Indies. Special Publications of the Cushman Laboratory for Foraminiferal Research, vol. 18, pp. 1–48.

Szczechura, J. and Pozaryska, K., 1974, Foraminiferida from the Paleocene of the Polish Carpathians (Babica Clays). Paleontologia Polonica, vol. 31, pp. 1–142.

Tjalsma, R.C. and Lohmann, G.P., 1983, Paleocene–Eocene bathyal and abyssal benthic foraminifera from the Atlantic Ocean. Micropaleontology, Special Publication, no. 4, pp. 1–90.

# *Bulimina rostrata*

## Brady, 1884

1. *Bulimina rostrata.* Ki Islands, Central Pacific Ocean; "Challenger" Station 191A, depth: 580 fathoms. Deposited in the micropaleontological collections of The Natural History Museum, London; NHMUK ZF1226. Syntype. Lateral view, NHMUK PM DI 0142, scale bar = 60 μm.

2. *Bulimina rostrata.* Ki Islands, Central Pacific Ocean; "Challenger" Station 191A, depth: 580 fathoms. Deposited in the micropaleontological collections of The Natural History Museum, London; NHMUK ZF1226. Syntype. Apertural view, NHMUK PM DI 0143, scale bar = 36 μm.

3. *Bulimina rostrata.* Ki Islands, Central Pacific Ocean; "Challenger" Station 191A, depth: 580 fathoms. Deposited in the micropaleontological collections of The Natural History Museum, London; NHMUK ZF1226. Syntype. Lateral view, NHMUK PM DI 0144, scale bar = 68 μm.

4. *Bulimina rostrata.* Ki Islands, Central Pacific Ocean; "Challenger" Station 191A, depth: 580 fathoms. Deposited in the micropaleontological collections of The Natural History Museum, London; NHMUK ZF1226. Syntype. Apertural view, NHMUK PM DI 0145, scale bar = 48 μm.

## Original Designation

*Bulimina rostrata* Brady, 1884, p. 408, pl. 51, figs. 14–15.

## Type Specimen

Housed in the micropaleontological collections of The Natural History Museum, London, UK, and registered in slide ZF1226.

## Type Level

Recent, surface sediments.

## Type Locality

Not designated. Localities given: four "Challenger" stations, off Cape Verde Islands, 1070 fathoms, north of Tristan da Cunha, 1825 fathoms, between the Cape of Good Hope and Kerguelen Island, 1570 fathoms, and off the Ki Islands, 580 fathoms.

## Description

Test forms an elongate, fusiform, triserial series; subcircular in cross-section, with a finely pointed apical end, and a rounded apertural end. The greatest breadth of the test is near the middle. Chambers are moderately inflated and sutures are usually indistinct, obscured by surface ornamentation. Chamber walls are calcareous, finely perforate, and ornamented by prominent, stout, raised, longitudinal costae. The primary aperture is a loop-shaped opening that extends up from the base of the final chamber and is bordered by a lip that merges with an internal toothplate.

## Synonyms and Citations

*Bulimina rostrata* Brady. –Barker, 1960, pl. 51, figs. 14–15. –Jones, 1994, p. 56, pl. 51, figs. 14–15.

## Remarks

This species appears to be morphologically close to *Bulimina alazanensis* Cushman. Comparison of the types is needed to elucidate whether they are conspecific. Unfortunately, the holotype of *Bulimina alazanensis* is lost.

## Biogeography

Worldwide. Recorded from the Atlantic and Pacific Oceans (Brady, 1884).

## Bathymetry

Middle to lower bathyal. Reported between 580 and 1570 fathoms at "Challenger" stations (Jones, 1994).

## Chronostratigraphy

Oligocene to Recent (Jones, 1994).

## References


Barker, R.W., 1960, Taxonomic notes on the species figured by H.B. Brady in his report on the foraminifera dredged by H.M.S. Challenger during the years 1873–1876. Society of Economic Paleontologists and Mineralogists Special Publication, vol. 9, pp. 1–238.

Brady, H.B., 1884, Report on the foraminifera dredged by H.M.S. Challenger during the years 1873–1876. In Brady, H.B., ed., Report on the Scientific Results of the Voyage of the H.M.S. Challenger during the years 1873–1876. Zoology, vol. 9, pp. 1–814.

Jones, R.W., 1994, The Challenger Foraminifera. Oxford: Oxford University Press, 149 pp.

# *Bulimina semicostata*

## Nuttall, 1930

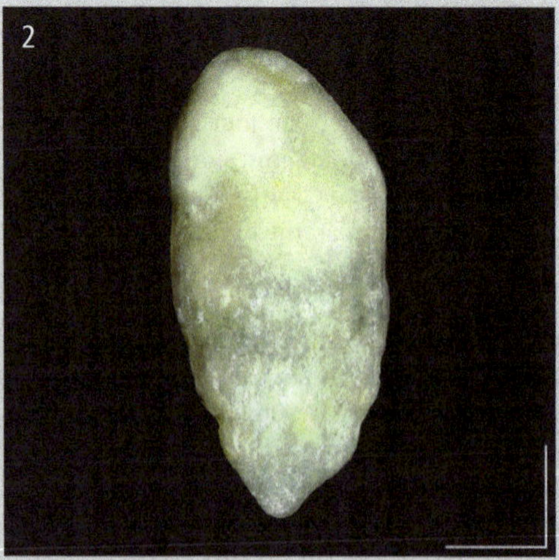

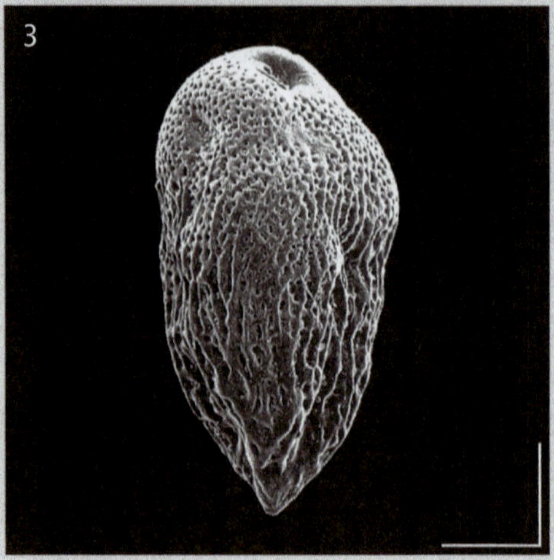

1. *Bulimina semicostata.* Aragon, Aroyo La Laja, 750 meters south and 70° west of La Antigua crossing, Mexico. Cotype. Lateral view, NHMUK PM DI 0146, scale bar = 112 μm.

2. *Bulimina semicostata.* Aragon, Aroyo La Laja, 750 meters south and 70° west of La Antigua crossing, Mexico. Cotype. Lateral view, NHMUK PM DI 0148, scale bar = 105 μm.

3. *Bulimina semicostata.* North Biscay, North Atlantic; DSDP Leg 48, Site 401, Core 10, Section 3, 98–102 cm. Lateral view, NHMUK PM DI 0147, scale bar = 101 μm.

## Original Designation

*Bulimina semicostata* Nuttall, 1930, p. 285, pl. 23, figs. 15–16.

## Type Specimen

Cotypes deposited in the collections of the United States National Museum, Washington, DC. Van Morkhoven et al. (1986) designated as lectotype CC 59482 (= Nuttall, 1930, pl. 23, fig. 16).

## Type Level

Not designated. Levels given: Upper Eocene, Chapapote Formation; Middle Eocene, Guyabal Formation; Lower Eocene, Aragon Formation.

## Type Locality

Arroyo La Laja, 720 meters south and 70° west of La Antigua crossing, Tampico region, Mexico.

## Description

Test forms a small, elongate, triserial series; tapered in outline, subcircular in cross-section with an acute to subacute initial portion and a rounded apertural end. The widest portion of this species' test is at the level of the penultimate chamber. The inflated, subspherical chambers increase gradually in size and are separated by narrow, depressed sutures, distinct in the final whorl, but obscured by surface ornamentation in earlier whorls. Fine, irregular, raised costae ornament the test, occasionally extending to the final whorl. Chamber walls are calcareous, and finely perforate except around the aperture and in some specimens over the final chambers. The primary aperture is an elongate, loop-shaped opening that extends from the base of the last chamber and is bordered by a lip that merges with an internal toothplate.

## Synonyms and Citations

*Bulimina pahiensis* Finlay, 1940, p. 455, pl. 64, figs. 87–88.

*Bulimina semicostata* Nuttall var. *lacrima* Mallory, 1959, p. 196, pl. 16, fig. 8.

*Bulimina semicostata* Nuttall. –Tjalsma and Lohmann, 1983, p. 25, pl. 13, figs. 1–3. –van Morkhoven et al., 1986, p. 279, pl. 93. –Bolli et al., 1994, p. 136, fig. 36.23. –Ortiz and Thomas, 2006, p. 114, pl. 4, fig. 14.

## Remarks

*Bulimina jarvisi* Cushman and Parker is larger and more slender, and has more inflated chambers, finer ornamentation, and a less triangular initial part than *Bulimina semicostata* Nuttall.

## Biogeography

Worldwide. See Observed Occurrences in van Morkhoven et al. (1986).

## Bathymetry

Lower bathyal and abyssal (van Morkhoven et al., 1986).

## Chronostratigraphy

Early Eocene (Ypresian) to Early Oligocene (Rupelian).

# References


Bolli, H.M., Beckmann, J.P., and Saunders, J.B., 1994, Benthic Foraminiferal Biostratigraphy of the South Caribbean Region. Cambridge: Cambridge University Press, 408 pp.

Finlay, H.J., 1940, New Zealand Foraminifera; key species in stratigraphy – No. 4. Transactions of the Royal Society of New Zealand, vol. 69, no. 4, pp. 448–472.

Mallory, V.S., 1959, Lower Tertiary Biostratigraphy of the California Coast Ranges. Tulsa, OK: American Association of Petroleum Geologists, 416 pp.

Nuttall, W.L.F., 1930, Eocene foraminifera from Mexico. Journal of Paleontology, vol. 4, pp. 271–293.

Ortiz, S. and Thomas, E., 2006, Lower–middle Eocene benthic foraminifera from the Fortuna Section (Betic Cordillera, southeastern Spain). Micropaleontology, vol. 52, pp. 97–150.

Tjalsma, R.C. and Lohmann, G.P., 1983, Paleocene–Eocene bathyal and abyssal benthic foraminifera from the Atlantic Ocean. Micropaleontology, Special Publication, no. 4, pp. 1–90.

van Morkhoven, F.P.C.M., Berggren, W.A., Edwards, A.S., et al., 1986, Cenozoic cosmopolitan deep-water benthic foraminifera. Bulletin des Centres de Recherches Exploration-Production Elf-Aquitaine, Mémoire, vol. 11, pp. 1–421.

# Bulimina taylorensis

## Cushman and Parker, 1935

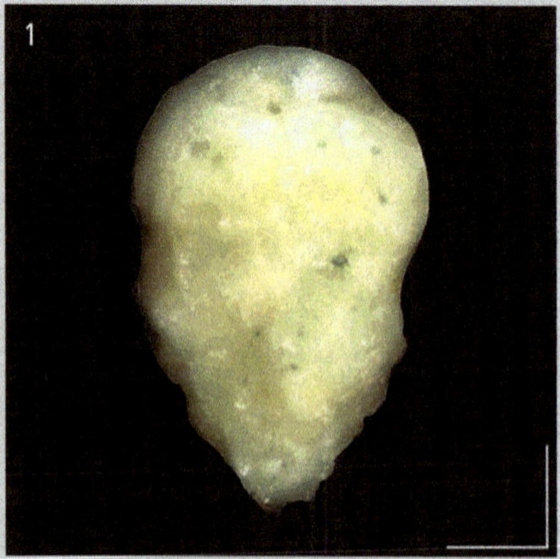

1. *Bulimina taylorensis*. Upper Taylor Branch of Kilkapoo Creek, Red River County, Texas; 1200 feet south of Public Rd, 1.8 miles morthwest of Ammona. Deposited in the collections of the United States National Museum, Washington, DC; No. 22367. Holotype. Lateral view, NHMUK PM DI 0149, scale bar = 122 μm.

### Original Designation

*Bulimina taylorensis* Cushman and Parker, 1935, p. 96, pl. 15, fig. 3.

### Type Specimen

Holotype (No. 22367) deposited in the collections of the United States National Museum, Washington, DC.

### Type Level

Cretaceous, Upper Taylor Formation.

### Type Locality

Branch of Kickapoo Creek, 1200 feet south of the public road, 1.8 miles northwest of Annona, Red River County, Texas, USA.

### Description

Test forms a small, elongate, triserial series; about one and a half to two times as long as broad and tapered in outline, with a subacute initial portion, a rounded apertural end, and a subcircular cross-section. The moderately inflated chambers increase rapidly in size and are separated by distinct, depressed sutures that are accentuated by a sharp undercutting below chambers in the whorl above. Chamber walls are calcareous, thin, finely perforate, and ornamented by irregular costae, giving a fluted appearance to the basal margin of chambers. The initial part of the test may bear one or more spines. The primary aperture is a loop-shaped opening that extends up from the base of the final chamber and is bordered by a lip that merges with an internal toothplate.

### Synonyms and Citations

*Bulimina taylorensis* Cushman and Parker. –Bolli et al., 1994, p. 136, fig. 36.27.

### Biogeography

Recorded from Texas (Cushman and Parker, 1935) and Trinidad (Bolli et al., 1994).

### Bathymetry

Bathyal.

### Chronostratigraphy

Late Cretaceous (Late Campanian to Maastrichtian).

## References

Bolli, H.M., Beckmann, J.P., and Saunders, J.B., 1994, Benthic Foraminiferal Biostratigraphy of the South Caribbean Region. Cambridge: Cambridge University Press, 408 pp.

Cushman, J.A. and Parker, F.L., 1935, Some American Cretaceous Buliminas. Contributions from the Cushman Laboratory for Foraminiferal Research, vol. 11, no. 4, pp. 96–101.

# Bulimina thanetensis

## Cushman and Parker, 1947

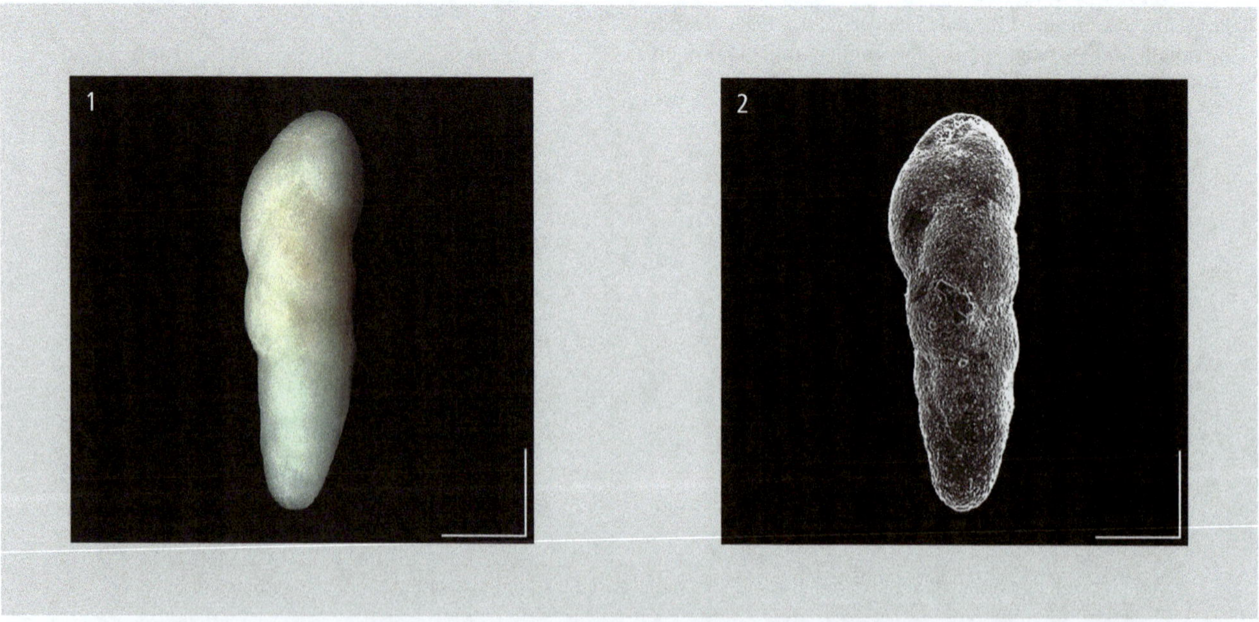

1. *Bulimina thanetensis.* Maud Rise, Weddell Sea, Antarctica; ODP Leg 113, Hole 690B, Core 4, Section 1, 120 cm. Lateral view, NHMUK PM DI 0150, scale bar = 72 μm.

2. *Bulimina thanetensis.* Maud Rise, Weddell Sea, Antarctica; ODP Leg 113, Hole 690B, Core 19, core catcher. Lateral view, NHMUK PM DI 0151, scale bar = 84 μm.

### Original Designation

*Bulimina thanetensis* Cushman and Parker, 1947, p. 92, pl. 21, fig. 26.

### Type Specimen

Holotype deposited in the collections of the United States National Museum, Washington, DC.

### Type Level

Eocene, Thanet Beds.

### Type Locality

Pegwell Bay, Kent, UK. Other locality given as Gotzreuther Graben, Germany.

### Description

Test small, elongate, slender, triserial, occasionally slightly twisted around the coiling axis, about three times as long as broad, slightly tapered in outline, with a subacute initial portion and a rounded apertural end. The early portion of the test exhibits a trigonal cross-section, becoming subcircular in the later portion. Slightly inflated, elongate chambers increase slowly in size, and are separated by distinct, slightly depressed sutures. Chamber walls are calcareous, thin, finely perforate, and smooth. The primary aperture is loop-shaped, extends up from the base of the final chamber, and is bordered by a lip that merges with an internal toothplate.

### Synonyms and Citations

*Bulimina thanetensis* Cushman and Parker, emend. Haynes, 1954, p. 186, textfigs. 1–6, 13–16; pl. 35, fig. 1.

*Bulimina thanetensis* Cushman and Parker. –Tjalsma and Lohmann, 1983, p. 26, pl. 12, fig. 5.

### Remarks

The holotype differs slightly from specimens illustrated by Haynes (1954) and from deep-sea forms by its more rounded, squatter chambers and less-prominent sutures.

### Biogeography

Mainly Atlantic. Reported from England (Cushman and Parker, 1947; Haynes, 1954) and the North and South Atlantic (Tjalsma and Lohmann, 1983).

### Bathymetry
Bathyal.

### Chronostratigraphy
Early Eocene (Ypresian) to Middle Eocene (Bartonian). Tjalsma and Lohmann (1983) reported a more extended Eocene range (Bartonian) in Atlantic deeper sites.

## References


Cushman, J.A. and Parker, F.L., 1947, Bulimina and related foraminiferal genera. Professional Paper of the United States Geological Survey, vol. 210-D, pp. 55–176.

Haynes, J., 1954, Taxonomic position of some British Palaeocene Buliminidae. Contributions to the Cushman Foundation of Foraminiferal Research, vol. 5, no. 4, pp. 185–191.

Tjalsma, R.C. and Lohmann, G.P., 1983, Paleocene–Eocene bathyal and abyssal benthic foraminifera from the Atlantic Ocean. Micropaleontology, Special Publication, no. 4, pp. 1–90.

# Bulimina trinitatensis

## Cushman and Jarvis, 1928

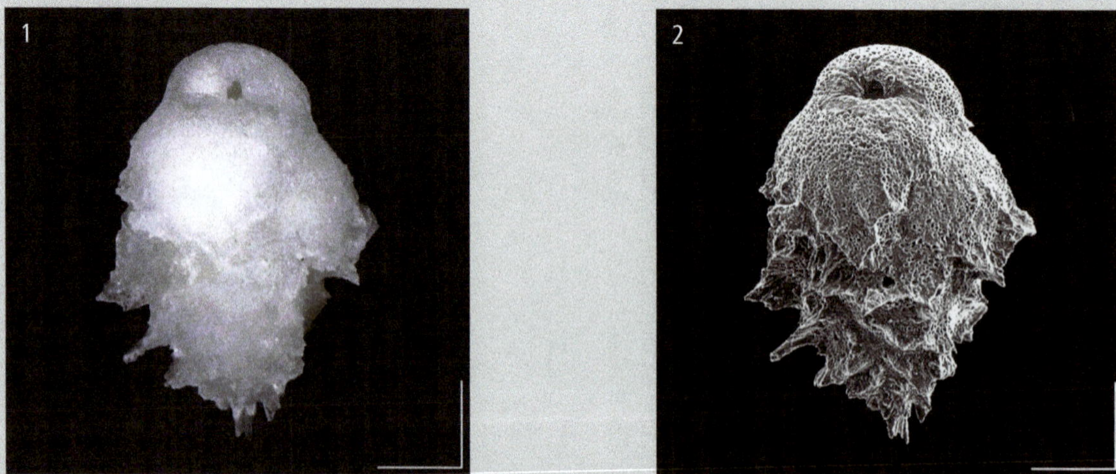


1. *Bulimina trinitatensis*. Shatsky Rise, West Central Pacific; DSDP Leg 86, Hole 577A, Core 10, Section 6, 50–57 cm. Lateral view, NHMUK PM DI 0152, scale bar = 102 μm.

2. *Bulimina trinitatensis*. Shatsky Rise, West Central Pacific; DSDP Leg 86, Hole 577A, Core 10, Section 6, 50–57 cm. Lateral view, NHMUK PM DI 0153, scale bar = 90 μm.

### Original Designation
*Bulimina trinitatensis* Cushman and Jarvis, 1928, p. 102, pl. 14, fig. 12.

### Type Specimen
Holotype (CC 9602) deposited in the collections of the United States National Museum, Washington, DC.

### Type Level
Paleocene (given as Upper Cretaceous by Cushman and Jarvis, 1928).

### Type Locality
Lizard Springs, southeastern Trinidad.

### Description
Test forms an elongate, stout, triserial series; slightly tapered in outline and subcircular in cross-section, with a subrounded initial portion and a rounded apertural end. The inflated chambers increase rapidly in size and are separated by distinct, depressed sutures. Chamber walls are calcareous and finely perforate. The base of chambers is ornamented by irregular reticulations ending in short spines and overhanging the sutures. The primary aperture is an elongate loop-shaped opening that extends from the base of the last chamber, and is bordered by a lip that merges with an internal toothplate.

### Synonyms and Citations
*Bulimina trinitatensis* Cushman and Jarvis, 1928. –Tjalsma and Lohmann, 1983, p. 7, pl. 3, figs. 3–4; pl. 14, fig. 1. –van Morkhoven et al., 1986, p. 299, pl. 98. –Thomas, 1990, pl. 2, fig. 7. –Bolli et al., 1994, p. 136, figs. 36.28–29. –Ortiz and Thomas, 2006, p. 115, pl. 4, figs. 17–18.

### Remarks
In older specimens, Tjalsma and Lohmann (1983) noted a tendency toward heavier ornamentation on the lower margins of chambers. Such specimens are close to the Late Cretaceous species *Bulimina taylorensis* described by Cushman and Parker.

According to van Morkhoven et al. (1986), *Bulimina trinitatensis* Cushman and Jarvis generally has five whorls, and is longer and more slender than the younger *Bulimina impedens* Parker and Bermúdez (three to four whorls).

### Biogeography

Worldwide. See Observed Occurrences in van Morkhoven et al. (1986).

### Bathymetry

Bathyal to abyssal (van Morkhoven et al., 1986). *Bulimina trinitatensis* appears to exhibit a shallower distribution pattern in the Eocene (Tjalsma and Lohmann, 1983).

### Chronostratigraphy

Early Paleocene (Danian) to Late Eocene (Priabonian).

## References


Bolli, H.M., Beckmann, J.P., and Saunders, J.B., 1994, Benthic Foraminiferal Biostratigraphy of the South Caribbean Region. Cambridge: Cambridge University Press, 408 pp.

Cushman, J.A. and Jarvis, P.W., 1928, Cretaceous foraminifera from Trinidad. Contributions from the Cushman Laboratory for Foraminiferal Research, vol. 4, pp. 85–103.

Ortiz, S. and Thomas, E., 2006, Lower-middle Eocene benthic foraminifera from the Fortuna Section (Betic Cordillera, southeastern Spain). Micropaleontology, vol. 52, pp. 97–150.

Thomas, E., 1990, Late Cretaceous through Neogene benthic foraminifers. In Barker, P.F., Kennett, J.P., et al., ed., Scientific Results of the Proceedings of the Ocean Drilling Program, vol. 113, pp. 571–594.

Tjalsma, R.C. and Lohmann, G.P., 1983, Paleocene–Eocene bathyal and abyssal benthic foraminifera from the Atlantic Ocean. Micropaleontology, Special Publication, no. 4, pp. 1–90.

van Morkhoven, F.P.C.M., Berggren, W.A., Edwards, A.S., et al., 1986, Cenozoic cosmopolitan deep-water benthic foraminifera. Bulletin des Centres de Recherches Exploration-Production Elf-Aquitaine, Mémoire, vol. 11, pp. 1–421.

# *Bulimina tuxpamensis*

## Cole, 1928

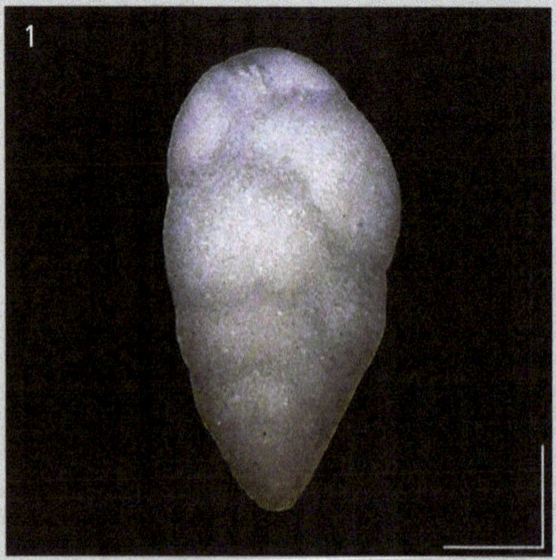


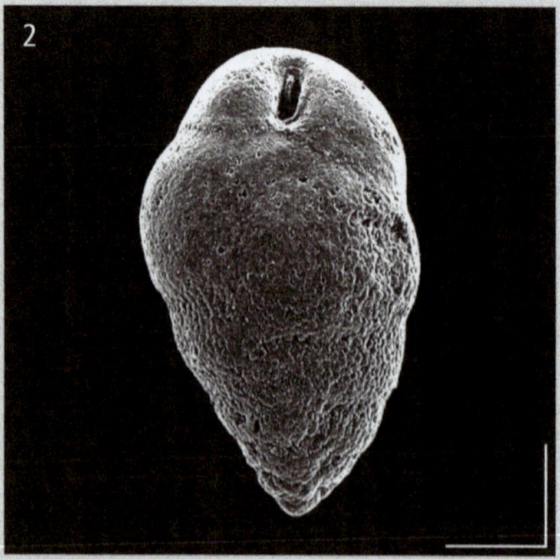


1. *Bulimina tuxpamensis.* Rio Grande Rise, South Atlantic; DSDP Leg 72, Hole 516F, Core 32, Section 6, 70–74 cm. Lateral view, NHMUK PM DI 0154, scale bar = 71 μm.

2. *Bulimina tuxpamensis.* Northeast Georgia Rise, South Atlantic; ODP Leg 114, Hole 698A, Core 5, Section 1, 68–72 cm. Lateral view, NHMUK PM DI 0155, scale bar = 76 μm.

### Original Designation

*Bulimina tuxpamensis* Cole, 1928, p. 212, pl. 32, fig. 23.

### Type Specimen

Holotype deposited in the Eocene collection of Prof. G.D. Harris, Cornell University, Ithaca, NY, USA.

### Type Level

Eocene, gray clay. Probably Jackson, Chapapote Formation.

### Type Locality

Huasteca Petroleum Company's well, Cerro Azul No. 75 at 1040 feet depth, Tampico, Mexico.

### Description

Test forms an elongate, stout, triserial series; tapered in outline and subcylindrical in cross-section with a subacute initial portion and a rounded apertural end. The moderately inflated chambers increase rapidly in size and are separated by distinct, relatively wide, depressed sutures. Chamber walls are calcareous, finely perforate, and unornamented. The primary aperture is a loop-shaped opening extending up from the base of the final chamber and bordered by a lip that merges with an internal toothplate.

### Synonyms and Citations

*Bulimina tuxpamensis* Cole. –Parker and Bermúdez, 1937, p. 513, pl. 58, figs. 1.
*Bulimina miolaevis* Finlay, 1940, p. 454, pl. 64, figs. 70–71.
?*Bulimina bradbury* Martin, 1943, p. 109, pl. 6, fig. 4.
*Bulimina tuxpamensis* Cole. –Tjalsma, 1983, p. 739, pl. 6, figs. 1–2. –Tjalsma and Lohmann, 1983, p. 26, pl. 12, figs. 3–4. –van Morkhoven et al., 1986, p.155, pl. 51A–B. –Robertson, 1998, p. 150, pl. 56, fig. 8.

### Remarks

Some specimens may have a granular appearance (van Morkhoven et al., 1986). *Bulimina bradbury* Martin, 1943 is morphologically close to *Bulimina tuxpamensis* Cole. The two species may be conspecific.

## Biogeography

Worldwide. Recorded from Mexico, Cuba, Trinidad, the Dominican Republic, the Pacific, the Indian Ocean, the North and South Atlantic, the Tasman Sea, the Coral Sea, and the Gulf of Mexico (van Morkhoven et al., 1986).

## Bathymetry

Bathyal to abyssal.

## Chronostratigraphy

Early Eocene (Ypresian) to Middle Miocene (Langhian). Doubtful reports of Serravallian occurrences (van Morkhoven et al., 1986).

# References


Cole, W.S., 1928, A foraminiferal fauna from the Chapapote formation in Mexico. Bulletin of American Paleontology, vol. 14, pp. 200–232.

Finlay, H.J., 1940, New Zealand foraminifera; key species in stratigraphy – No. 4. Transactions of the Royal Society of New Zealand, vol. 69, no. 4, pp. 448–472.

Martin, L.T., 1943, Eocene foraminifera from the Type Lodo Formation, Fresno County, California. Stanford University Publications, Geological Sciences, vol. 3, no. 3, pp. 1–35.

Parker, F.L. and Bermúdez, P.J., 1937, Eocene species of the genera Bulimina and Buliminella from Cuba. Journal of Paleontology, vol. 11, pp. 513–516.

Robertson, B.E., 1998, Systematics and paleoecology of the benthic foraminiferida from the Buff Bay section, Miocene of Jamaica. Micropaleontology, vol. 44, supplement 2, pp. 1–266.

Tjalsma, R.C., 1983, Eocene to Miocene benthic foraminifera from DSDP site 516, Rio Grande Rise, South Atlantic. In Barker, P.F., Carlson, R.L., Johnson, D.A., et al., ed., Initial Reports of the Deep Sea Drilling Project, Washington (U.S. Government Printing Office), vol. 72, pp. 731–755.

Tjalsma, R.C. and Lohmann, G.P., 1983, Paleocene–Eocene bathyal and abyssal benthic foraminifera from the Atlantic Ocean. Micropaleontology, Special Publication, no. 4, pp. 1–90.

van Morkhoven, F.P.C.M., Berggren, W.A., Edwards, A.S., et al., 1986, Cenozoic cosmopolitan deep-water benthic foraminifera. Bulletin des Centres de Recherches Exploration-Production Elf-Aquitaine, Mémoire, vol. 11, pp. 1–421.

# *Bulimina velascoensis*

## (Cushman), 1925

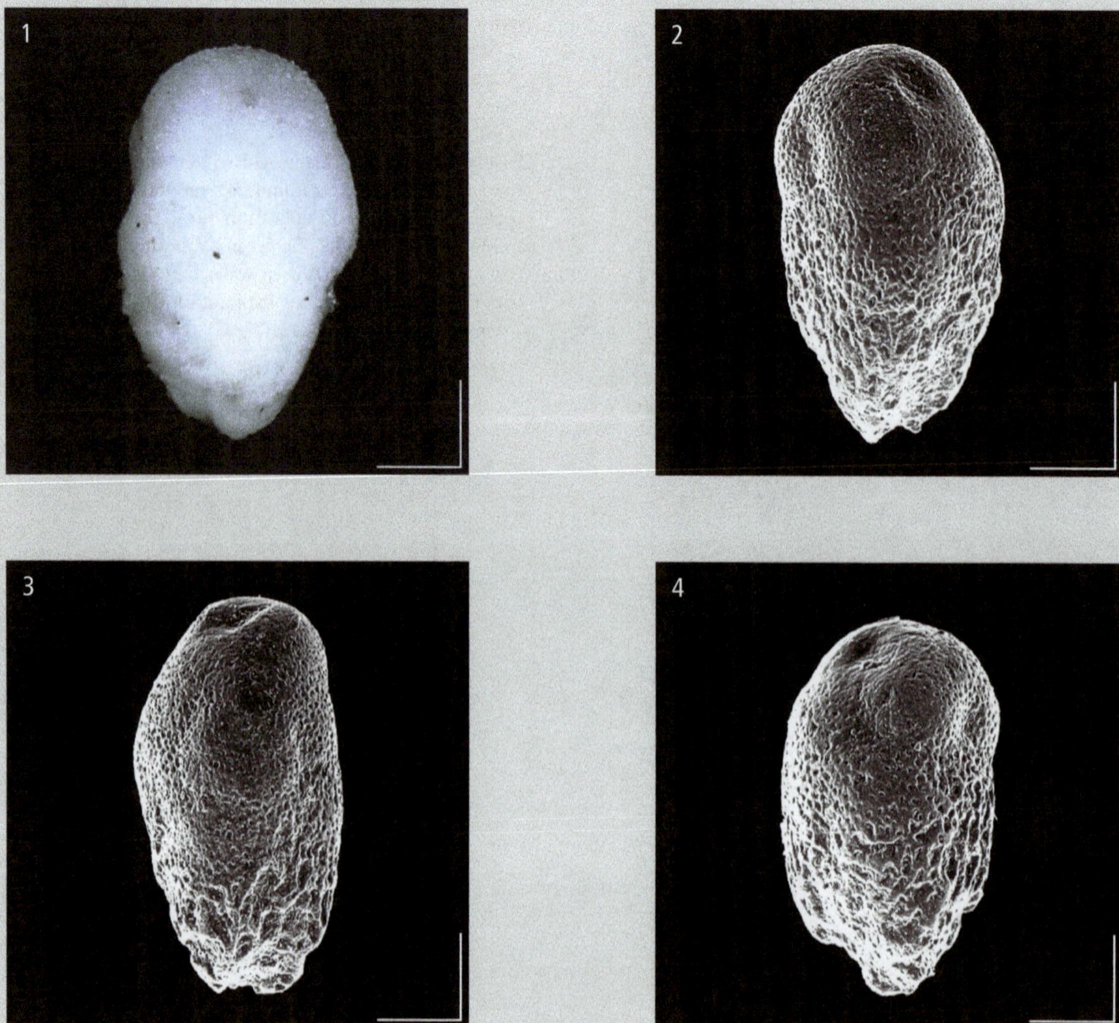

1. *Bulimina velascoensis.* Allison Guyot, Mid-Pacific Mountains, Northwest Pacific; ODP Leg 143, Hole 865B, Core 15, Section 3, 52–57 cm. Lateral view, NHMUK PM DI 0156, scale bar = 79 μm.

2. *Bulimina velascoensis.* Velasco Formation, Mexico; 0.5 km west of Velasco Station. Specimen from the White (1929) Collection. Deposited in the collections of the American Museum of Natural History, no. 46757. Lateral view, NHMUK PM DI 0157, scale bar = 75 μm.

3. *Bulimina velascoensis.* Velasco Formation, Mexico; 0.5 km west of Velasco Station. Specimen from the White (1929) Collection. Deposited in the collections of the American Museum of Natural History, no. 46757. Lateral view, NHMUK PM DI 0158, scale bar = 101 μm.

4. *Bulimina velascoensis.* Velasco Formation, Mexico; 0.5 km west of Velasco Station. Specimen from the White (1929) Collection. Deposited in the collections of the American Museum of Natural History, no. 46757. Lateral view, NHMUK PM DI 0159, scale bar = 78 μm.

## Original Designation

*Gaudryina velascoensis* Cushman, 1925, p. 20, pl. 3, fig. 7.

## Type Specimen

Holotype (No. 4349) deposited in the collections of the United States National Museum, Washington, DC.

## Type Level

Paleocene Velasco Shale. Originally given as Upper Cretaceous by Cushman (1925).

## Type Locality

Tamalte Arroyo, Hacienda El Limón, State of San Luis Potosi, Mexico.

## Description

Test forms an initially triserial series and an elongate, quadrangular, biserial, final portion. The chambers are generally indistinct in the early portion, but inflated and distinct in the biserial portion and separated by slightly depressed sutures. Chamber walls are calcareous, coarsely perforate, and finely striate in the early portion, and smooth in the final portion. The primary aperture is an elongate, loop-shaped opening extending up from the base of the final chamber and bordered by a thin lip that merges with an internal toothplate.

## Synonyms and Citations

*Bulimina rugifera* Glaessner, 1937, p. 372, pl. 2, fig. 19.

*Bulimina velascoensis* (Cushman). –Tjalsma and Lohmann, 1983, p. 8, pl. 3, fig. 2. –van Morkhoven et al., 1986, p. 335, pl. 109.

## Biogeography

Worldwide. See Observed Occurrences in van Morkhoven et al. (1986).

## Bathymetry

Lower bathyal to abyssal. Reported from depths between 1000 and 4000 meters (Tjalsma and Lohmann, 1983).

## Chronostratigraphy

Early Paleocene (Danian) to Early Eocene (earliest Ypresian). Doubtful in Biozone P6a (van Morkhoven et al., 1986).

# References


Cushman, J.A., 1925, Some new foraminifera from the Velasco Shale of Mexico. Contributions from the Cushman Laboratory for Foraminiferal Research, vol. 1, no. 1, pp. 18–23.

Glaessner, M.F., 1937, Studien über Foraminiferen aus der Kreide und dem Tertiär des Kaukasus; 1. Die Foraminiferen der ältesten Tertiärschichten des Nordwest-Kaukas. Problemy Paleontologii, Paleontologichesjaya Laboratoriya Moskovskogo Gosudarstvennogo Universiteta, vol. 2–3, pp. 349–408.

Tjalsma, R.C. and Lohmann, G.P., 1983, Paleocene–Eocene bathyal and abyssal benthic foraminifera from the Atlantic Ocean. Micropaleontology, Special Publication, no. 4, pp. 1–90.

van Morkhoven, F.P.C.M., Berggren, W.A., Edwards, A.S., et al., 1986, Cenozoic cosmopolitan deep-water benthic foraminifera. Bulletin des Centres de Recherches Exploration-Production Elf-Aquitaine, Mémoire, vol. 11, pp. 1–421.

White, M.P., 1929, Some index foraminifera of the Tampico Embayment area of Mexico (Part III). Journal of Paleontology, vol. 3, no. 1, pp. 30–58.

# *Buliminella beaumonti*

## Cushman and Renz, 1946

1. *Buliminella beaumonti*. Shatsky Rise, West Central Pacific; DSDP Leg 86, Site 577, Core 9, Section 2, 63–70 cm. Lateral view, NHMUK PM DI 0160, scale bar = 61 μm.

### Original Designation

*Buliminella beaumonti* Cushman and Renz, 1946, p. 36, pl. 6, fig. 7.

### Type Specimen

Holotype deposited in the collections of the United States National Museum, Washington, DC.

### Type Level

Paleocene, Lizard Springs Marl. Originally given as Upper Cretaceous by Cushman and Renz (1946).

### Type Locality

Racine of the Ampelu River, a small tributary of the Ortoire River system, Lizard Springs area, southeastern Trinidad.

### Description

Test small, elongate, high trochospire; slightly longer than broad, tapered in outline with a subacute initial portion, a broadly rounded apertural end, and a subcircular cross-section. The inflated chambers (four to a whorl) increase rapidly in size, and are separated by curved, slightly depressed sutures. The base of the chambers is ornamented by crenulations forming a characteristic backward-projecting lobe pattern. Chamber walls are calcareous, finely perforate, and smooth. The primary aperture is a loop in the depressed face of the final chamber, broadest at the upper end, with a high rim and a simple internal toothplate.

### Synonyms and Citations

*Praebulimina beaumonti* Cushman and Renz. –Proto Decima and Bolli, 1978, p. 795, pl. 1, fig. 19.

*Buliminella beaumonti* Cushman and Renz. –Tjalsma and Lohmann, 1983, p. 9, pl. 3, fig. 6.

### Remarks

Bolli et al. (1994) considered *Buliminella beaumonti* Cushman and Renz and *Buliminella grata* Parker and Bermúdez to be synonyms. Van Morkhoven et al. (1986) tentatively included *Buliminella beaumonti* in the synonymy of *Buliminella grata*, extending the lower stratigraphic range of *Buliminella grata* from Early Eocene (P6b) to Early Paleocene (P1).

Tjalsma and Lohmann (1983, pl. 3, figs. 5, 7) illustrated Late Paleocene forms, which they considered to be intermediate between *Buliminella beaumonti* and *Buliminella grata*. These authors considered that *Buliminella grata* differs from *Buliminella beaumonti* mainly by its shorter and squatter test.

### Biogeography

Mainly Atlantic. Recorded from the Caribbean (Cushman and Renz, 1946), Austria (Hillebrandt, 1962), the South Atlantic

(Proto Decima and Bolli, 1978), and the North and South Atlantic (Tjalsma and Lohmann, 1983).

### Bathymetry

Bathyal.

### Chronostratigraphy

Paleocene (Danian) to Middle Eocene (Lutetian).

## References


Bolli, H.M., Beckmann, J.P., and Saunders, J.B., 1994, Benthic Foraminiferal Biostratigraphy of the South Caribbean Region. Cambridge: Cambridge University Press, 408 pp.

Cushman, J.A. and Renz, H.H., 1946, The foraminiferal fauna of the Lizard Springs Formation of Trinidad, British West Indies. Special Publications of the Cushman Laboratory for Foraminiferal Research, vol. 18, pp. 1–48.

Hillebrandt, A. von, 1962, Das Paleozaen und seine Foraminiferenfauna im Becken von Reichenhall und Salzburg. Abhandlungen, Bayerische Akademie der Wissenschaften, Mathematisch-Naturwissenschaftliche Klasse, N.F., no. 108, pp. 9–182.

Proto Decima, F. and Bolli, H.M., 1978, Southeast Atlantic Leg 40 Paleogene benthic foraminifers. In Bolli, H.M., Ryan, W.B.F., et al., ed., Initial Reports of the Deep Sea Drilling Project, Washington (U.S. Government Printing Office), vol. 40, pp. 783–809.

Tjalsma, R.C. and Lohmann, G.P., 1983, Paleocene–Eocene bathyal and abyssal benthic foraminifera from the Atlantic Ocean. Micropaleontology, Special Publication, no. 4, pp. 1–90.

van Morkhoven, F.P.C.M., Berggren, W.A., Edwards, A.S., et al., 1986, Cenozoic cosmopolitan deep-water benthic foraminifera. Bulletin des Centres de Recherches Exploration-Production Elf-Aquitaine, Mémoire, vol. 11, pp. 1–421.

# Buliminella grata

## Parker and Bermúdez, 1937

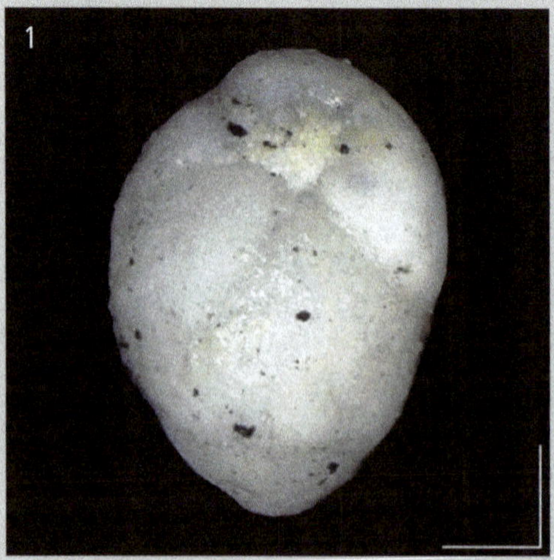

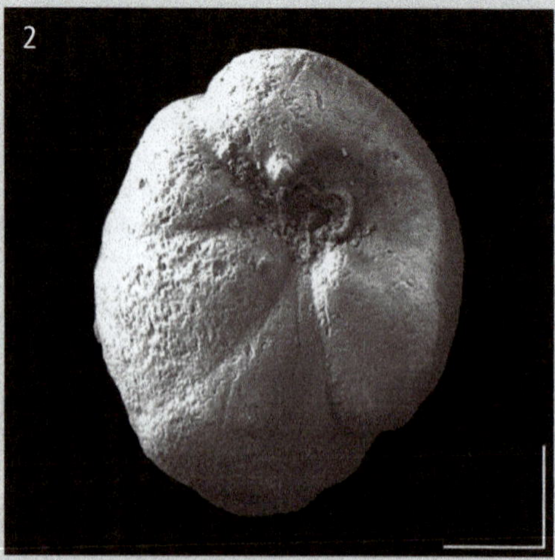

1. *Buliminella grata.* Loma Principe, Havana, Cuba. Deposited in the collections of the United States National Museum, Washington, DC; CC 23341. Lateral view, NHMUK PM DI 0161, scale bar = 89 μm.

2. *Buliminella grata.* North Biscay, North Atlantic; DSDP Leg 48, Site 401, Core 10, Section 3, 98–102 cm. Lateral view, NHMUK PM DI 0162, scale bar = 52 μm.

### Original Designation

*Buliminella grata* Parker and Bermúdez, 1937, p. 515, pl. 59, figs. 6a–c.

### Type Specimen

Holotype (No. 23340) deposited in the collections of the United States National Museum, Washington, DC.

### Type Level

Eocene.

### Type Locality

Loma Principe, cut between F and Avenida de los Presidentes Streets, 20 meters west of José M. Gómez monument, Havana, Cuba.

### Description

Test is a small, squat, elongate, high trochospire; slightly longer than broad, strongly tapered in outline, and subcircular in cross-section, with a subacute initial portion, and a rounded apertural end. The inflated, subspherical chambers (four to a whorl) increase rapidly in size and are separated by curved, flush or slightly depressed sutures. Chamber bases are crenulated, giving the test a distinctive scalloped appearance. Chamber walls are calcareous, finely perforate, and smooth. The primary aperture is a loop in the depressed face of the final chamber, broadest at the upper end, a with high rim and simple internal toothplate. A pattern of surficial creases radiates from the aperture across the face of the final chambers.

### Synonyms and Citations

*Bulimina ovata* d'Orbigny. –Douglas, 1973, p. 634, pl. 7, figs. 6–7.

*Buliminella grata* Parker and Bermúdez. –Tjalsma and Lohmann, 1983, p. 26, pl. 12, figs. 7a–b. –van Morkhoven et al., 1986, p. 163, pl. 54. –Miller and Katz, 1987, p. 125, pl. 2, fig. 4. –Bolli et al., p. 137, fig. 37.3.

### Remarks

Bolli et al. (1994) considered *Buliminella beaumonti* Cushman and Renz and *Buliminella grata* Parker and Bermúdez to be synonyms. Van Morkhoven et al. (1986) tentatively included *Buliminella beaumonti* in the synonymy of *Buliminella grata*, extending the lower stratigraphic range of *Buliminella grata* from Early Eocene (Biozone P6b) to Early Paleocene (Biozone P1).

Tjalsma and Lohmann (1983, pl. 3, figs. 5, 7) illustrated Late Paleocene forms, which they considered to be intermediate between *Buliminella beaumonti* and *Buliminella grata*. These authors considered that *Buliminella grata* differs from *Buliminella beaumonti* mainly by its shorter and squatter test.

### Biogeography

Worldwide. Recorded from the Caribbean, the Gulf of Mexico, the North and South Atlantic, the Pacific, and Europe (van Morkhoven et al., 1986).

### Bathymetry

Bathyal. Van Morkhoven et al. (1986) and Tjalsma and Lohmann (1983) recorded maximum abundances of *Buliminella grata* at depths of approximately 1–2 kilometers in Eocene deposits.

### Chronostratigraphy

Early Eocene (Ypresian) to Middle Miocene (Serravallian).

## References


Bolli, H.M., Beckmann, J.P., and Saunders, J.B., 1994, Benthic Foraminiferal Biostratigraphy of the South Caribbean Region. Cambridge: Cambridge University Press, 408 pp.

Douglas, R.G., 1973, Benthonic foraminiferal biostratigraphy in the central North Pacific, Leg 17, Deep Sea Drilling Project. In Winterer, E.L., Ewing, J.I., et al., ed., Initial Reports of the Deep Sea Drilling Project, Washington (U.S. Government Printing Office), vol. 17, pp. 607–671.

Miller, K.G. and Katz, M.E., 1987, Oligocene to Miocene benthic foraminiferal and abyssal circulation changes in the north Atlantic. Micropaleontology, vol. 33, no. 2, pp. 97–149.

Parker, F.L. and Bermúdez, P.J., 1937, Eocene species of the genera Bulimina and Buliminella from Cuba. Journal of Paleontology, vol. 11, pp. 513–516.

Tjalsma, R.C. and Lohmann, G.P., 1983, Paleocene–Eocene bathyal and abyssal benthic foraminifera from the Atlantic Ocean. Micropaleontology, Special Publication, no. 4, pp. 1–90.

van Morkhoven, F.P.C.M., Berggren, W.A., Edwards, A.S., et al., 1986, Cenozoic cosmopolitan deep-water benthic foraminifera. Bulletin des Centres de Recherches Exploration-Production Elf-Aquitaine, Mémoire, vol. 11, pp. 1–421.

# Buzasina galeata

## (Brady), 1881

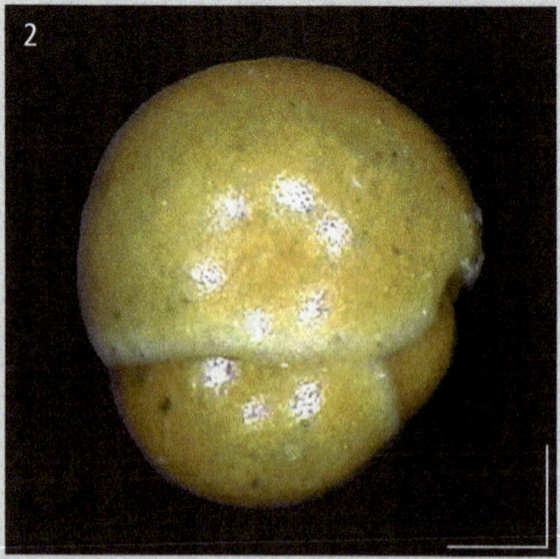

1. *Buzasina galeata*. South Atlantic; "Challenger" Station 346, depth: 2350 fathoms. Deposited in the micropaleontological collections of The Natural History Museum, London; NHMUK ZF2501. Syntype. Apertural view, NHMUK PM DI 0163, scale bar = 125 μm.

2. *Buzasina galeata*. South Atlantic; "Challenger" Station 346, depth: 2350 fathoms. Deposited in the micropaleontological collections of The Natural History Museum, London; NHMUK ZF2501. Syntype. Lateral view, NHMUK PM DI 0164, scale bar = 123 μm.

### Original Designation

*Trochammina galeata* Brady, 1881, p. 52 (type description), illustrated in Brady, 1884, pl. 40, figs. 19–23.

### Type Specimen

Housed in the micropaleontological collections of The Natural History Museum, London, UK, and registered in slides ZF2501–ZF2502.

### Type Level

Recent, surface sediments.

### Type Locality

Not designated. Figured specimens from "Challenger" Station 346, latitude 2° 42′ S, longitude 14° 41′ W, in the South Atlantic Ocean, north of Ascension Island, 2350 fathoms (figs. 19, 20, 23); and from "Challenger" Station 332, latitude 37° 29′ S, longitude 27° 31′ W, in the middle of the South Atlantic Ocean, 2200 fathoms.

### Description

Test forms a small, globular planispire with a rounded periphery. The three to four strongly inflated, overlapping chambers increase rapidly in size and are separated by thin, depressed sutures. The final chamber represents more than half the test. Chamber walls are thin, finely agglutinated, and well cemented. The primary aperture is an areal, equatorial slit bordered by a narrow lip.

### Synonyms and Citations

*Trochammina galeata* Brady. –Brady, 1884, p. 344, pl. 40, figs. 19–23.

*Cystammina galeata* (Brady). –Barker, 1960, pl. 40, figs. 19–23.

*Buzasina galeata* (Brady). –Jones, 1994, p. 45, pl. 40, figs. 19–23.

### Biogeography

Worldwide.

### Bathymetry

Lower bathyal to abyssal. Reported between 1825 and 2750 fathoms at "Challenger " stations (Jones, 1994).

### Chronostratigraphy

Recent (Brady's specimens). However, Charnock and Jones (1990) and Jones (1994) included Late Cretaceous and Paleogene forms (including *Labrospira pacifica* Krasheninnikov) into the synonymy of *Buzasina galeata* (Brady), extending its range into the Late Cretaceous.

## References


Barker, R.W., 1960, Taxonomic notes on the species figured by H.B. Brady in his report on the foraminifera dredged by H.M.S. Challenger during the years 1873–1876. Society of Economic Paleontologists and Mineralogists Special Publication, vol. 9, pp. 1–238.

Brady, H.B., 1881, Notes on some of the reticularian Rhizopoda of the Challenger Expedition, Part III. Quarterly Journal of Microscopical Science (New Series), vol. 21, pp. 31–71.

Brady, H.B., 1884, Report on the foraminifera dredged by H.M.S. Challenger during the years 1873–1876. In Brady, H.B., ed., Report on the Scientific Results of the Voyage of the H.M.S. Challenger during the years 1873–1876. Zoology, vol. 9, pp. 1–814.

Charnock, M.A. and Jones, R.W., 1990, Agglutinated foraminifera from the Paleogene of the North Sea. In Hemleben, C., Kaminski, M.A., Kuhnt, W., and Scott, D.B., ed., Paleoecology, Biostratigraphy, Paleoceanography and Taxonomy of Agglutinated Foraminifera, NATO ASI Series, C: Mathematical and Physical Sciences, Dordrecht: Kluwer Academic, pp. 139–244.

Jones, R.W., 1994, The Challenger Foraminifera. Oxford: Oxford University Press, 149 pp.

# *Cancris auriculus*

## (Fichtel and Moll), 1798

1. *Cancris auriculus.* Off South Africa, Atlantic; "Challenger" Station 142, depth: 150 fathoms. Deposited in the micropaleontological collections of The Natural History Museum, London; NHMUK ZF2230. Spiral view, NHMUK PM DI 0165, scale bar = 228 μm.

2. *Cancris auriculus.* Off South Africa, Atlantic; "Challenger" Station 142, depth: 150 fathoms. Deposited in the micropaleontological collections of The Natural History Museum, London; NHMUK ZF2230. Apertural view, NHMUK PM DI 0166, scale bar = 198 μm.

3. *Cancris auriculus.* Off South Africa, Atlantic; "Challenger" Station 142, depth: 150 fathoms. Deposited in the micropaleontological collections of The Natural History Museum, London; NHMUK ZF2230. Umbilical view, NHMUK PM DI 0167, scale bar = 201 μm.

## Original Designation

*Nautilus auricula* Fichtel and Moll, 1798, p. 108, pl. 20, figs. a–f.

## Type Specimen

Lectotype designated by Rögl and Hansen (1984) and deposited in the collections of the Natural History Museum, Vienna, Austria (MI-564).

## Type Level

Recent by designation of lectotype.

## Type Locality

Mediterranean Sea by designation of lectotype.

## Description

Test forms a low trochospire; ovate in outline and unequally biconvex in cross-section with the umbilical side more convex than the spiral side. Test length is approximately one and a quarter times the maximum width. The periphery is acute with a narrow, thin keel, except in the early portion, where the periphery is broadly rounded. Approximately seven chambers are present in the final whorl, increasing gradually in size, except for the final chamber, which is greatly enlarged and inflated. Sutures are thin and slightly depressed. Chamber walls are calcareous, smooth, and finely perforate. The primary aperture is an interiomarginal umbilical–extraumbilical slit, covered with an umbilical flap.

## Synonyms and Citations

*Pulvulina oblonga* (Williamson). –Brady, 1884, p. 688, pl. 106, fig. 4.

*Cancris auriculus* (Fichtel and Moll). –Cushman and Todd, 1942, p. 74, pl. 18, figs. 1–11; pl. 23, fig. 6. –Rögl and Hansen, 1984, p. 647, pl. 26, figs. 3–8. –Jones, 1994, p. 105, pl. 106, fig. 4. –Milker and Schmiedl, 2012, p. 93, figs. 21.14–15.

## Remarks

Type species of the genus *Cancris* de Montfort, 1808.

Rögl and Hansen (1984) designated as lectotype the single specimen representing the variety β in the collection of Fichtel and Moll. These authors considered the four specimens representing the variety α in the collection to be conspecific with the lectotype. The paratypes have a more elongate outline and a stronger overlap of chambers on the umbilical side.

*Cancris auriculus* (Fichtel and Moll) shows close similarity to *Rotalina oblonga* Williamson, but it has a more robust, inflated test and a wider apertural area.

## Biogeography

Worldwide. Cosmopolitan in low to moderate latitudes.

## Bathymetry

Littoral to middle bathyal (500 fathoms, Jones, 1994).

## Chronostratigraphy

Miocene to Recent (Jones, 1994).

# References


Brady, H.B., 1884, Report on the foraminifera dredged by H.M.S. Challenger during the years 1873–1876. In Brady, H.B., ed., Report on the Scientific Results of the Voyage of the H.M.S. Challenger during the years 1873–1876. Zoology, vol. 9, pp. 1–814.

Cushman, J.A. and Todd, R., 1942, The genus Cancris and its species. Contributions from the Cushman Laboratory for Foraminiferal Research, vol. 18, no. 4, pp. 72–94.

Fichtel, L. von and Moll, J.P.C. von, 1798, Testacea microscopica aliaque minuta ex generibus Argonauta et Nautilus ad naturam delineata et descripta. Microscopische und andere kleine Schalthiere aus den Geschlechtern Argonaute und Schiffer, nach der Natur gezeichnet und beschrieben. Vienna: Camesina, 124 pp.

Jones, R.W., 1994, The Challenger Foraminifera. Oxford: Oxford University Press, 149 pp.

Milker, Y. and Schmiedl, G., 2012, A taxonomic guide to modern benthic shelf foraminifera of the western Mediterranean Sea. Palaeontologia Electronica, vol. 15, Issue 2;16A, 134 pp.

Rögl, F. and Hansen, J.H., 1984, Foraminifera described by Fichtel and Moll in 1798; A revision of Testacea Microscopica. Appendix Testacea Microscopica alique minuta ex Generibus Argonauta et Nautilus. Reprint of original plates. Neue Denkschriften des Naturhistorischen Museum in Wien, vol. 3, pp. 1–143.

# *Cancris nuttalli*

## (Palmer and Bermúdez), 1936

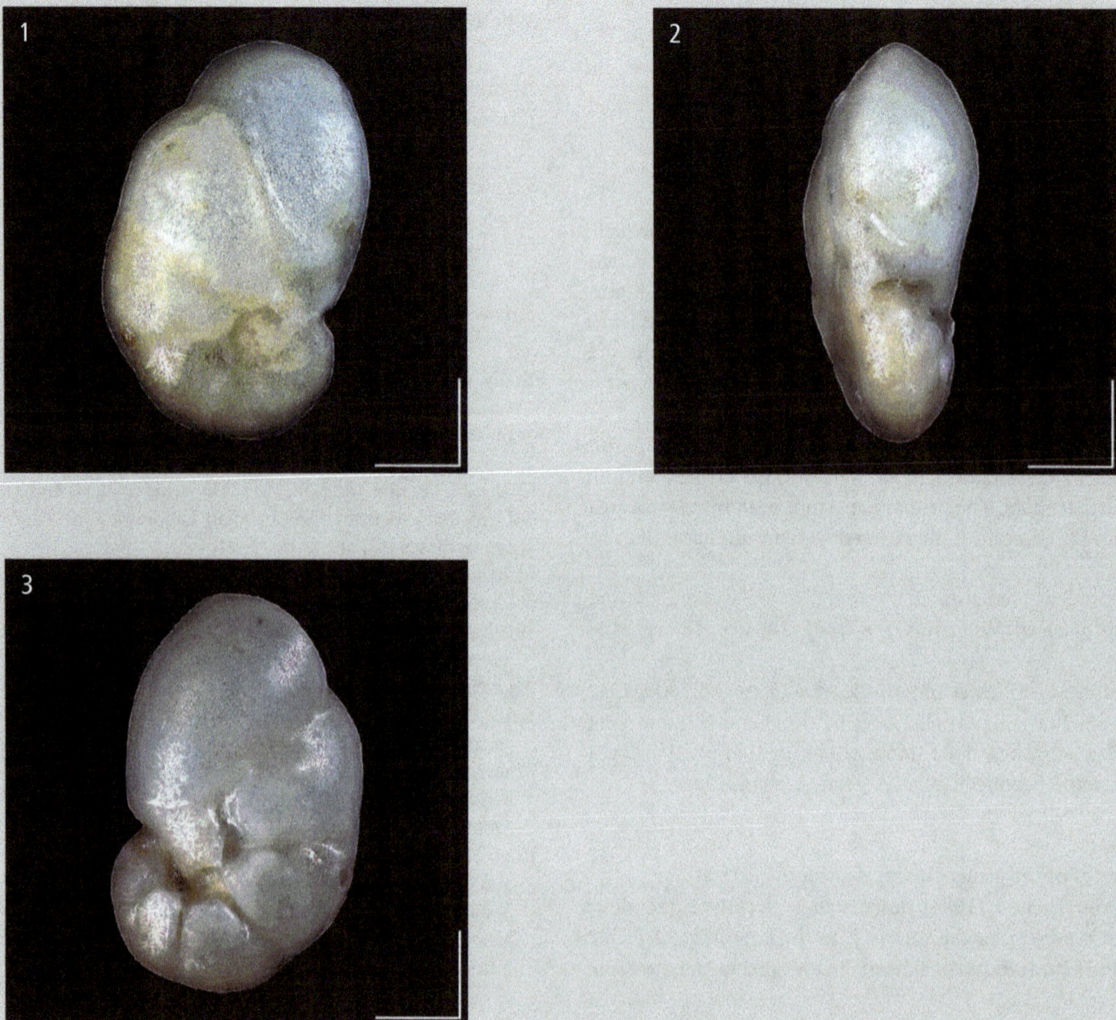

1. *Cancris nuttalli.* Gulf of Mexico; Eureka Core E68-151A. Deposited in the collections of the United States National Museum, Washington, DC. Specimen illustrated in van Morkhoven et al. 1986, Ref. no. 38. Spiral view, NHMUK PM DI 0168, scale bar = 186 μm.

2. *Cancris nuttalli.* Gulf of Mexico; Eureka Core E68-151A. Deposited in the collections of the United States National Museum, Washington, DC. Specimen illustrated in van Morkhoven et al. 1986, Ref. no. 38. Apertural view, NHMUK PM DI 0169, scale bar = 187 μm.

3. *Cancris nuttalli.* Gulf of Mexico; Eureka Core E68-151A. Deposited in the collections of the United States National Museum, Washington, DC. Specimen illustrated in van Morkhoven et al. 1986, Ref. no. 38. Umbilical view, NHMUK PM DI 0170, scale bar = 192 μm.

## Original Designation

*Valvulineria nuttalli* Palmer and Bermúdez, 1936, p. 300, pl. 19, figs. 3–5.

## Type Specimen

Holotype (No. C-41) deposited in the Dorothy K. Palmer collections at the Paleontological Research Institution, Ithaca, NY, USA.

## Type Level

Upper Oligocene (van Morkhoven et al., 1986), originally given as Lower Oligocene, soft, buff marl.

## Type Locality

Shallow-water well on Finca Adelina, approximately 9.5 kilometers east of Colon, Matanzas Province, and 0.5 kilometer north of the Central Highway, Cuba.

## Description

Test forms a large trochospire; ovate in outline and suboval in cross-section with a slightly more convex umbilical side and a subrounded to subacute periphery. Approximately six moderately inflated chambers are present in the last whorl, each increasing rapidly in size (final chamber occupies nearly one-third of the test), and separated from one another by gently curved, flush sutures on the spiral side and by radial, slightly depressed sutures on the umbilical side. Chamber walls are calcareous, finely perforate, and smooth, with an imperforate area above the primary aperture, extending into the large, umbilical flap that covers the umbilicus. The primary aperture is a slit at the base of the last chamber extending from the periphery to the umbilicus and partly covered by the umbilical flap.

## Synonyms (following van Morkhoven et al. 1986) and Citations

*Cancris sagra* (d'Orbigny). –Nuttall, 1932, p. 27, pl. 6, figs. 6–7.

*Valvulineria inaequalis* (d'Orbigny). –Coryell and Rivero, 1940, p. 338, pl. 43, fig. 26.

*Cancris mexicanus* Cushman and Todd, 1942, p. 93, pl. 24, figs. 8–11.

*Valvulineria nuttalli* Palmer and Bermúdez. –Bermúdez, 1949, pl. 18, figs. 25–27.

*Baggina notoensis* Asano, 1953, p. 10, pl. 3, fig. 17.

*Cancris nuttalli* (Palmer and Bermúdez). –van Morkhoven et al., 1986, p. 120, pl. 38. –Katz and Miller, 1993, pl. 6, fig. 5.

## Biogeography

Worldwide. Recorded from Puerto Rico, the Dominican Republic, the North and South Atlantic, Europe, the eastern Pacific, and the Gulf of Mexico (van Morkhoven et al., 1986).

## Bathymetry

Primarily upper to middle bathyal.

## Chronostratigraphy

Early Oligocene (Rupelian) to Late Miocene (Tortonian). Doubtful in the late Tortonian to Early Pliocene (van Morkhoven et al., 1986).

# References


Asano, K., 1953, Miocene foraminifera from the Noto Peninsula, Ishikawa Prefecture. Short Papers, Institute of Geology and Paleontology, Tohoku University, vol. 5, pp. 1–21.

Bermúdez, P.J., 1949, Tertiary smaller foraminifera of the Dominican Republic. Special Publications of the Cushman Laboratory for Foraminiferal Research, no. 25, pp. 1–322.

Coryell, H.N. and Rivero, F.C., 1940, A Miocene microfauna of Haiti. Journal of Paleontology, vol. 14, pp. 324–344.

Cushman, J.A. and Todd, R., 1942, The genus Cancris and its species. Contributions from the Cushman Laboratory for Foraminiferal Research, vol. 18, no. 4, pp. 72–94.

Katz, M.E. and Miller, K.G., 1993, Latest Oligocene to Earliest Pliocene benthic foraminiferal biofacies of the northeastern Gulf of Mexico. Micropaleontology, vol. 39, no. 4, pp. 367–403.

Nuttall, W.L.F., 1932, Lower Oligocene foraminifera from Mexico. Journal of Paleontology, vol. 6, pp. 3–35.

Palmer, D.K. and Bermúdez, P.J., 1936, An Oligocene foraminiferal fauna from Cuba; Part 2. Memorias de la Sociedad Cubana de Historia Natural "Filipe Poey", vol. 10, no. 5, pp. 273–317.

van Morkhoven, F.P.C.M., Berggren, W.A., Edwards, A.S., et al., 1986, Cenozoic cosmopolitan deep-water benthic foraminifera. Bulletin des Centres de Recherches Exploration-Production Elf-Aquitaine, Mémoire, vol. 11, pp. 1–421.

# *Cassidulina teretis*

## Tappan, 1951

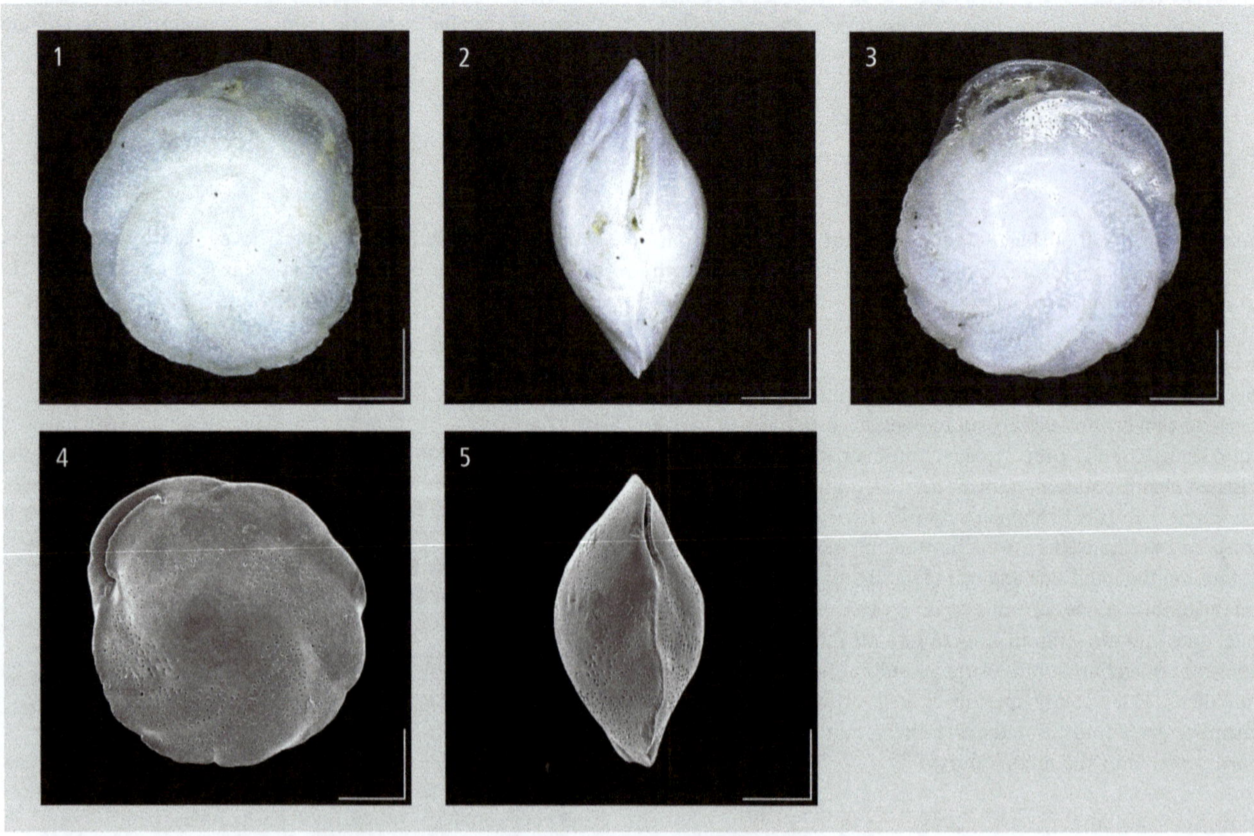

1. *Cassidulina teretis.* West of Ireland, North Atlantic Ocean; "Porcupine" Station 11, depth: 1630 fathoms. Deposited in the micropaleontological collections of The Natural History Museum, London; NHMUK ZF1261. Umbilical view, NHMUK PM DI 0171, scale bar = 88 μm.

2. *Cassidulina teretis.* West of Ireland, North Atlantic Ocean; "Porcupine" Station 11, depth: 1630 fathoms. Deposited in the micropaleontological collections of The Natural History Museum, London; NHMUK ZF1261. Edge view, NHMUK PM DI 0172, scale bar = 88 μm.

3. *Cassidulina teretis.* West of Ireland, North Atlantic Ocean; "Porcupine" Station 11, depth: 1630 fathoms. Deposited in the micropaleontological collections of The Natural History Museum, London; NHMUK ZF1261. Apertural view, NHMUK PM DI 0173, scale bar = 84 μm.

4. *Cassidulina teretis.* West of Ireland, North Atlantic Ocean; "Porcupine" Station 11, depth: 1630 fathoms. Deposited in the micropaleontological collections of The Natural History Museum, London; NHMUK ZF1261. Apertural view, NHMUK PM DI 0174, scale bar = 99 μm.

5. *Cassidulina teretis.* West of Ireland, North Atlantic Ocean; "Porcupine" Station 11, depth: 1630 fathoms. Deposited in the micropaleontological collections of The Natural History Museum, London; NHMUK ZF1261. Edge view, NHMUK PM DI 0175, scale bar = 101 μm.

## Original Designation

*Cassidulina teretis* Tappan, 1951, pl. 1, fig. 30.

## Type Specimen

Holotype and paratypes deposited in the collections of the United States National Museum, Washington, DC; holotype: USNM 560409.

## Type Level

Pleistocene, Gubik Formation.

## Type Locality

Core at 100–105 feet in Point Barrow Core Test No. 1, north-northeast of Barrow village, northern Alaska.

## Description

Test forms a lenticular planispiral, biconvex in cross-section, with a clear umbilical boss on each side and an acute, keeled periphery. Approximately eight to ten moderately inflated, biserial chambers are separated by distinct, gently curved, flush sutures. Chambers extend from the umbilical boss on one side, across the peripheral keel and approximately halfway to the umbilical boss of the opposite side. Chamber walls are calcareous, optically granular, coarsely perforate, and smooth. The primary aperture is a narrow, arched slit, extending from the base of the final chamber to the peripheral keel.

## Synonyms and Citations

*Cassidulina laevigata* d'Orbigny. –Brady, 1884, pl. 54, fig. 1 (not figs. 2–3).

*Cassidulina teretis* Tappan. –Barker, 1960, pl. 54, fig. 1. –Jones, 1994, p. 59, pl. 54, fig. 1.

## Remarks

Feyling-Hanssen and Buzas (1976) emended the definition of *Cassidulina*, and transferred one of the paratypes of *Cassidulina teretis* (USNM 560413) to *Islandiella helenae* n. sp., on the basis of the radial wall structure and characteristic apertural features. According to these authors, *Islandiella helenae* has a relatively short and broad apertural slit with a free apertural tongue, whereas *Cassidulina teretis* has a narrow apertural slit with a flap paralleling the periphery of the test. Feyling-Hanssen and Buzas (1976) also referred specimens from Recent Arctic waters off Point Barrow (illustrated by Loeblich and Tappan (1953) in pl. 24, figs. 3–4) to *Islandiella helenae*, and designated them as paratype material for this species.

This species differs from *Cassidulina limbata* Cushman and Hughes by its wider chambers, which are ovate in outline as seen in side view, rather than narrow and band-like, and by its more elongate, slit-like aperture.

## Biogeography

Worldwide. Common in Arctic and Boreal seas.

## Bathymetry

Lower bathyal. Reported from 1630 fathoms at "Porcupine" Station 11, west of Ireland, North Atlantic (Jones, 1994).

## Chronostratigraphy

Pliocene to Recent (Jones, 1994).

# References


Barker, R.W., 1960, Taxonomic notes on the species figured by H.B. Brady in his report on the foraminifera dredged by H.M.S. Challenger during the years 1873–1876. Society of Economic Paleontologists and Mineralogists Special Publication, vol. 9, pp. 1–238.

Brady, H.B., 1884, Report on the foraminifera dredged by H.M.S. Challenger during the years 1873–1876. In Brady, H.B., ed., Report on the Scientific Results of the Voyage of the H.M.S. Challenger during the years 1873–1876. Zoology, vol. 9, pp. 1–814.

Feyling-Hanssen, R.W. and Buzas, M.A., 1976, Emendation of Cassidulina and Islandiella helenae new species. Journal of Foraminiferal Research, vol. 6, pp. 154–158.

Jones, R.W., 1994, The Challenger Foraminifera. Oxford: Oxford University Press, 149 pp.

Loeblich, A.R. Jr and Tappan, H., 1953, Studies of Arctic Foraminifera. Smithsonian Miscellaneous Collections, vol. 121, no. 7, pp. 1–150.

Tappan, H., 1951, Northern Alaska index foraminifera. Contributions from the Cushman Foundation for Foraminiferal Research, vol. 2, pt 1, pp. 1–8.

# Cassidulinoides parkerianus

## (Brady), 1881

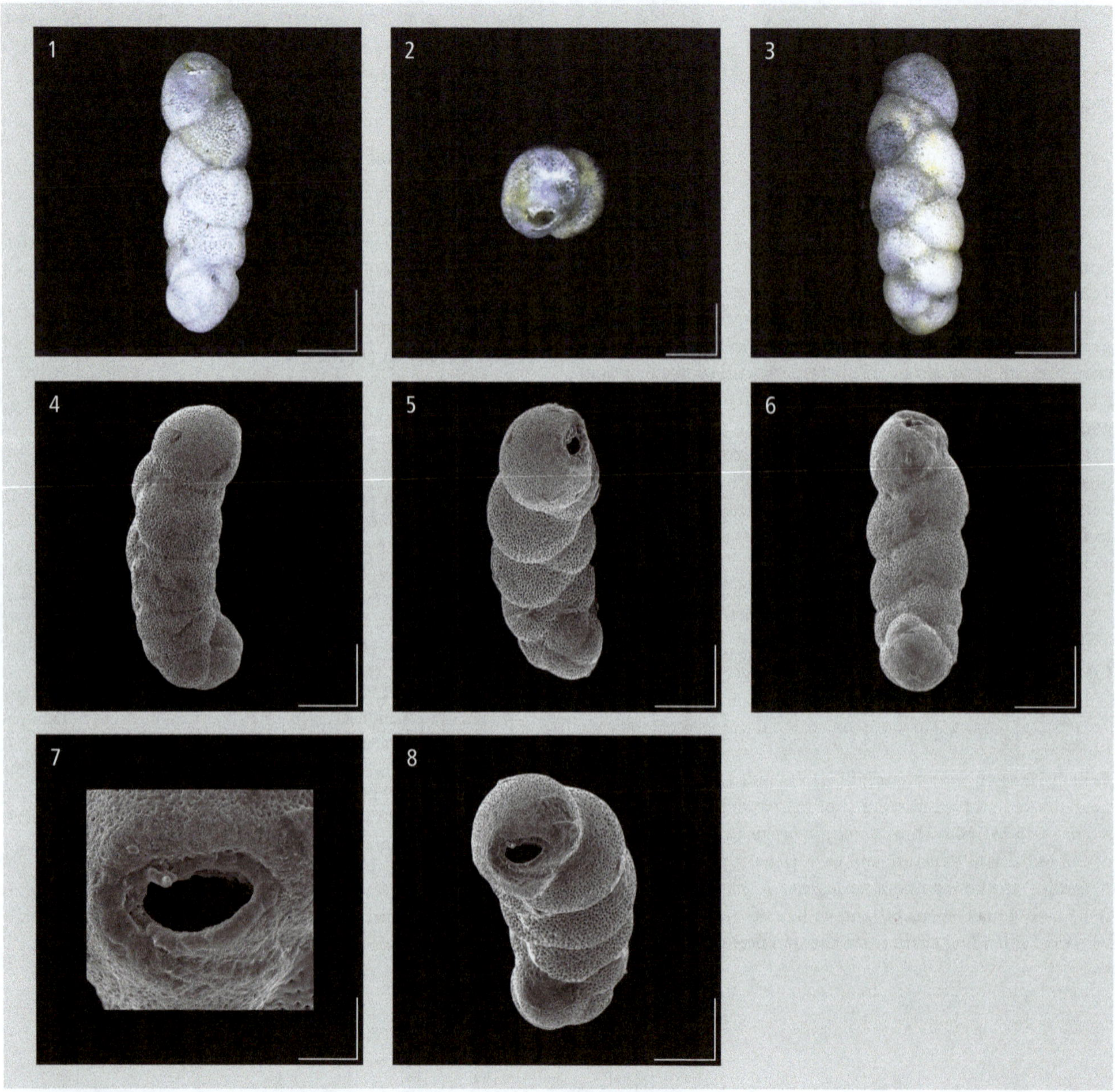


*Please see opposite for captions*

## Original Designation

*Cassidulina parkeriana* Brady, 1881, p. 59 (type description), illustrated in Brady, 1884, pl. 54, figs. 11–16.

## Type Specimen

Figured specimens are housed in the micropaleontological collections of The Natural History Museum, London, UK, and are registered in slides ZF1264–ZF1266.

## Type Level

Recent, surface sediments.

## Type Locality

Not designated. Localities given "Challenger" Station 304 (latitude 46° 53′ S, longitude 75° 11′ W, west of Patagonia, East Pacific), "Challenger" Station 305 (latitude 47° 48′ S, longitude 74° 48′ W, west of Patagonia, East Pacific), and "Challenger" Station 308 (latitude 50° 10′ S, longitude 74° 42′ W, Gulf of Penas, Patagonia, East Pacific).

## Description

Test elongate, initially planispiral, becoming rectilinear and cylindrical with a subcircular cross-section and a rounded periphery. The biserially arranged, moderately inflated chambers exhibit a small triangular overlap onto the opposite side at the periphery, and are separated by distinct, slightly depressed sutures. Chamber walls are calcareous, perforate, and smooth. The oval or loop-shaped primary aperture is terminal with a small toothplate.

## Synonyms and Citations

*Cassidulina parkeriana* Brady. –Brady, 1884, p. 432, pl. 54, figs. 11–16.

*Cassidulinoides parkerianus* (Brady). –Loeblich and Tappan, 1987, p. 504, pl. 555, figs. 10–13. –Jones, 1994, p. 60, pl. 54, figs. 11–16.

## Remarks

Type species of the genus *Cassidulinoides* Cushman, 1927.

## Biogeography

Worldwide. Cosmopolitan in low to moderate latitudes.

## Bathymetry

Upper bathyal. South Pacific 45–175 fathoms (Jones, 1994).

## Chronostratigraphy

Late Miocene to Recent.

## References


Brady, H.B., 1881, Notes on some of the reticularian Rhizopoda of the Challenger Expedition, Part III. Quarterly Journal of Microscopical Science (New Series), vol. 21, pp. 31–71.

Brady, H.B., 1884, Report on the foraminifera dredged by H.M.S. Challenger during the years 1873–1876. In Brady, H.B., ed., Report on the Scientific Results of the Voyage of the H.M.S. Challenger during the years 1873–1876. Zoology, vol. 9, pp. 1–814.

Jones, R.W., 1994, The Challenger Foraminifera. Oxford: Oxford University Press, 149 pp.

Loeblich, A.R. Jr and Tappan, H., 1987, Foraminiferal Genera and Their Classification. New York: Van Nostrand Reinhold, 970 pp.


1. *Cassidulinoides parkerianus*. West of Patagonia, East Pacific Ocean; "Challenger" Station 308, depth: 175 fathoms. Deposited in the micropaleontological collections of The Natural History Museum, London; NHMUK ZF1264. Syntype. Lateral view, NHMUK PM DI 0176, scale bar = 88 μm.

2. *Cassidulinoides parkerianus*. West of Patagonia, East Pacific Ocean; "Challenger" Station 308, depth: 175 fathoms. Deposited in the micropaleontological collections of The Natural History Museum, London; NHMUK ZF1264. Syntype. Apertural view, NHMUK PM DI 0177, scale bar = 31 μm.

3. *Cassidulinoides parkerianus*. West of Patagonia, East Pacific Ocean; "Challenger" Station 308, depth: 175 fathoms. Deposited in the micropaleontological collections of The Natural History Museum, London; NHMUK ZF1264. Syntype. Lateral view, NHMUK PM DI 0178, scale bar = 88 μm.

4. *Cassidulinoides parkerianus*. West of Patagonia, East Pacific Ocean; "Challenger" Station 308, depth: 175 fathoms. Deposited in the micropaleontological collections of The Natural History Museum, London; NHMUK ZF1264. Syntype. Lateral view, NHMUK PM DI 0179, scale bar = 92 μm.

5. *Cassidulinoides parkerianus*. West of Patagonia, East Pacific Ocean; "Challenger" Station 308, depth: 175 fathoms. Deposited in the micropaleontological collections of The Natural History Museum, London; NHMUK ZF1264. Syntype. Lateral view, NHMUK PM DI 0180, scale bar = 81 μm.

6. *Cassidulinoides parkerianus*. West of Patagonia, East Pacific Ocean; "Challenger" Station 308, depth: 175 fathoms. Deposited in the micropaleontological collections of The Natural History Museum, London; NHMUK ZF1264. Syntype. Lateral view, NHMUK PM DI 0181, scale bar = 94 μm.

7. *Cassidulinoides parkerianus*. West of Patagonia, East Pacific Ocean; "Challenger" Station 308, depth: 175 fathoms. Deposited in the micropaleontological collections of The Natural History Museum, London; NHMUK ZF1264. Syntype. Apertural detail, NHMUK PM DI 0182, scale bar = 15 μm.

8. *Cassidulinoides parkerianus*. West of Patagonia, East Pacific Ocean; "Challenger" Station 308, depth: 175 fathoms. Deposited in the micropaleontological collections of The Natural History Museum, London; NHMUK ZF1264. Syntype. Lateral view, NHMUK PM DI 0183, scale bar = 64 μm.

# *Caudammina excelsa*

## (Dylàzanka), 1923

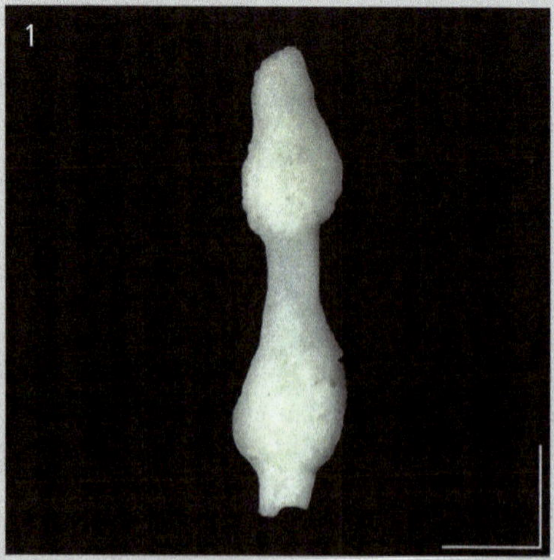

1. *Caudammina excelsa.* Węglówko near Krosno, Poland; Subsilesian Unit, 6h, 5t, 384.6–391.6 meters, greenish marls. Deposited in the Geroch Collection, Jagiellonian University, Kraków, Poland. Lateral view, NHMUK PM DI 0184, scale bar = 134 μm.

2. *Caudammina excelsa.* Siary near Gorlice, Poland; Ropianka beds, Magura Unit, 5.7.91-10, gray shale. Deposited in the Geroch Collection, Jagiellonian University, Kraków, Poland. Lateral view, NHMUK PM DI 0185, scale bar = 80 μm.

### Original Designation

*Hyperammina excelsa* Dylàzanka, 1923, p. 66, pl. 1, fig. 3.

### Type Specimen

Lectotype designated by Kaminski and Geroch (1993, pl. 17, fig. 3), and deposited in the micropaleontological collections of the Jagiellonian University, Kraków (M. Dylàzanka Collection no. UJ=133-P, 4/21).

### Type Level

Campanian to Maastrichtian, Inoceramian Beds of the Magura Unit, Polish Carpathians.

### Type Locality

Described from a quarry in the village of Szymbark, near Gorlice, Poland.

### Description

Elongated, uniserial, rectilinear test with two or more pear-shaped chambers. Chambers, broad at the prolocular end and strongly tapering in the upper part, are connected by thick stolons, about half the width of the chambers. Chamber walls are thick, finely agglutinated, and smoothly finished. The primary aperture is terminal and round.

### Synonyms and Citations

*Hormosina excelsa* (Dylàzanka). –Geroch and Nowak, 1984, pl. 1, fig. 8; pl. 5, figs. 23–24. –Morgiel and Olszewska, p. 9, pl. 2, fig. 6. –Kuhnt and Kaminski, 1990, p. 472, pl. 1, figs. f–i.

*Hyperammina excelsa* Dylàzanka. –Kaminski and Geroch, 1993, p. 281, pl. 17, figs. 1–4b.

*Caudammina excelsa* (Dylàzanka). –Kaminski and Gradstein, 2005, pp. 230–232, fig. 40, pl. 40.

### Remarks

*Caudammina excelsa* (Dylàzanka) differs from *Caudammina ovuloides* (Grzybowski) in possessing more slender, elongate chambers and more robust stolons connecting chambers (Kaminski and Gradstein, 2005). Paleocene specimens tend to be larger than Upper Cretaceous specimens (Kaminski and Gradstein, 2005).

## Biogeography

Worldwide. See also Observed Occurrences in Kaminski and Gradstein (2005).

## Bathymetry

Lower bathyal to abyssal. Generally sub-CCD.

## Chronostratigraphy

Cretaceous (Turonian) to Early Eocene (Lutetian). Geroch and Nowak (1984) reported its first appearance in the *Uvigerinammina jankoi* Biozone. Morgiel and Olszewska (1981) reported *Caudammina excelsa* from the Eocene of the Moroccan Rif.

# References


Dylàzanka, M., 1923, Warstwy inoceramowe z lomu w Szymbarku kolo Gorlic. Rocznik Polskiego Towarzystwa Geologicznego w Krakowie, vol. 1, pp. 36–81.

Geroch, S. and Nowak, W., 1984, Proposal of zonation for the late Tithonian–late Eocene, based upon arenaceous foraminifera from the outer Carpathians, Poland. In Oertli, H., ed., Benthos '83, 2nd International Symposium on Benthic Foraminifera, Pau (France), April 11–15, 1983. Elf Aquitaine, ESSO REP and TOTAL CFP, Pau and Bordeaux, pp. 225–239.

Kaminski, M.A. and Geroch, S., 1993, A revision of foraminiferal species in the Grzybowski Collection. In Kaminski, M.A., Geroch, S., and Kaminski, D.G., ed., The Origins of Applied Micropalaeontology: The School of Jósef Grzybowski (with a taxonomic revision of the Grzyboswski Collection). Grzybowski Foundation Special Publication, no. 1, pp. 239–330.

Kaminski, M.A. and Gradstein, F.M., 2005, Atlas of Palaeogene Cosmopolitan Deep-Water Agglutinated Foraminifera. Grzybowski Foundation Special Publication, no. 10, 547 + vii pp.

Kuhnt, W. and Kaminski, M.A., 1990, Paleoecology of Late Cretaceous to Paleocene deep-water agglutinated foraminifera from the North Atlantic and western Tethys. In Hemleben, C., Kaminski, M.A., Kuhnt, W., and Scott, D.B., ed., Paleoecology, Biostratigraphy, Paleoceanography and Taxonomy of Agglutinated Foraminifera. NATO ASI Series, Series C: Mathematical and Physical Sciences. Dordrecht: Kluwer Academic, vol. 327, pp. 433–505.

Morgiel, J. and Olszewska, B., 1981, Biostratigraphy of the Polish External Carpathians based on agglutinated foraminifera. Micropaleontology, vol. 27, no. 1, pp. 1–30.

# Caudammina gigantea

## (Geroch), 1960

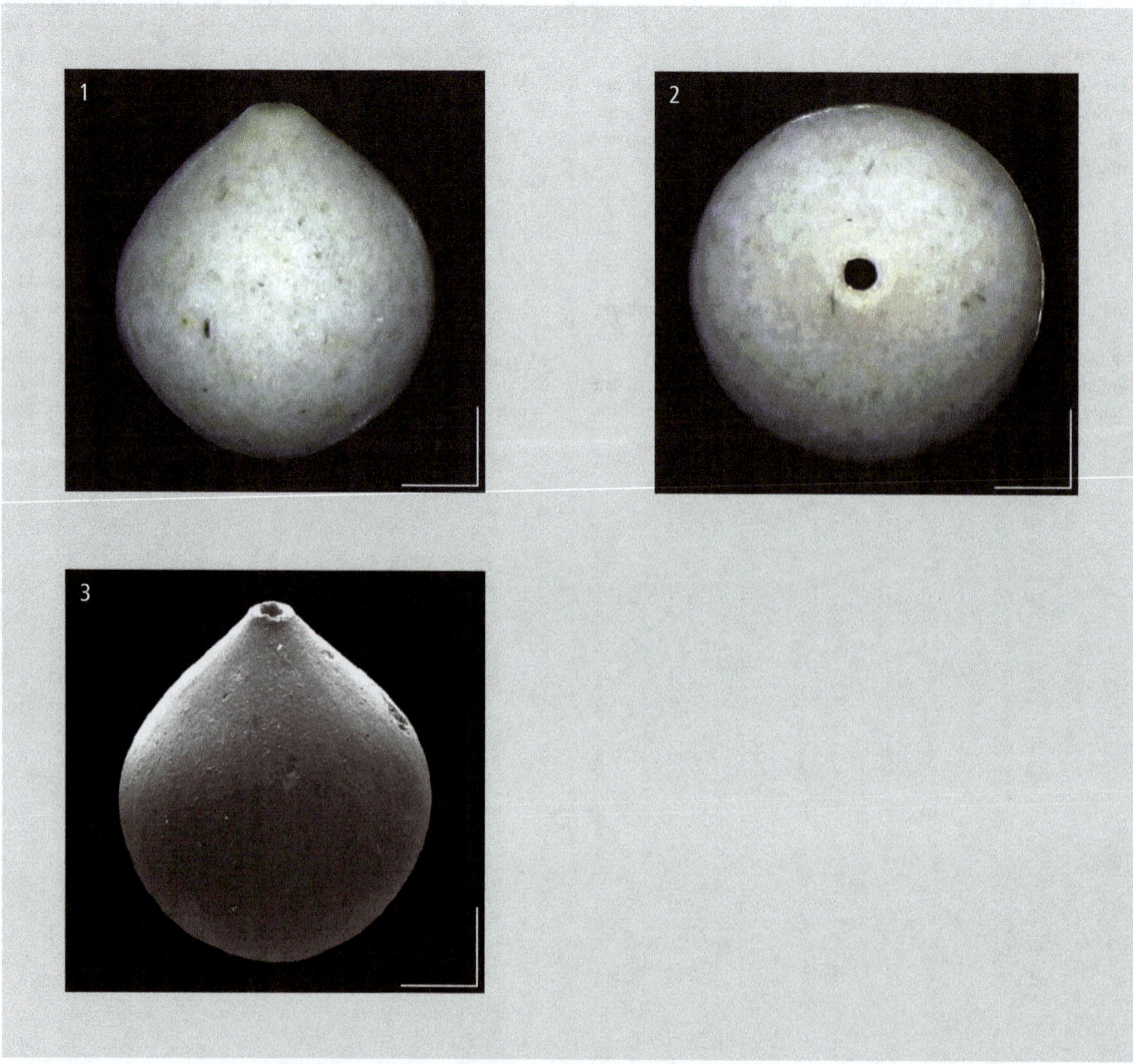

1. *Caudammina gigantea.* Vigo Seamount, North Atlantic; DSDP Leg 47, Hole 398D, Core 48, Section 5, 112–115 cm. Lateral view, NHMUK PM DI 0186, scale bar = 110 μm.

2. *Caudammina gigantea.* Vigo Seamount, North Atlantic; DSDP Leg 47, Hole 398D, Core 48, Section 5, 112–115 cm. Apertural view, NHMUK PM DI 0187, scale bar = 97 μm.

3. *Caudammina gigantea.* Galicia Bank, Northeast Atlantic; ODP Leg 103, Hole 641A, Core 1, core catcher. Lateral view, NHMUK PM DI 0188, scale bar = 98 μm.

## Original Designation

*Hormosina ovulum* (Grzybowski) var. *gigantea* Geroch, 1960, p. 43, pl. 2, figs. 18–19; pl. 11, fig. 6.

## Type Specimen

Depository not given.

## Type Level

Not designated. One figured specimen (fig. 18) from the Upper Cretaceous, Senonian, Lower Istebna Beds; other figured specimens (fig. 19) from beds transitional from the Godula Beds (Aptian–Albian to Senonian) to the Lower Istebna Beds. Also occurs in the Upper Istebna Beds (uppermost Cretaceous? to Paleocene) and Ciezkowice Beds (Paleocene).

## Type Locality

Not designated. Figured specimens from Rudnik near Wadowice (fig. 18), Czarne (fig. 19), and along the Czarna Wiselka stream (fig. 6), all in the Silesian Beskids, Poland.

## Description

Test composed of large, spherical chambers arranged in a rectilinear or arcuate series but typically found as single chamber fragments. Chamber size exceeds 400 µm in width. Chamber wall is multilayered in contrast to stolon connections that are thin, consisting mostly of a single layer often of coarse quartz grains. The difference in wall structure between the chambers and stolons probably accounts for the isolated occurrence of single chamber fragments. The primary aperture is terminal and produced on a tubular neck. Chamber walls are thick, finely agglutinated, and smoothly cemented with organic cement.

## Synonyms and Citations

*Hormosina ovulum* (Grzybowski). –Geroch and Gradzinski, 1955, pl. 5, fig. 3.

*Hormosina ovulum gigantea* Geroch. –Geroch and Nowak, 1984, pl. 1, fig. 20; pl. 5, figs. 15–16. –Kuhnt and Kaminski, 1990, p. 474, pl. 1, figs. a–b.

*Hormosina gigantea* Geroch. –Moullade et al., 1988, p. 365, pl. 2, fig. 11.

## Remarks

We follow Morgiel and Olszewska (1981), Moullade et al. (1988), and Kaminski and Gradstein (2005) in assigning the larger morphotype in the *Caudammina ovula* group to a separate species.

## Biogeography

Worldwide.

## Bathymetry

Bathyal to abyssal. Common in flysch-type assemblages.

## Chronostratigraphy

Late Cretaceous (Campanian to Maastrichtian) in the North Atlantic and Labrador (Kuhnt and Kaminski, 1990). In the Carpathian flysch, the first occurrence of *Caudammina gigantea* marks the base of the Early Campanian to Maastrichtian *Hormosina ovulum gigantea* Biozone of Geroch and Nowak (1984).

## References


Geroch, S., 1960, Microfaunal assemblages from the Cretaceous and Paleogene Silesian Unit in the Beskid Śląski Mts. (western Carpathians). Biuletyn Instytutu Geologicznego, vol. 153, pp. 7–138.

Geroch, S. and Gradzinski, R., 1955, Stratigrafia serii podsląskiej Żywieckiego okna tektonicznego. Rocznik Polskiego Towarzystwa Geologicznego, vol. 24, pp. 3–62.

Geroch, S. and Nowak, W., 1984, Proposal of zonation for the late Tithonian–late Eocene, based upon arenaceous foraminifera from the outer Carpathians, Poland. In Oertli, H., ed., Benthos '83, 2nd International Symposium on Benthic Foraminifera, Pau (France), April 11–15, 1983. Elf Aquitaine, ESSO REP and TOTAL CFP, Pau and Bordeaux, pp. 225–239.

Kaminski, M.A. and Gradstein, F.M., 2005, Atlas of Palaeogene Cosmopolitan Deep-Water Agglutinated Foraminifera. Grzybowski Foundation Special Publication, no. 10, 547 + vii pp.

Kuhnt, W. and Kaminski, M.A., 1990, Paleoecology of Late Cretaceous to Paleocene deep-water agglutinated foraminifera from the North Atlantic and western Tethys. In Hemleben, C., Kaminski, M.A., Kuhnt, W., and Scott, D.B., ed., Paleoecology, Biostratigraphy, Paleoceanography and Taxonomy of Agglutinated Foraminifera. NATO ASI Series, Series C: Mathematical and Physical Sciences. Dordrecht: Kluwer Academic, vol. 327, pp. 433–505.

Morgiel, J. and Olszewska, B., 1981, Biostratigraphy of the Polish External Carpathians based on agglutinated foraminifera. Micropaleontology, vol. 27, no. 1, pp. 1–30.

Moullade, M., Kuhnt, W., and Thurow, J., 1988, Agglutinated benthic foraminifers from Upper Cretaceous variegated clays of the North Atlantic Ocean (DSDP Leg 93 and ODP Leg 103). In Boillot, G., Winterer, E.L., et al., ed. Scientific Results of the Proceedings of the Ocean Drilling Program, vol. 103, pp. 349–377.

# Caudammina ovula

## (Grzybowski), 1896 emend. Geroch, 1960

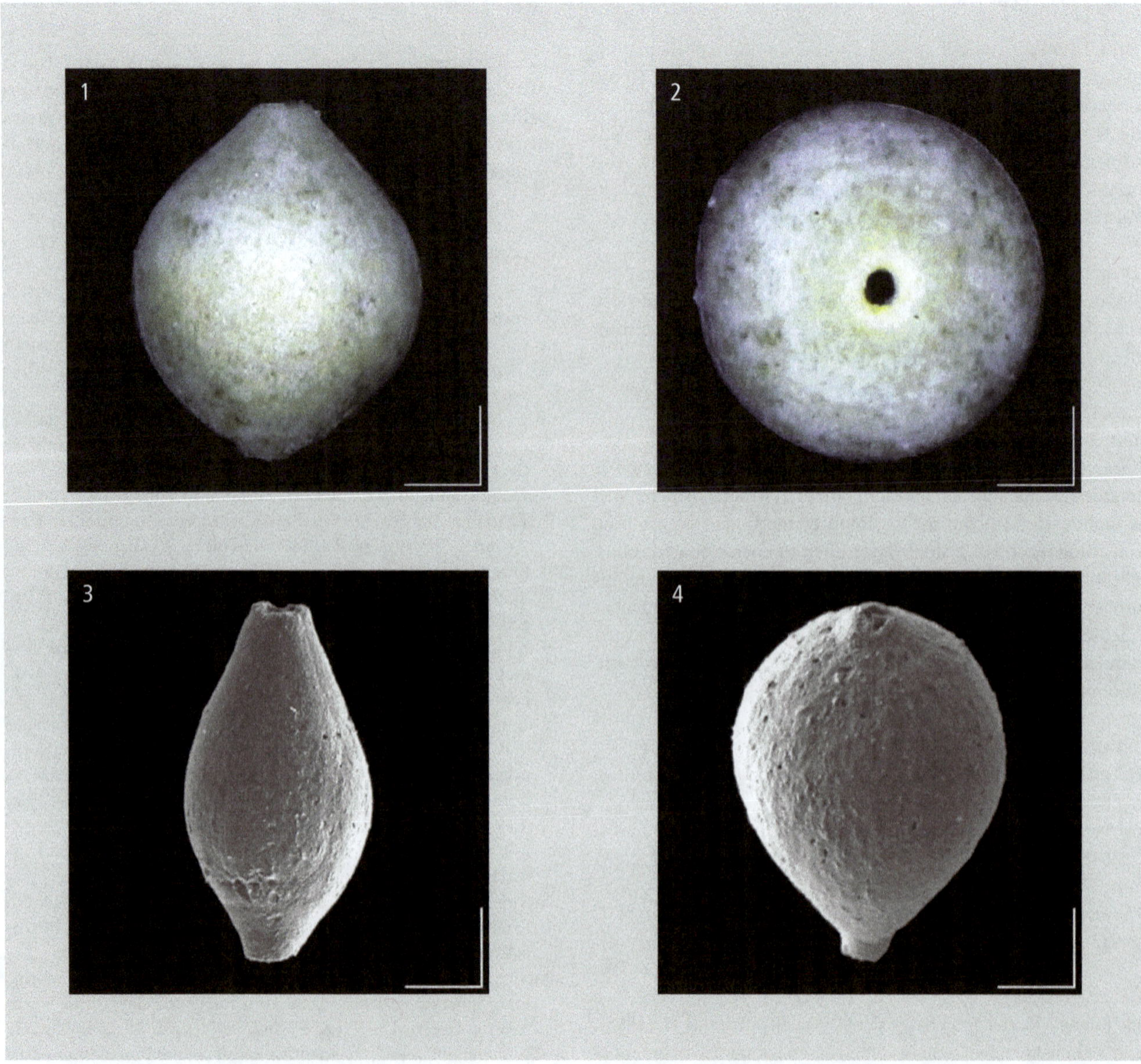


1. *Caudammina ovula.* Vigo Seamount, North Atlantic; DSDP Leg 47, Hole 398D, Core 48, Section 5, 112–115 cm. Lateral view, NHMUK PM DI 0189, scale bar = 67 µm.

2. *Caudammina ovula.* Vigo Seamount, North Atlantic; DSDP Leg 47, Hole 398D, Core 48, Section 5, 112–115 cm. Apertural view, NHMUK PM DI 0190, scale bar = 81 µm.

3. *Caudammina ovula.* Galicia Bank, Northeast Atlantic; ODP Leg 103, Hole 641A, Core 2, core catcher. Lateral view, NHMUK PM DI 0191, scale bar = 52 µm.

4. *Caudammina ovula.* Galicia Bank, Northeast Atlantic; ODP Leg 103, Hole 641A, Core 2, core catcher. Lateral view, NHMUK PM DI 0192, scale bar = 42 µm.

## Original Designation

*Reophax ovulum* Grzybowski, 1896, p. 276, pl. 8, figs. 19–21.

## Type Specimen

Depository not given.

## Type Level

Upper Cretaceous (Campanian) red clay.

## Type Locality

Wadowice, Poland.

## Description

Test composed of ovate chambers arranged in a rectilinear or arcuate series, but usually found as single chamber fragments. Chamber wall is multilayered, in contrast to stolon connections, which are thin, consisting mostly of a single layer often of coarse quartz grains. Differences in wall structure between chambers and stolons probably accounts for the isolated occurrence of single chamber fragments. The primary aperture is terminal, and produced on a tubular neck. Chamber walls are thick, finely agglutinated, and smoothly cemented with organic cement.

## Synonyms and Citations

*Hormosina ovulum* (Grzybowski). –Krasheninnikov, 1974, p. 644, pl. 7, fig. 12. –Moullade et al., 1988, p. 364, pl. 2, figs. 7–9.

*Hormosina ovulum ovulum* (Grzybowski). –Kaminski et al., 1988, p. 186, pl. 2, fig. 10. –Kuhnt and Kaminski, 1990, p. 474, pl. 1, figs. c–e. –Wightman and Kuhnt, 1992, p. 257, pl. 2, fig. 2.

*Hormosinella ovulum* (Grzybowski). –Bolli et al., 1994, p. 72, figs. 19.17–19.

*Caudammina ovula* (Grzybowski) emend. Geroch, 1960. –Kaminski and Gradstein, 2005, p. 233–236, fig. 41, pl. 41.

## Remarks

*Caudammina ovula* (Grzybowski) is similar to *Caudammina gigantea* (Geroch) in its very smooth surface and narrow connections between the chambers, but has a smaller size (200–400 µm) and more elongated chambers. *Caudammina gigantea* also has a more restricted stratigraphic range (Geroch, 1960; Geroch and Nowak, 1984; Moullade et al., 1988; Kaminski and Gradstein, 2005). Extremely elongated forms show similarity to *Caudammina ovuloides* (Grzybowski), but can be distinguished from this species by their narrow connections between the chambers and a more symmetrical egg-shape. See Remarks in Kaminski and Gradstein (2005).

## Biogeography

Worldwide. See also Observed Occurrences in Kaminski and Gradstein (2005).

## Bathymetry

Bathyal to abyssal. Common in flysch-type assemblages.

## Chronostratigraphy

Early Cretaceous (Hauterivian) to Middle Eocene (Kaminski and Gradstein, 2005).

# References


Bolli, H.M., Beckmann, J.P., and Saunders, J.B., 1994, Benthic Foraminiferal Biostratigraphy of the South Caribbean Region. Cambridge: Cambridge University Press, 408 pp.

Geroch, S., 1960, Microfaunal assemblages from the Cretaceous and Paleogene Silesian Unit in the Beskid Śląski Mts. (western Carpathians). Biuletyn Instytutu Geologicznego, vol. 153, pp. 7–138.

Geroch, S. and Nowak, W., 1984, Proposal of zonation for the late Tithonian–late Eocene, based upon arenaceous foraminifera from the outer Carpathians, Poland. In Oertli, H., ed., Benthos '83, 2nd International Symposium on Benthic Foraminifera, Pau (France), April 11–15, 1983. Elf Aquitaine, ESSO REP and TOTAL CFP, Pau and Bordeaux, pp. 225–239.

Grzybowski, J., 1896, Otwornice czerwonych iłów z Wadowic. Rozprawy Wydziału Matematyczno-Przyrodniczego, Akademia Umiejętności w Krakowie, serya 2, vol. 30, pp. 261–308.

Kaminski, M.A. and Gradstein, F.M., 2005, Atlas of Palaeogene Cosmopolitan Deep-Water Agglutinated Foraminifera. Grzybowski Foundation Special Publication, no. 10, 547 + vii pp.

Kaminski, M.A., Gradstein, F.M., Berggren, W.A., Geroch, S., and Beckmann, J.P., 1988, Agglutinated foraminiferal assemblages from Trinidad: taxonomy, stratigraphy and paleobathymetry. In Gradstein, F.M. and Rögl, F., ed., Second International Workshop on Agglutinated Foraminifera, Vienna 1986, Abhandlungen der Geologischen Bundesanstalt, vol. 41, pp. 155–228.

Krasheninnikov, V.A., 1974, Upper Cretaceous benthonic agglutinated foraminifera, Leg 27 of the Deep Sea Drilling Project. In Veevers, J.J., Heirtzler, J.R., et al., ed., Initial Reports of the Deep Sea Drilling Project, Washington (U.S. Government Printing Office), vol. 27, pp. 531–662.

Kuhnt, W. and Kaminski, M.A., 1990, Paleoecology of Late Cretaceous to Paleocene deep-water agglutinated foraminifera from the North Atlantic and western Tethys. In Hemleben, C., Kaminski, M.A., Kuhnt, W., and Scott, D.B., ed., Paleoecology, Biostratigraphy, Paleoceanography and Taxonomy of Agglutinated Foraminifera. NATO ASI Series, Series C: Mathematical and Physical Sciences. Dordrecht: Kluwer Academic, vol. 327, pp. 433–505.

Moullade, M., Kuhnt, W., and Thurow, J., 1988, Agglutinated benthic foraminifers from Upper Cretaceous variegated clays of the North Atlantic Ocean (DSDP Leg 93 and ODP Leg 103). In Boillot, G., Winterer, E.L., et al., ed. Scientific Results of the Proceedings of the Ocean Drilling Program, vol. 103, pp. 349–377.

Wightman, W.G. and Kuhnt, W., 1992, Biostratigraphy and Paleoecology of late Cretaceous abyssal agglutinated foraminifers from the western Pacific Ocean (Deep Sea Drilling Project Holes 196A and 198A and Ocean Drilling Program Holes 800A and 801A). In Larson, R.L., Lancelot, Y., et al., ed., Scientific Results of the Proceedings of the Ocean Drilling Program, vol. 129, pp. 247–264.

# Chilostomella oolina

## Schwager, 1878

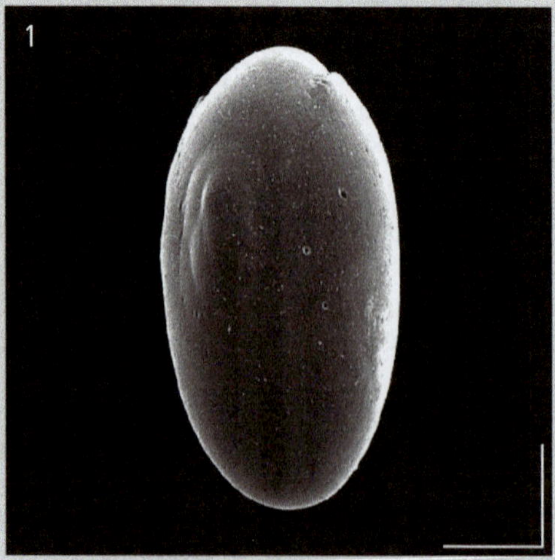

1. *Chilostomella oolina.* North Atlantic; 12347-2, 260–262.5 cm. Deposited in the Lutze Collection no. Z2132, Institute of Geosciences, Kiel University, Germany. Lateral view, NHMUK PM DI 0193, scale bar = 56 μm.

2. *Chilostomella oolina.* North Atlantic; 12347-2, 260–262.5 cm. Deposited in the Lutze Collection no. Z2132, Institute of Geosciences, Kiel University, Germany. Lateral view, NHMUK PM DI 0193, scale bar = 87 μm.

### Original Designation

*Chilostomella oolina* Schwager, 1878, p. 527, pl. 1, fig. 16.

### Type Specimen

Depository not given.

### Type Level

Miocene, Tortonian tuff.

### Type Locality

Streto, presso Girgenti, Sicily, Italy.

### Description

Test ovate, elongate, rounded, planispiral, involute. There are two strongly embracing chambers per whorl, the last one occupying more than half of the test and the initial ones often in an oblique position relative to the long axis of the test. Chamber walls are thin, smooth, transparent, shiny, and finely perforate. The aperture is a narrow, interiomarginal slit, situated approximately two-thirds of the way along the length of the test and sometimes bordered by a small lip.

### Synonyms and Citations

*Chilostomella ovoidea* Reuss. –Brady, 1884, p. 436, pl. 55, figs. 12–14, 17–18. –Barker, 1960, p. 112, pl. 55, figs. 12–14, 17–18.

*Chilostomella oolina* Schwager. –Hermelin, 1989, p. 76, pl. 14, fig. 5. –van Marle, 1991, p. 128, pl. 10, figs. 12–13. –Loeblich and Tappan, 1994, p. 160, pl. 349, figs. 12–13. –Jones, 1994, p. 61, pl. 55, figs. 12–14, 17–18.

### Remarks

*Chilostomella oolina* Schwager closely resembles *Chilostomella ovoidea* Reuss. However, the latter species is generally broader and more ovate.

### Biogeography

Worldwide. Originally described from Sicily, Italy. Recorded from the Gulf of Mexico (Pflum and Frerichs, 1976), the continental margin of New Zealand (Lewis, 1979), eastern Indonesia (van Marle, 1991), the Sahul Shelf and Timor Sea (Loeblich and Tappan, 1994), and the Pacific Ocean (Jones, 1994).

### Bathymetry

Bathyal. Corliss (1985) reported that *Chilostomella oolina* was a deep infaunal species, living at depths deeper than 6 cm within the sediment and tolerant of low oxygenation.

### Chronostratigraphy

Miocene to Recent (Jones, 1994). Early Miocene (Biozone M5) to Recent (van Marle, 1991).

## References


Barker, R.W., 1960, Taxonomic notes on the species figured by H.B. Brady in his report on the foraminifera dredged by H.M.S. Challenger during the years 1873–1876. Society of Economic Paleontologists and Mineralogists Special Publication, vol. 9, pp. 1–238.

Brady, H.B., 1884, Report on the foraminifera dredged by H.M.S. Challenger during the years 1873–1876. In Brady, H.B., ed., Report on the Scientific Results of the Voyage of the H.M.S. Challenger during the years 1873–1876. Zoology, vol. 9, pp. 1–814.

Corliss, B.H., 1985, Microhabitats of benthic foraminifera within deep-sea sediments. Nature, vol. 314, pp. 435–438.

Hermelin, J.O.R., 1989, Pliocene benthic foraminifera from the Ontong-Java Plateau (Western Equatorial Pacific Ocean): faunal response to changing paleoenvironment. Special Publication of the Cushman Foundation for Foraminiferal Research, vol. 26, pp. 1–143.

Jones, R.W., 1994, The Challenger Foraminifera. Oxford: Oxford University Press, 149 pp.

Lewis, K.B., 1979, Foraminifera on the continental shelf and slope off southern Hawke's Bay. Memoirs of the New Zealand Oceanographic Institute, vol. 84, pp. 1–45.

Loeblich, A.R. Jr and Tappan, H., 1994, Foraminifera of the Sahul Shelf and Timor Sea. Special Publications of the Cushman Laboratory for Foraminiferal Research, vol. 31, pp. 1–661.

Pflum, C.E. and Frerichs, W.E., 1976, Gulf of Mexico deep-water foraminifers. Cushman Foundation, Special Publication, vol. 14, pp. 1–125.

Schwager, C., 1878, Nota su alcuni Foraminiferi nuovi del Tubo di Stretto presso Girgenti. Bolletino R. Comitato Geologico d'Italia, vol. 8, pp. 519–529.

van Marle, L.J., 1991, Eastern Indonesian Late Cenozoic smaller benthic foraminifera. Verhandelingen der Koninklijke Nederlanse Akademie van Wetenschappen, Afdeeling Natuurkunde, Eerste Reeks, vol. 34, pp. 1–328.

# *Chrysalidinella dimorpha*

## (Brady), 1881

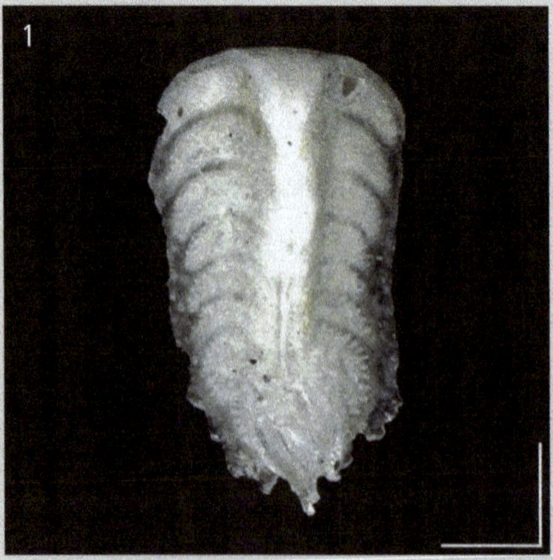


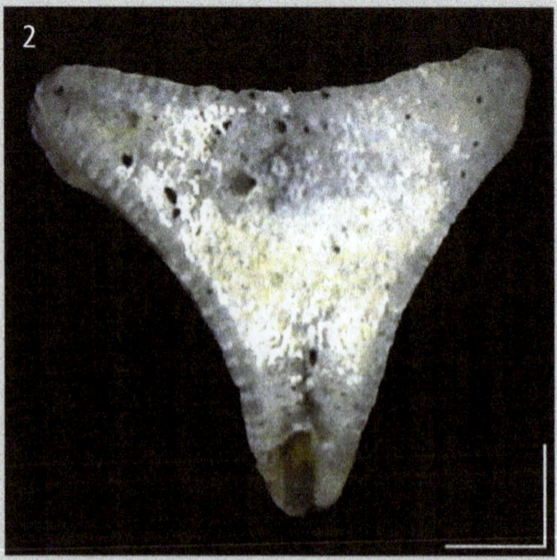


1. *Chrysalidinella dimorpha.* Hong Kong Harbour, Pacific; "Challenger" Station 205A, depth: 7 fathoms. Deposited in the micropaleontological collections of The Natural History Museum, London; NHMUK ZF1273. Lateral view, NHMUK PM DI 0194, scale bar = 106 μm.

2. *Chrysalidinella dimorpha.* Hong Kong Harbour, Pacific; "Challenger" Station 205A, depth: 7 fathoms. Deposited in the micropaleontological collections of The Natural History Museum, London; NHMUK ZF1273. Apertural view, NHMUK PM DI 0195, scale bar = 65 μm.

### Original Designation

*Verneuilina (Chrysalidina) dimorpha* Brady, 1881, p. 54 (type description), illustrated as *Chrysalidina dimorpha* in Brady, 1884, pl. 46, figs. 20–21.

### Type Specimen

Housed in the micropaleontological collections of The Natural History Museum, London, UK, and registered in slide ZF1273.

### Type Level

Recent.

### Type Locality

Figured specimens from "Challenger" Station 205A, Hong Kong Harbour, in 7 fathoms (fig. 20); and from "Challenger" Station 260A, off reefs Honolulu, Sandwich Islands, Hawaii, in 40 fathoms (fig. 21).

### Description

Test elongate, rectilinear, uniserial, initially pluriserial in the microspheric generation, and triangular in cross-section. The broad and low triangular chambers increase gradually in size, and are separated by distinct, curved sutures. The outer edges of the chambers may bear faint keels. Chamber walls are calcareous, and perforate with coarse pores in regular rows. The primary aperture is not clearly defined, but the apex of the chambers is pierced by irregular round openings.

### Synonyms and Citations

*Chrysalidina dimorpha* Brady, 1884, p. 388, pl. 46, figs. 20–21.

*Chrysalidinella dimorpha* (Brady). –Barker, 1960, pl. 46, figs. 20–21. –Loeblich and Tappan, 1987, p. 527, pl. 575, figs. 3–5. –Revets, 1991, p. 10, pl. 4, figs. 4–6. –Jones, 1994, p. 51, pl. 46, figs. 20–21.

### Remarks

Type species of the genus *Chrysalidinella* Schubert, 1908. Revets (1991) revised the description of the genus *Chrysalidinella.*

### Biogeography

Worldwide. Tropical areas in Atlantic and Pacific Oceans.

### Bathymetry

Neritic to upper bathyal. Reported between 7 and 155 fathoms at "Challenger" stations (Jones, 1994).

### Chronostratigraphy

Late Miocene to Recent (Jones, 1994).

## References


Barker, R.W., 1960, Taxonomic notes on the species figured by H.B. Brady in his report on the foraminifera dredged by H.M.S. Challenger during the years 1873–1876. Society of Economic Paleontologists and Mineralogists Special Publication, vol. 9, pp. 1–238.

Brady, H.B., 1881, Notes on some of the reticularian Rhizopoda of the Challenger Expedition, Part III. Quarterly Journal of Microscopical Science (New Series), vol. 21, pp. 31–71.

Brady, H.B., 1884, Report on the foraminifera dredged by H.M.S. Challenger during the years 1873–1876. In Brady, H.B., ed., Report on the Scientific Results of the Voyage of the H.M.S. Challenger during the years 1873–1876. Zoology, vol. 9, pp. 1–814.

Jones, R.W., 1994, The Challenger Foraminifera. Oxford: Oxford University Press, 149 pp.

Loeblich, A.R. Jr and Tappan, H., 1987, Foraminiferal Genera and Their Classification. New York: Van Nostrand Reinhold, 970 pp.

Revets, S., 1991, The generic revision of the reussellids (Foraminiferida). Journal of Micropalaeontology, vol. 10, pp. 1–15.

# *Cibicides lobatulus*

## (Walker and Jacob), 1798

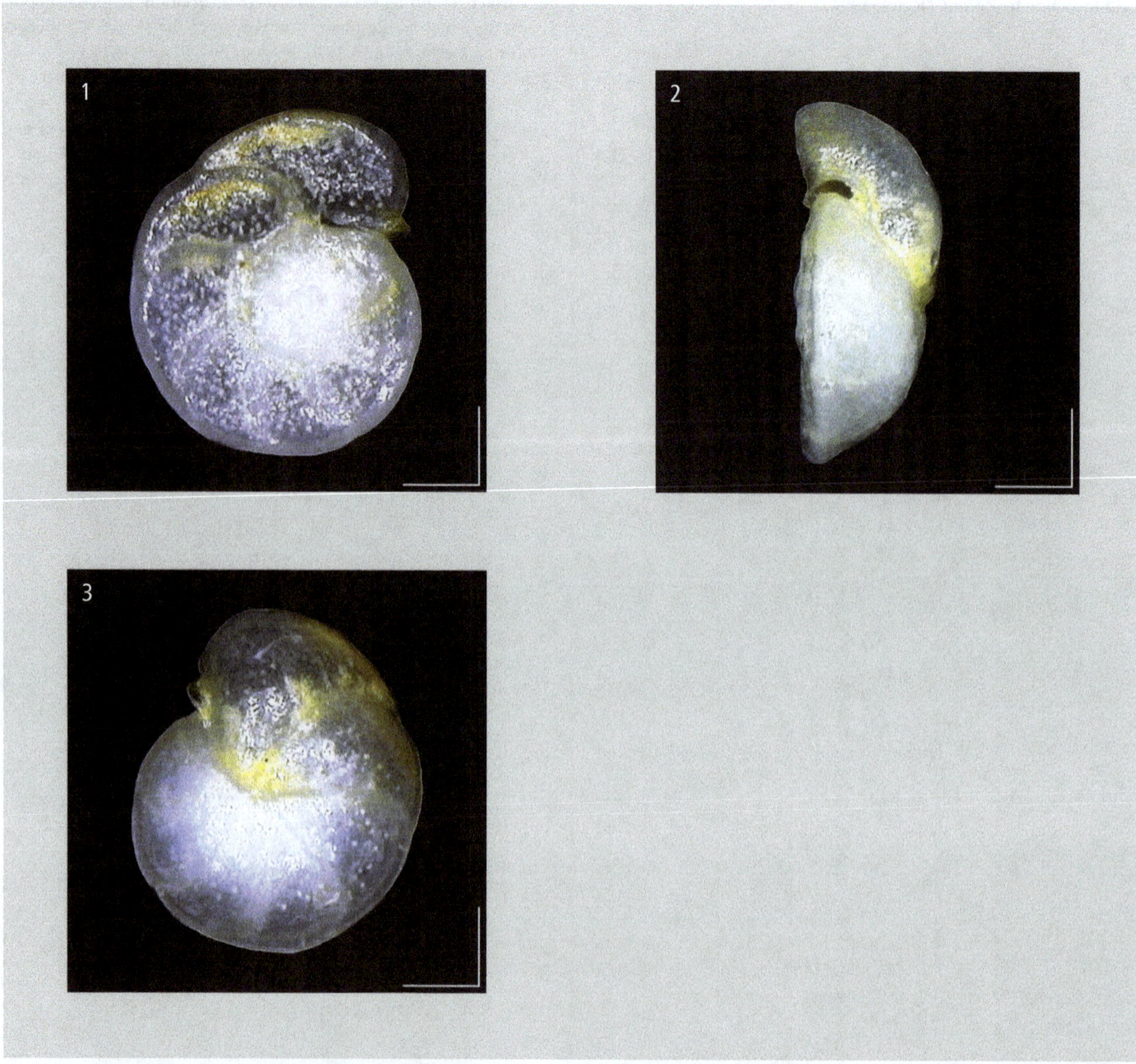


1. *Cibicides lobatulus*. Sunten sands, Boxgrove excavation, Southern England. J.E. Whittaker Collection. Spiral view, NHMUK PM DI 0196, scale bar = 65 μm.

2. *Cibicides lobatulus*. Sunten sands, Boxgrove excavation, Southern England. J.E. Whittaker Collection. Apertural view, NHMUK PM DI 0197, scale bar = 62 μm.

3. *Cibicides lobatulus*. Sunten sands, Boxgrove excavation, Southern England. J.E. Whittaker Collection. Umbilical view, NHMUK PM DI 0198, scale bar = 65 μm.

## Original Designation

*Nautilus lobatulus* Walker and Jacob, 1798, p. 642, pl. 14, fig. 36.

## Type Specimen

Depository not given.

## Type Level

Recent, shore sand.

## Type Locality

Whitestable, Kent, England.

## Description

Test forms a low trochospire; slightly lobulate (last chambers) in outline, and planoconvex in cross-section, with an involute, convex umbilical side, a flattened or slightly convex, evolute spiral side, and a subacute, imperforate periphery. Seven to nine moderately inflated chambers increase gradually in size and are separated by broad, curved, flush or slightly depressed sutures. Chamber walls are calcareous, thick, and coarsely perforate on both sides of the test. The primary aperture is a narrow equatorial slit bordered by a thin lip, extending on the spiral side.

## Synonyms and Citations

*Truncatulina lobatula* (Walker and Jacob. –Brady, 1884, p. 60, pl. 92, fig. 10; pl. 93, figs. 1, 4–5, pl. 115, figs. 4–5.

*Cibicides lobatulus* (Walker and Jacob). –Barker, 1960, pl. 92, fig. 10; pl. 93, figs. 1, 4–5; pl. 115, figs. 4–5. –Jones, 1994, p. 97, pl. 92, fig. 10; pl. 93, figs. 1, 4–5; p. 114, pl. 115, figs. 4–5.

*Lobatula lobatula* (Walker and Jacob). –Milker and Schmiedl, 2012, p. 107, figs. 24.17–20.

## Remarks

*Cibicides lobatulus* provides a reliable record of bottom-water $\delta^{18}O$ and $\delta^{13}C$ values in shallow to intermediate water masses, where *Planulina wuellerstorfi* is generally absent (Weinelt et al., 2000).

## Biogeography

Worldwide.

## Bathymetry

Neritic to bathyal. *Cibicides lobatulus* is an epibenthic species that often occurs in relatively shallow water (less than 1000 meters) over a wide range of organic export flux rates. This species appears to be closely linked to near-bottom currents for its nutritional needs (Altenbach et al., 1999). Reported from shallow water (11 fathoms) to 3000 fathoms at "Challenger" stations (Jones, 1994).

## Chronostratigraphy

Middle Miocene to Recent (Jones, 1994).

# References

Altenbach, A.V., Pflaumann, U., Schiebel, R., Thies, A., Timm, S., and Trauth, M., 1999, Scaling percentages and distributional patterns of benthic foraminifera with flux rates of organic carbon. Journal of Foraminiferal Research, vol. 29, pp. 173–185.

Barker, R.W., 1960, Taxonomic notes on the species figured by H.B. Brady in his report on the foraminifera dredged by H.M.S. Challenger during the years 1873–1876. Society of Economic Paleontologists and Mineralogists Special Publication, vol. 9, pp. 1–238.

Brady, H.B., 1884, Report on the foraminifera dredged by H.M.S. Challenger during the years 1873–1876. In Brady, H.B., ed., Report on the Scientific Results of the Voyage of the H.M.S. Challenger during the years 1873–1876. Zoology, vol. 9, pp. 1–814.

Jones, R.W., 1994, The Challenger Foraminifera. Oxford: Oxford University Press, 149 pp.

Milker, Y. and Schmiedl, G., 2012, A taxonomic guide to modern benthic shelf foraminifera of the western Mediterranean Sea. Palaeontologia Electronica, vol. 15, Issue 2;16A, 134 pp.

Walker, G. and Jacob, E., 1798, Descriptions of Serpula (Lagena) sulcata and Nautilus lobatulus. In Adams, E., ed., Essays on the microscope, 2nd edn with considerable additions and improvements by F. Kanmacher. London: Dillon & Keating, pp. 634 and 642.

Weinelt, M., Kuhnt, W., Sarnthein, M., Altenbach, A., Costello O., Erlenkeuser, H., Matthiessen, J., Pflaumann, U., Simstich, J., Struck, U., Thies, A., Trauth, M., and Vogelsang, E., 2000, Paleoceanographic proxies in the Northern North Atlantic. In Schäfer, P., Schlüter, M., Schröder-Ritzrau, W., and Thiede, J., ed., The Northern North Atlantic: A Changing Environment. New York: Springer-Verlag, pp. 319–352.

# Cibicides refulgens

## de Montfort, 1808

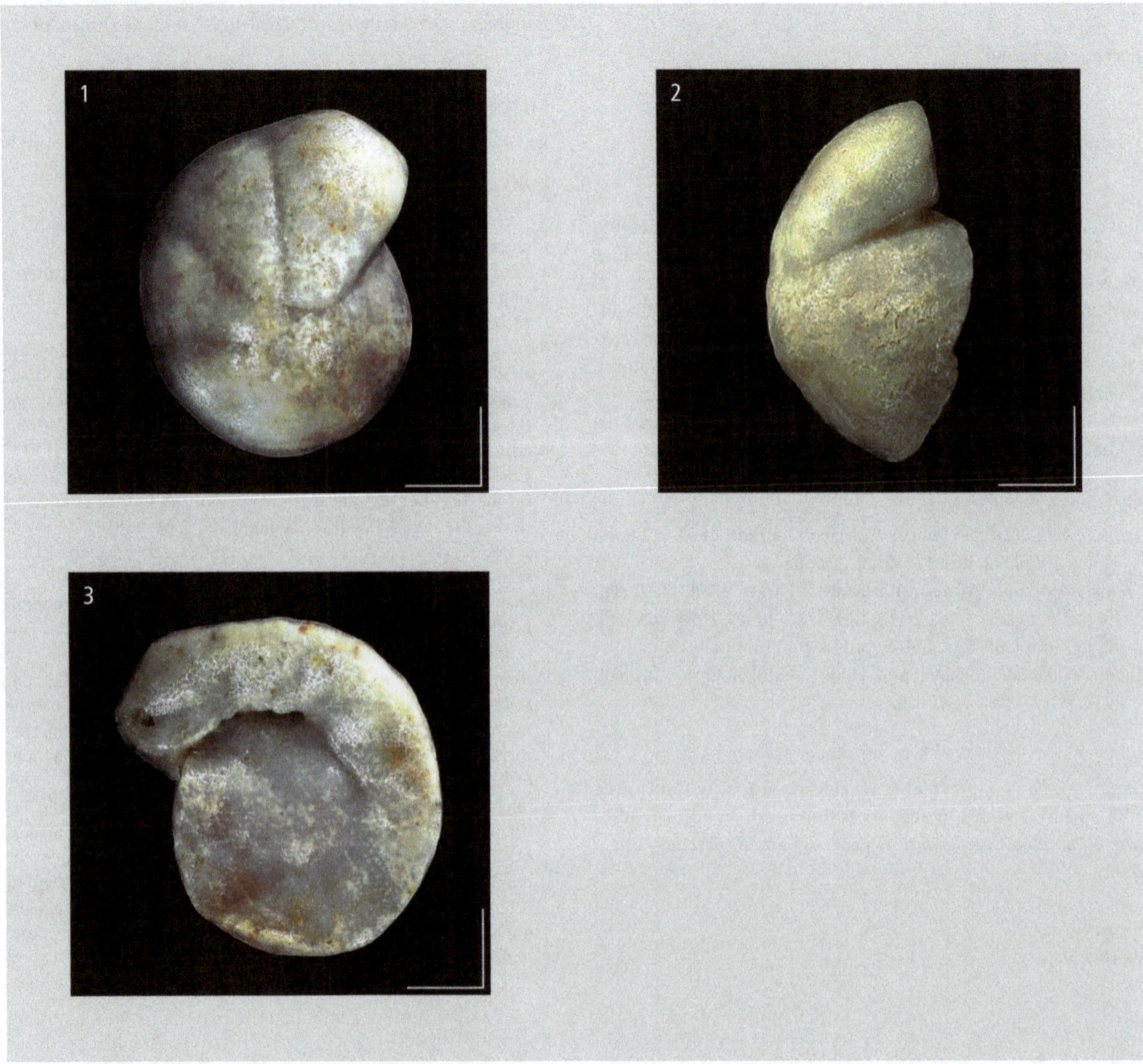


1. *Cibicides refulgens.* West Africa; "Meteor" Cruise M25. Deposited in the Lutze Collection (West Africa), no. 172. Institute of Geosciences, Kiel University, Germany. Umbilical view, NHMUK PM DI 0199, scale bar = 192 μm.

2. *Cibicides refulgens.* West Africa; "Meteor" Cruise M25. Deposited in the Lutze Collection (West Africa), no. 172. Institute of Geosciences, Kiel University, Germany. Apertural view, NHMUK PM DI 0200, scale bar = 201 μm.

3. *Cibicides refulgens.* West Africa; "Meteor" Cruise M25. Deposited in the Lutze Collection (West Africa), no. 172. Institute of Geosciences, Kiel University, Germany. Spiral view, NHMUK PM DI 0201, scale bar = 201 μm.

## Original Designation

*Cibicides refulgens* de Montfort, 1808, pp. 122–123.

## Type Specimen

Not designated.

## Type Level

Not designated. Levels given: fossil and Recent.

## Type Locality

Not designated. Localities given: near Livourne and in the Sienna area, Tuscany, Italy.

## Description

Test forms a compressed trochospire; planoconvex in cross-section, with an involute, convex umbilical side, an evolute, flattened spiral side, and an angular, keeled, imperforate periphery. The seven or eight inflated, crescentic chambers increase gradually in size and are separated by curved, flush, or slightly raised limbate sutures on the spiral side, and by sigmoid, slightly depressed sutures on the umbilical side. A shallow umbilicus is obscured by an umbonal boss. Chamber walls are calcareous, smooth, and coarsely perforate on the spiral side, and more finely perforate on the umbilical side. The primary aperture is a narrow equatorial slit bordered by a thin lip, continuing on the spiral side as a simple slit in the spiral suture.

## Synonyms and Citations

*Truncatulina refulgens* (de Montfort). –Brady, 1884, p. 659, pl. 92, figs. 7–9.

*Cibicides refulgens* de Montfort. –Barker, 1960, pl. 92, figs. 7–9. –Loeblich and Tappan, 1987, p. 582, pl. 634, figs. 1–3. –van Marle, 1991, p. 200, pl. 21, figs. 15–16; pl. 22, fig. 1. –Loeblich and Tappan, 1994, p. 149, pl. 318, figs. 7–9. –Jones, 1994, p. 97, pl. 92, figs. 7–9. –Milker and Schmiedl, 2012, p. 107, figs. 24.14–16.

## Remarks

Type species of the genus *Cibicides* de Montfort, 1808.

## Biogeography

Worldwide. Atlantic, Pacific (Jones, 1994). Eastern Indonesia (van Marle, 1991).

## Bathymetry

Middle neritic to bathyal. Reported between 50 and 2400 fathoms at "Challenger" stations (Jones, 1994).

## Chronostratigraphy

Late Miocene to Recent (van Marle, 1991; Jones, 1994).

# References


Barker, R.W., 1960, Taxonomic notes on the species figured by H.B. Brady in his report on the foraminifera dredged by H.M.S. Challenger during the years 1873–1876. Society of Economic Paleontologists and Mineralogists Special Publication, vol. 9, pp. 1–238.

Brady, H.B., 1884, Report on the foraminifera dredged by H.M.S. Challenger during the years 1873–1876. In Brady, H.B., ed., Report on the Scientific Results of the Voyage of the H.M.S. Challenger during the years 1873–1876. Zoology, vol. 9, pp. 1–814.

de Montfort, P.D., 1808, Conchyliologie systématique et classification méthodique des coquilles. Paris: F. Schoell, vol. 1, lxxxvii + 409 pp.

Jones, R.W., 1994, The Challenger Foraminifera. Oxford: Oxford University Press, 149 pp.

Loeblich, A.R. Jr and Tappan, H., 1987, Foraminiferal Genera and Their Classification. New York: Van Nostrand Reinhold, 970 pp.

Loeblich, A.R. Jr and Tappan, H., 1994, Foraminifera of the Sahul Shelf and Timor Sea. Special Publications of the Cushman Laboratory for Foraminiferal Research, vol. 31, pp. 1–661.

Milker, Y. and Schmiedl, G., 2012, A taxonomic guide to modern benthic shelf foraminifera of the western Mediterranean Sea. Palaeontologia Electronica, vol. 15, Issue 2;16A, 134 pp.

van Marle, L.J., 1991, Eastern Indonesian Late Cenozoic smaller benthic foraminifera. Verhandelingen der Koninklijke Nederlanse Akademie van Wetenschappen, Afdeeling Natuurkunde, Eerste Reeks, vol. 34, pp. 1–328.

# *Cibicidina walli*

## Bandy, 1949

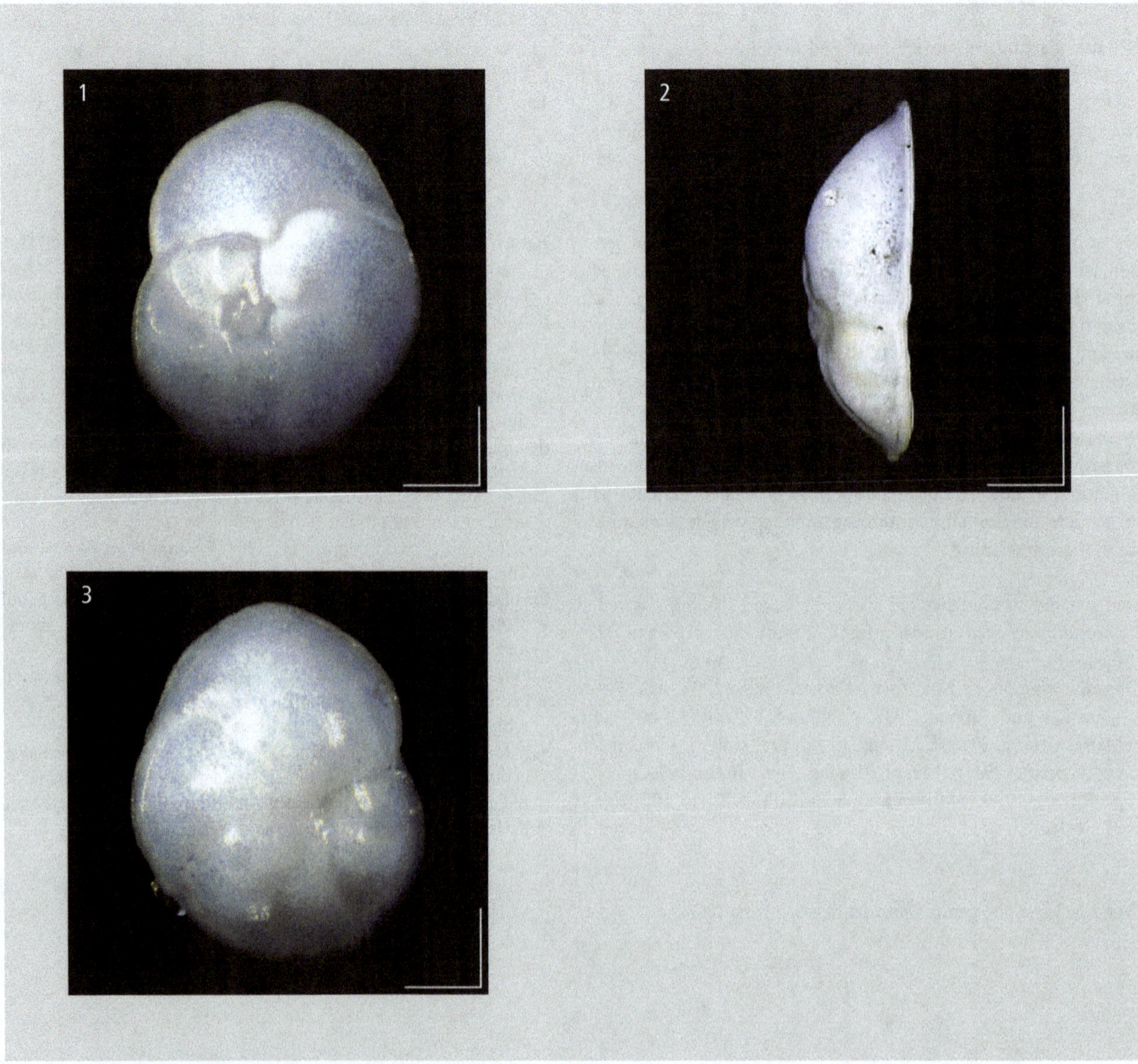


1. *Cibicidina walli.* Gulf of Mexico; Eureka Core E68, Section 151A, depth: 4608 feet. Spiral view, NHMUK PM DI 0202, scale bar = 102 μm.
2. *Cibicidina walli.* Gulf of Mexico; Eureka Core E68, Section 151A, depth: 4608 feet. Apertural view, NHMUK PM DI 0203, scale bar = 103 μm.
3. *Cibicidina walli.* Gulf of Mexico; Eureka Core E68, Section 151A, depth: 4608 feet. Umbilical view, NHMUK PM DI 0204, scale bar = 101 μm.

## Original Designation

*Cibicidina walli* Bandy, 1949, p. 95, pl. 15, fig. 5.

## Type Specimen

Holotype (No. 4855) deposited in the collections of the Paleontological Laboratory, Indiana University, Bloomington, Indiana, USA.

## Type Level

Upper Eocene, upper part of the Jackson Formation, Zone B; bluish, glauconitic marl with phosphatic nodules, about 5 feet below the top of the Jackson Formation.

## Type Locality

Locality 58 in the creek bed on the south side, along the upper course of Little Stave Creek, about 150 feet upstream from a falls and about 8675 feet upstream from the base of the exposed section, which begins at the intersection of the creek with the Jackson fault, near the northeast corner of Sec. 30, T. 7 N, R 2 E, about 2–2.5 miles north of the town of Jackson, Clarke County, Alabama, USA.

## Description

Test forms a trochospire; slightly lobulate in outline, concavo-convex in cross-section, with a concave, partially evolute, spiral side, a convex, involute umbilical side, and a slightly flattened periphery. The periphery is acute and keeled. The seven to eight moderately inflated chambers increase rapidly in size and are separated by slightly depressed sutures that are strongly curved on the spiral side and more gently curved on the umbilical side. Chamber walls are calcareous, finely perforate, and smooth on both sides of the test. The primary aperture is an equatorial arch, extending at the base of the last whorl to the center of the spiral side.

## Synonyms and Citations

?*Cibicides delicatulus* LeRoy, 1944, p. 42, pl. 2, figs. 19–21.

*Cibicidina walli* Bandy. –van Morkhoven et al., 1986, p. 178, pl. 60. –Loeblich and Tappan, 1987, p. 582, pl. 634, figs. 13–15.

## Remarks

Type species of the genus *Cibicidina* Bandy, 1949.

## Biogeography

Worldwide. Recorded from the North and South Atlantic, the eastern Pacific, and the Gulf of Mexico (van Morkhoven et al., 1986).

## Bathymetry

Lower neritic to bathyal (van Morkhoven et al., 1986).

## Chronostratigraphy

Late Eocene (Priabonian) to Middle Miocene (Serravallian).

## References


Bandy, O.L., 1949, Eocene and Oligocene foraminifera from Little Stave Creek, Clarke County: Alabama. Bulletins of American Paleontology, vol. 32, pp. 5–206.

LeRoy, L.W., 1944, Miocene foraminifera from Sumatra and Java, Netherlands East Indies; Part 1. Miocene Foraminifera of Central Sumatra, Netherlands East Indies. Quarterly of the Colorado School of Mines, vol. 39, no. 3, pp. 8–69.

Loeblich, A.R. Jr and Tappan, H., 1987, Foraminiferal Genera and Their Classification. New York: Van Nostrand Reinhold, 970 pp.

van Morkhoven, F.P.C.M., Berggren, W.A., Edwards, A.S., et al., 1986, Cenozoic cosmopolitan deep-water benthic foraminifera. Bulletin des Centres de Recherches Exploration-Production Elf-Aquitaine, Mémoire, vol. 11, pp. 1–421.

# Cibicidoides alazanensis

## (Nuttall), 1932

1. *Cibicidoides alazanensis.* Western Asumcion, State of Vera Cruz, Mexico. Deposited in the collections of the United States National Museum, Washington, DC; CC 16462. Paratype. Umbilical view, NHMUK PM DI 0205, scale bar = 194 μm.

2. *Cibicidoides alazanensis.* Western Asumcion, State of Vera Cruz, Mexico. Deposited in the collections of the United States National Museum, Washington, DC; CC 16462. Paratype. Spiral view, NHMUK PM DI 0206, scale bar = 208 μm.

### Original Designation

*Anomalina alazanensis* Nuttall, 1932, p. 31, pl. 8, figs. 5–7.

### Type Specimen

Nuttall (1932) used three specimens for illustration: the holotype (CC 16460) for the apertural view and the paratypes (CC 16461 and CC 16462) for the umbilical and spiral sides. All specimens are deposited in the collections of the United States National Museum, Washington, DC.

### Type Level

Lower Oligocene, upper Alazan Shale.

### Type Locality

A strip about 4 kilometers wide, bounded to the southwest by Tumbadero Hacienda, to the north by Temapache, and to the south by the Tuxpam River, Vera Cruz, Mexico.

### Description

Test forms a compressed trochospire; planoconvex in cross-section, with a partially evolute, convex umbilical side, an evolute, flattened spiral side, and a rounded, imperforate periphery. The fifteen to seventeen moderately inflated chambers in the last whorl increase gradually in size and are separated by curved, flush or slightly raised, limbate sutures. A shallow umbilicus is obscured by an umbonal boss. Chamber walls are calcareous, coarsely perforate, and unornamented. The primary aperture is a narrow equatorial slit bordered by a thin lip and extends onto the spiral side.

### Synonyms and Citations

*Cibicidoides alazanensis* (Nuttall). –van Morkhoven et al., 1986, p. 201, pl. 68.

*Anomalinoides alazanensis* (Nuttall).–Bolli et al., 1994, p. 373, figs. 59.5a–c, 6–9, 51, 79.20a–b, 73.–Ortiz and Thomas, 2006, p. 111, pl. 3, fig. 3.

### Biogeography

Worldwide. Cuba, Puerto Rico, Dominican Republic, Ecuador, Trinidad, Venezuela, Gulf of Mexico, North and South Atlantic, Pacific, and Europe (van Morkhoven et al., 1986).

### Bathymetry

Primarily upper to middle bathyal, rarely found in outer neritic deposits (van Morkhoven et al., 1986).

### Chronostratigraphy

Oligocene to Middle Miocene (van Morkoven et al., 1986). Early to Middle Eocene in the Fortuna Section (Betic Cordillera, southeastern Spain (Ortiz and Thomas, 2006)).

## References


Bolli, H.M., Beckmann, J.P., and Saunders, J.B., 1994, Benthic Foraminiferal Biostratigraphy of the South Caribbean Region. Cambridge: Cambridge University Press, 408 pp.

Nuttall, W.L.F., 1932, Lower Oligocene foraminifera from Mexico. Journal of Paleontology, vol. 6, pp. 3–35.

Ortiz, S. and Thomas, E., 2006, Lower–middle Eocene benthic foraminifera from the Fortuna Section (Betic Cordillera, southeastern Spain). Micropaleontology, vol. 52, pp. 97–150.

van Morkhoven, F.P.C.M., Berggren, W.A., Edwards, A.S., et al., 1986, Cenozoic cosmopolitan deep-water benthic foraminifera. Bulletin des Centres de Recherches Exploration-Production Elf-Aquitaine, Mémoire, vol. 11, pp. 1–421.

# Cibicidoides barnetti

## (Bermúdez), 1949

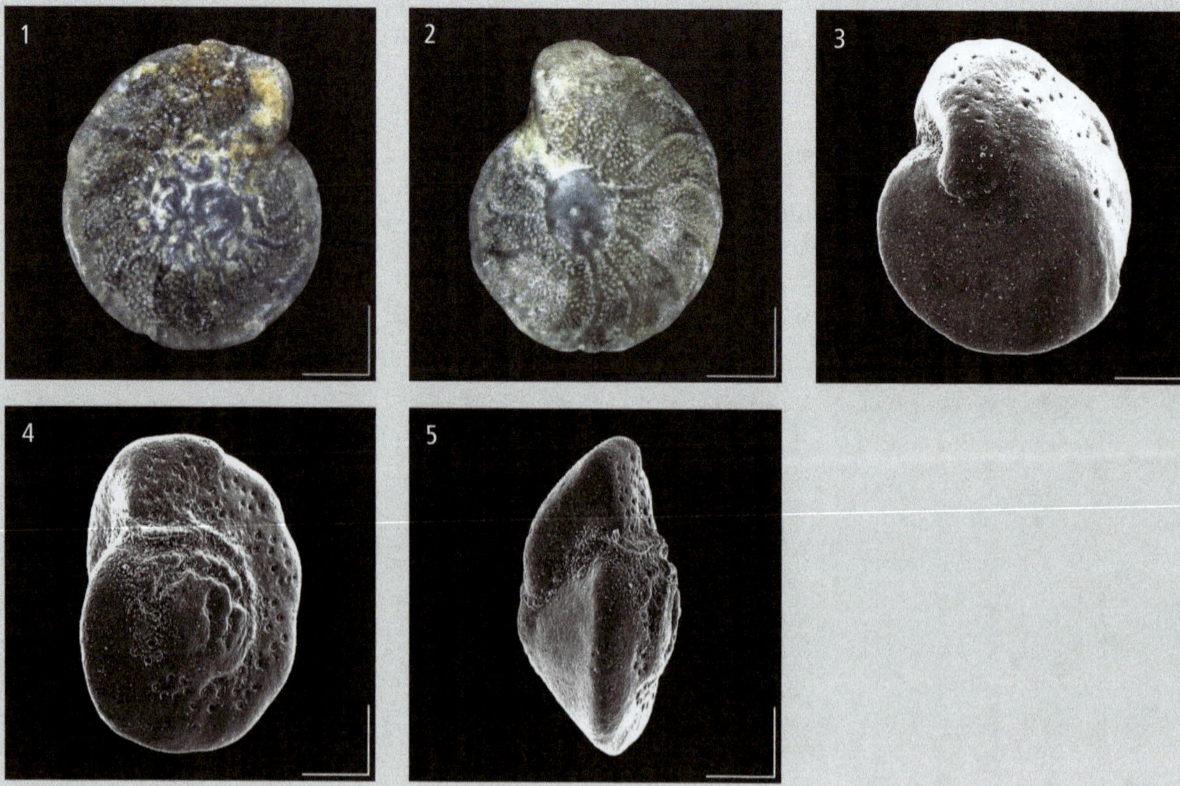


1. *Cibicidoides barnetti.* Sombrerito Formation, Benefactor Province, Cañada Gerinaldo, Dominican Republic. Deposited in the collections of the United States National Museum, Washington, DC; USNM 62546. Holotype. Spiral view, NHMUK PM DI 0207, scale bar = 220 μm.

2. *Cibicidoides barnetti.* Sombrerito Formation, Benefactor Province, Cañada Gerinaldo, Dominican Republic. Deposited in the collections of the United States National Museum, Washington, DC; USNM 62546. Holotype. Umbilical view, NHMUK PM DI 0208, scale bar = 216 μm.

3. *Cibicidoides barnetti.* Northeast continental slope of the South China Sea; ODP Leg 184, Hole 1148A, Core 33, Section 3, 70–72 cm. Umbilical view, NHMUK PM DI 0209, scale bar = 132 μm.

4. *Cibicidoides barnetti.* Northeast continental slope of the South China Sea; ODP Leg 184, Hole 1148A, Core 33, Section 3, 70–72 cm. Spiral view, NHMUK PM DI 0210, scale bar = 135 μm.

5. *Cibicidoides barnetti.* Northeast continental slope of the South China Sea; ODP Leg 184, Hole 1148A, Core 33, Section 3, 70–72 cm. Apertural view, NHMUK PM DI 0211, scale bar = 134 μm.

## Original Designation

*Cibicides barnetti* Bermúdez, 1949, p. 295, pl. 24, figs. 7–9.

## Type Specimen

Holotype (USNM 62546) and paratypes deposited in the collections of the United States National Museum, Washington, DC.

## Type Level

Lower Miocene (van Morkhoven et al., 1986), tan, massive, fossiliferous clay shale, Sombrerito Formation. Found abundantly only in shale samples from the upper part of the Sombrerito Formation. Given as Middle Oligocene by Bermúdez (1949).

## Type Locality

Sample H-1066 from the Cañada Gerinaldo, Benefactor Province, Dominican Republic.

## Description

Test forms a trochospire; biconvex in cross-section, with a partially evolute, more convex umbilical side, and an acute, imperforate periphery. The (approximately) fourteen moderately inflated chambers in the last whorl increase gradually in size and are separated by curved, raised, limbate sutures extending along the periphery to form a "pseudokeel". On the umbilical side the sutures converge toward a large, glossy, clear umbo; on the spiral side the early sutures coalesce to form irregular seams. Chamber walls are calcareous, thick, and uniformly perforate on both sides of the test. The primary aperture is a narrow, equatorial slit bordered by a thin lip and extends onto the spiral side.

## Synonyms and Citations

*Cibicidoides barnetti* (Bermúdez). –van Morkhoven et al., 1986, p. 187, pl. 63. –Katz and Miller, 1993, pl. 2, fig. 1. –Kuhnt et al., 2002, p. 141, pl. 7, figs. 14–18.

## Biogeography

Worldwide. Recorded in Barbados, Gulf of Mexico, the Pacific and Atlantic Oceans (van Morkhoven et al., 1986), the South China Sea ODP Leg 184 (Kuhnt et al., 2002).

## Bathymetry

Bathyal.

## Chronostratigraphy

Middle Eocene (Bartonian) to Middle Miocene (Serravallian).

# References


Bermúdez, P.J., 1949, Tertiary smaller foraminifera of the Dominican Republic. Special Publications of the Cushman Laboratory for Foraminiferal Research, no. 25, pp. 1–322.

Katz, M.E. and Miller, K.G., 1993, Latest Oligocene to Earliest Pliocene benthic foraminiferal biofacies of the northeastern Gulf of Mexico. Micropaleontology, vol. 39, no. 4, pp. 367–403.

Kuhnt, W., Holbourn, A.E., and Zhao, Q., 2002, The early history of the South China Sea: Evolution of Oligocene–Miocene deep water environments. Revue de Micropaléontologie, vol. 45, pp. 99–159.

van Morkhoven, F.P.C.M., Berggren, W.A., Edwards, A.S., et al., 1986, Cenozoic cosmopolitan deep-water benthic foraminifera. Bulletin des Centres de Recherches Exploration-Production Elf-Aquitaine, Mémoire, vol. 11, pp. 1–421.

# *Cibicidoides bradyi*

## (Trauth) 1918

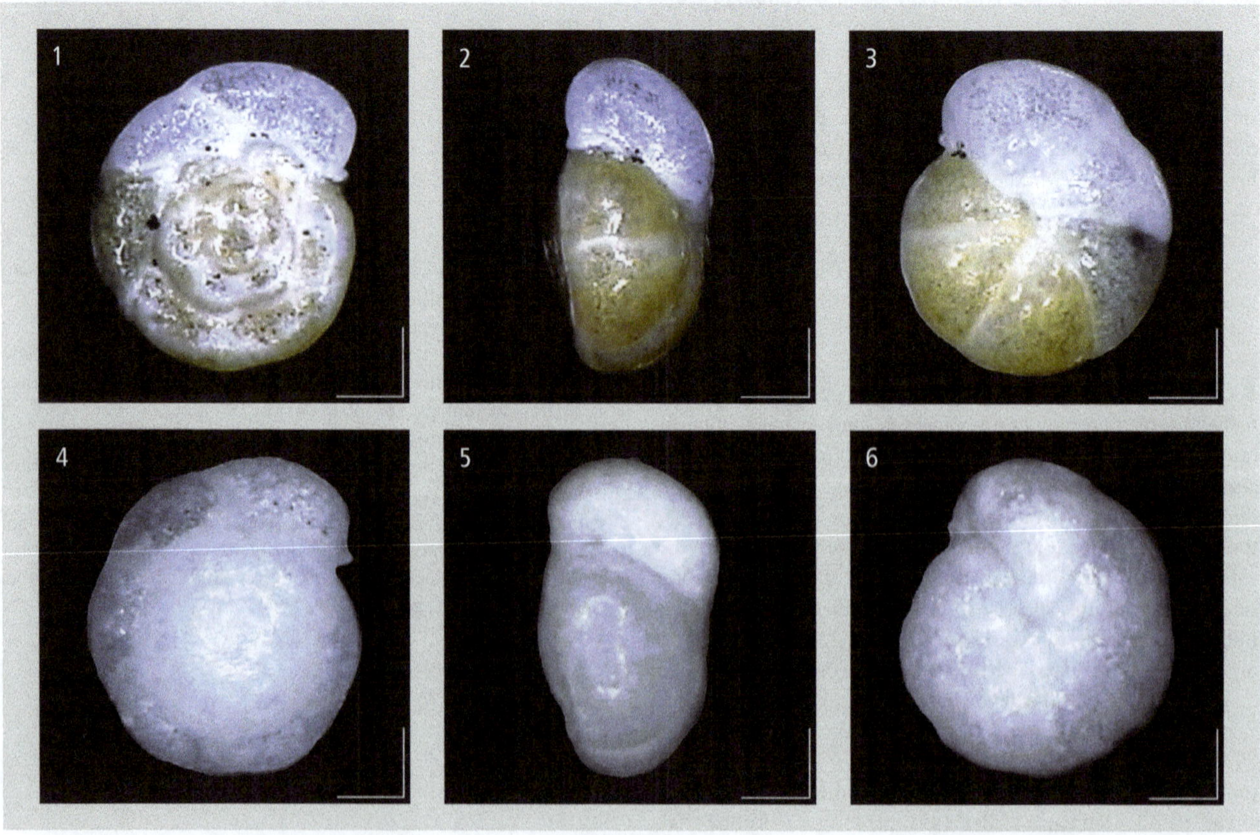


1. *Cibicidoides bradyi.* South Atlantic Ocean; "Challenger" Station 323, depth: 1900 fathoms. Deposited in the micropaleontological collections of The Natural History Museum, London; NHMUK ZF2523. Spiral view, NHMUK PM DI 0212, scale bar = 72 µm.

2. *Cibicidoides bradyi.* South Atlantic Ocean; "Challenger" Station 323, depth: 1900 fathoms. Deposited in the micropaleontological collections of The Natural History Museum, London; NHMUK ZF2523. Apertural view, NHMUK PM DI 0213, scale bar = 73 µm.

3. *Cibicidoides bradyi.* South Atlantic Ocean; "Challenger" Station 323, depth: 1900 fathoms. Deposited in the micropaleontological collections of The Natural History Museum, London; NHMUK ZF2523. Umbilical view, NHMUK PM DI 0214, scale bar = 72 µm.

4. *Cibicidoides bradyi.* Meteor Rise, South Atlantic; ODP Leg 114, Hole 703A, Core 10, Section 5, 88–92 cm. Spiral view, NHMUK PM DI 0215, scale bar = 64 µm.

5. *Cibicidoides bradyi.* Meteor Rise, South Atlantic; ODP Leg 114, Hole 703A, Core 10, Section 5, 88–92 cm. Apertural view, NHMUK PM DI 0216, scale bar = 56 µm.

6. *Cibicidoides bradyi.* Meteor Rise, South Atlantic; ODP Leg 114, Hole 703A, Core 10, Section 5, 88–92 cm. Umbilical view, NHMUK PM DI 0217, scale bar = 66 µm.

## Original Designation

*Truncatulina bradyi* Trauth, 1918, p. 235, illustrated with figures for *Truncatulina dutemplei* (d'Orbigny) from Brady, 1884, pl. 95, fig. 5 and from Egger, 1893, pl. 16, figs. 22–23, 30, 54–56.

## Type Specimen

Specimen illustrated by Brady (1884) is housed in the micropaleontological collections of The Natural History Museum, London, UK.

## Type Level

Recent, surface sediments.

## Type Locality

Not designated. Localities given: "Challenger" Station 323, latitude 35° 39′ S, longitude 50° 47′ W, South Atlantic, 1900 fathoms and north of Papua, Pacific Ocean, 1070 fathoms (Brady, 1884, p. 665); "Gazelle" Stations: Western Africa, 677 meters; Kerguelen Islands, 104 meters; New Amsterdam, 2624 meters; Mauritius, 137 meters; between New Amsterdam and Australia, 398 meters; Western Australia, 1188 meters; Fiji Islands, 2432 meters; between Fiji and South America, 3658 meters; between Montevideo and Tristan da Cunha, 3950 meters (Egger, 1893).

## Description

Test forms a trochospire; unequally biconvex in cross-section, with an involute umbilical side and an evolute, more convex spiral side with an imperforate, rounded periphery. The nine to ten inflated chambers in the last whorl increase gradually in size and are separated by radial sutures on the umbilical side, and oblique or slightly curved sutures on the spiral side. On the umbilical side, the sutures converge toward an open umbilicus that may be covered by a clear umbilical plug. Chamber walls are calcareous, coarsely perforate on the spiral side and smooth, finely perforate on the umbilical side. The primary aperture is a narrow equatorial slit, bordered by a thin lip, and extends onto the spiral side.

## Synonyms (following van Morkhoven et al., 1986) and Citations

*Truncatulina dutemplei* (d'Orbigny). –Brady, 1884, p. 665, pl. 95, fig. 5. –Egger, 1893, p. 208 (400), pl. 16, figs. 22–23, 30, 54–56.

*Pulvinulina umbonata* (Reuss) var. *multisepta* Koch, 1926, p. 749, 737, pl. 25.

*Cibicides hyalina* Hofker, 1951, p. 359, figs. 244–245.

*Cibicides bradyi* (Trauth). –Barker, 1960, pl. 95, fig. 5.

*Cibicidoides bradyi* (Trauth). –Pflum and Frerichs, 1976, pl. 3, figs. 6–7.

*Cibicidoides haitiensis* (Coryell and Rivero). –Tjalsma and Lohmann, 1983, p. 26, pl. 17, fig. 6. –Miller, 1983, p. 433, pl. 2, fig. 5.

*Cibicidoides bradyi* (Trauth). –van Morkhoven et al., 1986, p. 100, pl. 30. –Miller and Katz, 1987, p. 126, pl. 7, fig. 2. –Kuhnt et al., 2002, p. 141, pl. 8. figs. 1–2. –Nomura, 1995, p. 272, pl. 1, figs. 1–2.

*Gyroidina bradyi* (Trauth). –Jones, 1994, p. 99, pl. 95, fig. 5.

## Remarks

*Cibicidoides bradyi* has a smaller test size, fewer chambers per whorl, and a more rounded periphery than *Cibicidoides robertsonianus*. *Cibicidoides bradyi* may be the ancestral form of *Cibicidoides robertsonianus*. Intergrades between these two species occur in the Miocene, when their stratigraphic ranges overlap (Kuhnt et al., 2002).

## Biogeography

Worldwide. Recorded from the Indian Ocean, the Pacific, the North and South Atlantic, the Mediterranean, the Gulf of Mexico, and Papua New Guinea (van Morkhoven et al., 1986), and the South China Sea (Kuhnt et al., 2002).

## Bathymetry

Bathyal to abyssal.

## Chronostratigraphy

Early Eocene (Ypresian) to Recent.

# References


Barker, R.W., 1960, Taxonomic notes on the species figured by H.B. Brady in his report on the foraminifera dredged by H.M.S. Challenger during the years 1873–1876. Society of Economic Paleontologists and Mineralogists Special Publication, vol. 9, pp. 1–238.

Brady, H.B., 1884, Report on the foraminifera dredged by H.M.S. Challenger during the years 1873–1876. In Brady, H.B., ed., Report on the Scientific Results of the Voyage of the H.M.S. Challenger during the years 1873–1876. Zoology, vol. 9, pp. 1–814.

Egger, J.G., 1893, Foraminiferen aus Meeresgrundproben, gelothet von 1874 bis 1876 von S.M. Sch. Gazelle. Abhandlungen der Bayerischen Akademie der Wissenschaften, München, Math.-Phys. Cl., vol. 18, no. 2, pp. 193–458.

Hofker, J., 1951, The Foraminifera of the "Siboga" Expedition; Part III: Siboga-Expeditie, Monographie IVa. Leiden: E.J. Brill, 513 pp.

Jones, R.W., 1994, The Challenger Foraminifera. Oxford: Oxford University Press, 149 pp.

Koch, R., 1926, Mitteltertiaere Foraminiferen aus Bulongan, Ost-Borneo. Eclogae Geologicae Helvetiae, vol. 19, no. 3, pp. 722–751.

Kuhnt, W., Holbourn, A.E., and Zhao, Q., 2002, The early history of the South China Sea: Evolution of Oligocene–Miocene deep water environments. Revue de Micropaléontologie, vol. 45, pp. 99–159.

Miller, K.G., 1983, Eocene–Oligocene paleoceanography of the deep Bay of Biscay: benthic foraminiferal evidence. Marine Micropaleontology, vol. 7, pp. 403–440.

Miller, K.G. and Katz, M.E., 1987, Oligocene to Miocene benthic foraminiferal and abyssal circulation changes in the north Atlantic. Micropaleontology, vol. 33, no. 2, pp. 97–149.

Nomura R., 1995, Paleogene to Neogene deep-sea paleoceanography in the eastern Indian Ocean: benthic foraminifera from ODP Sites 747, 757 and 758. Micropaleontology, vol. 41, pp. 251–290.

Pflum, C.E. and Frerichs, W.E., 1976, Gulf of Mexico deep-water foraminifers. Cushman Foundation, Special Publication, vol. 14, pp. 1–125.

Tjalsma, R.C. and Lohmann, G.P., 1983, Paleocene–Eocene bathyal and abyssal benthic foraminifera from the Atlantic Ocean. Micropaleontology, Special Publication, no. 4, pp. 1–90.

Trauth, F., 1918, Das Eozanvorkommen bei Radstadt im Pongou und seine Beziehungen zu den gleichalterigen Ablagerungen bei Kirchberg am Wechsel und Wimpassing am Leithagebirge. Denkschriften der Kaiserlichen Akademie der Wissenschaften Wien, Mathematisch-Naturwissenschaftliche Classe, vol. 95, pp. 171–278.

van Morkhoven, F.P.C.M., Berggren, W.A., Edwards, A.S., et al., 1986, Cenozoic cosmopolitan deep-water benthic foraminifera. Bulletin des Centres de Recherches Exploration-Production Elf-Aquitaine, Mémoire, vol. 11, pp. 1–421.

# *Cibicidoides cicatricosus*

## (Schwager), 1866

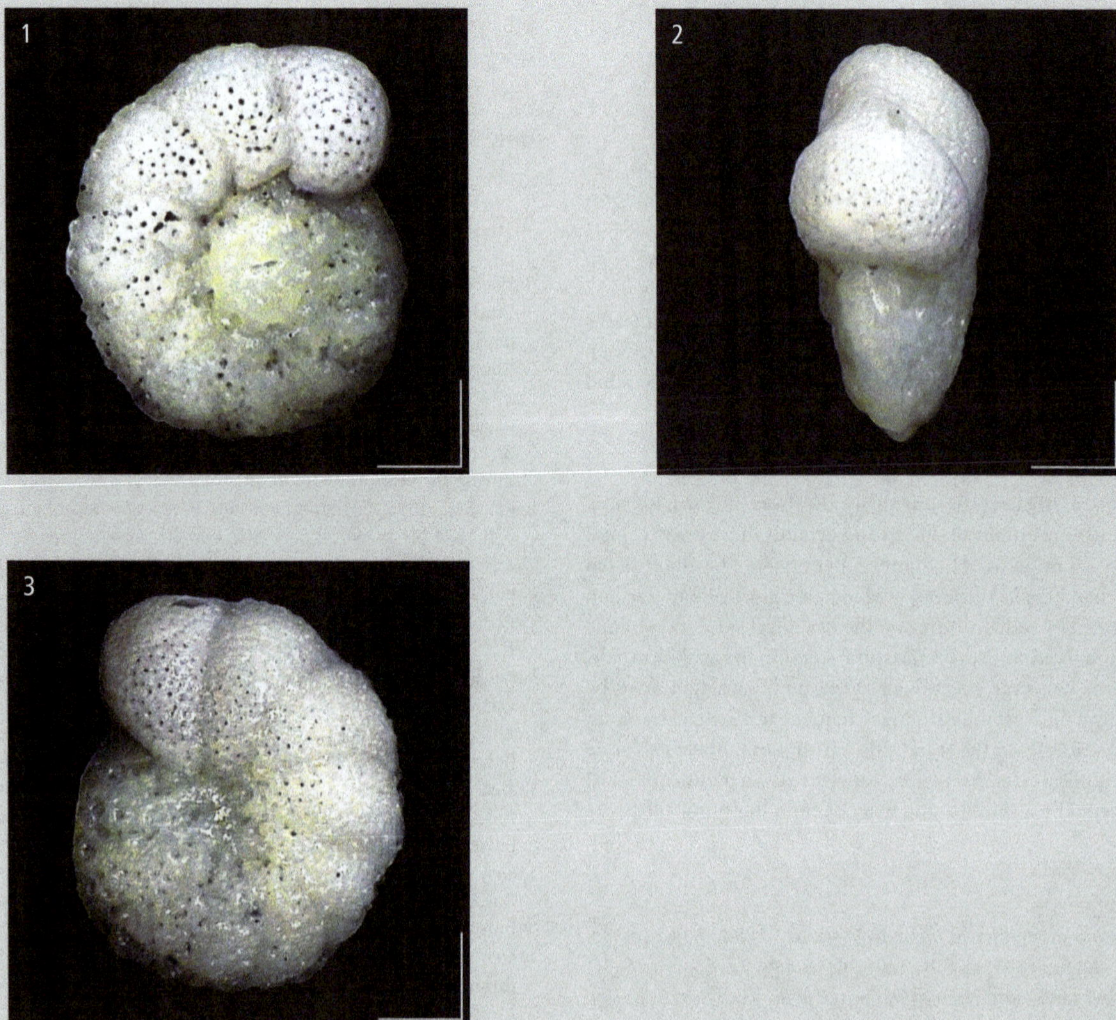

1. *Cibicidoides cicatricosus*. Gulf of Mexico; Eureka Core E68-151A, depth: 4510 feet. Spiral view, NHMUK PM DI 0218, scale bar = 123 μm.
2. *Cibicidoides cicatricosus*. Gulf of Mexico; Eureka Core E68-151A, depth: 4510 feet. Apertural view, NHMUK PM DI 0219, scale bar = 110 μm.
3. *Cibicidoides cicatricosus*. Gulf of Mexico; Eureka Core E68-151A, depth: 4510 feet. Umbilical view, NHMUK PM DI 0220, scale bar = 111 μm.

## Original Designation

*Anomalina cicatricosa* Schwager, 1866, p. 260, pl. 7, figs. 4, 108.

## Type Specimen

Depository not given. Neotype (Mf 12) designated by Srinivasan and Sharma (1980) and deposited in the Micropalaeontology Section, Department of Geology, Banaras Hindu University, Banaras, India.

## Type Level

Jüngere Tertiärschichten. Untere und obere Thone [Younger Tertiary Strata, lower and upper clay].

## Type Locality

Car Nicobar, Andaman Sea.

## Description

Test forms a trochospire; planoconvex in cross-section, with a tightly coiled, convex umbilical side, a partially evolute, slightly concave spiral side, and an imperforate, rounded periphery. The ten to twelve moderately inflated chambers in the last whorl increase gradually in size and are separated by thick, curved, limbate sutures that coalesce in the early part of the test, obscuring the central area. In the final part of the test the sutures are thinner, slightly depressed, and not limbate. Chamber walls are calcareous and coarsely perforate on both sides of the test. The primary aperture is a narrow equatorial slit bordered by a distinct lip, and extends onto the spiral side.

## Synonyms and Citations

*Truncatulina akneriana* (d'Orbigny). –Brady, 1884, pl. 94, fig. 8.

*Cibicides cicatricosus* (Schwager). –Barker, 1960, pl. 94, fig. 8. –Srinivasan and Sharma, 1980, p. 56, pl. 7, figs. 27–29.

*Cibicidoides cicatricosus* (Schwager). –van Morkhoven et al., 1986, p. 53, pl. 16. –Katz and Miller, 1993, pl. 3, fig. 2. –Jones, 1994, p. 98, pl. 94, fig. 8. –Robertson, 1998, p. 200, pl. 78, fig. 1.

## Remarks

Revised stratigraphy at the Eureka boreholes in the Gulf of Mexico (Katz and Miller, 1993) has extended the first occurrence of this species to Biozone M2, in contrast to the Biozone M5 occurrence reported by van Morkhoven et al. (1986).

## Biogeography

Worldwide. See Observed Occurrences in van Morkhoven et al. (1986).

## Bathymetry

Bathyal to abyssal.

## Chronostratigraphy

Early Miocene to Recent.

## References


Barker, R.W., 1960, Taxonomic notes on the species figured by H.B. Brady in his report on the foraminifera dredged by H.M.S. Challenger during the years 1873–1876. Society of Economic Paleontologists and Mineralogists Special Publication, vol. 9, pp. 1–238.

Brady, H.B., 1884, Report on the foraminifera dredged by H.M.S. Challenger during the years 1873–1876. In Brady, H.B., ed., Report on the Scientific Results of the Voyage of the H.M.S. Challenger during the years 1873–1876. Zoology, vol. 9, pp. 1–814.

Jones, R.W., 1994, The Challenger Foraminifera. Oxford: Oxford University Press, 149 pp.

Katz, M.E. and Miller, K.G., 1993, Latest Oligocene to Earliest Pliocene benthic foraminiferal biofacies of the northeastern Gulf of Mexico. Micropaleontology, vol. 39, no. 4, pp. 367–403.

Robertson, B.E., 1998, Systematics and paleoecology of the benthic foraminiferida from the Buff Bay section, Miocene of Jamaica. Micropaleontology, vol. 44, supplement 2, pp. 1–266.

Schwager, C., 1866, Fossile Foraminiferen von Kar Nikobar, Reise der Österreichischen Fregatte Novara um Erde in den Jahren 1857, 1858, 1859 unten den Befehlen des Commodore B. Von Wüllerstorf-Urbair. Geologischer Theil, vol. 2, no. 1, Geologische Beobachtung, no. 2, Paläontologische Mittheilung, pp. 187–268.

Srinivasan, M.S. and Sharma, V., 1980, Schwager's Car Nicobar Foraminifera in the Reports of the Novara Expedition – A Revision. New Delhi: Today & Tomorrow's Printers and Publishers, 83 pp.

van Morkhoven, F.P.C.M., Berggren, W.A., Edwards, A.S., et al., 1986, Cenozoic cosmopolitan deep-water benthic foraminifera. Bulletin des Centres de Recherches Exploration-Production Elf-Aquitaine, Mémoire, vol. 11, pp. 1–421.

# *Cibicidoides compressus*

## (Cushman and Renz), 1941

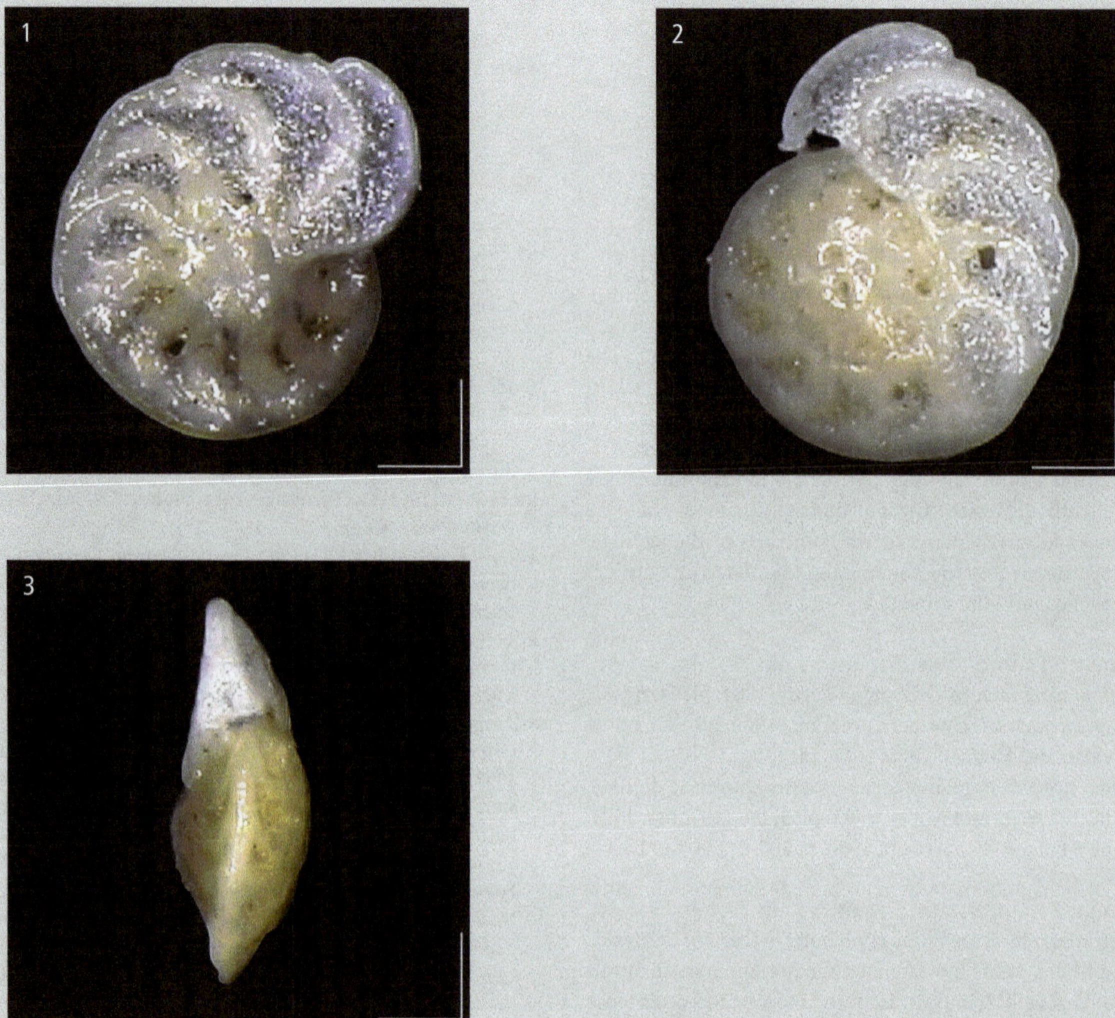


1. *Cibicidoides compressus*. Gulf of Mexico; Eureka Core E68-136, depth: 2677 feet. Umbilical view, NHMUK PM DI 0221, scale bar = 75 μm.
2. *Cibicidoides compressus*. Gulf of Mexico; Eureka Core E68-136, depth: 2677 feet. Spiral view, NHMUK PM DI 0222, scale bar = 82 μm.
3. *Cibicidoides compressus*. Gulf of Mexico; Eureka Core E68-136, depth: 2677 feet. Apertural view, NHMUK PM DI 0223, scale bar = 83 μm.

## Original Designation

*Cibicides floridanus* (Cushman) var. *compressa* Cushman and Renz, 1941, p. 26, pl. 4, fig. 9.

## Type Specimen

Holotype (CC 35882) deposited in the collections of the United States National Museum, Washington, DC.

## Type Level

Upper Agua Salada Formation; Zone of *Marginulina basispinosa* and *Robulus senni*. Given as Lower to Middle Miocene by Cushman and Renz (1941).

## Type Locality

Sample L.P. 167, from Isidro, 33.2 kilometers east (79° 25′) of Pueblo Píritu, District Zamora, State of Falcón, Venezuela.

## Description

Test forms a compressed trochospire; biconvex in cross-section, with an involute umbilical side, an evolute spiral side, and an acute, imperforate periphery. The ten to twelve slightly inflated chambers in the last whorl increase gradually in size and are separated by strongly curved, thick, limbate sutures. These sutures coalesce in the center of the spiral side, concealing earlier whorls. Chamber walls are calcareous and coarsely perforate on both sides of the test. The primary aperture is a narrow equatorial slit bordered by a thin lip, and extends onto the spiral side.

## Synonyms and Citations

*Cibicides floridanus* (Cushman) var. *compressa* Cushman and Renz. –Renz, 1948, p. 127, pl. 10, fig. 9.

*Cibicidoides compressus* (Cushman and Renz). –van Morkhoven et al., 1986, p. 137, pl. 44. –Katz and Miller, 1993, pl. 2, fig. 2.

## Remarks

*Cibicidoides compressus* (Cushman and Renz) differs from *Cibicidoides pachyderma* (Rzehak) by its more compressed test and coarser perforations on both sides of the test.

Revised stratigraphy at the Eureka boreholes in the Gulf of Mexico (Katz and Miller, 1993) has extended the first occurrence of this species to Biozone M1, rather than the Biozone M2 occurrence reported by van Morkhoven et al. (1986).

## Biogeography

Costa Rica, Barbados, Trinidad, and Venezuela (van Morkhoven et al., 1986).

## Bathymetry

Upper to middle bathyal. Predominantly upper bathyal (Katz and Miller, 1993).

## Chronostratigraphy

Miocene (Aquitanian to Tortonian).

## References


Cushman, J.A. and Renz, H.H., 1941, New Oligocene–Miocene foraminifera from Venezuela. Contributions from the Cushman Laboratory for Foraminiferal Research, vol. 17, pp. 1–27.

Katz, M.E. and Miller, K.G., 1993, Latest Oligocene to Earliest Pliocene benthic foraminiferal biofacies of the northeastern Gulf of Mexico. Micropaleontology, vol. 39, no. 4, pp. 367–403.

Renz, H.H., 1948, Stratigraphy and fauna of the Agua Salada group, State of Falcón, Venezuela. Memoirs of the Geological Society of America, vol. 32, pp. 1–219.

van Morkhoven, F.P.C.M., Berggren, W.A., Edwards, A.S., et al., 1986, Cenozoic cosmopolitan deep-water benthic foraminifera. Bulletin des Centres de Recherches Exploration-Production Elf-Aquitaine, Mémoire, vol. 11, pp. 1–421.

# *Cibicidoides crebbsi*

**(Hedberg), 1937**

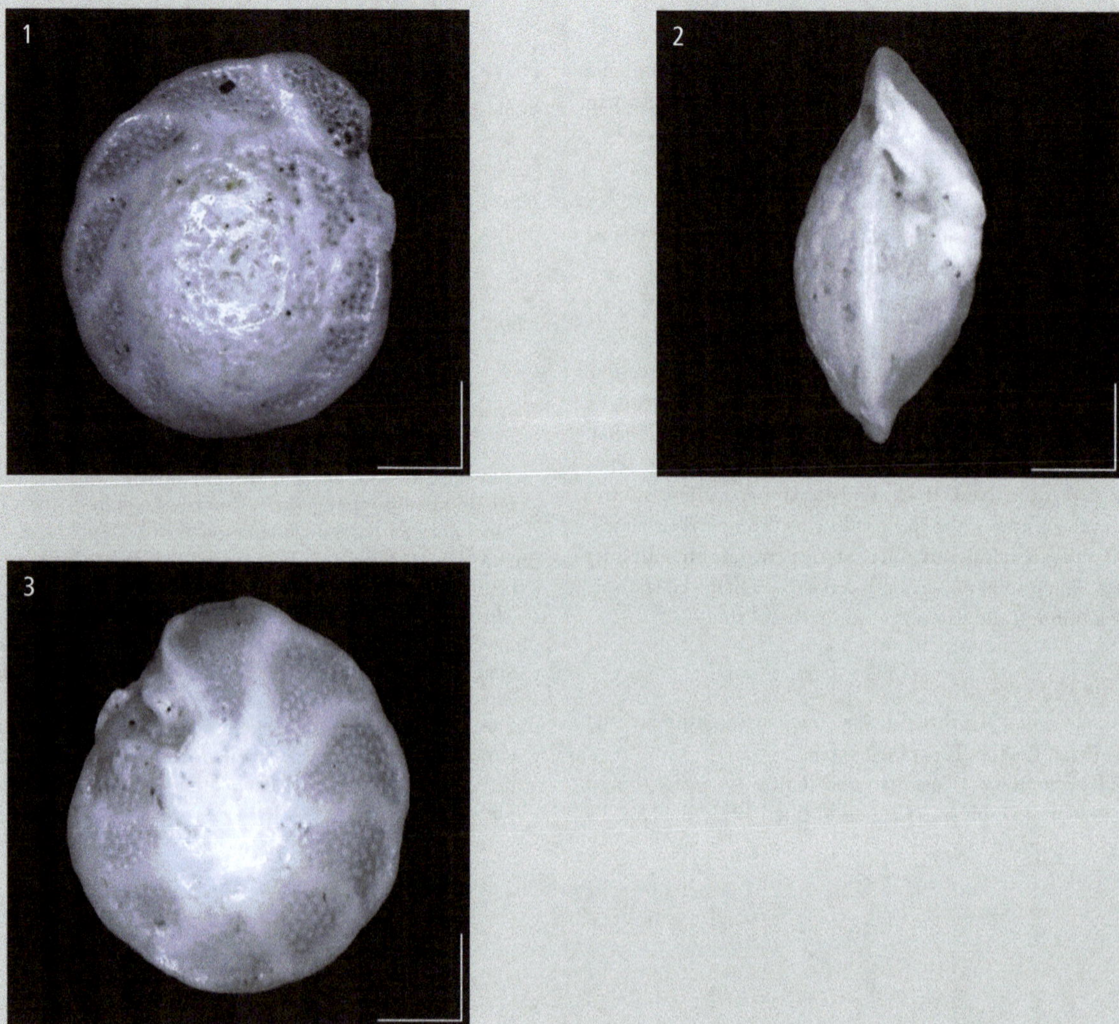

1. *Cibicidoides crebbsi.* Gulf of Mexico; Eureka Core E66-73, depth: 3438 feet. Spiral view, NHMUK PM DI 0224, scale bar = 42 μm.
2. *Cibicidoides crebbsi.* Gulf of Mexico; Eureka Core E66-73, depth of 3438 feet. Apertural view, NHMUK PM DI 0225, scale bar = 39 μm.
3. *Cibicidoides crebbsi.* Gulf of Mexico; Eureka Core E66-73, depth: 3438 feet. Umbilical view, NHMUK PM DI 0226, scale bar = 42 μm.

## Original Designation

*Eponides crebbsi* Hedberg, 1937, p. 679, pl. 92, fig. 1.

## Type Specimen

Holotype (CC 23607) deposited in the collections of the United States National Museum, Washington, DC.

## Type Level

Lower Miocene (van Morkhoven et al., 1986), Carapita Formation, middle zone. Originally given as Middle Tertiary by Hedberg (1937).

## Type Locality

North of the town of Santa Inés on Quebrada Carapita, a small tributary of the Rio Querecal, District of Libertad, State of Anzoategui, northeastern Venezuela.

## Description

Test forms a trochospire; unequally biconvex in cross-section, with an involute, convex umbilical side, an evolute, less convex, spiral side, and an imperforate, subacute periphery. The ten to thirteen moderately inflated chambers in the last whorl increase gradually in size and are separated by broad, strongly sinuous, kinked sutures on the umbilical side, and by broad, straight sutures, tangential to the periphery, on the spiral side. On the umbilical side, the sutures converge toward a glossy boss; on the spiral side, the sutures merge, obscuring early chambers. Chamber walls are calcareous, and coarsely perforate on both sides of the test. The primary aperture is a narrow equatorial slit bordered by a thin lip, and extends onto the spiral side.

## Synonyms and Citations

*Truncatulina floridana* Cushman. –Nuttall, 1928, p. 98, pl. 7, figs. 14, 16.

*Cibicidoides crebbsi* (Hedberg). –van Morkhoven et al., 1986, p. 139 pl. 45. –Katz and Miller, 1993, pl. 2, fig. 5.

## Remarks

*Cibicidoides crebbsi* (Hedberg) differs from *Cibicidoides matanzasensis* (Hadley) by its less compressed, less involute shape, and lack of beaded ornamentation over the central whorl on the spiral side. *Cibicidoides matanzasensis* is also distinguished by its sutures, which are curved and limbate on the spiral side (van Morkhoven et al., 1986).

Re-study of the Eureka boreholes in the Gulf of Mexico (Katz and Miller, 1993) extended the stratigraphic top of this species with rare occurrences into Biozone PL1.

## Biogeography

Recorded from Venezuela, the Dominican Republic, Trinidad, Libya, and the Gulf of Mexico (van Morkhoven et al., 1986).

## Bathymetry

Upper to middle bathyal (Katz and Miller, 1993).

## Chronostratigraphy

Early Oligocene (Rupelian) to Early Pliocene (Zanclean).

# References


Hedberg, H.D., 1937, Foraminifera of the middle Tertiary Carapitana formation of northeastern Venezuela. Journal of Paleontology, vol. 11, pp. 661–697.

Katz, M.E. and Miller, K.G., 1993, Latest Oligocene to Earliest Pliocene benthic foraminiferal biofacies of the northeastern Gulf of Mexico. Micropaleontology, vol. 39, no. 4, pp. 367–403.

Nuttall, W.L.F., 1928, Tertiary foraminifera from the Naparima region of Trinidad (British West Indies). Quarterly Journal of the Geological Society of London, vol. 84, pp. 57–117.

van Morkhoven, F.P.C.M., Berggren, W.A., Edwards, A.S., et al., 1986, Cenozoic cosmopolitan deep-water benthic foraminifera. Bulletin des Centres de Recherches Exploration-Production Elf-Aquitaine, Mémoire, vol. 11, pp. 1–421.

# Cibicidoides dohmi

## (Bermúdez), 1949

1. *Cibicidoides dohmi.* Arroyo Blanco Formation, Higuerito Member, Azua Province, Arroyo Blanco, Dominican Republic. Deposited in the collections of the United States National Museum, Washington, DC; CC 63103. Holotype. Spiral view, NHMUK PM DI 0227, scale bar = 168 μm.

2. *Cibicidoides dohmi.* Arroyo Blanco Formation, Higuerito Member, Azua Province, Arroyo Blanco, Dominican Republic. Deposited in the collections of the United States National Museum, Washington, DC; CC 63103. Holotype. Umbilical view, NHMUK PM DI 0228, scale bar = 169 μm.

### Original Designation

*Cibicides dohmi* Bermúdez, 1949, p. 297, pl. 24, figs. 25–27.

### Type Specimen

Holotype (CC 63103) and paratypes deposited in the collections of the United States National Museum, Washington, DC.

### Type Level

Lower Miocene, Arroyo Blanco Formation, Higuerito Member (a deep-water zone); greenish-gray to blue, calcareous clay with macrofossils.

### Type Locality

Sample H-15702 from the Arroyo Blanco, approximately 1.5 kilometers west of the Arroyo Palmarito and 2.5 kilometers east of the Rio Yaque del Sur, Azua Province, Dominican Republic.

### Description

Test forms a stout trochospire; nearly circular in outline and unequally biconvex in cross-section, with a nearly flat spiral side, strongly convex umbilical side, and a rounded, imperforate periphery. The (approximately) fourteen inflated chambers in the last whorl increase gradually in size, and are separated by thick, raised, slightly oblique or curved, limbate sutures on the umbilical side. On the spiral side, a characteristic clear umbo is surrounded by a narrow depression followed by a ring of coalescing, thickened, limbate sutures that become narrower toward the periphery. Chamber walls are calcareous, thick, and coarsely perforate on both sides of the test. The primary aperture is a narrow equatorial slit bordered by a thin lip, and extends onto the spiral side.

### Synonyms and Citations

*Cibicidoides dohmi* (Bermúdez). –van Morkhoven et al., 1986, p. 195, pl. 66.

### Remarks

According to van Morkhoven et al. (1986), *Cibicidoides dohmi* (Bermúdez) differs from *Cibicidoides hunteri* (Bermúdez) by its less convex umbilical side, the greater number of chambers in the final whorl, and the higher density of perforations. *Cibicidoides hunteri* has thin, depressed, straight sutures on the umbilical side, in contrast to *Cibicidoides dohmi*, which has thick, limbate, raised, curved sutures. On the spiral side, *Cibicidoides hunteri* has strongly limbate, oblique sutures curving away from a limbate spiral suture, whereas *Cibicidoides dohmi* has thick, limbate, raised, slightly oblique sutures, which fuse around a depressed spiral suture.

### Biogeography
Worldwide. Widely distributed in the Pacific, Atlantic, Philippine Sea, and South Ocean (van Morkhoven et al., 1986).

### Bathymetry
Lower bathyal to abyssal (van Morkhoven et al., 1986).

### Chronostratigraphy
Late Oligocene (Chattian) to Middle Miocene (Langhian).

## References


Bermúdez, P.J., 1949, Tertiary smaller foraminifera of the Dominican Republic. Special Publications of the Cushman Laboratory for Foraminiferal Research, no. 25, pp. 1–322.

van Morkhoven, F.P.C.M., Berggren, W.A., Edwards, A.S., et al., 1986, Cenozoic cosmopolitan deep-water benthic foraminifera. Bulletin des Centres de Recherches Exploration-Production Elf-Aquitaine, Mémoire, vol. 11, pp. 1–421.

# *Cibicidoides eocaenus*

## (Gümbel), 1868

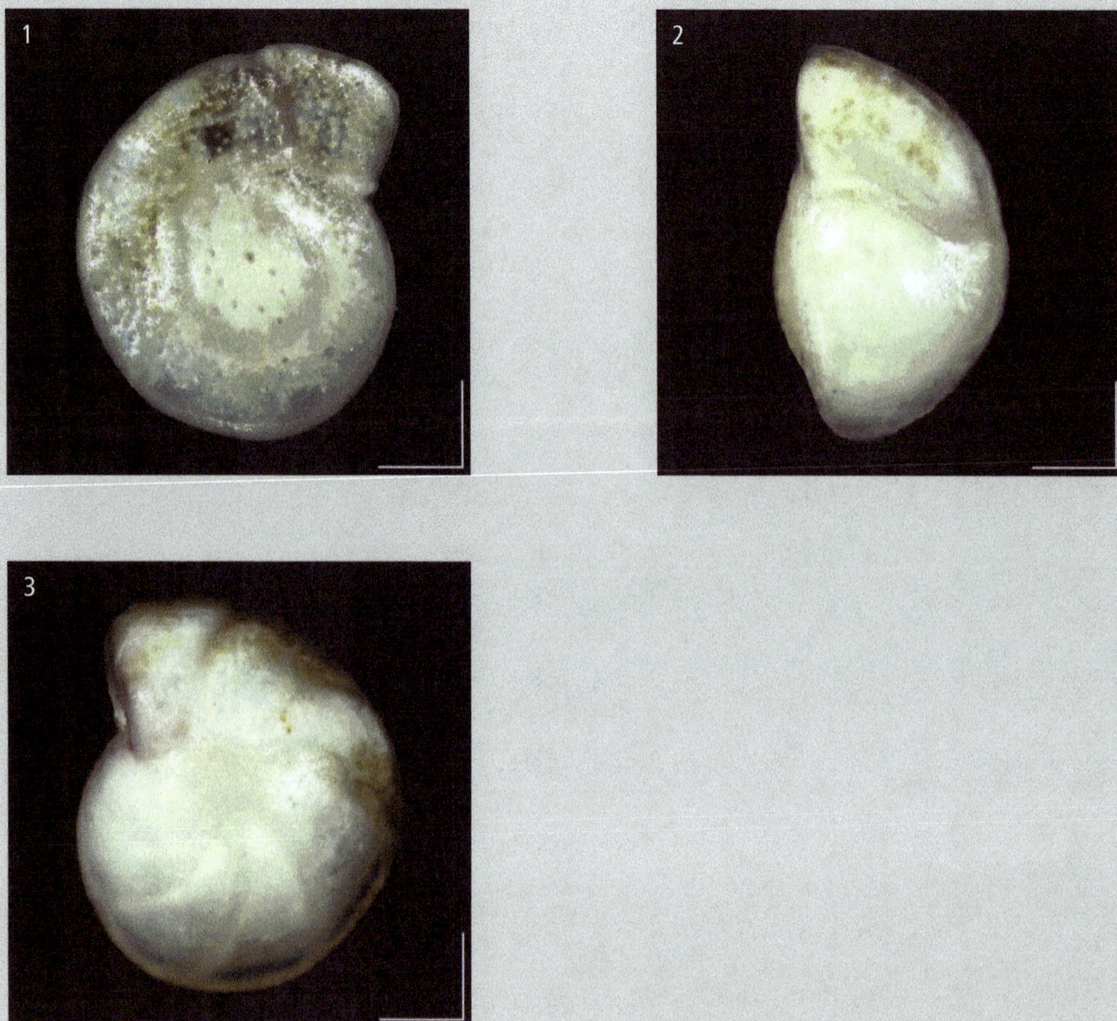


1. *Cibicidoides eocaenus*. North Biscay, North Atlantic; DSDP Leg 48, Hole 402A, Core 1, Section 3, 51–55 cm. Spiral view, NHMUK PM DI 0229, scale bar = 161 μm.

2. *Cibicidoides eocaenus*. North Biscay, North Atlantic; DSDP Leg 48, Hole 402A, Core 1, Section 3, 51–55 cm. Apertural view, NHMUK PM DI 0230, scale bar = 161 μm.

3. *Cibicidoides eocaenus*. North Biscay, North Atlantic; DSDP Leg 48, Hole 402A, Core 1, Section 3, 51–55 cm. Umbilical view, NHMUK PM DI 0231, scale bar = 160 μm.

## Original Designation
*Rotalia eocaena* Gümbel, 1868, p. 650, pl. 2, figs. 87a–b.

## Type Specimen
Depository not given.

## Type Level
Lower Eocene, Nummulitenkalk [Nummulitic limestone].

## Type Locality
Not designated. Localities given: in dem Westdistrikte: in der Umgebung des Grünten und bei Sonthofen; ostwärst am Rande der Alpen: in der Gegend von Tötz; am Ostrande des Innthaals: bei Neubeuern und Sinning; in Traungebiete: von Bergen bis Achthal, besonders am Kressenberge; im Kessel der Saalach: bei Reichenhall und am Fusse des Unterberges bis gegen Hallthurm, bayerische Alpen, Deutschland. [in the western district: vicinity of the Grünten and close to Sonthofen; eastwards along the mragin of the Alps: in the region around Tölz, at the eastern margin of the Inn valley: close to Neubeuern and Sinning, in the Traun area: from Bergen to the Aach valley, especially at the Kressenberg; in the depression of the Saalach; close to Reichenhalls at the foot of the Untersberg to Hallthurm, Bavarian Alps, Germany].

## Description
Test forms a small trochospire; planoconvex to biconvex in cross-section, with a partially evolute, more convex, spiral side, a less convex, evolute, umbilical side, and a subrounded, imperforate periphery. The twelve to fifteen inflated chambers in the last whorl increase gradually in size, and are separated by curved, limbate sutures that are wide and clear on the umbilical side, but coalesce in the central area on the spiral side. The test possesses a distinct spiral suture line and a prominent, clear, umbilical umbo. Chamber walls are calcareous, and coarsely perforate on both sides of the test. The primary aperture is a narrow equatorial slit bordered by a thin lip, and extends onto the spiral side.

## Synonyms and Citations
*?Rotalia propinqua* Reuss, 1856, p. 241, pl. 4, fig. 53.
*Cibicides tuxpamensis* Cole, 1928, p. 219, pl. 1, figs. 2–3, pl. 3, figs. 5–6.
*Cibicides perlucida* Nuttall, 1932, p. 33, pl. 8, figs. 10–12.
*Cibicidoides tuxpamensis* (Cole). –Tjalsma and Lohmann, 1983, p. 28, pl. 18, figs. 3–4; pl. 22, figs. 1–3. –Tjalsma, 1983, p. 742, pl. 9, figs. 4–5. –Miller et al., 1985, pl. 1, figs. 1–3.
*Cibicidoides eocaenus* (Gümbel). –van Morkhoven et al., 1986, p. 256, pl. 86A–C. –Miller and Katz, 1987, p. 126, pl. 8, fig. 3. –Nomura, 1995, p. 272, pl. 1, fig. 4. –Ortiz and Thomas, 2006, p. 115, pl. 5, fig. 2.

## Remarks
Van Morkhoven et al. (1986) considered *Cibicidoides tuxpamensis* (Cole) and *Cibicidoides perlucidus* (Nuttall) to be ecophenotypic variants of *Cibicidoides eocaenus* (Gümbel), differing mainly by their broader periphery (*Cibicidoides tuxpamensis*) and sharper periphery (*Cibicidoides perlucidus*). *Cibicidoides eocaenus* is also probably conspecific with *Rotalia propinqua* Reuss: see detailed discussion in van Morkhoven et al. (1986).

Revised stratigraphy of the Eureka boreholes in the Gulf of Mexico (Katz and Miller, 1993) has extended the last occurrence of this species to Biozone M2, in contrast to the Biozone M1 occurrence reported by van Morkhoven et al. (1986).

## Biogeography
Worldwide. Recorded in Europe, Cuba, Puerto Rico, Barbados, Venezuela, Ecuador, the Dominican Republic, Mexico, Trinidad, Oregon, the North and South Atlantic, and the Pacific (van Morkhoven et al., 1986).

## Bathymetry
Primarily bathyal, also outer neritic and abyssal (van Morkhoven et al., 1986). Common from middle bathyal to lower abyssal paleodepths in the Early Oligocene, rare at depths shallower than 3000 meters by the Late Oligocene.

## Chronostratigraphy
Early Eocene (Ypresian) to Early Miocene (Burdigalian).

# References


Cole, W.S., 1928, A foraminiferal fauna from the Chapapote formation in Mexico. Bulletin of American Paleontology, vol. 14, pp. 200–232.

Gümbel, C.W., 1868, Beitrage zur Foraminiferenfauna der nordalpinen Eocängebilde. Abhandlungen der K. Bayerischen Akademie der Wissenschaften, cl. 2 (1870), vol. 10, no. 2, pp. 581–730.

Miller, K.G. and Katz, M.E., 1987, Oligocene to Miocene benthic foraminiferal and abyssal circulation changes in the north Atlantic. Micropaleontology, vol. 33, no. 2, pp. 97–149.

Miller, K.G., Curry, W.B., and Ostermann, D.R., 1985, Late Paleogene (Eocene to Oligocene) benthic foraminiferal oceanography of the Goban Spur region, Deep Sea Drilling Project Leg 80. In de Graciansky, P.C., Poag, C.W., et al., ed., Initial Reports of the Deep Sea Drilling Project, Washington (U.S. Government Printing Office), vol. 80, pp. 505–538.

Nomura R., 1995, Paleogene to Neogene deep-sea paleoceanography in the eastern Indian Ocean: benthic foraminifera from ODP Sites 747, 757 and 758. Micropaleontology, vol. 41, pp. 251–290.

Nuttall, W.L.F., 1932, Lower Oligocene foraminifera from Mexico. Journal of Paleontology, vol. 6, pp. 3–35.

Ortiz, S. and Thomas, E., 2006, Lower-middle Eocene benthic foraminifera from the Fortuna Section (Betic Cordillera, southeastern Spain). Micropaleontology, vol. 52, pp. 97–150.

Reuss, A.E., 1856, Beitraege zur Charakteristik der Tertiaerschichten des noerdlichen und mittleren Deutschlands. Sitzungsberichte der Kaiserlichen Akademie der Wissenschaften in Wien, Mathematisch-Naturwissenschaftliche Classe, vol. 18, no. 2, pp. 197–273.

Tjalsma, R.C., 1983, Eocene to Miocene benthic foraminifera from DSDP site 516, Rio Grande Rise, South Atlantic. In Barker, P.F., Carlson, R.L., Johnson, D.A. et al., ed., Initial Reports of the Deep Sea Drilling Project, Washington (U.S. Government Printing Office), vol. 72, pp. 731–755.

Tjalsma, R.C. and Lohmann, G.P., 1983, Paleocene–Eocene bathyal and abyssal benthic foraminifera from the Atlantic Ocean. Micropaleontology, Special Publication, no. 4, pp. 1–90.

van Morkhoven, F.P.C.M., Berggren, W.A., Edwards, A.S., et al., 1986, Cenozoic cosmopolitan deep-water benthic foraminifera. Bulletin des Centres de Recherches Exploration-Production Elf-Aquitaine, Mémoire, vol. 11, pp. 1–421.

# *Cibicidoides grimsdalei*

## (Nuttall), 1930

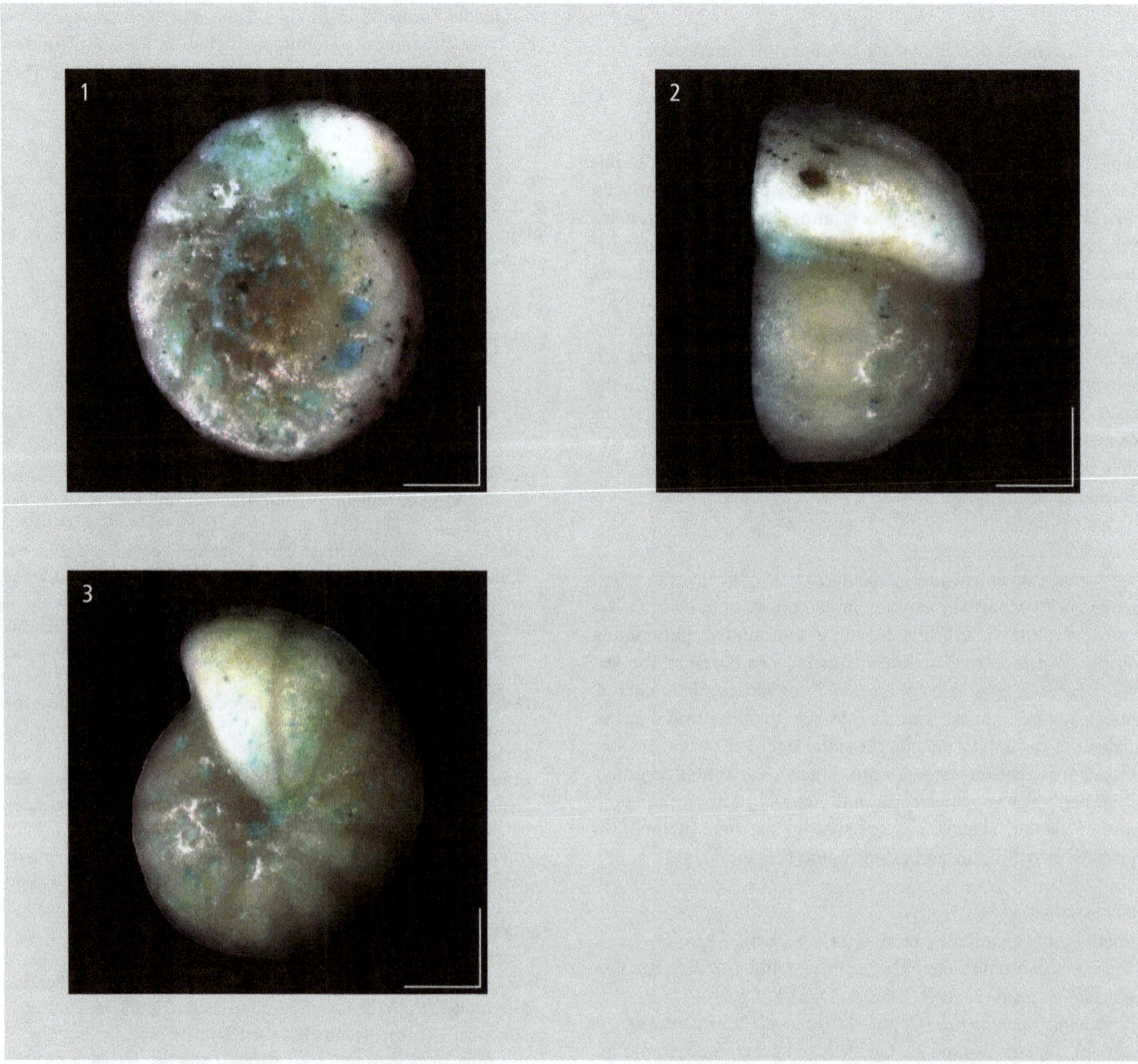


1. *Cibicidoides grimsdalei.* Rio La Puerta, 200 meters upstream from La Antigua crossing, Aragon, Mexico. Deposited in the collections of the United States National Museum, Washington, DC. Cotype. Spiral view, NHMUK PM DI 0232, scale bar = 215 $\mu$m.

2. *Cibicidoides grimsdalei.* Rio La Puerta, 200 meters upstream from La Antigua crossing, Aragon, Mexico. Deposited in the collections of the United States National Museum, Washington, DC. Cotype. Apertural view, NHMUK PM DI 0233, scale bar = 199 $\mu$m.

3. *Cibicidoides grimsdalei.* Rio La Puerta, 200 meters upstream from La Antigua crossing, Aragon, Mexico. Deposited in the collections of the United States National Museum, Washington, DC. Cotype. Umbilical view, NHMUK PM DI 0234, scale bar = 202 $\mu$m.

## Original Designation

*Cibicides grimsdalei* Nuttall, 1930, p. 291, pl. 25, figs. 7–8, 11.

## Type Specimen

Cotypes (CC 59518 and CC 59519) deposited in the collections of the United States National Museum, Washington, DC. Van Morkhoven et al. (1986) designated as lectotype CC 59518 (= Nuttall, 1930, pl. 25, fig. 7).

## Type Level

Lower Eocene, Aragon Formation.

## Type Locality

Not designated. Localities given: Rio la Puerta, 200 meters upstream from la Antigua crossing (figs. 7 and 11) and 1670 meters north 70° west of la Antigua crossing (fig. 8), Tampico region, Mexico.

## Description

Test forms a small, compact trochospire; planoconvex in cross-section, with an involute, strongly convex umbilical side, a flattened, evolute, spiral side, and a rounded, imperforate periphery. The ten to twelve inflated chambers in the last whorl increase gradually in size and are separated by gently curved, limbate sutures on the spiral side, and by flush, limbate sutures on the umbilical side. Chamber walls are calcareous, thick, and coarsely perforate on both sides of the test. On the spiral side, the intercameral sutures coalesce with the spiral suture and, together with the pattern of coarse perforations, give the test its distinctive pitted appearance. The primary aperture is a narrow equatorial arch bordered by a thin lip, and extends onto the spiral side.

## Synonyms and Citations

*Cibicidoides grimsdalei* (Nuttall). –Douglas, 1973, pl. 15, figs. 1–6; pl. 25, figs. 13–14. –Schnitker, 1979, pl. 11, figs. 1–3. –Tjalsma and Lohmann, 1983, p. 26, pl. 18, fig. 2; pl. 22, fig. 6 (not fig. 7). –Miller, 1983, p. 433, pl. 2, fig. 8. –Miller et al., 1985, pl. 1, figs. 7–9. –van Morkhoven et al., 1986, p. 247, pl. 83A–B. –Miller and Katz, 1987, p. 126, pl. 8, fig. 1. –Ortiz and Thomas, 2006, p. 116, pl. 5, fig. 3.

## Remarks

*Cibicidoides grimsdalei* (Nuttall) can be distinguished from *Cibicidoides perlucidus* (Nuttall) by its pitted and coarsely perforate spiral side (Douglas, 1973).

*Cibicidoides grimsdalei* differs from *Cibicidoides mexicanus* (Nuttall) by its pock-marked spiral side, generally higher domed umbilical side, and rounder periphery (M. Katz, pers. comm. 1998).

## Biogeography

Worldwide. Originally described from Mexico, *Cibicidoides grimsdalei* Nuttall has been recorded from Trinidad, Barbados, the Caribbean, the North and South Atlantic, California, the Gulf of Mexico, the Labrador Sea, Europe, the Pacific, the Tasman Sea, and the Coral Sea (van Morkhoven et al., 1986).

## Bathymetry

Lower bathyal to abyssal.

## Chronostratigraphy

Early Eocene (Ypresian) to Early Miocene (Aquitanian).

# References


Douglas, R.G., 1973, Benthonic foraminiferal biostratigraphy in the central North Pacific, Leg 17, Deep Sea Drilling Project. In Winterer, E.L., Ewing, J.I., et al., ed., Initial Reports of the Deep Sea Drilling Project, Washington (U.S. Government Printing Office), vol. 17, pp. 607–671.

Miller, K.G., 1983, Eocene–Oligocene paleoceanography of the deep Bay of Biscay: benthic foraminiferal evidence. Marine Micropaleontology, vol. 7, pp. 403–440.

Miller, K.G. and Katz, M.E., 1987, Oligocene to Miocene benthic foraminiferal and abyssal circulation changes in the north Atlantic. Micropaleontology, vol. 33, no. 2, pp. 97–149.

Miller, K.G., Curry, W.B., and Ostermann, D.R., 1985, Late Paleogene (Eocene to Oligocene) benthic foraminiferal oceanography of the Goban Spur region, Deep Sea Drilling Project Leg 80. In de Graciansky, P.C., Poag, C.W., et al., ed., Initial Reports of the Deep Sea Drilling Project, Washington (U.S. Government Printing Office), vol. 80, pp. 505–538.

Nuttall, W.L.F., 1930, Eocene foraminifera from Mexico. Journal of Paleontology, vol. 4, pp. 271–293.

Ortiz, S. and Thomas, E., 2006, Lower-middle Eocene benthic foraminifera from the Fortuna Section (Betic Cordillera, southeastern Spain). Micropaleontology, vol. 52, pp. 97–150.

Schnitker, D., 1979, Cenozoic deep water foraminifers, Bay of Biscay. In Montadert, L., Roberts, D.G., et al., ed. Initial Reports of the Deep Sea Drilling Project, Washington (U.S. Government Printing Office), vol. 48, pp. 377–413.

Tjalsma, R.C. and Lohmann, G.P., 1983, Paleocene–Eocene bathyal and abyssal benthic foraminifera from the Atlantic Ocean. Micropaleontology, Special Publication, no. 4, pp. 1–90.

van Morkhoven, F.P.C.M., Berggren, W.A., Edwards, A.S., et al., 1986, Cenozoic cosmopolitan deep-water benthic foraminifera. Bulletin des Centres de Recherches Exploration-Production Elf-Aquitaine, Mémoire, vol. 11, pp. 1–421.

# *Cibicidoides grosseperforatus*

## van Morkhoven and Berggren, 1986

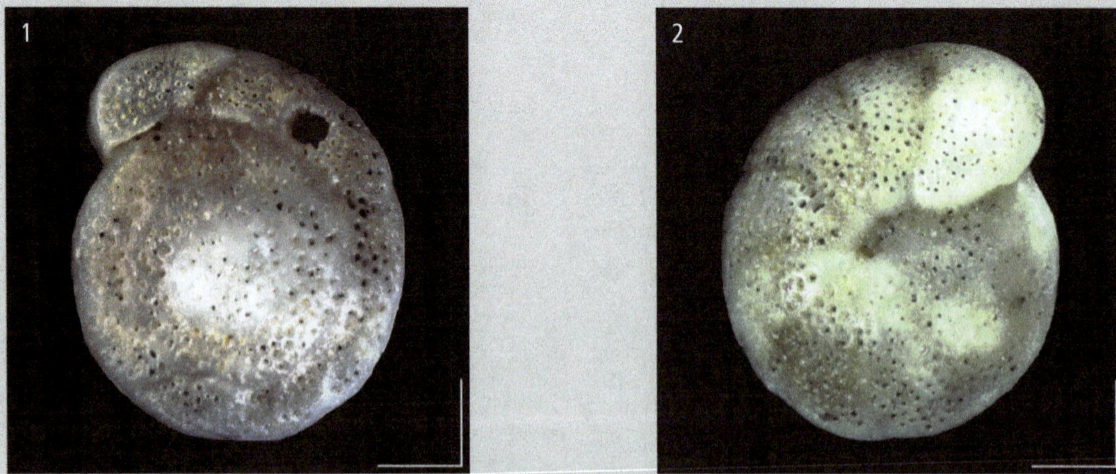


1. *Cibicidoides grosseperforatus.* Gulf of Mexico; Eureka Core E66-40D, depth: 826 meters. Deposited in the collections of the United States National Museum, Washington, DC; USNM No. 406404. Holotype. Spiral view, NHMUK PM DI 0235, scale bar = 286 μm.

2. *Cibicidoides grosseperforatus.* Gulf of Mexico; Eureka Core E66-40D, depth: 826 meters. Deposited in the collections of the United States National Museum, Washington, DC; USNM No. 406404. Holotype. Umbilical view, NHMUK PM DI 0236, scale bar = 256 μm.

### Original Designation

*Cibicidoides grosseperforatus* van Morkhoven and Berggren, 1986. In van Morkhoven et al., 1986, p. 103, textfig. 1a–c, pl. 31, figs. 1–2.

### Type Specimen

Holotype (USNM 406404) and several paratypes deposited in the collections of the United States National Museum, Washington, DC.

### Type Level

Middle Pliocene. Specimens figured in pl. 31, figs. 1–2 are from the Upper Miocene.

### Type Locality

A depth of 826 meters in Shell Oil Company's Eureka Core E66–40, in water depth of 2547 feet, latitude 25° 59′ N, longitude 95° 47′ W, Gulf of Mexico.

### Description

Test forms a large trochospire; unequally biconvex in cross-section, with an involute, more convex umbilical side (slightly depressed at the umbilicus), an evolute, less convex spiral side, and a subacute, imperforate periphery. The ten to eleven moderately inflated chambers in the final whorl increase gradually in size, and are separated by gently curved, flush, thickened sutures, becoming slightly depressed in the final chambers. Chamber walls are calcareous and very coarsely perforate on both sides of the test. The primary aperture is a narrow equatorial arch bordered by a thin lip, and extends onto the spiral side.

### Synonyms

None verified.

### Biogeography

Worldwide. Recorded from the Gulf of Mexico and the North and South Atlantic (van Morkhoven, 1981; van Morkhoven et al., 1986).

### Bathymetry

Bathyal (van Morkhoven et al., 1986).

### Chronostratigraphy

Late Miocene (Tortonian) to Late Pliocene (Gelasian) (van Morkhoven et al., 1986).

## References


van Morkhoven, F.P.C.M., 1981, Cosmopolitan Tertiary bathyal benthic foraminifera. Transactions of the Gulf Coast Association of Geological Societies, Supplement 31, 445 abstracts with accompanying range chart.

van Morkhoven, F.P.C.M., Berggren, W.A., Edwards, A.S., et al., 1986, Cenozoic cosmopolitan deep-water benthic foraminifera. Bulletin des Centres de Recherches Exploration-Production Elf-Aquitaine, Mémoire, vol. 11, pp. 1–421.

# *Cibicidoides guazumalensis*

## (Bermúdez), 1949

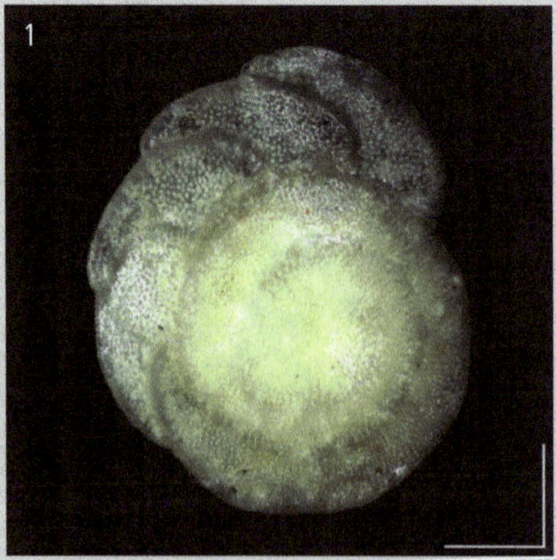

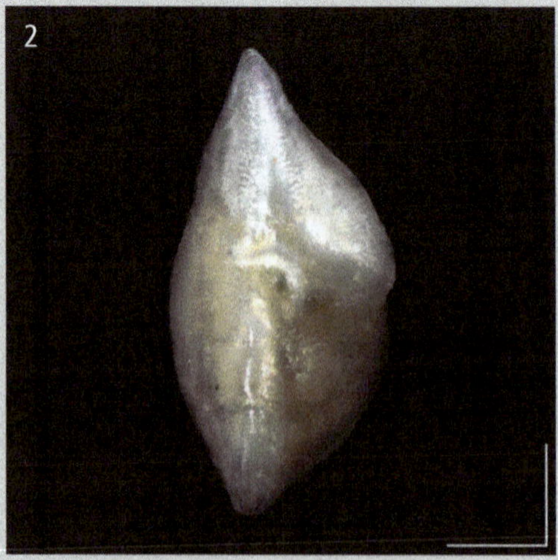

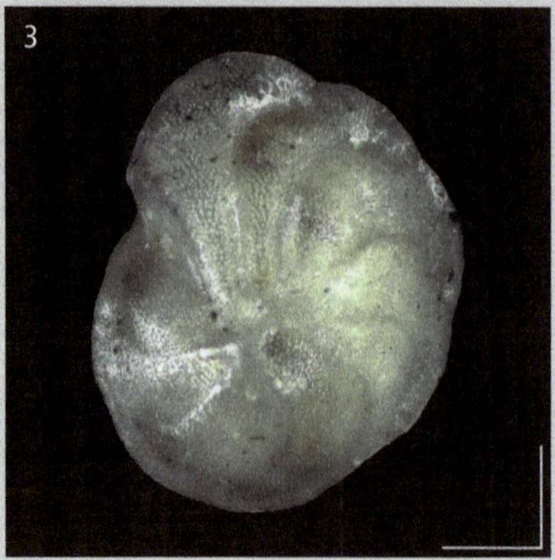

1. *Cibicidoides guazumalensis.* Thomonde Formation, Benefactor Province, Dominican Republic; 1600 meters south and 23° east, where Pio Guasimal crosses highway. Deposited in the collections of the United States National Museum, Washington, DC. Paratype. Spiral view, NHMUK PM DI 0237, scale bar = 82 μm.

2. *Cibicidoides guazumalensis.* Thomonde Formation, Benefactor Province, Dominican Republic; 1600 meters south and 23° east, where Pio Guasimal crosses highway. Deposited in the collections of the United States National Museum, Washington, DC. Paratype. Apertural view, NHMUK PM DI 0238, scale bar = 82 μm.

3. *Cibicidoides guazumalensis.* Thomonde Formation, Benefactor Province, Dominican Republic; 1600 meters south and 23° east, where Pio Guasimal crosses highway. Deposited in the collections of the United States National Museum, Washington, DC. Paratype. Umbilical view, NHMUK PM DI 0239, scale bar = 82 μm.

## Original Designation

*Cibicides subtenuissimus* (Nuttall) var. *guazumalensis* Bermúdez, 1949, p. 306, pl. 25, figs. 43–45.

## Type Specimen

Holotype (CC 63077) deposited in the collections of the United States National Museum, Washington, DC.

## Type Level

Middle Miocene (van Morkhoven et al., 1986), gray, calcareous clay shale. Originally given as Upper Oligocene by Bermúdez (1949).

## Type Locality

Sample H-20465, from along the Rio Guazumal, 1600 meters south and 23° east of the point where it crosses the highway, Comendador area, Benefactor Province, Dominican Republic. Found only in the Comendador area.

## Description

Test forms an unequally biconvex trochospire with a partially evolute, more convex umbilical side, and an acute, imperforate periphery. The ultimate chamber appears sharply truncated in edge view. The eight to ten chambers in the last whorl increase gradually in size, and are separated by sinuous sutures extending along the periphery. Chamber walls are calcareous, finely and densely perforate on both sides of the test. The primary aperture is a narrow equatorial slit bordered by a thin lip, and extends onto the spiral side.

## Synonyms and Citations

*Cibicidoides guazumalensis* (Bermúdez). –van Morkhoven et al., 1986, p. 143, pl. 46.

## Remarks

*Cibicidoides guazumalensis* (Bermúdez) differs from *Planulina subtenuissima* (Nuttall) by the greater degree of convexity of the umbilical side in relation to the compressed area at the periphery. It differs from *Cibicidoides subspiratus* (Nuttall) by its very sinuous sutures (van Morkhoven et al., 1986).

## Biogeography

Originally described from the Dominican Republic; also recorded in the Aquitaine Basin of France, and Cuba (van Morkhoven et al., 1986).

## Bathymetry

Lower neritic to upper bathyal (van Morkhoven et al., 1986).

## Chronostratigraphy

Early Miocene (Burdigalian) to Late Miocene (Tortonian).

## References


Bermúdez, P.J., 1949, Tertiary smaller foraminifera of the Dominican Republic. Special Publications of the Cushman Laboratory for Foraminiferal Research, no. 25, pp. 1–322.

van Morkhoven, F.P.C.M., Berggren, W.A., Edwards, A.S., et al., 1986, Cenozoic cosmopolitan deep-water benthic foraminifera. Bulletin des Centres de Recherches Exploration-Production Elf-Aquitaine, Mémoire, vol. 11, pp. 1–421.

# *Cibicidoides havanensis*

## (Cushman and Bermúdez), 1937

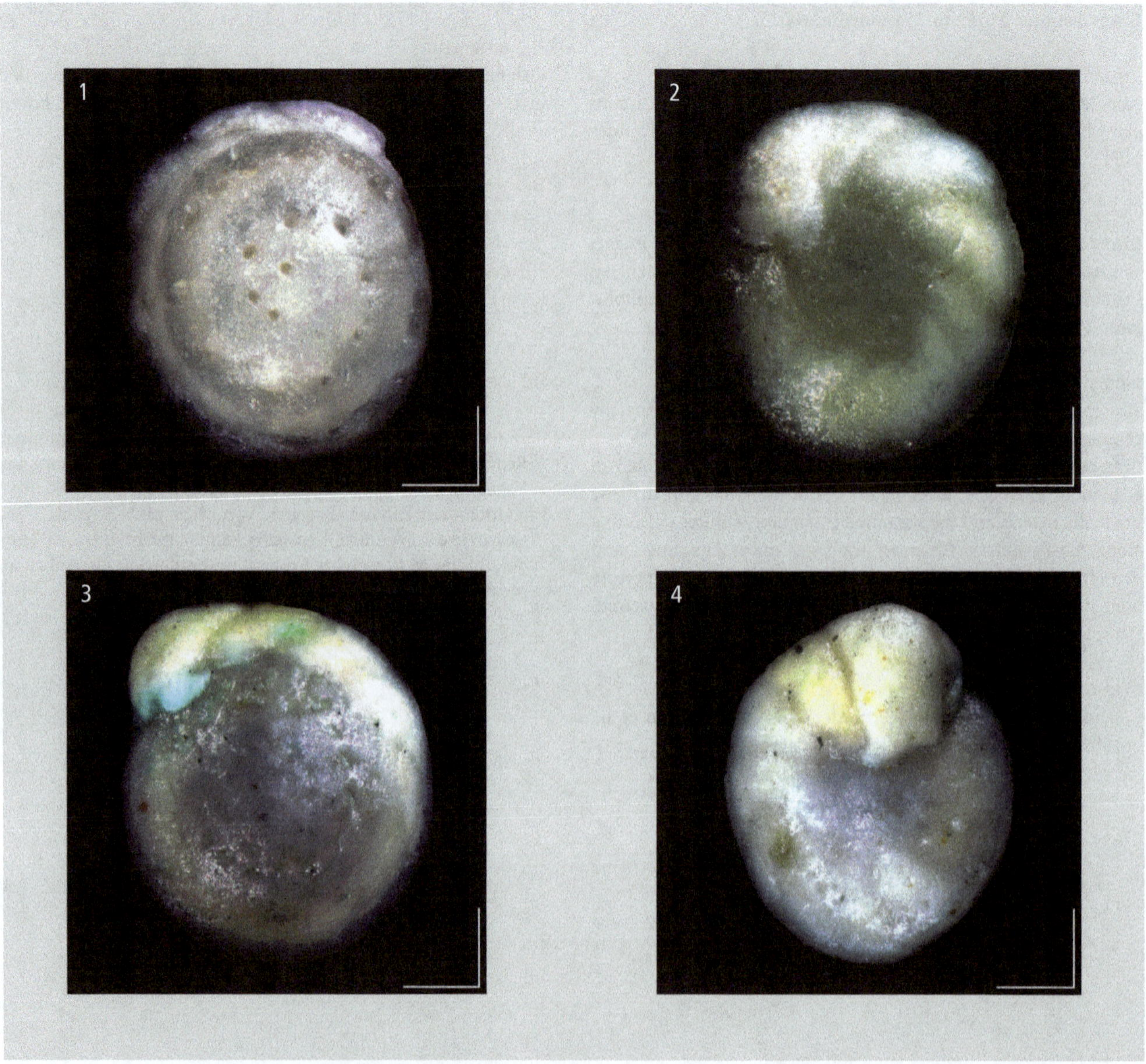

1. *Cibicidoides havanensis.* Teja "Cuba" Arroyo Naranjo (lower beds), Havana Province, Cuba. Deposited in the collections of the United States National Museum, Washington, DC; CC 23427. Holotype. Spiral view, NHMUK PM DI 0242, scale bar = 145 μm.

2. *Cibicidoides havanensis.* Teja "Cuba" Arroyo Naranjo (lower beds), Havana Province, Cuba. Deposited in the collections of the United States National Museum, Washington, DC; CC 23427. Holotype. Umbilical view, NHMUK PM DI 0241, scale bar = 145 μm.

3. *Cibicidoides havanensis.* North Biscay, North Atlantic; DSDP Leg 48, Hole 400A, Core 47, Section 6, 75–79 cm. Spiral view, NHMUK PM DI 0240, scale bar = 132 μm.

4. *Cibicidoides havanensis.* North Biscay, North Atlantic; DSDP Leg 48, Hole 400A, Core 47, Section 6, 75–79 cm. Umbilical view, NHMUK PM DI 0243, scale bar = 142 μm.

## Original Designation

*Cibicides havanensis* Cushman and Bermúdez, 1937, p. 28, pl. 3, figs. 1–3.

## Type Specimen

Holotype (CC 23427) deposited in the collections of the United States National Museum, Washington, DC.

## Type Level

Eocene.

## Type Locality

Bermúdez Station 235C, Tejar "Cuba," Arroyo Naranjo (lower beds), Havana Province, Cuba.

## Description

Test forms a trochospire; unequally biconvex in cross-section, with a partially evolute, convex umbilical side, a more strongly convex, evolute spiral side, and a subacute, imperforate periphery. The (approximately) twelve moderately inflated chambers in the last whorl increase gradually in size, and are separated by gently curved, flush sutures on the umbilical side, and by strongly oblique, limbate sutures on the spiral side. The sutures on the umbilical side converge toward a thick plug of secondary shell growth that obscures the umbilicus. Chamber walls are calcareous, thick, and coarsely perforate, particularly along the spiral suture. The primary aperture is a narrow equatorial slit bordered by a thin lip, and extends onto the spiral side.

## Synonyms and Citations

*Cibicidoides havanensis* (Cushman and Bermúdez). –Tjalsma and Lohmann, 1983, p. 27, pl. 22, fig. 4. –Miller, 1983, p. 433, pl. 2, figs. 9–10. –Miller et al., 1985, pl. 2, figs. 7–9. –van Morkhoven et al., 1986, p.189, pl. 64A–B. –Miller and Katz, 1987, p. 128, pl. 7, fig. 5. –Nomura, 1995, p. 272, pl. 1, figs. 1–2. –Robertson, 1998, p. 204, pl. 80, fig. 1. –Kuhnt et al., 2002, p. 141, pl. 8, figs. 3–11.

## Remarks

*Cibicidoides havanensis* (Cushman and Bermúdez) grades into *Cibicidoides mundulus* (Brady, Parker, and Jones) and *Cibicidoides praemundulus* Berggren and Miller (M. Katz, pers. comm. 1998).

## Biogeography

Worldwide. Recorded from Pacific and Atlantic DSDP and ODP sites and from Cuba, Guam, Barbados, and Morocco (van Morkhoven et al., 1986). Also reported from the South China Sea (Kuhnt et al., 2002).

## Bathymetry

Primarily abyssal. Also found up to lower middle bathyal depths (van Morkhoven et al., 1986).

## Chronostratigraphy

Early Eocene (Ypresian) to Middle Miocene (Serravallian).

# References


Cushman, J.A. and Bermúdez, P.J., 1937, Further new species of foraminifera from the Eocene of Cuba. Contributions from the Cushman Laboratory for Foraminiferal Research, vol. 13, pp. 1–29.

Kuhnt, W., Holbourn, A.E., and Zhao, Q., 2002, The early history of the South China Sea: Evolution of Oligocene–Miocene deep water environments. Revue de Micropaléontologie, vol. 45, pp. 99–159.

Miller, K.G., 1983, Eocene–Oligocene paleoceanography of the deep Bay of Biscay: benthic foraminiferal evidence. Marine Micropaleontology, vol. 7, pp. 403–440.

Miller, K.G. and Katz, M.E., 1987, Oligocene to Miocene benthic foraminiferal and abyssal circulation changes in the north Atlantic. Micropaleontology, vol. 33, no. 2, pp. 97–149.

Miller, K.G., Curry, W.B., and Ostermann, D.R., 1985, Late Paleogene (Eocene to Oligocene) benthic foraminiferal oceanography of the Goban Spur region, Deep Sea Drilling Project Leg 80. In de Graciansky, P.C., Poag, C.W., et al., ed., Initial Reports of the Deep Sea Drilling Project, Washington (U.S. Government Printing Office), vol. 80, pp. 505–538.

Nomura R., 1995, Paleogene to Neogene deep-sea paleoceanography in the eastern Indian Ocean: benthic foraminifera from ODP Sites 747, 757 and 758. Micropaleontology, vol. 41, pp. 251–290.

Robertson, B.E., 1998, Systematics and paleoecology of the benthic foraminiferida from the Buff Bay section, Miocene of Jamaica. Micropaleontology, vol. 44, supplement 2, pp. 1–266.

Tjalsma, R.C. and Lohmann, G.P., 1983, Paleocene–Eocene bathyal and abyssal benthic foraminifera from the Atlantic Ocean. Micropaleontology, Special Publication, no. 4, pp. 1–90.

van Morkhoven, F.P.C.M., Berggren, W.A., Edwards, A.S., et al., 1986, Cenozoic cosmopolitan deep-water benthic foraminifera. Bulletin des Centres de Recherches Exploration-Production Elf-Aquitaine, Mémoire, vol. 11, pp. 1–421.

# *Cibicidoides hyphalus*

**(Fisher), 1969**

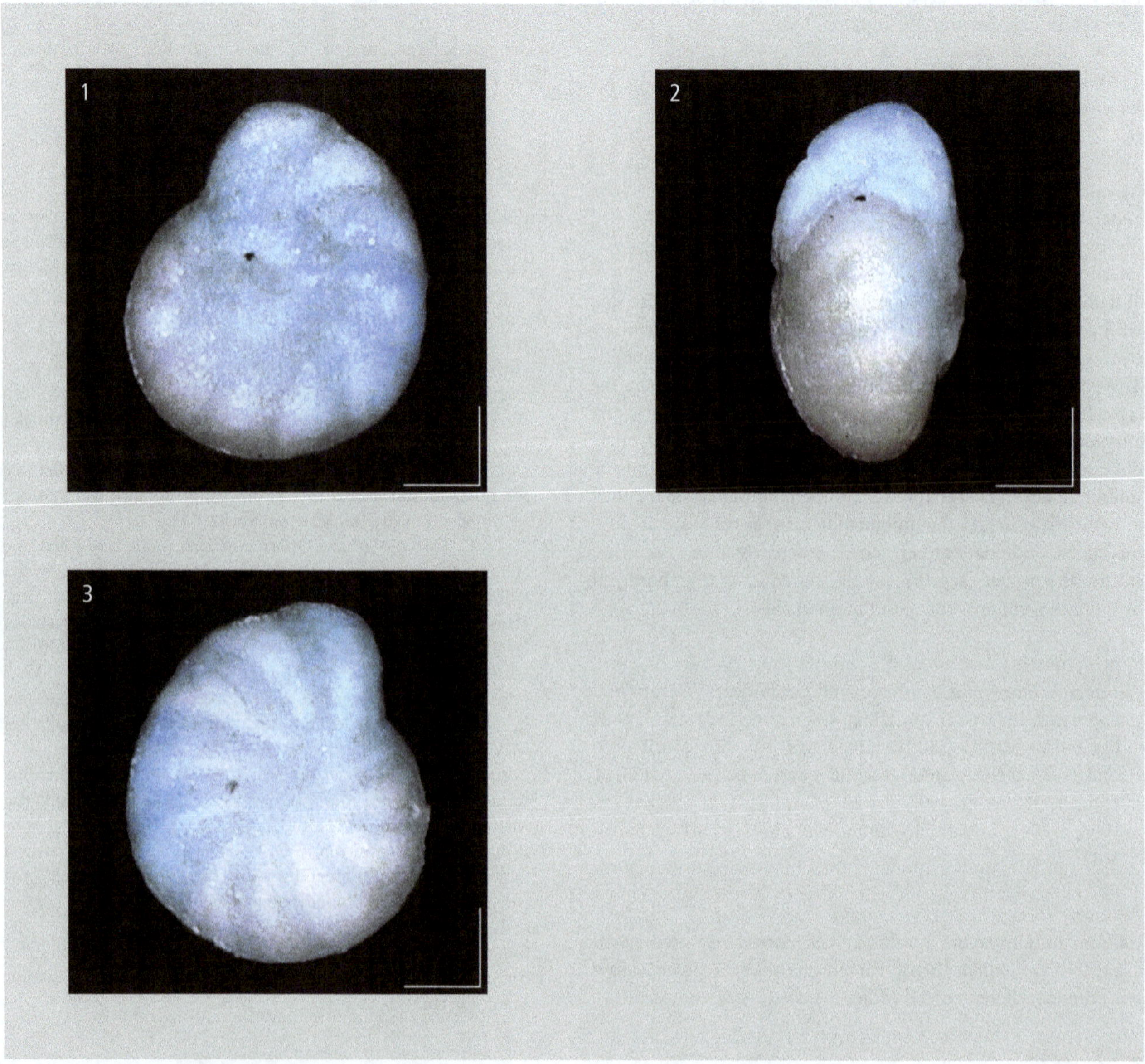


1. *Cibicidoides hyphalus*. Northeast Georgia Rise, South Atlantic; ODP Leg 114, Hole 700B, Core 30, Section 1, 80–84 cm. Spiral view, NHMUK PM DI 0244, scale bar = 46 μm.

2. *Cibicidoides hyphalus*. Northeast Georgia Rise, South Atlantic; ODP Leg 114, Hole 700B, Core 30, Section 1, 80–84 cm. Apertural view, NHMUK PM DI 0245, scale bar = 45 μm.

3. *Cibicidoides hyphalus*. Northeast Georgia Rise, South Atlantic; ODP Leg 114, Hole 700B, Core 30, Section 1, 80–84 cm. Umbilical view, NHMUK PM DI 0246, scale bar = 62 μm.

## Original Designation

*Anomalinoides hyphalus* Fisher, 1969, p. 197, textfig. 3a–c.

## Type Specimen

Holotype (No. 1210) deposited in the collections of the Sedgwick Museum, Cambridge, UK.

## Type Level

Upper Cretaceous (Maastrichtian).

## Type Locality

Galicia Bank seamount, latitude 42° 36′ N, longitude 11° 35′ W, off the west coast of Spain, chalk fragments dredged from 650–700 meters.

## Description

Test forms a trochospire; unequally biconvex, with an involute to partially evolute, more convex umbilical side characterized by a clear umbo, an evolute, slightly convex spiral side with a plug of secondary shell growth, and an imperforate, subacute to rounded periphery. The ten to twelve moderately inflated chambers in the last whorl increase gradually in size and are separated by strongly curved, limbate sutures, flush on the umbilical side and raised on the spiral side. The final whorl is slightly elevated and partly overlaps preceding whorls on the spiral side. Chamber walls are calcareous, coarsely perforate on the spiral side, and finely perforate on the umbilical side. The primary aperture is a narrow equatorial slit bordered by a thin lip, and extends onto the spiral side.

## Synonyms (following van Morkhoven et al., 1986) and Citations

*Planulina dayi* White var. –White, 1928, p. 302, pl. 41, fig. 5.

*Gavelinella velascoensis* (Cushman). –Hillebrandt, 1962, p. 102, pl. 8, fig. 4.

*?Cibicides commatus* Morozova. –Szczechura and Pozaryska, 1974, p. 91, pl. 21, figs. 4–5.

*Cibicidoides hyphalus* (Fisher). –van Morkhoven et al., 1986, p. 359, pl. 116.

## Remarks

*Cibicidoides velascoensis* (Cushman) and *Cibicidoides hyphalus* (Fisher) are morphologically close taxa that share the same stratigraphic range. However, *Cibicidoides velascoensis* differs in its much larger diameter, more strongly convex umbilical side, and the smaller number of chambers in its last whorl (van Morkhoven et al., 1986).

## Biogeography

Worldwide. Recorded from the Gulf of Mexico, the Caribbean, the South Atlantic, the Indian Ocean, Europe, and Russia (van Morkhoven et al., 1986).

## Bathymetry

Lower neritic to abyssal (van Morkhoven et al., 1986).

## Chronostratigraphy

Late Cretaceous (Maastrichtian) to Early Eocene (Ypresian).

# References


Fisher, M.J., 1969, Benthonic foraminifera from the Maastrichtian chalk of Galicia Bank, west of Spain. Palaeontology, vol. 12, no. 2, pp. 189–200.

Hillebrandt, A. von, 1962, Das Paleozaen und seine Foraminiferenfauna im Becken von Reichenhall und Salzburg. Abhandlungen, Bayerische Akademie der Wissenschaften, Mathematisch-Naturwissenschaftliche Klasse, N.F., no. 108, pp. 9–182.

Szczechura, J. and Pozaryska, K., 1974, Foraminiferida from the Paleocene of the Polish Carpathians (Babica Clays). Paleontologica Polonica, vol. 31, pp. 1–142.

van Morkhoven, F.P.C.M., Berggren, W.A., Edwards, A.S., et al., 1986, Cenozoic cosmopolitan deep-water benthic foraminifera. Bulletin des Centres de Recherches Exploration-Production Elf-Aquitaine, Mémoire, vol. 11, pp. 1–421.

White, M.P., 1928, Some index foraminifera of the Tampico Embayment area of Mexico (Part II). Journal of Paleontology, vol. 2, pp. 280–317.

# *Cibicidoides incrassatus*

## (Fichtel and Moll), 1798

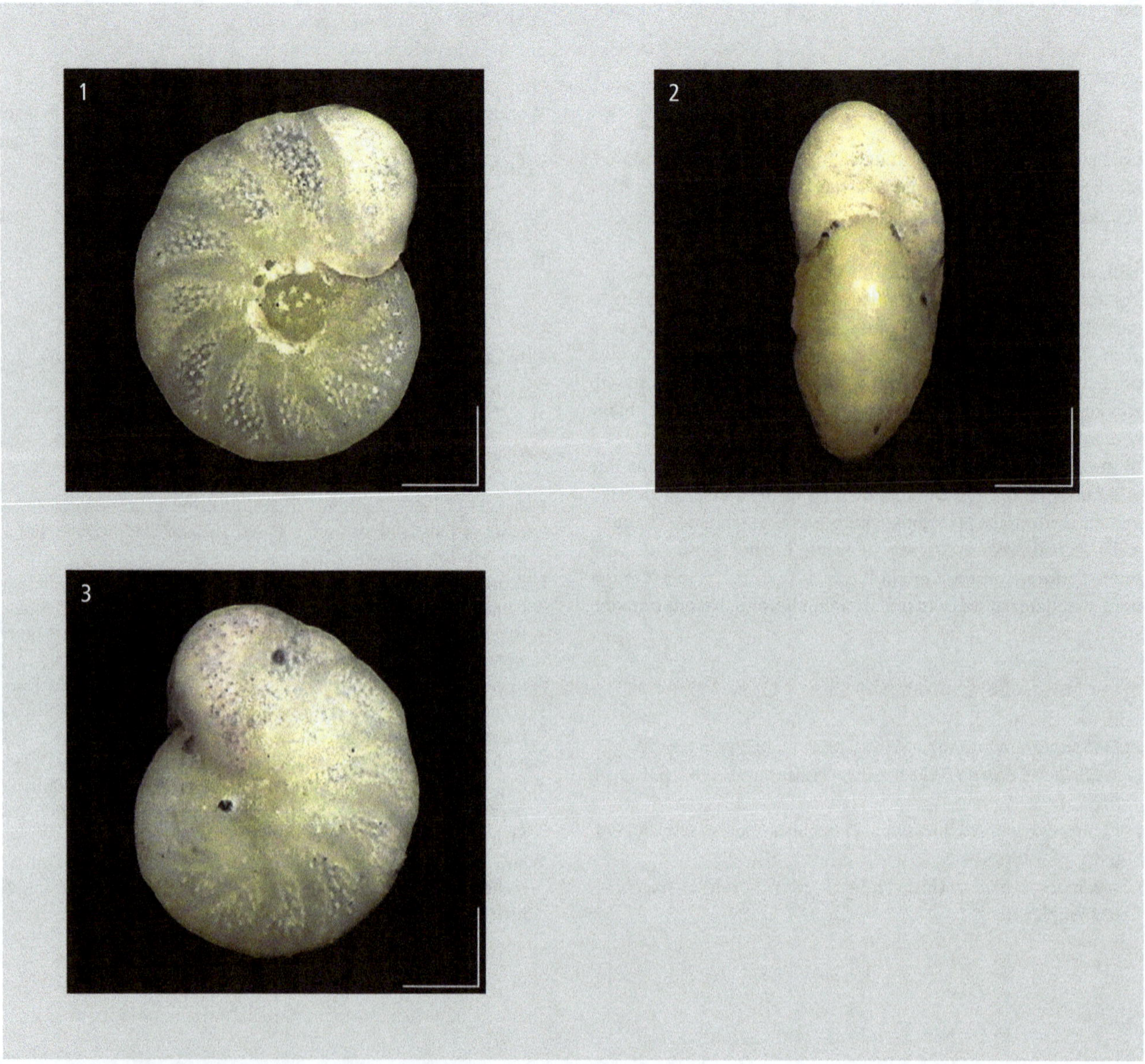


1. *Cibicidoides incrassatus*. Santaren Channel, west of the Great Bahaman Bank; ODP Leg 166, Hole 1006A, Core 55, Section 1, 30–35 cm. Umbilical view, NHMUK PM DI 0247, scale bar = 102 μm.

2. *Cibicidoides incrassatus*. Santaren Channel, west of the Great Bahaman Bank; ODP Leg 166, Hole 1006A, Core 55, Section 1, 30–35 cm. Apertural view, NHMUK PM DI 0248, scale bar = 102 μm.

3. *Cibicidoides incrassatus*. Santaren Channel, west of the Great Bahaman Bank; ODP Leg 166, Hole 1006A, Core 55, Section 1, 30–35 cm. Spiral view, NHMUK PM DI 0249, scale bar = 105 μm.

### Original Designation

*Nautilus incrassatus* Fichtel and Moll, 1798, p. 38, pl. 4, figs. a–c.

### Type Specimen

Depository of holotype not given. Lectotype (MI-473) and paratype (MI-474) designated by Rögl and Hansen (1984) and deposited in the collections of the Museum of Natural History, Vienna, Austria.

### Type Level

Recent, surface sediments.

### Type Locality

Portoferraio, Island of Elba, Mediterranean Sea.

### Description

Test forms a trochospire; unequally biconvex to planoconvex in cross-section, with an involute, more convex umbilical side, a flatter, evolute spiral side, and a subacute to subrounded, imperforate periphery. The (approximately) thirteen moderately inflated chambers in the last whorl increase gradually in size, and are separated by broad, slightly curved, limbate sutures. Sutures converge toward a central umbo on the umbilical side. On the spiral side, a large, flat, glassy plug is prominent. Chamber walls are calcareous, thick, and coarsely perforate on both sides of the test. The primary aperture is a narrow equatorial slit bordered by a thin lip, and extends onto the spiral side.

### Synonyms and Citations

*Truncatulina nucleata* Seguenza, 1880, p. 64, pl. 7, fig. 8.
*Truncatulina trinitatensis* Nuttall, 1928, p. 97, pl. 7, figs. 3–6.
*Cibicides repettoensis* LeRoy, 1941, p. 622, pl. 84, figs. 1–4.
*Cibicides corpulentus* Phleger and Parker, 1952, p. 31, pl. 17, figs. 1–4.
*Anomalinoides incrassatus* (Fichtel and Moll). –Rögl and Hansen, 1984, p. 36, pl. 8, figs. 4–6.
*Cibicidoides incrassatus* (Fichtel and Moll). –van Morkhoven et al., 1986, p. 83, pl. 25A–C.

### Remarks

See detailed Remarks in van Morkhoven et al. (1986).

### Biogeography

Worldwide. Reported under various names (see synonymy) from the Caribbean Region, Mexico, Venezuela, the North and South Atlantic, the Pacific, Morocco, and Europe (van Morkhoven et al., 1986).

### Bathymetry

Lower neritic to bathyal. More common in outer neritic to upper bathyal deposits (van Morkhoven et al., 1986).

### Chronostratigraphy

Early Oligocene (Rupelian) to Recent. Van Morkhoven et al. (1986) recorded a disjoint stratigraphic range for this species: Early Oligocene (Biozone P19) to Pleistocene (Biozone PT1) in the Caribbean region and western Atlantic. Doubtful in Early Oligocene Biozone P18. Middle Miocene to Late Pliocene and possibly basal Pleistocene in Europe and the Mediterranean Region.

## References


Fichtel, L. von and Moll, J.P.C. von, 1798, Testacea microscopica aliaque minuta ex generibus Argonauta et Nautilus ad naturam delineata et descripta. Microscopische und andere kleine Schalthiere aus den Geschlechtern Argonaute und Schiffer, nach der Natur gezeichnet und beschrieben. Vienna: Camesina, 124 pp.

LeRoy, L.W., 1941, A new species of Cibicides from the lower Pliocene (Repetto Formation) of Southern California. Journal of Paleontology, vol. 15, no. 6, pp. 622–623.

Nuttall, W.L.F., 1928, Tertiary foraminifera from the Naparima region of Trinidad (British West Indies). Quarterly Journal of the Geological Society of London, vol. 84, pp. 57–117.

Phleger, F.B. and Parker, F.L., 1952, New names for northwestern Gulf of Mexico foraminifera. Contributions from the Cushman Laboratory for Foraminiferal Research, vol. 3, no. 1, p. 14.

Rögl, F. and Hansen, J.H., 1984, Foraminifera described by Fichtel and Moll in 1798; A revision of Testacea Microscopica. Appendix Testacea Microscopica alique minuta ex Generibus Argonauta et Nautilus. Reprint of original plates. Neue Denkschriften des Naturhistorischen Museum in Wien, vol. 3, pp. 1–143.

Seguenza, G., 1880, Le formazioni terziarie nella provincia di Reggio (Calabria). Reale Accademia Nazionale dei Lincei, Classe di Scienze Fisiche, Matematiche e Naturali, Memorie, ser. 3, vol. 6, pp. 3–446.

van Morkhoven, F.P.C.M., Berggren, W.A., Edwards, A.S., et al., 1986, Cenozoic cosmopolitan deep-water benthic foraminifera. Bulletin des Centres de Recherches Exploration-Production Elf-Aquitaine, Mémoire, vol. 11, pp. 1–421.

# *Cibicidoides lamontdohertyi*

## Miller and Katz, 1987

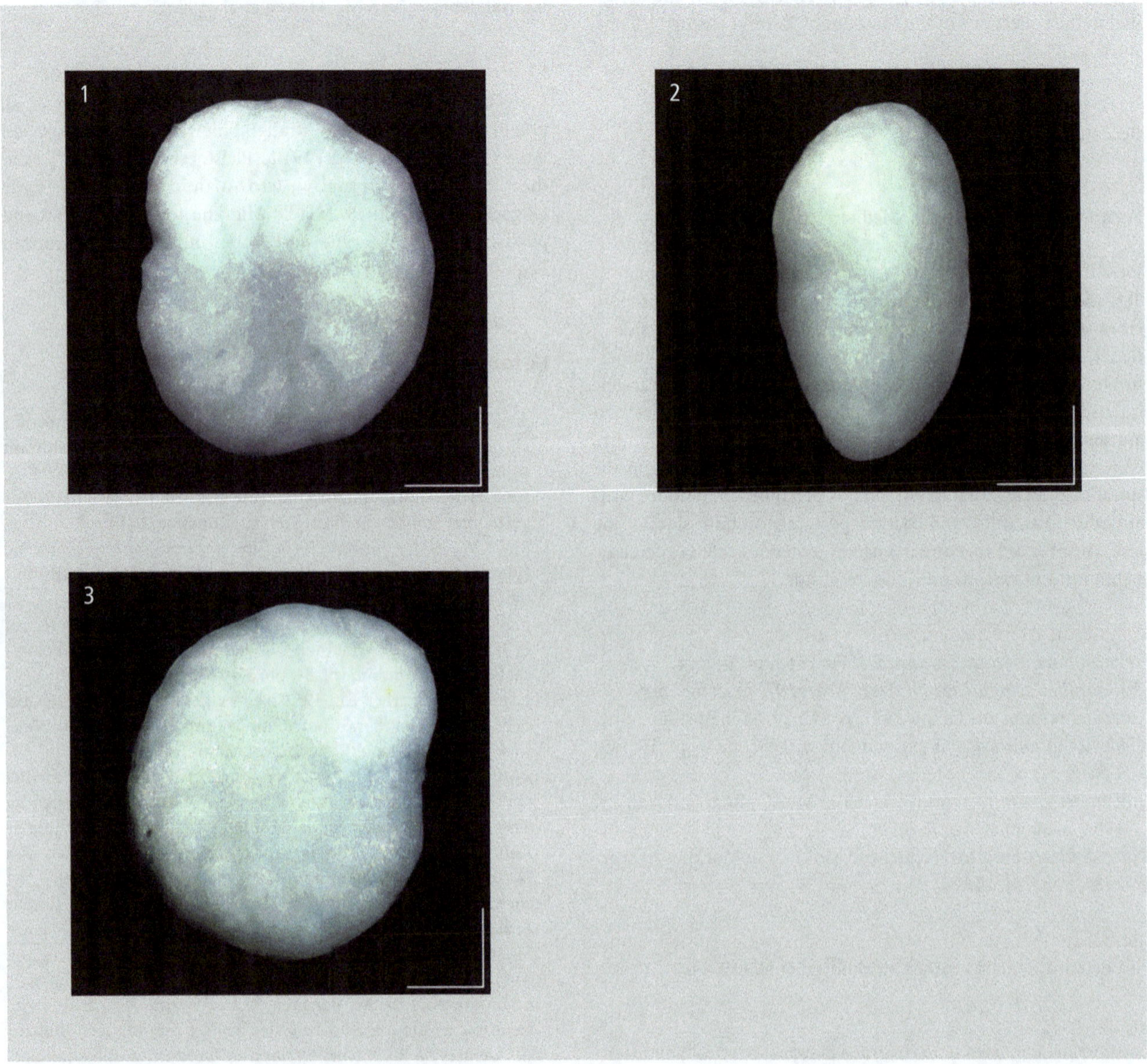


1. *Cibicidoides lamontdohertyi*. North Biscay, North Atlantic; DSDP Leg 48, Hole 400A, Core 46, Section 3, 146–150 cm. Umbilical view, NHMUK PM DI 0252, scale bar = 51 μm.

2. *Cibicidoides lamontdohertyi*. North Biscay, North Atlantic; DSDP Leg 48, Hole 400A, Core 46, Section 3, 146–150 cm. Apertural view, NHMUK PM DI 0251, scale bar = 49 μm.

3. *Cibicidoides lamontdohertyi*. North Biscay, North Atlantic; DSDP Leg 48, Hole 400A, Core 46, Section 3, 146–150 cm. Spiral view, NHMUK PM DI 0250, scale bar = 49 μm.

## Original Designation

*Cibicidoides lamontdohertyi* Miller and Katz, 1987, p. 130, pl. 9, figs. 1a–3c.

## Type Specimen

Depository not given.

## Type Level

Upper Oligocene, probably Zone NP24, Zone P21 (Miller, 1983).

## Type Locality

DSDP Leg 12, Site 119, Sample 15-CC, Bay of Biscay, North Atlantic, latitude 45° 01.9′ N, longitude 07° 58.5′ W, 4447 meters present water depth, approximately 4000 meters Oligocene depth.

## Description

Test forms a small trochospire; lobulate in outline (last chambers), biconvex to unequally biconvex in cross-section, with a partially evolute umbilical side, an evolute spiral side, and a subacute, imperforate periphery. The (approximately) eleven moderately inflated chambers in the last whorl increase gradually in size, and are separated by strongly curved sutures in the earlier part of the test. These sutures become more radial in the later part of the test, terminating along an umbilical–extraumbilical line, as the last chambers tend to uncoil and become more triangular in shape. Chamber walls are coarsely perforate on the spiral side, and more finely perforate on the umbilical side. The primary aperture is a narrow equatorial slit bordered by a thin lip, and extends onto the spiral side.

## Synonyms and Citations

*Cibicidoides lamontdohertyi* Miller and Katz, 1987. –Katz et al., 2003, p. 32, pl. 4, fig. 1.

## Remarks

*Cibicidoides lamontdohertyi* Miller and Katz differs from *Cibicidoides mundulus* (Brady, Parker, and Jones) and *Cibicidoides praemundulus* Berggren and Miller by its flatter test and smaller size (Miller and Katz, 1987).

## Biogeography

Common or abundant throughout the Atlantic.

## Bathymetry

Abyssal.

## Chronostratigraphy

Late Eocene (Priabonian) to Late Miocene (Messinian).

# References


Katz, M.E., Tjalsma, R.C., and Miller, K.G., 2003, Oligocene bathyal to abyssal benthic foraminifera of the Atlantic Ocean. Micropaleontology, vol. 49, supplement 2, pp. 1–45.

Miller, K.G., 1983, Eocene–Oligocene paleoceanography of the deep Bay of Biscay: benthic foraminiferal evidence. Marine Micropaleontology, vol. 7, pp. 403–440.

Miller, K.G. and Katz, M.E., 1987, Oligocene to Miocene benthic foraminiferal and abyssal circulation changes in the north Atlantic. Micropaleontology, vol. 33, no. 2, pp. 97–149.

# *Cibicidoides laurisae*

## (Mallory), 1959

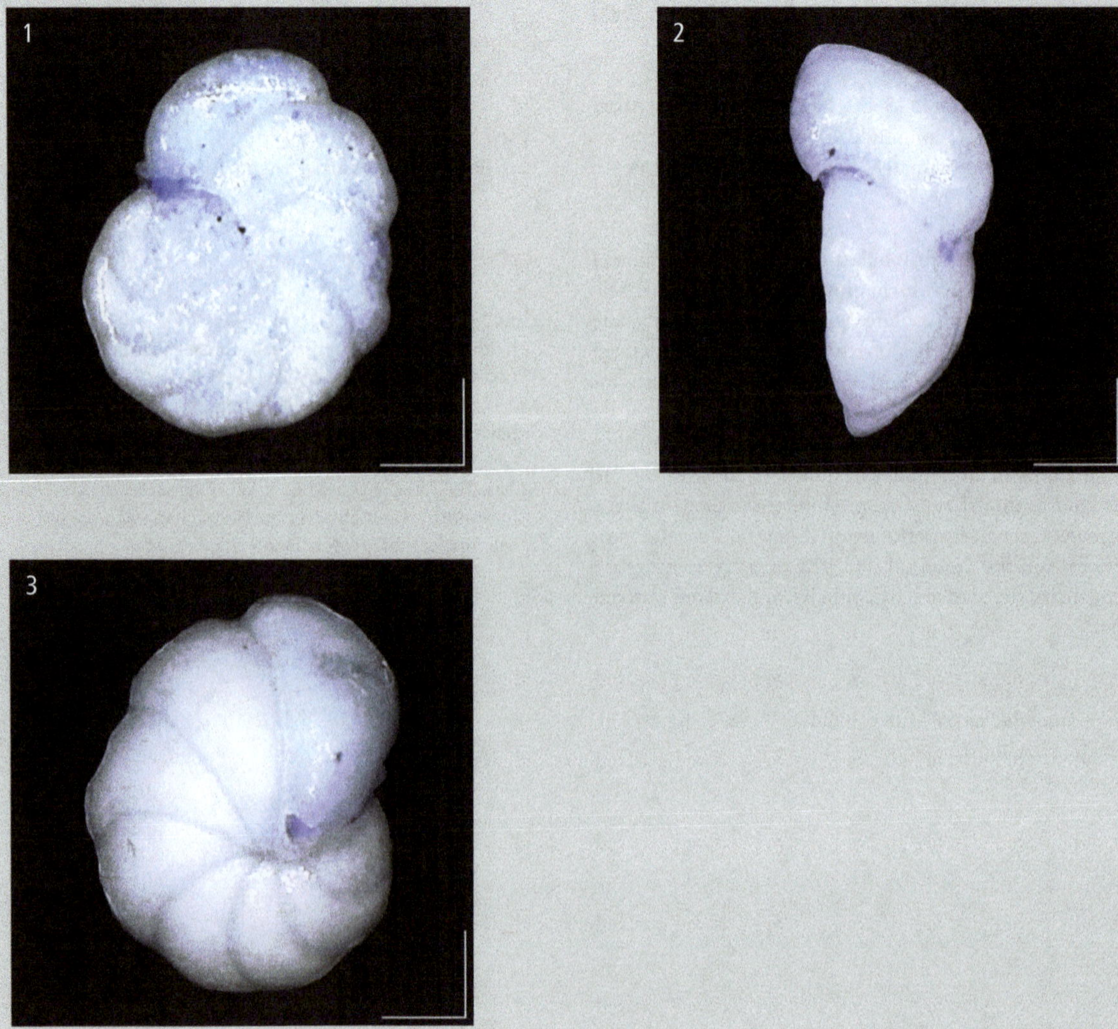

1. *Cibicidoides laurisae*. Goban Spur, North Atlantic; DSDP Leg 80, Hole 549A, Core 14, Section 1, 116–121 cm. Spiral view, NHMUK PM DI 0253, scale bar = 52 μm.

2. *Cibicidoides laurisae*. Goban Spur, North Atlantic; DSDP Leg 80, Hole 549A, Core 14, Section 1, 116–121 cm. Apertural view, NHMUK PM DI 0254, scale bar = 53 μm.

3. *Cibicidoides laurisae*. Goban Spur, North Atlantic; DSDP Leg 80, Hole 549A, Core 14, Section 1, 116–121 cm. Umbilical view, NHMUK PM DI 0255, scale bar = 52 μm.

## Original Designation

*Cibicides laurisae* Mallory, 1959, p. 267, pl. 24, figs. 8a–c.

## Type Specimen

Holotype (No. 42040) and unfigured paratypes (Nos 42041–42043) deposited in the collections of the Micropalaeontology Museum, Department of Paleontology, University of California, Berkeley, USA.

## Type Level

Upper Eocene.

## Type Locality

Holotype and unfigured paratypes from the Devils Den area of northwestern Kern County, California, USA. Found also at Cantua Creek (Kreyenhagen), at the north limb of the Devils Kitchen syncline (San Emigdio), in the Santa Rosa Hills (Cozy Dell), and at Cañada del Cojo (Cozy Dell), all in California.

## Description

Test forms a small trochospire; planoconvex or concavo-convex in cross-section, with an involute, strongly convex, umbilical side, a flattened or slightly concave, evolute, spiral side, and an acute, imperforate periphery. The seven to nine chambers in the last whorl increase rapidly in size and in convexity on the umbilical side, and are separated by slightly depressed, strongly curved, limbate sutures. The ultimate chamber is considerably more convex and inflated than previous chambers. The umbilicus may be filled by a plug of clear material. Chamber walls are calcareous and finely perforate on both sides of the test. The primary aperture is a narrow equatorial slit bordered by a thin lip, and extends onto the spiral side.

## Synonyms and Citations

?*Cibicides dampelae* Bykova and Khramaya in Vasilenko, 1954, p. 124, pl. 17, figs. 6–7.

*Cibicidoides* aff. *laurisae* (Mallory). –Tjalsma and Lohmann, 1983, p. 27, pl. 17, fig. 2.

*Cibicidoides laurisae* (Mallory). –Miller et al., 1985, pl. 2, figs. 4–6. –van Morkhoven et al., 1986, p. 253, pl. 85. –Katz et al., 2003, p. 32, pl. 2, fig. 2.

## Remarks

Tjalsma and Lohmann (1983) reported as *Cibicidoides* aff. *laurisae* forms with a greater number of coarse pores on the spiral side, a more convex umbilical side, less strongly curved sutures, and without an umbilical infilling.

## Biogeography

Worldwide. Recorded from the Atlantic, the Pacific, and the Gulf of Mexico (van Morkhoven et al., 1986). The form *Cibicidoides* aff. *laurisae* (Mallory) was recorded by Tjalsma and Lohmann (1983) from the Atlantic and Gulf of Mexico.

## Bathymetry

Middle bathyal to abyssal (van Morkhoven et al., 1986).

## Chronostratigraphy

Middle Eocene (Lutetian) to Early Miocene (Aquitanian).

# References

Katz, M.E., Tjalsma, R.C., and Miller, K.G., 2003, Oligocene bathyal to abyssal benthic foraminifera of the Atlantic Ocean. Micropaleontology, vol. 49, supplement 2, pp. 1–45.

Mallory, V.S., 1959, Lower Tertiary Biostratigraphy of the California Coast Ranges. Tulsa, OK: American Association of Petroleum Geologists, 416 pp.

Miller, K.G., Curry, W.B., and Ostermann, D.R., 1985, Late Paleogene (Eocene to Oligocene) benthic foraminiferal oceanography of the Goban Spur region, Deep Sea Drilling Project Leg 80. In de Graciansky, P.C., Poag, C.W., et al., ed., Initial Reports of the Deep Sea Drilling Project, Washington (U.S. Government Printing Office), vol. 80, pp. 505–538.

Tjalsma, R.C. and Lohmann, G.P., 1983, Paleocene–Eocene bathyal and abyssal benthic foraminifera from the Atlantic Ocean. Micropaleontology, Special Publication, no. 4, pp. 1–90.

van Morkhoven, F.P.C.M., Berggren, W.A., Edwards, A.S., et al., 1986, Cenozoic cosmopolitan deep-water benthic foraminifera. Bulletin des Centres de Recherches Exploration-Production Elf-Aquitaine, Mémoire, vol. 11, pp. 1–421.

Vasilenko, V.P., 1954, Fossil foraminifera of the USSR. Anomalinidae. Trudy Vsesoyuznogo Neftyanogo Nauchno-Issledovatel'skogo, Geologo-razvedochnogo Instituta (VNIGRI) (New Series), vol. 80, pp. 1–282.

# *Cibicidoides matanzasensis*

## (Hadley), 1934

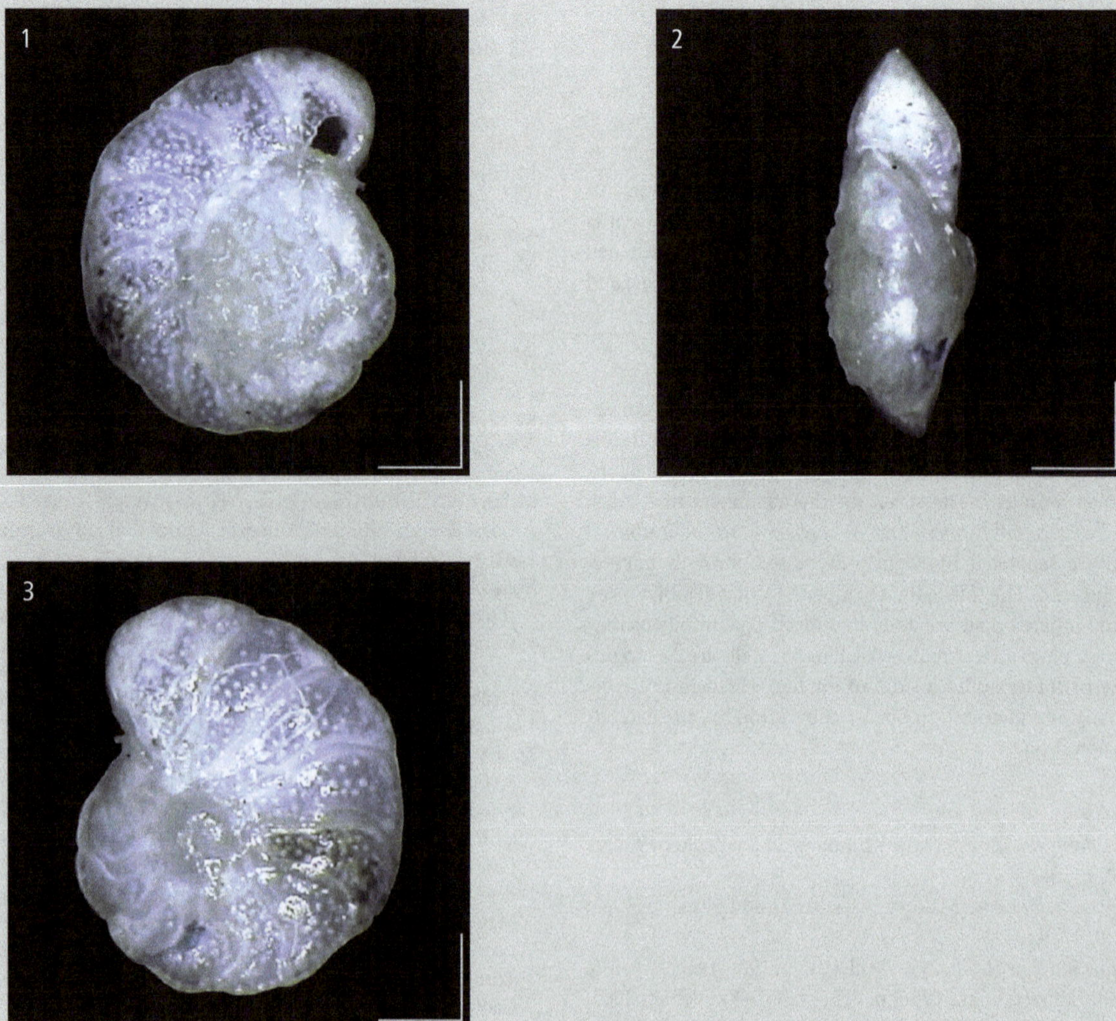


1. *Cibicidoides matanzasensis.* Central Great Barrier Reef, Grafton Passage, Northeast Australian Margin, tropical Southwest Pacific; ODP Leg 133, Hole 820A, Core 3, core catcher. Spiral view, NHMUK PM DI 0256, scale bar = 132 μm.

2. *Cibicidoides matanzasensis.* Central Great Barrier Reef, Grafton Passage, Northeast Australian Margin, tropical Southwest Pacific; ODP Leg 133, Hole 820A, Core 3, core catcher. Apertural view, NHMUK PM DI 0257, scale bar = 133 μm.

3. *Cibicidoides matanzasensis.* Central Great Barrier Reef, Grafton Passage, Northeast Australian Margin, tropical Southwest Pacific; ODP Leg 133, Hole 820A, Core 3, core catcher. Umbilical view, NHMUK PM DI 0258, scale bar = 132 μm.

## Original Designation

*Planulina matanzasensis* Hadley, 1934, p. 27, pl. 4, figs. 1–3.

## Type Specimen

Deposited in the collections of the Paleontological Research Institution, Ithaca, NY, USA.

## Type Level

Middle Miocene (van Morkhoven et al., 1986), basal marl. Originally given as "probably Oligocene" by Hadley (1934).

## Type Locality

Station 0–2 Yumuri River Gorge near the town of Matanzas, Matanzas Province, Cuba.

## Description

Test forms a trochospire; unequally biconvex in cross-section, with a partially evolute, more convex umbilical side and an acute, imperforate periphery. The eleven to twelve slightly inflated chambers in the last whorl increase gradually in size, and are separated by thick, raised, strongly curved, limbate sutures extending along the periphery to form a "pseudokeel". These spiral-side sutures fuse in the umbilical region to form a bead-like ornament over the central part of the test. Sutures on the umbilical side coalesce to form a distinct umbilical plug. Chamber walls are calcareous, thick, and coarsely perforate on both sides of the test. The primary aperture is a narrow equatorial slit bordered by a thin lip, and extends onto the spiral side.

## Synonyms (following van Morkhoven et al., 1986) and Citations

*Pulvinulina partschiana* (d'Orbigny). –MacFayden, 1930, p. 102, pl. 4, fig. 15.
*Cibicides dorsopustulosus* LeRoy, 1939, p. 268, pl. 1, figs. 1–3.
*Cibicides matanzasensis* (Hadley). –Renz, 1948, p. 129, pl. 11, figs. 12. –Bermúdez, 1949, p. 302, pl. 24, figs. 4–6.
*Cibicides cravenensis* Gibson, 1983, p. 395, pl. 14, figs. 1–12.
*Cibicidoides matanzasensis* (Hadley). –van Morkhoven et al., 1986, p.158, pl. 52.

## Remarks

Often occurs with *Cibicidoides crebbsi* (Hedberg) in upper bathyal assemblages of Early–Middle Miocene age (van Morkhoven et al., 1986).

Van Morkhoven et al. (1986) noted that Hadley's original description does not correspond with paratypes from the Middle Miocene of Cuba (CC 22374) now in the collections of the United States National Museum. These paratypes are umbilico-convex, and have chambers of a different shape than the form recognized by most authors as *Cibicidoides matanzasensis*.

## Biogeography

Worldwide. Recorded from the Caribbean, the North and South Atlantic, Venezuela, and the Gulf of Mexico (van Morkhoven et al., 1986). *Cibicides dorsopustulosus* was described from the Miocene of Sumatra by LeRoy (1939).

## Bathymetry

Primarily middle neritic to middle bathyal. Rarely, deeper bathyal (Katz and Miller, 1993).

## Chronostratigraphy

Early Miocene (Aquitanian) to Middle Miocene (Serravallian).

# References


Bermúdez, P.J., 1949, Tertiary smaller foraminifera of the Dominican Republic. Special Publications of the Cushman Laboratory for Foraminiferal Research, no. 25, pp. 1–322.

Gibson, T.G., 1983, Key foraminifera from upper Oligocene to lower Pleistocene strata of the Central Atlantic Coastal Plain. Smithsonian Contributions to Paleobiology, vol. 53, pp. 355–453.

Hadley, W.H., 1934, Some Tertiary foraminifera from the north coast of Cuba. Bulletin of American Paleontology, vol. 20, pp. 1–40.

Katz, M.E. and Miller, K.G., 1993, Latest Oligocene to Earliest Pliocene benthic foraminiferal biofacies of the northeastern Gulf of Mexico. Micropaleontology, vol. 39, no. 4, pp. 367–403.

LeRoy, L.W., 1939, Some small foraminifera, ostracoda and otoliths from the Neogene ("Miocene") of the Rokan-Tapanoeli area, Central Sumatra. Natuurkundig Tijdschrift voor Nederlandsch-Indië, vol. 99, no. 6, pp. 215–296.

MacFadyen, W.A., 1930, Miocene foraminifera from the Clysmic area of Egypt and Sinai. Cairo: Geological Survey of Egypt, 149 pp.

Renz, H.H., 1948, Stratigraphy and fauna of the Agua Salada group, State of Falcón, Venezuela. Memoirs of the Geological Society of America, vol. 32, pp. 1–219.

van Morkhoven, F.P.C.M., Berggren, W.A., Edwards, A.S., et al., 1986, Cenozoic cosmopolitan deep-water benthic foraminifera. Bulletin des Centres de Recherches Exploration-Production Elf-Aquitaine, Mémoire, vol. 11, pp. 1–421.

# *Cibicidoides mexicanus*

## (Nuttall), 1932

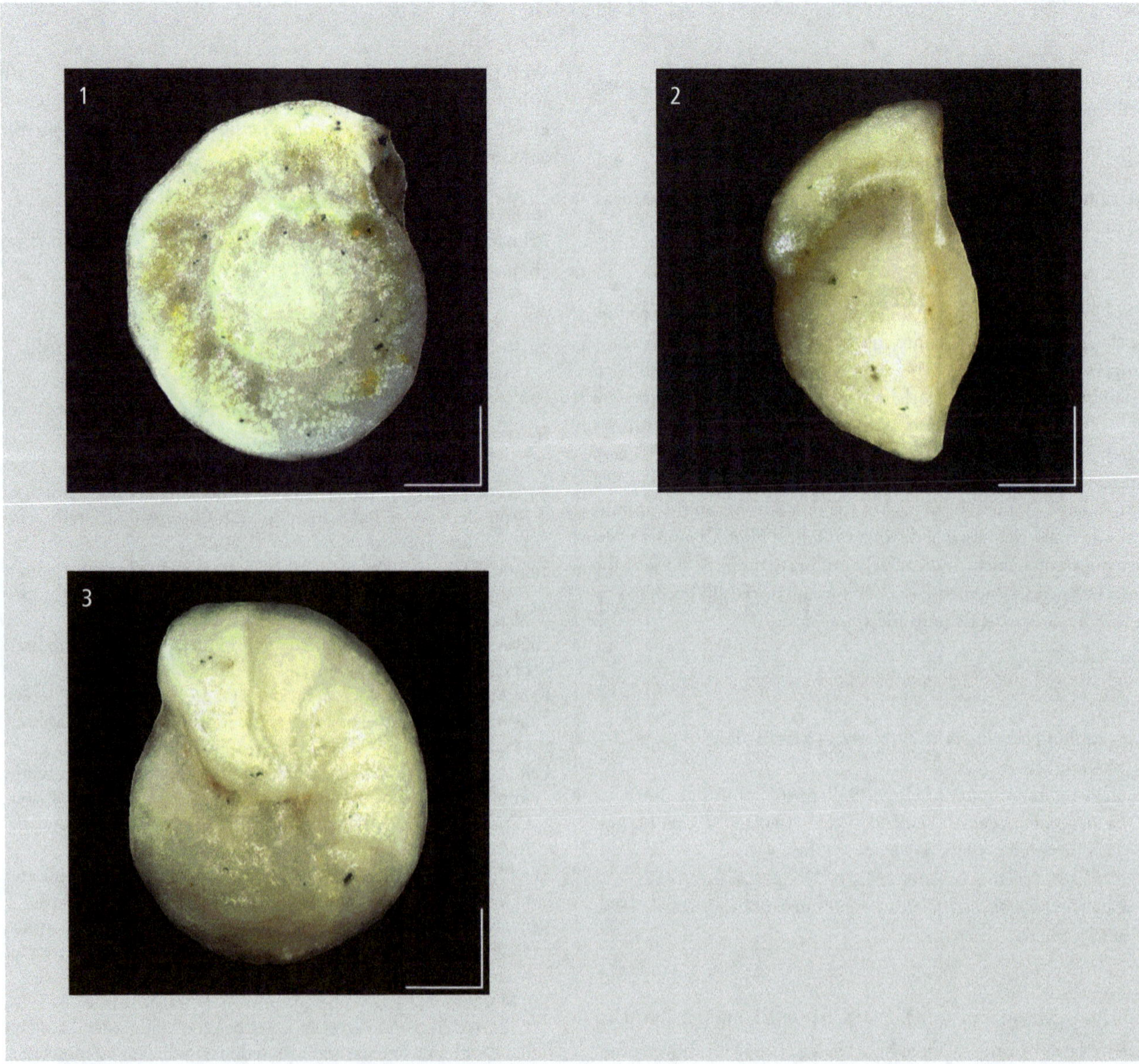


1. *Cibicidoides mexicanus.* Near San Miguel village, Hola, Palma Real, State of Vera Cruz, Mexico. Deposited in the collections of the United States National Museum, Washington, DC; CC 16465. Holotype. Spiral view, NHMUK PM DI 0259, scale bar = 170 μm.

2. *Cibicidoides mexicanus.* Near San Miguel village, Hola, Palma Real, State of Vera Cruz, Mexico. Deposited in the collections of the United States National Museum, Washington, DC; CC 16465. Holotype. Apertural view, NHMUK PM DI 0260, scale bar = 142 μm.

3. *Cibicidoides mexicanus.* Near San Miguel village, Hola, Palma Real, State of Vera Cruz, Mexico. Deposited in the collections of the United States National Museum, Washington, DC; CC 16465. Holotype. Umbilical view, NHMUK PM DI 0261, scale bar = 162 μm.

## Original Designation

*Cibicides mexicana* Nuttall, 1932, p. 33, pl. 9, figs. 7–9.

## Type Specimen

Holotype (CC 16465) deposited in the collections of the United States National Museum, Washington, DC.

## Type Level

Lower Oligocene, Lower Alazan Shale.

## Type Locality

San Miguel village, situated just south of the boundary of Palma Real and Llano Grande, Tampico Embayment, Vera Cruz, Mexico.

## Description

Test forms a trochospire; planoconvex in cross-section, with a flattened spiral side, a convex umbilical side, a small, shallow umbilicus, and a subacute, imperforate periphery. The (approximately) ten inflated chambers in the last whorl are separated by flush or slightly raised, curved, limbate sutures. On the spiral side, a thickened, translucent area over the early whorls is surrounded by a prominent, depressed spiral suture, and often protrudes above the flattened spiral side. Chamber walls are calcareous, and coarsely perforate on both sides of the test. The primary aperture is a narrow equatorial slit bordered by a thin lip, and extends onto the spiral side.

## Synonyms (following van Morkhoven et al., 1986) and Citations

*Truncatulina dutemplei* (d'Orbigny). –Reuss, 1866, p. 160, pl. 4, fig. 16.

*Heterolepa bullata* Franzenau, 1884, p. 217, pl. 5, figs. 5, 7, 9, 11.

*?Heterolepa costata* Franzenau, 1884, p. 216, pl. 5, figs. 2, 8.

*Heterolepa praecincta* Franzenau, 1884, p. 216, pl. 5, figs. 4, 6, 10.

*?Heterolepa simplex* Franzenau, 1884, p. 215, pl. 5, fig. 1.

*Cibicides pippeni* Cushman and Garret, 1938, p. 64, pl. 11, fig. 2.

*Cibicides dalmatina* van Bellen, 1941, p. 1002, fig. 30.

*Cibicidoides mexicanus* (Nuttall). –Tjalsma, 1983, p. 742, pl. 3, fig. 5; pl. 9, figs. 1–2. –van Morkhoven et al., 1986, p. 222, pl. 76. –Miller and Katz, 1987, p. 130, pl. 8, fig. 2.

*Heterolepa mexicana* (Nuttall). –Bolli et al., 1994, p. 376, figs. 60.1–5, 89.6.

## Remarks

The initial whorls of *Cibicidoides mexicanus* (Nuttall) are almost always obscured by translucent shell material. *Cibicidoides mexicanus* can be distinguished from *Cibicidoides grimsdalei* (Nuttall) by its generally less inflated umbilical side and its more acute, sharper-edged periphery. It differs from *Heterelopa dutemplei* (d'Orbigny) by the chamber shape and arrangement visible on the spiral side, although the umbilical sides of these two species are nearly identical with their pinwheel sutural patterns. Furthermore, there is almost no overlap in the biostratigraphic ranges of *Cibicidoides mexicanus* and *Heterelopa dutemplei* (M. Katz, pers. comm. 1998).

See also Remarks in van Morkhoven et al. (1986) for detailed taxonomic discussion.

## Biogeography

Worldwide. Recorded from Cuba, Puerto Rico, Trinidad, the Dominican Republic, Barbados, Ecuador, Europe, the North and South Atlantic, the eastern Pacific, and the Gulf of Mexico (van Morkhoven et al. 1986).

## Bathymetry

Middle to lower bathyal. The *Heterolepa simplex–costata* group, considered as shallow-water morphotypes of *Cibicidoides mexicanus*, occur in middle neritic to upper bathyal deposits (van Morkhoven et al., 1986).

## Chronostratigraphy

Late Eocene (Priabonian) to Early Miocene (Burdigalian).

# References


Bolli, H.M., Beckmann, J.P., and Saunders, J.B., 1994, Benthic Foraminiferal Biostratigraphy of the South Caribbean Region. Cambridge: Cambridge University Press, 408 pp.

Cushman, J.A. and Garret, J.B., 1938, Three new Rotaliform foraminifera from the Lower Oligocene and Upper Eocene of Alabama. Contributions from the Cushman Laboratory for Foraminiferal Research, vol. 14, no. 3, pp. 62–66.

d'Orbigny, A.D., 1846, Foraminifères fossiles du Bassin Tertiaire de Vienne (Autriche). Paris: Gide et Compe, xxxvii + 312 pp.

Franzenau, A., 1884, Heterolepa, egy uj genus a foraminiferak rendjében. Természetrajzi Füzetek, Budapest, vol. 8, pp. 181–184 (214–217).

Miller, K.G. and Katz, M.E., 1987, Oligocene to Miocene benthic foraminiferal and abyssal circulation changes in the north Atlantic. Micropaleontology, vol. 33, no. 2, pp. 97–149.

Nuttall, W.L.F., 1932, Lower Oligocene foraminifera from Mexico. Journal of Paleontology, vol. 6, pp. 3–35.

Reuss, A.E., 1866, Die Foraminiferen, Anthozoen und Bryozoen des deutschen Septarienthones. Denkschriften der Kaiserlichen Akademie der Wissenschaften, Mathematisch-Naturwissenschaftliche Classe, vol. 25, no. 1, pp. 117–214.

Tjalsma, R.C., 1983, Eocene to Miocene benthic foraminifera from DSDP site 516, Rio Grande Rise, South Atlantic. In Barker, P.F., Carlson, R.L., Johnson, D.A., et al., ed., Initial Reports of the Deep Sea Drilling Project, Washington (U.S. Government Printing Office), vol. 72, pp. 731–755.

van Bellen, R.C., 1941, Some Eocene foraminifera from the neighborhood of Ricice near Imotski, E. Dalmatia, Yugoslavia. Proceedings of the Koninklijke Nederlandse Akademie van Wetenschappen, vol. 44, no. 9, pp. 996–1005.

van Morkhoven, F.P.C.M., Berggren, W.A., Edwards, A.S., et al., 1986, Cenozoic cosmopolitan deep-water benthic foraminifera. Bulletin des Centres de Recherches Exploration-Production Elf-Aquitaine, Mémoire, vol. 11, pp. 1–421.

# *Cibicidoides micrus*

## (Bermúdez), 1949

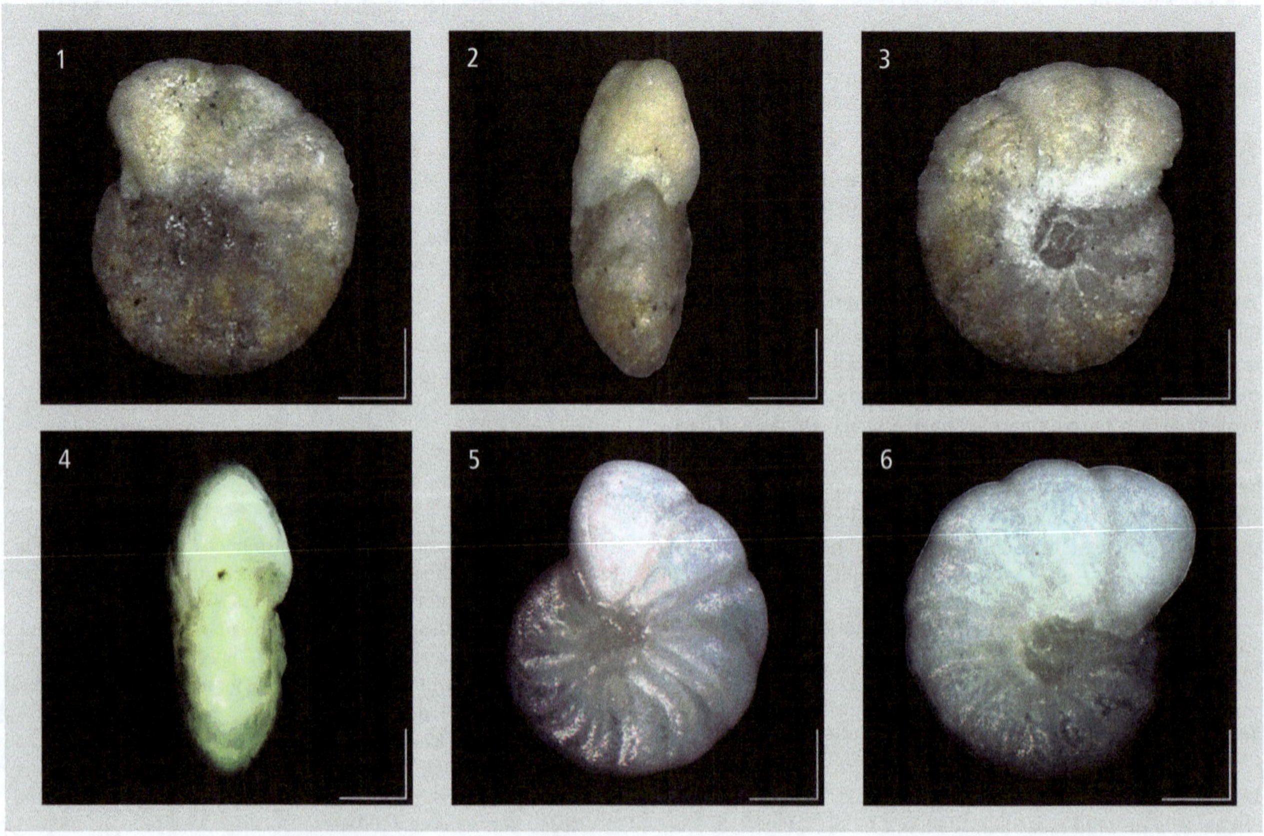


1. *Cibicidoides micrus*. Abuillot Formation, Abuillot River Traverse, Central Plain, Haiti. Deposited in the collections of the United States National Museum, Washington, DC; CC 62431. Holotype. Spiral view, NHMUK PM DI 0262, scale bar = 72 μm.

2. *Cibicidoides micrus*. Abuillot Formation, Abuillot River Traverse, Central Plain, Haiti. Deposited in the collections of the United States National Museum, Washington, DC; CC 62431. Holotype. Apertual view, NHMUK PM DI 0263, scale bar = 72 μm.

3. *Cibicidoides micrus*. Abuillot Formation, Abuillot River Traverse, Central Plain, Haiti. Deposited in the collections of the United States National Museum, Washington, DC; CC 62431. Holotype. Umbilical view, NHMUK PM DI 0264, scale bar = 72 μm.

4. *Cibicidoides micrus*. North Biscay, North Atlantic; DSDP Leg 48, Hole 402A, Core 2, Section 1, 18–22 cm. Apertural view, NHMUK PM DI 0265, scale bar = 68 μm.

5. *Cibicidoides micrus*. North Biscay, North Atlantic; DSDP Leg 48, Hole 402A, Core 2, Section 1, 18–22 cm. Umbilical view, NHMUK PM DI 0266, scale bar = 69 μm.

6. *Cibicidoides micrus*. North Biscay, North Atlantic; DSDP Leg 48, Hole 402A, Core 2, Section 1, 18–22 cm. Umbilical view, NHMUK PM DI 0267, scale bar = 68 μm.

## Original Designation

*Cibicides micrus* Bermúdez, 1949, p. 302, pl. 24, figs. 34–36.

## Type Specimen

Holotype (No. 62431) and paratypes deposited in the collections of the United States National Museum, Washington, DC. Topotypes deposited in the collections of the Standard Oil Company of Cuba at the University of Havana; the University of Santo Domingo, Ciudad Trudillo, Dominican Republic; the Dorothy K. Palmer Collection at the Paleontological Research Institution, Ithaca, NY, USA; and in the author's personal collection, now housed in the Paleontological Laboratory of INTEVEP, Los Teques (suburb of Caracas), Venezuela.

## Type Level

Lower Eocene, Abuillot Formation; highly calcareous, slightly sandy shale.

## Type Locality

Sample H-843, from a cut in a bluff along the Abuillot River, a confluent of the Canot River, Hinche area, Central Plain, Haiti (Type Locality of the Abuillot Formation).

## Description

Test forms a lenticular trochospire; subcircular in outline and unequally biconvex with a slightly convex, evolute spiral side, a more flattened, partially evolute umbilical side, and a subacute, imperforate periphery. The (approximately) twelve slightly inflated chambers in the last whorl increase gradually in size and are separated by distinct, slightly curved sutures. Spiral sutures are markedly depressed and the umbilicus is closed. Chamber walls are thick, finely perforate, and smooth. The primary aperture is a narrow equatorial slit bordered by a thin lip, and extends onto the spiral side.

## Synonyms

None verified.

## Remarks

Specimens illustrated by van Morkhoven et al. (1986) were incorrectly identified as *Cibicidoides micrus* (Bermúdez). They are topotypes of *Cocoarota cocoaensis* (Cushman), the type species of the genus *Cocoarota* Loeblich and Tappan 1986 (see Loeblich and Tappan, 1987, p. 637; Revets, 1996, p. 74).

Morphotypes from lower bathyal depths are generally larger, whereas abyssal forms are more rounded and distinctly evolute on the spiral side (Tjalsma and Lohmann, 1983).

## Biogeography

Worldwide. Found in the North and South Atlantic, Pacific, and Indian Oceans.

## Bathymetry

Bathyal to abyssal.

## Chronostratigraphy

Early Eocene (Ypresian) to Late Oligocene (Chattian).

## References


Bermúdez, P.J., 1949, Tertiary smaller foraminifera of the Dominican Republic. Special Publications of the Cushman Laboratory for Foraminiferal Research, no. 25, pp. 1–322.

Loeblich, A.R. and Tappan, H., 1986, Some new and revised genera and families of hyaline calcareous Foraminiferida (Protozoa). Transactions of the American Microscopical Society, vol. 105, pp. 239–265.

Loeblich, A.R. and Tappan, H., 1987, Foraminiferal Genera and Their Classification. New York: Van Nostrand Reinhold, 970 pp.

Revets, S.A. 1996, The generic revision of five families of Rotaliine Foraminifera. Special Publications of the Cushman Foundation for Foraminiferal Research, vol. 34, pp. 1–108.

Tjalsma, R.C. and Lohmann, G.P. 1983, Paleocene–Eocene bathyal and abyssal benthic foraminifera from the Atlantic Ocean. Micropaleontology Special Publication, vol. 4, pp. 1–90.

van Morkhoven, F.P.C.M., Berggren, W.A., Edwards, A.S., et al., 1986, Cenozoic cosmopolitan deep-water benthic foraminifera. Bulletin des Centres Recherches Exploration-Production Elf-Aquitaine, Memoir, vol. 11, pp. 1–423.

# Cibicidoides mundulus

## (Brady, Parker, and Jones), 1888

1. *Cibicidoides mundulus.* "Meteor" Cruise off West Africa; M25. Deposited in the Lutze Collection (West Africa), no. 198. Institute of Geosciences, Kiel University, Germany. Spiral view, NHMUK PM DI 0268, scale bar = 129 μm.

2. *Cibicidoides mundulus.* "Meteor" Cruise off West Africa; M25. Deposited in the Lutze Collection (West Africa), no. 198. Institute of Geosciences, Kiel University, Germany. Apertural view, NHMUK PM DI 0269, scale bar = 129 μm.

3. *Cibicidoides mundulus.* "Meteor" Cruise off West Africa; M25. Deposited in the Lutze Collection (West Africa), no. 198. Institute of Geosciences, Kiel University, Germany. Umbilical view, NHMUK PM DI 0270, scale bar = 130 μm.

4. *Cibicidoides mundulus.* Abrahlos Bank, off the coast of Brazil, South America. "Plumper" Station 4, latitude 22° 54′ S, longitude 40° 37′ W, depth: 260 fathoms. Deposited in the micropaleontological collections of The Natural History Museum, London; NHMUK ZF3585. Lectotype. Umbilical view, NHMUK PM DI 0271, scale bar = 130 μm.

## Original Designation

*Truncatulina mundula* Brady, Parker, and Jones, 1888, p. 228, pl. 45, fig. 25a–c.

## Type Specimen

Depository not given. Lectotype (ZF3585) designated by Loeblich and Tappan (1955), and housed in the micropaleontological collections of The Natural History Museum, London, UK.

## Type Level

Recent, olive-brown clay.

## Type Locality

"Plumper" Station 4, latitude 22° 54′ S, longitude 40° 37′ W, over Abrolhos Bank, off the coast of Brazil; 260 fathoms.

## Description

Test forms a biumbonate trochospire; subcircular in outline and nearly biconvex in cross-section with an involute, slightly more convex umbilical side, an evolute spiral side, and a subacute, imperforate periphery. The ten to twelve moderately inflated chambers in the last whorl increase gradually in size and are separated by flush, curved sutures. Chamber walls are calcareous and finely perforate on the umbilical side, and coarsely perforate on the spiral side. A coil of coarse perforations is visible through the clear umbo on the spiral side. The primary aperture is a narrow equatorial slit bordered by a thin lip, and extends onto the spiral side.

## Synonyms and Citations

*Truncatulina* sp. near *T. haidingerii*. –Brady, 1884, pl. 95, fig. 6.
*Cibicides* sp. 2 Phleger and Parker, 1951, p. 32, pl. 18, figs. 1–2.
*Cibicides kullenbergi* Parker. In Phleger et al., 1953, p. 49, pl. 11, figs. 7–8.
*Cibicidoides mundulus* (Brady, Parker, and Jones). –Loeblich and Tappan, 1955, p. 25, fig. 4a–c. –Barker, 1960, pl. 95, fig. 6. –van Morkhoven et al., 1986, p. 65, pl. 21. –Miller and Katz, 1987, p. 130, pl. 7, fig. 3. –Loeblich and Tappan, 1987, p. 572, pl. 626, figs. 1–3. –Jones, 1994, p. 99, pl. 95, fig. 6. –Revets, 1996, p. 65, pl. 1, figs. 9–12.

## Remarks

Type species of the genus *Cibicidoides* Thalmann, 1939.

We follow van Morkhoven et al. (1986) in considering *Cibicidoides kullenbergi* (Parker) to be a junior synonym of *Cibicidoides mundulus* (Brady, Parker, and Jones). The original figure shows that *Cibicidoides kullenbergi* has strongly curved sutures on the umbilical side, in contrast to *Cibicidoides mundulus*, which has straighter sutures. However, direct comparison between the lectotype of *Cibicidoides mundulus* and the holotype of *Cibicidoides kullenbergi* revealed that *Cibicidoides kullenbergi* also had straight sutures on the umbilical side. The two taxa exhibit a similar range of variability in the degree of curvature of their sutures and in the size of their test and cannot, therefore, be distinguished as separate species on the basis of these criteria.

## Biogeography

Worldwide.

## Bathymetry

Bathyal to abyssal.

## Chronostratigraphy

Early Oligocene (Rupelian) to Recent.

## References


Barker, R.W., 1960, Taxonomic notes on the species figured by H.B. Brady in his report on the foraminifera dredged by H.M.S. Challenger during the years 1873–1876. Society of Economic Paleontologists and Mineralogists Special Publication, vol. 9, pp. 1–238.

Brady, H.B., 1884, Report on the foraminifera dredged by H.M.S. Challenger during the years 1873–1876. In Brady, H.B., ed., Report on the Scientific Results of the Voyage of the H.M.S. Challenger during the years 1873–1876. Zoology, vol. 9, pp. 1–814.

Brady, H.B., Parker, W.K., and Jones, T.R., 1888, On some foraminifera from the Abrolhos Bank. Transactions of the Zoological Society of London, vol. 12 (1890), pp. 211–239.

Jones, R.W., 1994, The Challenger Foraminifera. Oxford: Oxford University Press, 149 pp.

Loeblich, A.R. and Tappan, H., 1955, Revision of some Recent foraminiferal genera. Smithsonian Miscellaneous Collections, vol. 128, pp. 1–37.

Loeblich, A.R. Jr and Tappan, H., 1987, Foraminiferal Genera and Their Classification. New York: Van Nostrand Reinhold, 970 pp.

Miller, K.G. and Katz, M.E., 1987, Oligocene to Miocene benthic foraminiferal and abyssal circulation changes in the north Atlantic. Micropaleontology, vol. 33, no. 2, pp. 97–149.

Phleger, F.B. and Parker, F.L., 1951, Ecology of foraminifera, northwest Gulf of Mexico; Part II – Foraminifera species. Memoirs of the Geological Society of America, vol. 46, pp. 1–64.

Phleger, F.B., Parker, F.L., and Pierson, J.F., 1953, North Atlantic foraminifera. Reports of the Swedish Deep Sea Expedition, 1947–1948, vol. 7, pp. 3–122.

Revets, S.A., 1996, The generic revision of five families of Rotaliine Foraminifera. Special Publications of the Cushman Foundation for Foraminiferal Research, vol. 34, pp. 1–108.

van Morkhoven, F.P.C.M., Berggren, W.A., Edwards, A.S., et al., 1986, Cenozoic cosmopolitan deep-water benthic foraminifera. Bulletin des Centres de Recherches Exploration-Production Elf-Aquitaine, Mémoire, vol. 11, pp. 1–421.


5. *Cibicidoides mundulus*. Abrahlos Bank, off the coast of Brazil, South America. "Plumper" Station 4, latitude 22° 54′ S, longitude 40° 37′ W, depth: 260 fathoms. Deposited in the micropaleontological collections of The Natural History Museum, London; NHMUK ZF3585. Lectotype. Apertural view. NHMUK PM DI 0272. scale bar = 129 μm.

6. *Cibicidoides mundulus*. Abrahlos Bank, off the coast of Brazil, South America. "Plumper" Station 4, latitude 22° 54′ S, longitude 40° 37′ W, depth: 260 fathoms. Deposited in the micropaleontological collections of The Natural History Museum, London; NHMUK ZF3585. Lectotype. Spiral view, NHMUK PM DI 0274, scale bar = 130 μm.

7. *Cibicidoides mundulus*. Central North Atlantic; Rehabeh Station 3-38, latitude 36° 34′ N, longitude 38° 15′ W, depth: 3109 meters. Deposited in the collections of the United States National Museum, Washington, DC. Paratype of *Cibicides kullenbergi* Parker. Spiral view, NHMUK PM DI 0273, scale bar = 129 μm.

8. *Cibicidoides mundulus*. Central North Atlantic; Rehabeh Station 3-38, latitude 36° 34′ N, longitude 38° 15′ W, depth: 3109 meters. Deposited in the collections of the United States National Museum, Washington, DC. Paratype of *Cibicides kullenbergi* Parker. Umbilical view, NHMUK PM DI 0275, scale bar = 130 μm.

# *Cibicidoides pachyderma*

**(Rzehak), 1886**

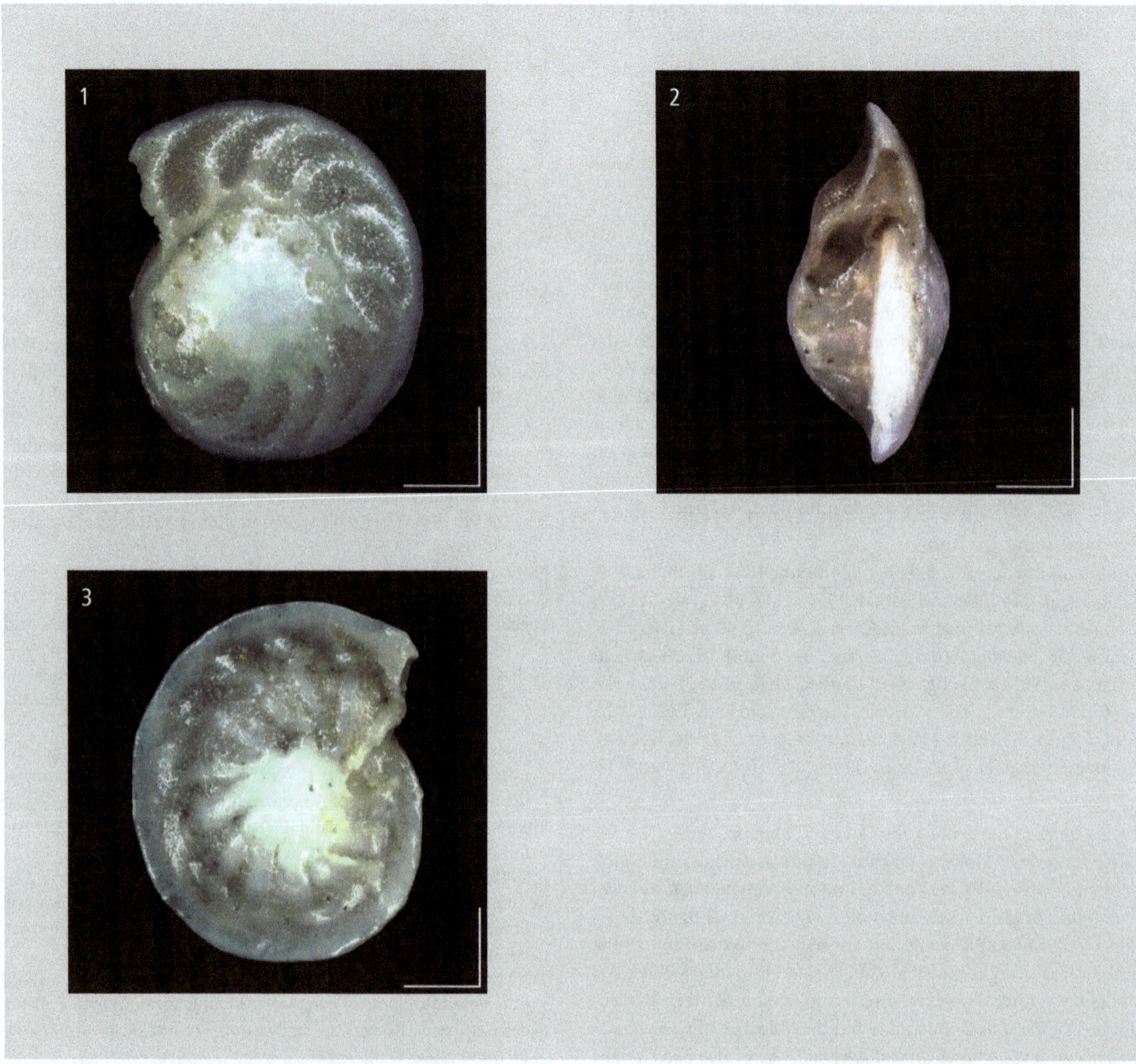


1. *Cibicidoides pachyderma*. Gulf of Mexico. Deposited in the R.W. Jones Collection. Spiral view, NHMUK PM DI 0276, scale bar = 215 μm.
2. *Cibicidoides pachyderma*. Gulf of Mexico. Deposited in the R.W. Jones Collection. Apertural view, NHMUK PM DI 0277, scale bar = 211 μm.
3. *Cibicidoides pachyderma*. Gulf of Mexico. Deposited in the R.W. Jones Collection. Umbilical view, NHMUK PM DI 0278, scale bar = 209 μm.

## Original Designation

*Truncatulina pachyderma* Rzehak, 1886, p. 87, pl. 1, fig. 5.

## Type Specimen

Holotype (missing). Lectotype designated by van Morkhoven et al. (1986), and deposited in the collections of the Natural History Museum, Vienna, Austria.

## Type Level

Miocene (Badenian), Tegel.

## Type Locality

Peterswald (Albrechts-Schacht, circa 150 meters deep), Umgebung von Mährisch-Ostrau (Czechoslovakia). Other localities given: Dombrau, Schlesien, Polnisch-Ostrau (Josefs-Schacht), Umgebung von Mährisch-Ostrau (Czechoslovakia).

## Description

Test forms a biumbonate trochospire; subcircular in outline and nearly biconvex in cross-section with an involute, slightly more convex umbilical side, an evolute spiral side, and a subacute periphery with a thick, imperforate peripheral band. The ten to twelve moderately inflated chambers in the last whorl increase gradually in size and are separated by thick sutures, flush or slightly depressed and gently curved on the umbilical side; flush, strongly curved and weakly limbate on the spiral side. On the umbilical side, the sutures converge toward a flat, central umbo, giving the test the characteristic profile of a flattened saucer in edge view. On the spiral side the early chambers are often obscured by extraneous carbonate overgrowth. Chamber walls are calcareous and finely perforate on the umbilical side, and coarsely perforate on the spiral side. The primary aperture is a narrow equatorial slit bordered by a thin lip, and extends onto the spiral side.

## Synonyms (following van Morkhoven et al., 1986) and Citations

*Truncatulina ungeriana* d'Orbigny. –Brady, 1884, p. 664, pl. 94, fig. 9.
*Cibicides pseudoungeriana* (Cushman). –Cushman, 1931, p. 123, pl. 22, figs. 3–4, 6 (not figs. 5, 7).
*Cibicides mediocris* Finlay, 1940, p. 464, pl. 67, figs. 198–199.
*Cibicides floridanus* (Cushman). –Galloway and Heminway, 1941, p. 392, pl. 23, fig. 2.
*Cibicides* aff. *floridanus* (Cushman). –Phleger and Parker, 1951, p. 30, pl. 16, fig. 1 (not figs. 2–4).
*Cibicides pseudoungerianus* (Cushman). –Barker, 1960, pl. 94, fig. 9.
*Cibicides cicatricosus* (Schwager). –Bock, 1971, p. 59.
*Cibicides* cf. *pseudoungerianus* (Cushman). –Pflum and Frerichs, 1976, p. 29, pl. 2, fig. 9; pl. 3, figs. 1–2.
*Cibicidoides* "*floridanus*" forma *bathyalis* Poag, 1981, p. 53, pl. 29, fig. 1; pl. 30, fig. 1.
*Cibicidoides* "*floridanus*" forma *sublittoralis* Poag, 1981, p. 53, pl. 29, fig. 2; pl. 30, fig. 2.
*Cibicidoides pachyderma* (Rzehak). –van Morkhoven et al., 1986, p. 68, pl. 22. –Katz and Miller, 1993, pl. 2, fig. 4. –Jones, 1994, p. 98, pl. 94, fig. 9.

## Remarks

See van Morkoven et al. (1986) for a detailed account of the complex taxonomic history of this taxon. Also see Remarks for *Cibicidoides mundulus* (Brady, Parker, and Jones).

## Biogeography

Worldwide.

## Bathymetry

Primarily upper bathyal.

## Chronostratigraphy

Early Oligocene (Rupelian) to Recent.

## References


Barker, R.W., 1960, Taxonomic notes on the species figured by H.B. Brady in his report on the foraminifera dredged by H.M.S. Challenger during the years 1873–1876. Society of Economic Paleontologists and Mineralogists Special Publication, vol. 9, pp. 1–238.

Bock, W.D., 1971, A handbook of benthonic foraminifera of Florida Bay and adjacent waters. In Jones, J.I. and Bock, W.D., ed., A Symposium of Recent South Florida Foraminifera. Memoir of the Miami Geological Society, vol. 1, pp. 1–72.

Brady, H.B., 1884, Report on the foraminifera dredged by H.M.S. Challenger during the years 1873–1876. In Brady, H.B., ed., Report on the Scientific Results of the Voyage of the H.M.S. Challenger during the years 1873–1876. Zoology, vol. 9, pp. 1–814.

Cushman, J.A., 1931, The foraminifera of the Atlantic Ocean; Part VIII – Rotaliidae, Amphisteginidae, Calcarinidae, Cymbaloporettidae, Globorotaliidae, Anomalinidae, Planorbulinidae, Rupertiidae and Homotremidae. Bulletin of the United States National Museum, vol. 104, no. 8, pp. 1–179.

Finlay, H.J., 1940, New Zealand Foraminifera; key species in stratigraphy – No. 4. Transactions of the Royal Society of New Zealand, vol. 69, no. 4, pp. 448–472.

Galloway, J.J. and Hemingway, C.E., 1941, The Tertiary Foraminifera of Porto Rico. New York Academy of Science, Scientific Survey of Porto Rico and the Virgin Islands, vol. 3, no. 4, pp. 275–491.

Jones, R.W., 1994, The Challenger Foraminifera. Oxford: Oxford University Press, 149 pp.

Katz, M.E. and Miller, K.G., 1993, Latest Oligocene to Earliest Pliocene benthic foraminiferal biofacies of the northeastern Gulf of Mexico. Micropaleontology, vol. 39, no. 4, pp. 367–403.

Pflum, C.E. and Frerichs, W.E., 1976, Gulf of Mexico deep-water foraminifers. Cushman Foundation, Special Publication, vol. 14, p. 1–125.

Phleger, F.B. and Parker, F.L., 1951, Ecology of foraminifera, northwest Gulf of Mexico; Part II – Foraminifera species. Memoirs of the Geological Society of America, vol. 46, pp. 1–64.

Poag, C.W., 1981, Ecologic Atlas of Benthic Foraminifera of the Gulf of Mexico. Stroudsburg, PA: Hutchinson Ross, 174 pp.

Rzehak, A., 1886, Die Foraminiferenfauna der Neogenformation der Umgebung von Mähr: Ostrau. Verhandlungen des Naturforschenden Vereins in Brünn, vol. 24 (1885), pp. 77–126.

van Morkhoven, F.P.C.M., Berggren, W.A., Edwards, A.S., et al., 1986, Cenozoic cosmopolitan deep-water benthic foraminifera. Bulletin des Centres de Recherches Exploration-Production Elf-Aquitaine, Mémoire, vol. 11, pp. 1–421.

# *Cibicidoides praemundulus*

## Berggren and Miller, 1986

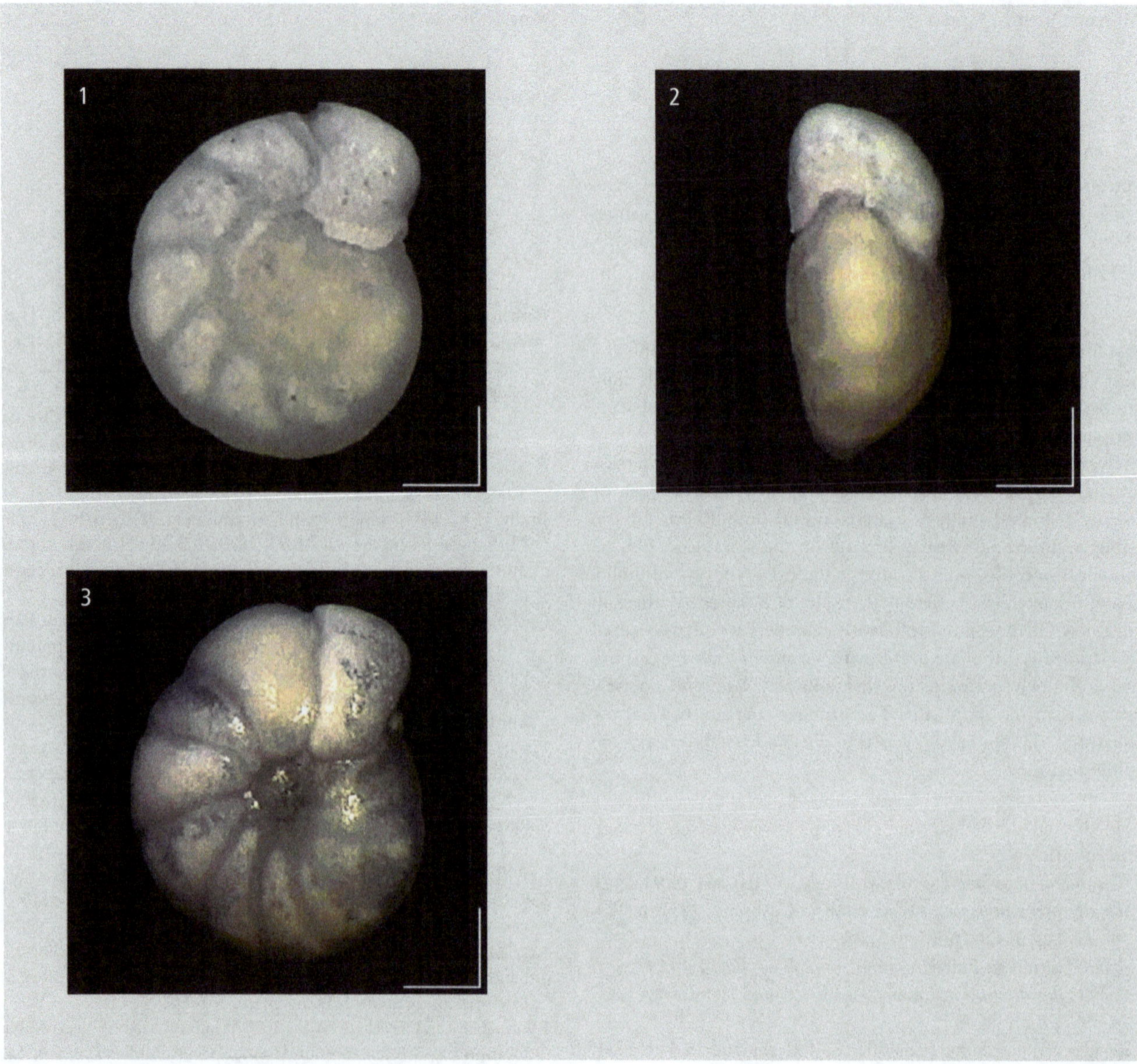


1. *Cibicidoides praemundulus.* Islas Orcadas Rise, South Atlantic; ODP Leg 114, Hole 702B, Core 3, Section 6, 68–72 cm. Spiral view, NHMUK PM DI 0279, scale bar = 112 μm.

2. *Cibicidoides praemundulus.* Islas Orcadas Rise, South Atlantic; ODP Leg 114, Hole 702B, Core 3, Section 6, 68–72 cm. Apertural view, NHMUK PM DI 0280, scale bar = 116 μm.

3. *Cibicidoides praemundulus.* Islas Orcadas Rise, South Atlantic; ODP Leg 114, Hole 702B, Core 3, Section 6, 68–72 cm. Umbilical view, NHMUK PM DI 0281, scale bar = 103 μm.

## Original Designation

*Cibicidoides praemundulus* Berggren and Miller, 1986. In van Morkhoven et al., 1986, p. 264, textfigs. 5–6, pl. 87.

## Type Specimen

Van Morkhoven et al. (1986) designated as holotype (USNM 406407) a specimen originally identified as *Cibicidoides ungerianus* (d'Orbigny) and figured in pl. 20, fig. 6 by Tjalsma and Lohmann (1983). Deposited in the collections of the United States National Museum, Washington, DC.

## Type Level

Upper Eocene.

## Type Locality

DSDP Leg 10, Site 94, latitude 24° 31.64′ N, longitude 88° 28.16′ W, Gulf of Mexico.

## Description

Test forms a biumbonate trochospire; subcircular in outline and nearly biconvex in cross-section with an involute, slightly more convex umbilical side, an evolute, convex spiral side, and a subacute, imperforate periphery. Ten to twelve moderately inflated chambers in the last whorl increase gradually in size, and are separated by flush, curved sutures. Sutures in the final whorl may be tangential to the spiral suture. Chamber walls are calcareous and finely perforate on the umbilical side with a distinctive, clear, glossy umbilical plug, and coarsely perforate on the spiral side. The primary aperture is a narrow equatorial slit bordered by a thin lip, and extends onto the spiral side.

## Synonyms and Citations

*Cibicidoides ungerianus* (d'Orbigny). –Tjalsma and Lohmann, 1983, p. 28, pl. 18, fig. 1; pl. 21, figs. 5–6. –Miller, 1983, p. 435, pl. 2, figs. 6–7.–Miller et al., 1985, pl. 1, figs. 4–6.

*Cibicidoides praemundulus* Berggren and Miller. –Nomura, 1995, p. 272, pl. 1, fig. 10. –Kuhnt et al., 2002, p. 141, pl. 8. figs. 12–17.

## Remarks

*Cibicidoides praemundulus* Berggren and Miller mainly differs from *Cibicidoides mundulus* (Brady, Parker, and Jones) by the distribution of pores on the spiral side. *Cibicidoides mundulus* has a coil of coarse perforations visible through the clear umbo on the spiral side. *Cibicidoides mundulus* intergrades with *Cibicidoides praemundulus* over the last two zones of the Oligocene (M. Katz, pers. comm. 1998).

## Biogeography

Worldwide. Recorded from Barbados, the Gulf of Mexico, the North and South Atlantic, and Europe (van Morkhoven et al., 1986).

## Bathymetry

Bathyal to abyssal.

## Chronostratigraphy

Early Eocene (Ypresian) to Early Miocene (Aquitanian).

# References


Kuhnt, W., Holbourn, A.E., and Zhao, Q., 2002, The early history of the South China Sea: Evolution of Oligocene–Miocene deep water environments. Revue de Micropaléontologie, vol. 45, pp. 99–159.

Miller, K.G., 1983, Eocene–Oligocene paleoceanography of the deep Bay of Biscay: benthic foraminiferal evidence. Marine Micropaleontology, vol. 7, pp. 403–440.

Miller, K.G., Curry, W.B., and Ostermann, D.R., 1985, Late Paleogene (Eocene to Oligocene) benthic foraminiferal oceanography of the Goban Spur region, Deep Sea Drilling Project Leg 80. In de Graciansky, P.C., Poag, C.W., et al., ed., Initial Reports of the Deep Sea Drilling Project, Washington (U.S. Government Printing Office), vol. 80, pp. 505–538.

Nomura, R., 1995, Paleogene to Neogene deep-sea paleoceanography in the eastern Indian Ocean: benthic foraminifera from ODP Sites 747, 757 and 758. Micropaleontology, vol. 41, pp. 251–290.

Tjalsma, R.C. and Lohmann, G.P., 1983, Paleocene–Eocene bathyal and abyssal benthic foraminifera from the Atlantic Ocean. Micropaleontology, Special Publication, no. 4, pp. 1–90.

van Morkhoven, F.P.C.M., Berggren, W.A., Edwards, A.S., et al., 1986, Cenozoic cosmopolitan deep-water benthic foraminifera. Bulletin des Centres de Recherches Exploration-Production Elf-Aquitaine, Mémoire, vol. 11, pp. 1–421.

# Cibicidoides robertsonianus

## (Brady), 1881

1. *Cibicidoides robertsonianus.* West Indies; "Challenger" Station 24, depth: 390 fathoms. Deposited in the micropaleontological collections of The Natural History Museum, London; NHMUK ZF2547. Spiral view, NHMUK PM DI 0282, scale bar = 276 μm.

2. *Cibicidoides robertsonianus.* West Indies; "Challenger" Station 24, depth: 390 fathoms. Deposited in the micropaleontological collections of The Natural History Museum, London; NHMUK ZF2547. Apertural view, NHMUK PM DI 0283, scale bar = 276 μm.

3. *Cibicidoides robertsonianus.* West Indies; "Challenger" Station 24, depth: 390 fathoms. Deposited in the micropaleontological collections of The Natural History Museum, London; NHMUK ZF2547. Umbilical view, NHMUK PM DI 0284, scale bar = 282 μm.

## Original Designation

*Planorbulina robertsoniana* Brady, 1881, p. 65 (type description), illustrated as *Truncatulina robertsoniana* in Brady, 1884, pl. 95, fig. 4.

## Type Specimen

Figured specimen deposited in the micropaleontological collections of The Natural History Museum, London, UK.

## Type Level

Recent, surface sediments.

## Type Locality

Not given. Figured specimen from "Challenger" Station 24, off Culebra Island, north of St Thomas, West Indies; 390 fathoms.

## Description

Test forms a trochospire; subcircular in outline and unequally biconvex in cross-section with a partially evolute, more convex umbilical side, an involute, convex umbilical side slightly depressed at the umbilicus, and a subacute, imperforate periphery. The thirteen to fourteen moderately inflated chambers in the last whorl (final chambers slightly more inflated) increase gradually in size and are separated by curved sutures; flush or slightly depressed on the umbilical side, thickened and flush on the spiral side. Sutures converge toward a large, glossy umbo on the umbilical side. Chamber walls are calcareous, thick, and smooth, coarsely perforate on the spiral side, and finely perforate on the umbilical side. Well-preserved specimens have distinctive rich brown-colored chambers visible through the translucent wall. The primary aperture is a narrow equatorial slit, bordered by a thin lip, and extends onto the spiral side.

## Synonyms and Citations

*Truncatulina robertsoniana* (Brady). –Brady, 1884, p. 664, pl. 95, fig. 4.

*Cibicides robertsonianus* var. *haitensis* Coryell and Rivero, 1940, p. 335, pl. 44, figs. 4–6.

*Gyroidina jarvisi* Cushman and Stainforth, 1945, p. 62, pl. 11, fig. 3.

*Cibicides robertsonianus* (Brady). –Barker, 1960, pl. 95, fig. 4. –Pflum and Frerichs, 1976, pl. 3, figs. 3–5.

*Cibicidoides robertsonianus* (Brady). –van Morkhoven et al., 1986, p. 41, pl. 11. –Miller and Katz, 1987, p. 132, pl. 7, fig. 1. –Jones, 1994, p. 99, pl. 95, fig. 4. –Robertson, 1998, p. 206, pl. 81, fig. 3.

## Remarks

*Cibicidoides robertsonianus* (Brady) differs from *Cibicidoides bradyi* (Trauth) by its larger test, and more angular periphery.

## Biogeography

Worldwide. Recorded from the Pacific, Tasman Sea, Atlantic, Mediterranean, and Gulf of Mexico (van Morkhoven et al., 1986).

## Bathymetry

Bathyal to abyssal.

## Chronostratigraphy

Middle Miocene (Langhian) to Recent.

# References


Barker, R.W., 1960, Taxonomic notes on the species figured by H.B. Brady in his report on the foraminifera dredged by H.M.S. Challenger during the years 1873–1876. Society of Economic Paleontologists and Mineralogists Special Publication, vol. 9, pp. 1–238.

Brady, H.B., 1881, Notes on some of the reticularian Rhizopoda of the Challenger Expedition, Part III. Quarterly Journal of Microscopical Science (New Series), vol. 21, pp. 31–71.

Brady, H.B., 1884, Report on the foraminifera dredged by H.M.S. Challenger during the years 1873–1876. In Brady, H.B., ed., Report on the Scientific Results of the Voyage of the H.M.S. Challenger during the years 1873–1876, Zoology, vol. 9, pp. 1–814.

Coryell, H.N. and Rivero, F.C., 1940, A Miocene microfauna of Haiti. Journal of Paleontology, vol. 14, pp. 324–344.

Cushman, J.A. and Stainforth, R.M., 1945, The foraminifera of the Cipero Marl Formation of Trinidad, British West Indies. Contributions from the Cushman Laboratory for Foraminiferal Research, vol. 14, pp. 3–75.

Jones, R.W., 1994, The Challenger Foraminifera. Oxford: Oxford University Press, 149 pp.

Miller, K.G. and Katz, M.E., 1987, Oligocene to Miocene benthic foraminiferal and abyssal circulation changes in the north Atlantic. Micropaleontology, vol. 33, no. 2, pp. 97–149.

Pflum, C.E. and Frerichs, W.E., 1976, Gulf of Mexico deep-water foraminifers. Cushman Foundation, Special Publication, vol. 14, pp. 1–125.

Robertson, B.E., 1998, Systematics and paleoecology of the benthic foraminiferida from the Buff Bay section, Miocene of Jamaica. Micropaleontology, vol. 44, supplement 2, pp. 1–266.

van Morkhoven, F.P.C.M., Berggren, W.A., Edwards, A.S., et al., 1986, Cenozoic cosmopolitan deep-water benthic foraminifera. Bulletin des Centres de Recherches Exploration-Production Elf-Aquitaine, Mémoire, vol. 11, pp. 1–42.

# *Cibicidoides subhaidingerii*

**(Parr), 1950**

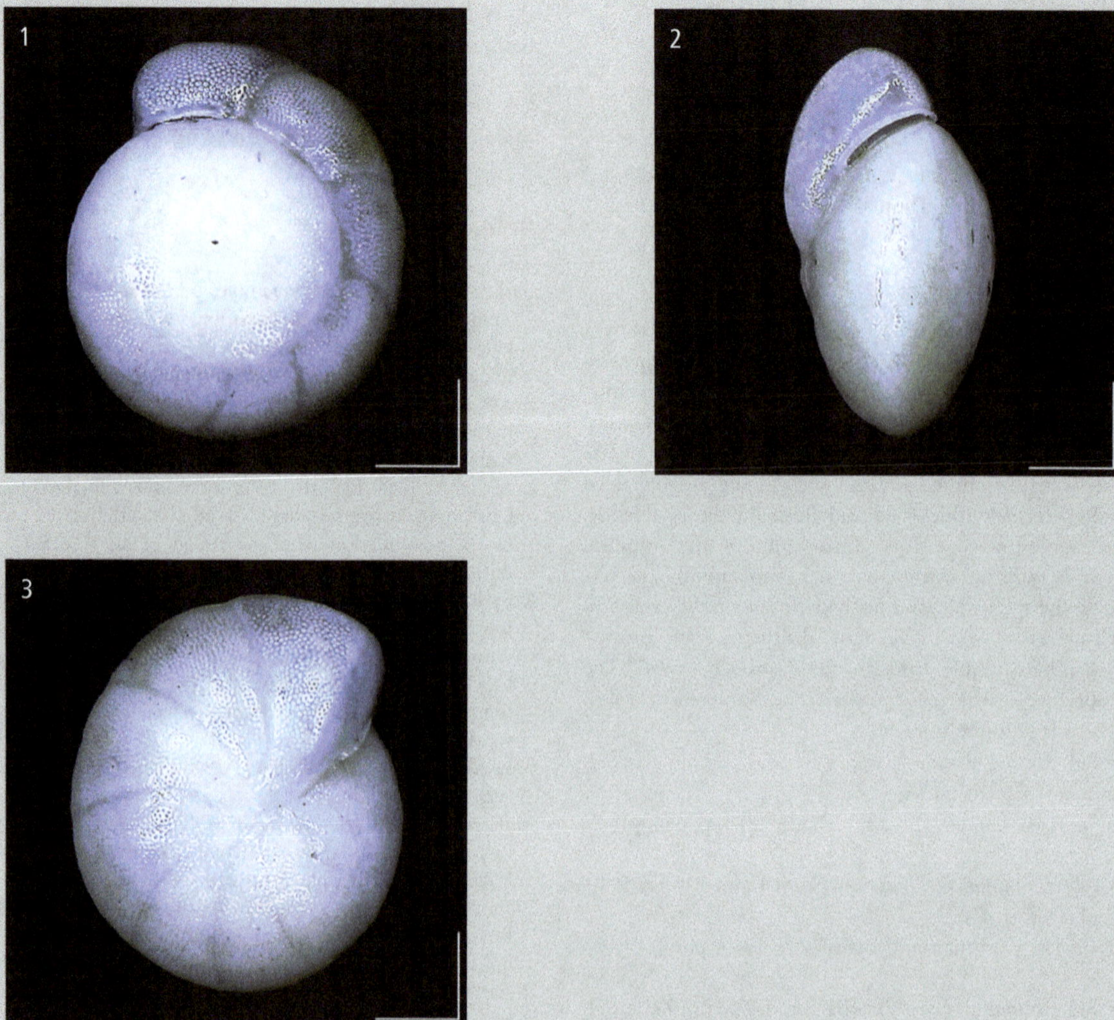


1. *Cibicidoides subhaidingerii*. Torres Strait, Pacific Ocean; "Challenger" Station 185, depth: 155 fathoms. Deposited in the micropaleontological collections of The Natural History Museum, London; NHMUK ZF2528. Spiral view, NHMUK PM DI 0285, scale bar = 243 μm.

2. *Cibicidoides subhaidingerii*. Torres Strait, Pacific Ocean; "Challenger" Station 185, depth: 155 fathoms. Deposited in the micropaleontological collections of The Natural History Museum, London; NHMUK ZF2528. Apertural view, NHMUK PM DI 0286, scale bar = 242 μm.

3. *Cibicidoides subhaidingerii*. Torres Strait, Pacific Ocean; "Challenger" Station 185, depth: 155 fathoms. Deposited in the micropaleontological collections of The Natural History Museum, London; NHMUK ZF2528. Umbilical view, NHMUK PM DI 0287, scale bar = 239 μm.

## Original Designation

*Cibicides subhaidingerii* Parr, 1950, p. 364, pl. 15, fig. 7.

## Type Specimen

Holotype and other specimens deposited at the University of Adelaide, Australia.

## Type Level

Recent, surface sediments.

## Type Locality

"Banzare" Station 115, latitude 41° 03′ S, longitude 148° 42′ E, off northeastern Tasmania, in 128 meters water depth (frequent). Other localities given: Station 113, off Maria Island, off the east coast of Tasmania, in 122–155 meters water depth, and Station 76, in the Indian Ocean, east of Albany, on the south coast of Western Australia, on the continental shelf, in 62 meters water depth.

## Description

Test forms a low trochospire; unequally biconvex in cross-section, with an involute, strongly convex umbilical side, a slightly convex, evolute spiral side, and a subrounded, imperforate periphery. The eight to ten moderately inflated chambers in the last whorl increase gradually in size and are separated by limbate, flush or slightly depressed sutures; radial, straight, or slightly curved on the umbilical side, curved on the spiral side. Chamber walls are calcareous, thick, and coarsely perforate on both sides of the test. The primary aperture is a narrow equatorial slit bordered by a thin lip, and extends onto the spiral side.

## Synonyms and Citations

*Truncatulina haidingerii* (d'Orbigny). –Brady, 1884, p. 663, pl. 95, fig. 7.

*Cibicides subhaidingerii* Parr. –Barker, 1960, pl. 95, fig. 7.

?*Cibicidoides yoitaensis* Matsunaga, 1963, p. 117, pl. 52, figs. 3a–c.

*Cibicidoides subhaidingerii* (Parr). –van Morkhoven et al., 1986, p. 95, pl. 28. –Jones, 1994, p. 99, pl. 95, fig. 7. –Robertson, 1998, p. 208, pl. 81, fig. 4.

## Remarks

Specimens may differ from the holotype by having a more acute periphery, thus resembling *Heterelopa dutemplei* (d'Orbigny). However, *Heterelopa dutemplei* has strongly curved, limbate umbilical sutures and lacks a large umbonal boss.

## Biogeography

Worldwide. Common along the east coast of Australia.

## Bathymetry

Neritic to lower bathyal (90–1775 fathoms, Jones, 1994).

## Chronostratigraphy

Late Eocene (Priabonian) to Recent.

## References


Barker, R.W., 1960, Taxonomic notes on the species figured by H.B. Brady in his report on the foraminifera dredged by H.M.S. Challenger during the years 1873–1876. Society of Economic Paleontologists and Mineralogists Special Publication, vol. 9, pp. 1–238.

Brady, H.B., 1884, Report on the foraminifera dredged by H.M.S. Challenger during the years 1873–1876. In Brady, H.B., ed., Report on the Scientific Results of the Voyage of the H.M.S. Challenger during the years 1873–1876. Zoology, vol. 9, pp. 1–814.

Jones, R.W., 1994, The Challenger Foraminifera. Oxford: Oxford University Press, 149 pp.

Matsunaga, T., 1963, Benthonic smaller foraminifera from the oil fields of northern Japan. Science Reports of the Tôhoku University, Sendai, Japan, ser. 2 (Geology), vol. 35, no. 2, pp. 67–122.

Parr, W.J., 1950, Foraminifera. Reports of the BANZ Antarctic Research Expedition 1929–1931, Series B (Zoology, Botany), vol. 5, no. 6, pp. 232–392.

Robertson, B.E., 1998, Systematics and paleoecology of the benthic foraminiferida from the Buff Bay section, Miocene of Jamaica. Micropaleontology, vol. 44, supplement 2, pp. 1–266.

van Morkhoven, F.P.C.M., Berggren, W.A., Edwards, A.S., et al., 1986, Cenozoic cosmopolitan deep-water benthic foraminifera. Bulletin des Centres de Recherches Exploration-Production Elf-Aquitaine, Mémoire, vol. 11, pp. 1–421.

# *Cibicidoides subspiratus*

## (Nuttall), 1930

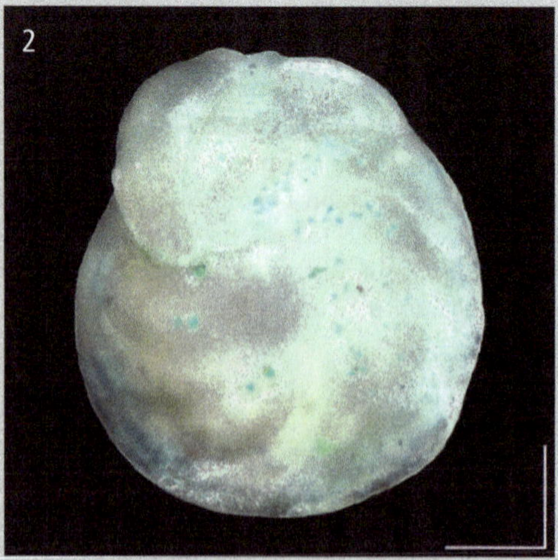

1. *Cibicidoides subspiratus.* Aragon, La Antigua, Mexico. Deposited in the collections of the United States National Museum, Washington, DC; CC 59520. Cotype. Spiral view, NHMUK PM DI 0288, scale bar = 123 µm.

2. *Cibicidoides subspiratus.* Aragon, La Antigua, Mexico. Deposited in the collections of the United States National Museum, Washington, DC; CC 59520. Cotype. Umbilical view, NHMUK PM DI 0289, scale bar = 133 µm.

### Original Designation

*Cibicides subspirata* Nuttall, 1930, p. 292, pl. 25, figs. 9, 10, 14.

### Type Specimen

Cotypes (CC 59520 and CC 59521) deposited in the collections of the United States National Museum, Washington, DC. Van Morkhoven et al. (1986) designated as lectotype CC 59521.

### Type Level

Lower Eocene, Aragon Formation. Other level: Lower Eocene Chicontepec Formation.

### Type Locality

La Antigua, Rio La Puerta, Tampico region, Mexico.

### Description

Test forms a trochospire; unequally biconvex in cross-section with an involute, strongly convex, umbilical side, an evolute, slightly convex spiral side, and a subacute, keeled periphery. The (approximately) ten moderately inflated chambers in the last whorl increase gradually in size, and are separated by strongly curved, slightly depressed sutures. The umbilical side has a distinct, clear, central boss. Early whorls are visible on the spiral side. Chamber walls are calcareous and finely perforate on both sides of the test. The primary aperture is a narrow equatorial arch bordered by a thin lip, and extends onto the spiral side.

### Synonyms and Citations

*Cibicides subspiratus* Nuttall var. *limbatus* Cita, 1950, p. 102, pl. 9, fig. 7.

*Cibicidoides subspiratus* (Nuttall). –van Morkhoven et al., 1986, p. 314, pl. 102. –Bolli et al., 1994. p. 367, fig. 56.7–8.

### Biogeography

Worldwide. Recorded from Cuba, Trinidad, Haiti, the Dominican Republic, the North and South Atlantic, the Caribbean, the Gulf of Mexico, and Europe (van Morkhoven et al., 1986).

### Bathymetry

Bathyal to abyssal (van Morkhoven et al., 1986).

### Chronostratigraphy

Early Eocene (Ypresian) to Middle Eocene (Bartonian).

## References


Bolli, H.M., Beckmann, J.P., and Saunders, J.B., 1994, Benthic Foraminiferal Biostratigraphy of the South Caribbean Region. Cambridge: Cambridge University Press, 408 pp.

Cita, M.B., 1950, L'Eocene della sponda occidentale del Lago di Garda; Parte 1 – Piccoli foraminiferi. Rivista Italiana Paleontologia e Stratigrafia, vol. 56, no. 3, pp. 81–113.

Nuttall, W.L.F., 1930, Eocene foraminifera from Mexico. Journal of Paleontology, vol. 4, pp. 271–293.

van Morkhoven, F.P.C.M., Berggren, W.A., Edwards, A.S., et al., 1986, Cenozoic cosmopolitan deep-water benthic foraminifera. Bulletin des Centres de Recherches Exploration-Production Elf-Aquitaine, Mémoire, vol. 11, pp. 1–421.

# Cibicidoides velascoensis

## (Cushman), 1925

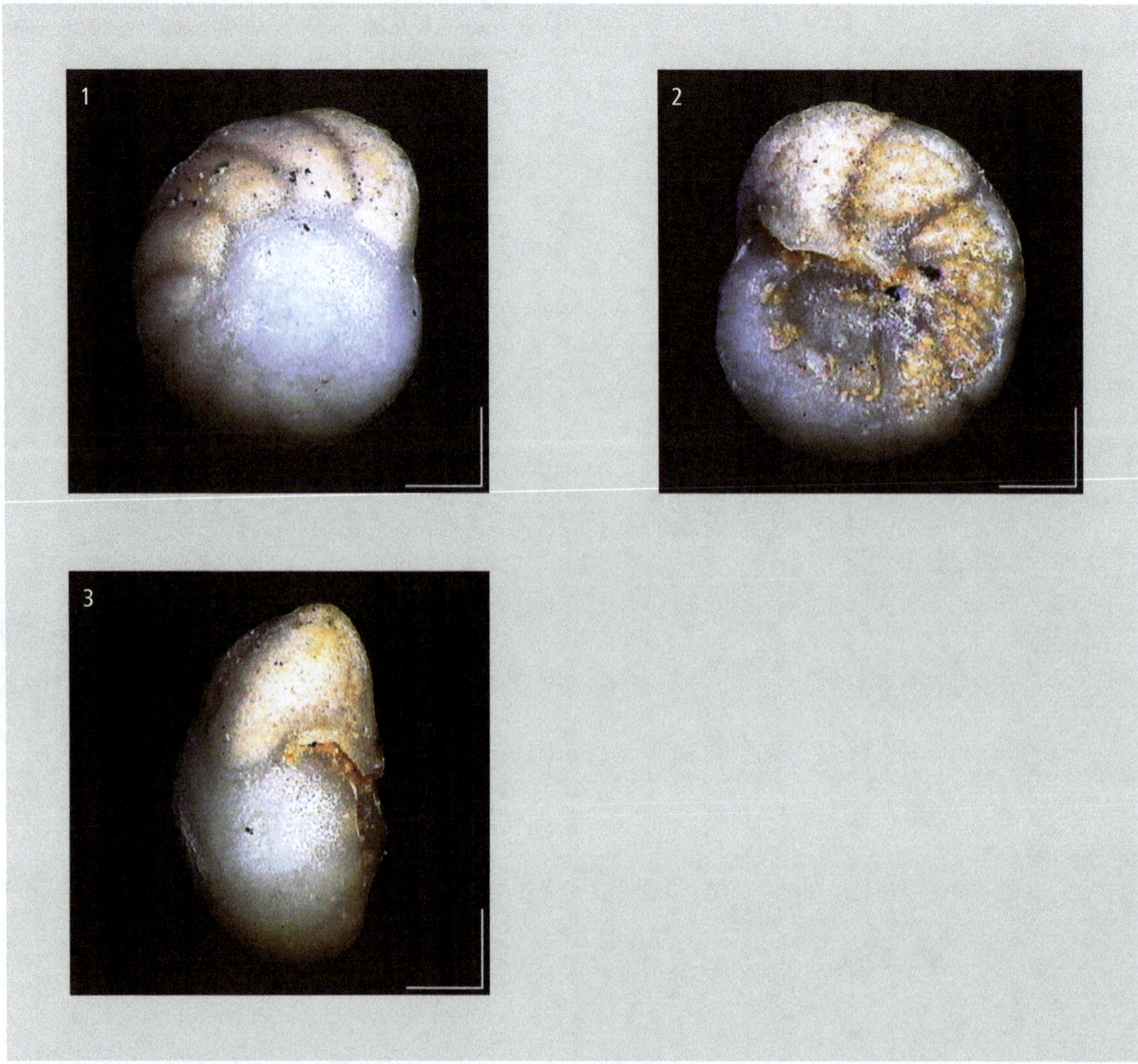


1. *Cibicidoides velascoensis*. Middle Velasco Formation, Top Red Bed, Tamaulipas, Mexico. Deposited in the micropaleontological collections of The Natural History Museum, London; NHMUK P25368. Spiral view, NHMUK PM DI 0290, scale bar = 120 μm.

2. *Cibicidoides velascoensis*. Middle Velasco Formation, Top Red Bed, Tamaulipas, Mexico. Deposited in the micropaleontological collections of The Natural History Museum, London; NHMUK P25368. Umbilical view, NHMUK PM DI 0291, scale bar = 120 μm.

3. *Cibicidoides velascoensis*. Middle Velasco Formation, Top Red Bed, Tamaulipas, Mexico. Deposited in the micropaleontological collections of The Natural History Museum, London; NHMUK P25368. Apertural view, NHMUK PM DI 0292, scale bar = 118 μm.

## Original Designation

*Anomalina velascoensis* Cushman, 1925, p. 21, pl. 3, fig. 3.

## Type Specimen

Holotype (No. 4345) deposited in the collections of the United States National Museum, Washington, DC.

## Type Level

Paleocene (van Morkhoven et al., 1986), Velasco Shale. Originally given as Upper Cretaceous by Cushman (1925).

## Type Locality

Tamalte Arroyo, Hacienda El Limón, State of San Luis Potosi, Mexico.

## Description

Test forms a trochospire; planoconvex in cross-section with a flattened, evolute, spiral side, a strongly convex, partially evolute, umbilical side, and a rounded, imperforate periphery. The nine to ten inflated chambers in the last whorl increase gradually in size and are separated by limbate, curved sutures that are raised on the spiral side but flush on the umbilical side. Chamber walls are calcareous and finely perforate on the umbilical side, and coarsely perforate on the spiral side. The central area of spiral side is thickened and generally slightly elevated. The primary aperture is a narrow equatorial slit bordered by a thin lip, and extends onto the spiral side.

## Synonyms and Citations

*Anomalina misrensis* Said and Kenawy, 1956, p. 153, pl. 6, fig. 22.
*Cibicidoides velascoensis* (Cushman). –van Morkhoven et al., 1986, p. 371, pl. 121. –Nomura, 1991, p. 22, pl. 3, fig. 5.

## Remarks

*Cibicidoides velascoensis* (Cushman) and *Cibicidoides hyphalus* (Fisher) share the same stratigraphic range. However, *Cibicidoides velascoensis* differs in its much larger diameter, more strongly convex umbilical side, and the smaller number of chambers in its last whorl (van Morkhoven et al., 1986).

## Biogeography

Worldwide. Recorded from the Caribbean, North and South Atlantic, Gulf of Mexico, California, Europe, Tunisia, and Egypt (van Morkhoven et al., 1986).

## Bathymetry

Bathyal to abyssal (van Morkhoven et al., 1986).

## Chronostratigraphy

Late Cretaceous (Campanian) to Early Eocene (Ypresian).

# References


Cushman, J.A., 1925, Some new foraminifera from the Velasco Shale of Mexico. Contributions from the Cushman Laboratory for Foraminiferal Research, vol. 1, no. 1, pp. 18–23.

Nomura, R., 1991, Paleoceanography of upper Maestrichtian to Eocene benthic foraminiferal assemblages at Sites 752, 753 and 754, eastern Indian Ocean. In Weissel, J., Price, J., Taylor, E., Alt, J. et al., ed., Scientific Results of the Proceedings of the Ocean Drilling Program, vol. 121, pp. 3–29.

Said, R. and Kenawy, A., 1956, Upper Cretaceous and Lower Tertiary foraminifera from northern Sinai, Egypt. Micropaleontology, vol. 2, no. 2, pp. 105–173.

van Morkhoven, F.P.C.M., Berggren, W.A., Edwards, A.S., et al., 1986, Cenozoic cosmopolitan deep-water benthic foraminifera. Bulletin des Centres de Recherches Exploration-Production Elf-Aquitaine, Mémoire, vol. 11, pp. 1–421.

# *Cibicorbis hitchcockae*

## (Galloway and Wissler), 1927

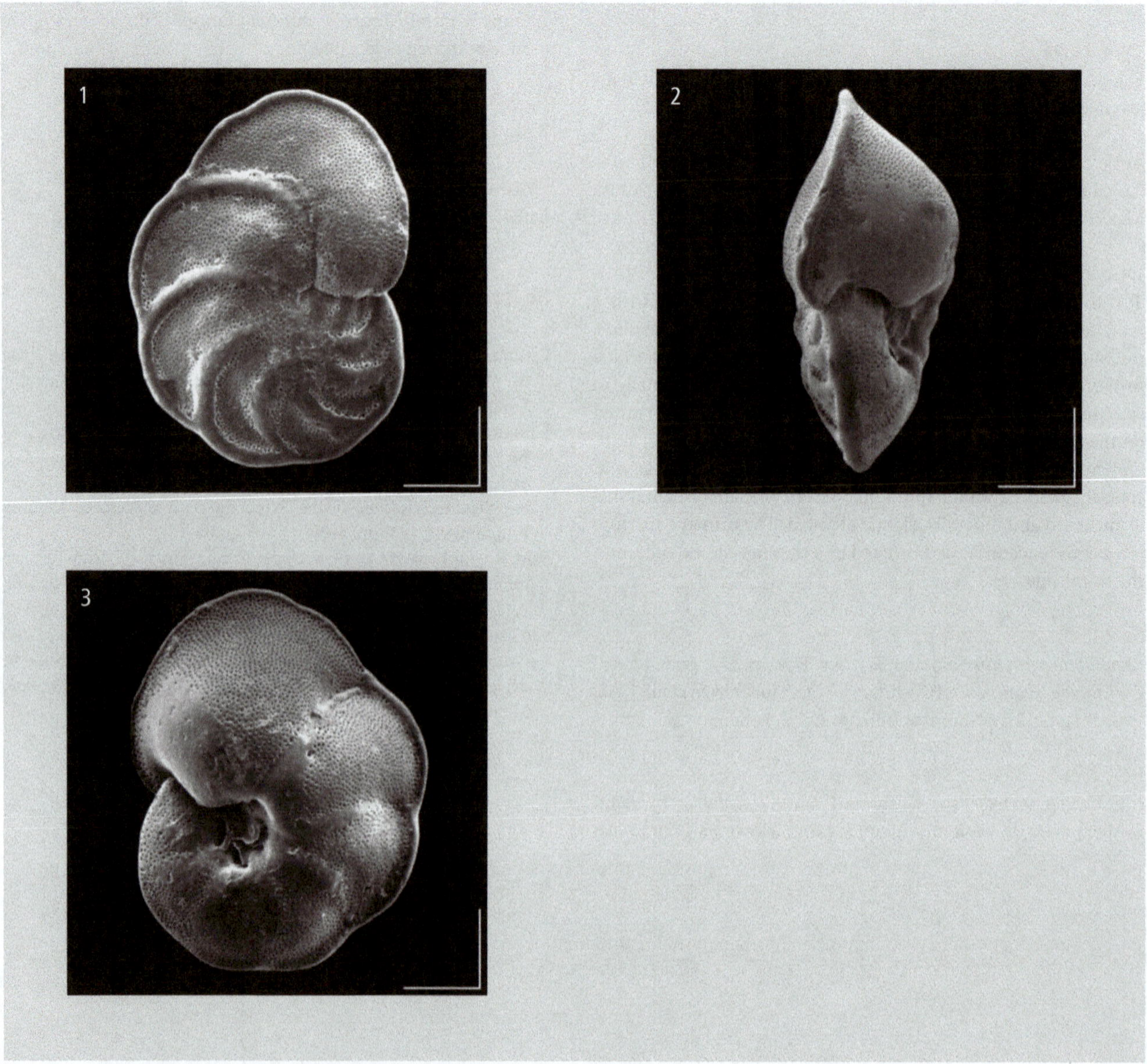

1. *Cibicorbis hitchcockae.* Upper Onzole Formation, Borbón Basin, Ecuador; Sample JW 325 (Esmeraldas Sheet). Deposited in the micropaleontological collections of The Natural History Museum, London; NHMUK P51846. Specimen illustrated in Whittaker, 1988, pl. 15, fig. 1. Spiral view, NHMUK PM DI 0293, scale bar = 133 μm.

2. *Cibicorbis hitchcockae.* Upper Onzole Formation, Borbón Basin, Ecuador; Sample JW 325 (Esmeraldas Sheet). Deposited in the micropaleontological collections of The Natural History Museum, London; NHMUK P51846. Specimen illustrated in Whittaker, 1988, pl. 15, fig. 1. Apertural view, NHMUK PM DI 0294, scale bar = 133 μm.

3. *Cibicorbis hitchcockae.* Upper Onzole Formation, Borbón Basin, Ecuador; Sample JW 325 (Esmeraldas Sheet). Deposited in the micropaleontological collections of The Natural History Museum, London; NHMUK P51846. Specimen illustrated in Whittaker, 1988, pl. 15, fig. 1. Umbilical view, NHMUK PM DI 0295, scale bar = 136 μm.

## Original Designation

*Rosalina hitchcockae* Galloway and Wissler, 1927, p. 62, pl. 10, fig. 2.

## Type Specimen

Deposited in Columbia University, New York, NY, USA, Paleontology Collection No. 19850.

## Type Level

Pleistocene, Lower San Pedro, lower bed, limestone.

## Type Locality

"Lomita Quarry" known as "Decomposed Marine Shell and Bone Quarry" on the north side of Palos Verdes Hills, about 20 miles south of Los Angeles and 2 miles south of the town of Lomita, California.

## Description

Test forms a trochospire; lobulate in outline in the final part of the test, planoconvex in cross-section, with a flattened or slightly concavo-convex, evolute spiral side and a convex, strongly flaring, involute, umbilical side. The periphery is carinate, acute on the spiral side, and sharply convex on the umbilical side. The (approximately) eight broad, low chambers rapidly increase in height and are strongly inflated on the umbilical side. The chambers are separated by thick, arcuate, raised sutures on the spiral side that merge to form the peripheral keel. On the umbilical side, the sutures are thick, straight, and raised or flush. Chamber walls are calcareous, thick, smooth, and finely perforate on both sides of the test. The primary aperture is an interiomarginal slit on the umbilical side, extending from the periphery to the umbilicus and partially covered by a very large, imperforate, imbricate umbilical flap, overlapping previous flaps and obscuring the umbilicus.

## Synonyms and Citations

*Cibicorbis herricki* Hadley, 1934, 26, pl. 5, figs. 1–3.

*Valvulineria collis* Coryell and Rivero, 1940, p. 338, pl. 43, fig. 24.

*Cibicides kugleri* Cushman and Renz, 1941, p. 27, pl. 4, fig. 11.

*Cibicorbis herricki* Hadley. –van Morkhoven et al., 1986, p. 107, pl. 33. –Loeblich and Tappan, 1987, p. 546, pl. 591, figs. 8–10. –Revets, 1996, p. 77, pl. 10, figs. 9–12. –Robertson, 1998, p. 188, pl. 70, fig. 2.

## Remarks

Type species of the genus *Cibicorbis* Hadley, 1934.

We follow Crouch and Poag (1979) and Whittaker (1988) in considering *Cibicorbis herricki* Hadley to be a junior synonym of *Cibicorbis hitchcockae* (Galloway and Wissler). Revets (1996, p. 77), however, regarded *Cibicorbis herricki* and *Cibicorbis hitchcockae* as distinct species, on account of morphological differences, the most obvious being the size of the umbilical flaps covering the umbilicus.

## Biogeography

May have a limited geographic distribution. Recorded from Cuba, the Dominican Republic, Mexico, Jamaica, Venezuela, and the Peru–Chile Trench area (van Morkhoven et al., 1986).

## Bathymetry

Lower neritic to upper bathyal (van Morkhoven et al., 1986).

## Chronostratigraphy

Middle Miocene (Langhian) to Recent.

# References


Coryell, H.N. and Rivero, F.C., 1940, A Miocene microfauna of Haiti. Journal of Paleontology, vol. 14, pp. 324–344.

Crouch, R.W. and Poag, C.W., 1979, Amphistegina gibbosa d'Orbigny from the California borderlands. Journal of Foraminiferal Research, vol. 9, no. 2, pp. 85–105.

Cushman, J.A. and Renz, H.H., 1941, New Oligocene–Miocene foraminifera from Venezuela. Contributions from the Cushman Laboratory for Foraminiferal Research, vol. 17, pp. 1–27.

Galloway, J.J. and Wissler, S.G., 1927, Pleistocene foraminifera from the Lomita Quarry, Palos Verdes Hills, California. Journal of Paleontology, vol. 1, pp. 35–87.

Hadley, W.H., 1934, Some Tertiary foraminifera from the north coast of Cuba. Bulletin of American Paleontology, vol. 20, pp. 1–40.

Loeblich, A.R. Jr and Tappan, H., 1987, Foraminiferal Genera and Their Classification. New York: Van Nostrand Reinhold, 970 pp.

Revets, S.A., 1996, The generic revision of five families of Rotaliine Foraminifera. Special Publications of the Cushman Foundation for Foraminiferal Research, vol. 34, pp. 1–108.

Robertson, B.E., 1998, Systematics and paleoecology of the benthic foraminiferida from the Buff Bay section, Miocene of Jamaica. Micropaleontology, vol. 44, supplement 2, pp. 1–266.

van Morkhoven, F.P.C.M., Berggren, W.A., Edwards, A.S., et al., 1986, Cenozoic cosmopolitan deep-water benthic foraminifera. Bulletin des Centres de Recherches Exploration-Production Elf-Aquitaine, Mémoire, vol. 11, pp. 1–421.

Whittaker, J.E., 1988, Benthic Foraminifera from Ecuador: Taxonomy and Distribution of Smaller Benthic Foraminifera from Coastal Ecuador (Late Oligocene–Late Pliocene). London: British Museum (Natural History), xi + 194 pp.

# Clavulinoides aspera

## (Cushman), 1926

1. *Clavulinoides aspera.* Velasco Formation, Hacienda El Limon, Vera Cruz, Mexico; Franco-Espanola Well 5686 feet, Marland Oil Co. Deposited in the collections of the United States National Museum, Washington, DC, CC 5154. Holotype. Lateral view, NHMUK PM DI 0296, scale bar = 211 μm.

### Original Designation

*Clavulina trilatera* Cushman var. *aspera* Cushman, 1926, p. 589, pl. 17, fig. 3.

### Type Specimen

Holotype (CC 5154) deposited in the collections of the United States National Museum, Washington, DC.

### Type Level

Paleocene, Velasco Shale. Originally given as Upper Cretaceous by Cushman (1926).

### Type Locality

Eastern part of the Hacienda El Limón, State of San Luis, Potosi, Mexico.

### Description

Test large and elongate, triangular in cross-section throughout, initially triserial but becoming uniserial. Chambers inflated, broad, and low in the uniserial portion, separated by flush sutures in the triserial portion and slightly depressed sutures in the uniserial portion. Chamber walls are very coarsely agglutinated and canaliculate with a rough finish. The primary aperture is circular and terminal, without a distinct tooth.

### Synonyms and Citations

*Clavulina trilatera* Cushman var. *aspera* Cushman. –Cushman and Jarvis, 1928, p. 93, pl. 13, fig. 5.

*Clavulinoides aspera* (Cushman). –Cushman and Renz, 1946, p. 21, pl. 2, fig. 23. –Kaminski et al., 1988, p. 194, pl. 8, fig. 13.

*Tritaxia aspera* (Cushman). –Tjalsma and Lohmann, 1983, p. 20, pl. 1, fig. 1.

*Clavulinoides asper* (Cushman). –Bolli et al., 1994, p. 95, figs. 25.8, 15–16.

### Biogeography

Worldwide.

### Bathymetry

Bathyal to abyssal.

### Chronostratigraphy

Early Paleocene (Danian) to Early Eocene (Ypresian). Bolli et al. (1994) reported a range from the Early Maastrichtian to Late Paleocene in Trinidad.

## References

Bolli, H.M., Beckmann, J.P., and Saunders, J.B., 1994, Benthic Foraminiferal Biostratigraphy of the South Caribbean Region. Cambridge: Cambridge University Press, 408 pp.

Cushman, J.A., 1926, The foraminifera of the Velasco Shale of the Tampico Embayment. Bulletin of the American Association of Petroleum Geologists, vol. 10, no. 6, pp. 581–612.

Cushman, J.A. and Jarvis, P.W., 1928, Cretaceous foraminifera from Trinidad. Contributions from the Cushman Laboratory for Foraminiferal Research, vol. 4, pp. 85–103.

Cushman, J.A. and Renz, H.H., 1946, The foraminiferal fauna of the Lizard Springs Formation of Trinidad, British West Indies. Special Publications of the Cushman Laboratory for Foraminiferal Research, vol. 18, pp. 1–48.

Kaminski, M.A., Gradstein, F.M., Berggren, W.A., Geroch, S., and Beckmann, J.P., 1988, Agglutinated foraminiferal assemblages from Trinidad: taxonomy, stratigraphy and paleobathymetry. In Gradstein, F.M. and Rögl, F., ed., Second International Workshop on Agglutinated Foraminifera, Vienna 1986, Abhandlungen der Geologischen Bundesanstalt, vol. 41, pp. 155–228.

Tjalsma, R.C. and Lohmann, G.P., 1983, Paleocene–Eocene bathyal and abyssal benthic foraminifera from the Atlantic Ocean. Micropaleontology, Special Publication, no. 4, pp. 1–90.

# *Conotrochammina voeringensis*

## Gradstein and Kaminski, 1997

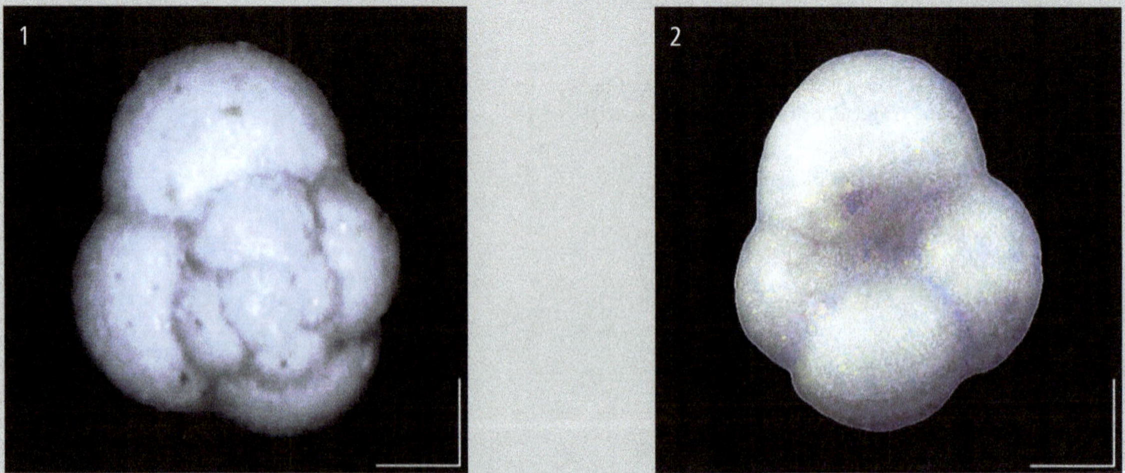


1. *Conotrochammina voeringensis*. Norwegian Sea; ODP Leg 104, Hole 643A, Core 57, Section 3, 83–88 cm. Deposited in the micropaleontological collections of The Natural History Museum, London; NHMUK PF62945. Holotype. Spiral view, NHMUK PM DI 0297, scale bar = 64 µm.

2. *Conotrochammina voeringensis*. Norwegian Sea; ODP Leg 104, Hole 643A, Core 57, Section 3, 83–88 cm. Deposited in the micropaleontological collections of The Natural History Museum, London; NHMUK PF62945. Holotype. Umbilical view, NHMUK PM DI 0298, scale bar = 66 µm.

### Original Designation

*Conotrochammina voeringensis* Gradstein and Kaminski, 1997, p. 224, figs. 9–10.

### Type Specimen

Holotype registered as NHMUK PF62945, and housed in the micropaleontological collections of The Natural History Museum, London, UK.

### Type Level

Lower Eocene.

### Type Locality

ODP Hole 643A, Voering Slope, Norwegian Sea, Samples 643A-57X-3, 83–88 cm.

### Description

Test forms a small, high trochospire with a lobulate outline, a rounded periphery, and a closed, slightly depressed umbilicus. Test consists of three or four whorls with four globular chambers per whorl, each separated by depressed sutures. On the spiral side sutures are often aligned and radiating outward. Chamber walls are medium to finely agglutinated and smoothly finished. The primary aperture is an oval, areal, extraumbilical opening, surrounded by a lip, and situated in the axis of trochospiral coiling.

### Synonyms and Citations

*Conotrochamina* sp. –Kaminski et al., 1990, p. 370, pl. 8, figs. 1–2.

*Trochammina* sp. 1. –Charnock and Jones, 1990, p. 187, pl. 11, figs. 4–6; pl. 22, fig. 5.

*Conotrochammina voeringensis* Gradstein and Kaminski. –Kaminski and Gradstein, 2005, pp. 460–462, fig. 113, pl. 113.

### Remarks

*Conotrochammina voeringensis* Gradstein and Kaminski differs from *Conotrochammina whangaia* Finlay by its smaller size, fine-grained wall, and closed umbilicus, and by the presence of a lip surrounding the aperture (Gradstein and Kaminski, 1997).

### Biogeography

Common in flysch-type agglutinated assemblages in the Paleogene of the North Sea (Charnock and Jones, 1990; Gradstein and Kaminski, 1997). See Observed Occurrences in Kaminski and Gradstein (2005).

### Bathymetry

Lower bathyal to abyssal (flysch assemblages).

### Chronostratigraphy

Late Cretaceous (Campanian) to Middle Eocene (Charnock and Jones, 1990). Extending into the Oligocene in the Norwegian Sea (Kaminski and Gradstein, 2005).

## References


Charnock, M.A. and Jones, R.W., 1990, Agglutinated foraminifera from the Paleogene of the North Sea. In Hemleben, C., Kaminski, M.A., Kuhnt, W., and Scott, D.B., ed., Paleoecology, Biostratigraphy, Paleoceanography and Taxonomy of Agglutinated Foraminifera, NATO ASI Series, C: Mathematical and Physical Sciences, Dordrecht: Kluwer Academic, pp. 139–244.

Gradstein, F.M. and Kaminski, M.A., 1997, New species of Paleogene deep-water agglutinated foraminifera from the North Sea and Norwegian Sea. Annales Societatis Geologorum Poloniae, vol. 67, pp. 217–229.

Kaminski, M.A. and Gradstein, F.M., 2005, Atlas of Palaeogene Cosmopolitan Deep-Water Agglutinated Foraminifera. Grzybowski Foundation Special Publication, no. 10, 547 + vii pp.

# Coryphostoma incrassata

## (Reuss), 1851

1. *Coryphostoma incrassata.* Gulf of Mexico; Eureka Core E68-139. Deposited in the collections of the United States National Museum, Washington, DC. Specimen illustrated in van Morkhoven et al., 1986, Ref. no. 126. Lateral view, NHMUK PM DI 0299, scale bar = 233 µm.

### Original Designation

*Bolivina incrassata* Reuss, 1851, p. 45, pl. 5, fig. 13.

### Type Specimen

Depository not given.

### Type Level

Upper Cretaceous marl, lower gray layers.

### Type Locality

Lemberg, East Galicia, Poland.

### Description

Test forms an elongate, compressed, biserial arrangement of chambers, about two and a half times as long as broad and tapered in outline, oval in cross-section, with a rounded initial portion, and with a subrounded periphery. The low, broad, slightly inflated chambers increase gradually in size and are separated by distinct, slightly depressed, strongly oblique sutures that are occasionally limbate. Chamber walls are calcareous, finely perforate, and smooth or ornamented by very fine striae. The primary aperture is loop-shaped, with an internal toothplate, and extends up from the base of the final chamber, progressively becoming terminal in adults.

### Synonyms and Citations

*Coryphostoma incrassata* (Reuss). –van Morkhoven et al., 1986, p. 382, pl. 126.

*Bolivina incrassata* Reuss. –Hart et al., 1989, p. 322, pl. 7.4, figs. 8–9.

*Coryphostoma incrassatum* (Reuss). –Bolli et al., 1994, p. 138, figs. 37.10–12.

### Remarks

*Coryphostoma incrassata* (Reuss) differs from *Coryphostoma midwayensis* (Cushman) by its larger, broader test. The two species also have different stratigraphic ranges: *Coryphostoma midwayensis* ranges from the Maastrichtian to the Early Eocene; in contrast, *Coryphostoma incrassata* ranges from the late Campanian to late Maastrichtian. The degree of taper is variable in megalospheric and microspheric forms.

### Biogeography

Worldwide.

### Bathymetry

Primarily bathyal. Also found in outer neritic deposits (van Morkhoven et al., 1986).

### Chronostratigraphy

Late Cretaceous (Campanian to Maastrichtian).

## References


Bolli, H.M., Beckmann, J.P., and Saunders, J.B., 1994, Benthic Foraminiferal Biostratigraphy of the South Caribbean Region. Cambridge: Cambridge University Press, 408 pp.

Hart, M.B., Bailey, H.W., Crittenden, S., Fletcher, B.N., Price, R.J., and Swiecicki, A., 1989, Cretaceous. In Jenkins, D.G., and Murray, J.W., ed., Stratigraphical Atlas of Fossil Foraminifera, 2nd edn. Chichester: Ellis Horwood, pp. 273–371.

Reuss, A.E., 1851, Die Foraminiferen und Entomostraceen des Kreidemergels von Lemberg. Naturwissenschaftliche Abhandlungen, Wien, Österreich vol. 4, pp. 17–52.

van Morkhoven, F.P.C.M., Berggren, W.A., Edwards, A.S., et al., 1986, Cenozoic cosmopolitan deep-water benthic foraminifera. Bulletin des Centres de Recherches Exploration-Production Elf-Aquitaine, Mémoire, vol. 11, pp. 1–421.

# Coryphostoma midwayensis

## (Cushman), 1936

1. *Coryphostoma midwayensis.* Midway Formantion, 4 miles southeast of Corsicana, Texas. Deposited in the collections of the United States National Museum, Washington, DC; CC 5335. Holotype. Lateral view, NHMUK PM DI 0300, scale bar = 88 µm.

### Original Designation

*Bolivina midwayensis* Cushman, 1936, p. 50, pl. 7, fig. 12.

### Type Specimen

Holotype (No. 5335) deposited in the collections of the United States National Museum, Washington, DC.

### Type Level

Paleocene (van Morkhoven et al., 1986), Midway. Originally given as Lower Eocene by Cushman (1936).

### Type Locality

Shallow ditch at road corner southeast of New Corsicana Reservoir, on road to Mildred, Navarro County, Texas.

### Description

Test forms an elongate, compressed, biserial series, narrow and slightly tapered in outline, suboval in cross-section with a subrounded periphery. The low, broad, slightly inflated chambers increase gradually in size, overlap slightly, and are separated by distinct, slightly depressed, strongly oblique sutures that form an angle of at least 45° with the test axis. Chamber walls are calcareous and finely perforate. The primary aperture is loop-shaped, with an internal toothplate, extending up from the base of the final chamber and becoming terminal in adults.

### Synonyms (following van Morkhoven et al., 1986) and Citations

?*Bolivina primatumida* White, 1929, p. 44, pl. 4, fig. 20.

*Loxostomum limonense* (Cushman). –Cushman and Renz, 1946, p. 39, pl. 6, fig. 23.

*Coryphostoma paleocenica* Akpati, 1966, p. 138, pl. 10, fig. 12.

*Coryphostoma midwayensis* (Cushman). –van Morkhoven et al., 1986, p. 364, pl. 118.

*Coryphostoma midwayense* (Cushman). –Bolli et al., 1994, p. 138, figs. 37.13–15.

### Remarks

*Coryphostoma incrassata* (Reuss) differs from *Coryphostoma midwayensis* (Cushman) by its larger, broader test. The two species also have different stratigraphic ranges: *Coryphostoma midwayensis* ranges from the Maastrichtian to the Early Eocene; in contrast, *Coryphostoma incrassata* ranges from the late Campanian to late Maastrichtian. The degree of taper is variable in megalospheric and microspheric forms.

### Biogeography

Worldwide.

### Bathymetry

Primarily found in lower neritic to bathyal deposits (van Morkhoven et al., 1986).

### Chronostratigraphy

Late Cretaceous (Maastrichtian) to Early Eocene (Ypresian).

## References


Akpati, B.N., 1966, Some Lower Paleocene foraminifera from the Anita Formation: Santa Barbara County, California. Contributions from the Cushman Laboratory for Foraminiferal Research, vol. 17, no. 4, pp. 135–139.

Bolli, H.M., Beckmann, J.P., and Saunders, J.B., 1994, Benthic Foraminiferal Biostratigraphy of the South Caribbean Region. Cambridge: Cambridge University Press, 408 pp.

Cushman, J.A., 1936, New genera and species of the families Verneuilinidae and Valvulinidae and of the subfamily Virgulininae. Contributions from the Cushman Laboratory for Foraminiferal Research, vol. 6, pp. 1–71.

Cushman, J.A. and Renz, H.H., 1946, The foraminiferal fauna of the Lizard Springs Formation of Trinidad, British West Indies. Special Publications of the Cushman Laboratory for Foraminiferal Research, vol. 18, pp. 1–48.

van Morkhoven, F.P.C.M., Berggren, W.A., Edwards, A.S., et al., 1986, Cenozoic cosmopolitan deep-water benthic foraminifera. Bulletin des Centres de Recherches Exploration-Production Elf-Aquitaine, Mémoire, vol. 11, pp. 1–421.

White, M.P., 1929, Some index foraminifera of the Tampico Embayment area of Mexico (Part III). Journal of Paleontology, vol. 3, no. 1, pp. 30–58.

# Cribrostomoides subglobosus

## (Cushman), 1910

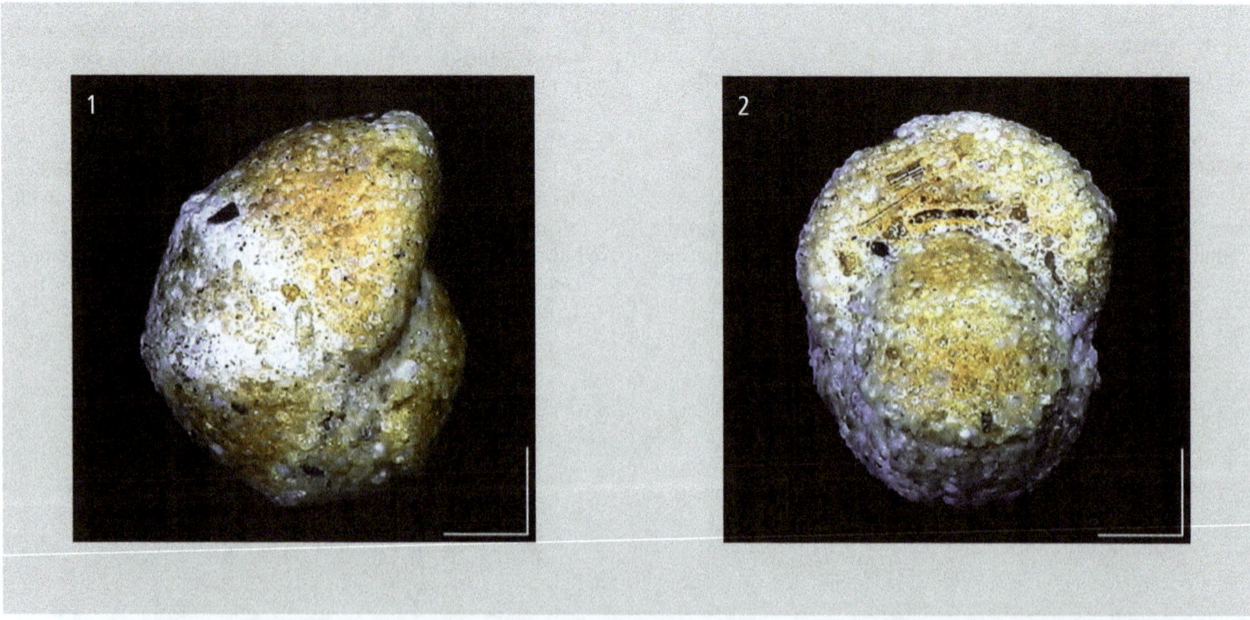


1. *Cribrostomoides subglobosus.* Off Culebra Island, West Indies; "Challenger" Station 24, depth: 390 fathoms. Deposited in the micropaleontological collections of The Natural History Museum, London; NHMUK ZF1542. Lateral view, NHMUK PM DI 0302, scale bar = 244 μm.

2. *Cribrostomoides subglobosus.* Off Culebra Island, West Indies; "Challenger" Station 24, depth: 390 fathoms. Deposited in the micropaleontological collections of The Natural History Museum, London; NHMUK ZF1542. Apertural view, NHMUK PM DI 0301, scale bar = 256 μm.

### Original Designation

*Haplophragmoides subglobosum* Cushman, 1910, p. 105, figs. 162–164.

### Type Specimen

Cushman's specimens are deposited in the collections of the United States National Museum, Washington, DC. Lectotype (fig. 164 in Cushman, 1910) designated by Jones et al. (1993) is registered as USNM 8219b.

### Type Level

Recent, surface sediments.

### Type Locality

North Pacific Ocean. Lectotype is from "Albatross" Station D3603 in the North Pacific, 1771 fathoms.

### Description

Test is involute, initially streptospiral, later planispiral with a subcircular outline and a rounded periphery. Chambers are low and broad, highly inflated, and separated by indistinct sutures. Chamber walls are agglutinated, non-calcareous, and usually smoothly finished. In early chambers the primary aperture is an oval to slit-like areal opening bordered by a narrow lip above the base of the last chamber. In later chambers the primary aperture may become more elongated and the lip may fuse to form multiple openings.

### Synonyms and Citations

*Lituola subglogosa* Sars, M., 1868 (1869), p. 250.

*Lituola subglobosa* Sars, G.O., 1871 (1872), p. 253.

*Haplophragmium latidorsatum* Bornemann. –Brady, 1884, p. 310, pl. 34, figs. 8–10 (not figs. 7, 14).

*Cribrostomoides subglobosus* (Cushman). –Charnock and Jones, 1990, p. 168, pl. 5, figs. 4–6; pl. 16, fig. 3. –Jones et al., 1993, figs. 1–2, pls. 1–2. –Jones, 1994, p. 40, pl. 34, figs. 8–10. –Kaminski and Gradstein, 2005, pp. 391–395, fig. 92, pl. 92.

### Remarks

Type species of the genus *Cribrostomoides* Cushman, 1910, emend. Jones et al. (1993). See detailed taxonomic history of *Cribrosto-*

*moides subglobosus* (Cushman) in Jones et al. (1993) and in Kaminski and Gradstein (2005).

### Biogeography

Worldwide. See Observed Occurrences in Kaminski and Gradstein (2005).

### Bathymetry

Bathyal to abyssal.

### Chronostratigraphy

Late Cretaceous to Recent.

## References


Bornemann, J.G., 1855, Die mikroskopische Fauna des Septarienthones von Hermsdorf bei Berlin. Zeitschrift der Deutschen Geologischen Gesellschaft, vol. 7, pp. 307–371.

Brady, H.B., 1884, Report on the foraminifera dredged by H.M.S. Challenger during the years 1873–1876. In Brady, H.B., ed., Report on the Scientific Results of the Voyage of the H.M.S. Challenger during the years 1873–1876. Zoology, vol. 9, pp. 1–814.

Charnock, M.A. and Jones, R.W., 1990, Agglutinated foraminifera from the Paleogene of the North Sea. In Hemleben, C., Kaminski, M.A., Kuhnt, W. and Scott, D.B., ed., Paleoecology, Biostratigraphy, Paleoceanography and Taxonomy of Agglutinated Foraminifera, NATO ASI Series, C: Mathematical and Physical Sciences, Dordrecht: Kluwer Academic, pp. 139–244.

Cushman, J.A., 1910, A monograph of the Foraminifera of the North Pacific Ocean. Part 1. Astrorhizidae and Lituolidae. Bulletin of the United States National Museum, vol. 71, no. 1, pp. 1–134.

Jones, R.W., 1994, The Challenger Foraminifera. Oxford: Oxford University Press, 149 pp.

Jones, R.W., Bender, H., Charnock, M.A., and Kaminski, M.A., 1993, Emendation of the foraminiferal genus Cribrostomoides Cushman, 1910, and its taxonomic implications. Journal of Micropalaeontology, vol. 12, pp. 181–193.

Kaminski, M.A. and Gradstein, F.M., 2005, Atlas of Palaeogene Cosmopolitan Deep-Water Agglutinated Foraminifera. Grzybowski Foundation Special Publication, no. 10, 547 + vii pp.

Sars, G.O., 1872, Undersøgelser over Hardangerfjordens Fauna. Forhandlinger i Videnskasselskabet i Kristiania, vol. 1871, pp. 246–255.

Sars, M., 1869, Fortsatte bemaerkninger over det dyriske livs udbredning i havets dybder. Forhandlinger i Videnskasselskabet i Kristiania, vol. 1868, pp. 246–275.

# Cyclammina cancellata

## Brady, 1879

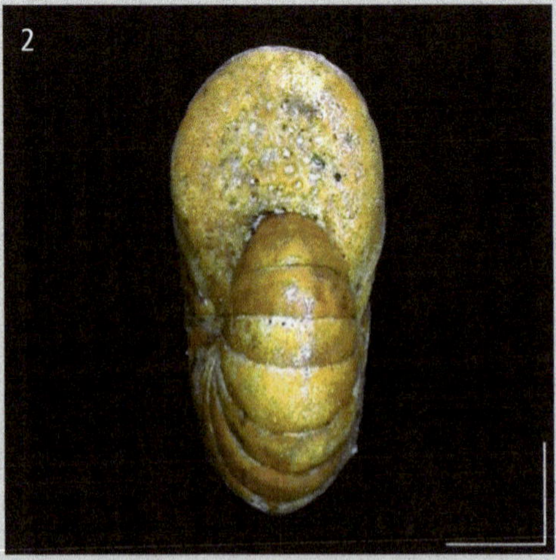


1. *Cyclammina cancellata.* Off New Zealand; "Challenger" Station 168, depth: 1100 fathoms. Deposited in the micropaleontological collections of The Natural History Museum, London; NHMUK1964.12.9.32 (ex NHMUK ZF1361). Lectotype. Lateral view, NHMUK PM DI 0303, scale bar = 614 μm.

2. *Cyclammina cancellata.* Off Culebra Island, West Indies; "Challenger" Station 24, depth: 390 fathoms. Deposited in the micropaleontological collections of The Natural History Museum, London; NHMUK ZF1360. Aperural view, NHMUK PM DI 0304, scale bar = 656 μm.

### Original Designation

*Cyclammina cancellata* Brady, 1879, p. 62 (type description), illustrated in Brady, 1884, pl. 37, figs. 8–16.

### Type Specimen

Figured specimens are housed in the micropaleontological collections of The Natural History Museum, London, UK. Lectotype was designated by Banner (1966) from slide ZF1361. Lectotype is the specimen illustrated by Brady (1884) in pl. 37, fig. 9, and is registered as NHMUK 1964.12.9.32.

### Type Level

Recent, surface sediments.

### Type Locality

Not designated. Lectotype (fig. 9 in Brady, 1884) is from "Challenger" Station 168, off the northeast coast of New Zealand (latitude 40° 28′ S, longitude 177° 43′ E, 1100 fathoms). Other localities given: "Porcupine" and "Valorous" dredgings in the North Atlantic; "Challenger" dredgings from off the Canary Islands and West Indies; from the South Pacific, off New Zealand, and from the Eastern Archipelago (Indonesia). Depths of soundings range from 350 to 1900 fathoms. Also "Challenger" Station 24, off Culebra Island, north of St Thomas, West Indies (390 fathoms).

### Description

Test forms a large, biconvex, involute planispire with a subcircular outline and a rounded to subacute periphery. The nine to sixteen slightly inflated chambers in the last whorl increase gradually in size and are separated by radial or sigmoidal, slightly depressed sutures. Chamber walls are finely agglutinated with a thin, imperforate, outer layer and a thick, alveolar, subepidermal meshwork. The primary aperture is multiple, consisting of an interiomarginal slit and a series of rounded pores scattered over the apertural face.

### Synonyms and Citations

*Cyclammina cancellata* Brady. –Brady, 1884, p. 351, pl. 37, figs. 8–16. –Barker, 1960, pl. 37, figs. 8–16. –Pflum and Frerichs, 1976, pl. 1, fig. 1. –Loeblich and Tappan, 1987, p. 105, pl. 107, figs. 2–6. –Hermelin, 1989, p. 29, pl. 1, fig. 1. –Jones, 1994, p.

43, pl. 37, figs. 8–16. –Kaminski and Gradstein, 2005, pp. 476–480, fig. 118, pl. 118.

### Remarks

Type species of the genus *Cyclammina* Brady, 1879. See Remarks in Kaminski and Gradstein (2005) for detailed taxonomic history of *Cyclammina cancellata* Brady.

### Biogeography

Worldwide.

### Bathymetry

Bathyal to abyssal. Reported between 250 and 1000 fathoms at "Challenger" stations (Jones, 1994).

### Chronostratigraphy

Late Eocene to Recent.

## References


Banner, F.T., 1966, The morphology, classification and stratigraphic value of the Spirocyclinidae. Voprosy Mikropaleontologii, vol. 10, pp. 210–224 [in Russian].

Barker, R.W., 1960, Taxonomic notes on the species figured by H.B. Brady in his report on the foraminifera dredged by H.M.S. Challenger during the years 1873–1876. Society of Economic Paleontologists and Mineralogists Special Publication, vol. 9, pp. 1–238.

Brady, H.B., 1879, Notes on some of the reticularian Rhizopoda of the Challenger Expedition, Part I. On new or little known Arenaceous types. Quarterly Journal of Microscopical Science (New Series), vol. 19, pp. 20–63.

Brady, H.B., 1884, Report on the foraminifera dredged by H.M.S. Challenger during the years 1873–1876. In Brady, H.B., ed., Report on the Scientific Results of the Voyage of the H.M.S. Challenger during the years 1873–1876. Zoology, vol. 9, pp. 1–814.

Hermelin, J.O.R., 1989, Pliocene benthic foraminifera from the Ontong-Java Plateau (Western Equatorial Pacific Ocean): faunal response to changing paleoenvironment. Special Publication of the Cushman Foundation for Foraminiferal Research, vol. 26, pp. 1–143.

Jones, R.W., 1994, The Challenger Foraminifera. Oxford: Oxford University Press, 149 pp.

Kaminski, M.A. and Gradstein, F.M., 2005, Atlas of Palaeogene Cosmopolitan Deep-Water Agglutinated Foraminifera. Grzybowski Foundation Special Publication, no. 10, 547 + vii pp.

Loeblich, A.R. Jr and Tappan, H., 1987, Foraminiferal Genera and Their Classification. New York: Van Nostrand Reinhold, 970 pp.

Pflum, C.E. and Frerichs, W.E., 1976, Gulf of Mexico deep-water foraminifers. Cushman Foundation, Special Publication, vol. 14, pp. 1–125.

# Cyclammina pusilla

## Brady, 1881

1. *Cyclammina pusilla*. South China Sea; Station 17955, latitude 14° 07.30′ N, longitude 112° 10.60′ E, water depth: 2404 meters, surface sample. Deposited in the South China Sea Collection, Institute of Geosciences, Kiel University, Germany. Lateral view, NHMUK PM DI 0305, scale bar = 217 μm.

### Original Designation

*Cyclammina pusilla* Brady, 1881, p. 53 (type description), illustrated in Brady, 1884, pl. 37, figs. 20–23.

### Type Specimen

Specimens illustrated by Brady (1884) are housed in the micropaleontological collections of The Natural History Museum, London, UK, and are registered in slide ZF1365.

### Type Level

Recent, surface sediments.

### Type Locality

Not designated. Figured specimens from "Challenger" Station 323, latitude 35° 39′ S, longitude 50° 47′ W, 1900 fathoms, east of Buenos Aires, Argentina, South Atlantic.

### Description

Test forms a biconvex, involute, slightly lobulate planispire with a subacute periphery. The numerous, slightly inflated chambers in the last whorl increase gradually in size and are separated by radial, slightly depressed sutures. Chamber walls are finely agglutinated, with thin, imperforate outer layers and thick, alveolar, subepidermal meshworks. The primary aperture is multiple, consisting of an interiomarginal slit and a series of rounded pores scattered over the apertural face.

### Synonyms and Citations

*Cyclammina pusilla* Brady. –Brady, 1884, pp. 353–354, pl. 37, figs. 20–23. Barker, 1960, pl. 37, figs. 20–23. –Schröder, 1986, p. 49, pl. 18, fig. 10. –Zheng, 1988, p. 74, pl. 21, figs. 3–4. –Jones, 1994, p. 43, pl. 37, figs. 20–23.

### Biogeography

Worldwide. Atlantic and Southern Ocean (Jones 1994).

### Bathymetry

Upper bathyal. Reported between 1675 and 1900 fathoms at "Challenger" stations in the South Atlantic and Southern Ocean (Jones, 1994).

### Chronostratigraphy

Early Miocene to Recent.

## References

Barker, R.W., 1960, Taxonomic notes on the species figured by H.B. Brady in his report on the foraminifera dredged by H.M.S. Challenger during the years 1873–1876. Society of Economic Paleontologists and Mineralogists Special Publication, vol. 9, pp. 1–238.

Brady, H.B., 1881, Notes on some of the reticularian Rhizopoda of the Challenger Expedition, Part III. Quarterly Journal of Microscopical Science (New Series), vol. 21, pp. 31–71.

Brady, H.B., 1884, Report on the foraminifera dredged by H.M.S. Challenger during the years 1873–1876. In Brady, H.B., ed., Report on the Scientific Results of the Voyage of the H.M.S. Challenger during the years 1873–1876. Zoology, vol. 9, pp. 1–814.

Jones, R.W., 1994, The Challenger Foraminifera. Oxford: Oxford University Press, 149 pp.

Schröder, C.J., 1986, Deep-Water Arenaceous Foraminifera in the Northwest Atlantic Ocean. Canadian Technical Report of Hydrography and Ocean Sciences No. 71. Atlantic Geoscience Centre, Bedford Institute of Oceanography, 191 pp.

Zheng, S.Y., 1988, The Agglutinated and Porcelaneous Foraminifera of the East China Sea. Beijing: China Ocean Press, 337 pp.

# Cystammina pauciloculata

## (Brady), 1879

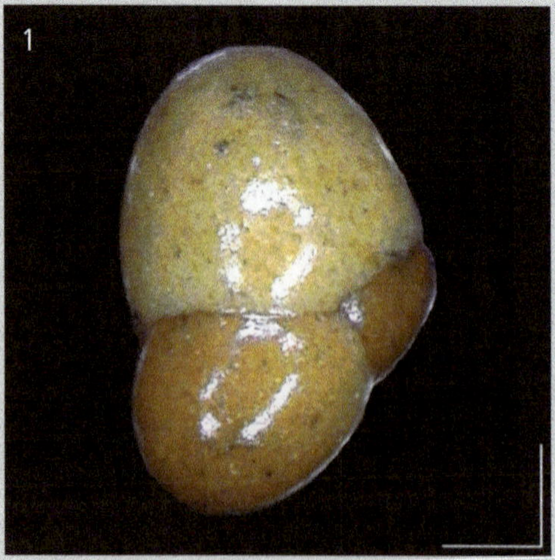

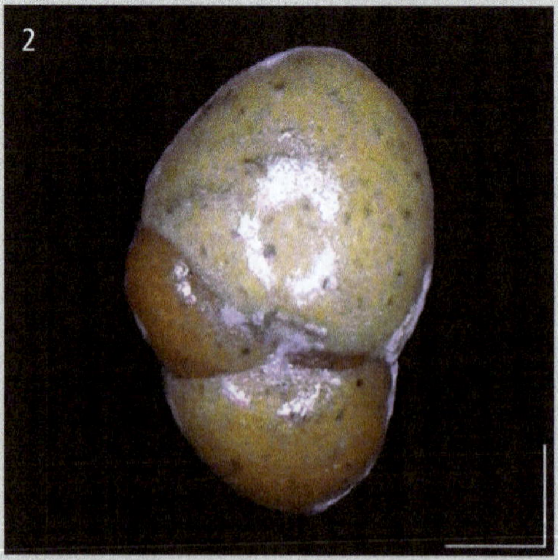

1. *Cystammina pauciloculata.* North of Juan Fernandez Island, South Pacific Ocean; "Challenger" Station 300, depth: 1375 fathoms. Deposited in the micropaleontological collections of The Natural History Museum, London; NHMUK ZF2508. Lectotype. Lateral view, NHMUK PM DI 0306, scale bar = 104 μm.

2. *Cystammina pauciloculata.* North of Juan Fernandez Island, South Pacific Ocean; "Challenger" Station 300, depth: 1375 fathoms. Deposited in the micropaleontological collections of The Natural History Museum, London; NHMUK ZF2508. Lectotype. Apertural view, NHMUK PM DI 0307, scale bar = 104 μm.

### Original Designation

*Trochammina pauciloculata* Brady, 1879, p. 58, pl. 5, figs. 13–14.

### Type Specimen

Lectotype (ZF2508) designated by Loeblich and Tappan (1987) and housed in the micropaleontological collections of The Natural History Museum, London, UK.

### Type Level

Recent, deep-sea ooze.

### Type Locality

Lectotype is from "Challenger" Station 300, north of Juan Fernandez Island, 1375 fathoms.

### Description

Test forms a small streptospire with a rounded periphery. The highly inflated chambers increase rapidly in size and are separated by fine, distinct, depressed sutures. Commonly only four or five chambers are visible externally. Chamber walls are thin and finely agglutinated with little cement, often reddish-brown to yellowish in color. The primary aperture is an elongate, areal slit above the suture of the final chamber.

### Synonyms and Citations

*Trochammina pauciloculata* Brady. –Brady, 1884, p. 344, pl. 41, fig. 1 (not fig. 2).

*Cystammina pauciloculata* (Brady). –Barker, 1960, pl. 41, fig. 1 (not fig. 2). –Loeblich and Tappan, 1987, p. 82, pl. 68, figs. 1–6. –Jones, 1994, p. 45, pl. 41, fig. 1. –Kaminski and Gradstein, 2005, pp. 380–382, fig. 88, pl. 88.

### Remarks

Type species of the genus *Cystammina* Neumayr, 1889.

The specimen illustrated in Brady (1884, pl. 41, fig. 2) as *Trochammina pauciloculata* Brady and as *Cystammina* sp. nov. in Jones (1994, pl. 41, fig. 2) differs in having an interiomarginal aperture. This specimen probably belongs to the genus *Ammosphaeroidina*.

## Biogeography

Worldwide. See Observed Occurrences in Kaminski and Gradstein (2005).

## Bathymetry

Bathyal to abyssal.

## Chronostratigraphy

Early Eocene (Ypresian) to Recent.

# References


Barker, R.W., 1960, Taxonomic notes on the species figured by H.B. Brady in his report on the foraminifera dredged by H.M.S. Challenger during the years 1873–1876. Society of Economic Paleontologists and Mineralogists Special Publication, vol. 9, pp. 1–238.

Brady, H.B., 1879, Notes on some of the reticularian Rhizopoda of the Challenger Expedition, Part I. On new or little known Arenaceous types. Quarterly Journal of Microscopical Science (New Series), vol. 19, pp. 20–63.

Brady, H.B., 1884, Report on the foraminifera dredged by H.M.S. Challenger during the years 1873–1876. In Brady, H.B., ed., Report on the Scientific Results of the Voyage of the H.M.S. Challenger during the years 1873–1876. Zoology, vol. 9, pp. 1–814.

Jones, R.W., 1994, The Challenger Foraminifera. Oxford: Oxford University Press, 149 pp.

Kaminski, M.A. and Gradstein, F.M., 2005, Atlas of Palaeogene Cosmopolitan Deep-Water Agglutinated Foraminifera. Grzybowski Foundation Special Publication, no. 10, 547 + vii pp.

Loeblich, A.R. Jr and Tappan, H., 1987, Foraminiferal Genera and Their Classification. New York: Van Nostrand Reinhold, 970 pp.

# *Cystammina sveni*

## Gradstein and Kaminski, 1997

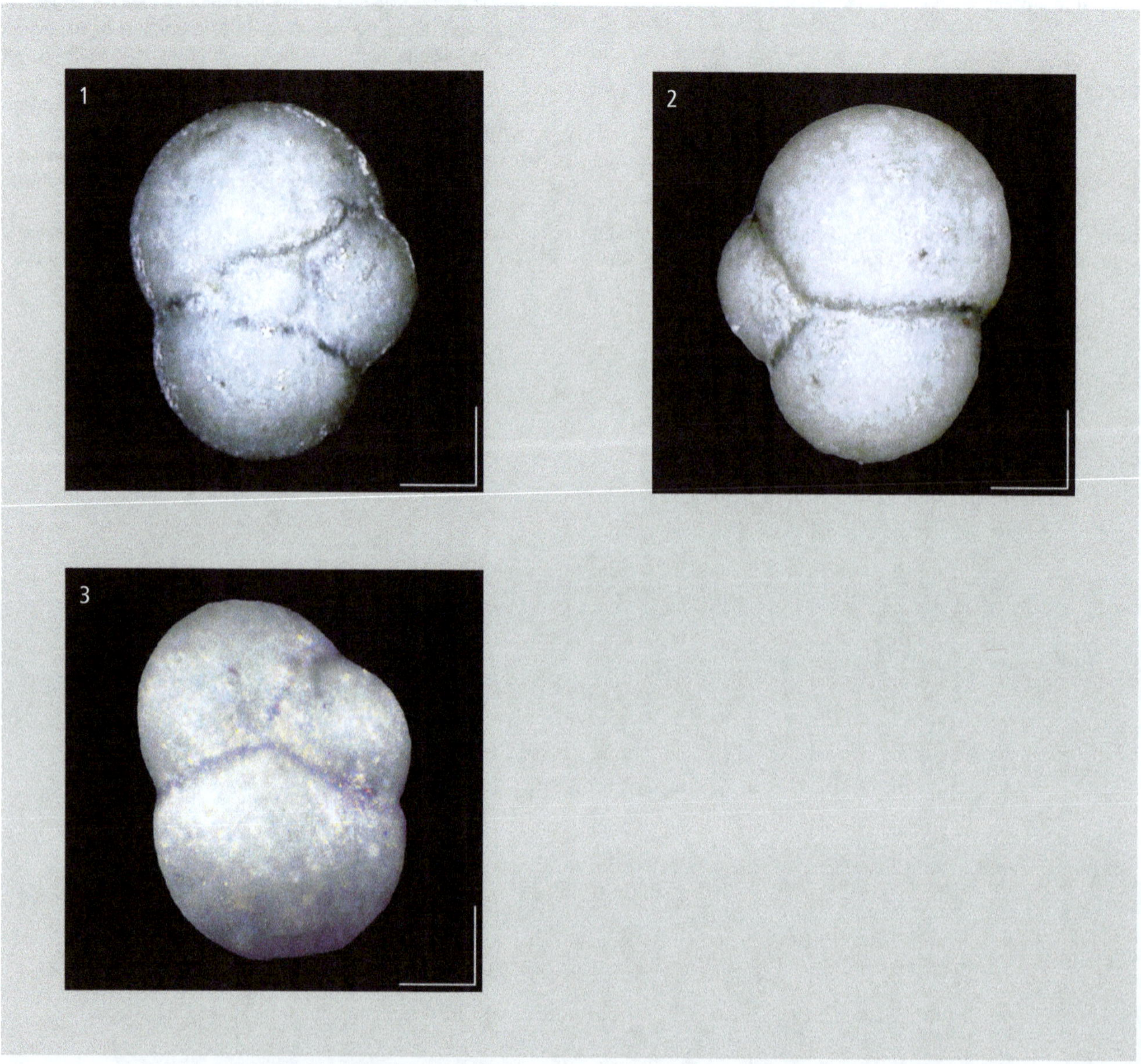


1. *Cystammina sveni*. Norwegian Sea; Saga Petroleum Well 6406/8-1, depth: 2560 meters. Deposited in the micropaleontological collections of The Natural History Museum, London; NHMUK PF67229. Paratype. Spiral view, NHMUK PM DI 0308, scale bar = 72 μm.

2. *Cystammina sveni*. Norwegian Sea; Saga Petroleum Well 6406/8-1, depth: 2560 meters. Deposited in the micropaleontological collections of The Natural History Museum, London; NHMUK PF67229. Paratype. Umbilical view, NHMUK PM DI 0309, scale bar = 73 μm.

3. *Cystammina sveni*. Norwegian Sea; Saga Petroleum Well 6406/8-1, depth: 2560 meters. Deposited in the micropaleontological collections of The Natural History Museum, London; NHMUK PF67229. Paratype. Umbilical view, NHMUK PM DI 0310, scale bar = 72 μm.

## Original Designation

*Cystammina sveni* Gradstein and Kaminski, 1997, p. 225, fig. 11, fig. 12 (1–6).

## Type Specimen

Holotype is housed in the micropaleontological collections of The Natural History Museum, London, UK, and is registered in slide PF62946. Additional paratypes are registered in slides PF67229–67230.

## Type Level

Paleocene, green clays.

## Type Locality

Elf Aquitaine well 6406/8–1, 2560 meters depth, offshore Norway.

## Description

The test is a compact, involute streptospire with an oval to spherical outline and a rounded periphery. The strongly embracing and rounded (not elongate) chambers increase rapidly in size, and are separated by depressed sutures. Two to three chambers are visible on one side and three or three to four chambers are visible on the other side. Chamber walls are finely agglutinated and smoothly finished. The primary aperture is an areal slit, often with upturned corners, and with an overhanging lip.

## Synonyms (following Kaminski and Gradstein, 2005) and Citations

*Praecystammina* aff. *globigerinaeformis* Krasheninnikov. –Gradstein and Berggren, 1981, p. 258–259, pl. 9, figs. 11–15.

*Cystammina globigerinaeformis* (Krasheninnikov). –Miller et al., 1982, p. 21, pl. 2, figs. 13, 17, 21.

*Ammosphaeroidina* sp. –Kaminski et al., 1989, pl. 5, fig. 4.

*Cystammina pauciloculata* (Brady). –Charnock and Jones, 1990, pl. 16, fig. 4a–c.

*Praecystammina* sp. –Kaminski et al., 1990, p. 370, pl. 7, figs. 4a, b, 5.

*Cystammina sveni* Gradstein and Kaminski. –Kaminski and Gradstein, 2005, pp. 383–385, fig. 89, pl. 89.

## Remarks

*Cystammina sveni* Gradstein and Kaminski differs from *Cystammina pauciloculata* (Brady) by its more tightly coiled test and less numerous, less elongated chambers in the last whorl. *Cystammina sveni* differs from *Praecystammina globigerinaeformis* Krasheninnikov by its more tightly coiled, less evolute test with fewer chambers in the last whorl, and by the shape of the aperture (Kaminski and Gradstein, 2005).

## Biogeography

North Sea, offshore Norway, Trinidad, Labrador Shelf, Alpine–Carpathian region, northern Spain, northern Morocco, and Atlantic Ocean (Gradstein and Kaminski, 1997).

## Bathymetry

Bathyal to abyssal.

## Chronostratigraphy

Late Cretaceous (Campanian) to Middle Eocene (Lutetian). Typical in Paleocene middle bathyal to abyssal sediments (Kaminski and Gradstein, 2005).

# References


Charnock, M.A. and Jones, R.W., 1990, Agglutinated foraminifera from the Palaeogene of the North Sea. In Hemleben, C., Kaminski, M.A., Kuhnt, W. and Scott, D.B., ed., Paleoecology, Biostratigraphy, Paleoceanography and Taxonomy of Agglutinated Foraminifera, NATO ASI Series, C: Mathematical and Physical Sciences, vol. 327. Dordrecht: Kluwer Academic, pp. 139–244.

Gradstein, F.M. and Berggren, W.A., 1981, Flysch-type agglutinated foraminifera and the Maestrichtian to Paleogene history of the Labrador and North Seas. Marine Micropaleontology, vol. 6, pp. 448–472.

Gradstein, F.M. and Kaminski, M.A., 1997, New species of Paleogene deep-water agglutinated foraminifera from the North Sea and Norwegian Sea. Annales Societatis Geologorum Poloniae, vol. 67, pp. 217–229.

Kaminski, M.A. and Gradstein, F.M., 2005, Atlas of Palaeogene Cosmopolitan Deep-Water Agglutinated Foraminifera. Grzybowski Foundation Special Publication, no. 10, 547 + vii pp.

Kaminski, M.A., Gradstein, F.M., and Berggren, W.A., 1989, Paeogene benthic foraminifer biostratigraphy and paleoecology at site 647, southern Labrador Sea. In Srivastava, S.P., Arthur, M.A., Clement, B., et al., ed., Scientific Results of the Proceedings of the Ocean Drilling Program, vol. 105, pp. 705–730.

Kaminski, M.A., Gradstein, F.M., Goll, R.M., and Greig, D., 1990, Biostratigraphy and Paleoecology of deep-water aggutinated foraminifera at ODP Site 643, Norwegian–Greenland Sea. In Hemleben, C., Kaminski, M.A., Kuhnt, W., and Scott, D.B., ed., Paleoecology, Biostratigraphy, Paleoceanography and Taxonomy of Agglutinated Foraminifera. NATO ASI Series, Series C: Mathematical and Physical Sciences, vol. 327. Dordrecht: Kluwer Academic, pp. 345–386.

Miller, K.G., Gradstein, F.M., and Berggren, W.A., 1982, Late Cretaceous to early Tertiary agglutinated benthic foraminifera in the Labrador Sea. Micropaleontology, vol. 28, pp. 1–30.

# *Discorbinella bertheloti*

## (d'Orbigny), 1839

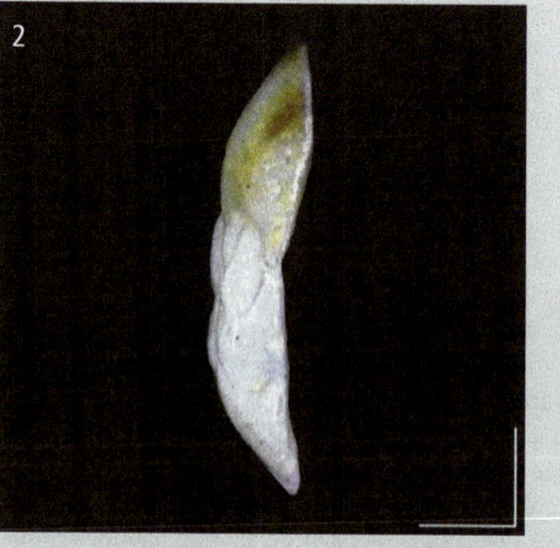

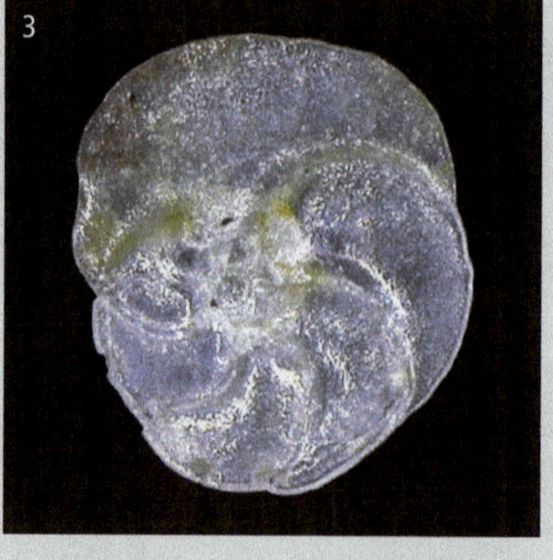

1. *Discorbinella bertheloti*. Philippines; "Challenger" Station 209, depth: 95–100 fathoms. Deposited in the micropaleontological collections of The Natural History Museum, London. NHMUK ZF1384. Umbilical view, NHMUK PM DI 0311, scale bar = 83 μm.

2. *Discorbinella bertheloti*. Philippines; "Challenger" Station 209, depth: 95–100 fathoms. Deposited in the micropaleontological collections of The Natural History Museum, London. NHMUK ZF1384. Apertural view, NHMUK PM DI 0312, scale bar = 82 μm.

3. *Discorbinella bertheloti*. Philippines; "Challenger" Station 209, depth: 95–100 fathoms. Deposited in the micropaleontological collections of The Natural History Museum, London. NHMUK ZF1384. Spiral view, NHMUK PM DI 0313, scale bar = 83 μm.

## Original Designation

*Rosalina bertheloti* d'Orbigny, 1839, p. 135, pl. 1, figs. 28–30.

## Type Specimen

Depository not given.

## Type Level

Recent, marine sands.

## Type Locality

Teneriffe, Canary Islands.

## Description

Test forms a low, compressed, planoconvex or concavo-convex trochospire; ovate in outline, with a convex spiral side, a flattened or slightly concave umbilical side, and an acute, keeled periphery. The slightly inflated, broad, low chambers increase rapidly in size and are separated by curved, slightly depressed sutures. Chambers on the umbilical side have projecting umbilical flaps. Chamber walls are calcareous, smooth, and finely perforate. The primary aperture is an interiomarginal arch at the periphery. Small secondary apertures are also present at the posterior end of the umbilical flaps.

## Synonyms and Citations

*Discorbina bertheloti* (d'Orbigny). –Brady, 1884, p. 650, pl. 89, figs. 10–12.

*Discorbis bertheloti* (d'Orbigny). –Cushman, 1921, pp. 305–306, pl. 59, figs. 1a–c. –Phleger and Parker, 1951, p. 20, pl. 10, figs. 17–19.

*Discopulvinulina bertheloti* (d'Orbigny). –Barker, 1960, pl. 89, figs. 10–12.

*Discorbinella bertheloti* (d'Orbigny). –Loeblich and Tappan, 1964, p. C575, textfig. 453, figs. 3a–b. –Belford, 1966, pp. 90–91, pl. 12, figs. 15–22, text-fig. 8:1–2, 9:1–2. –Hermelin, 1989, p. 67, pl. 12, figs. 9–10. –Jones, 1994, p. 95, pl. 89, figs. 10–12.

## Biogeography

Worldwide.

## Bathymetry

Neritic to middle bathyal. Generally less than 500 fathoms (Jones, 1994).

## Chronostratigraphy

Middle? Miocene to Recent (Jones, 1994).

# References


Barker, R.W., 1960, Taxonomic notes on the species figured by H.B. Brady in his report on the foraminifera dredged by H.M.S. Challenger during the years 1873–1876. Society of Economic Paleontologists and Mineralogists Special Publication, vol. 9, pp. 1–238.

Belford, D.J., 1966, Miocene and Pliocene smaller foraminifera from Papua and New Guinea. Australia Bureau of Mineral Resources Geology and Geophysics Bulletin, vol. 79, pp. 1–306.

Brady, H.B., 1884, Report on the foraminifera dredged by H.M.S. Challenger during the years 1873–1876. In Brady, H.B., ed., Report on the Scientific Results of the Voyage of the H.M.S. Challenger during the years 1873–1876. Zoology, vol. 9, pp. 1–814.

Cushman, J.A., 1921, Foraminifera of the Phillipine and adjacent seas. Bulletin of the United States National Museum, vol. 100, pp. 1–608.

d'Orbigny, A.D., 1839, Foraminifères des Îles Canaries. In Barker-Webb, P. and Berthelot, S., ed., Historie Naturelle des Îles Canaries, vol. 2, part 2: Paris (Béthune), pp. 119–146.

Hermelin, J.O.R., 1989, Pliocene benthic foraminifera from the Ontong-Java Plateau (Western Equatorial Pacific Ocean): faunal response to changing paleoenvironment. Special Publication of the Cushman Foundation for Foraminiferal Research, vol. 26, pp. 1–143.

Jones, R.W., 1994, The Challenger Foraminifera. Oxford: Oxford University Press, 149 pp.

Loeblich, A.R. Jr and Tappan, H., 1964, Sarcodina, chiefly "Thecamoebians" and Foraminiferida. In Moore, R.C., ed., Treatise on Invertebrate Paleontology, Part C, Protista 2 (2 vols). Lawrence, KS: Geological Society of America and University of Kansas Press, pp. 1–900.

Phleger, F.B. and Parker, F.L., 1951, Ecology of foraminifera, northwest Gulf of Mexico; Part II – Foraminifera species. Memoirs of the Geological Society of America, vol. 46, pp. 1–64.

# *Eggerella bradyi*

## (Cushman), 1911

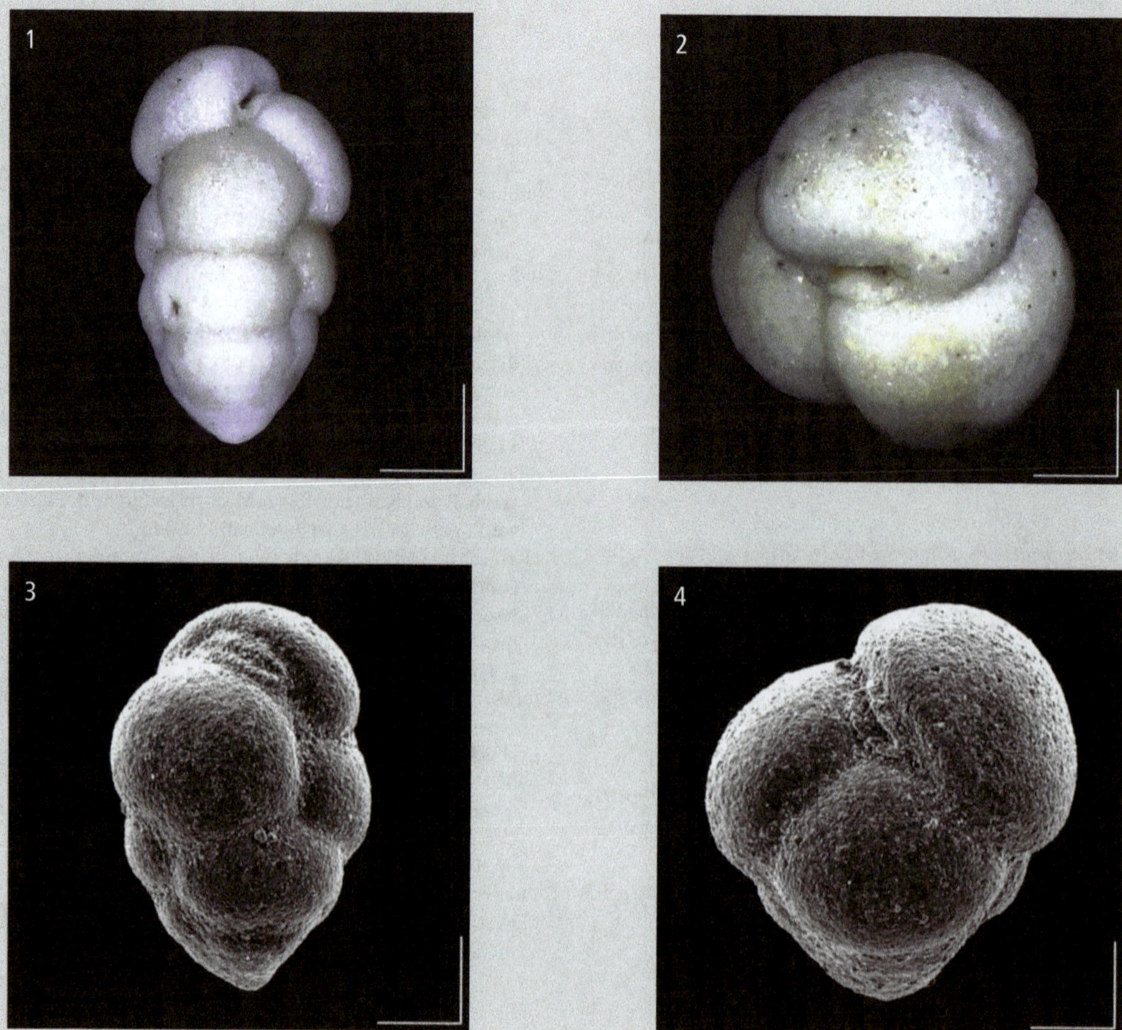


1. *Eggerella bradyi.* Off Sombrero Island, West Indies; "Challenger" Station 23, depth: 450 fathoms. Deposited in the micropaleontological collections of The Natural History Museum, London; NHMUK ZF2605. Lateral view, NHMUK PM DI 0314, scale bar = 228 μm.

2. *Eggerella bradyi.* Off Sombrero Island, West Indies; "Challenger" Station 23, depth: 450 fathoms. Deposited in the micropaleontological collections of The Natural History Museum, London; NHMUK ZF2605. Apertural view, NHMUK PM DI 0315, scale bar = 143 μm.

3. *Eggerella bradyi.* Northeast continental slope of the South China Sea; ODP Leg 184, Hole 1148A, Core 74, Section 3, 70–72 cm. Lateral view, NHMUK PM DI 0316, scale bar = 89 μm.

4. *Eggerella bradyi.* Northeast continental slope of the South China Sea; ODP Leg 184, Hole 1148A, Core 74, Section 3, 70–72 cm. Apertural view, NHMUK PM DI 0317, scale bar = 68 μm.

### Original Designation

*Verneuilina bradyi* Cushman, 1911, p. 54, fig. 87.

### Type Specimen

Deposited in the collections of the United States National Museum, Washington, DC.

### Type Level

Recent, surface sediments.

### Type Locality

Not designated. Localities mentioned are in the North Pacific Ocean, over 1000 fathoms.

### Description

Test nearly conical, oval in cross-section, initially trochospiral, later (in adult portion) reduced to three chambers per whorl. The highly inflated chambers increase rapidly in size and are separated by thin, depressed sutures. Chamber walls are finely agglutinated, canaliculate, and smoothly cemented with calcareous cement. The primary aperture is an elongate slit at the base of the final chamber, bordered by a narrow lip.

### Synonyms and Citations

*Verneuilina pygmaea* (Egger). –Brady, 1884, p. 385, pl. 47, figs. 4–7.

*Eggerella bradyi* (Cushman). –Phleger and Parker, 1951, p. 6, pl. 3, figs. 1–2. –Barker, 1960, pl. 47, figs. 4–7. –Jones, 1994, pp. 51–52, pl. 47, figs. 4–7. –Robertson, 1998, p. 24, pl. 3, fig. 4. –Kuhnt et al., 2002, p. 124, pl. 5. figs. 5–7.

### Remarks

Type species of the genus *Eggerella* Cushman, 1933.

### Biogeography

Worldwide.

### Bathymetry

Bathyal to abyssal. Recorded between 129 and 3125 fathoms at "Challenger" stations (Jones, 1994). Middle to lower bathyal on the Sunda Shelf of the South China Sea (Szarek et al., 2009).

### Chronostratigraphy

Late Eocene to Recent.

## References


Barker, R.W., 1960, Taxonomic notes on the species figured by H.B. Brady in his report on the foraminifera dredged by H.M.S. Challenger during the years 1873–1876. Society of Economic Paleontologists and Mineralogists Special Publication, vol. 9, pp. 1–238.

Brady, H.B., 1884, Report on the foraminifera dredged by H.M.S. Challenger during the years 1873–1876. In Brady, H.B., ed., Report on the Scientific Results of the Voyage of the H.M.S. Challenger during the years 1873–1876. Zoology, vol. 9, pp. 1–814.

Cushman, J.A., 1911, A monograph of the Foraminifera of the North Pacific Ocean. Part II. Textulariidae. Bulletin of the United States National Museum, vol. 71, no. 2, pp. 1–108.

Jones, R.W., 1994, The Challenger Foraminifera. Oxford: Oxford University Press, 149 pp.

Kuhnt, W., Holbourn, A.E., and Zhao, Q., 2002, The early history of the South China Sea: Evolution of Oligocene–Miocene deep water environments. Revue de Micropaléontologie, vol. 45, pp. 99–159.

Phleger, F.B. and Parker, F.L., 1951, Ecology of foraminifera, northwest Gulf of Mexico; Part II – Foraminifera species. Memoirs of the Geological Society of America, vol. 46, pp. 1–64.

Robertson, B.E., 1998, Systematics and paleoecology of the benthic foraminiferida from the Buff Bay section, Miocene of Jamaica. Micropaleontology, vol. 44, supplement 2, pp. 1–266.

Szarek, R., Kuhnt, W., Kawamura, H., and Nishi, H., 2009, Distribution of recent benthic foraminifera along the continental slope of the Sunda Shelf (South China Sea). Marine Micropaleontology, vol. 71, pp. 41–59.

# *Ehrenbergina hystrix*

## Brady, 1881

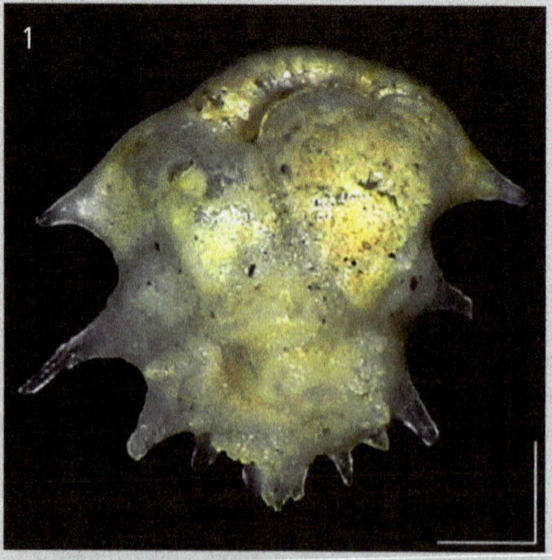


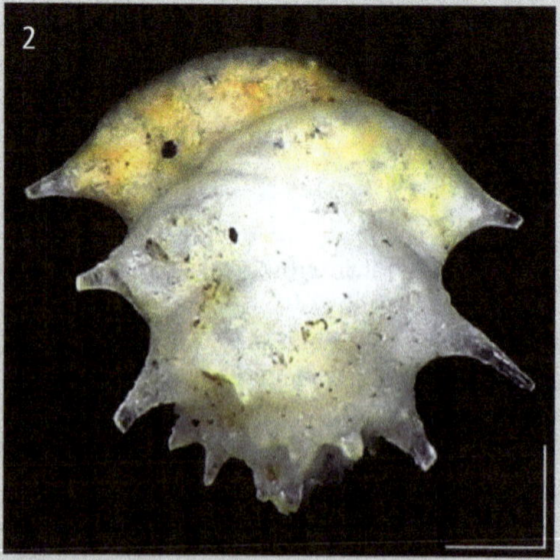


1. *Ehrenbergina hystrix.* South Pacific Ocean; "Challenger" Station 280, depth: 1940 fathoms. Deposited in the micropaleontological collections of The Natural History Museum, London; NHMUK ZF1434. Syntype. Lateral view, NHMUK PM DI 0318, scale bar = 143 μm.

2. *Ehrenbergina hystrix.* South Pacific Ocean; "Challenger" Station 280, depth: 1940 fathoms. Deposited in the micropaleontological collections of The Natural History Museum, London; NHMUK ZF1434. Syntype. Lateral view, NHMUK PM DI 0319, scale bar = 138 μm.

### Original Designation

*Ehrenbergina hystrix* Brady, 1881, p. 60 (type description), illustrated in Brady, 1884, pl. 55, figs. 8–11.

### Type Specimen

Figured specimens are housed in the micropaleontological collections of The Natural History Museum, London, UK, and are registered in slide ZF1434.

### Type Level

Recent, surface sediments.

### Type Locality

Not designated. Syntypes are from "Challenger" Station 280 in the South Pacific, 1940 fathoms.

### Description

Test forms an elongate initial planispire consisting of biserially arranged chambers, later becoming uncoiled. The broad and low, overlapping chambers are separated by curved, slightly depressed sutures. Chamber walls are calcareous and perforate with lateral spines that are most prominent in the early part of the test. The primary aperture is large, oval, and terminal, with a small toothplate.

### Synonyms and Citations

*Ehrenbergina hystrix* Brady. –Brady, 1884, p. 434, pl. 55, figs. 8–11. –Barker, 1960, pl. 55, figs. 8–11. –van Marle, 1991, p. 121, pl. 9, fig. 7. –Jones, 1994, p. 61, pl. 55, figs. 8–11.

*Reissia hystrix* (Brady). –Loeblich and Tappan, 1987, p. 508, pl. 561, figs. 1–4.

### Remarks

Type species of the genus *Reissia* Loeblich and Tappan, 1964, regarded by Jones (1994) as a junior synonym of *Ehrenbergina* Reuss, 1850.

### Biogeography

Worldwide. Cosmopolitan in low to moderate latitudes.

### Bathymetry

Bathyal to abyssal. Highest frequency between 914 and 1564 meters in eastern Indonesia (van Marle, 1991). Lower bathyal to upper abyssal (3000–4500 meters, Loeblich and Tappan, 1987).

### Chronostratigraphy

Pliocene to Recent in eastern Indonesia (van Marle, 1991). Pleistocene to Recent (Jones, 1994).

## References


Barker, R.W., 1960, Taxonomic notes on the species figured by H.B. Brady in his report on the foraminifera dredged by H.M.S. Challenger during the years 1873–1876. Society of Economic Paleontologists.

Brady, H.B., 1881, Notes on some of the reticularian Rhizopoda of the Challenger Expedition, Part III. Quarterly Journal of Microscopical Science (New Series), vol. 21, pp. 31–71.

Brady, H.B., 1884, Report on the foraminifera dredged by H.M.S. Challenger during the years 1873–1876. In Brady, H.B., ed., Report on the Scientific Results of the Voyage of the H.M.S. Challenger during the years 1873–1876. Zoology, vol. 9, pp. 1–814.

Jones, R.W., 1994, The Challenger Foraminifera. Oxford: Oxford University Press, 149 pp.

Loeblich, A.R. Jr and Tappan, H., 1987, Foraminiferal Genera and Their Classification. New York: Van Nostrand Reinhold, 970 pp.

van Marle, L.J., 1991, Eastern Indonesian Late Cenozoic smaller benthic foraminifera. Verhandelingen der Koninklijke Nederlanse Akademie van Wetenschappen, Afdeeling Natuurkunde, Eerste Reeks, vol. 34, pp. 1–328.

# *Elphidium crispum*

## (Linnaeus), 1758

1. *Elphidium crispum*. Southern England; Boxgrove excavation. J.E. Whittaker Collection. Lateral view, NHMUK PM DI 0320, scale bar = 134 μm.
2. *Elphidium crispum*. Southern England; Boxgrove excavation. J.E. Whittaker Collection. Apertural view, NHMUK PM DI 0321, scale bar = 187 μm.
3. *Elphidium crispum*. Southern England; Boxgrove excavation. J.E. Whittaker Collection. Lateral view, NHMUK PM DI 0322, scale bar = 135 μm.

## Original Designation

*Nautilus crispus* Linnaeus, 1758, p. 709, No. 236 (fide Ellis and Messina).

## Type Specimen

Five Recent specimens; one sectioned and illustrated by Fichtel and Moll (1798) on pl. 5, fig. 1-b. Deposited in the collections of the Natural History Museum, Vienna, Austria (1812-IX-7. Inv. No. MI-475, 476/1–3, 477).

## Type Level

Recent, surface sediments.

## Type Locality

Mediterranean and Adriatic Seas.

## Description

Test forms a lenticular, biumbonate planispire; subcircular in outline, with a subacute, keeled periphery. Chambers are generally numerous, increasing gradually in size and separated by prominent, raised, curved sutures. Chamber walls are calcareous, finely perforate, and ornamented by external ponticuli spanning the sutures. The primary aperture is an interiomarginal opening. Additional areal openings may also be present.

## Synonyms and Citations

*Nautilus crispus* Linnaeus. –Fichtel and Moll, 1798, p. 40 (partim), pl. 5, figs. a–b.

*Themeon rigatus* de Montfort, 1808, p. 203 (partim), textfig. on p. 202.

*Polystomella crispa* (Linnaeus). –Lamarck, 1822, p. 625. –Brady, 1884, p. 736, pl. 110, figs. 6–7.

*Elphidium crispum* (Linnaeus). –Cushman and Grant, 1927, p. 73, pl. 7, fig. 3a–b. –Barker, 1960, pl. 110, figs. 6–7. –Rögl and Hansen, 1984, p. 37, pl. 8, fig. 7, textfig. 10. –Jones, 1994, p. 109, pl. 110, figs. 6–7.

## Remarks

Rögl and Hansen (1984) pointed out that material in the Fichtel and Moll Collection under the name *Nautilus crispus* includes two different morphotypes, as originally indicated by Fichtel and Moll. According to Rögl and Hansen (1984), the Recent specimens agree well with the species concept of *Elphidium crispum* (Linnaeus) sensu stricto.

## Biogeography

Worldwide.

## Bathymetry

Neritic to middle bathyal (to 355 fathoms) according to Jones (1994).

## Chronostratigraphy

Miocene to Recent (Jones, 1994).

# References


Barker, R.W., 1960, Taxonomic notes on the species figured by H.B. Brady in his report on the foraminifera dredged by H.M.S. Challenger during the years 1873–1876. Society of Economic Paleontologists and Mineralogists Special Publication, vol. 9, pp. 1–238.

Brady, H.B., 1884, Report on the foraminifera dredged by H.M.S. Challenger during the years 1873–1876. In Brady, H.B., ed., Report on the Scientific Results of the Voyage of the H.M.S. Challenger during the years 1873–1876. Zoology, vol. 9, pp. 1–814.

Cushman, J.A. and Grant, U.S., 1927, Late Tertiary and Quaternary Elphidium of the West Coast of North America. Transactions of the San Diego Society of Natural History, vol. 5, pp. 69–82.

de Montfort, P.D., 1808, Conchyliologie systématique et classification méthodique des coquilles. Paris: F. Schoell, vol. 1, lxxxvii + 409 pp.

Fichtel, L. von and Moll, J.P.C. von, 1798, Testacea microscopica aliaque minuta ex generibus Argonauta et Nautilus ad naturam delineata et descripta. (Microscopische und andere kleine Schalthiere aus den Geschlechtern Argonaute und Schiffer, nach der Natur gezeichnet und beschrieben). Vienna: Camesina, 124 pp.

Jones, R.W., 1994, The Challenger Foraminifera. Oxford: Oxford University Press, 149 pp.

Lamarck, J.B., 1822, Histoire naturelle des animaux sans vertèbres, vol. 7. Paris: L'auteur, 711 pp.

Linnaeus, C., 1758, Systema Naturae, 10th edn. Lipsiae (Leipzig): G. Engelman, vol. 1, 824 pp.

Rögl, F. and Hansen, J.H., 1984, Foraminifera described by Fichtel and Moll in 1798; A revision of Testacea Microscopica. Appendix Testacea Microscopica alique minuta ex Generibus Argonauta et Nautilus. Reprint of original plates. Neue Denkschriften des Naturhistorischen Museum in Wien, vol. 3, pp. 1–143.

# *Elphidium macellum*

## (Fichtel and Moll), 1798

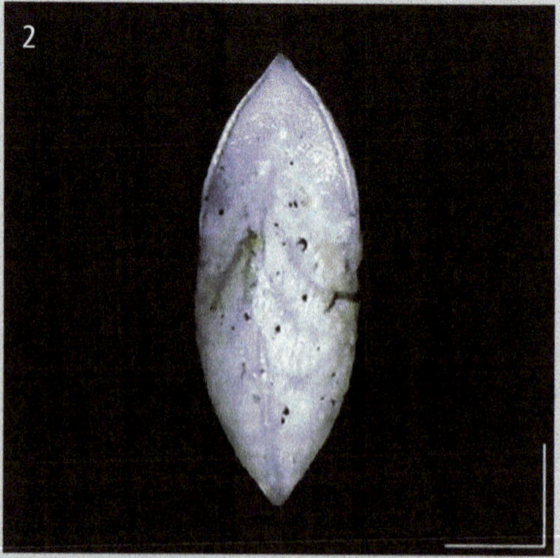


1. *Elphidium macellum.* Cardigan Bay, Wales; Sample Station 742 (Haynes 1973). Deposited in the micropaleontological collections of The Natural History Museum, London. NHMUK 1970.11.26.409. Lateral view, NHMUK PM DI 0323, scale bar = 124 μm.

2. *Elphidium macellum.* Cardigan Bay, Wales; Sample Station 742 (Haynes 1973). Deposited in the micropaleontological collections of The Natural History Museum, London. NHMUK 1970.11.26.409. Apertural view, NHMUK PM DI 0324, scale bar = 128 μm.

### Original Designation

*Nautilus macellus* var. β Fichtel and Moll, 1798, p. 68, pl. 10, figs. h–k.

### Type Specimen

Lectotype designated by Rögl and Hansen (1984) and deposited in the collections of the Natural History Museum, Vienna, Austria (MI-481).

### Type Level

Recent, "Zoophyten-Concretionen" (coralligene facies, shelf sediments with calcareous algae and bryozoans).

### Type Locality

Mediterranean Sea.

### Description

Test forms a lenticular, compressed planispire; subcircular in outline with a subacute, keeled periphery, occasionally spinose. Chambers are generally numerous (17.5 in the lectotype), increasing gradually in size, and separated by prominent, raised, curved sutures. Chamber walls are calcareous, finely perforate, and ornamented by external ponticuli spanning the sutures. The primary aperture is an interiomarginal opening. Additional areal openings may be present.

### Synonyms and Citations

?*Nautilus strigilatus* var. α Fichtel and Moll, 1798, p. 49, pl. 5, figs. c–e.

*Nautilus strigilatus* var. β Fichtel and Moll, 1798, p. 50, pl. 5, figs. f–g.

*Vorticialis marginata* Lamarck, 1816, p. 14, pl. 470, fig. 3 a–b.

*Polystomella strigilata* Fichtel. –d'Orbigny, 1826, p. 284, No. 4.

*Polystomella macella* (Fichtel and Moll). –Brady, 1884, p. 737, pl. 110, figs. 8, 11 (not figs. 9–10).

*Elphidium macellum* (Fichtel and Moll). –Barker, 1960, p. 109, pl. 110, figs. 8, 11. –Loeblich and Tappan, 1987, p. 674, pl. 786, figs. 6–7; pl. 789, figs. 1–5. –Rögl and Hansen, 1984, p. 39, pl. 9, figs. 3–4, textfig. 12. –Jones, 1994, p. 109, pl. 110, figs. 8, 11.

### Remarks

The variety β of *Nautilus strigilatus* Fichtel and Moll, 1798, was designated as type species of the genus *Sporilus* de Montfort, 1808, subsequently synonymized with the genus *Elphidium* de Montfort, 1808. Thus, *Elphidium macellum* is the type species of the genus *Elphidium* de Montfort, 1808.

According to Rögl and Hansen (1984), the development of the spinose keel is highly variable. Populations from the Adriatic coast studied by these authors show all variations ranging from non-spinose to having well-defined spines.

### Biogeography

Worldwide.

### Bathymetry

Neritic to bathyal (Jones, 1994).

### Chronostratigraphy

Miocene to Recent (Jones, 1994).

## References


Barker, R.W., 1960, Taxonomic notes on the species figured by H.B. Brady in his report on the foraminifera dredged by H.M.S. Challenger during the years 1873–1876. Society of Economic Paleontologists and Mineralogists Special Publication, vol. 9, pp. 1–238.

Brady, H.B., 1884, Report on the foraminifera dredged by H.M.S. Challenger during the years 1873–1876. In Brady, H.B., ed., Report on the Scientific Results of the Voyage of the H.M.S. Challenger during the years 1873–1876. Zoology, vol. 9, pp. 1–814.

d'Orbigny, A., 1826, Tableau méthodique de la classe des Céphalopodes. Annales des Sciences Naturelles, vol. 7, pp. 245–314.

Fichtel, L. von and Moll, J.P.C. von, 1798, Testacea microscopica aliaque minuta ex generibus Argonauta et Nautilus ad naturam delineata et descripta. (Microscopische und andere kleine Schalthiere aus den Geschlechtern Argonaute und Schiffer, nach der Natur gezeichnet und beschrieben). Vienna: Camesina, 124 pp.

Haynes, J.R., 1973, Cardigan Bay Recent Foraminifera. (Cruises of the R.V. Antur, 1962–1964). Bulletin of the British Museum (Natural History) Zoological Supplement, no. 4, pp. 1–245.

Jones, R.W., 1994, The Challenger Foraminifera. Oxford: Oxford University Press, 149 pp.

Lamarck, J.B., 1816, Tableau encyclopédique et méthodique des trois règnes de la nature, partie 23 – Mollusques et polypes divers. Paris: Mme V. Agasse, pp. 1–16.

Loeblich, A.R. Jr and Tappan, H., 1987, Foraminiferal Genera and Their Classification. New York: Van Nostrand Reinhold, 970 pp.

Rögl, F. and Hansen, J.H., 1984, Foraminifera described by Fichtel and Moll in 1798; A revision of Testacea Microscopica. Appendix Testacea Microscopica alique minuta ex Generibus Argonauta et Nautilus. Reprint of original plates. Neue Denkschriften des Naturhistorischen Museum in Wien, vol. 3, pp. 1–143.

# *Epistominella exigua*

## (Brady), 1884

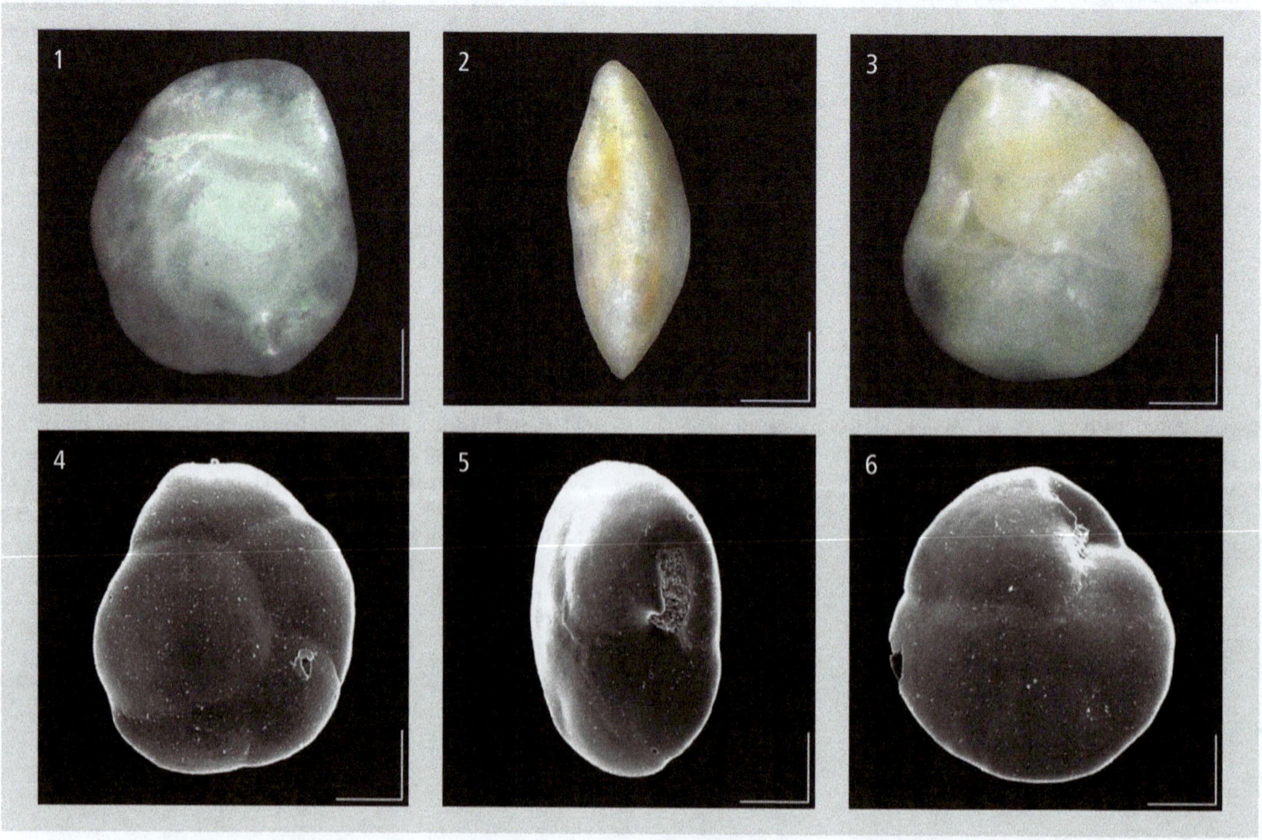

1. *Epistominella exigua*. Southern Ocean; "Challenger" Station 160, depth: 2600 fathoms. Deposited in the micropaleontological collections of The Natural History Museum, London; NHMUK ZF2214. Spiral view, NHMUK PM DI 0325, scale bar = 63 μm.

2. *Epistominella exigua*. Southern Ocean; "Challenger" Station 160, depth: 2600 fathoms. Deposited in the micropaleontological collections of The Natural History Museum, London; NHMUK ZF2214. Apertural view, NHMUK PM DI 0326, scale bar = 67 μm.

3. *Epistominella exigua*. Southern Ocean; "Challenger" Station 160, depth: 2600 fathoms. Deposited in the micropaleontological collections of The Natural History Museum, London; NHMUK ZF2214. Umbilical view, NHMUK PM DI 0327, scale bar = 68 μm.

4. *Epistominella exigua*. Scott Plateau, off Northwest Australia; Piston Core MD01-2378, Section 23, 105–106 cm. Deposited in the micropaleontological collections, Institute of Geosciences, Kiel University, Germany. Spiral view, NHMUK PM DI 0328, scale bar = 60 μm.

5. *Epistominella exigua*. Scott Plateau, off Northwest Australia; Piston Core MD01-2378, Section 23, 105–106 cm. Deposited in the micropaleontological collections, Institute of Geosciences, Kiel University, Germany. Apertural view, NHMUK PM DI 0329, scale bar = 53 μm.

6. *Epistominella exigua*. Scott Plateau, off Northwest Australia; Piston Core MD01-2378, Section 23, 105–106 cm. Deposited in the micropaleontological collections, Institute of Geosciences, Kiel University, Germany. Umbilical view, NHMUK PM DI 0330, scale bar = 59 μm.

## Original Designation

*Pulvinulina exigua* Brady, 1884, p. 696, pl. 103, figs. 13–14.

## Type Specimen

Figured specimens are housed in the micropaleontological collections of The Natural History Museum, London, UK, and are registered in slides ZF2214 and ZF2215.

## Type Level

Recent, surface sediments.

## Type Locality

Not designated. Localities given: twelve stations in the North Atlantic (64–2740 fathoms); four in the South Atlantic (1025–2475 fathoms); three in the Southern Ocean (1300–2600 fathoms); ten in the South Pacific (129–2350 fathoms); and five in the North Pacific (15–2300 fathoms).

## Description

Test forms a small trochospire; slightly lobulate in outline and unequally biconvex in cross-section, with an evolute, slightly convex spiral side, an involute, convex to nearly conical umbilical side, and an acute, lobulate periphery. The five moderately inflated chambers in the last whorl increase gradually in size, and are separated by flush, oblique, thickened sutures on the spiral side, and by curved, slightly depressed sutures on the umbilical side. Chamber walls are calcareous, finely perforate, and smooth. The primary aperture is an interiomarginal slit extending up the face of the final chamber on the umbilical side.

## Synonyms and Citations

*Pseudoparrella exigua* (Brady). –Phleger and Parker, 1951, p. 28, pl. 15, fig. 6 (not fig. 7). –Loeblich and Tappan, 1994, p. 146, pl. 307, figs. 1–7.

*Pulvinulinella exigua* (Brady). –Hofker, 1951, p. 322, textfigs. 219–221.

*Epistominella exigua* (Brady). –Barker, 1960, pl. 103, figs. 13–14. –Miller and Katz, 1987, p. 132, pl. 5, figs. 6a–b. –Hermelin, 1989, p. 67. –Thomas, 1990, p. 590. –Nomura, 1995, p. 276, pl. 3, fig. 3. –Katz et al., 2003, p. 35, pl. 2, fig. 3

*Alabaminoides exiguus* (Brady). –Jones, 1994, p. 103, pl. 103, figs. 13–14.

## Remarks

In Recent sediments the initial chambers characteristically have a brownish color. This species is a useful indicator of pulsed organic matter paleofluxes.

## Biogeography

Worldwide but not reported from the Mediterranean (Murray, 1991). Reported from the Timor Sea (Loeblich and Tappan, 1994), Weddell Sea, Antarctic (Thomas, 1990), North Atlantic (Miller and Katz, 1987), Indian Ocean, Atlantic, Pacific, and Gulf of Mexico (see Hermelin, 1989).

## Bathymetry

Bathyal to abyssal. Predominantly lower bathyal to abyssal. Deep water (Brady, 1884), lower bathyal (Thomas, 1990), abyssal, greater than 3.5 kilometers (Miller and Katz, 1987), reported from 15–2600 fathoms at "Challenger" stations (Jones, 1994).

Characteristically occurs in areas of seasonally pulsed phytodetritus flux, where abundant food is episodically available (Gooday, 1988, 1996; Smart et al., 1994).

## Chronostratigraphy

Middle Eocene (Lutetian) to Recent. Middle Eocene to Late Eocene (rare), topmost Eocene to Pleistocene (common) according to Thomas (1990). Rare in Early Oligocene, common to abundant in Late Oligocene to Miocene (Miller and Katz, 1987).

# References


Barker, R.W., 1960, Taxonomic notes on the species figured by H.B. Brady in his report on the foraminifera dredged by H.M.S. Challenger during the years 1873–1876. Society of Economic Paleontologists and Mineralogists Special Publication, vol. 9, pp. 1–238.

Brady, H.B., 1884, Report on the foraminifera dredged by H.M.S. Challenger during the years 1873–1876. In Brady, H.B., ed., Report on the Scientific Results of the Voyage of the H.M.S. Challenger during the years 1873–1876. Zoology, vol. 9, pp. 1–814.

Gooday, A.J., 1988, A response by benthic foraminifera to the deposition of phytodetritus in the deep-sea. Nature, vol. 332, pp. 70–73.

Gooday, A.J., 1996, Epifaunal and shallow infaunal foraminiferal communities at three abyssal NE Atlantic sites subject to differing phytodetritus input regimes. Deep-Sea Research, vol. 43, p. 1395–1421.

Hermelin, J.O.R., 1989, Pliocene benthic foraminifera from the Ontong-Java Plateau (Western Equatorial Pacific Ocean): faunal response to changing paleoenvironment. Special Publication of the Cushman Foundation for Foraminiferal Research, vol. 26, pp. 1–143.

Hofker, J., 1951, The Foraminifera of the "Siboga" Expedition. Part III: Siboga-Expeditie, Monographie IVa. Leiden: E.J. Brill, 513 pp.

Jones, R.W., 1994, The Challenger Foraminifera. Oxford: Oxford University Press, 149 pp.

Katz, M.E., Tjalsma, R.C., and Miller, K.G., 2003, Oligocene bathyal to abyssal benthic foraminifera of the Atlantic Ocean. Micropaleontology, vol. 49, supplement 2, pp. 1–45.

Loeblich, A.R. Jr and Tappan, H., 1994, Foraminifera of the Sahul Shelf and Timor Sea. Special Publications of the Cushman Laboratory for Foraminiferal Research, vol. 31, pp. 1–661.

Miller, K.G. and Katz, M.E., 1987, Oligocene to Miocene benthic foraminiferal and abyssal circulation changes in the north Atlantic. Micropaleontology, vol. 33, no. 2, pp. 97–149.

Murray, J.W., 1991, Ecology and Palaeoecology of Benthic Foraminifera. New York: John Wiley & Sons, 397 pp.

Nomura, R., 1995, Paleogene to Neogene deep-sea paleoceanography in the eastern Indian Ocean: benthic foraminifera from ODP Sites 747, 757 and 758. Micropaleontology, vol. 41, pp. 251–290.

Phleger, F.B. and Parker, F.L., 1951, Ecology of foraminifera, northwest Gulf of Mexico; Part II – Foraminifera species. Memoirs of the Geological Society of America, vol. 46, pp. 1–64.

Smart, C.W., King, S.C., Gooday, A.J., Murray, J.W., and Tomas, E. 1994, A benthic foraminiferal proxy of pulsed organic matter paleofluxes. Marine Micropaleontology, vol. 23, pp. 89–99.

Thomas, E., 1990, Late Cretaceous through Neogene benthic foraminifers. In Barker, P.F., Kennett, J.P., et al., ed., Scientific Results of the Proceedings of the Ocean Drilling Program, vol. 113, pp. 571–594.

# Eratidus foliaceus

## (Brady), 1881

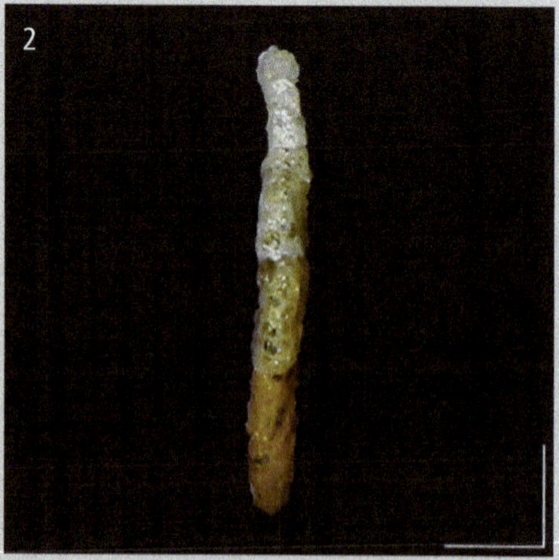

1. *Eratidus foliaceus.* South Atlantic; "Challenger" Station 323, depth: 1900 fathoms. Deposited in the micropaleontological collections of The Natural History Museum, London; NHMUK ZF1533. Lateral view, NHMUK PM DI 0331, scale bar = 59 µm.

2. *Eratidus foliaceus.* South Atlantic; "Challenger" Station 323, depth: 1900 fathoms. Deposited in the micropaleontological collections of The Natural History Museum, London; NHMUK ZF1535. Edge view, NHMUK PM DI 0332, scale bar = 59 µm.

### Original Designation

*Lituola (Haplophragmium) foliaceum* Brady, 1881, p. 50 (type description), illustrated as *Haplophragmium foliaceum* in Brady, 1884, pl. 33, figs. 20–25.

### Type Specimen

Figured specimens are housed in the micropaleontological collections of The Natural History Museum, London, UK, and are registered in slides ZF1533 and ZF1535.

### Type Level

Recent, surface sediments.

### Type Locality

Not designated. Figured specimens from "Challenger" Station 323, latitude 35° 39′ S, longitude 50° 47′ W, 1900 fathoms, east of Buenos Aires, Argentina, South Atlantic.

### Description

Test forms an initially strongly compressed planispire, later uncoiling to become rectilinear. The slightly inflated, numerous, broad and low chambers increase rapidly in size and are separated by distinct, slightly depressed sutures. Chamber walls are thin, and coarsely agglutinated with little cement. The primary aperture is an elongate, terminal slit.

### Synonyms and Citations

*Haplophragmium foliaceum* (Brady). –Brady, 1884, p. 304, pl. 33, figs. 20–25.

*Ammomarginulina foliaceus* (Brady). –Barker, 1960, pl. 33, figs. 20–25.

*Eratidus foliaceus* (Brady). –Loeblich and Tappan, 1987, p. 75, pl. 59, figs. 1–3. –Jones, 1994, p. 40, pl. 33, figs. 20–25.

### Remarks

Type species of the genus *Eratidus* Saidova, 1975.

### Biogeography

Worldwide.

### Bathymetry

Bathyal to abyssal.

### Chronostratigraphy

Middle Eocene to Recent (Jones, 1994).

## References

Barker, R.W., 1960, Taxonomic notes on the species figured by H.B. Brady in his report on the foraminifera dredged by H.M.S. Challenger during the years 1873–1876. Society of Economic Paleontologists and Mineralogists Special Publication, vol. 9, pp. 1–238.

Brady, H.B., 1881, Notes on some of the reticularian Rhizopoda of the Challenger Expedition, Part III. Quarterly Journal of Microscopical Science (New Series), vol. 21, pp. 31–71.

Brady, H.B., 1884, Report on the foraminifera dredged by H.M.S. Challenger during the years 1873–1876. In Brady, H.B., ed., Report on the Scientific Results of the Voyage of the H.M.S. Challenger during the years 1873–1876. Zoology, vol. 9, pp. 1–814.

Jones, R.W., 1994, The Challenger Foraminifera. Oxford: Oxford University Press, 149 pp.

Loeblich, A.R. Jr and Tappan, H., 1987, Foraminiferal Genera and Their Classification. New York: Van Nostrand Reinhold, 970 pp.

# Eubuliminella exilis

## (Brady), 1884

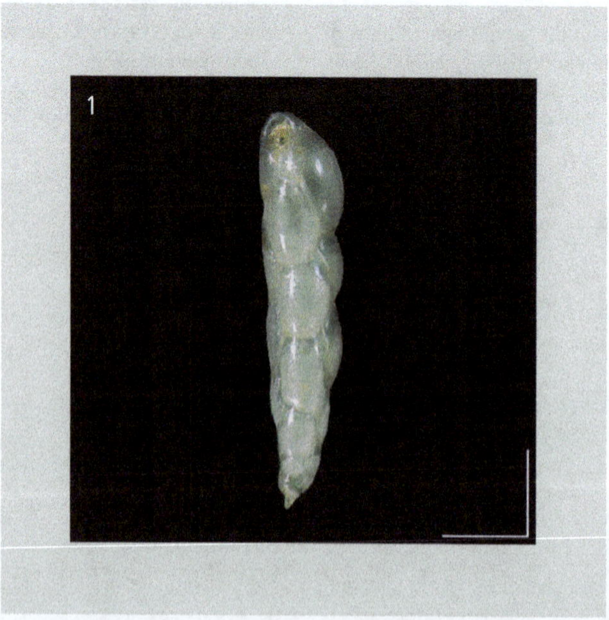

1. *Eubuliminella exilis*. Gulf of Mexico; Eureka Core E68-139. Deposited in the collections of the United States National Museum, Washington, DC. Specimen illustrated in van Morkhoven et al., 1986, Ref. no. 4. Lateral view, NHMUK PM DI 0333, scale bar = 126 μm.

### Original Designation

*Bulimina elegans* d'Orbigny var. *exilis* Brady, 1884, p. 399, pl. 50, figs. 5–6.

### Type Specimen

Lectotype (ZF4980) and paralectotypes (ZF1212 and ZF1213) are housed in the micropaleontological collections of The Natural History Museum, London, UK.

### Type Level

Recent, surface sediments.

### Type Locality

Localities given: 1000–1500 fathoms in the North Atlantic and in the North and South Pacific (350–800 fathoms).

### Description

Test forms a medium-sized, elongate, high trochospire in the earlier part (four chambers per whorl) reducing to a triserial, very slender (about five times as long as broad) tapered form in the later portion. The test is subcircular in cross-section, with a subacute initial portion bearing a spine and a subacute or subrounded apertural end. The inflated, ovate chambers are arranged in a straight or slightly oblique series, increasing rapidly in height (last whorl represents one-third of test length), and are separated by distinct, curved, depressed sutures. Chamber walls are calcareous, finely perforate, and smooth. The primary aperture is almost in the coiling axis, with prominent crista and a short toothplate attached to the lateral chamber wall.

### Synonyms and Citations

*Buliminella subfusiformis* var. *tenuata* Cushman, 1927, p. 149, pl. 2, fig. 9.

*Buliminella subfusiformis* (Cushman). –Cushman and Moyer, 1930, p. 56, pl. 7, fig. 20.

*Bulimina exilis* Brady. –Cushman and Parker, 1947, p. 123, pl. 28, figs. 27–28. –Barker, 1960, pl. 50, figs. 5–6. –van Morkhoven et al., 1986, p. 24, pl. 4. –van Marle, 1991, p. 86.

*Bulimina subfusiformis* var. *tenuata* (Cushman). –Cushman and Parker, 1947, p. 124, pl. 28, fig. 29.

*Eubuliminella exilis* (Brady). –Revets, 1993, p. 141, pl. 1, figs. 1–7. –Jones, 1994, p. 54, pl. 50, figs. 5–6.

### Remarks

Type species of the genus *Eubuliminella* Revets, 1993.

*Eubuliminella exilis* (Brady) is associated with sapropels and diatomites, formed during conditions of oxygen depletion in the Pliocene of Crete (Jonkers, 1984). According to den Dulk et al. (2000) it is a deep infaunal taxon, characteristic in oxygen-depleted sediments from the oxygen minimum zone.

### Biogeography

Worldwide. Recorded from Brazil, California, the North and South Atlantic, the Pacific, the eastern Mediterranean, and the Gulf of Mexico (van Morkhoven et al., 1986).

### Bathymetry

Bathyal. Van Morkhoven et al. (1986) reported highest abundance between 500 and 1000 meters.

### Chronostratigraphy

Late Miocene (Messinian) to Recent.

## References


Barker, R.W., 1960, Taxonomic notes on the species figured by H.B. Brady in his report on the foraminifera dredged by H.M.S. Challenger during the years 1873–1876. Society of Economic Paleontologists and Mineralogists Special Publication, vol. 9, pp. 1–238.

Brady, H.B., 1884, Report on the foraminifera dredged by H.M.S. Challenger during the years 1873–1876. In Brady, H.B., ed., Report on the Scientific Results of the Voyage of the H.M.S. Challenger during the years 1873–1876. Zoology, vol. 9, pp. 1–814.

Cushman, J.A., 1927, Recent foraminifera from off the West coast of America. Bulletin of the Scripps Institute of Oceanography, Technical Series, vol. 1, pp. 119–188.

Cushman, J.A. and Moyer, D.A., 1930, Some Recent foraminifera from off San Pedro. Contributions from the Cushman Laboratory for Foraminiferal Research, vol. 13, pp. 36–40.

Cushman, J.A. and Parker, F.L., 1947, Bulimina and related foraminiferal genera. Professional Paper of the United States Geological Survey, vol. 210-D, pp. 55–176.

den Dulk, M., Reichart, G.J., van Heyst, S., Zachariasse, W.J., and van der Zwaan, G.J., 2000. Benthic foraminifera as proxies of organic matter flux and bottom water oxygenation? A case history from the northern Arabian Sea. Palaeogeography, Palaeoclimatology, Palaeoecology, vol. 161, pp. 337–359.

Jones, R.W., 1994, The Challenger Foraminifera. Oxford: Oxford University Press, 149 pp.

Jonkers, H.A., 1984, Pliocene benthonic foraminifera from homogeneous and laminated marls on Crete. Utrecht Micropaleontological Bulletins, vol. 31, pp. 1–179.

Revets, S.A., 1993, The revision of the genus Buliminellita Cushman and Stainforth, 1947, and Eubuliminella gen. nov. Journal of Foraminiferal Research, vol. 23, no. 2, pp. 141–151.

van Marle, L.J., 1991, Eastern Indonesian Late Cenozoic smaller benthic foraminifera. Verhandelingen der Koninklijke Nederlanse Akademie van Wetenschappen, Afdeeling Natuurkunde, Eerste Reeks, vol. 34, pp. 1–328.

van Morkhoven, F.P.C.M., Berggren, W.A., Edwards, A.S., et al., 1986, Cenozoic cosmopolitan deep-water benthic foraminifera. Bulletin des Centres de Recherches Exploration-Production Elf-Aquitaine, Mémoire, vol. 11, pp. 1–421.

# *Evolutinella vocontianus*

## (Moullade), 1966

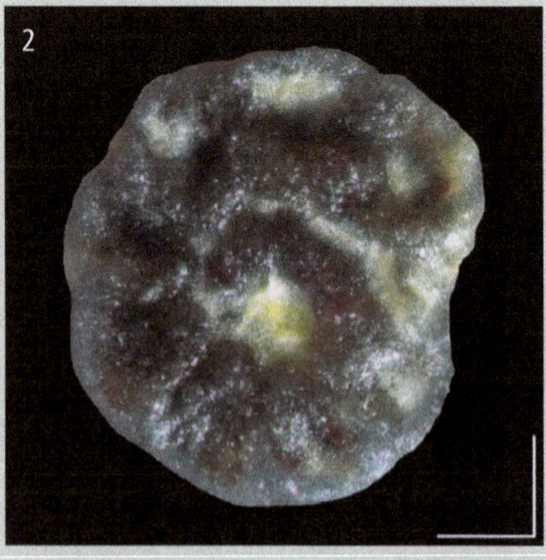

1. *Evolutinella vocontianus.* Cluse du Roubion Section, near Crupies (Drôme), Vocontian Trough, France. Deposited in the Moullade Collection, Nice, France; AA401. Umbilical view, NHMUK PM DI 0334, scale bar = 264 μm.

2. *Evolutinella vocontianus.* Cluse du Roubion Section, near Crupies (Drôme), Vocontian Trough, France. Deposited in the Moullade Collection, Nice, France; AA401. Umbilical view, NHMUK PM DI 0335, scale bar = 243 μm.

### Original Designation

*Haplophragmoides vocontianus* Moullade, 1966, p. 16, pl. 1, figs. 1–2.

### Type Specimen

Holotype figured in Moullade (1966) deposited, in the collections of the Département des Sciences de la Terre, University of Lyon, France (No. 50 020).

### Type Level

Lower Cretaceous (Hauterivian).

### Type Locality

Sisteron (Vocontian Trough, South France). Cutting in valley of Durance River.

### Description

Test planispiral, evolute, compressed, circular or slightly lobulate in outline, with a subacute periphery. Approximately nine to eleven chambers in the last whorl, slightly inflated, generally depressed in the central part, increase slowly in size, and are separated by fine, straight, radial, slightly depressed sutures. Chamber walls are finely agglutinated, and smoothly finished. The primary aperture is an interiomarginal slit.

### Synonyms

None verified.

### Remarks

We refer this species to the genus *Evolutinella* on account of its distinctly evolute test. It differs from *Haplophragmoides neocomianus* Chapman by its more evolute test.

### Biogeography
Tethys.

### Bathymetry
Bathyal to abyssal.

### Chronostratigraphy
Early Cretaceous (Hauterivian – Barremian).

## Reference

Moullade, M., 1966, Étude stratigraphique et microplaéontologique du Crétacé Inférieur de la "Fosse Vocontienne". Documents des Laboratoires de Géologie de la Faculté des Sciences de Lyon, vol. 15. no. 1–2, pp. 1–369.

# Favocassidulina favus

## (Brady), 1877

1. *Favocassidulina favus*. North of Juan Fernandez, Pacific Ocean; "Challenger" Station 300, depth: 1375 fathoms. Deposited in the micropaleontological collections of The Natural History Museum, London; NHMUK ZF2216. Lateral view, NHMUK PM DI 0336, scale bar = 167 μm.

2. *Favocassidulina favus*. North of Juan Fernandez, Pacific Ocean; "Challenger" Station 300, depth: 1375 fathoms. Deposited in the micropaleontological collections of The Natural History Museum, London; NHMUK ZF2216. Apertural view, NHMUK PM DI 0337, scale bar = 168 μm.

### Original Designation

*Pulvinulina favus* Brady, 1877, p. 535 (type description), illustrated in Brady, 1884, pl. 104, figs. 12–16.

### Type Specimen

Figured specimens are housed in the micropaleontological collections of The Natural History Museum, London, UK, and are registered in slides ZF2216 and ZF2217.

### Type Level

Recent, surface sediments.

### Type Locality

In a fragment of rock found only on the beach after an earthquake, in one district on the east side of New Ireland, Bismarck Archipelago, Territory of New Guinea. Brady's figured specimens are from "Challenger" Station 300, latitude 33° 42′ S, longitude 78° 18′ W, north of Juan Fernandez Island, South Pacific, 1375 fathoms; and "Challenger" Station 283 latitude 26° 90′ S, longitude 145° 17′ W, north of Juan Fernandez Island, South Pacific Ocean.

### Description

Test forms a lenticular planispire composed of biserial chambers, biconvex in cross-section, and with a subacute periphery. The moderately inflated chambers extend to the umbilicus on opposite sides. Chamber walls are calcareous, finely perforate, and exhibit a honeycomb-like secondary ornamentation that covers the entire test, except for the aperture. The primary aperture is a narrow, curved slit parallel to the chamber margin, with a narrow lip and a toothplate at the posterior margin.

### Synonyms and Citations

*Pulvinulina favus* Brady. –Brady, 1884, p. 701, pl. 104, figs. 12–16.

*Cassidulina* sp. 2. Phleger et al., 1953, pp. 45–46, p. 10, figs. 8–9.

*Favocassidulina favus* (Brady). –Barker, 1960, p. 214, pl. 104, figs. 12–16. –Belford, 1966, pp. 145–146, pl. 26, figs. 28–31, textfig. 18:6. –Douglas, 1973, pl. 6, figs. 4–5. –Boltovskoy, 1978, pl. 4, fig. 20. –Burke, 1981, pl. 3, figs. 3–4. –Loeblich and Tappan, 1987, p. 505, pl. 556, figs. 1–6. –Hermelin, 1989, p. 74, pl. 13, figs. 14, 17. –Jones, 1994, p. 104, pl. 104, figs. 12–16. –Nomura, 1995, pl. 2, fig. 34.

### Remarks
Type species of the genus *Favocassidulina* Loeblich and Tappan, 1957.

### Biogeography
Worldwide.

### Bathymetry
Bathyal to abyssal. Reported between 1375 and 2600 fathoms at "Challenger" stations (Jones, 1994).

### Chronostratigraphy
Middle Miocene to Recent (Jones, 1994).

## References


Barker, R.W., 1960, Taxonomic notes on the species figured by H.B. Brady in his report on the foraminifera dredged by H.M.S. Challenger during the years 1873–1876. Society of Economic Paleontologists and Mineralogists Special Publication, vol. 9, pp. 1–238.

Belford, D.J., 1966, Miocene and Pliocene smaller foraminifera from Papua and New Guinea. Australia Bureau of Mineral Resources Geology and Geophysics Bulletin, vol. 79, pp. 1–306.

Boltovskoy, E., 1978, Late Cenozoic benthonic foraminifera of the Ninetyeast Ridge (Indian Ocean). Marine Geology, vol. 26, pp. 139–175.

Brady, H.B., 1877, Supplementary note on the foraminifera of the Chalk(?) of the New Britain Group. Geological Magazine (New Series), vol. 4, pp. 534–536.

Brady, H.B., 1884, Report on the foraminifera dredged by H.M.S. Challenger during the years 1873–1876. In Brady, H.B., ed., Report on the Scientific Results of the Voyage of the H.M.S. Challenger during the years 1873–1876. Zoology, vol. 9, pp. 1–814.

Burke, S.C., 1981, Recent benthic foraminifera of the Ontong Java Plateau. Journal of Foraminiferal Research, vol. 11, pp. 1–19.

Douglas, R.G., 1973, Benthonic foraminiferal biostratigraphy in the central North Pacific, Leg 17, Deep Sea Drilling Project. In Winterer, E.L., Ewing, J.I., et al., ed., Initial Reports of the Deep Sea Drilling Project, Washington (U.S. Government Printing Office), vol. 17, pp. 607–671.

Hermelin, J.O.R., 1989, Pliocene benthic foraminifera from the Ontong-Java Plateau (Western Equatorial Pacific Ocean): faunal response to changing paleoenvironment. Special Publication of the Cushman Foundation for Foraminiferal Research, vol. 26, pp. 1–143.

Jones, R.W., 1994, The Challenger Foraminifera. Oxford: Oxford University Press, 149 pp.

Loeblich, A.R. Jr and Tappan, H., 1987, Foraminiferal Genera and Their Classification. New York: Van Nostrand Reinhold, 970 pp.

Nomura, R., 1995, Paleogene to Neogene deep-sea paleoceanography in the eastern Indian Ocean: benthic foraminifera from ODP Sites 747, 757 and 758. Micropaleontology, vol. 41, pp. 251–290.

Phleger, F.B., Parker, F.L., and Pierson, J.F., 1953, North Atlantic foraminifera. Reports of the Swedish Deep-Sea Expedition, 1947–1948, vol. 7, pp. 3–122.

# Fissurina orbignyana

## Seguenza, 1862

1. *Fissurina orbignyana.* Prince Edward Island, South Pacific Ocean; "Challenger" Station 145, depth: 50–150 fathoms. Deposited in the micropaleontological collections of The Natural History Museum, London; NHMUK ZF1726. Lateral view, NHMUK PM DI 0338, scale bar = 78 μm.

### Original Designation

*Fissurina (Fissurine) orbignyana* Seguenza, 1862, p. 66, pl. 2, figs. 25–26.

### Type Specimen

Depository not given. Brady's figured specimen is housed in the micropaleontological collections of The Natural History Museum, London, UK, and is registered in slide ZF1726.

### Type Level

Upper Miocene, marl.

### Type Locality

Rometta, distretto di Messina, Sicily.

### Description

Test unilocular, ovate in outline, biconvex in cross-section with a keeled periphery. Chamber walls are calcareous, finely perforate, and smooth. The primary aperture is a terminal slit with an entosolenian tube.

### Synonyms and Citations

*Lagena orbignyana* (Seguenza). –Brady, 1884, p. 484, pl. 59, fig. 18 (not figs. 20, 24–26).

*Fissurina orbignyana* Seguenza. –Barker, 1960, pl. 59, fig. 18 (not figs. 20, 26). –Jones, 1994, p. 68, pl. 59, fig. 18.

### Biogeography

Worldwide.

### Bathymetry

Neritic. Reported from 50–150 fathoms at "Challenger" Station 145, Prince Edward Island, South Pacific (Jones, 1994).

### Chronostratigraphy

Miocene to Recent (Jones, 1994).

## References

Barker, R.W., 1960, Taxonomic notes on the species figured by H.B. Brady in his report on the foraminifera dredged by H.M.S. Challenger during the years 1873–1876. Society of Economic Paleontologists and Mineralogists Special Publication, vol. 9, pp. 1–238.

Brady, H.B., 1884, Report on the foraminifera dredged by H.M.S. Challenger during the years 1873–1876. In Brady, H.B., ed., Report on the Scientific Results of the Voyage of the H.M.S. Challenger during the years 1873–1876. Zoology, vol. 9, pp. 1–814.

Jones, R.W., 1994, The Challenger Foraminifera. Oxford: Oxford University Press, 149 pp.

Seguenza, G., 1862, Prime recerche intorno ai rizopodi fossili delle argille Pleistoceniche dei dintorni di Catania. Accademia Gioenia di Scienze Naturali di Catania, Atti, ser. 2, vol. 18, pp. 84–126.

# *Fissurina seminiformis*

## (Schwager), 1866

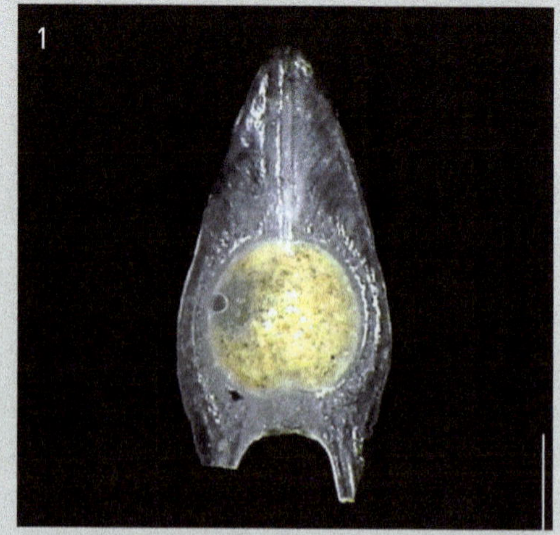


1. *Fissurina seminiformis*. South Pacific Ocean; "Challenger" Station 283, depth: 2075 fathoms. Deposited in the micropaleontological collections of The Natural History Museum, London; NHMUK ZF1750. Lateral view, NHMUK PM DI 0418, scale bar = 160 μm.

### Original Designation

*Lagena seminiformis* Schwager, 1866, p. 208, pl. 5, fig. 21.

### Type Specimen

Depository not given. Neotype (Mf 19) designated by Srinivasan and Sharma (1980) deposited in the Micropalaeontology Section, Department of Geology, Banaras Hindu University, Banaras, India. Brady's figured specimens are housed in the micropaleontological collections of The Natural History Museum, London, UK, and are registered in slides ZF1747, ZF1749, and ZF1750.

### Type Level

Lower Pliocene (Srinivasan and Sharma, 1980), Untere und obere Thone. Given as Lower Tertiary by Schwager (1866).

### Type Locality

Car Nicobar, Andaman Sea.

### Description

Test unilocular with a subspherical central body encircled by a wide keel, extending to the apertural neck and prolonged by two spines at the basal end. Chamber walls are calcareous, finely perforate, and smooth. The primary aperture is terminal, round, and at the end of a long tubular neck.

### Synonyms and Citations

*Lagena seminiformis* Schwager. –Brady, 1884, p. 478, pl. 59, figs. 28–30.

*Fissurina seminiformis* (Schwager). –Barker, 1960, pl. 59, figs. 28–30. –Srinavasan and Sharma, 1980, 42, pl. 3, figs. 19–20. –Hermelin, 1989, p. 49, pl. 7, fig. 7. –Jones, 1994, p. 68, pl. 59, figs. 28–30.

### Remarks

Occasionally referred to the genus *Lagnea* Popescu, 1983 on account of the complex tubule structures in the lateral carina.

### Biogeography

Worldwide.

### Bathymetry

Bathyal to abyssal. Reported between 1000 and 2350 fathoms at "Challenger" stations (Jones, 1994).

### Chronostratigraphy

Miocene to Recent (Jones, 1994).

## References


Barker, R.W., 1960, Taxonomic notes on the species figured by H.B. Brady in his report on the foraminifera dredged by H.M.S. Challenger during the years 1873–1876. Society of Economic Paleontologists and Mineralogists Special Publication, vol. 9, pp. 1–238.

Brady, H.B., 1884, Report on the foraminifera dredged by H.M.S. Challenger during the years 1873–1876. In Brady, H.B., ed., Report on the Scientific Results of the Voyage of the H.M.S. Challenger during the years 1873–1876. Zoology, vol. 9, pp. 1–814.

Hermelin, J.O.R., 1989, Pliocene benthic foraminifera from the Ontong-Java Plateau (Western Equatorial Pacific Ocean): faunal response to changing paleoenvironment. Special Publication of the Cushman Foundation for Foraminiferal Research, vol. 26, pp. 1–143.

Jones, R.W., 1994, The Challenger Foraminifera. Oxford: Oxford University Press, 149 pp.

Schwager, C., 1866, Fossile Foraminiferen von Kar Nikobar, Reise der Österreichischen Fregatte Novara um Erde in den Jahren 1857, 1858, 1859 unten den Befehlen des Commodore B. Von Wüllerstorf-Urbair. Geologischer Theil, vol. 2, no. 1, Geologische Beobachtung, no. 2, Paläontologische Mitteilung, pp. 187–268.

Srinivasan, M.S. and Sharma, V., 1980, Schwager's Car Nicobar Foraminifera in the Reports of the Novara Expedition – A Revision. New Delhi: Today & Tomorrow's Printers and Publishers, 83 pp.

# *Frondicularia saggitula*

## van den Broeck, 1876

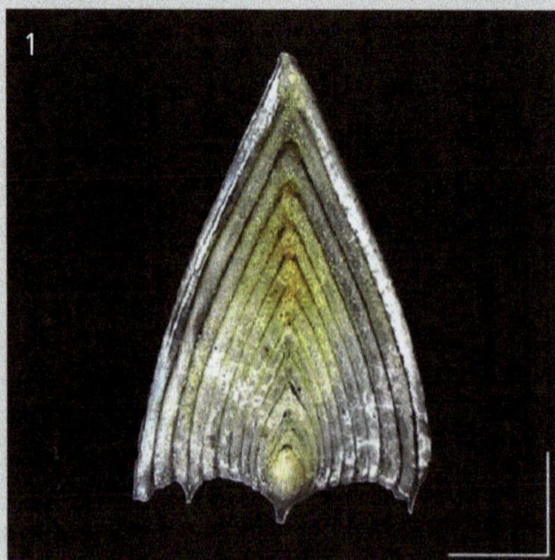
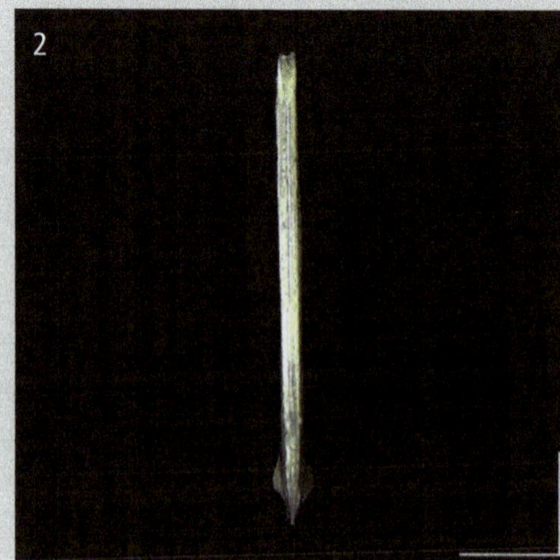

1. *Frondicularia saggitula*. Off Bermuda; "Challenger" Station 33, depth: 435 fathoms. Deposited in the micropaleontological collections of The Natural History Museum, London; NHMUK ZF1439. Lateral view, NHMUK PM DI 0339, scale bar = 843 μm.

2. *Frondicularia saggitula*. Off Bermuda; "Challenger" Station 33, depth: 435 fathoms. Deposited in the micropaleontological collections of The Natural History Museum, London; NHMUK ZF1439. Lateral view, NHMUK PM DI 0340, scale bar = 822 μm.

### Original Designation
*Frondicularia alata* d'Orbigny var. *saggitula* van den Broeck, 1876, p. 113, pl. 2, figs. 12, 14.

### Type Specimen
Depository not given. Brady's figured specimens are housed in the micropaleontological collections of The Natural History Museum, London, UK, and are registered in slides 1959.5.5.341, ZF1439, and ZF1440.

### Type Level
Recent, surface sediments.

### Type Locality
100 fathoms water depth, near Barbados, Lesser Antilles.

### Description
Test exhibits an elongate, strongly compressed, lanceolate form. The broad, low, equitant chambers increase gradually in size and are separated by distinct, depressed sutures, highly arched or angled in the center of the test. The initial chamber and chamber ends often bear a sharp spine. Chamber walls are calcareous, optically radial, perforate, and smooth. The primary aperture is terminal, radiate, and may be slightly produced.

### Synonyms and Citations
*Frondicularia alata* d'Orbigny. –Brady, 1884, p. 522, pl. 65, figs. 20–23, pl. 66, figs. 3–5.

*Frondicularia saggitula* (van den Broeck). –Barker, 1960, pl. 65, figs. 20–23, pl. 66, figs. 3–5. –Jones, 1994, p. 77, pl. 65, figs. 20–23; p. 78, pl. 66, figs. 3–5.

### Biogeography
Worldwide.

### Bathymetry
Lower neritic to middle bathyal. Reported between 84 and 435 fathoms at "Challenger" stations (Jones, 1994).

### Chronostratigraphy
Miocene to Recent.

## References

Barker, R.W., 1960, Taxonomic notes on the species figured by H.B. Brady in his report on the foraminifera dredged by H.M.S. Challenger during the years 1873–1876. Society of Economic Paleontologists and Mineralogists Special Publication, vol. 9, pp. 1–238.

Brady, H.B., 1884, Report on the foraminifera dredged by H.M.S. Challenger during the years 1873–1876. In Brady, H.B., ed., Report on the Scientific Results of the Voyage of the H.M.S. Challenger during the years 1873–1876. Zoology, vol. 9, pp. 1–814.

Jones, R.W., 1994, The Challenger Foraminifera. Oxford: Oxford University Press, 149 pp.

van den Broeck, E., 1876, Étude sur les foraminifères de la Barbade (Antilles) recueillis par L. Agassiz précédée de quelques considérations sur la classification et la nomenclature des Foraminifères. Annales de la Société Belge de Microscopie (Mémoires), vol. 1 (for 1876), pp. 55–152.

# *Fursenkoina bradyi*

## (Cushman), 1922

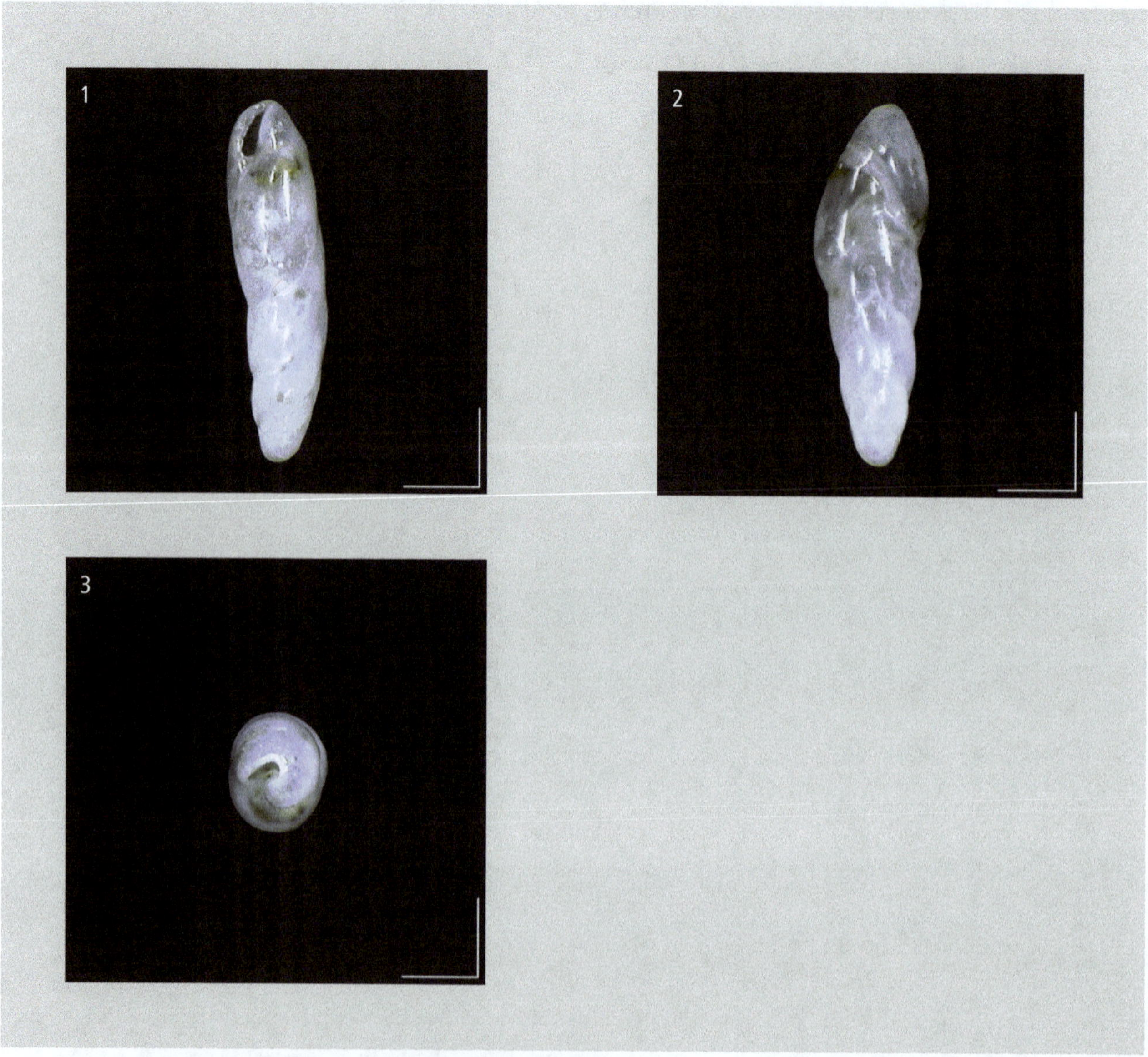


1. *Fursenkoina bradyi.* Off Tahiti; "Challenger" Station 279C, depth: 620 fathoms. Deposited in the micropaleontological collections of The Natural History Museum, London; NHMUK ZF2621. Lateral view, NHMUK PM DI 0341, scale bar = 149 μm.

2. *Fursenkoina bradyi.* Off Tahiti; "Challenger" Station 279C, depth: 620 fathoms. Deposited in the micropaleontological collections of The Natural History Museum, London; NHMUK ZF2621. Lateral view, NHMUK PM DI 0342, scale bar = 124 μm.

3. *Fursenkoina bradyi.* Off Tahiti; "Challenger" Station 279C, depth: 620 fathoms. Deposited in the micropaleontological collections of The Natural History Museum, London; NHMUK ZF2621. Apertural view, NHMUK PM DI 0343, scale bar = 52 μm.

## Original Designation

*Virgulina bradyi* Cushman, 1922, p. 115, pl. 24, fig. 1.

## Type Specimen

Deposited in the collections at the United States National Museum, Washington, DC (Nos 16283–16286, 16287).

## Type Level

Recent, surface sediments.

## Type Locality

"Albatross" Station D2568 (latitude 39° 15′ 00″ N; longitude 68° 08′ 00″ W) in 1781 fathoms, southeast of Nantucket, off the northeast coast of the USA.

## Description

Test exhibits an elongate, twisted, biserial, very slightly compressed fusiform shape with a blunt initial portion, a rounded apertural end, and a subcircular cross-section. The moderately inflated chambers are separated by distinct, curved, depressed sutures. Chamber walls are calcareous, smooth, and very finely perforate. The primary aperture is an elongate oval slit with a toothplate, extending up the face of the final chamber.

## Synonyms and Citations

*Virgulina subsquammosa* Egger. –Brady, 1884, p. 415, pl. 52, fig. 9 (not figs. 7–8, 10–11).

*Virgulina bradyi* Cushmann. –Barker, 1960, pl. 52, fig. 9.

*Fursenkoina bradyi* (Cushman). –Thomas, 1985, p. 676, pl. 3, fig. 7. –Jones, 1994, p. 57, pl. 52, fig. 9.

## Biogeography

Worldwide.

## Bathymetry

Upper bathyal. Reported from 620 fathoms at "Challenger" Station 279C, off Tahiti (Jones, 1994).

## Chronostratigraphy

Oligocene to Recent (Jones, 1994).

# References


Barker, R.W., 1960, Taxonomic notes on the species figured by H.B. Brady in his report on the foraminifera dredged by H.M.S. Challenger during the years 1873–1876. Society of Economic Paleontologists and Mineralogists Special Publication, vol. 9, pp. 1–238.

Brady, H.B., 1884, Report on the foraminifera dredged by H.M.S. Challenger during the years 1873–1876. In Brady, H.B., ed., Report on the Scientific Results of the Voyage of the H.M.S. Challenger during the years 1873–1876. Zoology, vol. 9, pp. 1–814.

Cushman, J.A., 1922, The foraminifera of the Atlantic Ocean; Part 3 – Textulariidae. Bulletin of the United States National Museum, vol. 104, pp. 1–149.

Jones, R.W., 1994, The Challenger Foraminifera. Oxford: Oxford University Press, 149 pp.

Thomas, E., 1985, Late Eocene to Recent deep-sea benthic foraminifers from the central equatorial Pacific Ocean. In Mayer, L., Theyer, F., et al., ed., Initial Reports of the Deep Sea Drilling Project, Washington (U.S. Government Printing Office), vol. 85, pp. 655–694.

# Fursenkoina complanata

## (Egger), 1893

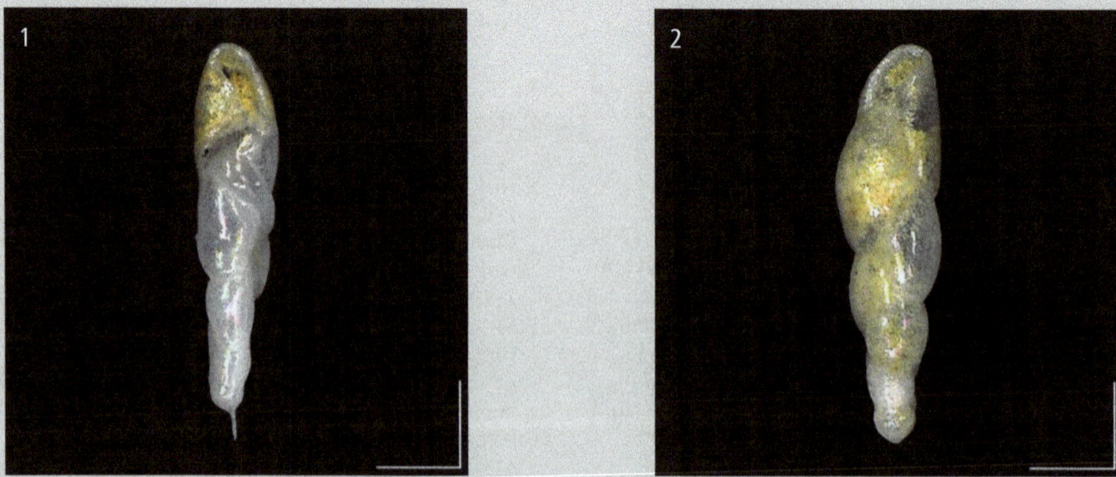


1. *Fursenkoina complanata*. South Pacific Ocean; "Challenger" Station 276, depth: 2350 fathoms. Deposited in the micropaleontological collections of The Natural History Museum, London; NHMUK ZF2618. Lateral view, NHMUK PM DI 0344, scale bar = 116 μm.

2. *Fursenkoina complanata*. South Pacific Ocean; "Challenger" Station 276, depth: 2350 fathoms. Deposited in the micropaleontological collections of The Natural History Museum, London; NHMUK ZF2618. Lateral view, NHMUK PM DI 0345, scale bar = 97 μm.

### Original Designation

*Virgulina schreibersiana* Czjzek var. *complanata* Egger, 1893, p. 292, pl. 8, figs. 91–92.

### Type Specimen

Depository not given.

### Type Level

Recent, surface sediments.

### Type Locality

Station 90: from 359 meters water depth, at 18° 52′ S, 116° 18′ E, north of Mermaid (Australia).

### Description

Test exhibits an elongate, biserial, slightly compressed, slightly twisted, fusiform shape with a pointed initial portion (often bearing a spine), a rounded apertural end, and a subcircular cross-section. The moderately inflated chambers increase rapidly in height and are separated by distinct, curved, depressed sutures. Chamber walls are calcareous, smooth, and finely perforate. The primary aperture is an elongate oval slit extending up the face of the final chamber, with a toothplate.

### Synonyms and Citations

*Virgulina schreibersiana* Czjzek. –Brady, 1884, p. 414, pl. 52, figs. 1–3.
*Virgulina davisi* Chapman and Parr. –Barker, 1960, pl. 52, figs. 1–3.
*Stainforthia complanata* (Cushman). –Thomas, 1985, p. 678, pl. 2, fig. 9.
*Fursenkoina complanata* (Egger). –Jones, 1994, p. 56, pl. 52, figs. 1–3.

### Biogeography

Worldwide.

### Bathymetry

Neritic to abyssal. Reported between 10 and 3000 fathoms at "Challenger" stations (Jones, 1994). Abyssal (Thomas, 1985).

### Chronostratigraphy

Pliocene to Recent. First appearance datum at 2.8 million years recorded in DSDP Site 573 (Equatorial Pacific) by Thomas (1985).

## References

Barker, R.W., 1960, Taxonomic notes on the species figured by H.B. Brady in his report on the foraminifera dredged by H.M.S. Challenger during the years 1873–1876. Society of Economic Paleontologists and Mineralogists Special Publication, vol. 9, pp. 1–238.

Brady, H.B., 1884, Report on the foraminifera dredged by H.M.S. Challenger during the years 1873–1876. In Brady, H.B., ed., Report on the Scientific Results of the Voyage of the H.M.S. Challenger during the years 1873–1876. Zoology, vol. 9, pp. 1–814.

Egger, J.G., 1893, Foraminiferen aus Meeresgrundproben, gelothet von 1874 bis 1876 von S.M. Sch. Gazelle. Abhandlungen der Bayerischen Akademie der Wissenschaften, München, Math.-Phys. Cl., vol. 18, no. 2, pp. 193–458.

Jones, R.W., 1994, The Challenger Foraminifera. Oxford: Oxford University Press, 149 pp.

Thomas, E., 1985, Late Eocene to Recent deep-sea benthic foraminifers from the central equatorial Pacific Ocean. In Mayer, L., Theyer, F., et al., ed., Initial Reports of the Deep Sea Drilling Project, Washington (U.S. Government Printing Office), vol. 85, pp. 655–694.

# Globobulimina pacifica

## Cushman, 1927

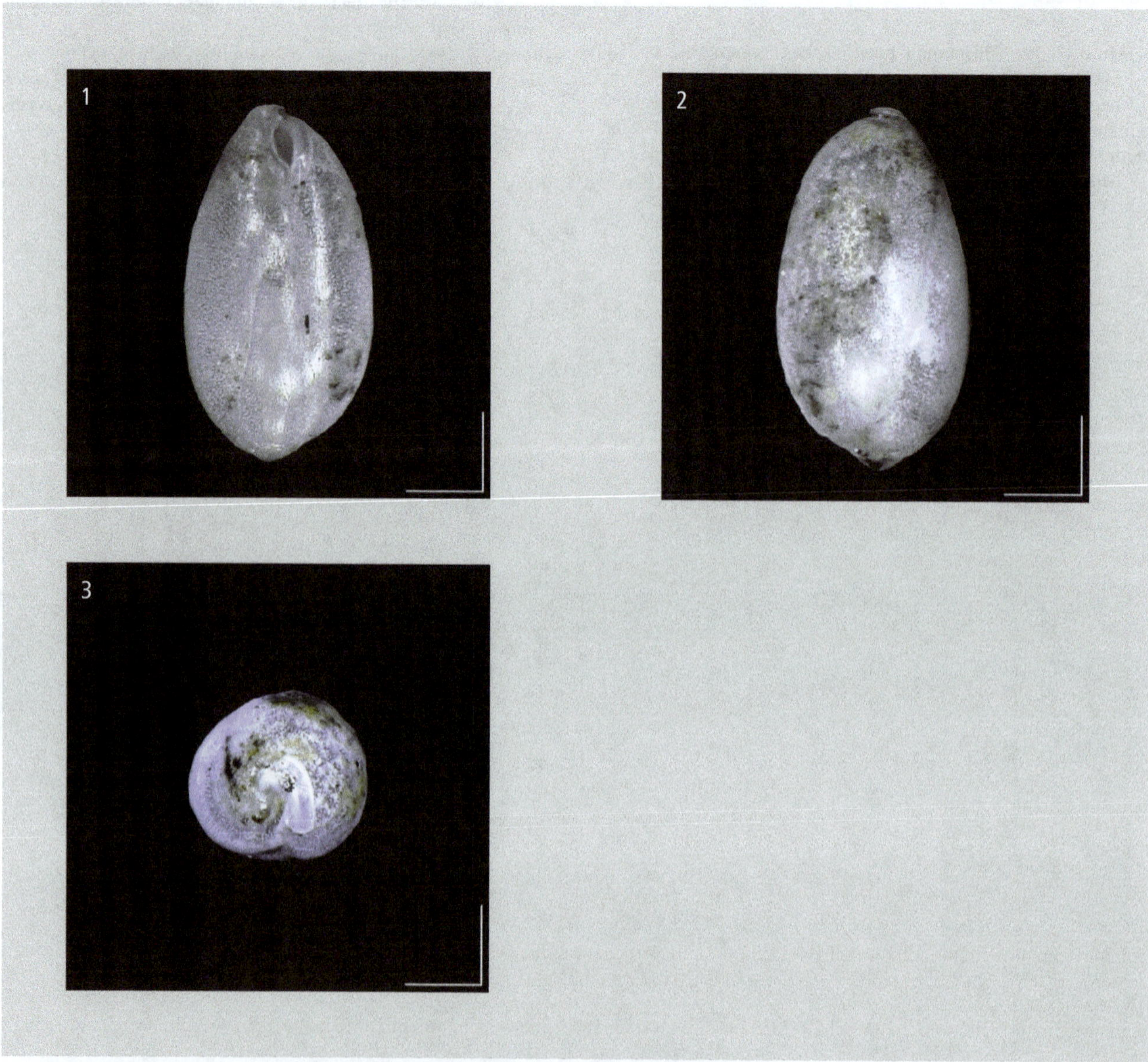

1. *Globobulimina pacifica*. Off the Azores, Atlantic Ocean; "Challenger" Station 75, depth: 450 fathoms. Deposited in the micropaleontological collections of The Natural History Museum, London; NHMUK ZF1222. Lateral view, NHMUK PM DI 0346, scale bar = 222 μm.

2. *Globobulimina pacifica*. Off the Azores, Atlantic Ocean; "Challenger" Station 75, depth: 450 fathoms. Deposited in the micropaleontological collections of The Natural History Museum, London; NHMUK ZF1222. Lateral view, NHMUK PM DI 0347, scale bar = 251 μm.

3. *Globobulimina pacifica*. Off the Azores, Atlantic Ocean; "Challenger" Station 75, depth: 450 fathoms. Deposited in the micropaleontological collections of The Natural History Museum, London; NHMUK ZF1222. Apertural view, NHMUK PM DI 0348, scale bar = 125 μm.

## Original Designation

*Globobulimina pacifica* Cushman, 1927, p. 67, pl. 14, fig. 12.

## Type Specimen

Deposited in the collections of the United States National Museum, Washington, DC. (No. 20285).

## Type Level

Recent, surface sediments.

## Type Locality

Eastern Pacific, latitude 17° 18′ N, longitude 102° 22′ W.

## Description

Test forms an elongate triserial series, ovate in outline and circular in cross-section, with the length equivalent to approximately one and a third the maximum width and the maximum width located in the middle portion of the test. The highly inflated, overlapping chambers increase in size rapidly, and are separated by thin, slightly depressed, oblique sutures that are nearly parallel to the margins of the test. Chamber walls are calcareous, smooth, and finely perforate. The primary aperture is a loop with an internal toothplate.

## Synonyms and Citations

*Bulimina pyrula* d'Orbigny. –Brady, 1884, p. 399, pl. 50, figs. 7–10.

*Globobulimina pacifica* Cushman?. –Barker, 1960, pl. 50, figs. 7–10.

*Globobulimina pacifica* Cushman. –Cushman and Jarvis, 1930, p. 362, pl. 33, fig. 6. –Bermúdez 1949, p. 185, pl. 12, fig. 14. –Loeblich and Tappan, 1987, p. 521, pl. 571, figs. 8–12. –van Marle, 1991, p. 90, pl. 5, figs. 11–12. –Jones, 1994, p. 54, pl. 50, figs. 7–10. –Robertson, 1998, p. 150, pl. 57, fig. 1. –Ortiz and Thomas, 2006, p. 118, pl. 6, figs. 8–11.

## Remarks

Type species of the genus *Globobulimina* Cushman, 1927.

*Globobulimina pacifica* is a deep infaunal taxon, characteristic in oxygen-depleted sediments from the oxygen minimum zone (den Dulk et al., 2000).

## Biogeography

Worldwide.

## Bathymetry

Middle neritic to bathyal. Reported from 129 fathoms (Central Pacific), 450 fathoms (Azores), and 1100 fathoms (off New Zealand) at "Challenger" stations (Jones, 1994).

## Chronostratigraphy

Miocene to Recent (Jones, 1994).

# References


Barker, R.W., 1960, Taxonomic notes on the species figured by H.B. Brady in his report on the foraminifera dredged by H.M.S. Challenger during the years 1873–1876. Society of Economic Paleontologists and Mineralogists Special Publication, vol. 9, pp. 1–238.

Bermúdez, P.J., 1949, Tertiary smaller foraminifera of the Dominican Republic. Special Publications of the Cushman Laboratory for Foraminiferal Research, no. 25, pp. 1–322.

Brady, H.B., 1884, Report on the foraminifera dredged by H.M.S. Challenger during the years 1873–1876. In Brady, H.B., ed., Report on the Scientific Results of the Voyage of the H.M.S. Challenger during the years 1873–1876. Zoology, vol. 9, pp. 1–814.

Cushman, J.A., 1927, An outline of a re-classification of the foraminifera. Contributions from the Cushman Laboratory for Foraminiferal Research, vol. 3, pp. 1–105.

Cushman, J.A. and Jarvis, P.W., 1930, Miocene foraminifera from Buff Bay, Jamaica. Journal of Paleontology, vol. 4, no. 4, pp. 353–368.

den Dulk, M., Reichart, G.J., van Heyst, S., Zachariasse, W.J., and van der Zwaan, G.J., 2000, Benthic foraminifera as proxies of organic matter flux and bottom water oxygenation? A case history from the northern Arabian Sea. Palaeogeography, Palaeoclimatology, Palaeoecology, vol. 161, pp. 337–359.

Jones, R.W., 1994, The Challenger Foraminifera. Oxford: Oxford University Press, 149 pp.

Loeblich, A.R. Jr and Tappan, H., 1987, Foraminiferal Genera and their Classification. New York: Van Nostrand Reinhold, 970 pp.

Ortiz, S. and Thomas, E., 2006, Lower-middle Eocene benthic foraminifera from the Fortuna Section (Betic Cordillera, southeastern Spain). Micropaleontology, vol. 52, pp. 97–150.

Robertson, B.E., 1998, Systematics and paleoecology of the benthic foraminiferida from the Buff Bay section, Miocene of Jamaica. Micropaleontology, vol. 44, supplement 2, pp. 1–266.

van Marle, L.J., 1991, Eastern Indonesian Late Cenozoic smaller benthic foraminifera. Verhandelingen der Koninklijke Nederlanse Akademie van Wetenschappen, Afdeeling Natuurkunde, Eerste Reeks, vol. 34, pp. 1–328.

# *Globocassidulina punctata*

## Berggren and Miller, 1986

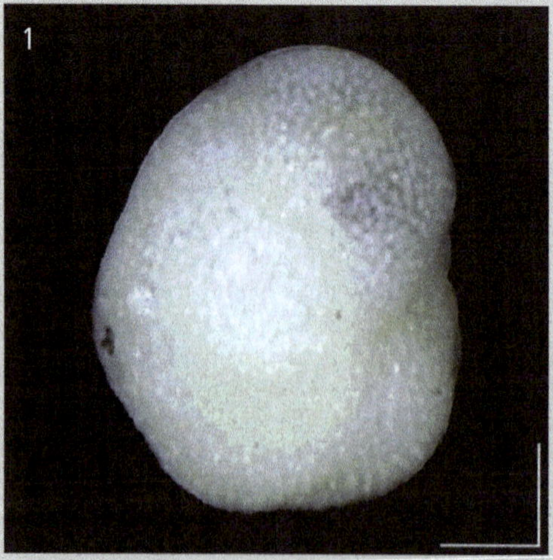


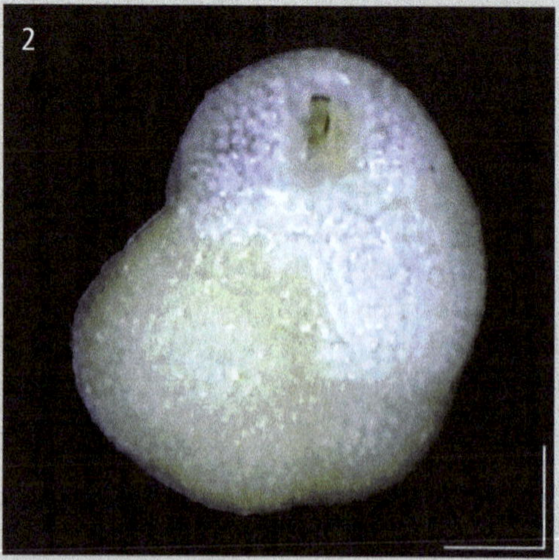


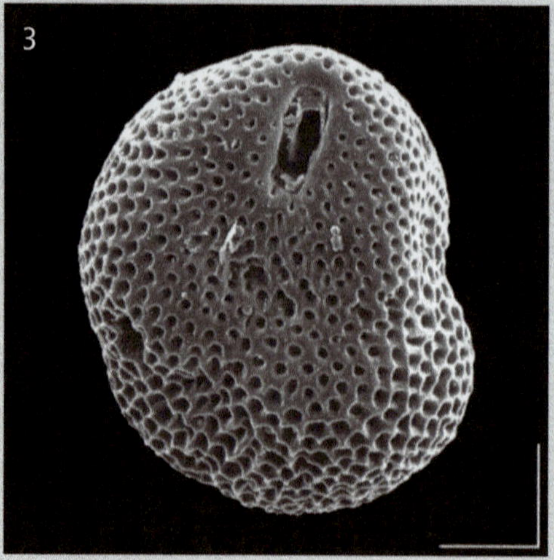


1. *Globocassidulina punctata.* Tosagua Formation, Manabi Basin, Ecuador; Sample CRB111A (Chone Sheet). Deposited in the micropaleontological collections of The Natural History Museum, London. NHMUK P51990. Lateral view, NHMUK PM DI 0349, scale bar = 47 μm.

2. *Globocassidulina punctata.* Tosagua Formation, Manabi Basin, Ecuador; Sample CRB111A (Chone Sheet). Deposited in the micropaleontological collections of The Natural History Museum, London. NHMUK P51990. Lateral view, NHMUK PM DI 0350, scale bar = 49 μm.

3. *Globocassidulina punctata.* Gulf of Mexico; Eureka Core E66-73, depth: 3258 feet. Lateral view, NHMUK PM DI 0351, scale bar = 64 μm.

### Original Designation

*Globocassidulina punctata* Berggren and Miller, 1986. In van Morkhoven et al., 1986, p. 118, textfig. 1, pl. 37.

### Type Specimen

Holotype (USNM 406405) and paratypes deposited in the collections of the United States National Museum, Washington, DC.

### Type Level

Middle Miocene, Biozone N11.

### Type Locality

Eureka Core E68–136, 747 meters, latitude 29° 17′ N, longitude 86° 55′ W, Gulf of Mexico.

### Description

Test forms a small biserial series, ovate in outline and with a broadly rounded initial portion, a tapered apertural end, and a rounded periphery. The enrolled chambers are separated by slightly depressed sutures. Chamber walls are calcareous and coarsely perforate, appearing strongly pitted. The primary aperture is a straight or curved slit extending up the apertural face, with a toothplate on the posterior side.

### Synonyms and Citations

*Globocassidulina punctata* Berggren and Miller. –Katz and Miller, 1993, pl. 5, fig. 4. –Robertson, 1998, p. 136, pl. 52, figs. 5–6.

### Remarks

High-resolution re-study of the Eureka boreholes in the Gulf of Mexico (Katz and Miller, 1993) extended the first occurrence of *Globocassidulina punctata* Berggren and Miller to Biozone M1, rather than the Biozone M3 reported by van Morkhoven et al. (1986). Its bathymetric range was also extended by Katz and Miller (1993) to the middle and lower bathyal zones, in contrast to the upper bathyal range given by van Morkhoven et al. (1986).

### Biogeography

Found in Cuba and the Gulf of Mexico (van Morkhoven et al., 1986). Recorded as *Cassidulina* sp. 1 in the North and South Atlantic (van Morkhoven, 1981).

### Bathymetry

Bathyal.

### Chronostratigraphy

Miocene (Aquitanian to Messinian).

## References


Katz, M.E. and Miller, K.G., 1993, Latest Oligocene to Earliest Pliocene benthic foraminiferal biofacies of the northeastern Gulf of Mexico. Micropaleontology, vol. 39, no. 4, pp. 367–403.

Robertson, B.E., 1998, Systematics and paleoecology of the benthic foraminiferida from the Buff Bay section, Miocene of Jamaica. Micropaleontology, vol. 44, supplement 2, pp. 1–266.

van Morkhoven, F.P.C.M., 1981, Cosmopolitan Tertiary bathyal benthic foraminifera. Transactions of the Gulf Coast Association of Geological Societies, Supplement 31, 445 abstracts with accompanying range chart.

van Morkhoven, F.P.C.M., Berggren, W.A., Edwards, A.S., et al., 1986, Cenozoic cosmopolitan deep-water benthic foraminifera. Bulletin des Centres de Recherches Exploration-Production Elf-Aquitaine, Mémoire, vol. 11, pp. 1–421.

# *Globocassidulina subglobosa*

## (Brady), 1881

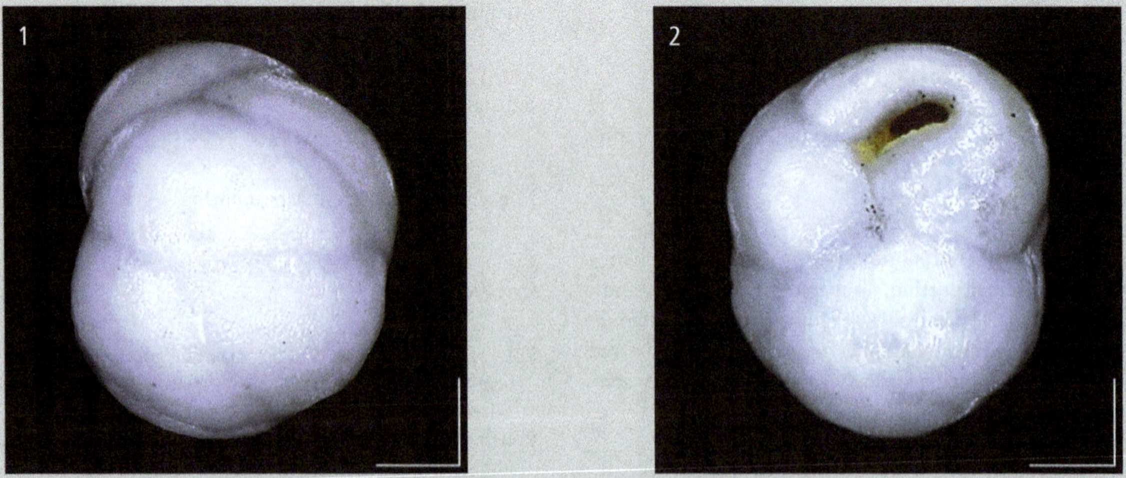

1. *Globocassidulina subglobosa.* Off Pernambuco, Atlantic Ocean; "Challenger" Station 120, depth: 675 fathoms. Deposited in the micropaleontological collections of The Natural History Museum, London; NHMUK ZF1267. Syntype. Lateral view, NHMUK PM DI 0352, scale bar = 146 μm.

2. *Globocassidulina subglobosa.* Off Pernambuco, Atlantic Ocean; "Challenger" Station 120, depth: 675 fathoms. Deposited in the micropaleontological collections of The Natural History Museum, London; NHMUK ZF1267. Syntype. Apertural view, NHMUK PM DI 0353, scale bar = 146 μm..

### Original Designation

*Cassidulina subglobosa* Brady, 1881, p. 60 (type description), illustrated in Brady, 1884, pl. 54, fig. 17.

### Type Specimen

Figured specimen is housed in the micropaleontological collections of The Natural History Museum, London, UK, and is registered in slide ZF1267.

### Type Level

Recent, surface sediments.

### Type Locality

Figured specimens from "Challenger" Station 120, latitude 8° 37′ S, longitude 34° 28′ W, off Pernambuco, Brazil, 675 fathoms.

### Description

Test forms a small, subglobular, biserial series, with a subcircular cross-section and a rounded periphery. The inflated chambers are separated by slightly depressed, curved sutures. Chamber walls are calcareous, finely perforate, and smooth. The primary aperture is a straight or curved slit extending up the apertural face, with a toothplate on the posterior side.

### Synonyms and Citations

*Cassidulina subglobosa* Brady. –Brady, 1884, p. 430, pl. 54, fig. 17. –Barker, 1960, pl. 54, fig. 17.

*Globocassidulina subglobosa* (Brady). –Tjalsma and Lohmann, 1983, p. 31, pl. 16, fig. 9. –Miller and Katz, 1987, p. 134, pl. 3, fig. 4. –Hermelin, 1989, p. 74. –Thomas, 1990, p. 590. –Jones, 1994, p. 60, pl. 54, fig. 17. –Robertson, 1998, p. 136, pl. 53, figs. 1–2. –Kuhnt et al., 2002, p. 144, pl. 10. figs. 3–5. –Milker and Schmiedl, 2012, p. 85, figs. 20.13–14.

### Biogeography

Worldwide (Jones, 1994); Weddell Sea, Antarctic (Thomas, 1990); North Atlantic (Miller and Katz, 1987); Indian Ocean, Atlantic, Pacific, Gulf of California, Gulf of Mexico, Mexico, Java, Okinawa,

Papua New Guinea, and Guam (see Hermelin, 1989); South China Sea (Kuhnt et al., 2002); Mediterranean (Milker and Schmiedl, 2012).

### Bathymetry

Primarily bathyal to abyssal. Reported between 12 and 2950 fathoms at "Challenger" stations (Jones, 1994); lower bathyal (Thomas, 1990; Katz and Miller, 1993); abyssal, greater than 3.5 kilometers (Miller and Katz, 1987).

### Chronostratigraphy

Middle Paleocene (Selandian) to Recent. Late Paleocene (Biozone P4) to Recent (Tjalsma and Lohmann, 1983); Oligocene to Recent (Jones, 1994); topmost Paleocene to Early Eocene (rare), topmost Eocene to Pleistocene (common) according to Thomas (1990); abundant in the Oligocene and Miocene (Miller and Katz, 1987); Late Miocene (Katz and Miller, 1993).

## References


Barker, R.W., 1960, Taxonomic notes on the species figured by H.B. Brady in his report on the foraminifera dredged by H.M.S. Challenger during the years 1873–1876. Society of Economic Paleontologists and Mineralogists Special Publication, vol. 9, pp. 1–238.

Brady, H.B., 1881, Notes on some of the reticularian Rhizopoda of the Challenger Expedition, Part III. Quarterly Journal of Microscopical Science (New Series), vol. 21, pp. 31–71.

Brady, H.B., 1884, Report on the foraminifera dredged by H.M.S. Challenger during the years 1873–1876. In Brady, H.B., ed., Report on the Scientific Results of the Voyage of the H.M.S. Challenger during the years 1873–1876. Zoology, vol. 9, pp. 1–814.

Hermelin, J.O.R., 1989, Pliocene benthic foraminifera from the Ontong-Java Plateau (Western Equatorial Pacific Ocean): faunal response to changing paleoenvironment. Special Publication of the Cushman Foundation for Foraminiferal Research, vol. 26, pp. 1–143.

Jones, R.W., 1994, The Challenger Foraminifera. Oxford: Oxford University Press, 149 pp.

Katz, M.E. and Miller, K.G., 1993, Latest Oligocene to Earliest Pliocene benthic foraminiferal biofacies of the northeastern Gulf of Mexico. Micropaleontology, vol. 39, no. 4, pp. 367–403.

Kuhnt, W., Holbourn, A.E., and Zhao, Q., 2002, The early history of the South China Sea: Evolution of Oligocene–Miocene deep water environments. Revue de Micropaléontologie, vol. 45, pp. 99–159.

Miller, K.G. and Katz, M.E., 1987, Oligocene to Miocene benthic foraminiferal and abyssal circulation changes in the north Atlantic. Micropaleontology, vol. 33, no. 2, pp. 97–149.

Milker, Y. and Schmiedl, G., 2012, A taxonomic guide to modern benthic shelf foraminifera of the western Mediterranean Sea. Palaeontologia Electronica, vol. 15, Issue 2;16A, 134 pp.

Robertson, B.E., 1998, Systematics and paleoecology of the benthic foraminiferida from the Buff Bay section, Miocene of Jamaica. Micropaleontology, vol. 44, supplement 2, pp. 1–266.

Thomas, E., 1990, Late Cretaceous through Neogene benthic foraminifers. In Barker, P.F., Kennett, J.P., et al., ed., Scientific Results of the Proceedings of the Ocean Drilling Program, vol. 113, pp. 571–594.

Tjalsma, R.C. and Lohmann, G.P., 1983, Paleocene–Eocene bathyal and abyssal benthic foraminifera from the Atlantic Ocean. Micropaleontology, Special Publication, no. 4, pp. 1–90.

# Globulina myristiformis

## (Williamson), 1858

1. *Globulina myristiformis*. Mounts Bay, Cornwall (dredged sample). Deposited in the micropaleontological collections of The Natural History Museum, London; NHMUK ZF2144. Lateral view, NHMUK PM DI 0354, scale bar = 126 μm.

2. *Globulina myristiformis*. Mounts Bay, Cornwall (dredged sample). Deposited in the micropaleontological collections of The Natural History Museum, London; NHMUK ZF2144. Apertural view, NHMUK PM DI 0355, scale bar = 105 μm.

### Original Designation

*Polymorphina myristiformis* Williamson, 1858, p. 73, pl. 6, figs. 156–157.

### Type Specimen

Figured specimens housed in the micropaleontological collections of The Natural History Museum, London, UK, and registered in slide ZF2144.

### Type Level

Recent, surface sediments.

### Type Locality

Not designated. Localities given: Oxwich, near Swansea; Manorbier; Cork; dredged sands of Arran; Shetland; Roundstone; Guernsey; Eddystone, British Isles.

### Description

Test globular, ovate, with a circular cross-section. The highly inflated, strongly overlapping chambers are separated by flush, oblique sutures. Chamber walls are calcareous, perforate, and ornamented by fine, irregular, raised costae. The primary aperture is terminal and radiate.

### Synonyms and Citations

*Polymorphina myristiformis* Williamson. –Brady, 1884, p. 571, pl. 73, figs. 9–10.
*Globulina gibba* var. (Williamson). –Barker, 1960, pl. 73, figs. 9–10.
*Globulina myristiformis* (Williamson). –Jones, 1994, p. 85, pl. 73, figs. 9–10. –Milker and Schmiedl, 2012, p. 75, Fig. 19.1.

### Biogeography

Atlantic (Jones 1994). Mediterranean (Milker and Schmiedl, 2012). Perhaps wider distribution.

### Bathymetry

Middle neritic to middle bathyal. Reported between 30 and 630 fathoms at "Challenger" stations (Jones, 1994).

### Chronostratigraphy

Miocene to Recent (Jones, 1994).

## References

Barker, R.W., 1960, Taxonomic notes on the species figured by H.B. Brady in his report on the foraminifera dredged by H.M.S. Challenger during the years 1873–1876. Society of Economic Paleontologists and Mineralogists Special Publication, vol. 9, pp. 1–238.

Brady, H.B., 1884, Report on the foraminifera dredged by H.M.S. Challenger during the years 1873–1876. In Brady, H.B., ed., Report on the Scientific Results of the Voyage of the H.M.S. Challenger during the years 1873–1876. Zoology, vol. 9, pp. 1–814.

Jones, R.W., 1994, The Challenger Foraminifera. Oxford: Oxford University Press, 149 pp.

Milker, Y. and Schmiedl, G., 2012, A taxonomic guide to modern benthic shelf foraminifera of the western Mediterranean Sea. Palaeontologia Electronica, vol. 15, Issue 2;16A, 134 pp.

Williamson, W.C., 1858, On the Recent Foraminifera of Great Britain. London: Ray Society, 107 pp.

# Glomospira charoides

## (Jones and Parker), 1860

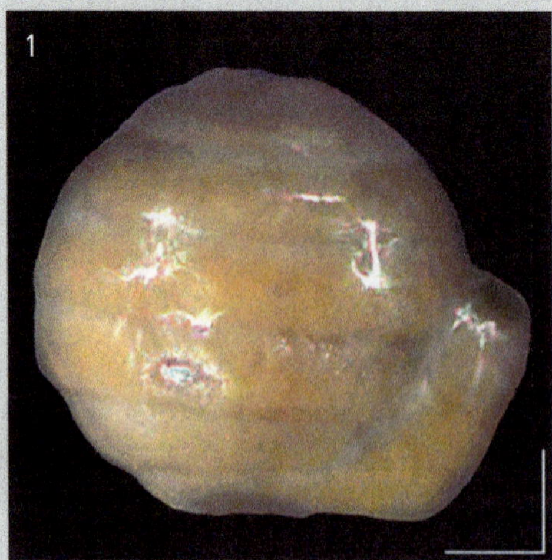

1. *Glomospira charoides*. Cape Demetri, latitude 36° 18′ N, longitude 13° 33′ E, off Gozo Island, Malta; HMS "Firebrand" Station, Parker Collection. Deposited in the micropaleontological collections of The Natural History Museum, London; NHMUK ZF4875. Lectotype. Lateral view, NHMUK PM DI 0356, scale bar = 73 μm.

### Original Designation

*Trochammina squamata* var. *charoides* Jones and Parker, 1860, p. 304 [no type figure given].

### Type Specimen

Deposited in the the micropaleontological collections of The Natural History Museum, London, UK, and registered in slides 1894.4.3.540, 1894.4.3.585, 1894.4.3.616, and 1894.4.3.624. The lectotype, designated by Berggren and Kaminski (1990, pl. 1, fig. 2), is the specimen illustrated by Carpenter et al. (1862, pl. 11, fig. 3). Originally in slide 1894.4.3.540, it has been assigned a new catalog number (ZF4875).

### Type Level

Recent, surface sediments.

### Type Locality

Not designated. Jones and Parker (1860) listed several locality stations in the Greek archipelago and off Crete. The lectotype specimen is from a dredge sample collected from a depth of 704 fathoms 33 miles southeast of Cape Demitri, Gozo Island, Malta (36° 18′ N, 13° 33′ E).

### Description

Test consists of a globular proloculus followed by a trochospirally enrolled, undivided, tubular second chamber coiled about a vertical axis. There are typically three layers of coils with six to seven whorls in the outer layer. Chamber walls are finely agglutinated, cemented with organic matter, with inner and outer organic linings, and a smooth finish. The primary aperture is at the open end of the tube.

### Synonyms and Citations

*Trochammina squamata* var. *charoides* Jones and Parker. –Carpenter, Parker, and Jones, 1862, p. 141, pl. 11, fig. 3.

*Ammodiscus charoides* (Jones and Parker). –Brady, 1884, p. 334, pl. 38, figs. 10–16.

*Glomospira charoides* (Jones and Parker) var. *corona* Cushman and Jarvis, 1928, p. 89, pl. 12, figs. 9–11.

*Usbekistania charoides antarctica* Saidova, 1975, p. 71, pl. 96, fig. 3.

*Usbekistania charoides profunda* Saidova, 1975, p. 71, pl. 18, fig. 5.

*Glomospira charoides* (Jones and Parker). –Barker, 1960, pl. 38, figs.10–16. –Berggren and Kaminski, 1990, pl. 1, fig. 2. –Holbourn and Kaminski, 1997, p. 36, pl. 5, figs. 7–11. –Kaminski and Gradstein, 2005, pp. 168–174, fig. 22, pl. 22.

*Glomospira charoides charoides* (Jones and Parker). –Weidich, 1990, p. 83, pl. 34, figs. 7, 11.

*Glomospira charoides corona* Cushman and Jarvis. –Weidich 1990, p. 83, pl. 34, figs. 8–9.

*Usbekistania charoides* (Jones and Parker). –Jones, 1994, p. 43, pl. 38, figs. 10–16.

### Remarks

Type species of the genus *Repmanina* Suleymanov, 1966, regarded as a junior synonym of *Usbekistania* Suleymanov 1960 by Jones (1994).

Numerous junior synonyms or subspecies of *Glomospira charoides* (Jones and Parker) have been described: see Remarks in Kaminski and Gradstein (2005) for detailed taxonomic history.

### Biogeography

Worldwide. See Observed Occurrences in Kaminski and Gradstein (2005).

### Bathymetry

Bathyal to abyssal in Cretaceous to Paleogene assemblages; may be restricted to the continental slope and rise in the modern Atlantic (Kaminski and Gradstein, 2005).

### Chronostratigraphy

Jurassic to Recent (Kaminski and Gradstein, 2005). *Glomospira charoides* (Jones and Parker) is the dominant species of *Glomospira* in Cretaceous and Paleogene abyssal assemblages and variations in the abundance of this species can be useful for paleoecology and biostratigraphy (Kaminski and Gradstein, 2005). In certain areas of the North Atlantic, as well as in the Mediterranean and Alpine flysch basins, a level containing increased proportions of *Glomospira* occurs in the upper Lower Eocene to lower Middle Eocene, in both clastic and pelagic, deep-sea sediments (Kaminski and Gradstein, 2005). In the North Atlantic ODP holes, this "*Glomospira* facies" is associated with siliceous microfossils and higher amounts of organic carbon, indicating higher organic productivity in the surface waters (Kaminski and Gradstein, 2005). This Early Eocene "*Glomospira* facies" is entirely comparable to the "Biofacies B" assemblages described from the lower Campanian by Kuhnt and Kaminski (1989), and from the Lower Cretaceous by Holbourn and Kaminski (1997).

## References


Barker, R.W., 1960, Taxonomic notes on the species figured by H.B. Brady in his report on the foraminifera dredged by H.M.S. Challenger during the years 1873–1876. Society of Economic Paleontologists and Mineralogists Special Publication, vol. 9, pp. 1–238.

Brady, H.B., 1884, Report on the foraminifera dredged by H.M.S. Challenger during the years 1873–1876. In Brady, H.B., ed., Report on the Scientific Results of the Voyage of the H.M.S. Challenger during the years 1873–1876. Zoology, vol. 9, pp. 1–814.

Berggren, W.A. and Kaminski, M.A., 1990, Abyssal agglutinates: back to basics. In Hemleben, C., Kaminski, M.A., Kuhnt, W., and Scott, D.B., ed., Paleoecology, Biostratigraphy, Paleoceanography and Taxonomy of Agglutinated Foraminifera. NATO ASI Series, Series C: Mathematical and Physical Sciences, vol. 327. Dordrecht: Kluwer Academic, pp. 53–76.

Carpenter, W.B., Parker, W.K., and Jones, J.P., 1862, Introduction to the Study of the Foraminifera. London: Ray Society, 319 pp.

Cushman, J.A. and Jarvis, P.W., 1928, Cretaceous foraminifera from Trinidad. Contributions from the Cushman Laboratory for Foraminiferal Research, vol. 4, pp. 85–103.

Holbourn, A.E.L. and Kaminski, M.A., 1997, Cretaceous deep-water benthic foraminifera of the Indian Ocean. Grzybowski Foundation Special Publication, no. 4, pp. 1–175.

Jones, R.W., 1994, The Challenger Foraminifera. Oxford: Oxford University Press, 149 pp.

Jones, J.P. and Parker, W.K., 1860, On the Rhizopodal fauna of the Mediterranean compared with that of the Italian and some other Tertiary deposits. Quarterly Journal of the Geological Society of London, vol. 16, pp. 292–307.

Kaminski, M.A. and Gradstein, F.M., 2005, Atlas of Palaeogene Cosmopolitan Deep-Water Agglutinated Foraminifera. Grzybowski Foundation Special Publication, no. 10, 547 + vii pp.

Kaminski, M.A., Gradstein, F.M., Goll, R.M., and Greig, D., 1990, Biostratigraphy and Paleoecology of deep-water aggutinated foraminifera at ODP Site 643, Norwegian–Greenland Sea. In Hemleben, C., Kaminski, M.A., Kuhnt, W., and Scott, D.B., ed., Paleoecology, Biostratigraphy, Paleoceanography and Taxonomy of Agglutinated Foraminifera. NATO ASI Series, Series C: Mathematical and Physical Sciences, vol. 327. Dordrecht: Kluwer Academic, pp. 345–386.

Kuhnt, W. and Kaminski, M.A., 1989, Upper Cretaceous deep-water agglutinated benthic foraminiferal assemblages from the western Mediterranean and adjacent areas. In Wiedmann, J., ed., Cretaceous of the western Tethys. Proceedings of the 3rd International Cretaceous Symposium, Tubingen, pp. 91–120.

Saidova, Kh.M., 1975, Bentosnye foraminifery Tikhogo Okeana, 3 vols. Moscow: Institut Okeanologii P.P. Shirshova, Akademiya Nauk SSSR, 875 pp.

Weidich, K., 1990, Die kalkalpine Unterkreide und ihre Foraminiferenfauna. Zitteliana, vol. 17, p. 1–312.

# Glomospira gordialis

## (Jones and Parker), 1860

1. *Glomospira gordialis.* Cape Demetri, latitude 36° 18′ N, longitude 13° 33′ E, off Gozo Island, Malta; HMS "Firebrand" Station, Parker Collection. Deposited in the micropaleontological collections of The Natural History Museum, London; NHMUK ZF4876. Lectotype. Lateral view, NHMUK PM DI 0360, scale bar = 128 μm.

### Original Designation

*Trochammina squamata* var. *gordialis* Jones and Parker, 1860, p. 304 [no type figure given].

### Type Specimen

First figured by Carpenter et al. (1862, pl. 11, fig. 4). Brady's figured specimens are housed in the micropaleontological collections of The Natural History Museum, London, UK. The lectotype (ZF4876) designated by Berggren and Kaminski (1990, p. 58, pl. 1, fig. 1) from the W.K. Parker Collection is from Slide 1894.4.3.540. Metatypes are housed in Slide 1894.4.5.709.

### Type Level

Recent.

### Type Locality

Not designated. The lectotype specimen is from a dredge sample collected from a depth of 704 fathoms 33 miles southeast of Cape Demitri, Gozo Island, Malta (36° 18′ N, 13° 33′ E).

### Description

Test consists of a globular proloculus followed by an undivided, tubular, second chamber, initially coiled in a high trochospiral manner around the proloculus, later coiled glomospirally about a common plane. The test is usually comprised of five to six whorls, with the terminal part of the tube often irregularly coiled. Chamber walls are finely agglutinated and cemented with organic matter with inner and outer organic linings and a smooth finish. The primary aperture is a round opening at the end of the tube.

### Synonyms and Citations

*Trochammina squamata* var. *gordialis* Jones and Parker. –Carpenter, Parker, and Jones, 1862, pl. 11, fig. 4.

*Ammodiscus gordialis* (Jones and Parker). –Brady, 1884, p. 333, pl. 38, figs. 7–9.

*Glomospira gordialis* (Jones and Parker). –Bartenstein and Brand, 1951, pl. 1, figs. 15–16. –Barker, 1960, pl. 38, figs. 7–9. –Berggren and Kaminski, 1990, p. 58, pl. 1, fig. 1. –Jones, 1994, p. 43, pl. 38, figs. 7–9. –Holbourn and Kaminski, 1997, p. 36, pl. 5, fig. 5. –Kaminski and Gradstein, 2005, pp. 181–185, fig. 25, pl. 25.

### Remarks

Type species of the genus *Glomospira* Rzehak, 1885.

*Glomospira gordialis* (Jones and Parker) differs from *Glomospirella diffundens* (Cushman and Renz) by its subcircular rather than elliptical shape and by having not more than one final planispiral whorl. See Remarks in Kaminski and Gradstein (2005) for detailed taxonomic history.

### Biogeography

Worldwide. Jones and Parker (1860) reported this taxon from the Indian Ocean and Arctic seas. Also see Observed Occurrences in Kaminski and Gradstein (2005).

### Bathymetry

Primarily bathyal to abyssal but may occur at shallower depths (Kaminski and Gradstein, 2005).

### Chronostratigraphy

Early Cretaceous to Recent.

## References


Barker, R.W., 1960, Taxonomic notes on the species figured by H.B. Brady in his report on the foraminifera dredged by H.M.S. Challenger during the years 1873–1876. Society of Economic Paleontologists and Mineralogists Special Publication, vol. 9, pp. 1–238.

Brady, H.B., 1884, Report on the foraminifera dredged by H.M.S. Challenger during the years 1873–1876. In Brady, H.B., ed., Report on the Scientific Results of the Voyage of the H.M.S. Challenger during the years 1873–1876. Zoology, vol. 9, pp. 1–814.

Bartenstein, H. and Brand, E., 1951, Mikropalaeontologische Untersuchungen zur Stratigraphie des nordwest-deutschen. Valendis Abhandlungen der Senckenbergischen Naturforschenden Gesellschaft, vol. 485, pp. 239–336.

Berggren, W.A. and Kaminski, M.A., 1990, Abyssal agglutinates: back to basics. In Hemleben, C., Kaminski, M.A., Kuhnt, W., and Scott, D.B., ed., Paleoecology, Biostratigraphy, Paleoceanography and Taxonomy of Agglutinated Foraminifera. NATO ASI Series, Series C: Mathematical and Physical Sciences, vol. 327. Dordrecht: Kluwer Academic, pp. 53–76.

Carpenter, W.B., Parker, W.K., and Jones, J.P., 1862, Introduction to the Study of the Foraminifera. London: Ray Society, 319 pp.

Holbourn, A.E.L. and Kaminski, M.A., 1997, Cretaceous deep-water benthic foraminifera of the Indian Ocean. Grzybowski Foundation Special Publication, no. 4, pp. 1–175.

Jones, R.W., 1994, The Challenger Foraminifera. Oxford: Oxford University Press, 149 pp.

Jones, J.P. and Parker, W.K., 1860, On the Rhizopodal fauna of the Mediterranean compared with that of the Italian and some other Tertiary deposits. Quarterly Journal of the Geological Society of London, vol. 16, pp. 292–307.

Kaminski, M.A. and Gradstein, F.M., 2005, Atlas of Palaeogene Cosmopolitan Deep-Water Agglutinated Foraminifera. Grzybowski Foundation Special Publication, no. 10, 547 + vii pp.

# Glomospirella diffundens

## (Cushman and Renz), 1946

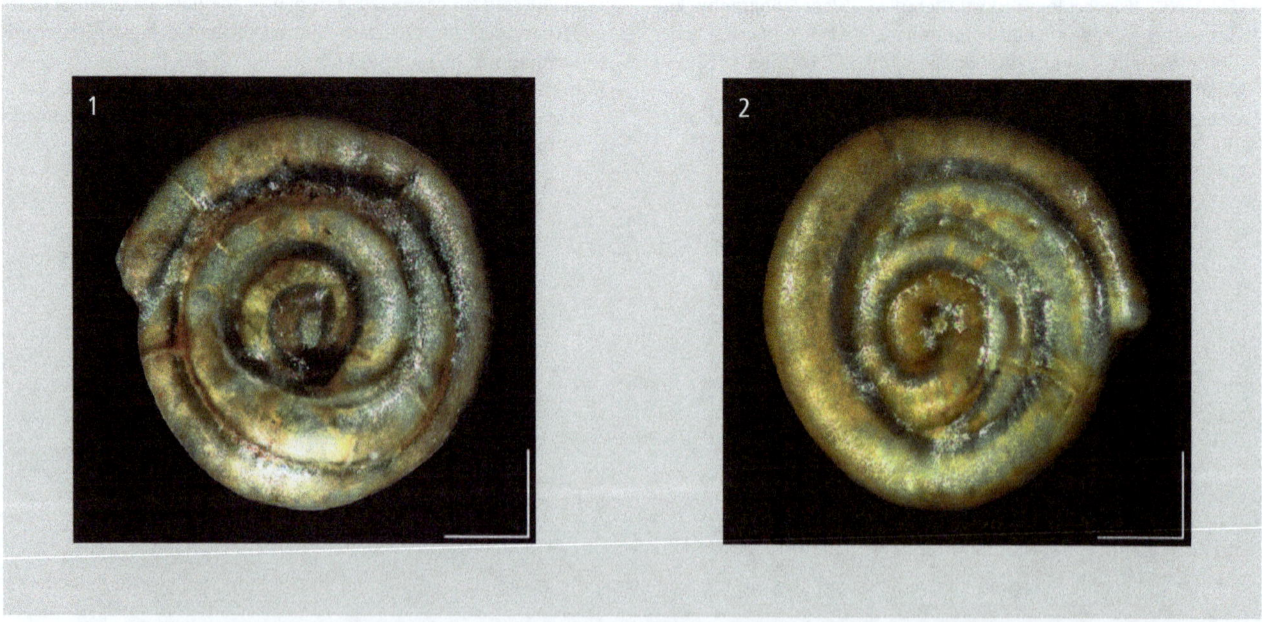


1. *Glomospirella diffundens*. Subsilesian Unit, gray-green shales, Radziechowy near Żywiec, Poland. Deposited in the Geroch Collection, Jagiellonian University, Kraków, Poland. Lateral view, NHMUK PM DI 0357, scale bar = 167 μm.

2. *Glomospirella diffundens*. Subsilesian Unit, gray-green shales, Radziechowy near Żywiec, Poland. Deposited in the Geroch Collection, Jagiellonian University, Kraków, Poland. Lateral view, NHMUK PM DI 0358, scale bar = 164 μm.

### Original Designation

*Glomospira gordialis* Jones and Parker var. *diffundens* Cushman and Renz, 1946, p. 15, pl. 1, fig. 30.

### Type Specimen

Holotype (CC 46488) and paratypes (CC 46489) deposited in the collections of the United States National Museum, Washington, DC.

### Type Level

Upper Paleocene.

### Type Locality

Lizards Springs Formation in Ravine Ampelu, Trinidad.

### Description

Test elliptical with a large, robust proloculus followed by an enrolled, undivided, broad, tubular second chamber glomospirally coiled about a common plane. The final two whorls show a characteristic tendency to become planispiral. Chamber walls are finely agglutinated, well cemented, and smoothly finished with organic cement. The primary aperture is round and situated at the end of the open tube.

### Synonyms and Citations

*Glomospira gordialis diffundens* Cushman and Renz. –Hemleben and Tröster, 1984, p. 519, pl. 1, fig. 21.

*Glomospira diffundens* Cushman and Renz. –Kaminski et al., 1988, p. 185, pl. 3, figs. 18–19. –Kaminski and Gradstein, 2005, pp. 175–177, fig. 23, pl. 23.

### Remarks

*Glomospirella diffundens* (Cushman and Renz) differs from *Glomospira gordialis* (Jones and Parker) by its elliptical rather than subcircular shape and by having more than one final planispiral whorl.

### Biogeography

Worldwide. See Observed Occurrences in Kaminski and Gradstein (2005).

### Bathymetry

Bathyal to abyssal.

### Chronostratigraphy

Late Cretaceous (Campanian) to Paleocene, possibly Eocene (Kaminski and Gradstein, 2005).

## References


Cushman, J.A. and Renz, H.H., 1946, The foraminiferal fauna of the Lizard Springs Formation of Trinidad, British West Indies. Special Publications of the Cushman Laboratory for Foraminiferal Research, vol. 18, pp. 1–48.

Hemleben, C. and Tröster, J., 1984, Campanian–Maestrichtian deep-water foraminifers from Hole 543A, Deep Sea Drilling Project. In Bij-Duval, B., Moore, J.C., et al., ed., Initial Reports of the Deep Sea Drilling Project, Washington (U.S. Government Printing Office), vol. 78A, pp. 509–532.

Kaminski, M.A. and Gradstein, F.M., 2005, Atlas of Palaeogene Cosmopolitan Deep-Water Agglutinated Foraminifera. Grzybowski Foundation Special Publication, no. 10, 547 + vii pp.

Kaminski, M.A., Gradstein, F.M., Berggren, W.A., Geroch, S., and Beckmann, J.P., 1988, Agglutinated foraminiferal assemblages from Trinidad: taxonomy, stratigraphy and paleobathymetry. In Gradstein, F.M. and Rögl, F., ed., Second International Workshop on Agglutinated Foraminifera, Vienna 1986. Abhandlungen der Geologischen Bundesanstalt, vol. 41, pp. 155–228.

# *Glomospirella gaultina*

## (Berthelin), 1880

1. *Glomospirella gaultina*. Barrow Group. Northern Exmouth Plateau, Carnarvon Basin, Western Australia; ODP Leg 122. R.W. Jones Collection. Lateral view, NHMUK PM DI 0359, scale bar = 82 μm.

### Original Designation
*Ammodiscus gaultinus* Berthelin, 1880, p. 19, pl. 1, fig. 3.

### Type Specimen
Holotype (Berthelin's fig. 3, Bartenstein's fig. 18), figured paratypes (Bartenstein's figs. 17, 19–20) and sixteen unfigured paratypes deposited in the collections of the Laboratoire de Paléontologie, Muséum d'Histoire Naturelle, Paris, France (tube no. 23.1).

### Type Level
Lower Cretaceous. Le Gault ou l'étage albien. L'assise marneuse brunâtre. [The Gault or the Albian stage. Marly brownish base].

### Type Locality
Montcley, on the left bank of the River Ognon, Doubs, France.

### Description
Test composed of a globular proloculus followed by a tubular second chamber, initially coiled trochospirally or glomospirally about a wobbly axis, later becoming planispiral (two or more planispiral whorls). Chamber walls are finely agglutinated and smoothly finished. The primary aperture is a round opening at the end of the tube.

### Synonyms and Citations
*Ammodiscus gaultinus* Berthelin emend. Bartenstein, 1954, p. 38, pl. 1, figs. 17–20.
*Glomospirella gaultina* (Berthelin). –Kuznetsova, 1974, pl. 1, figs. 2a, b. –Riegraf and Luterbacher, 1989, pl. 1, figs. 16–17.

### Remarks
Berthelin's species *Ammodiscus gaultinus* was designated as the type species of *Mjatliukaena* Suleimanov, 1969, distinguished by its siliceous, cryptocrystalline wall structure. As this is a diagenetic feature, Loeblich and Tappan (1987) placed this genus in the synonymy of *Glomospirella*.

### Biogeography
Worldwide.

### Bathymetry
Bathyal to abyssal.

### Chronostratigraphy
Early Cretaceous (Berriasian to Albian?). Probably also occurs in the Jurassic.

## References

Bartenstein, H., 1954, Revision von Berthelin's Mémoire 1880 über die Alb-Foraminiferen von Montcley. Senckenbergiana Lethaea, vol. 35, no. 1/2, pp. 37–50.

Berthelin, G., 1880, Mémoire sur les foraminifères fossiles de l'étage albien de Montcley (Doubs). Mémoires de la Société géologique de France, ser. 31, no. 5, pp. 1–84.

Kuznetsova, K.I., 1974, Distribution of benthonic Foraminifera in Upper Jurassic and Lower Cretaceous Deposits at Site 261, DSDP Leg 27, in the Eastern Indian Ocean. In Veevers, J.J., Heirtzler, J.R., et al., ed, Initial Reports of the Deep Sea Drilling Project, Washington (U.S. Government Printing Office), vol. 27, pp. 673–681.

Loeblich, A.R. Jr and Tappan, H., 1987, Foraminiferal Genera and Their Classification. New York: Van Nostrand Reinhold, 970 pp.

Riegraf, W. and Luterbacher, H., 1989, Benthonische Foraminiferen aus de Unterkreide des Deep Sea Drilling Project (Leg 1-79). Geologische Rundschau, vol. 78, no. 3, pp. 1063–1120.

# *Gyroidinoides quadratus*

## (Cushman and Church), 1929

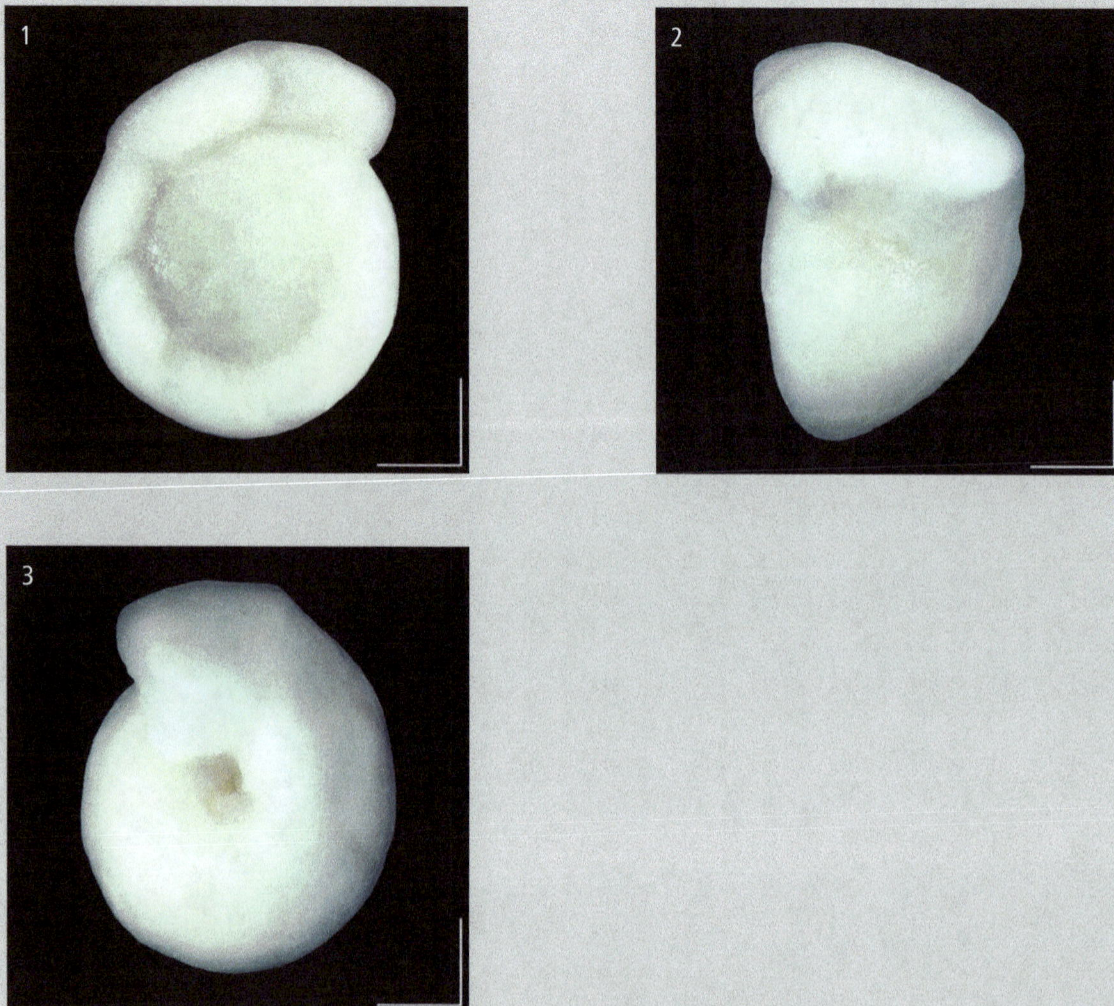


1. *Gyroidinoides quadratus*. Shatsky Rise, West Central Pacific; DSDP Leg 86, Hole 577A, Core 10, Section 6, 50–57 cm. Spiral view, NHMUK PM DI 0361, scale bar = 190 μm.

2. *Gyroidinoides quadratus*. Shatsky Rise, West Central Pacific; DSDP Leg 86, Hole 577A, Core 10, Section 6, 50–57 cm. Apertual view, NHMUK PM DI 0362, scale bar = 190 μm.

3. *Gyroidinoides quadratus*. Shatsky Rise, West Central Pacific; DSDP Leg 86, Hole 577A, Core 10, Section 6, 50–57 cm. Umbilical view, NHMUK PM DI 0363, scale bar = 190 μm.

## Original Designation

*Gyroidina quadrata* Cushman and Church, 1929, p. 516, pl. 41, figs. 7–9.

## Type Specimen

Holotype (No. 4754) deposited in the California Academy of Sciences Museum, San Francisco, California, USA.

## Type Level

Upper Cretaceous.

## Type Locality

California Academy of Sciences Loc. No. 1421, California Northern Petroleum Company Well No. 19. Sec. 2, T. 21S., R. 14E., M.D.M., depth 1135 feet, near Coalinga, Fresno County, California, USA.

## Description

Test forms a small trochospire; lobulate in outline, planoconvex to subquadrate in cross-section, with a slightly concave, evolute spiral side, a strongly convex umbilical side, and a broad, rounded periphery. The (approximately) six to eight moderately inflated chambers in the final whorl increase gradually in size, and are separated by depressed, slightly curved sutures that are slightly limbate on the spiral side and radiate on the umbilical side. Chamber walls are calcareous, finely perforate, and smooth. The primary aperture is a low interiomarginal slit extending from the periphery to the open umbilicus, which is partially obscured by umbilical flaps extending at the base of the chambers. Previous apertures may remain open in the umbilicus.

## Synonyms and Citations

*Gyroidinoides quadratus* (Cushman and Church). –Nyong and Olsson, 1984, p. 474, pl. 7, figs. 5–6. –Thomas, 1990, p. 590. –Bolli et al., 1994, p. 160, figs. 45.19–21.

## Biogeography

Worldwide. California (Cushman and Church, 1929); North American Basin and Atlantic Ocean (Nyong and Olsson, 1984); Weddell Sea, Antarctic (Thomas, 1990); Trinidad (Bolli et al., 1994).

## Bathymetry

Bathyal. Middle bathyal (Nyong and Olsson, 1984), lower bathyal (Thomas, 1990).

## Chronostratigraphy

Late Cretaceous (Turonian) to Early Paleocene (Danian). Campanian to early Maastrichtian (Nyong and Olsson, 1984). Early Maastrichtian to early Paleocene (Thomas, 1990). Turonian to late Maastrichtian (Bolli et al., 1994).

## References


Bolli, H.M., Beckmann, J.P., and Saunders, J.B., 1994, Benthic Foraminiferal Biostratigraphy of the South Caribbean Region. Cambridge: Cambridge University Press, 408 pp.

Cushman, J.A. and Church, C.C., 1929, Some Upper Cretaceous foraminifera from near Coalinga, California. Proceedings of the California Academy of (Natural) Sciences, ser. 4, vol. 18, no. 16, pp. 497–530.

Nyong, E.E. and Olsson, R.K., 1984, A paleoslope model for Campanian to lower Maestrichtian foraminifera of the North American Basin and adjacent continental margin. Marine Micropaleontology, vol. 8, pp. 437–477.

Thomas, E., 1990, Late Cretaceous through Neogene benthic foraminifers. In Barker, P.F., Kennett, J.P., et al., ed., Scientific Results of the Proceedings of the Ocean Drilling Program, vol. 113, pp. 571–594.

# *Gyroidinoides soldanii*

## (d'Orbigny), 1826

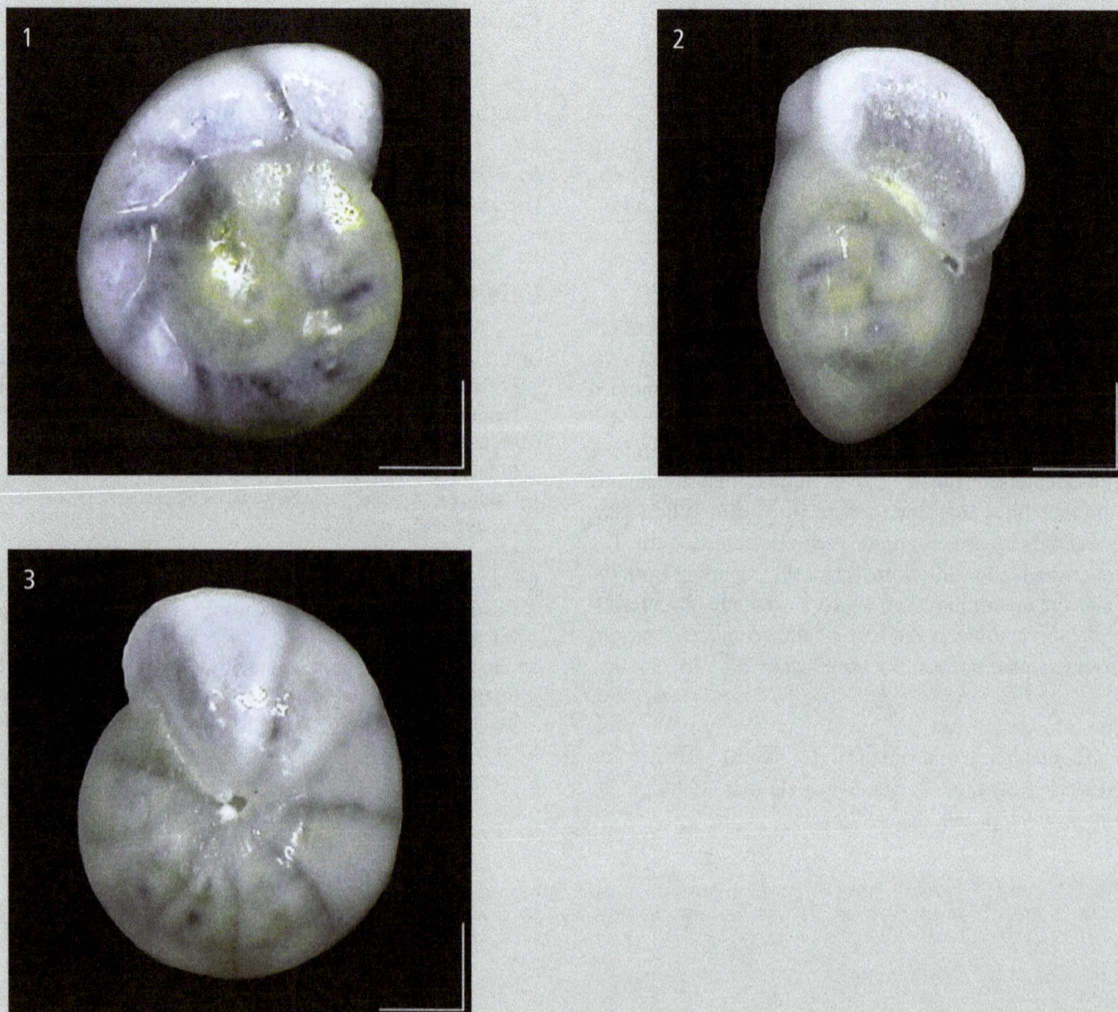


1. *Gyroidinoides soldanii.* South Pacific Ocean; "Challenger" Station 302, depth: 1450 fathoms. Deposited in the micropaleontological collections of The Natural History Museum, London; NHMUK ZF2330. Spiral view, NHMUK PM DI 0364, scale bar = 139 μm.

2. *Gyroidinoides soldanii.* South Pacific Ocean; "Challenger" Station 302, depth: 1450 fathoms. Deposited in the micropaleontological collections of The Natural History Museum, London; NHMUK ZF2330. Apertural view, NHMUK PM DI 0365, scale bar = 135 μm.

3. *Gyroidinoides soldanii.* South Pacific Ocean; "Challenger" Station 302, depth: 1450 fathoms. Deposited in the micropaleontological collections of The Natural History Museum, London; NHMUK ZF2330. Umbilical view, NHMUK PM DI 0366, scale bar = 135 μm.

## Original Designation

*Gyroïdina soldanii* d'Orbigny, 1826, p. 278, Model No. 36, illustrated in d'Orbigny, 1846, pl. 8, figs. 10–12.

## Type Specimen

Deposited in the collections of the Laboratoire de Paléontologie, Muséum d'Histoire Naturelle, Paris, France.

## Type Level

Recent, surface sediments.

## Type Locality

Adriatic Sea, near Rimini, Italy.

## Description

Test forms a planoconvex trochospire with a flattened, evolute spiral side, a strongly convex, involute, umbilical side, and a rounded periphery. The highly inflated chambers increase gradually in size, and are separated by slightly depressed sutures that are curved on the spiral side and straight on the umbilical side. Chamber walls are calcareous, smooth, and finely perforate. The primary aperture is a narrow, interiomarginal slit, extending from the umbilicus to the periphery. This aperture may be hidden by a broad crescentic flap, covering the umbilicus including previous flaps.

## Synonyms and Citations

*Gyroïdina soldanii* d'Orbigny. –d'Orbigny, 1846, pl. 8, figs. 10–12. –Parker et al., 1871, p. 176, pl. 12, fig. 151. –Papp and Schmid, 1985, p. 60, pl. 50, figs. 4–9.

*Gyroidinoides soldanii* (d'Orbigny). –Jones, 1994, p. 106, pl. 107, figs. 6–7. –Milker and Schmiedl, 2012, p. 116, figs. 26.20–21.

## Remarks

Papp and Schmid (1985, pl. 50, figs. 4–9) illustrated specimens from material collected in Nussdorf by Hauer, and previously described by d'Orbigny (1846).

## Biogeography

Worldwide.

## Bathymetry

Primarily bathyal. Reported between 300 and 2000 fathoms at "Challenger" stations (Jones, 1994).

## Chronostratigraphy

Oligocene to Recent (Jones, 1994).

# References

d'Orbigny, A., 1826, Tableau méthodique de la classe des Céphalopodes. Annales des Sciences Naturelles, vol. 7, pp. 245–314.

d'Orbigny, A.D., 1846, Foraminifères fossiles du Bassin Tertiaire de Vienne (Autriche). Paris: Gide et Compe, xxxvii + 312 pp.

Jones, R.W., 1994, The Challenger Foraminifera. Oxford: Oxford University Press, 149 pp.

Milker, Y. and Schmiedl, G., 2012, A taxonomic guide to modern benthic shelf foraminifera of the western Mediterranean Sea. Palaeontologia Electronica, vol. 15, Issue 2;16A, 134 pp.

Papp, A. and Schmid, M.E., 1985, Die fossilen Foraminiferen des tertiaeren Beckens von Wien: Revision der Monographie Alcide d'Orbigny (1846). Abhandlungen der Geologischen Bundesanstalt, vol. 37, pp. 1–311.

Parker, W.K., Jones, T.R., and Brady, H.B., 1871, On the nomenclature of the foraminifera. Part XIV – The species enumerated by d'Orbigny in the "Annales des Sciences Naturelles" 1826. Annals and Magazine of Natural History, Series 4, vol. 7 (continued), – The species founded upon the figures in Soldani's "Testaceographia ac Zoophytographia", Annals and Magazine of Natural History, Series 4, vol. 8, pp. 145–179, 238–266.

# *Hanzawaia ammophila*

## (Gümbel), 1868

1. *Hanzawaia ammophila.* North Biscay, North Atlantic; DSDP Leg 48, Hole 402A, Core 1, Section 3, 51–55 cm. Spiral view, NHMUK PM DI 0367, scale bar = 126 µm.

2. *Hanzawaia ammophila.* North Biscay, North Atlantic; DSDP Leg 48, Hole 402A, Core 1, Section 3, 51–55 cm. Apertural view, NHMUK PM DI 0368, scale bar = 126 µm.

3. *Hanzawaia ammophila.* North Biscay, North Atlantic; DSDP Leg 48, Hole 402A, Core 1, Section 3, 51–55 cm. Umbilical view, NHMUK PM DI 0369, scale bar = 128 µm.

## Original Designation

*Rotalia ammophila* Gümbel, 1868, p. 652, pl. 2, figs. 90a–b.

## Type Specimen

Original material deposited in the collections of the Institute of Historical Geology and Paleontology of the University, Munich, was destroyed during the Second World War. Comparative material from the collections of C. Schwager is available.

## Type Level

Upper Eocene (van Morkhoven et al., 1986). Given as Lower Eocene by Gümbel (1868).

Nummulitic limestone.

## Type Locality

Not designated. Various localities given in Bavarian Alps, Germany.

## Description

Test forms a small trochospire; planoconvex in cross-section with a flattened, partially evolute, spiral side, a convex, partially evolute, umbilical side (with clear central boss), and a subrounded, imperforate periphery. The (approximately) fifteen narrow, curved, moderately inflated chambers in the last whorl increase rapidly in size and are separated by strongly curved, limbate sutures that are raised on the spiral side and depressed on the umbilical side. Chamber walls are calcareous, smooth, and finely perforate on both sides of the test. The primary aperture is an interiomarginal equatorial slit bordered by a lip that extends on the spiral side under septal flaps. Fused flaps form an imperforate band with relict apertures on the spiral side (approximately four are typically visible).

## Synonyms and Citations

*Truncatulina scarenaensis* Hantken, 1883, p. 151, pl. 2, fig. 6

*Cibicides cushmani* Nuttall, 1930, p. 291, pl. 25, figs. 3, 5–6.

*Hanzawaia cushmani* (Nuttall). –Miller, 1983, p. 437, pl. 1, fig. 12. –Tjalsma and Lohmann, 1983, p. 32, pl. 17, fig. 1.

*Hanzawaia ammophila* (Gümbel). –van Morkhoven et al., 1986, p.168, pl. 56. –Miller and Katz, 1987, p. 134, pl. 6, fig. 3. –Ortiz and Thomas, 2006, p. 120, pl. 7, figs. 5–6.

## Biogeography

Worldwide. Recorded from Europe, Crimea, North Caucasus, Turkmenia, the Pacific, the Tasman Sea, the Coral Sea, the Chatham Rise, the Gulf of Mexico, Barbados, the North and South Atlantic, California, Mexico, Cuba, Haiti, and the Dominican Republic (van Morkhoven et al. 1986).

## Bathymetry

Primarily lower neritic to upper bathyal according to van Morkhoven et al. (1986). Found at middle bathyal to upper abyssal depths (600–3000 meters) in Oligocene samples from the Atlantic (M. Katz, pers. comm. 1998).

## Chronostratigraphy

Early Eocene (Ypresian) to Middle Miocene (Serravallian).

# References


Gümbel, C.W., 1868, Beitrage zur Foraminiferenfauna der nordalpinen Eocängebilde. Abhandlungen der K. Bayerischen Akademie der Wissenschaften, cl. 2 (1870), vol. 10, no. 2, pp. 581–730.

Hantken, M., 1883, Die Clavulina szabói-Schichten im Gebiete der Euganeen und der Meerealpen und die Cretacische Scaglia in den Euganeen. Mathematische und Naturwissenschaftliche Berichte aus Ungarn, Budapest, vol. 2 (1883–4), pp. 121–169.

Miller, K.G., 1983, Eocene–Oligocene paleoceanography of the deep Bay of Biscay: benthic foraminiferal evidence. Marine Micropaleontology, vol. 7, pp. 403–440.

Miller, K.G. and Katz, M.E., 1987, Oligocene to Miocene benthic foraminiferal and abyssal circulation changes in the north Atlantic. Micropaleontology, vol. 33, no. 2, pp. 97–149.

Nuttall, W.L.F., 1930, Eocene foraminifera from Mexico. Journal of Paleontology, vol. 4, pp. 271–293.

Ortiz, S. and Thomas, E., 2006, Lower-middle Eocene benthic foraminifera from the Fortuna Section (Betic Cordillera, southeastern Spain). Micropaleontology, vol. 52, pp. 97–150.

Tjalsma, R.C. and Lohmann, G.P., 1983, Paleocene–Eocene bathyal and abyssal benthic foraminifera from the Atlantic Ocean. Micropaleontology, Special Publication, no. 4, pp. 1–90.

van Morkhoven, F.P.C.M., Berggren, W.A., Edwards, A.S., et al., 1986, Cenozoic cosmopolitan deep-water benthic foraminifera. Bulletin des Centres de Recherches Exploration-Production Elf-Aquitaine, Mémoire, vol. 11, pp. 1–421.

# *Hanzawaia caribaea*

## (Cushman and Bermúdez), 1936

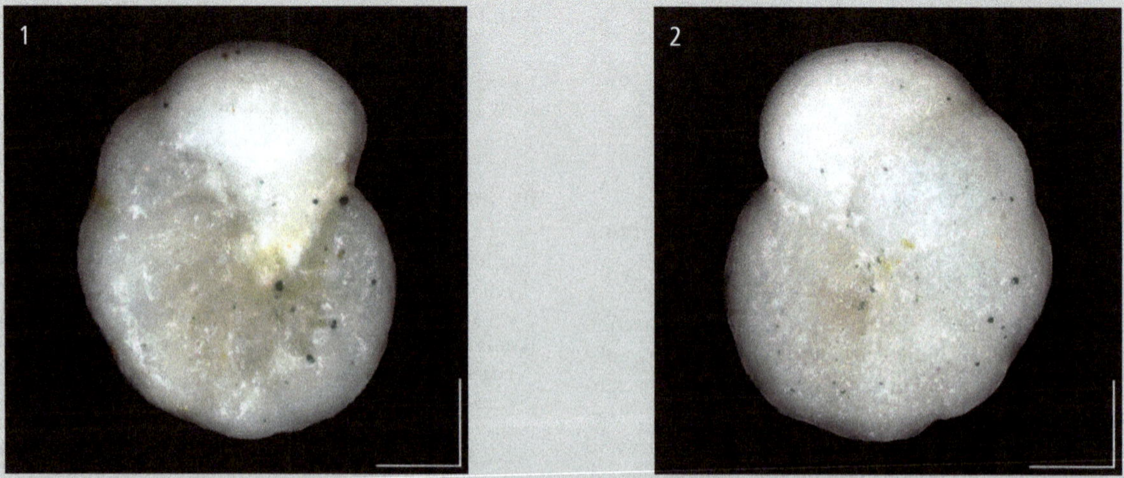


1. *Hanzawaia caribaea.* Lower Principle Formation, under the library of Havana University, Cuba. Deposited in the collections of the United States National Museum, Washington, DC; CC 23305. Holotype. Umbilical view, NHMUK PM DI 0370, scale bar = 126 μm.

2. *Hanzawaia caribaea.* Lower Principle Formation, under the library of Havana University, Cuba. Deposited in the collections of the United States National Museum, Washington, DC; CC 23305. Holotype. Spiral view, NHMUK PM DI 0371, scale bar = 126 μm.

### Original Designation
*Anomalina crassisepta* Cushman and Siegfus var. *caribaea* Cushman and Bermúdez, 1936, p. 62, pl. 11, figs. 12–14.

### Type Specimen
Holotype (CC 23305) deposited in the collections of the United States National Museum, Washington, DC.

### Type Level
Eocene, Lower Principe Formation.

### Type Locality
Bermúdez Station 257, under the library of Havana University, Havana, Cuba.

### Description
Test thick, trochospiral, planoconvex in cross-section, with a flattened, partially evolute spiral side and convex involute, umbilical side, slightly depressed in the center. The periphery is broadly rounded and angled on the spiral side. Eight to ten inflated chambers in the last whorl, increasing regularly in size, are separated by strongly curved sutures, broad, raised, and limbate on the spiral side, and flush on the umbilical side. Chamber walls are calcareous and finely perforate on both sides of the test. The primary aperture is an interiomarginal equatorial slit bordered by a lip, which extends on the spiral side under septal flaps. Fused flaps form imperforate band with relict apertures.

### Synonyms and Citations
*Anomalina crassisepta* Cushman and Siegfus. –Berggren and Aubert, 1983, pl. 5, figs. 4–6.
*Gavelinella caribaea* (Cushman and Bermúdez). –Tjalsma and Lohmann, 1983, p. 31, pl. 16, fig. 8.
*Hanzawaia caribaea* (Cushman and Bermúdez). –van Morkhoven et al., 1986, p. 303, pl. 99.

### Biogeography
Worldwide. Recorded from the North and South Atlantic, the eastern Pacific, and the Gulf of Mexico (van Morkhoven et al., 1986).

### Bathymetry
Bathyal with upper depth limit of about 500 meters (Berggren and Aubert, 1983).

### Chronostratigraphy

Early Eocene (Ypresian) to Middle Eocene (Bartonian).

## References


Berggren, W.A. and Aubert, J., 1983, Paleogene benthonic foraminiferal biostratigraphy and bathymetry of the Central Coast Ranges of California. In Brabb, E.E., ed., Studies in Tertiary stratigraphy of the California Coast Ranges. Professional Paper of the United States Geological Survey, vol. 1213, pp. 4–21.

Cushman, J.A. and Bermúdez, P.J., 1936, Additional new species of foraminifera and a new genus from the Eocene of Cuba. Contributions from the Cushman Laboratory for Foraminiferal Research, vol. 12, no. 3, pp. 55–63.

Cushman, J.A. and Siegfus, S.S., 1935, New species of foraminifera from the Kreyenhagen Shale of Fresno County, California. Contributions from the Cushman Laboratory for Foraminiferal Research, vol. 11, no. 4, pp. 90–95.

Tjalsma, R.C. and Lohmann, G.P., 1983, Paleocene–Eocene bathyal and abyssal benthic foraminifera from the Atlantic Ocean. Micropaleontology, Special Publication, no. 4, pp. 1–90.

van Morkhoven, F.P.C.M., Berggren, W.A., Edwards, A.S., et al., 1986, Cenozoic cosmopolitan deep-water benthic foraminifera. Bulletin des Centres de Recherches Exploration-Production Elf-Aquitaine, Mémoire, vol. 11, pp. 1–421.

# *Hanzawaia mantaensis*

## (Galloway and Morrey), 1929

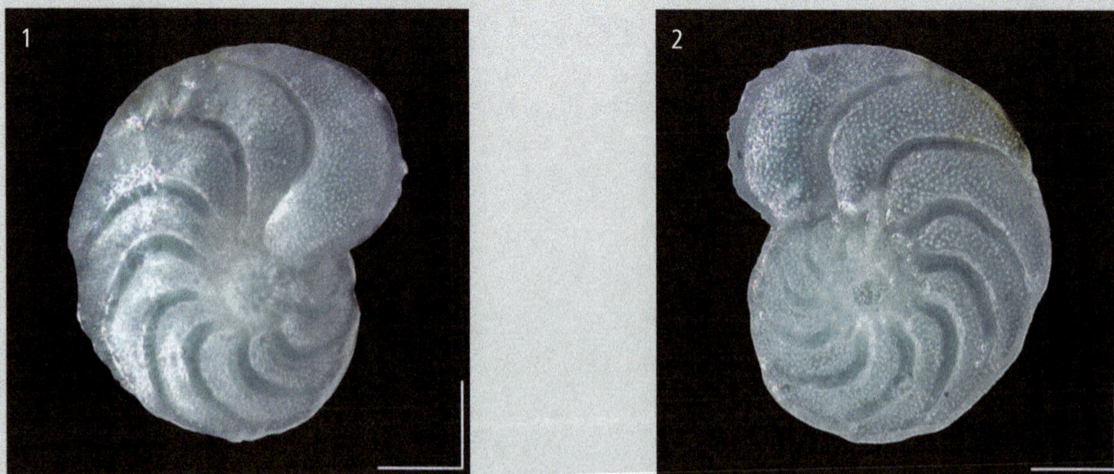


1. *Hanzawaia mantaensis.* Gulf of Mexico; Eureka Core E68-136, depth: 2627 feet. Umbilical view, NHMUK PM DI 0372, scale bar = 128 μm.
2. *Hanzawaia mantaensis.* Gulf of Mexico; Eureka Core E68-136, depth: 2627 feet. Spiral view, NHMUK PM DI 0373, scale bar = 128 μm.

### Original Designation

*Anomalina mantaensis* Galloway and Morrey, 1929, p. 28, pl. 4, fig. 5.

### Type Specimen

Holotype (No. 19954) now in the collections of the Department of Invertebrate Paleontology at the American Museum of Natural History, New York.

### Type Level

?Middle Miocene (van Morkhoven et al., 1986). Given as Upper Eocene by Galloway and Morrey (1929).

### Type Locality

Manta, Ecuador.

### Description

Test forms a trochospire, planoconvex in cross-section, with a flattened, partially evolute spiral side and slightly convex, involute, umbilical side with clear central boss. The periphery is acute and keeled. Approximately ten narrow, curved, slightly inflated chambers in the last whorl, increasing regularly in size, are separated by strongly curved, limbate sutures, raised on the spiral side, depressed on the umbilical side. Chamber walls are calcareous and finely perforate on both sides of the test. The primary aperture is an interiomarginal equatorial slit bordered by a lip, extending on the spiral side under the septal flaps. Fused flaps form an imperforate band with relict apertures (approximately four are visible) on the spiral side.

### Synonyms and Citations

*Truncatulina americana* Cushman. –Cushman, 1918, p. 68, pl. 23, fig. 2.

*Discorbis* sp. Cushman, 1923, p. 39, pl. 5, figs. 5–6.

*Cibicides mantaensis* (Galloway and Morrey). –Renz, 1948, p. 128, pl. 11, fig. 8. –Bermúdez, 1949, p. 302, pl. 25, figs. 22–24.

*Hanzawaia mantaensis* (Galloway and Morrey). –van Morkhoven et al., 1986, p.105, pl. 32. –Katz and Miller, 1993, pl. 5, fig. 5. –Robertson, 1998, p. 247, pl. 101, figs. 3–4.

### Biogeography

Worldwide. Recorded from Trinidad, the Dominican Republic, Venezuela, the North and South Atlantic, the eastern Pacific, and the Gulf of Mexico (van Morkhoven et al., 1986).

### Bathymetry

Primarily upper to middle bathyal. Occasionally outer neritic and lower bathyal (van Morkhoven et al., 1986).

### Chronostratigraphy

Oligocene to Late Pliocene (Gelasian).

## References


Bermúdez, P.J., 1949, Tertiary smaller foraminifera of the Dominican Republic. Special Publications of the Cushman Laboratory for Foraminiferal Research, no. 25, p. 1–322.

Cushman, J.A., 1918, The smaller fossil foraminifera of the Panama Canal Biozone. Bulletin of the United States National Museum, vol. 103, pp. 45–87.

Cushman, J.A., 1923, The foraminifera of the Vicksburg Group. Professional Paper of the United States Geological Survey, vol. 133, pp. 11–71.

Galloway, J.J. and Morrey, M., 1929, A lower Tertiary foraminiferal fauna from Manta, Ecuador. Bulletins of American Paleontology, vol. 15, pp. 7–56.

Katz, M.E. and Miller, K.G., 1993, Latest Oligocene to Earliest Pliocene benthic foraminiferal biofacies of the northeastern Gulf of Mexico. Micropaleontology, vol. 39, no. 4, pp. 367–403.

Renz, H.H., 1948, Stratigraphy and fauna of the Agua Salada group, State of Falcón, Venezuela. Memoirs of the Geological Society of America, vol. 32, pp. 1–219.

Robertson, B.E., 1998, Systematics and paleoecology of the benthic foraminiferida from the Buff Bay section, Miocene of Jamaica. Micropaleontology, vol. 44, supplement 2, pp. 1–266.

van Morkhoven, F.P.C.M., Berggren, W.A., Edwards, A.S., et al., 1986, Cenozoic cosmopolitan deep-water benthic foraminifera. Bulletin des Centres de Recherches Exploration-Production Elf-Aquitaine, Mémoire, vol. 11, pp. 1–421.

# *Haplophragmoides constrictus*

## Krasheninnikov, 1973

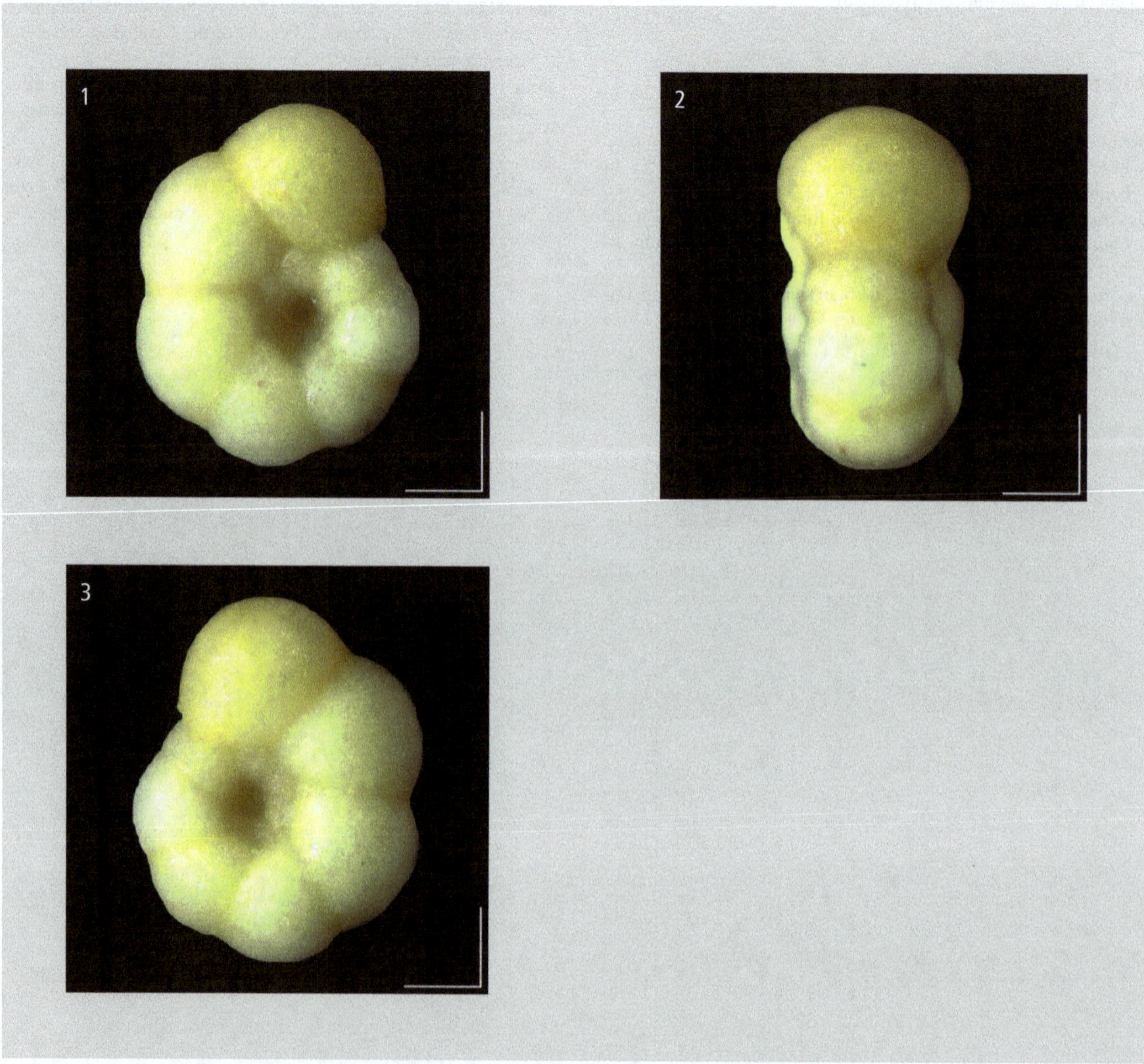


1. *Haplophragmoides constrictus*. Pigafetta Basin, West Pacific; ODP Leg 129, Hole 800A, Core 5, Section 1, 15–20 cm. Umbilical view, NHMUK PM DI 0374, scale bar = 128 µm.

2. *Haplophragmoides constrictus*. Pigafetta Basin, West Pacific; ODP Leg 129, Hole 800A, Core 5, Section 1, 15–20 cm. Apertural view, NHMUK PM DI 0375, scale bar = 128 µm.

3. *Haplophragmoides constrictus*. Pigafetta Basin, West Pacific; ODP Leg 129, Hole 800A, Core 5, Section 1, 15–20 cm. Umbilical view, NHMUK PM DI 0376, scale bar = 131 µm.

## Original Designation

*Haplophragmoides constrictus* Krasheninnikov, 1973, p. 207, pl. 1, fig. 5.

## Type Specimen

Holotype (No. 4011/4) deposited in the collections of the Micropaleontological Laboratory, Geological Institute, Russian Academy of Sciences, Moscow.

## Type Level

Upper Cretaceous, Santonian to Campanian.

## Type Locality

Holotype from DSDP Leg 20, Hole 196, Core 2, Section 1, 30–32 cm interval; 6194 meters depth on the abyssal floor, west of the Izu-Bonin Trench, latitude 30° 06.97′ N, longitude 148° 34.49′ E.

## Description

Test small to medium-sized, planispiral, flattened, slightly evolute, slightly lobulate in outline, oval in cross-section with rounded periphery. Chambers of previous whorls are partly visible through the broad, shallow umbilicus. Lateral sides are slightly concave to flat. Approximately six to seven inflated, broad, low trapeziform chambers in the last whorl, increasing slowly in size, are separated by straight, depressed sutures. Chamber walls are finely agglutinated and non-calcareous, with a smooth, glassy finish. The primary aperture is an elongate equatorial slit at the base of the last chamber.

## Synonyms and Citations

*Haplophragmoides constrictus* Krasheninnikov. –Krasheninnikov, 1974, p. 635, pl. 1, fig. 3. –Moullade et al., 1988, p. 364, pl. 4, figs. 7–9. –Kuhnt and Moullade, 1991, pl. 5, figs. D–E.

## Remarks

Useful stratigraphic marker and bathymetric indicator for Late Cretaceous abyssal paleoenvironments.

## Biogeography

Worldwide. Recorded in the Pacific Ocean (Krasheninnikov, 1973), the Indian Ocean (Krasheninnikov, 1974), and the North Atlantic (Moullade et al., 1988).

## Bathymetry

Abyssal, below CCD.

## Chronostratigraphy

Santonian to Campanian in the Pacific Ocean (Krasheninnikov, 1973), in the Indian Ocean (Krasheninnikov, 1974), and in the North Atlantic (Moullade et al., 1988).

# References


Krasheninnikov, V.A., 1973, Cretaceous benthonic foraminifera, Leg 20, Deep Sea Drilling Project. In Heezen, B.C,, MacGregor, I.D., et al., ed., Initial Reports of the Deep Sea Drilling Project, Washington (U.S. Government Printing Office), vol. 20, pp. 205–221.

Krasheninnikov, V.A., 1974, Upper Cretaceous benthonic agglutinated foraminifera, Leg 27 of the Deep Sea Drilling Project. In Veevers, J.J., Heirtzler, J.R., et al., ed., Initial Reports of the Deep Sea Drilling Project, Washington (U.S. Government Printing Office), vol. 27, pp. 531–662.

Kuhnt, W. and Moullade, M., 1991, Quantitative analysis of upper Cretaceous abyssal agglutinated foraminiferal distribution in the North Atlantic – paleoceanographic implications. Revue de Micropaléontologie, vol. 34, no. 4, pp. 313–349.

Moullade, M., Kuhnt, W., and Thurow, J., 1988, Agglutinated benthic foraminifers from Upper Cretaceous variegated clays of the North Atlantic Ocean (DSDP Leg 93 and ODP Leg 103). In Boillot, G., Winterer, E.L., et al., ed., Scientific Results of the Proceedings of the Ocean Drilling Program, vol. 103, pp. 349–377.

# *Haplophragmoides multicamerus*

## Krasheninnikov, 1973

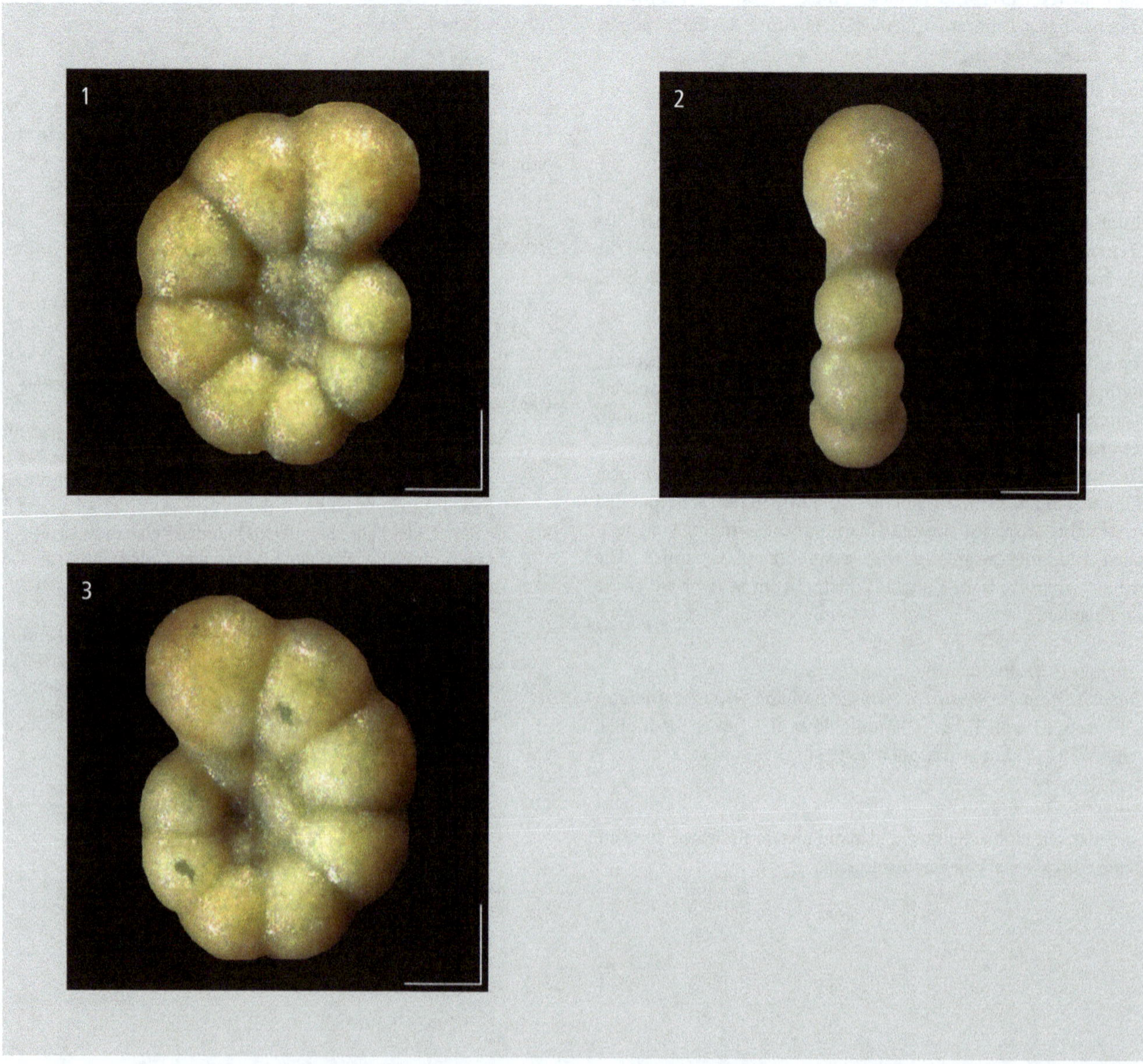


1. *Haplophragmoides multicamerus*. East of Izu-Bonin Trench, North Pacific; DSDP Leg 20, Site 196, Core 2, Section 1, 1–4 cm. Umbilical view, NHMUK PM DI 0377, scale bar = 134 μm.

2. *Haplophragmoides multicamerus*. East of Izu-Bonin Trench, North Pacific; DSDP Leg 20, Site 196, Core 2, Section 1, 1–4 cm. Apertural view, NHMUK PM DI 0378, scale bar = 127 μm.

3. *Haplophragmoides multicamerus*. East of Izu-Bonin Trench, North Pacific; DSDP Leg 20, Site 196, Core 2, Section 1, 1–4 cm. Umbilical view, NHMUK PM DI 0379, scale bar = 125 μm.

## Original Designation

*Haplophragmoides multicamerus*, Krasheninnikov, 1973, p. 207, pl. 1, figs. 1–2.

## Type Specimen

Holotype (No. 4011/1) deposited in the collections of the Micropaleontological Laboratory, Geological Institute, Russian Academy of Sciences, Moscow.

## Type Level

Upper Cretaceous, Santonian to Campanian.

## Type Locality

Holotype from DSDP Leg 20, Hole 196, Core 2, Section 1, 38–40 cm interval; 6194 meters depth on the abyssal floor, west of the Izu-Bonin Trench, latitude 30° 06.97′ N, longitude 148° 34.49′ E.

## Description

Test small, planispiral, flattened, evolute, slightly lobulate in outline, oval in cross-section with rounded periphery. Chambers of previous whorls are partly visible through the wide, shallow umbilicus. Lateral sides are slightly concave to flat. Eight to eleven inflated, trapeziform chambers in the last whorl, increasing relatively rapidly in height, are separated by straight, depressed sutures. Chamber walls are finely agglutinated with organic cement and show a smooth, glassy finish. The primary aperture is an elongate equatorial slit at the base of the last chamber.

## Synonyms and Citations

*Haplophragmoides multicamerus* Krasheninnikov. –Krasheninnikov, 1974, p. 635, pl. 1, fig. 4. –Moullade et al., 1988, p. 364, pl. 5, figs. 1–2.

## Remarks

Useful stratigraphic marker and bathymetric indicator for Late Cretaceous abyssal paleoenvironments.

## Biogeography

Worldwide. Recorded in the Pacific Ocean (Krasheninnikov, 1973; Wightman and Kuhnt, 1992), the Indian Ocean (Krasheninnikov, 1974), and the North Atlantic (Moullade et al., 1988).

## Bathymetry

Abyssal, below CCD.

## Chronostratigraphy

Late Cretaceous. Campanian in the Pacific Ocean (Krasheninnikov, 1973), Santonian to Campanian in the Indian Ocean (Krasheninnikov, 1974), Santonian to Maastrichtian in the North Atlantic (Moullade et al., 1988).

# References


Krasheninnikov, V.A., 1973, Cretaceous benthonic foraminifera, Leg 20, Deep Sea Drilling Project. In Heezen, B.C., MacGregor, I.D., et al., ed., Initial Reports of the Deep Sea Drilling Project, Washington (U.S. Government Printing Office), vol. 20, pp. 205–221.

Krasheninnikov, V.A., 1974, Upper Cretaceous benthonic agglutinated foraminifera, Leg 27 of the Deep Sea Drilling Project. In Veevers, J.J., Heirtzler, J.R., et al., ed., Initial Reports of the Deep Sea Drilling Project, Washington (U.S. Government Printing Office), vol. 27, pp. 531–662.

Moullade, M., Kuhnt, W., and Thurow, J., 1988, Agglutinated benthic foraminifers from Upper Cretaceous variegated clays of the North Atlantic Ocean (DSDP Leg 93 and ODP Leg 103). In Boillot, G., Winterer, E.L., et al., ed., Scientific Results of the Proceedings of the Ocean Drilling Program, vol. 103, pp. 349–377.

Wightman, W.G. and Kuhnt, W., 1992, Biostratigraphy and paleoecology of Late Cretaceous abyssal agglutinated foraminifers from the western Pacific Ocean (Deep Sea Drilling Project Holes 196A and 198A and Ocean Drilling Program Holes 800A and 801A). In Larson, R.L., Lancelot, Y. et al., ed., Scientific Results of the Proceedings of the Ocean Drilling Program, vol. 129, pp. 247–264.

# *Haplophragmoides pervagatus*

## Krasheninnikov, 1973

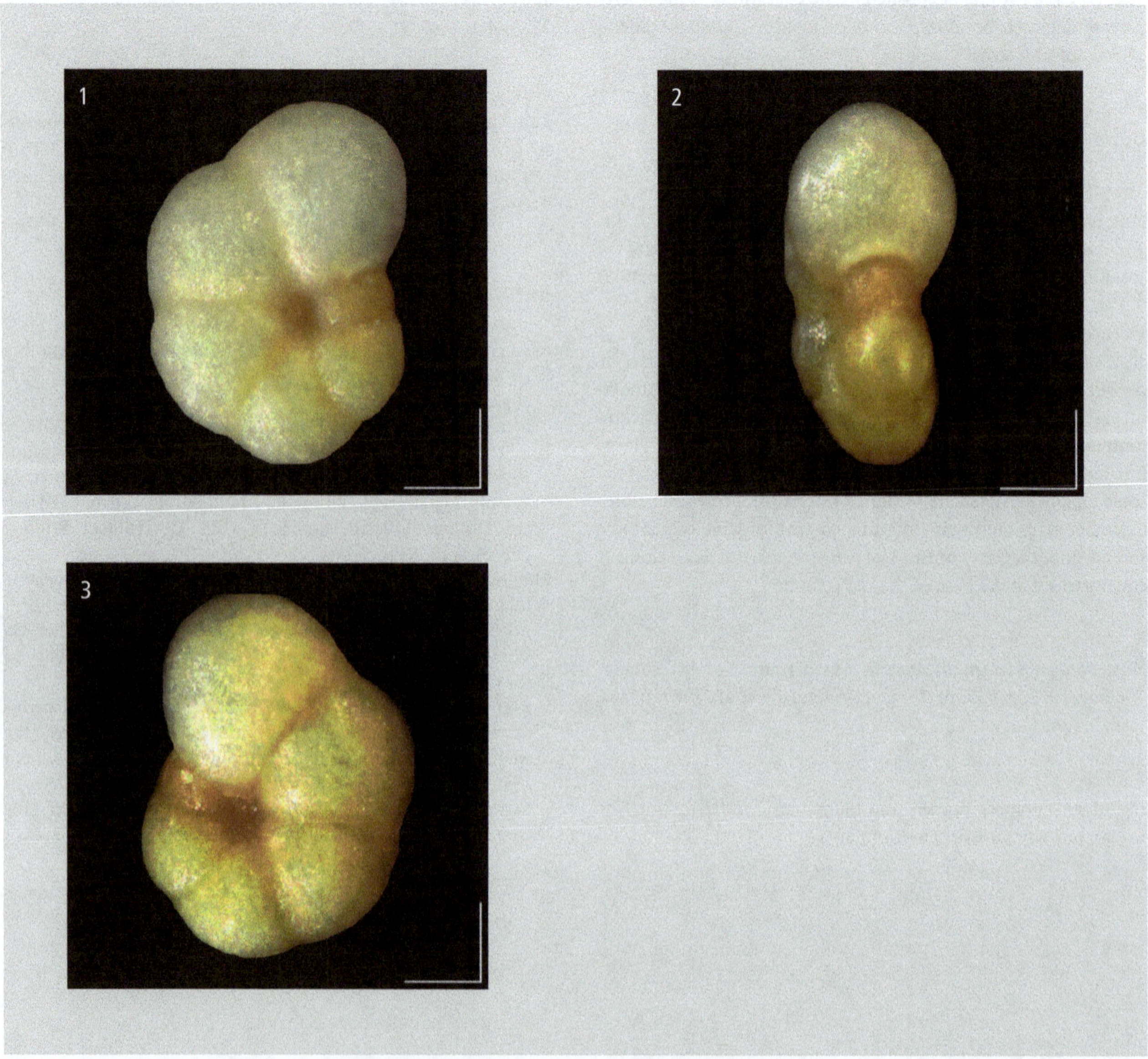


1. *Haplophragmoides pervagatus*. Continental Rise, North American Basin, North Atlantic; DSDP Leg 93, Hole 603B, Core 30, Section 3, 88–91 cm. Umbilical view, NHMUK PM DI 0380, scale bar = 120 μm.

2. *Haplophragmoides pervagatus*. Continental Rise, North American Basin, North Atlantic; DSDP Leg 93, Hole 603B, Core 30, Section 3, 88–91 cm. Apertural view, NHMUK PM DI 0381, scale bar = 120 μm.

3. *Haplophragmoides pervagatus*. Continental Rise, North American Basin, North Atlantic; DSDP Leg 93, Hole 603B, Core 30, Section 3, 88–91 cm. Umbilical view, NHMUK PM DI 0382, scale bar = 123 μm.

## Original Designation
*Haplophragmoides pervagatus* Krasheninnikov, 1973, p. 208, pl. 1, fig. 7.

## Type Specimen
Holotype (No. 4011/7) deposited in the collections of the Micropaleontological Laboratory, Geological Institute, Russian Academy of Sciences, Moscow.

## Type Level
Upper Cretaceous, Santonian to Campanian.

## Type Locality
Holotype from DSDP Leg 20, Hole 196, Core 2, Section 1, 64–66 cm interval; 6194 meters depth on the abyssal floor, west of the Izu-Bonin Trench, latitude 30° 06.97′ N, longitude 148° 34.49′ E.

## Description
Test small, planispiral, flattened, slightly evolute, slightly lobulate in outline, oval in cross-section with rounded periphery. Previous chambers of penultimate whorl are partly visible through the small, open umbilicus. Lateral sides are slightly concave, almost flat. Approximately six inflated, trapeziform chambers in the last whorl, increasing gradually in size, are separated by straight, depressed sutures. Chamber walls are finely agglutinated with organic cement and show a smooth, glassy finish. The primary aperture is an elongate equatorial slit at the base of the last chamber.

## Synonyms
None verified.

## Remarks
Useful stratigraphic marker and bathymetric indicator for Late Cretaceous abyssal paleoenvironments.

## Biogeography
Worldwide. Recorded in the Pacific Ocean (Krasheninnikov, 1973), as *Haplophragmoides perexplicatus* Krasheninnikov in the North Atlantic (Moullade et al., 1988), and as *Haplophragmoides* ex gr. *perexplicatus-constrictus* in the Pacific Ocean (Wightman and Kuhnt, 1992).

## Bathymetry
Abyssal, below CCD.

## Chronostratigraphy
Late Cretaceous (Santonian to Campanian). Campanian in the Pacific Ocean (Krasheninnikov, 1973), Santonian to Campanian in the Indian Ocean (Krasheninnikov, 1974), and in the Atlantic Ocean (Moullade et al., 1988; Wightman and Kuhnt, 1992).

# References


Krasheninnikov, V.A., 1973, Cretaceous benthonic foraminifera, Leg 20, Deep Sea Drilling Project. In Heezen, B.C., MacGregor, I.D., et al., ed., Initial Reports of the Deep Sea Drilling Project, Washington (U.S. Government Printing Office), vol. 20, pp. 205–221.

Moullade, M., Kuhnt, W., and Thurow, J., 1988, Agglutinated benthic foraminifers from Upper Cretaceous variegated clays of the North Atlantic Ocean (DSDP Leg 93 and ODP Leg 103). In Boillot, G., Winterer, E.L., et al., ed., Scientific Results of the Proceedings of the Ocean Drilling Program, vol. 103, pp. 349–377.

Wightman, W.G. and Kuhnt, W., 1992, Biostratigraphy and Paleoecology of Late Cretaceous abyssal agglutinated foraminifers from the western Pacific Ocean (Deep Sea Drilling Project Holes 196A and 198A and Ocean Drilling Program Holes 800A and 801A). In Larson, R.L., Lancelot, Y., et al., ed., Scientific Results of the Proceedings of the Ocean Drilling Program, vol. 129, pp. 247–264.

# *Haplophragmoides walteri*

## (Grzybowski), 1898

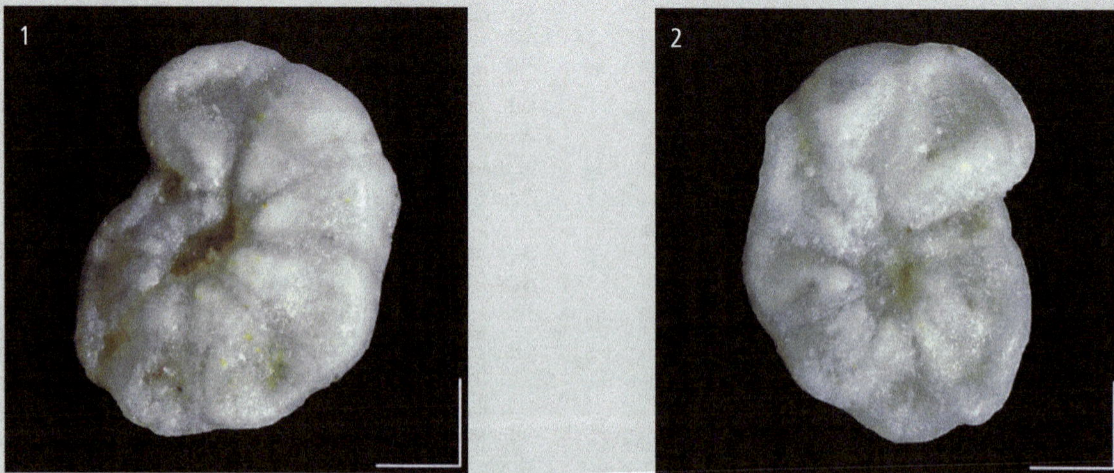


1. *Haplophragmoides walteri.* Silesian Unit, Karaś stream 87, Biecz near Gorlice, Poland. Deposited in the Geroch Collection, Jagiellonian University, Kraków, Poland. Oblique umbilical view, NHMUK PM DI 0383, scale bar = 361 μm.

2. *Haplophragmoides walteri.* Silesian Unit, Karaś stream 87, Biecz near Gorlice, Poland. Deposited in the Geroch Collection, Jagiellonian University, Kraków, Poland. Oblique umbilical view, NHMUK PM DI 0384, scale bar = 364 μm.

### Original Designation

*Trochammina walteri* Grzybowski, 1898, p. 290, pl. 11, fig. 31.

### Type Specimen

Syntypes deposited in the Grzybowski Collection at the Jagiellonian University, Kraków, Poland. As all syntypes are juveniles or deformed, a neotype was designated by Kaminski et al. (1993, pl. 10, fig. 4a–c) from another of Grzybowski's localities.

### Type Level

Paleogene, Silesian Unit of the Polish Carpathians.

### Type Locality

Vicinity of Krosno, Poland. Syntypes are from the Potok H-23 well, 327 meters, and the Potok H-34 well, 15 meters and 45 meters, Potok, Poland. Neotype is from outcrop on the bank of the Wislok River, near the Smierdziaczka Stream, Slasczka area, district of Krosno, Poland.

### Description

Test planispiral, involute, compressed, circular or slightly lobulate in outline, oval in cross-section with a small, open umbilicus and acute periphery. Approximately seven to nine slightly inflated, triangular chambers in the last whorl, increasing rapidly in size, are separated by straight to sigmoid, slightly depressed sutures. Chamber walls are non-calcareous, finely agglutinated, and finely finished. The primary aperture is an interiomarginal slit.

### Synonyms and Citations

*Trochammina tenuissima* Grzybowski, 1898, p. 290, pl. 11, fig. 30.
*Haplophragmoides grzybowskii* Mjatliuk, 1950, p. 268, pl. 1, figs. 9–10b.
*Haplophragmoides walteri* (Grzybowski). –Morgiel and Olszewska, 1981, p. 10, pl. 3, fig. 15. –Charnock and Jones, 1990, p. 171, pl. 6, figs. 3–4; pl. 17, fig. 2. –Kaminski and Geroch, 1993, p. 263, pl. 10, figs. 3a–7c. –Gradstein et al., 1994, pl. 7, figs. 1–5. –Kaminski and Gradstein, 2005, pp. 363–366, fig. 83, pl. 83.

### Remarks

Kaminski and Gradstein (2005) reported differences between specimens from Poland and Texas. Typical specimens of *Haplophragmoides walteri* (Grzybowski) from the Eocene of Poland are large, involute, have more than eight chambers, and occur in deep-water deposits. A different morphotype occurs in shallower environments (upper bathyal to neritic) such as the Navarro Formation of Texas (Cushman and Waters, 1927) and in the Eocene of the Labrador Margin. This morphotype differs from the typical form in its evolute coiling, coarser wall, more poorly sorted grains, and more excavate chambers. Specimens of *Haplophragmoides walteri* from older strata (Upper Cretaceous to Paleocene) tend to be smaller and possess fewer chambers than the Eocene variety. Kuhnt (1990) and Kuhnt and Kaminski (1990) referred this form to "*Haplophragmoides* cf. *walteri*".

### Biogeography

Worldwide. See Observed Occurrences in Kaminski and Gradstein (2005).

### Bathymetry

Bathyal to abyssal.

### Chronostratigraphy

Late Cretaceous to Late Eocene. Large, typical forms first occur in the Late Paleocene (Kaminski and Gradstein, 2005).

## References


Charnock, M.A. and Jones, R.W., 1990, Agglutinated foraminifera from the Palaeogene of the North Sea. In Hemleben, C., Kaminski, M.A., Kuhnt, W., and Scott, D.B., ed., Paleoecology, Biostratigraphy, Paleoceanography and Taxonomy of Agglutinated Foraminifera, NATO ASI Series, C: Mathematical and Physical Sciences. Dordrecht: Kluwer Academic, pp. 139–244.

Cushman, J.A. and Waters, J.A., 1927, Some arenaceous foraminifera from the Upper Cretaceous of Texas. Contributions from the Cushman Laboratory for Foraminiferal Research, vol. 2, pp. 81–85.

Gradstein, F.M., Agterberg, F.P., Ogg, J.G., Hardenbol, J., van Veen, P., Thierry, J., and Huang, Z., 1994, A Mesozoic time scale. Journal of Geophysical Research, vol. 99, B12, pp. 24051–24074.

Grzybowski, J., 1898, Otwornice pokładów naftonosnych okolicy Krosna. Rozprawy Wydziału Matematyczno-Przyrodniczego, Akademia Umiejętności w Krakowie, serya 2, vol. 33, pp. 257–305.

Kaminski, M.A. and Geroch, S., 1993, A revision of foraminiferal species in the Grzybowski collection. In Kaminski, M.A., Geroch, S., and Kaminski, D.G., ed., The Origins of Applied Micropalaeontology: The School of Jósef Grzybowski (with a taxonomic revision of the Grzybowski Collection). Grzybowski Foundation Special Publication, no. 1, pp. 239–330.

Kaminski, M.A. and Gradstein, F.M., 2005, Atlas of Palaeogene Cosmopolitan Deep-Water Agglutinated Foraminifera. Grzybowski Foundation Special Publication, no. 10, 547 + vii pp.

Kuhnt, W., 1990, Agglutinated foraminifera of western Mediterranean Upper Cretaceous pelagic limestones (Umbrian Apennines, Italy, and Betic Cordillera, Southern Spain). Micropaleontology, vol. 36, no. 4, pp. 297–330.

Kuhnt, W. and Kaminski, M.A., 1990, Paleoecology of Late Cretaceous to Paleocene deep-water agglutinated foraminifera from the North Atlantic and western Tethys. In Hemleben, C., Kaminski, M.A., Kuhnt, W., and Scott, D.B., ed., Paleoecology, Biostratigraphy, Paleoceanography and Taxonomy of Agglutinated Foraminifera. NATO ASI Series, Series C: Mathematical and Physical Sciences. Dordrecht: Kluwer Academic, vol. 327, pp. 433–505.

Mjatliuk, E.V., 1950, Stratigrafiya flishevykh osadkov Severnykh Karpat w svete dannykh fauny foraminifer. Trudy Vsesoyuznogo Nauchno-Issledovatel'skogo Geologorazvedochnogo Instituta VNIGRI (New Series), vol. 51, pp. 225–287.

Morgiel, J. and Olszewska, B., 1981, Biostratigraphy of the Polish External Carpathians based on agglutinated foraminifera. Micropaleontology, vol. 27, no. 1, pp. 1–30.

# *Heterolepa dutemplei*

## (d'Orbigny), 1846

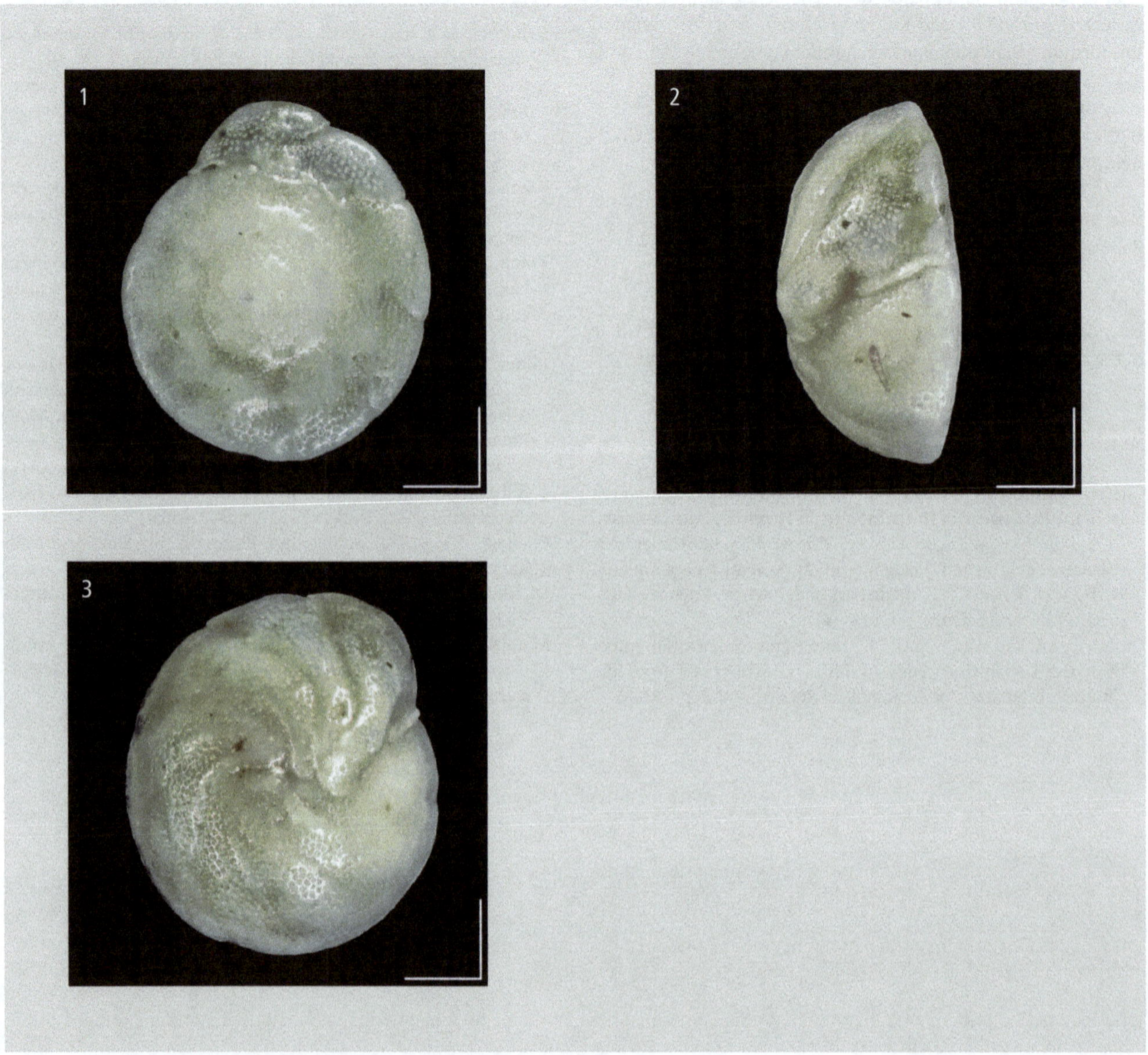


1. *Heterolepa dutemplei.* Queensland Trough, Northeast Australian Margin; ODP Leg 133, Hole 813A, Core 9, Section 4, 71–76 cm. Spiral view, NHMUK PM DI 0385, scale bar = 106 μm.

2. *Heterolepa dutemplei.* Queensland Trough, Northeast Australian Margin; ODP Leg 133, Hole 813A, Core 9, Section 4, 71–76 cm. Apertural view, NHMUK PM DI 0386, scale bar = 109 μm.

3. *Heterolepa dutemplei.* Queensland Trough, Northeast Australian Margin; ODP Leg 133, Hole 813A, Core 9, Section 4, 71–76 cm. Umbilical view, NHMUK PM DI 0387, scale bar = 110 μm.

### Original Designation

*Rotalina dutemplei* d'Orbigny, 1846, p. 157, pl. 8, figs. 19–21.

### Type Specimen

Depository not given. Lectotype (GBA 1981/03/240) designated by Papp and Schmid, 1985, pl. 52, figs. 1–6) and deposited in the collections of the Austrian Geological Survey, Vienna, Austria.

### Type Level

Middle Miocene (van Morkhoven et al., 1986). Originally given as Tertiary by d'Orbigny (1846).

### Type Locality

Near the village of Nussdorf, in the Vienna Basin, north of Vienna, Austria.

### Description

Test low trochospiral, slightly lobulate (last chambers) in outline, unequally biconvex in cross-section, with involute, strongly convex umbilical side, flattened or slightly convex, evolute spiral side, and subacute, imperforate periphery. Seven to ten moderately inflated chambers in the last whorl, increasing gradually in size, are separated by broad, flush or slightly depressed sutures, tangential and forming a distinctive pinwheel pattern on the spiral side, gently curved on the umbilical side. Chamber walls are calcareous, thick, and coarsely perforate on both sides of the test. The primary aperture is an interiomarginal slit bordered by a low lip, extending on the spiral side to about half of the chamber, partially hidden under the continuing apertural lip.

### Synonyms and Citations

*?Rotalia haidingerii* d'Orbigny, 1846, p. 154, pl. 8, figs. 7–9.

*?Rotalina kalembergensis* d'Orbigny, 1846, p. 151, pl. 7, figs. 19–21.

*Heterolepa simplex* Franzenau, 1884, p. 215, pl. 5, fig. 1.

*Heterolepa dutemplei* (d'Orbigny). –Papp and Schmid, 1985, p. 61, pl. 52, figs. 1–6. –Loeblich and Tappan, 1987, p. 632, pl. 709, figs. 1–8. –Revets, 1996, p. 67, pl. 2, figs. 9–12.

*Cibicidoides dutemplei* (d'Orbigny). –van Morkhoven et al., 1986, p. 112, pl. 35.

### Remarks

Type species of the genus *Heterolepa* Franzenau, 1884.

*Heterolepa dutemplei* (d'Orbigny) closely resembles *Cibicidoides mexicanus* (Nuttall), from which it may descend, and it may be difficult to separate the two species (van Morkhoven et al., 1986). It also resembles *Cibicidoides subhaidingerii* (Parr). However, *Heterelopa dutemplei* has strongly curved, limbate umbilical sutures and lacks a large umbonal boss. *Heterelopa dutemplei* also has a shallower bathymetric range than *Cibicidoides subhaidingerii.*

### Biogeography

Worldwide. Widely distributed in western Europe, the Mediterranean, and North Africa (van Morkhoven et al., 1986). Recorded from the northeastern Australian margin by Katz and Miller (1993).

### Bathymetry

Lower neritic to upper bathyal.

### Chronostratigraphy

Early Miocene to Pliocene (van Morkhoven et al., 1986). Pleistocene specimens identical to the Pliocene forms were found in ODP Site 812 on the northeastern Australian margin (Katz and Miller, 1993). *Heterolepa dutemplei* may have persisted in the South Pacific/Australian region through the Pleistocene.

## References


d'Orbigny, A.D., 1846, Foraminifères fossiles du Bassin Tertiaire de Vienne (Autriche). Paris: Gide et Compe, xxxvii + 312 pp.

Franzenau, A., 1884, Heterolepa, egy uj genus a foraminiferak rendjében. Természetrajzi Füzetek, Budapest, vol. 8, pp. 181–184 (214–217).

Katz, M.E. and Miller, K.G., 1993. Neogene subsidence along the Northeast Australian margin: benthic foraminiferal evidence. In Davies, P.J., McKenzie, J.A., Julson, A., et al., ed., Scientific Results of the Proceedings of the Ocean Drilling Program, vol. 133, pp. 75–92.

Loeblich, A.R. Jr and Tappan, H., 1987, Foraminiferal Genera and Their Classification. New York: Van Nostrand Reinhold, 970 pp.

Papp, A. and Schmid, M.E., 1985, Die fossilen Foraminiferen des tertiaeren Beckens von Wien: Revision der Monographie Alcide d'Orbigny (1846). Abhandlungen der Geologischen Bundesanstalt, vol. 37, pp. 1–311.

Revets, S.A., 1996, The generic revision of five families of Rotaliine Foraminifera. Special Publications of the Cushman Foundation for Foraminiferal Research, vol. 34, pp. 1–108.

van Morkhoven, F.P.C.M., Berggren, W.A., Edwards, A.S., et al., 1986, Cenozoic cosmopolitan deep-water benthic foraminifera. Bulletin des Centres de Recherches Exploration-Production Elf-Aquitaine, Mémoire, vol. 11, pp. 1–421.

# *Hippocrepina depressa*

## Vasicek, 1947

1. *Hippocrepina depressa*. Silesian Unit, MK 260, Black shales, Verovice Beds, Wilkowisko, east of Myślenice, Poland. Deposited in the Geroch Collection, Jagiellonian University, Kraków, Poland. Lateral view, NHMUK PM DI 0388, scale bar = 260 µm.

### Original Designation

*Hippocrepina depressa* Vacisek, 1947, p. 243, pl. 1, figs. 1–2.

### Type Specimen

Depository not given.

### Type Level

Cretaceous, lower part of the Magura flysch.

### Type Locality

Exploratory boring no. 12R, near Hluk, 5 miles southeast of Uherske Hradiste, on the Morava River about 40 miles east-southeast of Brno, central Moravia, Czech Republic.

### Description

Small, elongate, laterally compressed, conical test, which may be arched and have annular constrictions at irregular intervals. Chamber walls are thick, finely agglutinated with fine to medium particles, and smoothly finished. The primary aperture is terminal and rounded.

### Synonyms and Citations

*Hippocrepina depressa* Vacisek. –Geroch, 1960, pl. 6, fig. 1. –Bartenstein, 1974, p. 684, pl. 2, figs. 23–26. –Kaminski et al., 1992, p. 254, pl. 1, figs. 7–8. –Reicherter et al., 1994, pl. 7A, figs. E–F. –Kuhnt, 1995, pl. 1, fig. 5. –Holbourn and Kaminski, 1997, p. 34, pl. 4, figs. 1–2.

*Jacullela depressa* (Vacisek). –Riegraf and Luterbacher, 1989, p. 1085, pl. 1, figs. 4–6.

### Biogeography

Worldwide.

### Bathymetry

Bathyal to abyssal. Common in flysch-type sediments.

### Chronostratigraphy

Cretaceous (Hauterivian to Cenomanian).

## References

Bartenstein, H., 1974, Upper Jurassic–Lower Cretaceous primitive arenaceous foraminifera from DSDP Sites 259 and 261, Eastern Indian Ocean. In Veevers, J.J., Heirtzler, J.R., et al., ed., Initial Repoprts of the Deep Sea Drilling Project, Washington (U.S. Government Printing Office), vol. 27, pp. 683–695.

Geroch, S., 1960, Microfaunal assemblages from the Cretaceous and Paleogene Silesian Unit in the Beskid Śląski Mts. (western Carpathians). Biuletyn Instytutu Geologicznego, vol. 153, pp. 7–138.

Holbourn, A.E.L. and Kaminski, M.A., 1997, Cretaceous deep-water benthic foraminifera of the Indian Ocean. Grzybowski Foundation Special Publication, no. 4, pp. 1–175.

Kaminski, M.A., Gradstein, F., and Geroch, S., 1992, Uppermost Jurassic to Lower Cretaceous deep-water benthic foraminiferal assemblages from Site 765 on the Argo Abyssal Plain. In Gradstein, F., Ludden, J.N., et al., ed., Scientific Results of the Proceedings of the Ocean Drilling Program, vol. 123, pp. 239–269.

Kuhnt, W., 1995, Deep-water agglutinated foraminifera from the Lower Cretaceous (Neocomian) "complex à Aptychus" Formation (Corridor de Boyar, Betic Cordillera, southern Spain). Journal of Micropalaeontology, vol. 14, no. 1, pp. 37–52.

Reicherter, K., Pletsch, T., Kuhnt, W., Manthey, J., Homeier, G., Wiedmann, J., and Thurow, J., 1994, Mid-Cretaceous paleogeography and paleoceanography of the Betic Seaway (Betic Cordillera, Spain). Palaeogeography, Palaeoclimatology, Palaeoecology, vol. 107, pp. 1–33.

Riegraf, W. and Luterbacher, H., 1989, Benthonische Foraminiferen aus de Unterkreide des "Deep Sea Drilling Project" (Leg 1–79). Geologische Rundschau, vol. 78, no. 3, pp. 1063–1120.

Vasicek, M., 1947, Poznámky k mikrobiostratigrafii magurského flyse na Morave. Vestník Státního Geologického Ústavu Ceskoslovenské Republiky, Praha, vol. 22, pp. 235–256.

# Hoeglundina elegans

## (d'Orbigny), 1826

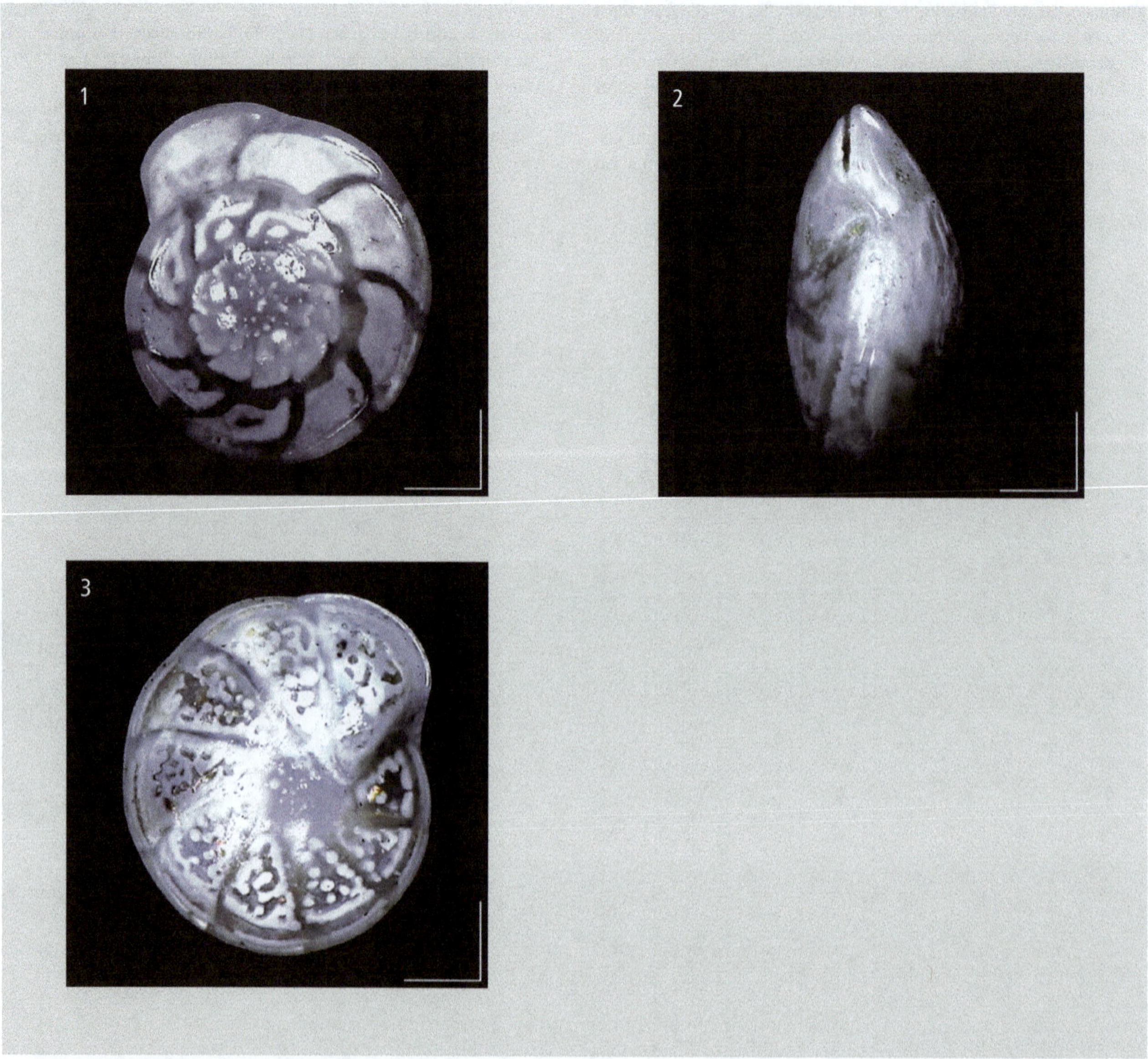

1. *Hoeglundina elegans.* West Indies; "Challenger" Station 24, depth: 390 fathoms. Deposited in the micropaleontological collections of The Natural History Museum, London; NHMUK ZF2212. Spiral view, NHMUK PM DI 0389, scale bar = 282 μm.

2. *Hoeglundina elegans.* West Indies; "Challenger" Station 24, depth: 390 fathoms. Deposited in the micropaleontological collections of The Natural History Museum, London; NHMUK ZF2212. Apertural view, NHMUK PM DI 0390, scale bar = 296 μm.

3. *Hoeglundina elegans.* West Indies; "Challenger" Station 24, depth: 390 fathoms. Deposited in the micropaleontological collections of The Natural History Museum, London; NHMUK ZF2212. Umbilical view, NHMUK PM DI 0391, scale bar = 261 μm.

## Original Designation

*Rotalia (Turbinulina) elegans* d'Orbigny, 1826, p. 276, Model No. 54, illustrated in Parker et al., 1871, pl. 12, fig. 142.

## Type Specimen

Deposited in the collections of the Laboratoire de Paléontologie, Muséum d'Histoire Naturelle, Paris, France (Terquem, 1878, p. 2).

## Type Level

Not designated. [Fossile (Soldani).]

## Type Locality

Not designated. [Coroncina, Italy (Soldani).]

## Description

Test forms a trochospire, subcircular in outline, biconvex in cross-section, with a subacute, keeled periphery. Approximately seven to eight chambers in the last whorl, increasing gradually in size, are separated by thick, flush or slightly depressed sutures, curved on the spiral side, straight and oblique on the umbilical side. Chamber walls are calcareous, aragonitic, smooth, finely perforate, and, in well-preserved specimens, opaque to transparent. The primary aperture is small, interiomarginal, and may be absent in some specimens. The supplementary aperture is an elongate lateromarginal slit, parallel to the peripheral keel and opening on the umbilical side.

## Synonyms and Citations

*Rotalia partschiana* d'Orbigny, 1846, p. 153, pl. 7, figs. 28–30; pl. 8, figs. 1–3.

*Rotalia flosculiformis* Schwager, 1866, p. 262, pl. 7, fig. 109.

*Epistomina bradyi* Galloway and Wissler, 1927, p. 60, pl. 10, fig. 1.

*Epistomina flinti* Galloway and Wissler, 1927, p. 61, pl. 9, fig. 16.

*Epistomina elegans* (d'Orbigny). –Cushman and Jarvis, 1930, p. 365, pl. 34, figs. 1a–c. –LeRoy, 1941, p. 40, pl. 1, figs. 5–7.

*Hoeglundina elegans* (d'Orbigny). –Bolli et al., 1994, p. 361, figs. 55.6–8, 9a–e. –Bermúdez, 1949, p. 250, pl. 17, figs. 34–36. –Phleger and Parker, 1951, p. 22, pl. 12, figs. 1a, b. –Belford, 1966, p. 190, pl. 36, figs. 8–13. –van Morkhoven et al., 1986, p. 97, pl. 29, figs. 1a, b, 2a, b. –Loeblich and Tappan, 1987, p. 446, pl. 478, figs. 1–5. –Jones, 1994, p. 104, pl. 105, figs. 3–6. –Robertson, 1998, p. 114, pl. 44, fig. 2. –Milker and Schmiedl, 2012, p. 79, figs. 19.15–16.

## Remarks

Type species of the genus *Hoeglundina* Brotzen, 1948.

## Biogeography

Worldwide, except in highest latitudes.

## Bathymetry

Neritic to bathyal (van Morkhoven et al., 1986).

## Chronostratigraphy

Late Eocene (Priabonian) to Recent.

# References


Belford, D.J., 1966, Miocene and Pliocene smaller foraminifera from Papua and New Guinea. Australia Bureau of Mineral Resources Geology and Geophysics Bulletin, vol. 79, pp. 1–306.

Bermúdez, P.J., 1949, Tertiary smaller foraminifera of the Dominican Republic. Special Publications of the Cushman Laboratory for Foraminiferal Research, no. 25, pp. 1–322.

Bolli, H.M., Beckmann, J.P., and Saunders, J.B., 1994, Benthic Foraminiferal Biostratigraphy of the South Caribbean Region. Cambridge: Cambridge University Press, 408 pp.

Cushman, J.A. and Jarvis, P.W., 1930, Miocene foraminifera from Buff Bay, Jamaica. Journal of Paleontology, vol. 4, no. 4, pp. 353–368.

d'Orbigny, A.D., 1846, Foraminifères fossiles du Bassin Tertiaire de Vienne (Autriche). Paris: Gide et Compe, xxxvii + 312 pp.

Galloway, J.J. and Wissler, S.G., 1927, Pleistocene foraminifera from the Lomita Quarry, Palos Verdes Hills, California. Journal of Paleontology, vol. 1, pp. 35–87.

Jones, R.W., 1994, The Challenger Foraminifera. Oxford: Oxford University Press, 149 pp.

LeRoy, L.W., 1941, A new species of Cibicides from the lower Pliocene (Repetto Formation) of Southern California. Journal of Paleontology, vol. 15, no. 6, pp. 622–623.

Loeblich, A.R. Jr and Tappan, H., 1987, Foraminiferal Genera and Their Classification. New York: Van Nostrand Reinhold, 970 pp.

Milker, Y. and Schmiedl, G., 2012, A taxonomic guide to modern benthic shelf foraminifera of the western Mediterranean Sea. Palaeontologia Electronica, vol. 15, Issue 2;16A, 134 pp.

Parker, W.K., Jones, T.R., and Brady, H.B., 1871, On the nomenclature of the foraminifera. Part XIV – The species enumerated by d'Orbigny in the "Annales des Sciences Naturelles" 1826. Annals and Magazine of Natural History, Series 4, vol. 7 (continued), – The species founded upon the figures in Soldani's "Testaceographia ac Zoophytographia", Annals and Magazine of Natural History, Series 4, vol. 8, pp. 145–179, 238–266.

Phleger, F.B. and Parker, F.L., 1951, Ecology of foraminifera, northwest Gulf of Mexico; Part II – Foraminifera species. Memoirs of the Geological Society of America, vol. 46, pp. 1–64.

Robertson, B.E., 1998, Systematics and paleoecology of the benthic foraminiferida from the Buff Bay section, Miocene of Jamaica. Micropaleontology, vol. 44, supplement 2, pp. 1–266.

Schwager, C., 1866, Fossile Foraminiferen von Kar Nikobar, Reise der Österreichischen Fregatte Novara um Erde in den Jahren 1857, 1858, 1859 unten den Befehlen des Commodore B. Von Wüllerstorf-Urbair. Geologischer Theil, vol. 2, no. 1, Geologische Beobachtung, no. 2, Paläontologische Mittheilung, pp. 187–268.

Terquem, O., 1878, Les Foraminifères et les Entomostracés–Ostracodes du Pliocène Supérieur de l'île de Rhodes. Mémoires de la Société géologique de France, ser. 3, vol. 1, pp.1–135.

van Morkhoven, F.P.C.M., Berggren, W.A., Edwards, A.S., et al., 1986, Cenozoic cosmopolitan deep-water benthic foraminifera. Bulletin des Centres de Recherches Exploration-Production Elf-Aquitaine, Mémoire, vol. 11, pp. 1–421.

# *Hormosina globulifera*

## Brady, 1879

1. *Hormosina globulifera*. Off northwestern Ireland, North Atlantic Ocean; "Porcupine" Station 23, depth: 630 fathoms. Deposited in the micropaleontological collections of The Natural History Museum, London; NHMUK ZF1584. Lectotype. Lateral view, NHMUK PM DI 0392, scale bar = 552 µm.

### Original Designation

*Hormosina globulifera* Brady, 1879, p. 60, pl. 4, figs. 4–5.

### Type Specimen

Figured specimens housed in the micropaleontological collections of The Natural History Museum, London, UK, and registered in slides ZF1581–ZF1584. Lectotype, designated by Kaminski and Cetean (2011), is the five-chambered specimen, illustrated by Brady (1884, pl. 39, fig. 4), and is preserved in the centre of slide ZF1584.

### Type Level

Recent.

### Type Locality

Not originally designated. Localities given: eight "Challenger" dredgings in North and South Atlantic and South Pacific, six locations over 100 fathoms (three of those over 2000 fathoms). Lectotype from "Porcupine" Station 23, off northwest Ireland, 630 fathoms.

### Description

Elongate, uniserial test, rectilinear or slightly arcuate, composed of large, globular chambers, increasing rapidly in size. Later chambers are slightly overlapping the previous ones. Chamber walls are finely agglutinated, of several layers, smoothly finished with abundant cement. The primary aperture is terminal, round, and at the end of a long, thin, tubular neck.

### Synonyms and Citations

*Hormosina globulifera* Brady.–Brady, 1884, p. 326, pl. 39, figs. 1–4, 6 (not fig. 5). –Barker, 1960, pl. 39, figs. 1–4, 6 (not fig. 5). –Hofker, 1972, p. 60, pl. 17, figs. 13–16. –Schröder, 1986, p. 41, pl. 13, figs. 1–3. –Loeblich and Tappan, 1987, p. 61, pl. 45, figs. 18–20. –Jones, 1994, p. 44, pl. 39, figs. 1–4, 6.

### Remarks

Type species of the genus *Hormosina* Brady, 1879.

Cushman and Renz (1946) described *Hormosina trinitatensis* as a separate species from *Hormosina globulifera* Brady on account of its smaller size, fewer chambers, coarser wall, and smaller apertural neck (see Remarks for *Hormosina trinitatensis* in Kaminski and Gradstein, 2005). Jones (1994) referred Brady's specimen in pl. 39, fig. 5 to *Ginesina* sp. nov.

### Biogeography

Worldwide.

### Bathymetry

Bathyal to abyssal.

### Chronostratigraphy

Late Paleocene to Recent (Jones, 1994).

## References


Barker, R.W., 1960, Taxonomic notes on the species figured by H.B. Brady in his report on the foraminifera dredged by H.M.S. Challenger during the years 1873–1876. Society of Economic Paleontologists and Mineralogists Special Publication, vol. 9, pp. 1–238.

Brady, H.B., 1879, Notes on some of the reticularian Rhizopoda of the Challenger Expedition, Part I. On new or little known Arenaceous types. Quarterly Journal of Microscopical Science (New Series), vol. 19, pp. 20–63.

Brady, H.B., 1884, Report on the foraminifera dredged by H.M.S. Challenger during the years 1873–1876. In Brady, H.B., ed., Report on the Scientific Results of the Voyage of the H.M.S. Challenger during the years 1873–1876. Zoology, vol. 9, pp. 1–814.

Cushman, J.A. and Renz, H.H., 1946, The foraminiferal fauna of the Lizard Springs Formation of Trinidad, British West Indies. Special Publications of the Cushman Laboratory for Foraminiferal Research, vol. 18, pp. 1–48.

Hofker, J., 1972, Primitive Agglutinated Foraminifera. Leiden: E.J. Brill, 95 pp.

Jones, R.W., 1994, The Challenger Foraminifera. Oxford: Oxford University Press, 149 pp.

Kaminski, M.A. and Cetean, C.G., 2011, Lectotypes of type species of Agglutinated Foraminiferal Genera in the Collections of the Natural History Museum, London. Part 2. Hippocrepinina and Hormosinina. In Kaminski, M.A. and Filipescu, S., ed., Proceedings of the Eighth International Workshop on Agglutinated Foraminifera, Grzybowski Foundation Special Publication, no. 16, pp. 61–69.

Kaminski, M.A. and Gradstein, F.M., 2005, Atlas of Palaeogene Cosmopolitan Deep-Water Agglutinated Foraminifera. Grzybowski Foundation Special Publication, no. 10, 547 + vii pp.

Loeblich, A.R. Jr and Tappan, H., 1987, Foraminiferal Genera and Their Classification. New York: Van Nostrand Reinhold, 970 pp.

Schröder, C.J., 1986, Deep-Water Arenaceous Foraminifera in the Northwest Atlantic Ocean. Canadian Technical Report of Hydrography and Ocean Sciences No. 71. Atlantic Geoscience Centre, Bedford Institute of Oceanography, 191 pp.

# *Hormosinella carpenteri*

## (Brady), 1881

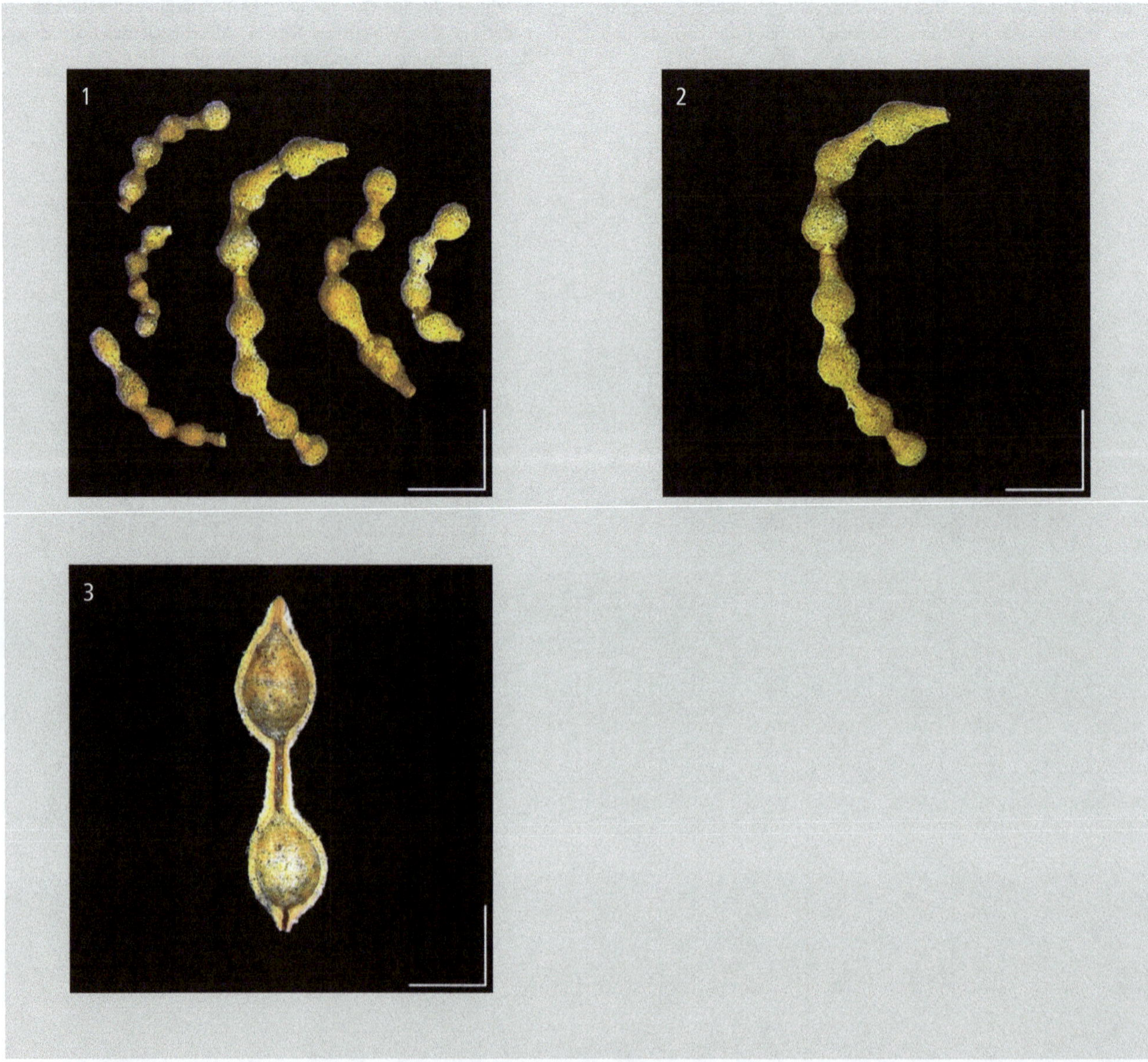


1. *Hormosinella carpenteri*. East of the Azores, Atlantic Ocean; "Challenger" Station 78, depth: 1000 fathoms. Deposited in the micropaleontological collections of The Natural History Museum, London; NHMUK ZF1579. View of slide NHMUK ZF1579 containing syntype specimens. Lateral view, NHMUK PM DI 0393, scale bar = 3.06 mm.

2. *Hormosinella carpenteri*. East of the Azores, Atlantic Ocean; "Challenger" Station 78, depth: 1000 fathoms. Deposited in the micropaleontological collections of The Natural History Museum, London; NHMUK ZF1579. Syntype. Lateral view, NHMUK PM DI 0395, scale bar = 2.76 mm.

3. *Hormosinella carpenteri*. Off the Azores, Atlantic Ocean; "Challenger" Station 76, depth 900 fathoms. Deposited in the micropaleontological collections of The Natural History Museum, London; cross-section view of a syntype specimen from slide NHMUK 1959.5.5.567-577. Lateral view, NHMUK PM DI 0394, scale bar = 1.21 mm.

## Original Designation

*Trochammina (Hormosina) carpenteri* Brady, 1881, p. 51 (type description), illustrated as *Hormosina carpenteri* in Brady, 1884, pl. 39, figs. 14–18.

## Type Specimen

Figured specimens housed in the micropaleontological collections of The Natural History Museum, London, UK, and registered in slides ZF1579 and 1959.5.5.567–577.

## Type Level

Recent.

## Type Locality

Off the coast of Borneo.

## Description

Elongate, uniserial test, rectilinear or arcuate, composed of ovate, inflated chambers, broader at the base, separated by elongate, fragile, stolon-like necks. Chamber walls are thin and agglutinated, consisting of a single layer of well-cemented medium or coarse grains. The primary aperture is terminal, and round at the open end of the neck.

## Synonyms and Citations

*Hormosina carpenteri* Brady, 1884, p. 327, pl. 39, figs. 14–18. –Barker, 1960, pl. 39, figs. 14–18.

*Hormosinella carpenteri* (Brady). –Jones, 1994, p. 44, pl. 39, figs. 14–18.

## Remarks

Specimens figured by Brady (1884) are from "Challenger" Stations 76 and 78, close to the Azores, Atlantic Ocean.

## Biogeography

Worldwide. Predominantly North Atlantic "Challenger" stations (Jones, 1994).

## Bathymetry

Bathyal (350–1940 fathoms, Jones, 1994).

## Chronostratigraphy

Late Paleocene to Recent (Jones, 1994).

# References


Barker, R.W., 1960, Taxonomic notes on the species figured by H.B. Brady in his report on the foraminifera dredged by H.M.S. Challenger during the years 1873–1876. Society of Economic Paleontologists and Mineralogists Special Publication, vol. 9, pp. 1–238.

Brady, H.B., 1881, Notes on some of the reticularian Rhizopoda of the Challenger Expedition, Part III. Quarterly Journal of Microscopical Science (New Series), vol. 21, pp. 31–71.

Brady, H.B., 1884, Report on the foraminifera dredged by H.M.S. Challenger during the years 1873–1876. In Brady, H.B., ed., Report on the Scientific Results of the Voyage of the H.M.S. Challenger during the years 1873–1876. Zoology, vol. 9, pp. 1–814.

Jones, R.W., 1994, The Challenger Foraminifera. Oxford: Oxford University Press, 149 pp.

# *Hormosinella distans*

## (Brady), 1881

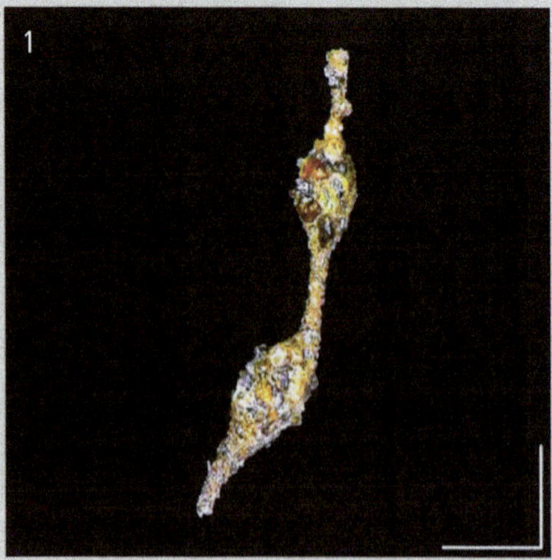

1. *Hormosinella distans.* North off Juan Fernandez Island, East Pacific Ocean Ocean; "Challenger" Station 300, depth: 1375 fathoms. Deposited in the micropaleontological collections of The Natural History Museum, London; NHMUK ZF2270. Syntype. Lateral view, NHMUK PM DI 0396, scale bar = 450 µm.

### Original Designation

*Lituola (Reophax) distans* Brady, 1881, p. 50 (type description), illustrated as *Reophax distans* in Brady, 1884, pl. 31, figs. 18–22.

### Type Specimen

Figured specimens are housed in the micropaleontological collections of The Natural History Museum, London, UK, and are registered in slides ZF2270–ZF2271. Specimens illustrated by Brady in pl. 31, figures 18, 19, 22 are from slide ZF2270; specimens corresponding to figures 20, 21 are from slide ZF2271. Lectotype (6th specimen in the 2nd row in slide ZF2271) designated by Kaminski and Gradstein (2005).

### Type Level

Recent.

### Type Locality

Not designated. Figured specimens from "Challenger" Station 300, (latitude 33° 42′ S, longitude 78° 18′ W), north of Juan Fernandez Island, South Pacific, 1375 fathoms.

### Description

Elongate, uniserial test, rectilinear or slightly arcuate, composed of subcylindrical or fusiform, inflated chambers, tapering at both ends and separated by elongate, narrow, stolon-like necks. Chamber walls are thin and agglutinated, consisting of a single layer of well-cemented medium or coarse grains. The primary aperture is terminal, and round at the open end of the neck.

### Synonyms and Citations

*Reophax distans.* –Brady, 1884, p. 296, pl. 31, figs. 18–22. –Barker, 1960, pl. 31, figs. 18–22.

*Reophax distans* Brady var. *delicatulus* Cushman, 1920, p. 13, pl. 4, fig. 2.

*Hormosinella distans* (Brady). –Shchedrina, 1969, p. 170. –Loeblich and Tappan, 1987, p. 57, pl. 44, figs. 6–9. –Jones, 1994, p. 38, pl. 31, figs. 18–22. –Kaminski and Gradstein, 2005, pp. 246–249, fig. 45, pl. 45.

*Hormosinella distans antarctica* Saidova, 1975, p. 53, pl. 10, fig. 5.

*Hormosinella distans tristus* Saidova, 1975, p. 53, pl. 10, fig. 4.

*Reophax distans* Brady. –Schröder, 1986, p. 44, pl. 16, figs. 3–5, 9.

### Remarks

Type species of the genus *Hormosinella* Shchedrina, 1969.

### Biogeography

Worldwide. See Observed Occurrences in Kaminski and Gradstein (2005).

### Bathymetry

Bathyal to abyssal. Reported between 355 and 2775 fathoms at "Challenger" stations (Jones, 1994).

### Chronostratigraphy

Late Cretaceous to Recent.

## References


Barker, R.W., 1960, Taxonomic notes on the species figured by H.B. Brady in his report on the foraminifera dredged by H.M.S. Challenger during the years 1873–1876. Society of Economic Paleontologists and Mineralogists Special Publication, vol. 9, pp. 1–238.

Brady, H.B., 1881, Notes on some of the reticularian Rhizopoda of the Challenger Expedition, Part III. Quarterly Journal of Microscopical Science (New Series), vol. 21, pp. 31–71.

Brady, H.B., 1884, Report on the foraminifera dredged by H.M.S. Challenger during the years 1873–1876. In Brady, H.B., ed., Report on the Scientific Results of the Voyage of the H.M.S. Challenger during the years 1873–1876, Zoology, vol. 9, pp. 1–814.

Cushman, J.A., 1920, The foraminifera of the Atlantic Ocean; Part 2 –Lituolidae. Bulletin of the United States National Museum, vol. 104, no. 2, pp. 1–111.

Jones, R.W., 1994, The Challenger Foraminifera. Oxford: Oxford University Press, 149 pp.

Kaminski, M.A. and Gradstein, F.M., 2005, Atlas of Palaeogene Cosmopolitan Deep-Water Agglutinated Foraminifera. Grzybowski Foundation Special Publication, no. 10, 547 + vii pp.

Loeblich, A.R. Jr and Tappan, H., 1987, Foraminiferal Genera and Their Classification. New York: Van Nostrand Reinhold, 970 pp.

Saidova, Kh.M., 1975, Bentosnye foraminifery Tikhogo Okeana, 3 vols. Moscow: Institut Okeanologii P.P. Shirshova, Akademiya Nauk SSSR, 875 pp.

Schröder, C.J., 1986, Deep-Water Arenaceous Foraminifera in the Northwest Atlantic Ocean: Canadian Technical Report of Hydrography and Ocean Sciences No. 71. Atlantic Geoscience Centre, Bedford Institute of Oceanography, 191 pp.

Shchedrina, Z.G., 1969, On some changes in the systematics of the families Astrorhizidae and Reophacidae (Foraminifera). Voprosy Mikropaleontologii, vol. 11, pp. 157–170.

# Hormosinelloides guttifer

## (Brady), 1881

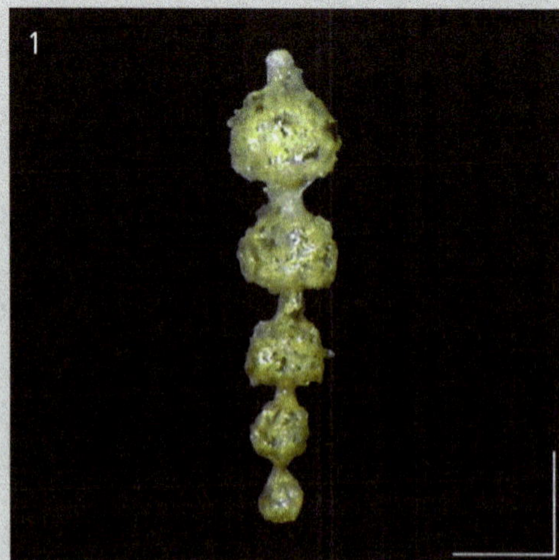


1. *Hormosinelloides guttifer.* South Atlantic Ocean; "Challenger" Station 323, depth: 1425 fathoms. Deposited in the micropaleontological collections of The Natural History Museum, London; NHMUK ZF2276. Syntype. Lateral view, NHMUK PM DI 0397, scale bar = 153 μm.

### Original Designation

*Reophax guttifera* Brady, 1881, p. 49 (type description), illustrated as *Reophax guttifera* in Brady, 1884, pl. 31, figs. 10–15.

### Type Specimen

Figured specimens are housed in the micropaleontological collections of The Natural History Museum, London, UK, and are registered in slide ZF2276. Lectotype (7th specimen in the third row in slide ZF2276) designated by Kaminski and Gradstein (2005).

### Type Level

Recent.

### Type Locality

"Challenger" Station 323, latitude 35° 39′ S, longitude 50° 47′ W, 1425 fathoms, east of Buenos Aires, Argentina, South Atlantic. Also recorded in "cold area of the Faröe Channel", 540 fathoms.

### Description

Test forms an elongate, uniserial series, rectilinear or slightly arcuate and composed of pyriform to nearly round chambers. Early chambers are appressed, increasing gradually in size. Later chambers are connected by narrow, stolon-like tubes. Chamber walls are thin and agglutinated with different types of grains held in a little organic cement. The primary aperture is a terminal, round opening at the end of a tubular neck.

### Synonyms and Citations

*Reophax guttifera* Brady. –Brady, 1884, p. 295, pl. 31, figs. 10–15.

*Reophax guttifer* Brady. –Barker, 1960, pl. 31, figs. 10–15.

*Hormosinella guttifera* (Brady). –Charnock and Jones, 1990, p. 164, pl. 4, fig. 12; pl. 15, fig. 12. –Jones, 1994, p. 38, pl. 31, figs. 10–15.

*Hormosinelloides guttifer* (Brady). –Kaminski and Gradstein, 2005, pp. 249–252, fig. 46, pl. 46.

### Remarks

Type species of the genus *Hormosinelloides* Zheng (2001), which differs from *Hormosinella* by having early chambers tightly appressed and later chambers connected by stolons.

### Biogeography

Worldwide. See Observed Occurrences in Kaminski and Gradstein (2005).

### Bathymetry

Bathyal to abyssal. Reported between 540 and 1900 fathoms at "Challenger" stations in the Atlantic Ocean (Jones, 1994).

### Chronostratigraphy

Late Cretaceous (Campanian) to Recent (Kaminski and Gradstein, 2005).

## References


Barker, R.W., 1960, Taxonomic notes on the species figured by H.B. Brady in his report on the foraminifera dredged by H.M.S. Challenger during the years 1873–1876. Society of Economic Paleontologists and Mineralogists Special Publication, vol. 9, pp. 1–238.

Brady, H.B., 1881, Notes on some of the reticularian Rhizopoda of the Challenger Expedition, Part III. Quarterly Journal of Microscopical Science (New Series), vol. 21, pp. 31–71.

Brady, H.B., 1884, Report on the foraminifera dredged by H.M.S. Challenger during the years 1873–1876. In Brady, H.B., ed., Report on the Scientific Results of the Voyage of the H.M.S. Challenger during the years 1873–1876. Zoology, vol. 9, pp. 1–814.

Charnock, M.A. and Jones, R.W., 1990, Agglutinated foraminifera from the Palaeogene of the North Sea. In Hemleben, C., Kaminski, M.A., Kuhnt, W., and Scott, D.B., ed., Paleoecology, Biostratigraphy, Paleoceanography and Taxonomy of Agglutinated Foraminifera, NATO ASI Series, C: Mathematical and Physical Sciences. Dordrecht: Kluwer Academic, pp. 139–244.

Jones, R.W., 1994, The Challenger Foraminifera. Oxford: Oxford University Press, 149 pp.

Kaminski, M.A. and Gradstein, F.M., 2005, Atlas of Paleogene Cosmopolitan Deep-Water Agglutinated Foraminifera. Grzybowski Foundation Special Publication, no. 10, 547 + vii pp.

# Hyalinea balthica

## (Schroeter), 1783

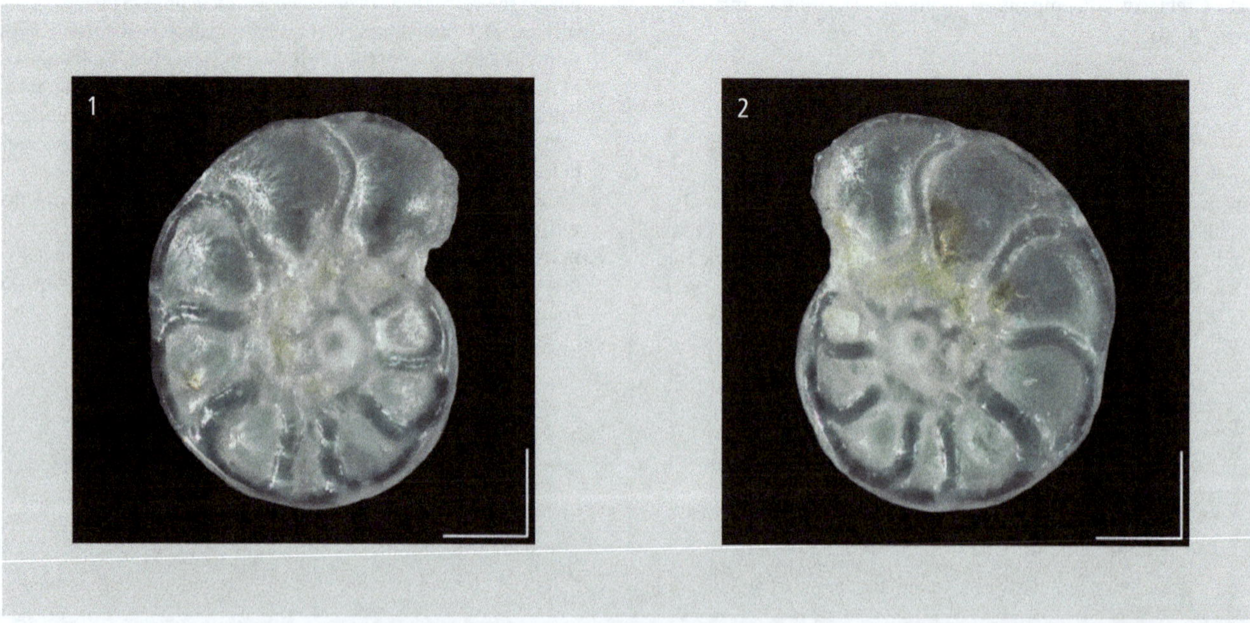


1. *Hyalinea balthica.* "Meteor" Cruise, West Africa; M25. Deposited in the Lutze Collection (West Africa), no. 170. Institute of Geosciences, Kiel University, Germany. Spiral view, NHMUK PM DI 0398, scale bar = 202 μm.

2. *Hyalinea balthica.* "Meteor" Cruise, West Africa; M25. Deposited in the Lutze Collection (West Africa), no. 170. Institute of Geosciences, Kiel University, Germany. Umbilical view, NHMUK PM DI 0399, scale bar = 202 μm.

### Original Designation
*Nautilus balthicus* Schroeter, 1783, p. 20, pl. 1, fig. 2.

### Type Specimen
Depository not given.

### Type Level
Recent.

### Type Locality
Baltic Sea.

### Description
Test discoidal, strongly compressed, very low trochospiral to nearly planispiral with flattened, semi-evolute umbilical and spiral sides and broad, imperforate, keeled periphery. Approximately eight to twelve slightly inflated chambers in the final whorl, increasing gradually in size, are separated by radial, slightly curved, prominent, limbate, raised sutures, which merge into the peripheral carina. Chamber walls are calcareous, finely perforate, and smooth, except for rare pustules in the umbilical area. The primary aperture is a low equatorial, interiomarginal arch, bordered above by a narrow lip, extending laterally below the folium on both sides of the test. Previous apertures remain open in the last few chambers before being closed by lamellar thickening.

### Synonyms and Citations
*Operculina complanata* (Defrance). –Parker and Jones, 1857, p. 285, pl. 11, figs. 3–4.
*Nonionina elegans* Williamson, 1858, p. 35, pl. 3, figs. 74–75.
*Operculina ammonoides* (Gronovius). –Brady, 1884, p. 745, pl. 112, figs. 1–2.
*Hyalinea balthica* (Schroeter). –Barker, pl. 112, figs. 1–2. –van Morkhoven et al., 1986, p. 21, pl. 3, figs. 1–3. –Loeblich and Tappan, 1987, p. 580, pl. 632, figs. 5–8. –Jones, 1994, p. 110, pl. 112, figs. 1–2. –Milker and Schmiedl, 2012, p. 104, figs. 24.1–2.

### Remarks
Type species of the genus *Hyalinea* Hofker, 1951.

### Biogeography
Worldwide. In the modern ocean, this species is most abundant in the cool water masses of the northeastern Atlantic; it also

occurs in the Barents Sea, Skagerrak, central North Sea, British Isles, Mediterranean, Italy, Pacific Ocean, and Indian Ocean (van Morkhoven et al., 1986; Loeblich and Tappan, 1987).

### Bathymetry

Primarily neritic to upper bathyal, but ranges down to about 1000 meters (van Morkhoven et al., 1986).

### Chronostratigraphy

Late Pliocene to Recent; doubtful occurrence in Middle Pliocene (van Morkhoven et al., 1986). Middle? Miocene to Recent (Jones, 1994).

## References


Barker, R.W., 1960, Taxonomic notes on the species figured by H.B. Brady in his report on the foraminifera dredged by H.M.S. Challenger during the years 1873–1876. Society of Economic Paleontologists and Mineralogists Special Publication, vol. 9, pp. 1–238.

Brady, H.B., 1884, Report on the foraminifera dredged by H.M.S. Challenger during the years 1873–1876. In Brady, H.B., ed., Report on the Scientific Results of the Voyage of the H.M.S. Challenger during the years 1873–1876. Zoology, vol. 9, pp. 1–814.

Jones, R.W., 1994, The Challenger Foraminifera. Oxford: Oxford University Press, 149 pp.

Loeblich, A.R. Jr and Tappan, H., 1987, Foraminiferal Genera and Their Classification. New York: Van Nostrand Reinhold, 970 pp.

Milker, Y. and Schmiedl, G., 2012, A taxonomic guide to modern benthic shelf foraminifera of the western Mediterranean Sea. Palaeontologia Electronica, vol. 15, Issue 2;16A, 134 pp.

Parker, W.K. and Jones, T.R., 1857, Description of some foraminifera from the coast of Norway. Annals and Magazine of Natural History, ser. 2, vol. 19, pp. 273–303.

Schroeter, J.S., 1783, Einleitung in die Conchylienkenntniss nach Linné vol. 1. Halle: J.J. Gebauer, 860 pp.

van Morkhoven, F.P.C.M., Berggren, W.A., Edwards, A.S., et al., 1986, Cenozoic cosmopolitan deep-water benthic foraminifera. Bulletin des Centres de Recherches Exploration-Production Elf-Aquitaine, Mémoire, vol. 11, pp. 1–421.

Williamson, W.C., 1858, On the Recent Foraminifera of Great Britain. London: Ray Society, 107 pp.

# Hyperammina elongata

**Brady, 1878**

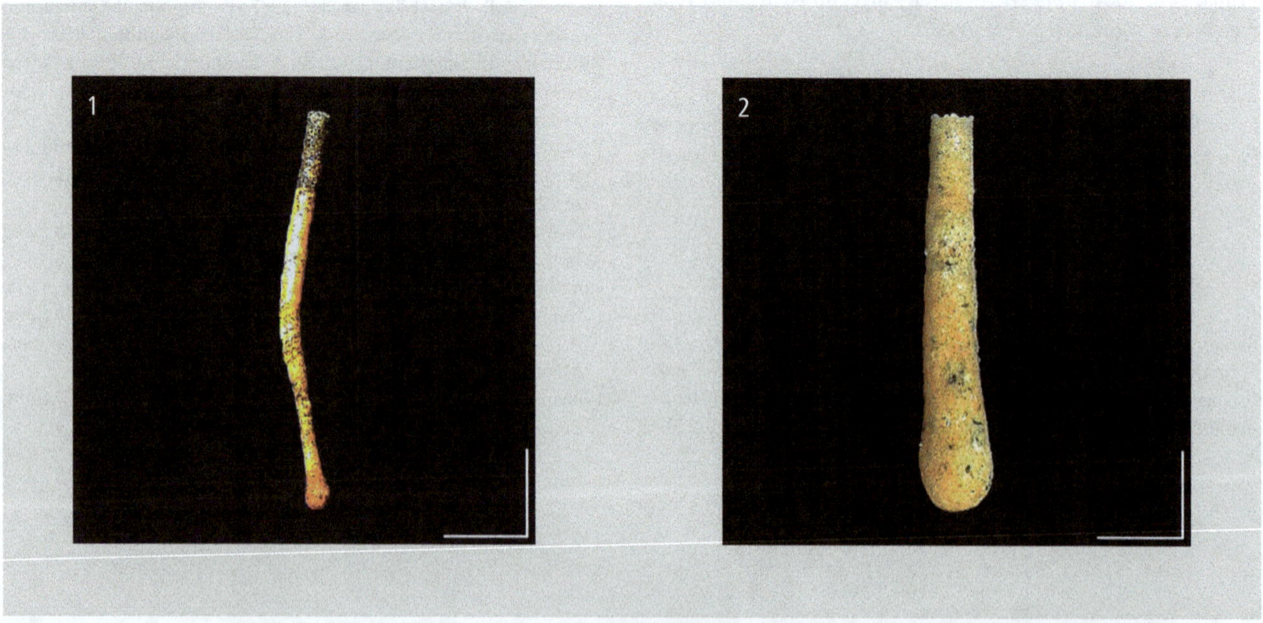


1. *Hyperammina elongata.* West Indies; "Challenger" Station 24, depth: 390 fathoms. Deposited in the micropaleontological collections of The Natural History Museum, London; NHMUK ZF1592. Lateral view, NHMUK PM DI 0400, scale bar = 441 μm.

2. *Hyperammina elongata.* South China Sea; Station 17956, latitude 13° 50.90′ N, longitude 112° 35.30′ E, water depth: 3387 meters, surface sample. Deposited in the South China Sea Collection, Institute of Geosciences, Kiel University, Germany. Lateral view, NHMUK PM DI 0401, scale bar = 208 μm.

## Original Designation
*Hyperammina elongata* Brady, 1878, p. 433, pl. 20, fig. 2.

## Type Specimen
Lectotype, designated by Loeblich and Tappan (1964), corresponds to specimen illustrated by Brady (1878, pl. 20, fig. 2). Lectotype is housed in the micropaleontological collections of The Natural History Museum, London, UK, and is registered in slide ZF3604. Specimen figured by Brady (1884, pl. 23, fig. 8) is registered in slide ZF1592.

## Type Level
Recent, surface sediments.

## Type Locality
North-Polar Expedition Station P, off Cape Frazer, 80 fathoms, latitude 79° 45′ N, Arctic Seas. Specimen figured by Brady (1884) is from "Challenger" Station 24, West Indies (390 fathoms).

## Description
Elongate test with large proloculus followed by undivided, tubular chamber, tapering slightly toward the aperture. Chamber walls are agglutinated with fine to medium particles and a variable amount of organic cement. Interior cavity of wall is smoothly finished. The primary aperture is terminal and rounded.

## Synonyms and Citations
*Hyperammina elongata* Brady. –Brady, 1884, p. 257, pl. 23, fig. 8. –Barker, 1960, pl. 23, fig. 8. –Loeblich and Tappan, 1987, p. 42, pl. 32, figs. 7–9. –Charnock and Jones, 1990, p. 150, pl. 1, fig. 13; pl. 13, fig. 14. –Jones, 1994, p. 33, pl. 23, fig. 8.

## Remarks
Type species of the genus *Hyperammina* Brady, 1878.

Specimens illustrated by Charnock and Jones (1990) are from the Paleogene of the North Sea.

## Biogeography
Worldwide.

## Bathymetry
Neritic to abyssal.

### Chronostratigraphy

Paleocene? to Recent.

## References

Barker, R.W., 1960, Taxonomic notes on the species figured by H.B. Brady in his report on the foraminifera dredged by H.M.S. Challenger during the years 1873–1876. Society of Economic Paleontologists and Mineralogists Special Publication, vol. 9, pp. 1–238.

Brady, H.B., 1878, On the Reticularian and Radiolarian Rhizopoda (Foraminifera and Polycystina) of the North Polar Expedition of 1875–76. Annals and Magazine of Natural History, ser. 5, no. 1, pp. 425–440.

Brady, H.B., 1884, Report on the foraminifera dredged by H.M.S. Challenger during the years 1873–1876. In Brady, H.B., ed., Report on the Scientific Results of the Voyage of the H.M.S. Challenger during the years 1873–1876. Zoology, vol. 9, pp. 1–814.

Charnock, M.A. and Jones, R.W., 1990, Agglutinated foraminifera from the Palaeogene of the North Sea. In Hemleben, C., Kaminski, M.A., Kuhnt, W. and Scott, D.B., ed., Paleoecology, Biostratigraphy, Paleoceanography and Taxonomy of Agglutinated Foraminifera, NATO ASI Series, C: Mathematical and Physical Sciences. Dordrecht: Kluwer Academic, pp. 139–244.

Jones, R.W., 1994, The Challenger Foraminifera. Oxford: Oxford University Press, 149 pp.

Loeblich, A.R. Jr and Tappan, H., 1964, Sarcodina, chiefly "Thecamoebians" and Foraminiferida. In Moore, R.C., ed., Treatise on Invertebrate Paleontology, Part C, Protista 2 (2 vols). Lawrence, KS: Geological Society of America and University of Kansas Press, pp. 1–900.

Loeblich, A.R. Jr and Tappan, H., 1987, Foraminiferal Genera and Their Classification. New York: Van Nostrand Reinhold, 970 pp.

# *Ioanella tumidula*

## (Brady), 1884

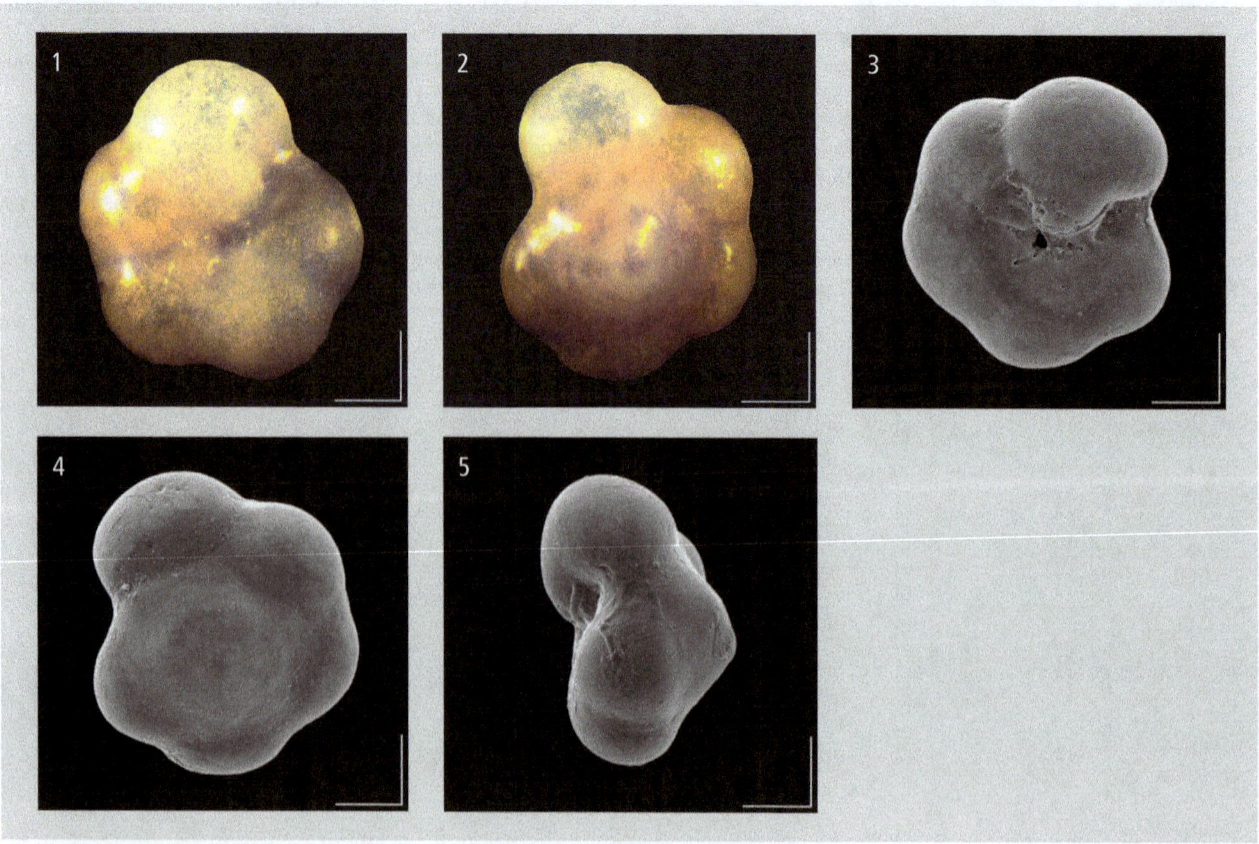

1. *Ioanella tumidula.* North Atlantic Ocean; "Challenger" Station 5, depth: 2740 fathoms. Deposited in the micropaleontological collections of The Natural History Museum, London; NHMUK ZF2553. Syntype. Umbilical view, NHMUK PM DI 0402, scale bar = 39 μm.

2. *Ioanella tumidula.* North Atlantic Ocean; "Challenger" Station 5, depth: 2740 fathoms. Deposited in the micropaleontological collections of The Natural History Museum, London; NHMUK ZF2553. Syntype. Spiral view, NHMUK PM DI 0403, scale bar = 39 μm.

3. *Ioanella tumidula.* North Atlantic Ocean; "Challenger" Station 5, depth: 2740 fathoms. Deposited in the micropaleontological collections of The Natural History Museum, London; NHMUK ZF2553. Syntype. Umbilical view, NHMUK PM DI 0404, scale bar = 44 μm.

4. *Ioanella tumidula.* North Atlantic Ocean; "Challenger" Station 5, depth: 2740 fathoms. Deposited in the micropaleontological collections of The Natural History Museum, London; NHMUK ZF2553. Syntype. Spiral view, NHMUK PM DI 0405, scale bar = 43 μm.

5. *Ioanella tumidula.* North Atlantic Ocean; "Challenger" Station 5, depth: 2740 fathoms. Deposited in the micropaleontological collections of The Natural History Museum, London; NHMUK ZF2553. Syntype. Edge view, NHMUK PM DI 0406, scale bar = 46 μm.

## Original Designation

*Truncatulina tumidula* Brady, 1884, p. 666, pl. 95, fig. 8.

## Type Specimen

Depository not designated. Figured specimen is housed in the micropaleontological collections of The Natural History Museum, London, UK, and is registered in slide ZF2553.

## Type Level

Recent, red clay.

## Type Locality

"Challenger" Station 5 (latitude 24° 20′ N, longitude 24° 28′ W), southwest of the Canaries, 2740 fathoms.

## Description

Test small, trochospiral, slightly lobulate in outline, unequally biconvex in cross-section with more convex spiral side and subrounded periphery. A deep, narrow umbilicus is present on the umbilical side. Approximately five to seven inflated chambers in the last whorl, increasing rapidly in size, are separated by radial, depressed sutures. Chamber walls are calcareous, smooth, and finely perforate. The primary aperture is an interiomarginal arch near the umbilicus with small secondary openings at the umbilical ends of the chambers.

## Synonyms and Citations

*Eponides tumidula* (Brady). –Barker, 1960, pl. 95, fig. 8.

*Ioanella tumidula* (Brady). –Loeblich and Tappan, 1987, p. 549, pl. 595, figs. 4–9. –Jones, 1994, p. 99, pl. 95, fig. 8.

## Remarks

Type species of the genus *Ioanella* Saidova, 1975.

## Biogeography

Worldwide (Murray, 1991).

## Bathymetry

Bathyal to abyssal.

## Chronostratigraphy

Pliocene? to Recent.

## References


Barker, R.W., 1960, Taxonomic notes on the species figured by H.B. Brady in his report on the foraminifera dredged by H.M.S. Challenger during the years 1873–1876. Society of Economic Paleontologists and Mineralogists Special Publication, vol. 9, pp. 1–238.

Brady, H.B., 1884, Report on the foraminifera dredged by H.M.S. Challenger during the years 1873–1876. In Brady, H.B., ed., Report on the Scientific Results of the Voyage of the H.M.S. Challenger during the years 1873–1876. Zoology, vol. 9, pp. 1–814.

Jones, R.W., 1994, The Challenger Foraminifera. Oxford: Oxford University Press, 149 pp.

Loeblich, A.R. Jr and Tappan, H., 1987, Foraminiferal Genera and Their Classification. New York: Van Nostrand Reinhold, 970 pp.

Murray, J.W., 1991, Ecology and Paleocecology of Benthic Foraminifera. New York: John Wiley & Sons, 397 pp.

# Jaculella acuta

## Brady, 1879

1. *Jaculella acuta.* Southeast of Pernambuco, Atlantic Ocean; "Challenger" Station 122, depth: 350 fathoms. Deposited in the micropaleontological collections of The Natural History Museum, London; NHMUK1966.2.11.5 (ex NHMUK ZF1602). Lectotype. Lateral view, NHMUK PM DI 0407, scale bar = 1.99 mm.

### Original Designation

*Jaculella acuta* Brady, 1879, p. 35, pl. 3, figs. 12–13.

### Type Specimen

Figured specimens are housed in the micropaleontological collections of The Natural History Museum, London, UK, and are registered in slides ZF1602 and ZF1603. A lectotype (1966.2.11.5) was designated by Loeblich and Tappan (1964) from slide ZF1602.

### Type Level

Recent, surface sediments.

### Type Locality

"Challenger" Station 122 (off the coast of Brazil, 350 fathoms), South Atlantic.

### Description

Test large, elongate, conical, and tapered consisting of proloculus and second non-septate chamber. Chamber walls are thick, solid, coarsely agglutinated with non-calcareous cement, and roughly finished. The primary aperture is a round, terminal opening at the slightly constricted open end of the tube.

### Synonyms and Citations

*Jaculella acuta* Brady. –Brady, 1884, p. 255, pl. 22, figs. 14–18. –Barker, 1960, pl. 22, figs. 14–18. –Loeblich and Tappan, 1987, p. 44, pl. 33, figs. 5–6. –Jones, 1994, p. 33, pl. 22, figs. 14–18.

### Remarks

Type species of the genus *Jaculella* Brady, 1879.

### Biogeography

Worldwide.

### Bathymetry

Neritic to abyssal. Reported between 60 and 2900 fathoms at "Challenger" stations (Jones, 1994).

### Chronostratigraphy

Early Miocene to Recent. *Jaculella acuta* Brady is mostly reported from Pliocene to Pleistocene sediments. Rögl (1976) recorded its occurrence in Lower Miocene sediments from the Southern Ocean.

## References


Barker, R.W., 1960, Taxonomic notes on the species figured by H.B. Brady in his report on the foraminifera dredged by H.M.S. Challenger during the years 1873–1876. Society of Economic Paleontologists and Mineralogists Special Publication, vol. 9, pp. 1–238.

Brady, H.B., 1879, Notes on some of the reticularian Rhizopoda of the Challenger Expedition, Part I. On new or little known Arenaceous types. Quarterly Journal of Microscopical Science (New Series), vol. 19, pp. 20–63.

Brady, H.B., 1884, Report on the foraminifera dredged by H.M.S. Challenger during the years 1873–1876. In Brady, H.B., ed., Report on the Scientific Results of the Voyage of the H.M.S. Challenger during the years 1873–1876. Zoology, vol. 9, pp. 1–814.

Jones, R.W., 1994, The Challenger Foraminifera. Oxford: Oxford University Press, 149 pp.

Loeblich, A.R. Jr and Tappan, H., 1964, Sarcodina, chiefly "Thecamoebians" and Foraminiferida. In Moore, R.C., ed., Treatise on Invertebrate Paleontology, Part C, Protista 2 (2 vols). Lawrence, KS: Geological Society of America and University of Kansas Press, pp. 1–900.

Loeblich, A.R. Jr and Tappan, H., 1987, Foraminiferal Genera and Their Classification. New York: Van Nostrand Reinhold, 970 pp.

Rögl, F., 1976, Late Cretaceous to Pleistocene foraminifera from the Southeast Pacific Basin, DSDP Leg 35. In Hollister, C.D., Craddock, C. et al., ed., Initial Reports of the Deep Sea Drilling Project, Washington (U.S. Government Printing Office), vol. 35, pp. 1–930.

# Kalamopsis grzybowskii

## (Dylàzanka), 1923

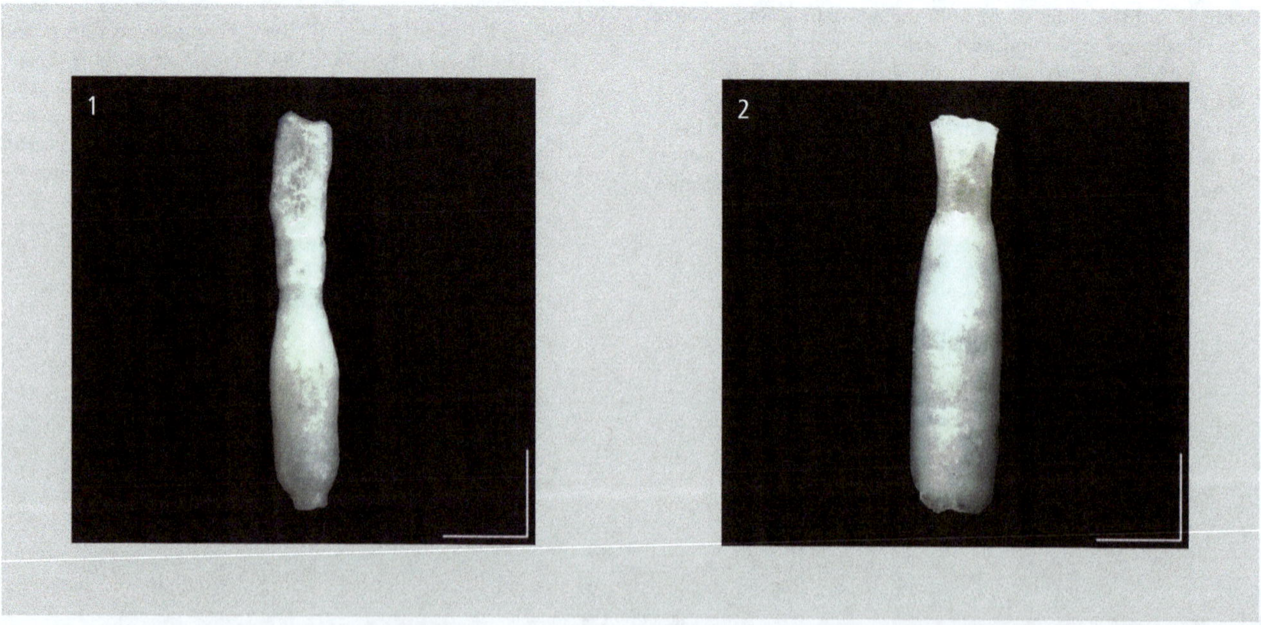


1. *Kalamopsis grzybowskii.* Subsilesian Unit (variegated marls), Poland; Węglówka borehole 5: 384.6–319.6 meters. Deposited in the Geroch Collection, Jagiellonian University, Kraków, Poland. Lateral view, NHMUK PM DI 0408, scale bar = 516 μm.

2. *Kalamopsis grzybowskii.* Subsilesian Unit (variegated marls), Poland; Węglówka borehole 5: 384.6–319.6 meters. Deposited in the Geroch Collection, Jagiellonian University, Kraków, Poland. Lateral view, NHMUK PM DI 0409, scale bar = 408 μm.

### Original Designation

*Hyperammina grzybowskii* Dylàzanka, 1923, p. 65 [no type figure given].

### Type Specimen

Dylàzanka did not provide a type figure, but referred to Grzybowski's drawing of *Hyperammina* sp. aff. *subnodosiformis* (Grzybowski, 1901, pl. 7, fig. 5). This specimen is not preserved in the Grzybowski Collection. A lectotype was designated by Berggren and Kaminski (1990, pl. 1, fig. 3), from unfigured specimens in the Dylàzanka Collection at the Jagiellonian University (No. UJ-133-P, 4/18a).

### Type Level

Upper Cretaceous, Inoceramian Beds of the Magura Unit, Polish Carpathians.

### Type Locality

Quarry in Szymbark, near Gorlice, Poland.

### Description

Cylindrical test comprised of a rounded proloculus followed by elongate, tubular chambers in a rectilinear series. Chambers are separated by internal partitions, which may correspond to points of constriction. Chamber walls are thin, finely agglutinated with organic cement, and finely finished, occasionally with some coarser grains. The primary aperture is simple and terminal.

### Synonyms and Citations

*Hyperammina grzybowskii* Dylàzanka. –Geroch and Gradzinski, 1955, pl. 5, fig. 1.

*Kalamopsis grzybowskii* (Dylàzanka). –Bartenstein, 1974, pl. 1, figs. 31–34; pl. 2, figs. 27–35. –Geroch and Nowak, 1984, pl. 1, figs. 2–3. –Riegraf and Luterbacher, 1989, pl. 1, fig. 4–5. –Kaminski and Geroch, 1993, p. 281, pl. 17, figs. 5–8. –Holbourn and Kaminski, 1997, p. 37, pl. 7, figs. 1–3. –Kaminski and Gradstein, 2005, pp. 252–255, fig. 47, pl. 47.

### Remarks

Mostly found as broken fragments.

### Biogeography

Worldwide. See Observed Occurrences in Kaminski and Gradstein (2005).

### Bathymetry

Bathyal to abyssal.

### Chronostratigraphy

Late Jurassic (Kimmeridgian) to Late Eocene (Kaminski and Gradstein, 2005). Riegraf and Luterbacher (1989) reported its range as Kimmeridgian to Eocene in the North Atlantic (ODP Sites 105, 367, 368, 370, and 398). Kuhnt et al. (2002) reported its occurrence in Oligocene sediments in the South China Sea (ODP Site 1148).

## References


Bartenstein, H., 1974, Upper Jurassic–Lower Cretaceous primitive arenaceous foraminifera from DSDP Sites 259 and 261, Eastern Indian Ocean. In Veevers, J.J., Heirtzler, J.R., et al., ed., Initial Repoprts of the Deep Sea Drilling Project, Washington (U.S. Government Printing Office), vol. 27, pp. 683–695.

Berggren, W.A. and Kaminski, M.A., 1990, Abyssal agglutinates: back to basics. In Hemleben, C., Kaminski, M.A., Kuhnt, W., and Scott, D.B., ed., Paleoecology, Biostratigraphy, Paleoceanography and Taxonomy of Agglutinated Foraminifera. NATO ASI Series, Series C: Mathematical and Physical Sciences, vol. 327. Dordrecht: Kluwer Academic, pp. 53–76.

Dylàzanka, M., 1923, Warstwy inoceramowe z lomu w Szymbarku kolo Gorlic. Rocznik Polskiego Towarzystwa Geologicznego w Krakowie, vol. 1, pp. 36–81.

Geroch, S. and Gradzinski, R., 1955, Stratigrafia serii podsląskiej Żywieckiego okna tektonicznego. Rocznik Polskiego Towarzystwa Geologicznego, vol. 24, pp. 3–62.

Geroch, S. and Nowak, W., 1984, Proposal of zonation for the late Tithonian–late Eocene, based upon arenaceous foraminifera from the outer Carpathians, Poland. In Oertli, H., ed., Benthos '83, 2nd International Symposium on Benthic Foraminifera, Pau (France), April 11–15, 1983. Elf Aquitaine, ESSO REP and TOTAL CFP, Pau and Bordeaux, p. 225–239.

Grzybowski, J., 1901, Otwornice warstw inoceramowych okolicy Gorlic. Rozprawy Wydziału Matematyczno-Przyrodniczego, Akademia Umiejętności w Krakowie, serya 2, vol. 41, pp. 219–286.

Holbourn, A.E.L. and Kaminski, M.A., 1997, Cretaceous deep-water benthic foraminifera of the Indian Ocean. Grzybowski Foundation Special Publication, no. 4, pp. 1–175.

Kaminski, M.A. and Geroch, S., 1993, A revision of foraminiferal species in the Grzybowski Collection. In Kaminski, M.A., Geroch, S., and Kaminski, D.G., ed., The Origins of Applied Micropalaeontology: The School of Jósef Grzybowski (with a taxonomic revision of the Grzyboswski Collection). Grzybowski Foundation Special Publication, no. 1, pp. 239–330.

Kaminski, M.A. and Gradstein, F.M., 2005, Atlas of Palaeogene Cosmopolitan Deep-Water Agglutinated Foraminifera. Grzybowski Foundation Special Publication, no. 10, 547 + vii pp.

Kuhnt, W., Holbourn, A., and Zhao, Q., 2002, The early history of the South China Sea: evolution of Oligocene–Miocene deep water environments. Revue de Micropaléontologie, vol. 45, no. 2, pp. 99–159.

Riegraf, W. and Luterbacher, H., 1989, Oberjura-Foraminiferen aus dem Nord- und Südatlantik (Deep Sea Drilling Project 1–79). Geologische Rundschau, vol. 78, no. 3, pp. 999–1045.

# *Karreriella bradyi*

## (Cushman), 1911

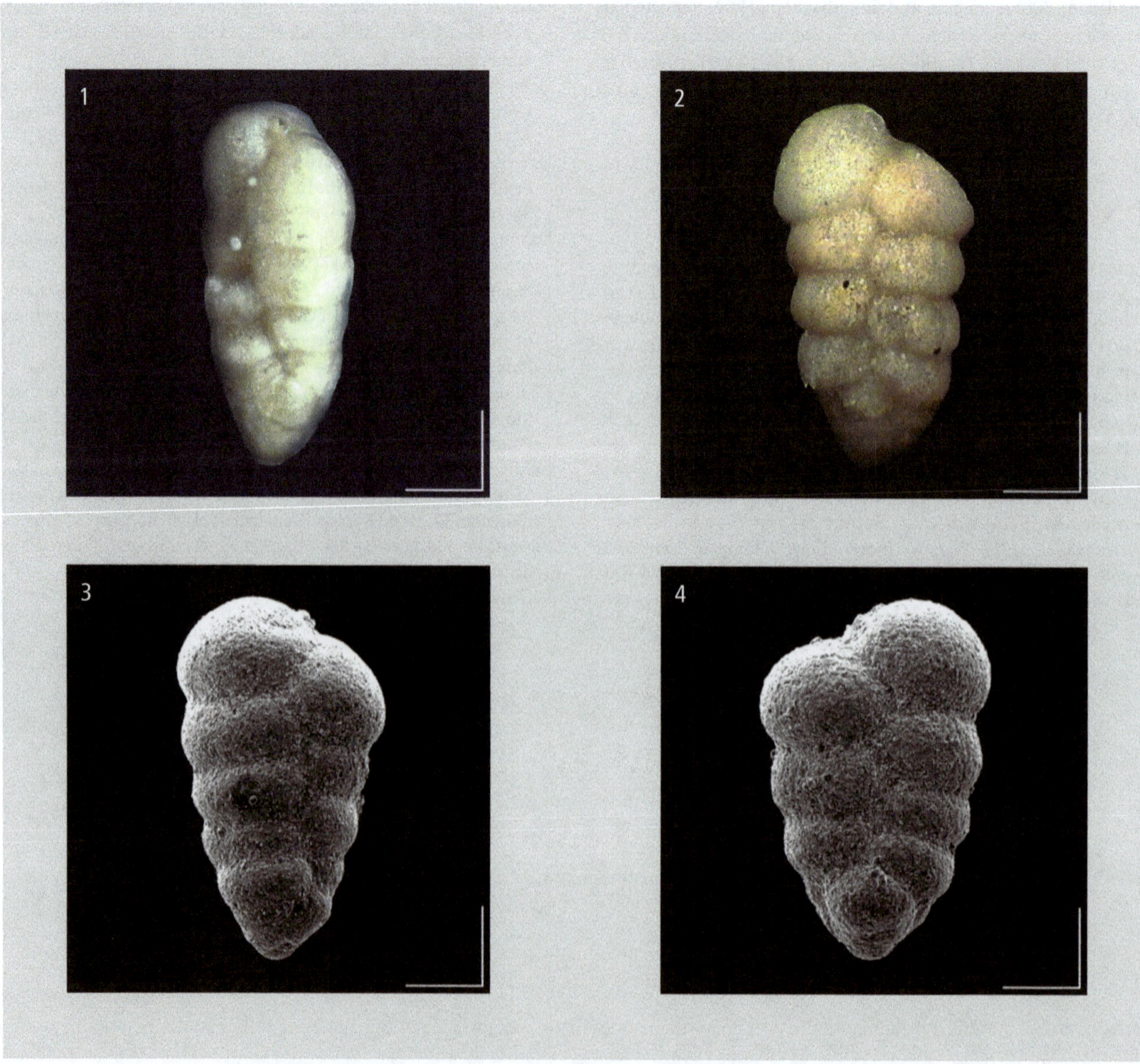


1. *Karreriella bradyi*. Off the Galapagos Islands; "Albatross" Station D2806; depth: 1379 fathoms. Deposited in the collections of the United States National Museum, Washington, DC; USNM 26307. Holotype. Lateral view, NHMUK PM DI 0410, scale bar = 158 μm.

2. *Karreriella bradyi*. Hebrides slope, Northeast Atlantic. Barra Fan, Seep Study. R.W. Jones Collection. Lateral view, NHMUK PM DI 0411, scale bar = 191 μm.

3. *Karreriella bradyi*. Northeast continental slope of the South China Sea; ODP Leg 184, Hole 1148A, Core 75, Section 3, 75–80 cm. Lateral view, NHMUK PM DI 0412, scale bar = 116 μm.

4. *Karreriella bradyi*. Northeast continental slope of the South China Sea; ODP Leg 184, Hole 1148A, Core 75, Section 3, 75–80 cm. Lateral view, NHMUK PM DI 0413, scale bar = 109 μm.

## Original Designation

*Gaudryina bradyi* Cushman, 1911, p. 67, fig. 107.

## Type Specimen

Holotype deposited in the collections of the United States National Museum, Washington, DC.

## Type Level

Recent.

## Type Locality

Not designated. Localities given: "Challenger" stations in 1850 and 2050 fathoms and "Albatross" stations in the vicinity of the Hawaian and Galapagos Islands, Guam, off Japan, in average depth of 1000 fathoms.

## Description

Test elongate, cylindrical, or slightly flaring in outline, trochospiral in early stage, with up to five chambers per whorl, later reduced to triserial and becoming biserial in adult portion. Chamber walls are finely agglutinated with calcareous cement. The aperture is areal, surrounded by a distinct lip. It is a rounded opening slightly above the base of the apertural face in the trochospiral stage, and is subterminal in the biserial adult.

## Synonyms and Citations

*Gaudryina pupoides* d'Orbigny. –Brady, 1884, p. 378, pl. 46, figs. 1–4.

*Gaudryina bradyi* Cushman. –Cushman, 1921, p. 149, pl. 29, fig. 3.

*Karreriella bradyi* (Cushman). –Barker, 1960, pl. 46, figs. 1–4. –Schröder, 1986, p. 55, pl. 22, figs. 8–9. –Zheng, 1988, p. 94, pl. 45, fig. 10; pl. 46, fig. 1; pl. 54, fig. 6; textfig. 21. –van Marle, 1991, p. 235, pl. 25, figs. 2–4. –Loeblich and Tappan, 1994, p. 25, pl. 30, figs. 8–16. –Jones, 1994, p. 50, pl. 46, figs. 1–4. –Robertson, 1998, p. 26, pl. 4, fig. 1. –Hayward, 2002, p. 288, pl. 2, fig. 1. –Kuhnt et al., 2002, p. 130, pl. 5. figs. 1–2, 14.

## Biogeography

Worldwide.

## Bathymetry

Bathyal to abyssal, above CCD. Typically occurs in deep-water, oligotrophic environments (Kuhnt et al., 1999).

## Chronostratigraphy

Paleocene to Recent.

## References


Barker, R.W., 1960, Taxonomic notes on the species figured by H.B. Brady in his report on the foraminifera dredged by H.M.S. Challenger during the years 1873–1876. Society of Economic Paleontologists and Mineralogists Special Publication, vol. 9, pp. 1–238.

Brady, H.B., 1884, Report on the foraminifera dredged by H.M.S. Challenger during the years 1873–1876. In Brady, H.B., ed., Report on the Scientific Results of the Voyage of the H.M.S. Challenger during the years 1873–1876. Zoology, vol. 9, pp. 1–814.

Cushman, J.A., 1911, A monograph of the Foraminifera of the North Pacific Ocean. Part II. Textulariidae. Bulletin of the United States National Museum, vol. 71, no. 2, pp. 1–108.

Cushman, J.A., 1921, Foraminifera of the Phillipine and adjacent seas. Bulletin of the United States National Museum, vol. 100, pp. 1–608.

Hayward, B., 2002, Late Pliocene to Middle Pleistocene extinctions of deep-sea benthic foraminifera ("Stilostomella extinction") in the southwest Pacific. Journal of Foraminiferal Research, vol. 32, pp. 274–307.

Jones, R.W., 1994, The Challenger Foraminifera. Oxford: Oxford University Press, 149 pp.

Loeblich, A.R. Jr and Tappan, H., 1994, Foraminifera of the Sahul Shelf and Timor Sea. Special Publications of the Cushman Laboratory for Foraminiferal Research, vol. 31, pp. 1–661.

Kuhnt, W., Hess, S., and Jian, Z., 1999, Quantitative composition of benthic foraminiferal assemblages as a proxy indicator for organic carbon flux rates in the South China Sea. In Sarnthein, M. and Wang, P.X., ed., Response of West Pacific Marginal Seas to Global Climate Change. Marine Geology, vol. 156, pp. 123–157.

Kuhnt, W., Holbourn, A.E., and Zhao, Q., 2002, The early history of the South China Sea: Evolution of Oligocene–Miocene deep water environments. Revue de Micropaléontologie, vol. 45, pp. 99–159.

Robertson, B.E., 1998, Systematics and paleoecology of the benthic foraminiferida from the Buff Bay section, Miocene of Jamaica. Micropaleontology, vol. 44, supplement 2, pp. 1–266.

Schröder, C.J., 1986, Deep-Water Arenaceous Foraminifera in the Northwest Atlantic Ocean. Canadian Technical Report of Hydrography and Ocean Sciences No. 71. Atlantic Geoscience Centre, Bedford Institute of Oceanography, 191 pp.

van Marle, L.J., 1991, Eastern Indonesian Late Cenozoic smaller benthic foraminifera. Verhandelingen der Koninklijke Nederlanse Akademie van Wetenschappen, Afdeeling Natuurkunde, Eerste Reeks, vol. 34, pp. 1–328.

Zheng, S.Y., 1988, The Agglutinated and Porcelaneous Foraminifera of the East China Sea. Beijing: China Ocean Press, 337 pp.

# Karreriella novangliae

## (Cushman), 1922

1. *Karreriella novangliae.* West of Ireland, North Atlantic Ocean; "Porcupine" Station 28, depth: 1215 fathoms. Deposited in the micropaleontological collections of The Natural History Museum, London; NHMUK ZF1448. Lateral view, NHMUK PM DI 0414, scale bar = 499 µm.

### Original Designation

*Gaudryina baccata* Schwager var. *novangliae* Cushman, 1922, no. 104, p. 76, pl. 13, fig. 4.

### Type Specimen

Deposited in the United States National Museum, Washington, DC, Catalog Nos 16701–16710 (type), 16711–16722.

### Type Level

Recent.

### Type Locality

"Albatross" Station D2105 (37° 50′ 00″ N; 73° 03′ 50″ W), 1395 fathoms, North Atlantic Ocean.

### Description

Test elongate, slightly compressed and flaring, initially trochospiral, becoming triserial, then biserial in the adult stage. Moderately inflated chambers, increasing gradually in size, are separated by distinct, depressed, curved sutures. Chamber walls are finely agglutinated with calcareous cement. The primary aperture is an elongate slit, close to the base of the final chamber, bordered by a lip.

### Synonyms and Citations

*Gaudryina baccata* Schwager. –Brady, 1884, p. 379, pl. 46, figs. 8–11.

*Meidamonella novangliae* (Cushman). –Loeblich and Tappan, 1987, p. 171, pl. 190, figs. 10–12.

*Karreriella novangliae* (Cushman). –Barker, 1960, pl. 46, figs. 8–11. –Jones, 1994, p. 51, pl. 46, figs. 8–11.

### Remarks

*Gaudryina baccata* Schwager var. *novangliae* Cushman, 1922, is the type species by original designation of the genus *Meidamonella* Loeblich and Tappan, 1986, regarded by Jones (1994) as a junior synonym of *Karreriella* Cushman, 1933.

Barker (1960) suggested that in view of the locality the specimen from "Challenger" Station 174C, Fiji (fig. 11) probably did not belong to the same species as the Atlantic specimens (figs. 8–10). However, Jones (1994) referred all figured specimens to *Karreriella novangliae* (Cushman).

### Biogeography

Worldwide.

### Bathymetry

Primarily bathyal. Reported between 210 and 2300 fathoms at "Challenger" stations (Jones, 1994).

### Chronostratigraphy

Pleistocene to Recent (Jones, 1994).

## References


Barker, R.W., 1960, Taxonomic notes on the species figured by H.B. Brady in his report on the foraminifera dredged by H.M.S. Challenger during the years 1873–1876. Society of Economic Paleontologists and Mineralogists Special Publication, vol. 9, pp. 1–238.

Brady, H.B., 1884, Report on the foraminifera dredged by H.M.S. Challenger during the years 1873–1876. In Brady, H.B., ed., Report on the Scientific Results of the Voyage of the H.M.S. Challenger during the years 1873–1876. Zoology, vol. 9, pp. 1–814.

Cushman, J.A., 1922, The foraminifera of the Atlantic Ocean; Part III – Textulariidae. Bulletin of the United States National Museum, vol. 104, pp. 1–149.

Jones, R.W., 1994, The Challenger Foraminifera. Oxford: Oxford University Press, 149 pp.

Loeblich, A.R. Jr and Tappan, H., 1987, Foraminiferal Genera and Their Classification. New York: Van Nostrand Reinhold, 970 pp.

# *Karrerulina conversa*

## (Grzybowski), 1901

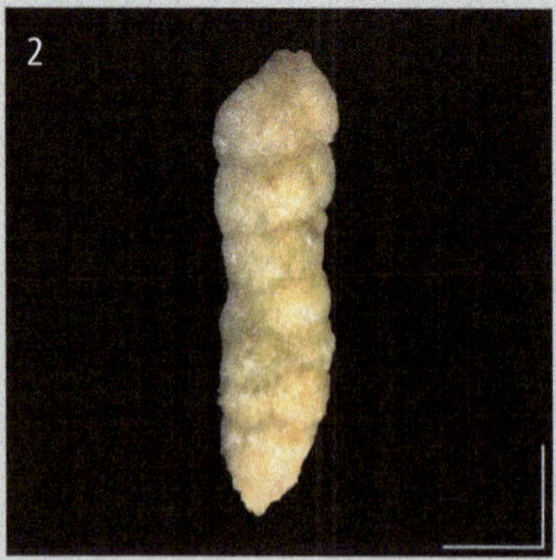


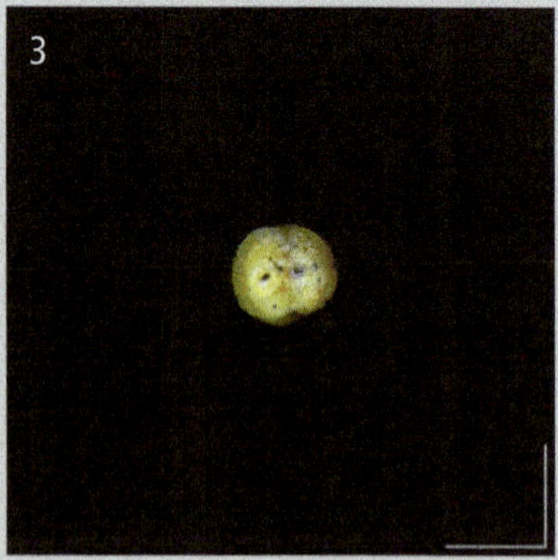


1. *Karrerulina conversa.* South Atlantic Ocean; "Challenger" Station 346, depth: 2350 fathoms. Deposited in the micropaleontological collections of The Natural History Museum, London; NHMUK ZF1461. Lateral view, NHMUK PM DI 0415, scale bar = 217 μm.

2. *Karrerulina conversa.* Carpathians, Poland. Deposited in the collections of the United States National Museum, Washington, DC. Lateral view, NHMUK PM DI 0416, scale bar = 210 μm.

3. *Karrerulina conversa.* South Atlantic Ocean; "Challenger" Station 346, depth: 2350 fathoms. Deposited in the micropaleontological collections of The Natural History Museum, London; NHMUK ZF1461. Apertural view, NHMUK PM DI 0417, scale bar = 217 μm.

### Original Designation

*Gaudryina conversa* Grzybowski, 1901, p. 285, pl. 7, figs. 15–16.

### Type Specimen

Not originally designated (see Kaminski and Gradstein, 2005). Lectotype (No. UJ-133-P, 2/105b) designated by Kaminski et al. (1993), and deposited in the Grzybowski Collection, Jagiellonian University, Kraków, Poland.

### Type Level

Eocene.

### Type Locality

Neotype locality: red clays in a stream bed in the village of Bodaki, Magura Unit of the Polish Carpathians, Poland (Kaminski and Gradstein, 2005).

### Description

Test elongate and initially trochospiral, then becoming triserial, and finally biserial. The trochospiral portion, which makes up about one-third of the test, has three to four whorls of four or five chambers and a circular cross-section. The triserial portion generally consists of one to two whorls or may be absent. The biserial portion is of variable length, somewhat flattened with subparallel or lobulate outline. Chambers increase rapidly in size and sutures are indistinct in the multiserial portion. Chambers are low, increasing slowly in size with distinct, fine, and depressed sutures in the biserial stage. The final chamber is often distinctly inflated. Chamber walls are agglutinated with fine to medium quartz grains and abundant silicified cement. The primary aperture is terminal, round, and at the end of a short neck.

### Synonyms and Citations

*Gaudryina apicularis* Cushman, 1911, p. 68, fig. 110.

*Karreriella aegra* Finlay, 1940, p. 451, pl. 62, figs. 21–22, 25–26.

*Plectina apicularis* (Cushman). –Phleger et al., 1953, p. 27, pl. 5, fig. 10.

*Karreriella indigena* Mjatliuk, 1970, p. 116, pl. 34, figs. 10–14.

*Plectina conversa* (Grzybowski). –Krasheninnikov and Pflaumann, 1978, p. 569, pl. 3, figs. 4a–b. –Moullade et al., 1988, p. 365, pl. 9, figs. 1–3.

*Plectina* aff. *conversa* (Grzybowski). –Hemleben and Tröster, 1984, p. 521, pl. 4, fig. 24.

*Karrerulina apicularis* (Cushman). –Loeblich and Tappan, 1987, p. 130, pl. 139, figs. 7–9.

*Karreriella conversa* (Grzybowski).–Kaminski et al., 1988, p. 196, pl. 9, figs. 17–18b.

*Gerochammina conversa* (Grzybowski). –Kaminski and Geroch, 1993, p. 279, pl. 13, figs. 5–11.

*Karrerulina conversa* (Grzybowski). –Charnock and Jones, 1990, p. 195, pl. 12, fig. 19; pl. 25, fig. 10. –Jones, 1994, p. 51, pl. 46, figs. 17–19. –Kaminski and Gradstein, 2005, pp. 468–472, fig. 116, pl. 116.

### Remarks

Type species of the genus *Karrerulina* Finlay, 1940, by synonymy with *Gaudryina apicularis* Cushman.

### Biogeography

Worldwide. See Observed Occurrences in Kaminski and Gradstein (2005).

### Bathymetry

Bathyal to abyssal. Infaunal species, generally occurring in oligotrophic deep-water settings with well-oxygenated bottom and interstitial waters (Kuhnt et al., 2000).

### Chronostratigraphy

Late Cretaceous to Recent. Predominantly Paleocene to Oligocene (Kaminski and Gradstein, 2005).

## References


Charnock, M.A. and Jones, R.W., 1990, Agglutinated foraminifera from the Palaeogene of the North Sea. In Hemleben, C., Kaminski, M.A., Kuhnt, W., and Scott, D.B., ed., Paleoecology, Biostratigraphy, Paleoceanography and Taxonomy of Agglutinated Foraminifera, NATO ASI Series, C: Mathematical and Physical Sciences Dordrecht: Kluwer Academic, pp. 139–244.

Cushman, J.A., 1911, A monograph of the Foraminifera of the North Pacific Ocean. Part II. Textulariidae. Bulletin of the United States National Museum, vol. 71, no. 2, pp. 1–108.

Finlay, H.J., 1940, New Zealand Foraminifera; key species in stratigraphy – No. 4. Transactions of the Royal Society of New Zealand, vol. 69, no. 4, pp. 448–472.

Grzybowski, J., 1901, Otwornice warstw inoceramowych okolicy Gorlic. Rozprawy Wydziału Matematyczno-Przyrodniczego, Akademia Umiejętności w Krakowie, serya 2, vol. 41, pp. 219–286.

Hemleben, C. and Tröster, J., 1984, Campanian–Maestrichtian deep-water foraminifers from Hole 543A, Deep Sea Drilling Project. In Bij-Duval, B., Moore, J.C., et al., ed., Initial Reports of the Deep Sea Drilling Project, Washington (U.S. Government Printing Office), vol. 78A, pp. 509–532.

Jones, R.W., 1994, The Challenger Foraminifera. Oxford: Oxford University Press, 149 pp.

Kaminski, M.A. and Geroch, S., 1993, A revision of foraminiferal species in the Grzybowski Collection. In Kaminski, M.A., Geroch, S., and Kaminski, D.G., ed., The Origins of Applied Micropalaeontology: The School of Jósef Grzybowski (with a taxonomic revision of the Grzyboswski Collection). Grzybowski Foundation Special Publication, no. 1, pp. 239–330.

Kaminski, M.A. and Gradstein, F.M., 2005, Atlas of Palaeogene Cosmopolitan Deep-Water Agglutinated Foraminifera. Grzybowski Foundation Special Publication, no. 10, 547 + vii pp.

Kaminski, M.A., Geroch, S., and Kaminski, D.G., 1993, The Origins of Applied Micropalaeontology: The School of Jósef Grzybowski (with a taxonomic revision of the Grzyboswski Collection). Grzybowski Foundation Special Publication, no. 1, pp. 1–336.

Kaminski, M.A., Gradstein, F.M., Berggren, W.A., Geroch, S., and Beckmann, J.P., 1988, Agglutinated foraminiferal assemblages from Trinidad: Taxonomy, Stratigraphy and Paleobathymetry. In Gradstein, F.M. and Rögl, F., ed., Second International Workshop on Agglutinated Foraminifera, Vienna 1986. Abhandlungen der Geologischen Bundesanstalt, vol. 41, pp. 155–228.

Krasheninnikov, V.A. and Pflaumann, U., 1978, Cretaceous agglutinated foraminifera of the Atlantic Ocean off West Africa (Leg 41, Deep Sea Drilling Project). In Lancelot, Y., Seibold, E., et al., ed., Initial Reports of the Deep Sea Drilling Project, Washington (U.S. Government Printing Office), vol. 41, pp. 565–580.

Kuhnt, W., Collins, C., and Scott, D.B., 2000, Deep water agglutinated foraminiferal assemblages across the Gulf Stream: distribution patterns and taphonomy. In Hart, M., Kaminski, M., and Smart, C.W., ed., Proceedings of the 5th international workshop on Agglutinated Foraminifera. Grzybowski Foundation Special Publication, no. 7, pp. 261–298.

Loeblich, A.R. Jr and Tappan, H., 1987, Foraminiferal Genera and Their Classification. New York: Van Nostrand Reinhold, 970 pp.

Mjatliuk, E.V., 1970, Foraminifery flishevykh otlozhenii vostochnykh Karpat (Mel – Paleogen). Trudy Vsesoyuznogo Nauchno-Issledovatel'skogo Geologorazvedochnogo Instituta VNIGRI (New Series), vol. 282, pp. 1–360.

Moullade, M., Kuhnt, W., and Thurow, J., 1988, Agglutinated benthic foraminifers from Upper Cretaceous variegated clays of the North Atlantic Ocean (DSDP Leg 93 and ODP Leg 103). In Boillot, G., Winterer, E.L., et al., ed., Scientific Results of the Proceedings of the Ocean Drilling Program, vol. 103, pp. 349–377.

Phleger, F.B., Parker, F.L., and Pierson, J.F., 1953, North Atlantic foraminifera. Reports of the Swedish Deep-Sea Expedition, 1947–1948, vol. 7, pp. 3–122.

# *Lagena sulcata*

## (Walker and Jacob), 1798

1. *Lagena sulcata*. Off Ki Islands, Central Pacific Ocean; "Challenger" Station 191A, depth: 580 fathoms. Deposited in the micropaleontological collections of The Natural History Museum, London; NHMUK ZF1785. Lateral view, NHMUK PM DI 0419, scale bar = 118 μm.

### Original Designation

*Serpula (Lagena) sulcata* Walker and Jacob, 1798, p. 634, pl. 14, fig. 5.

### Type Specimen

Specimens illustrated by Brady (1884) are housed in the micropaleontological collections of The Natural History Museum, London, UK, and are registered in slides ZF1783, ZF1785, ZF1786, and ZF1787.

### Type Level

Recent, shore sand.

### Type Locality

Not designated. Localities given: Sandwich, Reculver, and Shepey [Sheppey] Island, Kent, England.

### Description

Test unilocular, globular, ovate in outline, circular in cross-section. Chamber walls are calcareous, ornamented with prominent longitudinal costae. The primary aperture is terminal, round, and at the end of a long or short neck.

### Synonyms and Citations

*Lagena sulcata* (Walker and Jacob). –Michael, 1967, p. 76, pl. 4, fig. 36. –Neagu, 1975, pl. 69, figs. 18–19. –Loeblich and Tappan, 1987, p. 415, pl. 455, figs. 12–13. –Weidich, 1990, p. 122, pl. 40, figs. 25–26; pl. 45, figs. 6, 14. –Jones, 1994, p. 64, pl. 57, figs. 23, 25–27, 33–34; p. 65, pl. 58, figs. 5–6, 10?, 18.

### Remarks

Specimens in the "Challenger" Collection were referred to several varietal forms by various authors including Brady (1884). However, Jones (1994) did not maintain any formal distinction, as there appears to be a complete intergradation between forms.

### Biogeography

Worldwide.

### Bathymetry

Neritic to bathyal. Reported between 20 and 2350 fathoms at "Challenger" stations in the Pacific (Jones, 1994).

### Chronostratigraphy

Early Cretaceous to Recent. Cretaceous (Neagu, 1975; Weidich, 1990; Holbourn and Kaminski, 1997). Miocene to Recent (Jones, 1994).

## References

Brady, H.B., 1884, Report on the foraminifera dredged by H.M.S. Challenger during the years 1873–1876. In Brady, H.B., ed., Report on the Scientific Results of the Voyage of the H.M.S. Challenger during the years 1873–1876. Zoology, vol. 9, pp. 1–814.

Holbourn, A.E.L. and Kaminski, M.A., 1997, Cretaceous deep-water benthic foraminifera of the Indian Ocean. Grzybowski Foundation Special Publication, no. 4, pp. 1–175.

Jones, R.W., 1994, The Challenger Foraminifera. Oxford: Oxford University Press, 149 pp.

Loeblich, A.R. Jr and Tappan, H., 1987, Foraminiferal Genera and Their Classification. New York: Van Nostrand Reinhold, 970 pp.

Michael, E., 1967, Die Mikrofauna der nordwestdeutschen Barreme. Teil I – Die Foraminiferen des NW-deutschen Barreme. Paleontographica Supplement, vol. 12, pp. 1–176.

Neagu, T., 1975, Monographie de la faune des Foraminifères Éocrétacés du couloir de Dîmbovicioara, de Codlea et des Monts Persani (Couches de Carhaga). Memorii Institutul de Geologie si Geofizica, Bucaresti, vol. 35, pp. 1–141.

Walker, G. and Jacob, E., 1798, Descriptions of Serpula (Lagena) sulcata and Nautilus lobatulus. In Adams, E., ed., Essays on the Microscope, 2nd edn with considerable additions and improvements by F. Kanmacher. London: Dillon & Keating, pp. 634 and 642.

Weidich, K., 1990, Die kalkalpine Unterkreide und ihre Foraminiferenfauna. Zitteliana, vol. 17, pp. 1–312.

# *Latibolivina subreticulata*

## (Parr) 1932

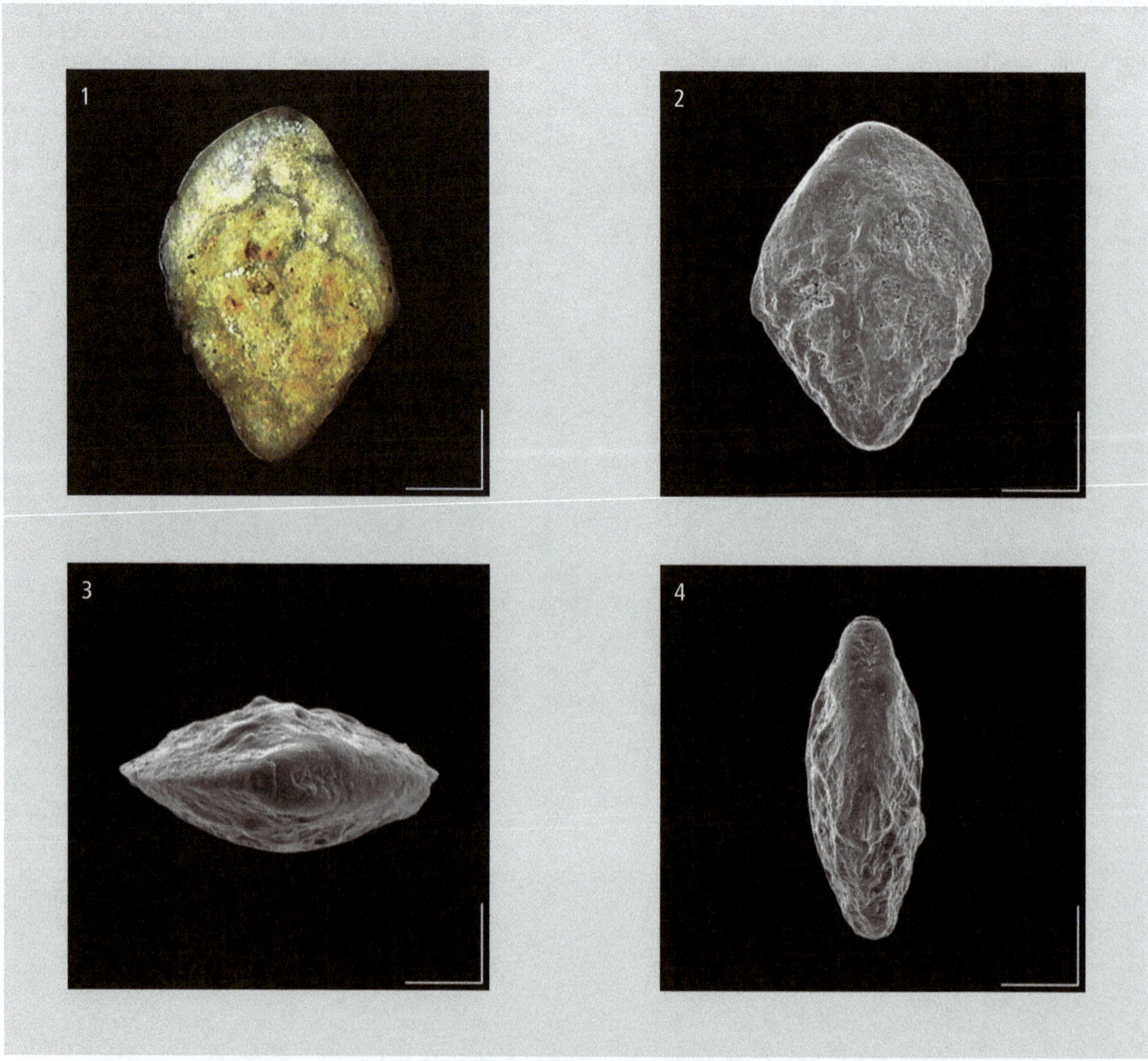


1. *Latibolivina subreticulata*. New Hebrides, South Pacific Ocean; "Challenger" Station 177, depth: 125 fathoms. Deposited in the micropaleontological collections of The Natural History Museum, London; NHMUK ZF1193. Lateral view, NHMUK PM DI 0420, scale bar = 64 μm.

2. *Latibolivina subreticulata*. New Hebrides, South Pacific Ocean; "Challenger" Station 177, depth: 125 fathoms. Deposited in the micropaleontological collections of The Natural History Museum, London; NHMUK ZF1193. Lateral view, NHMUK PM DI 0421, scale bar = 71 μm.

3. *Latibolivina subreticulata*. New Hebrides, South Pacific Ocean; "Challenger" Station 177, depth: 125 fathoms. Deposited in the micropaleontological collections of The Natural History Museum, London; NHMUK ZF1193. Apertural view, NHMUK PM DI 0422, scale bar = 53 μm.

4. *Latibolivina subreticulata*. New Hebrides, South Pacific Ocean; "Challenger" Station 177, depth: 125 fathoms. Deposited in the micropaleontological collections of The Natural History Museum, London; NHMUK ZF1193. Edge view, NHMUK PM DI 0423, scale bar = 71 μm.

## Original Designation
*Bolivina subreticulata* Parr, 1932, p. 12, pl. 1, fig. 21.

## Type Specimen
Holotype deposited in the Parr Collection. Brady's figured specimens are housed in the micropaleontological collections of The Natural History Museum, London, UK, and are registered in slide ZF1193.

## Type Level
Recent, surface sediments.

## Type Locality
"Challenger" Station 185, off Raine Island, Torres Strait, 155 fathoms. Other locality: Station 6, Point Lonsdale, Victoria, Australia. Brady's figured specimens are from "Challenger" Station 177, New Hebrides, South Pacific.

## Description
Test is elongate, biserial, rhomboidal, strongly compressed, and elliptical in cross-section, with an angular periphery. Slightly inflated, broad and low chambers, increasing slowly in height, are separated by curved, raised sutures, prolonged by short, irregular ribs forming a reticulate pattern in the final portion of the test. Wall is calcareous and finely perforate. Aperture is loop-shaped, bordered by a lip, with an internal toothplate.

## Synonyms and Citations
*Bolivina reticulata* Hantken. –Brady, 1884, p. 426, pl. 53, figs. 30–31.
*Bolivina subreticulata* Parr. –Barker, 1960, pl. 53, figs. 30–31.
*Brizalina subreticulata* (Parr). –Jones, 1994, p. 59, pl. 53, figs. 30–31. –van Marle, 1991, p. 173, pl. 18, figs. 1–3.
*Latibolivina subreticulata* (Parr). –Loeblich and Tappan, 1994, p. 112, pl. 217, figs. 1–11.

## Biogeography
Worldwide? Primarily recorded in the tropical Pacific and Indian Oceans.

## Bathymetry
Neritic to bathyal. Reported between 130 and 1570 fathoms at "Challenger" stations in the South Pacific and Southern Ocean (Jones, 1994).

## Chronostratigraphy
Miocene to Recent (Jones, 1994). Early Miocene to Recent (van Marle, 1991).

# References


Barker, R.W., 1960, Taxonomic notes on the species figured by H.B. Brady in his report on the foraminifera dredged by H.M.S. Challenger during the years 1873–1876. Society of Economic Paleontologists and Mineralogists Special Publication, vol. 9, pp. 1–238.

Brady, H.B., 1884, Report on the foraminifera dredged by H.M.S. Challenger during the years 1873–1876. In Brady, H.B., ed., Report on the Scientific Results of the Voyage of the H.M.S. Challenger during the years 1873–1876. Zoology, vol. 9, pp. 1–814.

Jones, R.W., 1994, The Challenger Foraminifera. Oxford: Oxford University Press, 149 pp.

Loeblich, A.R. Jr and Tappan, H., 1994, Foraminifera of the Sahul Shelf and Timor Sea. Special Publications of the Cushman Laboratory for Foraminiferal Research, vol. 31, pp. 1–661.

Parr, W.J., 1932, Victorian and south Australian shallow-water foraminifera. Part I. Proceedings of the Royal Society of Victoria, vol. 44, pp. 1–14.

van Marle, L.J., 1991, Eastern Indonesian Late Cenozoic smaller benthic foraminifera. Verhandelingen der Koninklijke Nederlanse Akademie van Wetenschappen, Afdeeling Natuurkunde, Eerste Reeks, vol. 34, pp. 1–328.

# *Laticarinina pauperata*

## (Parker and Jones), 1865

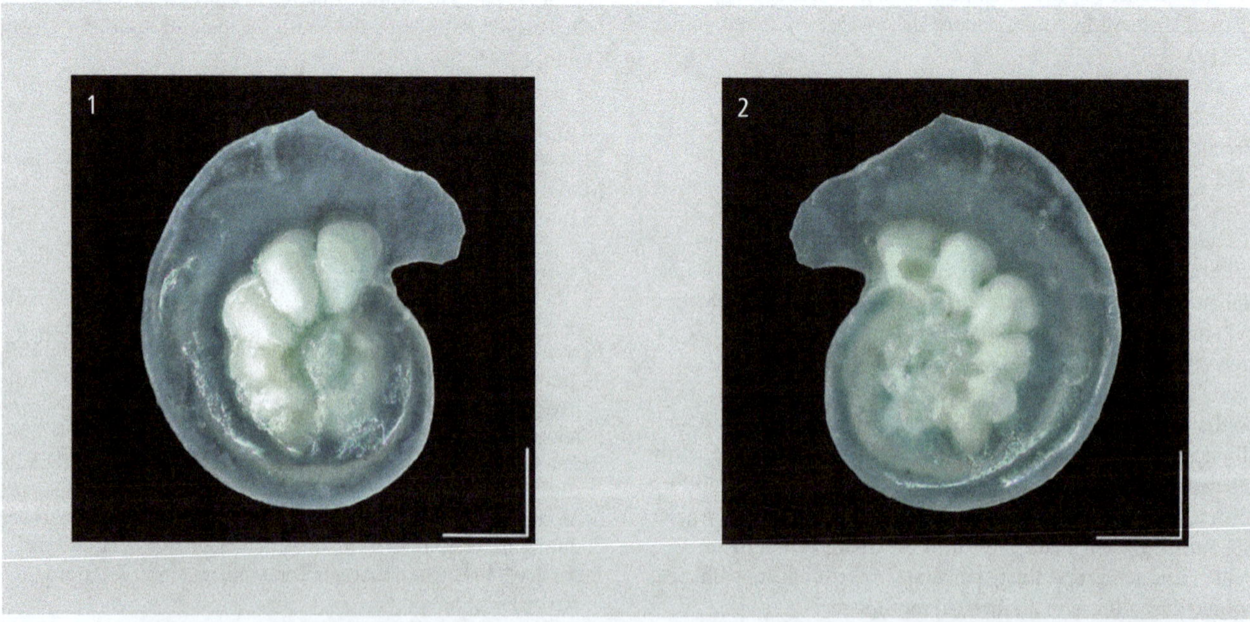

1. *Laticarinina pauperata*. South of Pico Fracture Zone, west flank of the Mid-Atlantic Ridge, North Atlantic; DSDP Leg 82, Site 558, Core 1, Section 2, 90–94 cm. Umbilical view, NHMUK PM DI 0424, scale bar = 142 μm.

2. *Laticarinina pauperata*. South of Pico Fracture Zone, west flank of the Mid-Atlantic Ridge, North Atlantic; DSDP Leg 82, Site 558, Core 1, Section 2, 90–94 cm. Spiral view, NHMUK PM DI 0425, scale bar = 140 μm.

### Original Designation

*Pulvinulina repanda* Fichtell and Moll var. *menardii* d'Orbigny, subvar. *pauperata* Parker and Jones, 1865, p. 395, pl. 16, figs. 50–51.

### Type Specimen

Depository not given.

### Type Level

Recent.

### Type Locality

Not designated. Localities given: 1450–2350 fathoms, between 52° 25′ N and 48° N, North Atlantic Ocean; from "Boral" tract toward Newfoundland Bank, latitude 50° 06′ N, longitude 48° 45′ W, North Atlantic; from "Celtic" tract, latitude 52° 16′ 30″ N, longitude 29° 28′ 30″ W, North Atlantic; 1080 fathoms, latitude 2° 20′ N, longitude 28° 45′ W, Equatorial Atlantic; 900 and 1120 fathoms, latitude 36° 58′ N, longitude 51° 49′ E, Indian Ocean.

### Description

Test strongly compressed, trochoid planoconvex in cross-section, with an evolute, flattened spiral side and an evolute, slightly convex umbilical side. Very broad, transparent, imperforate, peripheral keel projecting beyond the last chamber and showing sinuous growth lines. Twelve to thirteen slightly inflated chambers in the last whorl in megalospheric forms and up to 22 chambers in microspheric forms. Chambers become loosely coiled, cuneiform on the umbilical side and reniform on the spiral side, and are separated by sutures, radial and slightly depressed on the umbilical side and indistinct on the spiral side. Chamber walls are calcareous, smooth, and finely perforate on both sides of the test. Earlier chambers often have a thin lining of chitin. The primary aperture is a subequatorial slit, or a supplementary opening is present beneath the posterior umbilical margin.

### Synonyms and Citations

*Pulvinulina pauperata* (Parker and Jones). –Brady, 1884, p. 696, pl. 104, figs. 3–11.

*Laticarinina bulbrooki* Cushman and Todd, 1942, p. 19, pl. 4, figs. 8–9.

*Laticarinina crassicarinata* Cushman and Todd, 1942, p. 18, pl. 4, figs. 11–12.

*Laticarinina pauperata* (Parker and Jones). –Bermúdez, 1949, p. 309, pl. 23, figs. 43–45. –Phleger et al., 1953, p. 49, pl. 11, figs. 5–6. –van Morkhoven et al., 1986, p. 89, pl. 26. –Miller and Katz, 1987, p. 134, pl. 3, fig. 7. –Loeblich and Tappan, 1987, p. 578, pl. 631, figs. 1–4. –Jones, p. 104, pl. 104, figs. 3–11. –Robertson, 1998, p. 210, pl. 84, figs. 1–2.

### Remarks

Type species of the genus *Laticarinina* Galloway and Wissler, 1927.

### Biogeography

Worldwide. Recorded from the North and South Atlantic, Gulf of Mexico, Caribbean, Pacific, Indian Ocean, Arabian, Sea, Tasman Sea, Ross Sea, Weddell Sea, Scotia Sea, Mediterranean, Europe, and India (van Morkhoven et al., 1986).

### Bathymetry

Bathyal to abyssal.

### Chronostratigraphy

Early Oligocene (Rupelian) to Recent. Doubtful in Oligocene Biozone P18 (van Morkhoven et al., 1986).

## References


Brady, H.B., 1884, Report on the foraminifera dredged by H.M.S. Challenger during the years 1873–1876. In Brady, H.B., ed., Report on the Scientific Results of the Voyage of the H.M.S. Challenger during the years 1873–1876. Zoology, vol. 9, pp. 1–814.

Bermúdez, P.J., 1949, Tertiary smaller foraminifera of the Dominican Republic. Special Publications of the Cushman Laboratory for Foraminiferal Research, no. 25, pp. 1–322.

Cushman, J.A. and Todd, R., 1942, The Recent and fossil species of Laticarinina. Contributions from the Cushman Laboratory for Foraminiferal Research, vol. 18, no. 1, pp. 14–20.

Jones, R.W., 1994, The Challenger Foraminifera. Oxford: Oxford University Press, 149 pp.

Loeblich, A.R. Jr and Tappan, H., 1987, Foraminiferal Genera and Their Classification. New York: Van Nostrand Reinhold, 970 pp.

Miller, K.G. and Katz, M.E., 1987, Oligocene to Miocene benthic foraminiferal and abyssal circulation changes in the North Atlantic. Micropaleontology, vol. 33, no. 2, pp. 97–149.

Parker, W.K. and Jones, T.R., 1865, On some foraminifera from the North Atlantic and Arctic Oceans, including Davis Straits and Baffin's Bay. Philosophical Transactions of the Royal Society, vol. 155, pp. 325–441.

Phleger, F.B., Parker, F.L., and Pierson, J.F., 1953, North Atlantic foraminifera. Reports of the Swedish Deep-Sea Expedition, 1947–1948, vol. 7, pp. 3–122.

Robertson, B.E., 1998, Systematics and paleoecology of the benthic foraminiferida from the Buff Bay section, Miocene of Jamaica. Micropaleontology, vol. 44, supplement 2, pp. 1–266.

van Morkhoven, F.P.C.M., Berggren, W.A., Edwards, A.S., et al., 1986, Cenozoic cosmopolitan deep-water benthic foraminifera. Bulletin des Centres de Recherches Exploration-Production Elf-Aquitaine, Mémoire, vol. 11, pp. 1–421.

# Lenticulina anaglypta

## (Loeblich and Tappan), 1987

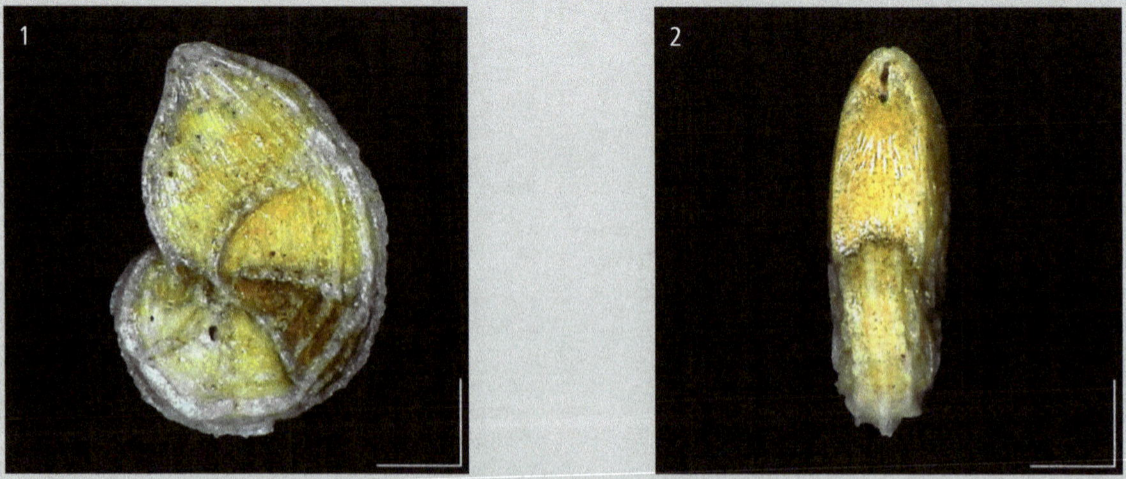


1. *Lenticulina anaglypta.* Off Raine Island, Torres Strait, Pacific Ocean. "Challenger" Station 185, depth: 155 fathoms. Deposited in the micropaleontological collections of The Natural History Museum, London; NHMUK ZF1313. Lateral view, NHMUK PM DI 0426, scale bar = 204 μm.

2. *Lenticulina anaglypta.* Off Raine Island, Torres Strait, Pacific Ocean. "Challenger" Station 185, depth: 155 fathoms. Deposited in the micropaleontological collections of The Natural History Museum, London; NHMUK ZF1313. Apertural view, NHMUK PM DI 0427, scale bar = 216 μm.

### Original Designation

*Nautilus costatus* Fichtel and Moll, 1798, p. 47, pl. 4, figs. g–i.

### Type Specimen

Holotype is the figured specimen, deposited in the Fichtel and Moll Collection, now in the micropaleontological collection, Department of Geology and Paleontology, Natural History Museum, Vienna, Austria (No. MI-480). Rögl and Hansen (1984) re-described and re-figured the holotype.

### Type Level

Recent.

### Type Locality

Moroccan Coast, Africa.

### Description

Test lenticular, planispiral, involute, subcircular in outline, compressed or biconvex in cross-section with keeled periphery. Moderately inflated chambers, increasing rapidly in size, are separated by slightly curved, depressed sutures. Chamber walls are calcareous, perforate, and ornamented by numerous prominent, raised costae, which may break up into rows or knobs. The primary aperture is radiate and terminal.

### Synonyms and Citations

*Robulina costata* (Fichtel and Moll). –d'Orbigny, 1826, p. 289, no. 13.

*Cristellaria costata* (Fichtel and Moll). –Brady, 1884, p. 555, pl. 71, fig. 9.

*Robulus costatus* (Fichtel and Moll). –Barker, 1960, pl. 71, fig. 9.

*Lenticulina costata* (Fichtel and Moll). –Rögl and Hansen, 1984, p. 38, pl. 9, figs. 1–2, textfig. 11.

*Spincterules anaglyptus* Loeblich and Tappan, 1987, p. 407, pl. 449, figs. 7–8.

*Lenticulina anaglypta* (Loeblich and Tappan). –Jones, 1994, p. 82, pl. 71, fig. 9.

### Remarks

*Nautilus costatus* Fichtel and Moll is a junior homonym of *Nautilus (Orthoceras) costatus* Batsch, 1791. It was renamed *Spincterules anaglyptus* by Loeblich and Tappan (1987). Jones (1994) regarded *Spincterules* de Montfort, 1808 as a junior synonym of *Lenticulina* Lamarck, 1804.

### Biogeography

Worldwide.

### Bathymetry

Neritic to bathyal.

### Chronostratigraphy

Miocene to Recent (Jones, 1994).

## References

Barker, R.W., 1960, Taxonomic notes on the species figured by H.B. Brady in his report on the foraminifera dredged by H.M.S. Challenger during the years 1873–1876. Society of Economic Paleontologists and Mineralogists Special Publication, vol. 9, pp. 1–238.

Brady, H.B., 1884, Report on the foraminifera dredged by H.M.S. Challenger during the years 1873–1876. In Brady, H.B., ed., Report on the Scientific Results of the Voyage of the H.M.S. Challenger during the years 1873–1876. Zoology, vol. 9, pp. 1–814.

d'Orbigny, A., 1826, Tableau méthodique de la classe des Céphalopodes. Annales des Sciences Naturelles, vol. 7, pp. 245–314.

Fichtel, L. von and Moll, J.P.C. von, 1798, Testacea microscopica aliaque minuta ex generibus Argonauta et Nautilus ad naturam delineata et descripta. (Microscopische und andere kleine Schalthiere aus den Geschlechtern Argonaute und Schiffer, nach der Natur gezeichnet und beschrieben). Vienna: Camesina, 124 pp.

Jones, R.W., 1994, The Challenger Foraminifera. Oxford: Oxford University Press, 149 pp.

Loeblich, A.R. Jr and Tappan, H., 1987, Foraminiferal Genera and Their Classification. New York: Van Nostrand Reinhold, 970 pp.

Rögl, F. and Hansen, J.H., 1984, Foraminifera described by Fichtel and Moll in 1798; A revision of Testacea Microscopica. Appendix Testacea Microscopica alique minuta ex Generibus Argonauta et Nautilus. Reprint of original plates. Neue Denkschriften des Naturhistorischen Museum in Wien, vol. 3, pp. 1–143.

# *Lenticulina convergens*

## (Bornemann), 1855

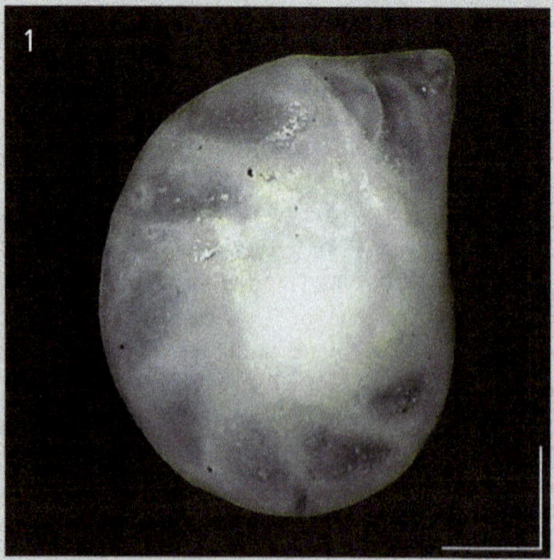

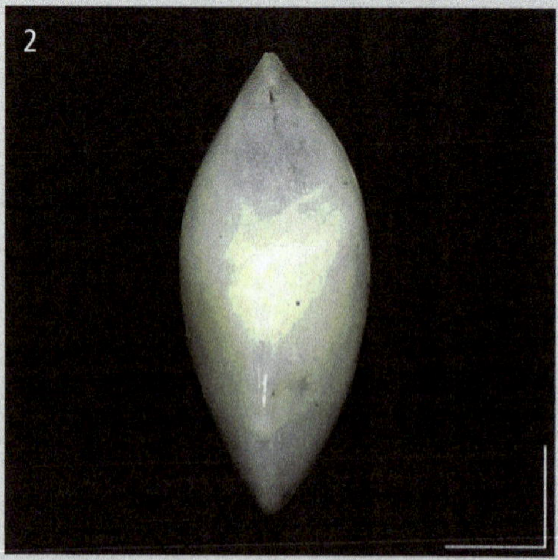

1. *Lenticulina convergens*. North Pacific Ocean; "Challenger" Station 224, depth: 1850 fathoms. Deposited in the micropaleontological collections of The Natural History Museum, London; NHMUK ZF1311. Umbilical view, NHMUK ZF DI 0369, scale bar = 175 μm.

2. *Lenticulina convergens*. North Pacific Ocean; "Challenger" Station 224, depth: 1850 fathoms. Deposited in the micropaleontological collections of The Natural History Museum, London; NHMUK ZF1311. Apertural view, NHMUK ZF DI 0370, scale bar = 187 μm.

### Original Designation

*Cristellaria convergens* Bornemann, 1855, p. 327, pl. 13, figs. 16–17.

### Type Specimen

Depository not given.

### Type Level

Oligocene, Septarienthon.

### Type Locality

Hermsdorf, near Berlin, Germany.

### Description

Test lenticular, planispiral, involute, subcircular in outline, biconvex in cross-section with subacute periphery. Numerous, moderately inflated chambers, increasing gradually in size, are separated by flush, oblique sutures fusing into a large, internal, umbilical boss. Chamber walls are calcareous, perforate, and smooth. The primary aperture is radiate and terminal.

### Synonyms and Citations

*Cristellaria convergens* Bornemann. –Brady, 1884, p. 646, pl. 69, figs. 6–7.

*Lenticulina convergens* (Bornemann). –Barker, 1960, pl. 69, figs. 6–7. –Hermelin, 1989, p. 44, pl. 5, figs. 7–8. –Jones, 1994, p. 80, pl. 69, figs. 6–7.

### Biogeography

Worldwide.

### Bathymetry

Neritic to abyssal. Reported between 16 and 2740 fathoms at "Challenger" stations (Jones, 1994).

### Chronostratigraphy

Oligocene to Recent.

## References

Barker, R.W., 1960, Taxonomic notes on the species figured by H.B. Brady in his report on the foraminifera dredged by H.M.S. Challenger during the years 1873–1876. Society of Economic Paleontologists and Mineralogists Special Publication, vol. 9, pp. 1–238.

Bornemann, J.G., 1855, Die mikroskopische Fauna des Septarienthones von Hermsdorf bei Berlin. Zeitschrift der Deutschen Geologischen Gesellschaft, vol. 7, pp. 307–371.

Brady, H.B., 1884, Report on the foraminifera dredged by H.M.S. Challenger during the years 1873–1876. In Brady, H.B., ed., Report on the Scientific Results of the Voyage of the H.M.S. Challenger during the years 1873–1876. Zoology, vol. 9, pp. 1–814.

Hermelin, J.O.R., 1989, Pliocene benthic foraminifera from the Ontong-Java Plateau (Western Equatorial Pacific Ocean): faunal response to changing paleoenvironment. Special Publication of the Cushman Foundation for Foraminiferal Research, vol. 26, pp. 1–143.

Jones, R.W., 1994, The Challenger Foraminifera. Oxford: Oxford University Press, 149 pp.

# Lenticulina gibba

## (d'Orbigny), 1839

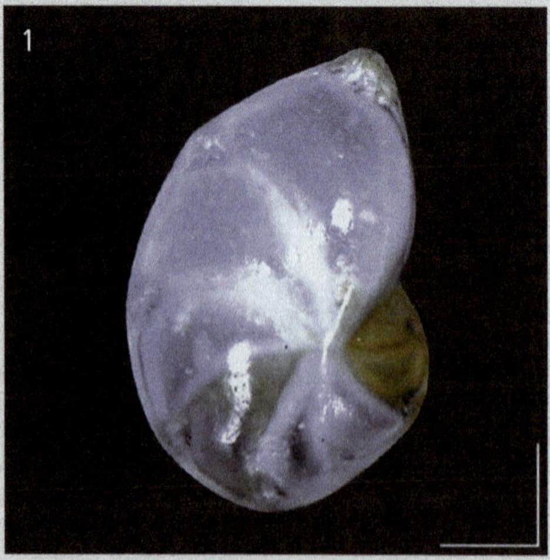

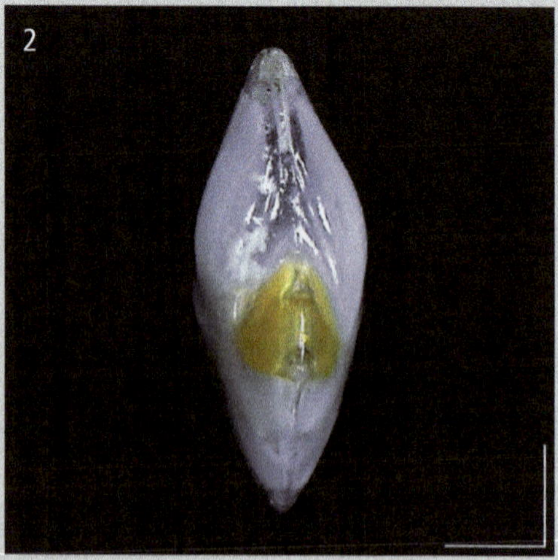

1. *Lenticulina gibba.* West Indies; "Challenger" Station 24, depth: 390 fathoms. Deposited in the micropaleontological collections of The Natural History Museum, London; NHMUK ZF1329. Umbilical view, NHMUK PM DI 0430, scale bar = 305 μm.

2. *Lenticulina gibba.* West Indies; "Challenger" Station 24, depth: 390 fathoms. Deposited in the micropaleontological collections of The Natural History Museum, London; NHMUK ZF1329. Apertural view, NHMUK PM DI 0431, scale bar = 308 μm.

### Original Designation
*Cristellaria gibba* d'Orbigny, 1839, p. 40, pl. 7, figs. 20–21.

### Type Specimen
Depository not given.

### Type Level
Recent, marine sands.

### Type Locality
Not designated. Localities given: Antilles Islands, Mediterranean.

### Description
Test involute, planispiral, elongate in outline, biconvex in cross-section with keeled, subacute periphery. Keel becomes narrower in later chambers. Approximately six to nine chambers in the final whorl, increasing gradually in size and tending slightly to uncoil, are separated by thin, flush or slightly raised, curved sutures. Chamber walls are calcareous, smooth, and finely perforate. The primary aperture is radiate and terminal.

### Synonyms and Citations
*Cristellaria gibba* d'Orbigny. –Brady, 1884, p. 546, pl. 69, figs. 8–9
*Robulus oblongus* Coryell and Rivero, 1940, p. 332, pl. 43, fig. 12.
*Robulus gibbus* (d'Orbigny). –Bermúdez, 1949, p. 126, pl. 7, figs. 53–54.
*Lenticulina gibba* (d'Orbigny). –Barker, 1960, pl. 69, figs. 8–9. –Jones, 1994, p. 81, pl. 69, figs. 8–9. –Robertson, 1998, p. 66, pl. 22, fig. 4.

### Biogeography
Worldwide.

### Bathymetry
Neritic to upper bathyal. Reported as less than 500 fathoms at "Challenger" stations (Jones, 1994).

### Chronostratigraphy
Early Miocene to Recent.

## References

Barker, R.W., 1960, Taxonomic notes on the species figured by H.B. Brady in his report on the foraminifera dredged by H.M.S. Challenger during the years 1873–1876. Society of Economic Paleontologists and Mineralogists Special Publication, vol. 9, pp. 1–238.

Bermúdez, P.J., 1949, Tertiary smaller foraminifera of the Dominican Republic. Special Publications of the Cushman Laboratory for Foraminiferal Research, no. 25, pp. 1–322.

Brady, H.B., 1884, Report on the foraminifera dredged by H.M.S. Challenger during the years 1873–1876. In Brady, H.B., ed., Report on the Scientific Results of the Voyage of the H.M.S. Challenger during the years 1873–1876. Zoology, vol. 9, pp. 1–814.

Coryell, H.N. and Rivero, F.C., 1940, A Miocene microfauna of Haiti. Journal of Paleontology, vol. 14, pp. 324–344.

d'Orbigny, A.D., 1839, Foraminifères. In Sagra, R. de la, ed., Histoire physique, politique et naturelle de l'île de Cuba. Paris: A. Bertrand, pp. 1–224, 12 pls.

Jones, R.W., 1994, The Challenger Foraminifera. Oxford: Oxford University Press, 149 pp.

Robertson, B.E., 1998, Systematics and paleoecology of the benthic foraminiferida from the Buff Bay section, Miocene of Jamaica. Micropaleontology, vol. 44, supplement 2, pp. 1–266.

# *Lenticulina iota*

## (Cushman), 1923

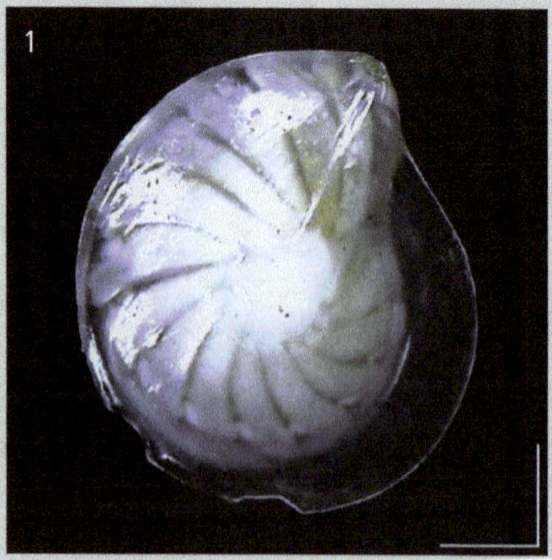

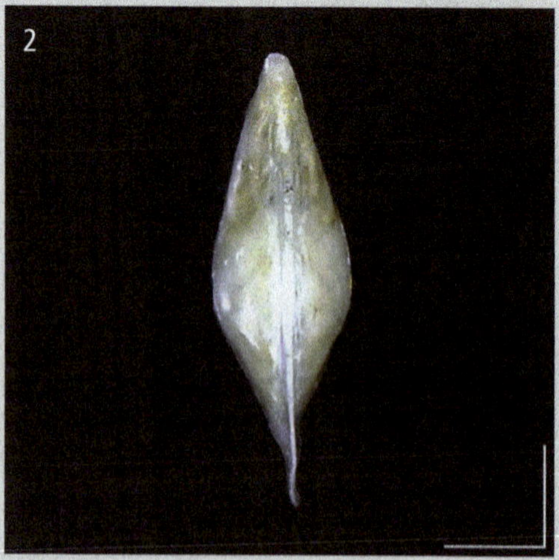

1. *Lenticulina iota.* Off Kandavu, Fiji; "Challenger" Station 174C, depth: 210 fathoms. Deposited in the micropaleontological collections of The Natural History Museum, London; NHMUK ZF1320. Umbilical view, NHMUK PM DI 0432, scale bar = 548 μm.

2. *Lenticulina iota.* Off Kandavu, Fiji; "Challenger" Station 174C, depth: 210 fathoms. Deposited in the micropaleontological collections of The Natural History Museum, London; NHMUK ZF1320. Apertural view, NHMUK PM DI 0433, scale bar = 544 μm.

### Original Designation

*Cristellaria iota* Cushman, 1923, p. 111, pl. 29, fig. 2; pl. 30, fig. 1.

### Type Specimen

Deposited in the United States National Museum, Washington, DC (Nos 18938, 18939, 18940).

### Type Level

Recent, surface sediments.

### Type Locality

"Albatross" Station D2399 (latitude 28° 44′ 00″ N, longitude 86° 18′ 00″ W), northern Gulf of Mexico.

### Description

Test large, involute, planispiral, compressed, subcircular in outline, and biconvex in cross-section, with a keeled, acute periphery and large clear internal umbos (covered by successive chamber lamellae). Keel initially broad, becomes narrower in later chambers. Approximately thirteen to fifteen chambers in the final whorl, increasing gradually in size, are separated by slightly limbate, curved sutures. Chamber walls are calcareous, smooth, or with slight transverse ridges, and finely perforate. The primary aperture is radiate and terminal.

### Synonyms and Citations

*Cristellaria cultrata* de Montfort. –Brady, 1884, p. 550, pl. 70, figs. 4–6.

*Lenticulina iota* (Cushman). –Barker, 1960, pl. 70, figs. 4–6. –Thomas, 1988, p. 74, pl. 1, fig. 10. –Jones, 1994, p. 81, pl. 70, figs. 4–6.

### Remarks

Thomas (1988, p. 74) emended the description of this taxon, including Neogene specimens from Canadian Atlantic margin wells with fewer chambers.

### Biogeography

Worldwide.

### Bathymetry

Neritic to upper bathyal.

### Chronostratigraphy

Miocene to Recent. Neogene in Canadian Atlantic margin wells (Thomas, 1988).

## References


Barker, R.W., 1960, Taxonomic notes on the species figured by H.B. Brady in his report on the foraminifera dredged by H.M.S. Challenger during the years 1873–1876. Society of Economic Paleontologists and Mineralogists Special Publication, vol. 9, pp. 1–238.

Brady, H.B., 1884, Report on the foraminifera dredged by H.M.S. Challenger during the years 1873–1876. In Brady, H.B., ed., Report on the Scientific Results of the Voyage of the H.M.S. Challenger during the years 1873–1876. Zoology, vol. 9, pp. 1–814.

Cushman, J.A., 1923, The foraminifera of the Atlantic Ocean; Part 4 – Lagenidae. Bulletin of the United States National Museum (US Nat. Mus. Bull.), vol. 104, no. 4, pp. 1–228.

Jones, R.W., 1994, The Challenger Foraminifera. Oxford: Oxford University Press, 149 pp.

Thomas, F.C., 1988, Taxonomy and stratigraphy of selected Cenozoic benthic foraminifera, Canadian Atlantic margin. Micropaleontology, vol. 34, no. 1, pp. 67–82.

# *Lenticulina muensteri*

## (Roemer), 1839

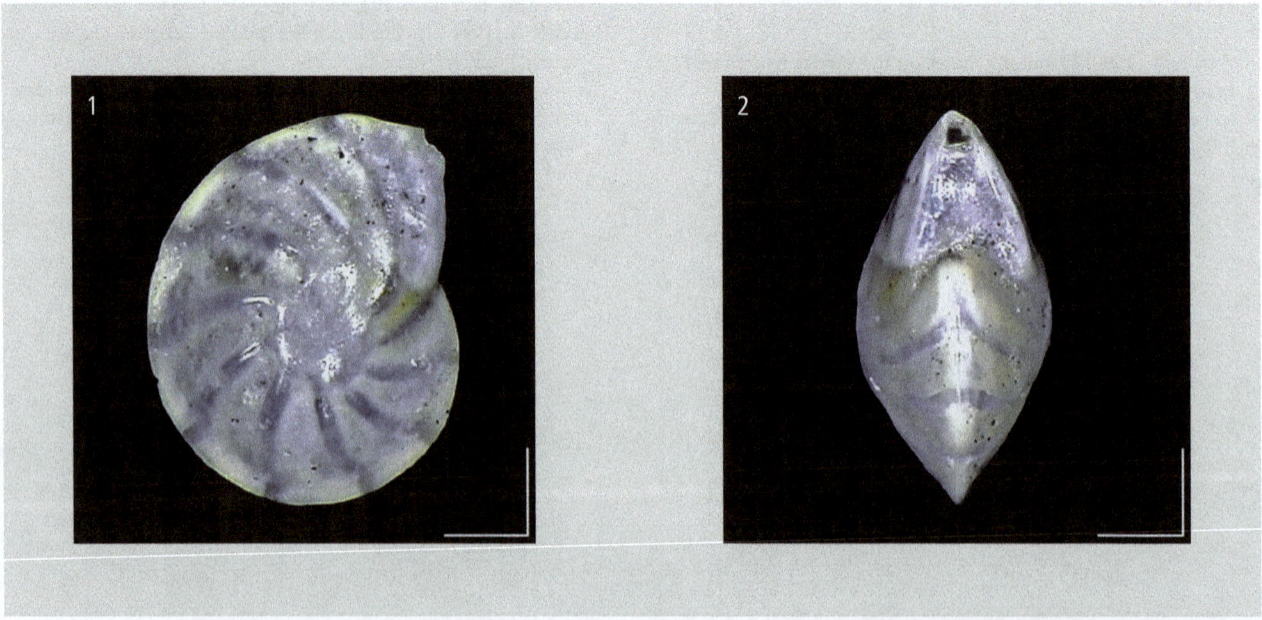


1. *Lenticulina muensteri*. Oxford Clay, Purton, Wiltshire, England. Deposited in the micropaleontological collections of The Natural History Museum, London; NHMUK P45240. Umbilical view, NHMUK PM DI 0434, scale bar = 158 μm.

2. *Lenticulina muensteri*. Oxford Clay, Purton, Wiltshire. Deposited in the micropaleontological collections of The Natural History Museum, London; NHMUK P45240. Apertural view, NHMUK PM DI 0435, scale bar = 174 μm.

### Original Designation
*Robulina muensteri* Roemer, 1839, p. 48, pl. 20, fig. 29.

### Type Specimen
Depository not given.

### Type Level
Cretaceous.

### Type Locality
Schöpenstedt, North Germany.

### Description
Test involute, lenticular or subcircular in outline, biconvex in cross-section with acute periphery. Moderately inflated chambers, increasing gradually in size, are separated by flush, or slightly raised, curved sutures, coalescing to form a slightly raised ring around an umbonal disc. Chamber walls are calcareous, perforate, and smooth. The primary aperture is radiate and terminal.

### Synonyms and Citations
*Lenticulina muensteri* (Roemer). –Jendryka-Fuglewicz, 1975, pls. 8–11, figs. 1–6; pl. 19; pl. 20, figs. 1–2. –Bartenstein and Bolli, 1986, pl. 4, figs. 25–26. –Meyn and Vespermann, 1994, pl. 23, figs. 12–17; pl. 24, figs. 1–17.

### Remarks
Jendryka-Fuglewicz (1975) distinguished six stratigraphically significant morphotypes in Jurassic to Albian deposits in Poland. Bartenstein and Bolli (1986) highlighted difficulties in separating *Lenticulina muensteri* (Roemer), *Lenticulina cultrata* (de Montfort), *Lenticulina rotulata* (Lamarck), *Lenticulina macrodisca* (Reuss), *Lenticulina roemeri* (Reuss), and *Lenticulina subalata* (Reuss) due to transitional forms. Meyn and Vespermann (1994) considered *Lenticulina ehrenbergi* (Roemer), *Lenticulina* crassa (Roemer), *Lenticulina impressa* (Reuss), and *Lenticulina macrodisca* (Reuss) to be synonyms of *Lenticulina muensteri* (Roemer).

### Biogeography
Worldwide.

### Bathymetry
Bathyal.

### Chronostratigraphy
Jurassic to Cretaceous.

## References

Bartenstein, H. and Bolli, H.M., 1986, The foraminifera of the Lower Cretaceous of Trinidad, W.I. Part 5: Maridale Formation, upper part; Hedbergella rohri zone. Eclogae Geologicae Helvetiae, vol. 7, no. 3, pp. 945–999.

Jendryka-Fuglewicz, B., 1975, Evolution of the Jurassic and Cretaceous smooth walled Lenticulina of Poland. Acta Palaeontologica Polonica, vol. 20, pp. 99–197.

Meyn, H. and Vespermann, J., 1994, Taxonomic revision of Lower Cretaceous Foraminifera described by Roemer (1839, 1841, 1842), Koch (1851), and Reuss (1863) from Lower Cretaceous Saxony. Senckenbergiana Lethaea, vol. 74, no. 1/2, pp. 49–272.

Roemer, F.A., 1839, Die Versteinerungen des norddeutschen Oolithen-Gebirges. Ein Nachtrag. Hannover: Hahn'schen Hofbuchhandlung, iv + 59 pp.

# *Lituotuba lituiformis*

## (Brady), 1879

1. *Lituotuba lituiformis.* Off Pernambuco, West Atlantic Ocean. "Challenger" Station 120, depth: 675 fathoms. Deposited in the micropaleontological collections of The Natural History Museum, London; NHMUK ZF2504. Syntype. Lateral view, NHMUK PM DI 0436, scale bar = 246 μm.

### Original Designation

*Trochammina lituiformis* Brady, 1879, p. 59, pl. 5, fig. 16.

### Type Specimen

Figured specimens are housed in the micropaleontological collections of The Natural History Museum, London, and are registered in slides ZF2503–ZF2505. Kaminski and Gradstein (2005) reported that the holotype is in slide ZF2505.

### Type Level

Recent.

### Type Locality

"Challenger" station 24, off Culebra Island, West Indies, 390 fathoms. Other figured specimens from stations off the Azores, 900 fathoms, and off Perambuco, Brazil, 675 fathoms.

### Description

Test elongate, irregularly coiled initially, becoming uncoiled and irregularly rectilinear. The coiled stage is undivided, later developing tubular chambers, separated by constrictions at irregular intervals in the rectilinear portion. Chamber walls are finely agglutinated, and smoothly finished with non-calcareous cement, often yellowish-brown in color. The primary aperture is terminal, round, and at the open end of the tube.

### Synonyms and Citations

*Trochammina lituiformis* Brady. –Brady, 1884, p. 342, pl. 40, figs. 4–7.
*Lituotuba lituiformis* (Brady). –Barker, 1960, pl. 40, figs. 4–7. –Loeblich and Tappan, 1987, p. 69, pl. 53, fig. 3. –Kaminski et al., 1988, p. 190, pl. 4, figs. 14–15. –Charnock and Jones, 1990, p. 160, pl. 3, fig. 11; pl. 15, fig. 1. –Jones, 1994, p. 44, pl. 40, figs. 4–7. –Kaminski and Gradstein, 2005, pp. 287–289, fig. 58, pl. 58.

### Remarks

Type species of the genus *Lituotuba* Rhumbler, 1895.

### Biogeography

Worldwide.

### Bathymetry

Bathyal to abyssal. Reported between 714 and 4000 meters in the North Atlantic (Charnock and Jones, 1990).

### Chronostratigraphy

Cretaceous to Recent (Kaminski and Gradstein, 2005).

## References


Barker, R.W., 1960, Taxonomic notes on the species figured by H.B. Brady in his report on the foraminifera dredged by H.M.S.

Challenger during the years 1873–1876. Society of Economic Paleontologists and Mineralogists Special Publication, vol. 9, pp. 1–238.

Brady, H.B., 1879, Notes on some of the reticularian Rhizopoda of the Challenger Expedition, Part I. On new or little known Arenaceous types. Quarterly Journal of Microscopical Science (New Series), vol. 19, pp. 20–63.

Brady, H.B., 1884, Report on the foraminifera dredged by H.M.S. Challenger during the years 1873–1876. In Brady, H.B., ed., Report on the Scientific Results of the Voyage of the H.M.S. Challenger during the years 1873–1876. Zoology, vol. 9, pp. 1–814.

Charnock, M.A. and Jones, R.W., 1990, Agglutinated foraminifera from the Palaeogene of the North Sea. In Hemleben, C., Kaminski, M.A., Kuhnt, W., and Scott, D.B., ed., Paleoecology, Biostratigraphy, Paleoceanography and Taxonomy of Agglutinated Foraminifera, NATO ASI Series, C: Mathematical and Physical Sciences. Dordrecht, Kluwer Academic, pp. 139–244.

Jones, R.W., 1994, The Challenger Foraminifera. Oxford: Oxford University Press, 149 pp.

Kaminski, M.A. and Gradstein, F.M., 2005, Atlas of Palaeogene Cosmopolitan Deep-Water Agglutinated Foraminifera. Grzybowski Foundation Special Publication, no. 10, 547 + vii pp.

Kaminski, M.A., Gradstein, F.M., Berggren, W.A., Geroch, S., and Beckmann, J.P., 1988, Agglutinated foraminiferal assemblages from Trinidad: taxonomy, stratigraphy and paleobathymetry. In Gradstein, F.M. and Rögl, F., ed., Second International Workshop on Agglutinated Foraminifera, Vienna 1986. Abhandlungen der Geologischen Bundesanstalt, vol. 41, pp. 155–228.

Loeblich, A.R. Jr and Tappan, H., 1987, Foraminiferal Genera and Their Classification. New York: Van Nostrand Reinhold, 970 pp.

# Loxostomina limbata

## (Brady), 1881

1. *Loxostomina limbata.* Admirality Islands, Pacific Ocean; "Challenger" Station 218A, depth: 16–25 fathoms. Deposited in the micropaleontological collections of The Natural History Museum, London; NHMUK ZF1183. Syntype. Lateral view, NHMUK PM DI 0437, scale bar = 146 μm.

2. *Loxostomina limbata.* Admirality Islands, Pacific Ocean; "Challenger" Station 218A, depth: 16–25 fathoms. Deposited in the micropaleontological collections of The Natural History Museum, London; NHMUK ZF1183. Syntype. Apertural view, NHMUK PM DI 0438, scale bar = 41 μm.

3. *Loxostomina limbata.* Admirality Islands, Pacific Ocean; "Challenger" Station 218A, depth: 16–25 fathoms. Deposited in the micropaleontological collections of The Natural History Museum, London; NHMUK ZF1183. Syntype. Lateral view, NHMUK PM DI 0439, scale bar = 123 μm.

### Original Designation
*Bulimina (Bolivina) limbata* Brady, 1881, p. 57 (type description), illustrated as *Bolivina limbata* in Brady, 1884, pl. 52, figs. 26–28.

### Type Specimen
Figured specimens are housed in the micropaleontological collections of The Natural History Museum, London, UK, and are registered in slides ZF1182–ZF1184.

### Type Level
Recent, surface sediments.

### Type Locality
Not designated. Figured specimens are from "Challenger" Station 260A off reefs Honolulu, Sandwich Islands, Hawaii, in 40 fathoms; "Challenger" Station 218A, Nares Harbour Admiralty Islands, in 16–25 fathoms; and "Challenger" Station 219, Nares Harbour, in 17 fathoms.

### Description
Test elongate, slightly compressed, fusiform or with subparallel sides, initially biserial, becoming uniserial, with oval cross-section. Chambers, increasing gradually in height, are separated by slightly depressed, overlapping sutures, oblique in the biserial part, straighter in the uniserial portion. Chamber walls are calcareous, perforate, and ornamented by fine, longitudinal costae. The primary aperture is terminal, elliptical, often slightly produced, with a lip and internal toothplate.

### Synonyms and Citations
*Bolivina limbata* Brady. –Brady, 1884, p. 419, pl. 52, figs. 26–28.

*Loxostomum limbatum* (Brady). –Barker, 1960, pl. 52, figs. 26–28.

*Loxostomella limbata* (Brady). –Saidova, 1975, p. pl. 86, fig. 4.

*Loxostomina limbata* (Brady). –Jones, 1994, p. 57, pl. 52, figs. 26–28.

### Remarks
Jones (1994) considered the genus *Loxostomella* Saidova, 1975, to be a junior synonym of *Loxostomina* Sellier de Civrieux, 1969.

### Biogeography
Worldwide.

### Bathymetry
Neritic to upper bathyal. Reported between 7 and 200 fathoms at "Challenger" stations (Jones, 1994).

### Chronostratigraphy
Middle Miocene to Recent (Jones, 1994).

## References


Barker, R.W., 1960, Taxonomic notes on the species figured by H.B. Brady in his report on the foraminifera dredged by H.M.S. Challenger during the years 1873–1876. Society of Economic Paleontologists and Mineralogists Special Publication, vol. 9, pp. 1–238.

Brady, H.B., 1881, Notes on some of the reticularian Rhizopoda of the Challenger Expedition, Part III. Quarterly Journal of Microscopical Science (New Series), vol. 21, pp. 31–71.

Brady, H.B., 1884, Report on the foraminifera dredged by H.M.S. Challenger during the years 1873–1876. In Brady, H.B., ed., Report on the Scientific Results of the Voyage of the H.M.S. Challenger during the years 1873–1876. Zoology, vol. 9, pp. 1–814.

Jones, R.W., 1994, The Challenger Foraminifera. Oxford: Oxford University Press, 149 pp.

Saidova, Kh.M., 1975, Bentosnye foraminifery Tikhogo Okeana, 3 vols. Moscow: Institut Okeanologii P.P. Shirshova, Akademiya Nauk SSSR, 875 pp.

# Loxostomoides applinae

## (Plummer), 1927

1. *Loxostomoides applinae*. Claypit, Mexia Brickworks, Mexia, Texas, USA. Deposited in the micropaleontological collections of The Natural History Museum, London; NHMUK PF53048. Lateral view, NHMUK PM DI 0440, scale bar = 103 μm.

### Original Designation

*Bolivina applini* Plummer, 1927, p. 69, pl. 4, fig. 1.

### Type Specimen

Cotypes (No. 33006) deposited in the Walker Museum at the University of Chicago, USA. Duplicate set deposited at the Bureau of Economic Geology, University of Texas, Austin, USA.

### Type Level

Paleocene. Lower part of the upper Midway Formation, clay above the Tehuacana Limestone horizon. Given as Eocene by Plummer (1927).

### Type Locality

Station 46, clay pit, Mexia Brick Works, about 1 mile west of the Town of Mexia, Limestone County, Texas.

### Description

Test initially biserial, becoming uniserial, elongate, slender, somewhat compressed, slightly tapered initially with oval cross-section. Broad and low chambers with slightly overlapping lower margins, increase relatively rapidly in height, and are separated by sutures, initially faint, becoming depressed and crenulate in the later portion. Chamber walls are calcareous, perforate, and ornamented by longitudinal striae in the early part of the test. The primary aperture is a loop, areal and slightly eccentric in the uniserial stage.

### Synonyms and Citations

*Loxostomoides applinae* (Plummer). –van Morkhoven et al., 1986, p. 327, pl. 106, figs. 1–3. –Loeblich and Tappan, 1987, p. 499, pl. 549, figs. 11–16.

### Remarks

Type species of the genus *Loxostomoides* Reiss, 1957.

*Loxostomoides applinae* (Plummer) is a useful index fossil for Paleocene strata, occurring less frequently in the Lower Eocene (van Morkhoven et al., 1986).

### Biogeography

Worldwide. See Observed Occurrences in van Morkhoven et al. (1986).

### Bathymetry

Neritic to upper bathyal (van Morkhoven et al., 1986).

### Chronostratigraphy

Early Paleocene to Early Eocene; doubtful occurrences in Early Eocene Biozones P7 through P9 (van Morkhoven et al., 1986).

## References

Loeblich, A.R. Jr and Tappan, H., 1987, Foraminiferal Genera and Their Classification. New York: Van Nostrand Reinhold, 970 pp.

Plummer, H.J., 1927, Foraminifera of the Midway Formation in Texas. Bulletin of the University of Texas Bureau of Economic Geology and Technology, vol. 2644, pp. 1–206.

van Morkhoven, F.P.C.M., Berggren, W.A., Edwards, A.S., et al., 1986, Cenozoic cosmopolitan deep-water benthic foraminifera. Bulletin des Centres de Recherches Exploration-Production Elf-Aquitaine, Mémoire, vol. 11, pp. 1–421.

# *Marginulina obesa*

## Cushman, 1923

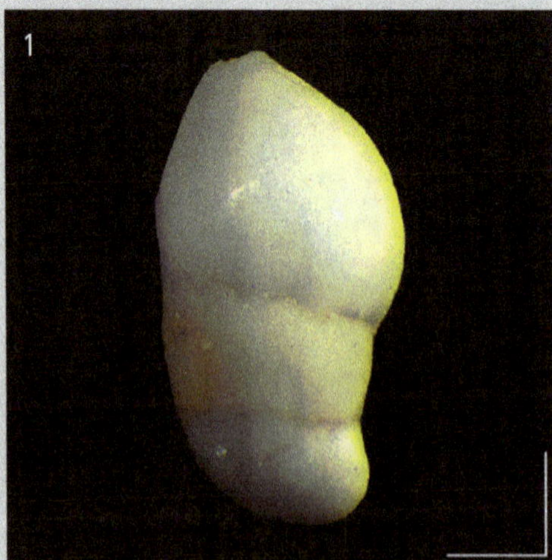


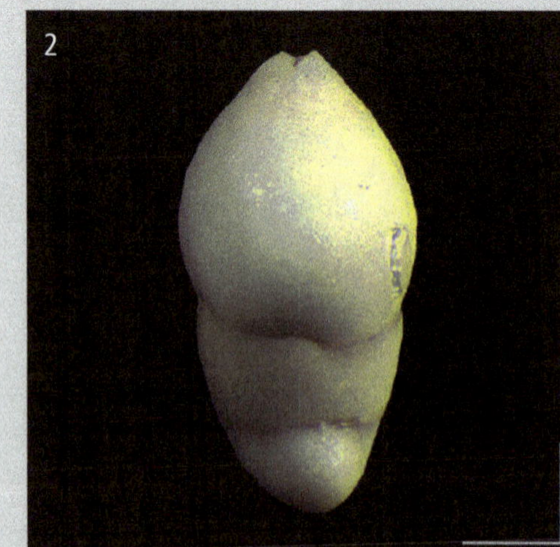


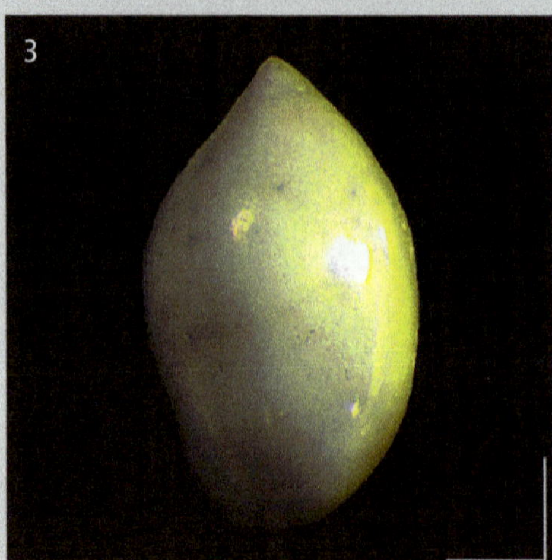


1. *Marginulina obesa.* North Pacific Ocean; "Challenger" Station 246, depth: 2050 fathoms. Deposited in the micropaleontological collections of The Natural History Museum, London; NHMUK ZF1809. Lateral view, NHMUK PM DI 0441, scale bar = 276 μm.

2. *Marginulina obesa.* North Pacific Ocean; "Challenger" Station 246, depth: 2050 fathoms. Deposited in the micropaleontological collections of The Natural History Museum, London; NHMUK ZF1809. Edge view, NHMUK PM DI 0442, scale bar = 207 μm.

3. *Marginulina obesa.* North Pacific Ocean; "Challenger" Station 246, depth: 2050 fathoms. Deposited in the micropaleontological collections of The Natural History Museum, London; NHMUK ZF1810. Lateral view, NHMUK PM DI 0443, scale bar = 244 μm.

## Original Designation

*Marginulina glabra* d'Orbigny var. *obesa* Cushman, 1923, p. 128, pl. 37, fig. 1.

## Type Specimen

Deposited in the United States National Museum, Washington, DC (Nos 17482–17488, 17489, 17490–17507).

## Type Level

Recent, surface sediments.

## Type Locality

"Albatross" Station D 2043 (39° 49′ 00″ N; 68° 28′ 30″ W) in 1467 fathoms, off the northeastern coast of the USA.

## Description

Test elongate, initially curved, later becoming rectilinear, circular in cross-section. Highly inflated chambers increasing rapidly in size, trigonal in umbilical view, are separated by straight, slightly depressed sutures. Chamber walls are calcareous, smooth, and finely perforate. The primary aperture is terminal and radiate.

## Synonyms and Citations

*Marginulina glabra* d'Orbigny. –Flint, 1899, p. 133, pl. 60, fig. 1 (in part).

*Marginulina glabra* d'Orbigny. –Brady, 1884, p. 527, pl. 65, figs. 5–6. –van Marle, 1991, p. 51, pl. 1, figs. 5–6.

*Marginulina obesa* Cushman. –Barker, 1960, pl. 65, figs. 5–6. –Jones, 1994, p. 77, pl. 65, figs. 5–6.

## Biogeography

Worldwide.

## Bathymetry

Neritic to abyssal. Reported between 15 and 2740 fathoms at "Challenger" stations (Jones, 1994).

## Chronostratigraphy

Late Miocene to Recent (van Marle, 1991; Jones, 1994).

# References


Barker, R.W., 1960, Taxonomic notes on the species figured by H.B. Brady in his report on the foraminifera dredged by H.M.S. Challenger during the years 1873–1876. Society of Economic Paleontologists and Mineralogists Special Publication, vol. 9, pp. 1–238.

Brady, H.B., 1884, Report on the foraminifera dredged by H.M.S. Challenger during the years 1873–1876. In Brady, H.B., ed., Report on the Scientific Results of the Voyage of the H.M.S. Challenger during the years 1873–1876. Zoology, vol. 9, pp. 1–814.

Cushman, J.A., 1923, The foraminifera of the Atlantic Ocean; Part 4 – Lagenidae. Bulletin of the United States National Museum (US Nat. Mus. Bull.), vol. 104, no. 4, pp. 1–228.

Flint, J.M., 1899, Recent foraminifera. A descriptive catalogue of specimens dredged by the US Fish Commmission Steamer Albatross. Report of the United States National Museum for 1897, pt 1, pp. 249–349.

Jones, R.W., 1994, The Challenger Foraminifera. Oxford: Oxford University Press, 149 pp.

van Marle, L.J., 1991, Eastern Indonesian Late Cenozoic smaller benthic foraminifera. Verhandelingen der Koninklijke Nederlanse Akademie van Wetenschappen, Afdeeling Natuurkunde, Eerste Reeks, vol. 34, pp. 1–328.

# *Marssonella oxycona*

## (Reuss), 1860

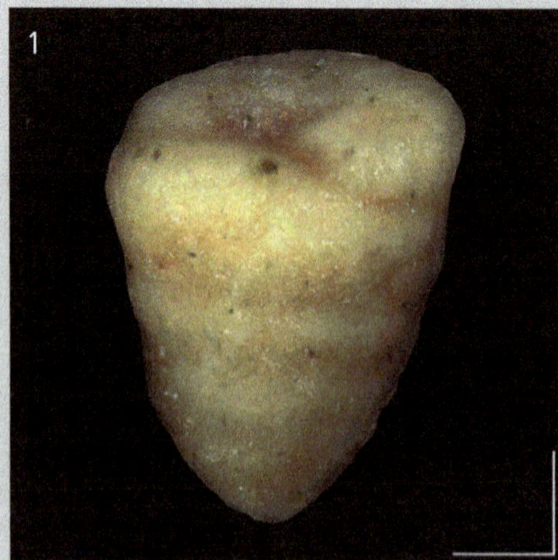


1. *Marssonella oxycona.* Zawadka near Wadowice, Poland; Subsilesian Unit, Sample J-B 4/88. Deposited in the Geroch Collection, Jagiellonian University, Kraków, Poland. Lateral view, NHMUK PM DI 0444, scale bar = 141 μm.

### Original Designation

*Gaudryina oxycona* Reuss, 1860, p. 229, pl. 12, fig. 3a–c.

### Type Specimen

Depository not given.

### Type Level

Not designated. Levels given: Cretaceous, Upper Senonian ("Mucronatenkreide"), marl, Lower Senonian ("Quadratenkreide"), Turonian and Cenomanian ("Pläner"), and uppermost Gault ("Minimuston"). Desai and Banner (1987) designated the *mucronata* chalk as neotype level.

### Type Locality

Not designated. Localities given: Hilgenberg near Hamm und Drensteinfurth ("Mucronatenkreide"), Hamm, Ostheide, der Rhyneberg ("Quadratenkreide"), Bergcapen, Wullen, Opherdicke und Ahaus (Pläner); Rheine ("Minimuston"), Haverlat und Fleischerkamp near Salzgitter (Cenomanian), Westphalen, Germany. Desai and Banner (1987) designated the *mucronata* chalk of Hildenberg was designated as the neotype locality.

### Description

Test forms a stout, initially trochospiral cone elongating in the later portion into a biserial series with a flat or slightly concave terminal face and a subcircular cross-section. The chambers in the initial trochospiral portion increase rapidly in size, giving the early part of the test a rounded appearance. Chambers in the biserial part increase rapidly in width and are separated by flush, indistinct sutures. Chamber walls are canaliculate and comprised mostly of calcareous particles with abundant calcareous cement. The terminal face is flattened or concave. The primary aperture is a low, narrow, interiomarginal slit in a slight re-entrant of the final chamber.

### Synonyms and Citations

*Dorothia oxycona* (Reuss). –Trujillo, 1960, p. 309, pl. 44, fig. 5. –Sliter, 1968, p. 50, pl. 3, fig. 13. –Geroch and Nowak, 1984, pl. 4, fig. 14.

*Marssonella oxycona* (Reuss). –Belford, 1960, p. 16, pl. 4, fig. 1. –Desai and Banner, 1987, p. 20, pl. 3, figs. a–i.

### Remarks

Desai and Banner (1987) highlighted significant structural differences in the *Dorothiidae*, and restricted the concept of *Marssonella oxycona* (Reuss) to Upper Cretaceous specimens, possessing canaliculate walls. Lower Cretaceous specimens with similar test form, coiling mode, and chamber shape, but with solid chamber walls, were referred to *Protomarssonella* Desai and Banner.

### Biogeography
Worldwide.

### Bathymetry
Bathyal to abyssal.

### Chronostratigraphy
Upper Cretaceous.

## References


Belford, D.J., 1960, Upper Cretaceous foraminifera from the Toolonga Calcilutite and Gingin Chalk, Western Australia. Bulletin of the Bureau of Mineral Resources, Geology and Geophysics, Australia, vol. 57, pp. 1–198.

Desai, D. and Banner, F.T., 1987, The evolution of Early Cretaceous Dorothiinae (Foraminiferid). Journal of Micropalaeontology, vol. 6, no. 2, pp. 13–27.

Geroch, S. and Nowak, W., 1984, Proposal of zonation for the late Tithonian–late Eocene, based upon arenaceous foraminifera from the outer Carpathians, Poland. In Oertli, H., ed., Benthos '83, 2nd International Symposium on Benthic Foraminifera, Pau (France), April 11–15, 1983. Elf Aquitaine, ESSO REP and TOTAL CFP, Pau and Bordeaux, pp. 225–239.

Reuss, A.E., 1860, Die Foraminiferen der westphälischen Kreideformation. Sitzungsberichte der Kaiserlichen Akademie der Wissenschaften in Wien, Mathematisch-Naturwissenschaftliche Classe, vol. 40, pp. 147–238.

Sliter, W.V., 1968, Upper Cretaceous foraminifera from southern California and northwestern Baja California, Mexico. Paleontological Contributions of the University of Kansas, no. 49, Protozoa, art. 7, pp. 1–141.

Trujillo, E.F., 1960, Upper Cretaceous foraminifera from near Redding, Shasta County, California. Journal of Paleontology, vol. 34, pp. 290–346.

# *Martinottiella communis*

## (d'Orbigny), 1826

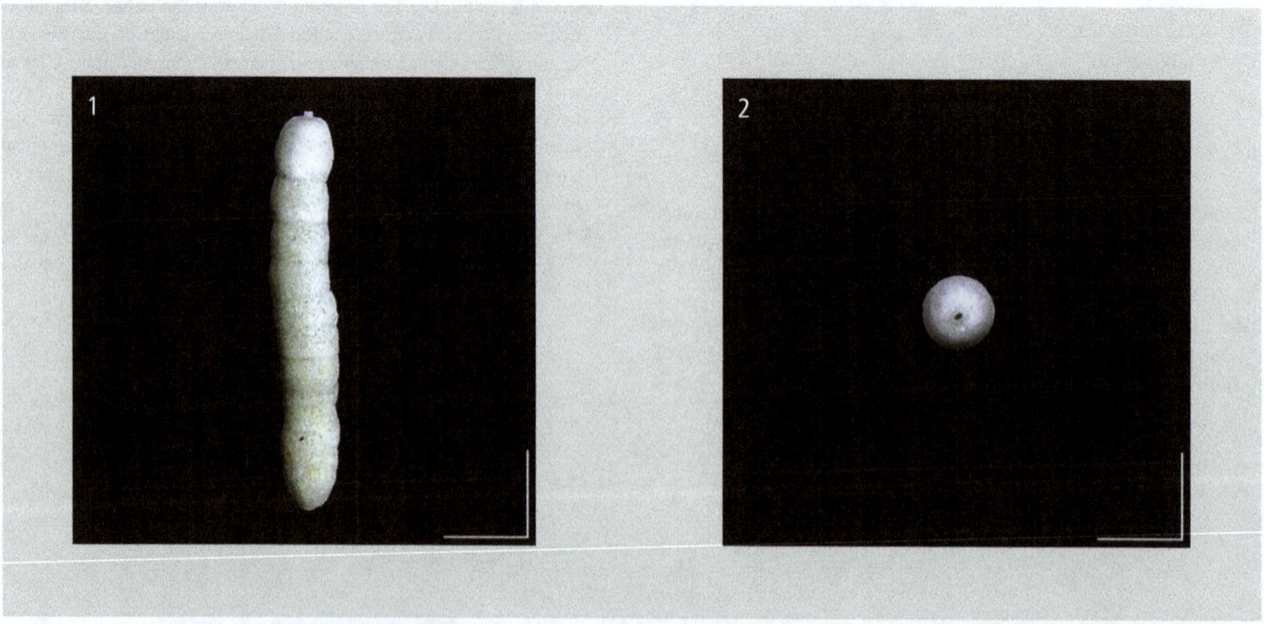


1. *Martinottiella communis.* Off Kanduva, Fiji, Pacific Ocean; "Challenger" Station 174C, depth: 210 fathoms. Deposited in the micropaleontological collections of The Natural History Museum, London; NHMUK ZF1277. Lateral view, NHMUK PM DI 0445, scale bar = 523 μm.

2. *Martinottiella communis.* Off Kanduva, Fiji, Pacific Ocean; "Challenger" Station 174C, depth: 210 fathoms. Deposited in the micropaleontological collections of The Natural History Museum, London; NHMUK ZF1277. Apertural view, NHMUK PM DI 0446, scale bar = 520 μm.

### Original Designation
*Clavulina communis* d'Orbigny, 1826, p. 102 [268], Model No. 4, illustrated in d'Orbigny, 1846, pl. 12, figs. 1–2.

### Type Specimen
Deposited in the collections of the Laboratoire de Paléontologie, Muséum d'Histoire Naturelle, Paris, France.

### Type Level
Tertiary, Tegel.

### Type Locality
Near the village of Nussdorf, north of Vienna, Austria.

### Description
Test forms an elongate, subcylindrical tube, initially trochospiral, later triserial, and rapidly reducing to uniserial, with circular cross-section. The chambers, usually indistinct in the early portion but numerous in the uniserial stage, increase gradually in size and are separated by slightly depressed sutures. Chamber walls are finely agglutinated and smooth. The primary aperture is terminal, round or slightly elliptical, bordered by a lip, and situated at the end of a short neck.

### Synonyms and Citations
*Clavulina communis* d'Orbigny. –d'Orbigny, 1846, p. 196, pl. 12, figs. 1–2. –Brady, 1884, p. 394, pl. 48, figs. 1–8 (not figs. 9–13).
*Martinottiella bradyana* Cushman. –Barker, 1960, pl. 48, figs. 1–2, 5.
*Martinottiella communis* (d'Orbigny). –Barker, 1960, pl. 48, figs. 3–4, 6–8. –Papp and Schmid, 1985, p. 74, pl. 66, figs. 88. –Loeblich and Tappan, 1987, p. 171, pl. 190, figs. 3–4. –Hermelin, 1989, p. 34, pl. 2, figs. 5–6. –Charnock and Jones, 1990, p. 192, pl. 12, figs. 10–11, pl. 25, fig. 2. –Jones, 1994, p. 52, pl. 48, figs. 1–2, ?3, 4–8.

### Remarks
Type species of the genus *Martinottiella* Cushman, 1933.

### Biogeography
Worldwide.

### Bathymetry

Bathyal. North Atlantic bathymetric range: 714 to 2938 meters (Charnock and Jones, 1990).

### Chronostratigraphy

Oligocene to Recent (Jones, 1994). Oligocene to Pliocene in the North Sea (Charnock and Jones, 1990).

## References


Barker, R.W., 1960, Taxonomic notes on the species figured by H.B. Brady in his report on the foraminifera dredged by H.M.S. Challenger during the years 1873–1876. Society of Economic Paleontologists and Mineralogists Special Publication, vol. 9, pp. 1–238.

Brady, H.B., 1884, Report on the foraminifera dredged by H.M.S. Challenger during the years 1873–1876. In Brady, H.B., ed., Report on the Scientific Results of the Voyage of the H.M.S. Challenger during the years 1873–1876. Zoology, vol. 9, pp. 1–814.

Charnock, M.A. and Jones, R.W., 1990, Agglutinated foraminifera from the Palaeogene of the North Sea. In Hemleben, C., Kaminski, M.A., Kuhnt, W., and Scott, D.B., ed., Paleoecology, Biostratigraphy, Paleoceanography and Taxonomy of Agglutinated Foraminifera, NATO ASI Series, C: Mathematical and Physical Sciences. Dordrecht: Kluwer Academic, pp. 139–244.

Cushman, J.A., 1933, Some new foraminiferal genera. Contributions from the Cushman Laboratory for Foraminiferal Research, vol. 9, pp. 32–38.

d'Orbigny, A., 1826, Tableau méthodique de la classe des Céphalopodes. Annales des Sciences Naturelles, vol. 7, pp. 245–314.

d'Orbigny, A.D., 1846, Foraminifères fossiles du Bassin Tertiaire de Vienne (Autriche). Paris: Gide et Compe, xxxvii + 312 pp.

Hermelin, J.O.R., 1989, Pliocene benthic foraminifera from the Ontong-Java Plateau (Western Equatorial Pacific Ocean): faunal response to changing paleoenvironment. Special Publication of the Cushman Foundation for Foraminiferal Research, vol. 26, pp. 1–143.

Jones, R.W., 1994, The Challenger Foraminifera. Oxford: Oxford University Press, 149 pp.

Loeblich, A.R. Jr and Tappan, H., 1987, Foraminiferal Genera and Their Classification. New York: Van Nostrand Reinhold, 970 pp.

Papp, A. and Schmid, M.E., 1985, Die fossilen Foraminiferen des tertiaeren Beckens von Wien: Revision der Monographie Alcide d'Orbigny (1846). Abhandlungen der Geologischen Bundesanstalt, vol. 37, pp. 1–311.

# *Matanzia bermudezi*

## Palmer, 1936

1. *Matanzia bermudezi*. Gulf of Mexico; Eureka Core E68-136. Deposited in the collections of the United States National Museum, Washington, DC. Specimen illustrated in van Morkhoven et al., 1986, Ref. no. 55. Lateral view, NHMUK PM DI 0447, scale bar = 281 μm.

### Original Designation

*Matanzia bermudezi* Palmer, 1936, p. 126, pl. 5, figs. 14–18.

### Type Specimen

Cotypes deposited in the Dorothy K. Palmer Collection at the Paleontological Research Institution, Ithaca, NY, USA.

### Type Level

Upper Oligocene, marl. Originally given as Lower Oligocene by Palmer (1936).

### Type Locality

Not designated. Localities given: shallow-water well on Finca Adelina, approximately 9.5 kilometers east of Colon and 0.5 kilometer north of the Carretera Central; Palmer Station FC-28, water well on the Batey of Central Alava, southeast of Banaglises, Matanzas Province, Cuba.

### Description

Test elongate with early trochospiral portion followed by a biserial stage representing approximately half of the test, cylindrical in outline, with a subacute initial portion, and a subcircular cross-section. The moderately inflated chambers are separated by indistinct sutures in the early trochospiral portion and by slightly depressed sutures in the biserial portion. Chambers are subdivided by up to ten interior, narrow, vertical partitions projecting inward from the outer wall. Chamber walls are finely agglutinated and smoothly finished with abundant calcareous cement. The primary aperture is a low arch at the base of the last chamber.

### Synonyms and Citations

*Matanzia bermudezi* Palmer. –van Morkhoven et al., 1986, p. 166, pl. 55. –Loeblich and Tappan, 1987, p. 145, pl. 153, figs. 4–7.

### Remarks

Type species of the genus *Matanzia* Palmer, 1936.

*Matanzia bermudezi* Palmer resembles *Gaudryina cylindrica* Nuttall, with which it is commonly associated in Cuba. However, *Gaudryina cylindrica* lacks the internal vertical partitions characteristic of the genus *Matanzia* (Palmer, 1936).

### Biogeography

Worldwide. Recorded from Cuba, the Dominican Republic, the eastern Pacific, the North and South Atlantic, and the Gulf of Mexico (van Morkhoven et al., 1986).

### Bathymetry

Bathyal.

### Chronostratigraphy

Late Oligocene (Chattian) to Middle Miocene (Serravallian).

## References


Loeblich, A.R. Jr and Tappan, H., 1987, Foraminiferal Genera and Their Classification. New York: Van Nostrand Reinhold, 970 pp.

Palmer, D.K., 1936, New genera and species of Cuban Oligocene Foraminifera. Memorias de la Sociedad Cubana de Historia Natural "Filipe Poey", vol. 10, no. 2, pp. 123–128.

van Morkhoven, F.P.C.M., Berggren, W.A., Edwards, A.S., et al., 1986, Cenozoic cosmopolitan deep-water benthic foraminifera. Bulletin des Centres de Recherches Exploration-Production Elf-Aquitaine, Mémoire, vol. 11, pp. 1–421.

# Melonis barleeanum

## (Williamson), 1858

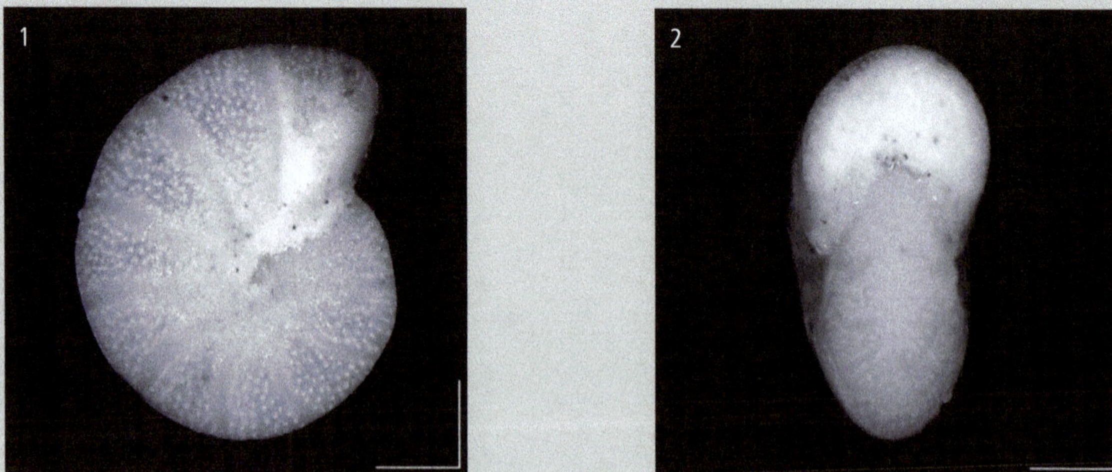

1. *Melonis barleeanum*. Great Bahama Bank; ODP Leg 166, Hole 1006A, Core 62, Section 5, 30–35 cm. Umbilical view, NHMUK PM DI 0448, scale bar = 80 µm.

2. *Melonis barleeanum*. Great Bahama Bank; ODP Leg 166, Hole 1006A, Core 62, Section 5, 30–35 cm. Apertural view, NHMUK PM DI 0448, scale bar = 82 µm.

### Original Designation
*Noniona barleeana* Williamson, 1858, p. 32, pl. 3, figs. 68–69.

### Type Specimen
Depository not given.

### Type Level
Recent.

### Type Locality
Not designated. Localities given: Scarborough, dredged sands of Skye, Stornoway and Shetland, British Isles.

### Description
Test forms a symmetrical, compressed, involute, biumbilicate planispire, with deep, open umbilici, subcircular in outline, and a broadly rounded periphery. The (approximately) ten to twelve moderately inflated chambers increase gradually in size and are separated by gently curved, flush or slightly depressed sutures that merge to form umbilical rings. Chamber walls are calcareous, smooth, and coarsely perforate, except for the imperforate apertural face. The primary aperture is a low interiomarginal, equatorial slit extending to the open umbilici, and bordered by a distinct lip.

### Synonyms and Citations
*Nonionina crassula* (Walker). –Parker and Jones, 1857, p. 14, pl. 11, figs. 5–6.

*Nonion barleeanum* (Williamson). –Cushman, 1930, p. 11, pl. 4, fig. 5.

*Melonis barleeanus* (Williamson). –Thomas, 1985, p. 677, pl. 12, fig. 3. –Loeblich and Tappan, 1994, p. 159, pl. 347, figs. 1–5.

*Melonis barleeanum* (Williamson). –Corliss, 1979, p. 10, pl. 5, figs. 78. –Loeblich and Tappan, 1987, pl. 696, figs. 5–6. –Miller and Katz, 1987, p. 136, pl. 4, fig. 5. –Hermelin, 1989, p. 88, pl. 17, fig. 12.

### Remarks
*Melonis barleeanum* (Williamson) is morphologically close to *Melonis affinis* (Reuss), and is considered by some authors to be a junior synonym. However, the two species have quite different type levels: Recent for *Melonis barleeanum* and Oligocene (originally given as Eocene) for *Melonis affinis*.

*Melonis barleeanum* (Williamson) is distinguished from *Melonis pompilioides* (Fichtel and Moll) and *Melonis sphaeroides* Voloshinova by its flatter test and generally more numerous chambers in the last whorl.

### Biogeography

Worldwide, except in highest latitudes. Central Equatorial Pacific (Thomas, 1985), Europe, Atlantic, Gulf of Mexico, Indian Ocean, Pacific (Hermelin, 1989), North Atlantic (Miller and Katz, 1987). Gulf of Mexico (Katz and Miller, 1993). Sahul Shelf and Timor Sea (Loeblich and Tappan, 1994).

### Bathymetry

Neritic to bathyal (Pflum and Frerichs, 1976). Bathyal (Katz and Miller, 1993).

### Chronostratigraphy

Oligocene to Recent. Early Miocene to Recent (Thomas, 1985). Rare in Oligocene, abundant in Middle and Late Miocene (Miller and Katz, 1987; Katz and Miller, 1993).

## References


Corliss, B.H., 1979, Taxonomy of Recent deep-sea benthonic foraminifera from the Southeast Indian Ocean. Micropaleontology, vol. 25, pp. 1–19.

Cushman, J.A., 1930, The foraminifera of the Atlantic Ocean, Part 7. Nonionidae, Camerinidae, Peneroplidae and Alveolinellidae. Bulletin of the United States National Museum, vol. 104, no. 7, pp. 1–79.

Hermelin, J.O.R., 1989, Pliocene benthic foraminifera from the Ontong-Java Plateau (Western Equatorial Pacific Ocean): faunal response to changing paleoenvironment. Special Publication of the Cushman Foundation for Foraminiferal Research, vol. 26, pp. 1–143.

Katz, M.E. and Miller, K.G., 1993, Latest Oligocene to Earliest Pliocene benthic foraminiferal biofacies of the northeastern Gulf of Mexico. Micropaleontology, vol. 39, no. 4, pp. 367–403.

Loeblich, A.R. Jr and Tappan, H., 1987, Foraminiferal Genera and Their Classification. New York: Van Nostrand Reinhold, 970 pp.

Loeblich, A.R. Jr and Tappan, H., 1994, Foraminifera of the Sahul Shelf and Timor Sea: Special Publications of the Cushman Laboratory for Foraminiferal Research, vol. 31, pp. 1–661.

Miller, K.G. and Katz, M.E., 1987, Oligocene to Miocene benthic foraminiferal and abyssal circulation changes in the North Atlantic. Micropaleontology, vol. 33, no. 2, pp. 97–149.

Parker, W.K. and Jones, T.R., 1857, Description of some foraminifera from the coast of Norway. Annals and Magazine of Natural History, ser. 2, vol. 19, pp. 273–303.

Pflum, C.E. and Frerichs, W.E., 1976, Gulf of Mexico deep-water foraminifers. Cushman Foundation, Special Publication, vol. 14, pp. 1–125.

Thomas, E., 1985, Late Eocene to Recent deep-sea benthic foraminifers from the central equatorial Pacific Ocean. In Mayer, L., Theyer, F., et al., ed., Initial Reports of the Deep Sea Drilling Project, Washington (U.S. Government Printing Office), vol. 85, pp. 655–694.

Williamson, W.C., 1858, On the Recent Foraminifera of Great Britain. London: Ray Society, 107 pp.

# Melonis pompilioides

## (Fichtel and Moll), 1798

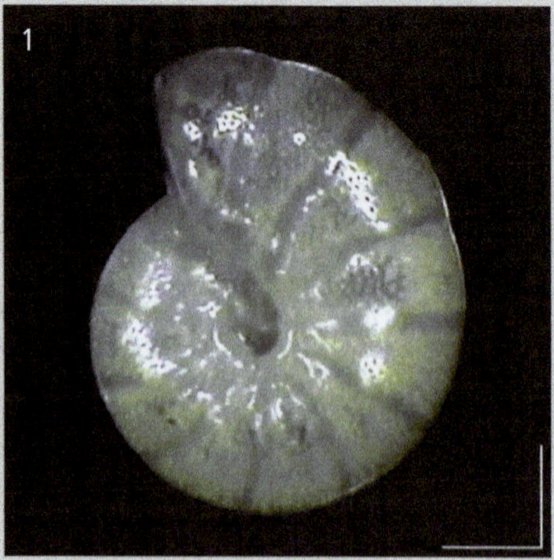


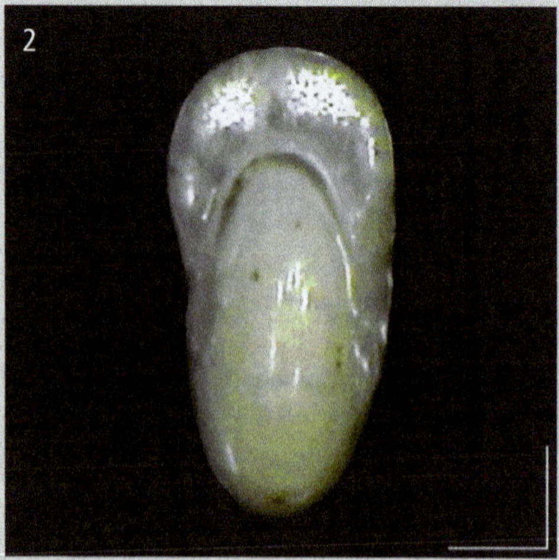


1. *Melonis pompilioides.* Great Sole Bank, Celtic Sea. Deposited in the micropaleontological collections of The Natural History Museum, London; NHMUK 1970.12.7.5. Umbilical view, NHMUK PM DI 0450, scale bar = 87 μm.

2. *Melonis pompilioides.* Great Sole Bank, Celtic Sea. Deposited in the micropaleontological collections of The Natural History Museum, London; NHMUK 1970.12.7.5. Apertural view, NHMUK PM DI 0451, scale bar = 89 μm.

### Original Designation

*Nautilus pompilioides* Fichtel and Moll, 1798, p. 31, pl. 2, figs. a–c.

### Type Specimen

Depository of holotype not given. Lectotype (MI-466) and paratypes (MI-467/1–5) designated by Rögl and Hansen (1984) are deposited in the collections of the Natural History Museum, Vienna, Austria.

### Type Level

Not designated. Levels given: Fossil and Recent.

### Type Locality

Not designated. Localities given: Coroncina, Tuscany, Italy (Fossil) and Mediterranean Sea (Recent). Lectotype from the Pliocene of Coroncina, Tuscany, Italy.

### Description

Test forms a symmetrical, involute, biumbilicate planispire with deep, open umbilici, subcircular in outline, and a broadly rounded periphery. The (approximately) eight to twelve strongly inflated, broad chambers increase relatively rapidly in width, and are separated by slightly curved, flush, or slightly depressed, limbate sutures that merge to form umbilical rings. Chamber walls are calcareous, smooth, and coarsely perforate, except for the imperforate apertural face. The primary aperture is a low interiomarginal, equatorial slit extending to the open umbilici, and bordered by a distinct lip.

### Synonyms and Citations

*Melonis etruscus* de Montfort, 1808, p. 67, fig. p. 66.

*Noniona soldanii* d'Orbigny, 1846, p. 109, pl. 5, figs. 15–16.

*Melonis pompilioides* (Fichtel and Moll). –van Morkhoven et al., 1986, p. 72, pl. 23A–C. –Loeblich and Tappan, 1987, p. 621, pl. 696, figs. 7–8. –Rögl and Hansen, 1984, p. 30, pl. 2, figs. 1–2; pl. 3, fig. 1.

### Remarks

Type species of the genus *Melonis* de Montfort, 1808. Loeblich and Tappan (1987) selected a shallow-water form from the Pliocene of Albania of Voloshinova (1958) to illustrate the original concept of *Melonis pompilioides*.

*Melonis pompilioides* (Fichtel and Moll) and *Melonis sphaeroides* Voloshinova are morphologically similar, bathymetrically segregated forms. *Melonis sphaeroides* is the deeper-water form, restricted to middle bathyal to abyssal depths, whereas *Melonis pompilioides* occurs primarily in outer neritic to middle bathyal depths. Voloshinova (1958) separated the North Atlantic specimens in Brady's Collection from neritic forms from Albania, on account of various morphological differences. According to Voloshinova (1958), *Melonis sphaeroides* has a more spherical test, a lesser number of chambers in the last whorl separated by completely flush sutures, and a more coarsely perforated test. Van Morkhoven et al. (1986) agreed that there were subtle morphological differences between the two taxa; however, these authors considered them to be bathymetrically controlled ecophenotypes rather than distinct species (see Remarks in van Morkhoven et al., 1986).

### Biogeography

Worldwide. See Observed Occurrences in van Morkhoven et al. (1986).

### Bathymetry

Lower neritic to middle bathyal.

### Chronostratigraphy

Late Oligocene? to Recent. The lower stratigraphic range of *Melonis pompilioides* is difficult to ascertain in view of its convoluted taxonomic history.

## References


de Montfort, P.D., 1808, Conchyliologie systématique et classification méthodique des coquilles. Paris: F. Schoell, vol. 1, lxxxvii + 409 pp.

d'Orbigny, A.D., 1846, Foraminifères fossiles du Bassin Tertiaire de Vienne (Autriche). Paris: Gide et Compe, xxxvii + 312 pp.

Fichtel, L. von and Moll, J.P.C. von, 1798, Testacea microscopica aliaque minuta ex generibus Argonauta et Nautilus ad naturam delineata et descripta. (Microscopische und andere kleine Schalthiere aus den Geschlechtern Argonaute und Schiffer, nach der Natur gezeichnet und beschrieben). Vienna: Camesina, 124 pp.

Loeblich, A.R. Jr and Tappan, H., 1987, Foraminiferal Genera and Their Classification. New York: Van Nostrand Reinhold, 970 pp.

Rögl, F. and Hansen, J.H., 1984, Foraminifera described by Fichtel and Moll in 1798; A revision of Testacea Microscopica. Appendix Microscopica alique minuta ex Generibus Argonauta et Nautilus. Reprint of original plates. Neue Denkschriften des Naturhistorischen Museum in Wien, vol. 3, pp. 1–143.

van Morkhoven, F.P.C.M., Berggren, W.A., Edwards, A.S., et al., 1986, Cenozoic cosmopolitan deep-water benthic foraminifera. Bulletin des Centres de Recherches Exploration-Production Elf-Aquitaine, Mémoire, vol. 11, pp. 1–421.

Voloshinova, N.A., 1958, On new systematics of the Nonionidae. Trudy Vsesoyuznogo Neftyanogo Nauchno-Issledovatel'skogo, Geologo-razvedochnogo Instituta (VNIGRI) (New Series), vol. 115, pp. 117–223.

# *Melonis sphaeroides*

## Voloshinova, 1958

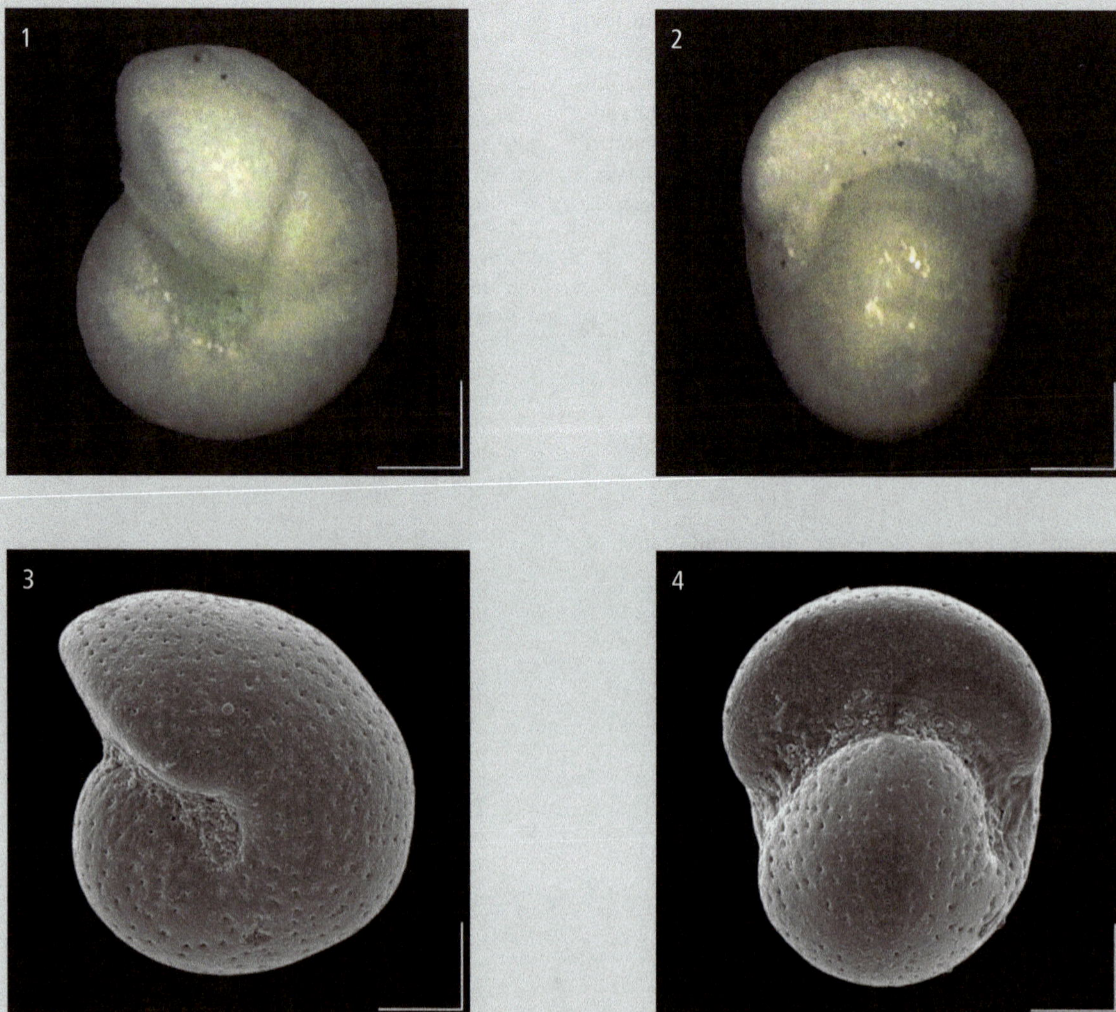


1. *Melonis sphaeroides*. South of Hayes Fracture Zone, West Flank of the Mid-Atlantic Ridge, North Atlantic; DSDP Leg 82, Site 563, Core 8, Section 2, 114–120 cm. Umbilical view, NHMUK PM DI 0452, scale bar = 83 μm.

2. *Melonis sphaeroides*. South of Hayes Fracture Zone, West Flank of the Mid-Atlantic Ridge, North Atlantic; DSDP Leg 82, Site 563, Core 8, Section 2, 114–120 cm. Apertural view, NHMUK PM DI 0453, scale bar = 87 μm.

3. *Melonis sphaeroides*. South of Hayes Fracture Zone, West Flank of the Mid-Atlantic Ridge, North Atlantic; DSDP Leg 82, Site 563, Core 8, Section 2, 114–120 cm. Umbilical view, NHMUK PM DI 0454, scale bar = 57 μm.

4. *Melonis sphaeroides*. South of Hayes Fracture Zone, West Flank of the Mid-Atlantic Ridge, North Atlantic; DSDP Leg 82, Site 563, Core 8, Section 2, 114–120 cm. Apertural view, NHMUK PM DI 0455, scale bar = 57 μm.

## Original Designation

*Melonis sphaeroides* Voloshinova, 1958, p. 153, pl. 3, fig. 10 after Brady, 1884.

## Type Specimen

Holotype was chosen by Voloshinova (1958) from specimens in slide ZF1995 referred to *Nonionina pompilioides* (Fichtel and Moll) by Brady (1884). Holotype (Brady's fig. 10) is housed in the micropaleontological collections of The Natural History Museum, London, UK.

## Type Level

Recent, surface sediments.

## Type Locality

"Porcupine" Station 20, West of Ireland, 1443 fathoms, in the eastern Atlantic.

## Description

Test forms a symmetrical, involute, markedly inflated, biumbilicate planispire, with deep, open umbilici, nearly circular in outline, and a broadly rounded periphery. The (approximately) seven to nine strongly inflated, broad chambers increase relatively rapidly in width and are separated by gently curved, flush sutures, often indistinct. Chamber walls are calcareous, smooth, and coarsely perforate except for the imperforate apertural face. The primary aperture is a low interiomarginal, equatorial slit extending to the open umbilici, bordered by a distinct lip and remaining open along the umbilical margin.

## Synonyms and Citations

*Nonionina pompilioides* (Fichtel and Moll). –Brady, 1884, p. 727, pl. 109, figs. 10–11.

*Nonion* (?) *pompilioides* (Fichtel and Moll). –Barker, 1960, pl. 109, figs. 10–11.

*Melonis pompilioides* (Fichtel and Moll) forma *sphaeroides* Voloshinova. –van Morkhoven et al., 1986, p. 76, pl. 23D–E.

*Melonis sphaeroides* Voloshinova. –Miller and Katz, 1987, p. 136, pl. 4, fig. 3.

*Melonis pompilioides* (Fichtel and Moll). –Jones, 1994, p. 108, pl. 109, figs. 10–11.

## Remarks

*Melonis pompilioides* (Fichtel and Moll) and *Melonis sphaeroides* Voloshinova are morphologically similar, bathymetrically segregated forms. *Melonis sphaeroides* is the deeper-water form, restricted to middle bathyal to abyssal depths, whereas *Melonis pompilioides* occurs primarily in outer neritic to middle bathyal depths. Voloshinova (1958) separated the North Atlantic specimens in Brady's Collection from neritic forms from Albania, on account of various morphological differences. According to Voloshinova (1958), *Melonis sphaeroides* has a more spherical test, a lesser number of chambers in the last whorl separated by completely flush sutures, and a more coarsely perforated test. Van Morkhoven et al. (1986) agreed that there were subtle morphological differences between the two taxa; however, these authors considered them to be bathymetrically controlled ecophenotypes rather than distinct species (see Remarks in van Morkhoven et al., 1986).

## Biogeography

Worldwide. See Observed Occurrences in van Morkhoven et al. (1986).

## Bathymetry

Lower bathyal to abyssal. Reported between 1000 and 2421 fathoms at "Challenger" stations (Jones, 1994).

## Chronostratigraphy

Late Oligocene (Chattian) to Recent.

# References


Barker, R.W., 1960, Taxonomic notes on the species figured by H.B. Brady in his report on the foraminifera dredged by H.M.S. Challenger during the years 1873–1876. Society of Economic Paleontologists and Mineralogists Special Publication, vol. 9, pp. 1–238.

Brady, H.B., 1884, Report on the foraminifera dredged by H.M.S. Challenger during the years 1873–1876. In Brady, H.B., ed., Report on the Scientific Results of the Voyage of the H.M.S. Challenger during the years 1873–1876. Zoology, vol. 9, pp. 1–814.

Jones, R.W., 1994, The Challenger Foraminifera. Oxford: Oxford University Press, 149 pp.

Miller, K.G. and Katz, M.E., 1987, Oligocene to Miocene benthic foraminiferal and abyssal circulation changes in the North Atlantic. Micropaleontology, vol. 33, no. 2, pp. 97–149.

van Morkhoven, F.P.C.M., Berggren, W.A., Edwards, A.S., et al., 1986, Cenozoic cosmopolitan deep-water benthic foraminifera. Bulletin des Centres de Recherches Exploration-Production Elf-Aquitaine, Mémoire, vol. 11, pp. 1–421.

Voloshinova, N.A., 1958, On new systematics of the Nonionidae. Trudy Vsesoyuznogo Neftyanogo Nauchno-Issledovatel'skogo, Geologo-razvedochnogo Instituta (VNIGRI) (New Series), vol. 115, pp. 117–223.

# *Neoconorbina terquemi*

## (Rzehak), 1888

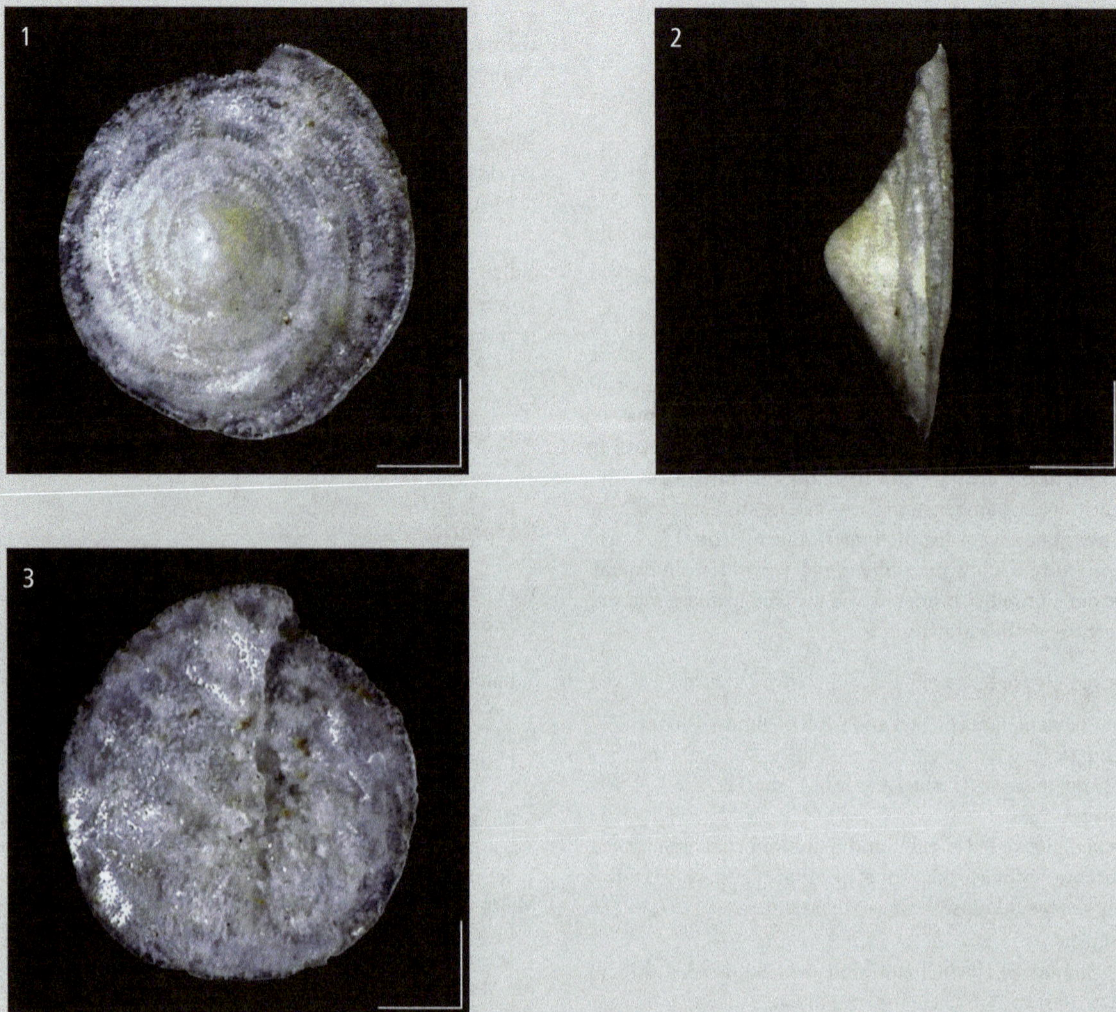


1. *Neoconorbina terquemi.* Off Raine Island, Torres Strait, Pacific Ocean; "Challenger" Station 184, depth: 155 fathoms. Deposited in the micropaleontological collections of The Natural History Museum, London; NHMUK ZF1402. Spiral view, NHMUK PM DI 0456, scale bar = 175 μm.

2. *Neoconorbina terquemi.* Off Raine Island, Torres Strait, Pacific Ocean; "Challenger" Station 184, depth: 155 fathoms. Deposited in the micropaleontological collections of The Natural History Museum, London; NHMUK ZF1402. Edge view, NHMUK PM DI 0457, scale bar = 171 μm.

3. *Neoconorbina terquemi.* Off Raine Island, Torres Strait, Pacific Ocean; "Challenger" Station 184, depth: 155 fathoms. Deposited in the micropaleontological collections of The Natural History Museum, London; NHMUK ZF1402. Umbilical view, NHMUK PM DI 0458, scale bar = 184 μm.

## Original Designation

*Discorbina terquemi* Rzehak, 1888, p. 228, illustrated as *Rosalina orbicularis* in Terquem, 1876 (non d'Orbigny, 1850), pl. 9, fig. 4.a.

## Type Specimen

Neotype selected by Lévy et al. (1975) from Dunkerque, France, was deposited in the collections of the Laboratoire de Paléontologie, Muséum d'Histoire Naturelle, Paris, France (No. FG449).

## Type Level

Upper Eocene, Bartonstufe, yellowish-gray, quartz-rich sandstone (Rzehak's specimens).

## Type Locality

Das Waschberg near Stockau, Austria (Rzehak's specimens).

## Description

Test forms a low trochospiral cone, circular in outline and concavo-convex in cross-section, with a convex, evolute spiral side, a concave, partially involute umbilical side, and an acute periphery. The broad, low, crescentic chambers (except for initial chambers) increase rapidly in breadth and are separated by thin, depressed, oblique sutures. The ultimate chamber is greatly enlarged, occupying one-third to one-half of the umbilical surface area. Umbilical extensions of chambers form folia. Chamber walls are calcareous, smooth, and finely perforate. The primary aperture is an umbilical–extraumbilical arch, partially covered by the folium.

## Synonyms and Citations

*Rosalina orbicularis* Terquem, 1876. (non d'Orbigny, 1850), p. 166, pl. 9, fig. 4.a.

*Discorbina terquemi* Rzehak, 1888, p. 228. New name for *Rosalina orbicularis* Terquem.

*Discorbina orbicularis* (Terquem) (non *Rosalina orbicularis* d'Orbigny, 1850). –Brady, 1884, p. 647, pl. 88, figs. 5–8.

*Neoconorbina terquemi* (Rzehak). –Barker, 1960, pl. 88, figs. 5–8. –Loeblich and Tappan, 1987, p. 560, pl. 609, figs. 8–10. –Jones, 1994, p. 94, pl. 88, figs. 5–8. –Robertson, 1998, p. 194, pl. 74, fig. 3. –Milker and Schmiedl, 2012, p. 97, figs. 22.5–6.

## Remarks

Type species of the genus *Neoconorbina* Hofker, 1951.

Rzehak (1888) renamed the species *Rosalina orbicularis* Terquem as *Discorbina terquemi*, as this species name was already occupied by *Rosalina orbicularis* d'Orbigny, 1850.

## Biogeography

Worldwide.

## Bathymetry

Neritic to middle bathyal. Reported between 2 and 435 fathoms at "Challenger" stations (Jones, 1994).

## Chronostratigraphy

Late Eocene to Recent.

# References


Barker, R.W., 1960, Taxonomic notes on the species figured by H.B. Brady in his report on the foraminifera dredged by H.M.S. Challenger during the years 1873–1876. Society of Economic Paleontologists and Mineralogists Special Publication, vol. 9, pp. 1–238.

Brady, H.B., 1884, Report on the foraminifera dredged by H.M.S. Challenger during the years 1873–1876. In Brady, H.B., ed., Report on the Scientific Results of the Voyage of the H.M.S. Challenger during the years 1873–1876. Zoology, vol. 9, pp. 1–814.

Hofker, J., 1951, The Foraminifera of the "Siboga" Expedition; Part III: Siboga-Expeditie, Monographie IVa. Leiden: E.J. Brill, 513 pp.

Jones, R.W., 1994, The Challenger Foraminifera. Oxford: Oxford University Press, 149 pp.

Lévy, A., Mathieu, R., Poignant, A., Rosset-Moulinier, M., and Rouvillois, A., 1975, Sur quelques foraminiféres actuels des plages de Dunkerque et des environs: néotypes et espéce nouvelle. Revue de Micropaléontologie, vol. 17, no. 4, pp. 171–181.

Loeblich, A.R. Jr and Tappan, H., 1987, Foraminiferal Genera and Their Classification. New York: Van Nostrand Reinhold, 970 pp.

Milker, Y. and Schmiedl, G., 2012, A taxonomic guide to modern benthic shelf foraminifera of the western Mediterranean Sea. Palaeontologia Electronica, vol. 15, Issue 2;16A, 134 pp.

Robertson, B.E., 1998, Systematics and paleoecology of the benthic foraminiferida from the Buff Bay section, Miocene of Jamaica. Micropaleontology, vol. 44, supplement 2, pp. 1–266.

Rzehak, A., 1888, Die Foraminiferen der Nummuliten-schichten des Waschberges und Michelsberges bei Stockerau in Nieder-Österreich. Verhandlungen der Geologischen Bundesanstalt, vol. 1888, pp. 226–229.

Terquem, O., 1876, Essai sur le classement des animaux qui vivent sur la plage et dans les environs de Dunquerque. Paris: Deuxième fascicule, pp. 55–100.

# Neoeponides auberii

## (d'Orbigny), 1839

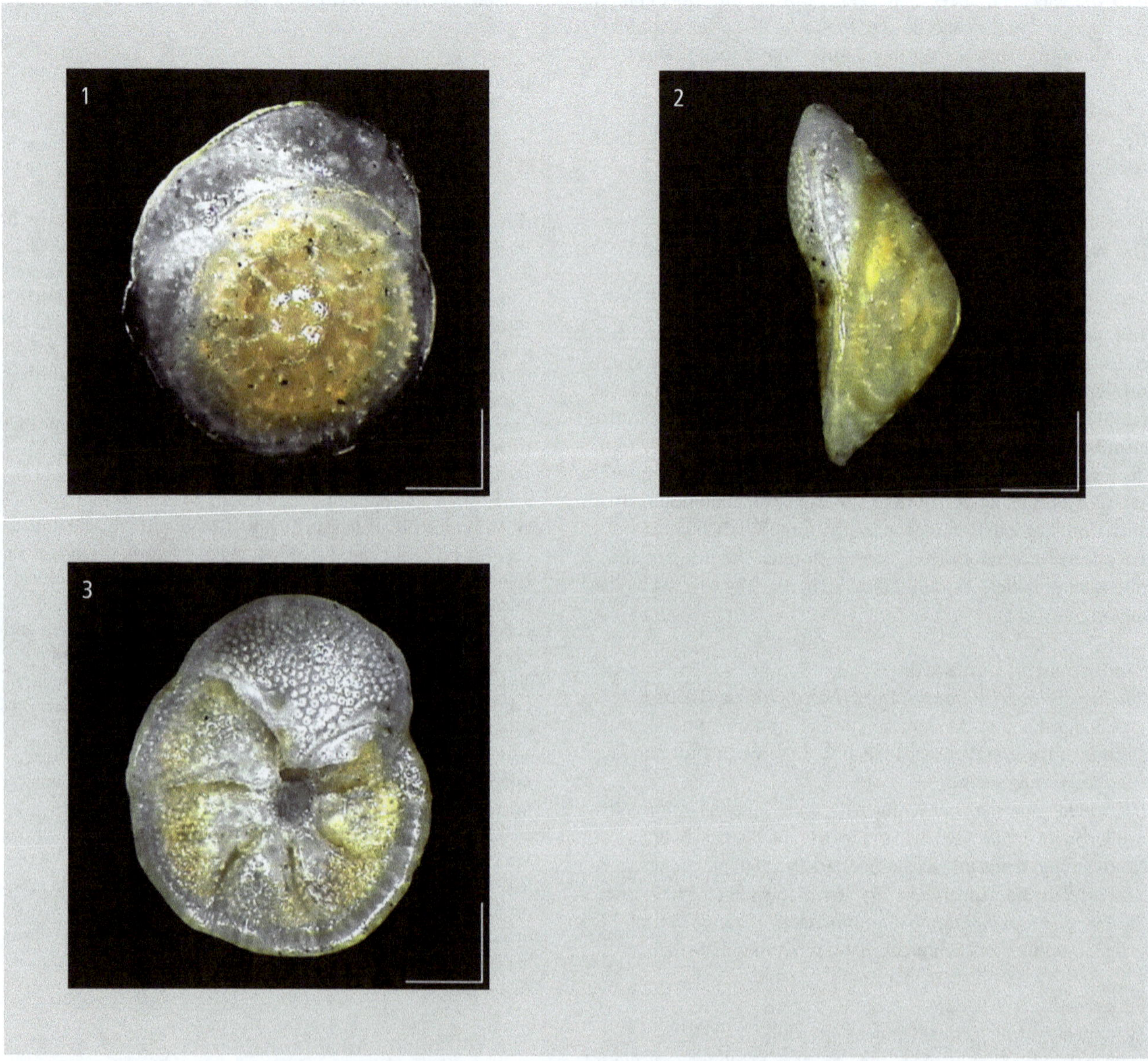

1. *Neoeponides auberii*. Off Cape Verde Islands, Atlantic Ocean; "Challenger" Station 352A, depth: 11 fathoms. Deposited in the micropaleontological collections of The Natural History Museum, London; NHMUK ZF1421. Spiral view, NHMUK PM DI 0459, scale bar = 142 μm.

2. *Neoeponides auberii*. Off Cape Verde Islands, Atlantic Ocean; "Challenger" Station 352A, depth: 11 fathoms. Deposited in the micropaleontological collections of The Natural History Museum, London; NHMUK ZF1421. Edge view, NHMUK PM DI 0460, scale bar = 165 μm.

3. *Neoeponides auberii*. Off Cape Verde Islands, Atlantic Ocean; "Challenger" Station 352A, depth: 11 fathoms. Deposited in the micropaleontological collections of The Natural History Museum, London; NHMUK ZF1421. Umbilical view, NHMUK PM DI 0461, scale bar = 141 μm.

## Original Designation

*Rosalina auberii* d'Orbigny, 1839, p. 94, pl. 4, figs. 5–8.

## Type Specimen

Depository not given.

## Type Level

Recent, marine sands.

## Type Locality

Not designated. Localities given: Cuba and Martinique.

## Description

Test forms a trochospire, subcircular in outline and planoconvex in cross-section with a flattened, involute, umbilical side, a conical, convex, spiral side, and an acute, keeled periphery. The (approximately) six crescentic, slightly inflated chambers in the last whorl increase gradually in size and are separated by flush, limbate sutures that are straight or slightly curved on the umbilical side and strongly curved, and slightly depressed on the spiral side. Chamber walls are calcareous, coarsely perforate, and smooth. The primary aperture is a low anteriomarginal arch extending from the periphery to the umbilicus.

## Synonyms and Citations

*Discorbis mira* Cushman, 1922, p. 39, pl. 6, figs. 10–11.
*Discorbina turbo* (d'Orbigny). –Brady, 1884, p. 642, pl. 87, fig. 8.
*Discorbina* (?) *mira* (Cushman). –Barker, 1960, pl. 87, fig. 8.
*Rotorbis auberi* (d'Orbigny). –Loeblich and Tappan, 1994, p. 137, pl. 278, figs. 1–11.
*Neoeponides auberii* (d'Orbigny). –Jones, 1994, p. 94, pl. 87, fig. 8.

## Remarks

Type species of the genus *Rotorbis* Sellier de Civrieux, 1977, regarded by Loeblich and Tappan (1987) as a junior synonym of *Neoeponides* Reiss, 1960.

## Biogeography

Worldwide.

## Bathymetry

Neritic to upper bathyal. Reported between 2 and 420 fathoms at "Challenger" stations (Jones, 1994) and between 24 and 44 meters in the Timor Sea (Loeblich and Tappan, 1994).

## Chronostratigraphy

Late Pliocene? to Recent.

## References


Barker, R.W., 1960, Taxonomic notes on the species figured by H.B. Brady in his report on the foraminifera dredged by H.M.S. Challenger during the years 1873–1876. Society of Economic Paleontologists and Mineralogists Special Publication, vol. 9, pp. 1–238.

Brady, H.B., 1884, Report on the foraminifera dredged by H.M.S. Challenger during the years 1873–1876. In Brady, H.B., ed., Report on the Scientific Results of the Voyage of the H.M.S. Challenger during the years 1873–1876. Zoology, vol. 9, pp. 1–814.

Cushman, J.A., 1922, Shallow-water foraminifera of the Tortugas region. Publications of the Carnegie Institution of Washington, no. 311, Department of Marine Biology, vol. 17, 85 pp.

d'Orbigny, A.D., 1839, Foraminifères. In Sagra, R. de la, ed., Histoire physique, politique et naturelle de l'île de Cuba. Paris: A. Bertrand, pp. 1–224, 12 pls.

Jones, R.W., 1994, The Challenger Foraminifera. Oxford: Oxford University Press, 149 pp.

Loeblich, A.R. Jr and Tappan, H., 1987, Foraminiferal Genera and Their Classification. New York: Van Nostrand Reinhold, 970 pp.

Loeblich, A.R. Jr and Tappan, H., 1994, Foraminifera of the Sahul Shelf and Timor Sea. Special Publications of the Cushman Laboratory for Foraminiferal Research, vol. 31, pp. 1–661.

# *Neoeponides hillebrandti*

## Fisher, 1969

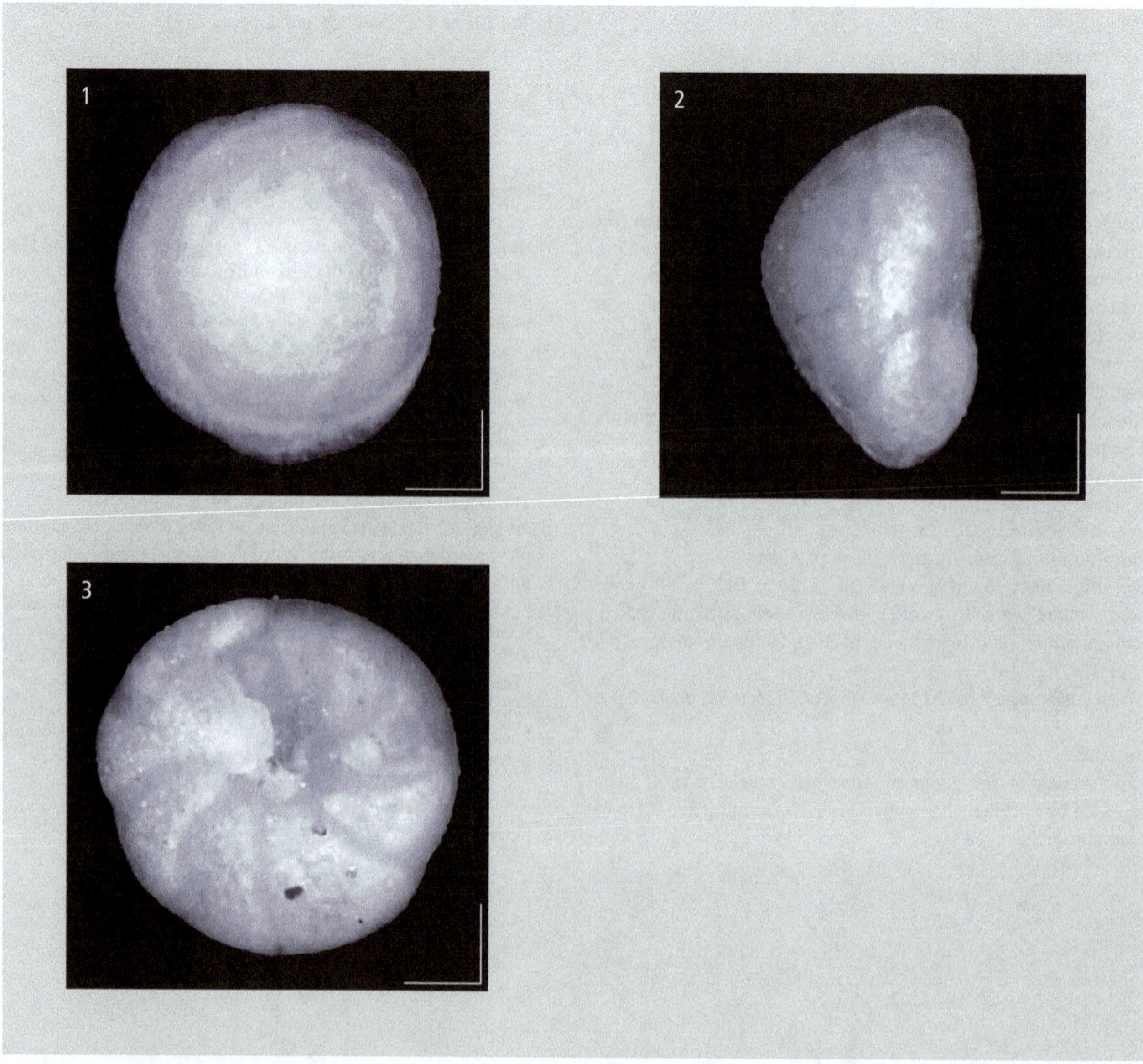


1. *Neoeponides hillebrandti*. Shatsky Rise, West Central Pacific; DSDP Leg 86, Site 577, Core 10, Section 1, 63–70 cm. Spiral view, NHMUK PM DI 0462, scale bar = 69 μm.

2. *Neoeponides hillebrandti*. Shatsky Rise, West Central Pacific; DSDP Leg 86, Site 577, Core 10, Section 1, 63–70 cm. Edge view, NHMUK PM DI 0463, scale bar = 67 μm.

3. *Neoeponides hillebrandti*. Shatsky Rise, West Central Pacific; DSDP Leg 86, Site 577, Core 10, Section 1, 63–70 cm. Umbilical view, NHMUK PM DI 0464, scale bar = 69 μm.

## Original Designation

*Neoeponides hillebrandti* Fisher, 1969, p. 196, illustrated as *Eponides whitei* in Hillebrandt, 1962 (non Brotzen, 1936), pl. 8, fig. 11.

## Type Specimen

Holotype originally deposited in the Bayerische Staatsammlung für Palaeontologie und Historische Geologie, Munich, Germany. Fisher's specimen deposited in the collections of the Sedgwick Museum, Cambridge, UK.

## Type Level

Paleocene, red marl facies of the Nierentaler beds (Hillebrandt's specimen). Upper Cretaceous (Fisher's specimen).

## Type Locality

Eitelgraben, on the north side of Untersberg, Salzburg Province, Austria (Hillebrandt's specimen). Fisher's specimen is from a chalk fragment dredged from a depth between 650 and 700 meters, Galicia Bank seamount, latitude 42° 36′ N, longitude 11° 35′ W, off the coast of Spain.

## Description

Test forms a trochospire; subcircular in outline and planoconvex in cross-section with a flattened, involute umbilical side, a conical, convex spiral side, and an acute periphery. The (approximately) ten crescentic, moderately inflated chambers in the last whorl increase gradually in size and are separated by flush, limbate, straight, or slightly curved sutures on the umbilical side and by strongly oblique, slightly depressed sutures on the spiral side. These sutures merge into a large umbilical boss on the umbilical side. Chamber walls are calcareous, perforate, and smooth. The primary aperture is a low anteriomarginal arch extending from the periphery to the umbilicus.

## Synonyms and Citations

*Rotalia* cf. *partschiana* (d'Orbigny). –White, 1928, p. 288, pl. 38, figs. 10a–c.

*Eponides whitei* Hillebrandt, 1962 (non Brotzen, 1936), p. 106, pl. 8, fig. 11.

*Neoeponides hillebrandti* Fisher, 1969, p. 196. New name for *Eponides whitei* Hillebrandt.

*Neoeponides hillebrandti* Fisher. –Tjalsma and Lohmann, 1983, p. 16, pl. 7, fig. 9. –Thomas, 1990, p. 590, pl. 2, figs. 3–4.

*Eponides* cf. *hillebrandti* (Fisher). –Bolli et al., 1994, p. 146, figs. 39.7–9.

## Remarks

Fisher (1969) renamed *Eponides whitei* Hillebrandt as *Neoeponides hillebrandti* because this species name was already occupied by *Eponides whitei* Brotzen, 1936.

Tjalsma and Lohmann (1983) found significant morphological variation in the height of the spiral side, the size of the umbilical boss, and the number of chambers in the last whorl (six to ten). Some specimens also have fairly depressed and curved sutures on the umbilical side.

Specimens referred to *Eponides* cf. *hillebrandti* (Fisher) by Bolli et al. (1994) are slightly smaller and possess fewer chambers in the last whorl (six to seven) than Hillebrandt's types.

## Biogeography

Worldwide. Reported from the Atlantic Ocean (Tjalsma and Lohmann, 1983), Austria, Mexico (Hillebrandt, 1962), Weddell Sea, Antarctic (Thomas, 1990).

## Bathymetry

Neritic to bathyal. Wide bathymetric range, but highest frequencies shift from deep sites during the Early Paleocene to shallow sites during the Late Paleocene (Tjalsma and Lohmann, 1983). Lower bathyal (Thomas, 1990).

## Chronostratigraphy

Late Cretaceous (Maastrichtian) to Late Paleocene (Thanetian). Rare to common in Biozones P1 to P5 (Tjalsma and Lohmann, 1983).

## References


Bolli, H.M., Beckmann, J.P., and Saunders, J.B., 1994, Benthic Foraminiferal Biostratigraphy of the South Caribbean Region. Cambridge: Cambridge University Press, 408 pp.

Brotzen, F., 1936. Foraminiferen aus dem schwedischen untersten Senon von Eriksdal in Schonen. Årsbok Sveriges geologiska Undersökning, vol. 30, no. 3, pp. 1–206.

Fisher, M.J., 1969, Benthonic foraminifera from the Maastrichtian chalk of Galicia Bank, west of Spain. Palaeontology, vol. 12, no. 2, pp. 189–200.

Hillebrandt, A. von, 1962, Das Paleozaen und seine Foraminiferenfauna im Becken von Reichenhall und Salzburg. Abhandlungen, Bayerische Akademie der Wissenschaften, Mathematisch-Naturwissenschaftliche Klasse, N.F., no. 108, pp. 9–182.

Thomas, E., 1990, Late Cretaceous through Neogene benthic foraminifers. In Barker, P.F., Kennett, J.P., et al., ed., Scientific Results of the Proceedings of the Ocean Drilling Program, vol. 113, pp. 571–594.

Tjalsma, R.C. and Lohmann, G.P., 1983, Paleocene–Eocene bathyal and abyssal benthic foraminifera from the Atlantic Ocean. Micropaleontology, Special Publication, no. 4, pp. 1–90.

White, M.P., 1928, Some index foraminifera of the Tampico Embayment area of Mexico (Part II). Journal of Paleontology, vol. 2, pp. 280–317.

# Neoflabellina jarvisi

## (Cushman), 1935

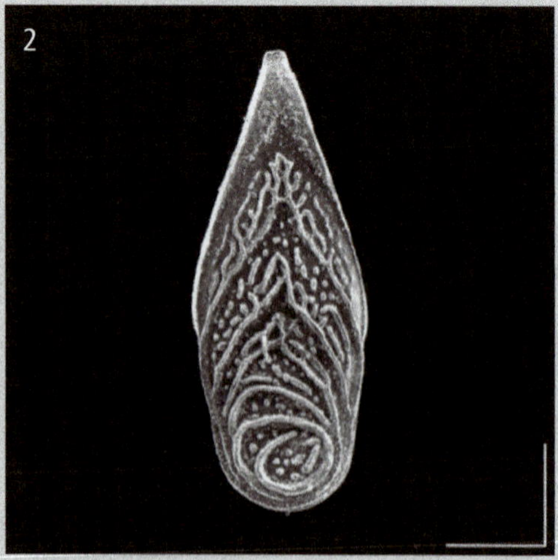


1. *Neoflabellina jarvisi.* Pit at Lizard Springs, near Guayaguayare, southeastern Trinidad. Deposited in the collections of the United States National Museum, Washington, DC, No. 15396. Holotype. Lateral view, NHMUK PM DI 0465, scale bar = 275 μm.

2. *Neoflabellina jarvisi.* Shatsky Rise, West Central Pacific; DSDP Leg 86, Site 577, Core 10, Section 4, 70–77 cm. Lateral view, NHMUK PM DI 0466, scale bar = 134 μm.

### Original Designation

*Flabellina jarvisi* Cushman, 1935, p. 85, pl. 13, figs. 7–8.

### Type Specimen

Holotype (No. 15396) deposited in the collections of the United States National Museum, Washington, DC.

### Type Level

Paleocene (van Morkhoven et al., 1986). Originally given as Upper Cretaceous by Cushman (1935).

### Type Locality

Pit at Lizard Springs, near Guayaguayare, southeastern Trinidad.

### Description

Test forms a large, elongate, compressed rhomboid with a truncate periphery. Chambers (up to ten) are narrow and increase slowly in size. Chambers are planispirally coiled initially, becoming rapidly uniserial, rectilinear, and chevron-shaped. Raised, plate-like sutures form semicircular loops at the apical end of each chamber. Chamber walls are calcareous, finely perforate, and papillate. The primary aperture is terminal and radiate.

### Synonyms and Citations

*Neoflabellina jarvisi* (Cushman). –Said and Kenawy, 1956, p. 135, pl. 2, fig. 42. –van Morkhoven et al., 1986, p. 361, pl. 117. –Bolli et al., 1994, p. 113, figs. 30.18–19, 25.

### Biogeography

Worldwide. Recorded from Mexico, the North and South Atlantic, the Caribbean, the Gulf of Mexico, the Pacific, Europe, and Egypt (van Morkhoven et al., 1986).

### Bathymetry

Lower neritic to bathyal.

### Chronostratigraphy

Late Cretaceous (Maastrichtian) to Late Paleocene (Thanetian).

## References

Bolli, H.M., Beckmann, J.P., and Saunders, J.B., 1994, Benthic Foraminiferal Biostratigraphy of the South Caribbean Region. Cambridge: Cambridge University Press, 408 pp.

Cushman, J.A., 1935, Notes on some American Cretaceous Flabellinas. Contributions from the Cushman Laboratory for Foraminiferal Research, vol. 11, no. 4, pp. 83–89.

Said, R. and Kenawy, A., 1956, Upper Cretaceous and Lower Tertiary foraminifera from northern Sinai, Egypt. Micropaleontology, vol. 2, no. 2, pp. 105–173.

van Morkhoven, F.P.C.M., Berggren, W.A., Edwards, A.S., et al., 1986, Cenozoic cosmopolitan deep-water benthic foraminifera. Bulletin des Centres de Recherches Exploration-Production Elf-Aquitaine, Mémoire, vol. 11, pp. 1–421.

# *Neolenticulina peregrina*

## (Schwager), 1866

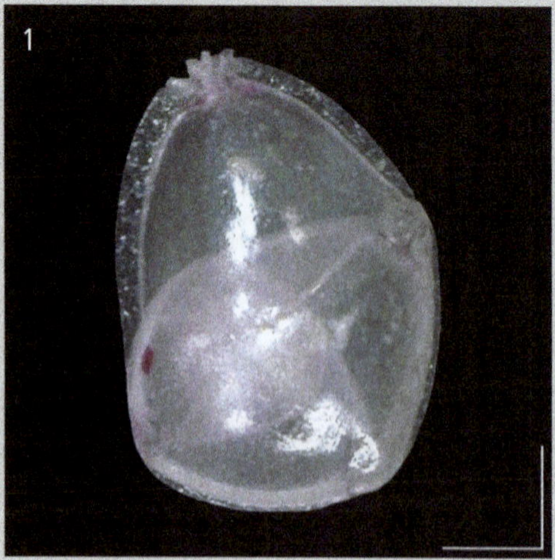


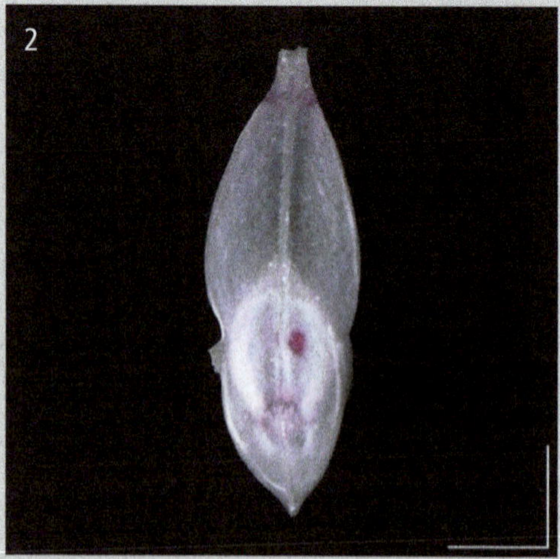


1. *Neolenticulina peregrina.* South China Sea; Station 17930, 1-17, latitude 20° 20.00′ N, longitude 115° 46.90′ E, water depth: 629 meters. Deposited in the South China Sea Collection, Institute of Geosciences, Kiel University, Germany. Lateral view, NHMUK PM DI 0467, scale bar = 106 μm.

2. *Neolenticulina peregrina.* South China Sea; Station 17930, 1-17, latitude 20° 20.00 N, longitude 115° 46.90 E, water depth: 629 meters. Deposited in the South China Sea Collection, Institute of Geosciences, Kiel University, Germany. Edge view, NHMUK PM DI 0468, scale bar = 105 μm.

### Original Designation

*Cristellaria peregrina* Schwager, 1866, p. 245, pl. 7, fig. 89.

### Type Specimen

Depository of holotype not given. Neotype designated by Srinivasan and Sharma (1980), and deposited in the Micropaleontology Section, Department of Geology, Banaras Hindu University, Banaras, India.

### Type Level

Lower to Middle Pliocene (Srinivasan and Sharma, 1980), Obere Thone. Given as Lower Tertiary by Schwager (1866).

### Type Locality

Car Nicobar, Andaman Sea.

### Description

Test forms a compressed, slightly evolute, biumbilical planispire; ovate in outline and elliptical in cross-section with a sharp, keeled periphery. The three to four slightly inflated chambers in the last whorl increase rapidly in size, and are separated by distinct, slightly curved, flush or slightly depressed sutures. Chamber walls are calcareous, finely perforate, and smooth. The primary aperture is terminal, centrally placed, and radiate.

### Synonyms and Citations

?*Cristellaria variabilis* Reuss, 1850, p. 369, pl. 46, figs. 15–16.

*Cristellaria variabilis* Reuss. –Brady, 1884, p. 541, pl. 68, figs. 11–16.

*Lenticulina peregrina* (Schwager). –Barker, 1960, pl. 68, figs. 11–16. –Srinivasan and Sharma, 1980, p. 34, pl. 6, fig. 26. –van Morkhoven et al., 1986, p. 92, pl. 27.

*Neolenticulina peregrina* (Schwager). –Loeblich and Tappan, 1987, p. 406, pl. 447, figs. 9–12, 16. –Loeblich and Tappan, 1994, p. 69, pl. 124, figs. 1–11. –Milker and Schmiedl, 2012, p. 73, fig. 18.21.

*Neolenticulina variabilis* (Reuss). –Jones, 1994, p. 80, pl. 68, figs. 11–16.

### Remarks

Srinivasan and Sharma (1980) noted that the specimens of *Cristellaria varibilis* Reuss figured by Brady (1884) exhibited the range of variation of their species. According to van Morkhoven et al. (1986), the specimens of *Cristellaria varibilis* figured by Reuss (1850, pl. 46, figs. 15–16) show close similarity to *Neolenticulina peregrina* (Schwager), and probably represent the same taxon. As the type material of Reuss was not available for study, these authors preferred to retain them as separate species. Jones (1994) regarded *Neolenticulina peregrina* as a junior synonym of *Cristellaria variabilis.*

### Biogeography

Worldwide. Recorded in the Andaman Sea, the Caribbean, the Pacific, the Gulf of Mexico, the Atlantic, and the Mediterranean (van Morkhoven et al., 1986).

### Bathymetry

Middle neritic to bathyal. Reported from depths between 22 meters (Vella, 1957) and 2092 meters (Pflum and Frerichs, 1976).

### Chronostratigraphy

Late Eocene (Priabonian) to Recent.

## References

Barker, R.W., 1960, Taxonomic notes on the species figured by H.B. Brady in his report on the foraminifera dredged by H.M.S. Challenger during the years 1873–1876. Society of Economic Paleontologists and Mineralogists Special Publication, vol. 9, pp. 1–238.

Brady, H.B., 1884, Report on the foraminifera dredged by H.M.S. Challenger during the years 1873–1876. In Brady, H.B., ed., Report on the Scientific Results of the Voyage of the H.M.S. Challenger during the years 1873–1876. Zoology, vol. 9, pp. 1–814.

Jones, R.W., 1994, The Challenger Foraminifera. Oxford: Oxford University Press, 149 pp.

Loeblich, A.R. Jr and Tappan, H., 1987, Foraminiferal Genera and Their Classification. New York: Van Nostrand Reinhold, 970 pp.

Loeblich, A.R. Jr and Tappan, H., 1994, Foraminifera of the Sahul Shelf and Timor Sea. Special Publications of the Cushman Laboratory for Foraminiferal Research, vol. 31, pp. 1–661.

Milker, Y. and Schmiedl, G., 2012, A taxonomic guide to modern benthic shelf foraminifera of the western Mediterranean Sea. Palaeontologia Electronica, vol. 15, Issue 2;16A, 134 pp.

Pflum, C.E. and Frerichs, W.E., 1976, Gulf of Mexico deep-water foraminifers. Cushman Foundation, Special Publication, vol. 14, pp. 1–125.

Reuss, A.E., 1850, Neue Foraminiferen aus den Schichten des österreichischen Tertiärbeckens. Denkschriften der Kaiserlichen Akademie der Wissenschaften, Mathematisch-Naturwissenschaftliche Classe, vol. 1, pp. 365–390.

Schwager, C., 1866, Fossile Foraminiferen von Kar Nikobar, Reise der Österreichischen Fregatte Novara um Erde in den Jahren 1857, 1858, 1859 unten den Befehlen des Commodore B. Von Wüllerstorf-Urbair. Geologischer Theil, vol. 2, no. 1, Geologische Beobachtung, no. 2, Paläontologische Mittheilung, pp. 187–268.

Srinivasan, M.S. and Sharma, V., 1980, Schwager's Car Nicobar Foraminifera in the Reports of the Novara Expedition – A Revision. New Delhi: Today & Tomorrow's Printers and Publishers, 83 pp.

van Morkhoven, F.P.C.M., Berggren, W.A., Edwards, A.S., et al., 1986, Cenozoic cosmopolitan deep-water benthic foraminifera. Bulletin des Centres de Recherches Exploration-Production Elf-Aquitaine, Mémoire, vol. 11, pp. 1–421.

Vella, P., 1957, Studies in New Zealand foraminifera. New Zealand Geological Survey, Paleontological Bulletin, no. 28, pp. 1–64.

# Nonion havanense

## Cushman and Bermúdez, 1937

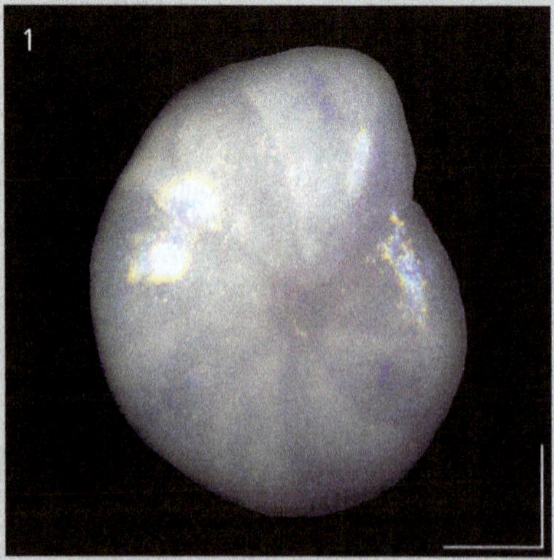

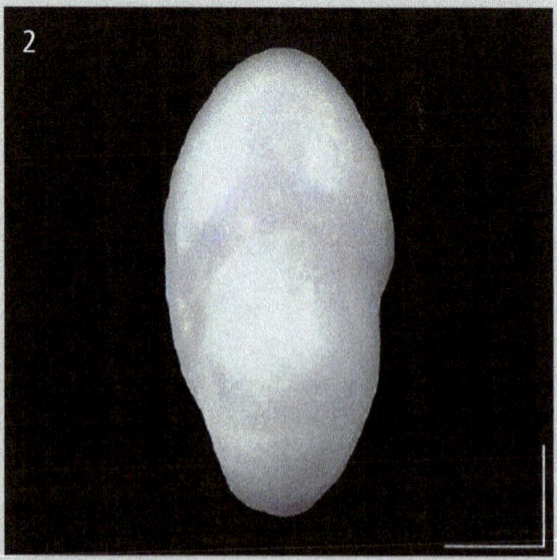

1. *Nonion havanense.* Islas Orcadas Rise, South Atlantic; ODP Leg 114, Hole 702A, Core 3, Section 6, 68–72 cm. Umbilical view, NHMUK PM DI 0469, scale bar = 69 μm.

2. *Nonion havanense.* Islas Orcadas Rise, South Atlantic; ODP Leg 114, Hole 702A, Core 3, Section 6, 68–72 cm. Apertural view, NHMUK PM DI 0470, scale bar = 69 μm.

### Original Designation

*Nonion havanense* Cushman and Bermúdez, 1937, p. 19, pl. 2, figs. 13–14.

### Type Specimen

Holotype (CC 23417) deposited in the collections of the United States National Museum, Washington, DC.

### Type Level

Eocene.

### Type Locality

Bermúdez Station 36, Tejar Consuelo (upper beds), Cerro, Havana, Cuba.

### Description

Test forms a slightly compressed, biumbilicate, involute or slightly evolute planispire; subcircular in outline and subrounded in cross-section. The eight to ten inflated chambers in the last whorl increase slowly in size, and are separated by slightly limbate, flush sutures. Chamber walls are calcareous, finely perforate, and smooth. The primary aperture is a low interiomarginal, equatorial slit, extending laterally nearly to the umbilicus.

### Synonyms and Citations

*Nonion havanense* Cushman and Bermúdez. –Tjalsma and Lohmann, 1983, p. 17, pl. 7, fig. 6. –Miller and Katz, 1987, p. 136, pl. 4, fig. 7. –Thomas, 1990, p. 590. –Bolli et al., 1994, p. 151, figs. 41.14–15. –Ortiz and Thomas, 2006, p. 123, pl. 9, figs. 5–6.

### Remarks

*Nonion havanense* Cushman and Bermúdez exhibits variation in the number of chambers, degree of lateral compression, curvature of sutures, and size (Tjalsma and Lohmann, 1983). Paleocene specimens are often larger than Eocene specimens, and have fewer chambers in the last whorl. *Nonion havanense* Cushman and Bermúdez differs from *Nonion advenum* (Cushman) by the less compressed shape of the test, less limbate sutures, and by the lack of a raised umbilical boss. It differs from *Nonion affine* (Reuss) by its more compressed test, lack of umbilical depression, and lower apertural face (Ortiz and Thomas, 2006).

### Biogeography

Worldwide. Weddell Sea, Antarctic (Thomas, 1990), North Atlantic (Miller and Katz, 1987).

### Bathymetry

Bathyal to abyssal. Wide bathymetric distribution (Tjalsma and Lohmann, 1983). Lower bathyal (Thomas, 1990). Abyssal, greater than 3.5 kilometers (Miller and Katz, 1987).

### Chronostratigraphy

Paleocene (Danian) to Miocene (Tortonian). Early Paleocene (Biozone P1) to Oligocene (Tjalsma and Lohmann, 1983). Abundant in the Early Oligocene, rare to common in the Miocene (Miller and Katz, 1987).

## References


Bolli, H.M., Beckmann, J.P., and Saunders, J.B., 1994, Benthic Foraminiferal Biostratigraphy of the South Caribbean Region. Cambridge: Cambridge University Press, 408 pp.

Cushman, J.A. and Bermúdez, P.J., 1937, Further new species of foraminifera from the Eocene of Cuba. Contributions from the Cushman Laboratory for Foraminiferal Research, vol. 13, pp. 1–29.

Miller, K.G. and Katz, M.E., 1987, Oligocene to Miocene benthic foraminiferal and abyssal circulation changes in the North Atlantic. Micropaleontology, vol. 33, no. 2, pp. 97–149.

Ortiz, S. and Thomas, E., 2006, Lower-middle Eocene benthic foraminifera from the Fortuna Section (Betic Cordillera, southeastern Spain). Micropaleontology, vol. 52, pp. 97–150.

Thomas, E., 1990, Late Cretaceous through Neogene benthic foraminifers. In Barker, P.F., Kennett, J.P., et al., ed., Scientific Results of the Proceedings of the Ocean Drilling Program, vol. 113, pp. 571–594.

Tjalsma, R.C. and Lohmann, G.P., 1983, Paleocene–Eocene bathyal and abyssal benthic foraminifera from the Atlantic Ocean. Micropaleontology, Special Publication, no. 4, pp. 1–90.

# *Nonionella turgida*

## (Williamson), 1858

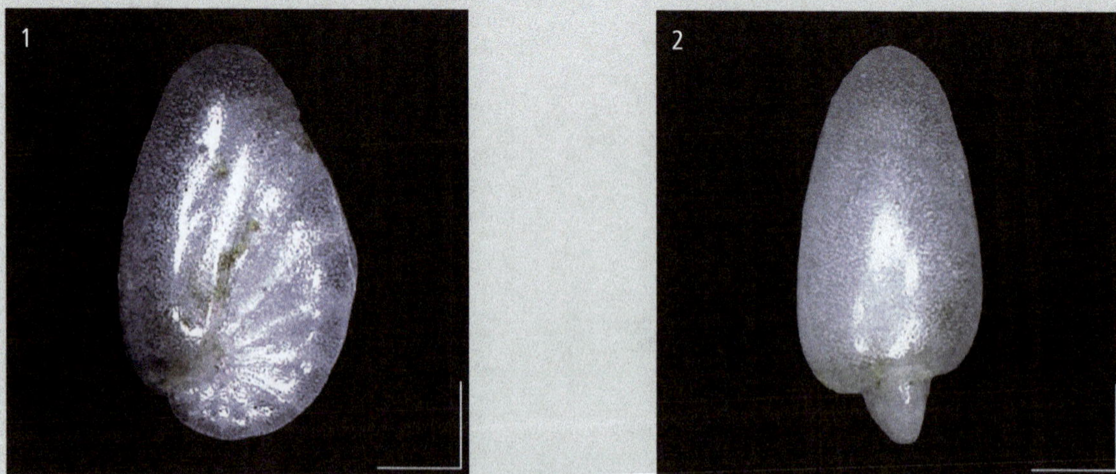

1. *Nonionella turgida*. East of the Shetland Islands, Atlantic Ocean; "Porcupine" Station 67-68, depth: 64–75 fathoms. Deposited in the micropaleontological collections of The Natural History Museum, London; NHMUK ZF2000. Lateral view, NHMUK PM DI 0471, scale bar = 82 μm.

2. *Nonionella turgida*. East of the Shetland Islands, Atlantic Ocean; "Porcupine" Station 67-68, depth: 64–75 fathoms. Deposited in the micropaleontological collections of The Natural History Museum, London; NHMUK ZF2000. Edge view, NHMUK PM DI 0472, scale bar = 80 μm.

### Original Designation
*Rotalina turgida* Williamson, 1858, p. 50, pl. 4, figs. 95–97.

### Type Specimen
Specimens illustrated by Brady (1884) are housed in the micropaleontological collections of The Natural History Museum, London, and are registered in slides ZF2000–ZF2001.

### Type Level
Recent.

### Type Locality
Not designated. Localities given: dredged sands of Skye, Arran, Shetland, and Whitehaven, British Isles.

### Description
Test forms a low, compressed trochospire, with a partially evolute spiral side, an involute umbilical side, and a subrounded periphery. The moderately inflated, broad, low chambers increase rapidly in size (giving the test a flaring appearance), and are separated by curved, depressed sutures. Chamber walls are calcareous, smooth, and finely perforate. The primary aperture is a small interiomarginal, near-equatorial arch, extending onto the umbilical side.

### Synonyms and Citations
*Nonionina turgida* (Williamson). –Brady, 1884, p. 731, pl. 109, figs. 17–19.

*Nonionella turgida* (Williamson). –Barker, 1960, pl. 109, figs. 17–19. –Jones, 1994, p. 108, pl. 109, figs. 17–19. –Mendes et al., 2012, p. 38, fig. 4.4. –Milker and Schmiedl, 2012, p. 113, figs. 26.1–6.

### Biogeography
Worldwide.

### Bathymetry
Neritic to bathyal. Reported between 11 and 1630 fathoms at "Challenger" stations by Jones (1994).

### Chronostratigraphy
Late Miocene to Recent.

## References


Barker, R.W., 1960, Taxonomic notes on the species figured by H.B. Brady in his report on the foraminifera dredged by H.M.S. Challenger during the years 1873–1876. Society of Economic Paleontologists and Mineralogists Special Publication, vol. 9, pp. 1–238.

Brady, H.B., 1884, Report on the foraminifera dredged by H.M.S. Challenger during the years 1873–1876. In Brady, H.B., ed., Report on the Scientific Results of the Voyage of the H.M.S. Challenger during the years 1873–1876. Zoology, vol. 9, pp. 1–814.

Jones, R.W., 1994, The Challenger Foraminifera. Oxford: Oxford University Press, 149 pp.

Mendes, I., Dias, J.A., Schönfeld, J., and Ferreira, O., 2012, Distribution of living foraminifera on the northern Gulf of Cadiz continental shelf. Journal of Foraminiferal Research, vol. 42, pp. 18–38.

Milker, Y. and Schmiedl, G., 2012, A taxonomic guide to modern benthic shelf foraminifera of the western Mediterranean Sea. Palaeontologia Electronica, vol. 15, Issue 2;16A, 134 pp.

Williamson, W.C., 1858, On the Recent Foraminifera of Great Britain. London: Ray Society, 107 pp.

# *Nonionellina labradorica*

## (Dawson), 1860

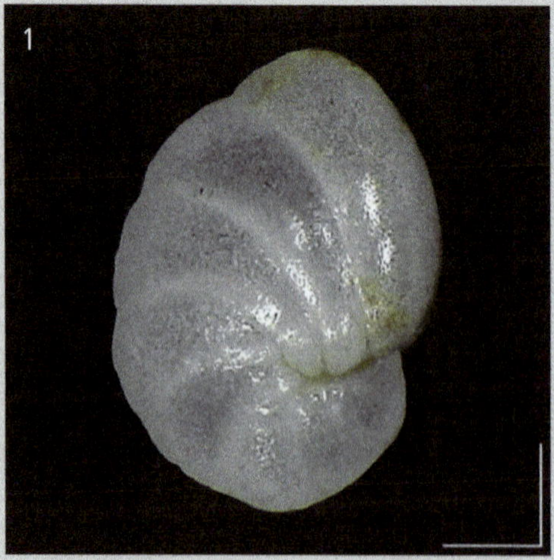



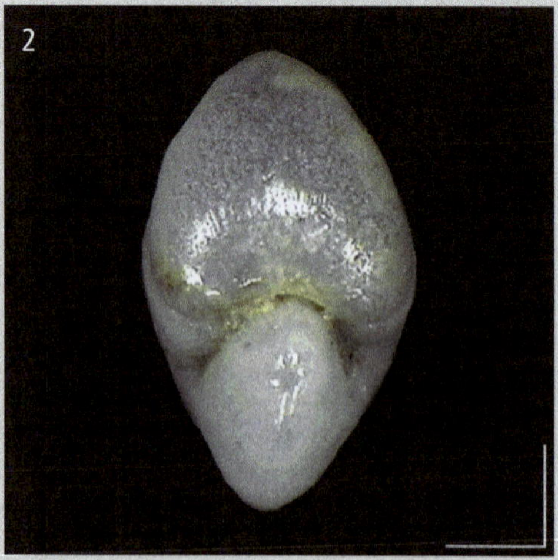



1. *Nonionellina labradorica*. Hebrides slope, Northeast Atlantic. Barra Fan Seep Study, R.W. Jones Collection. Lateral view, NHMUK PM DI 0473, scale bar = 152 μm.

2. *Nonionellina labradorica*. Hebrides slope, Northeast Atlantic. Barra Fan Seep Study, R.W. Jones Collection. Apertural view, NHMUK PM DI 0474, scale bar = 142 μm.

### Original Designation

*Nonionina labradorica* Dawson, 1860, pp. 191–192, fig. 4.

### Type Specimen

No repository given.

### Type Level

Not designated. Levels given: Tertiary, clay, and Recent.

### Type Locality

Not designated. Localities given: Vicinity of Tertiary Bay on the coast of Labrador (Tertiary) and off the coast of Anticosti from depths between 144 and 313 fathoms, Gulf of St Lawrence, Quebec, Canada (Recent).

### Description

Test forms an involute, biconvex, biumbilicate planispire with an ovate, slightly lobulate outline, a subacute periphery, and a broad, triangular apertural face. The (approximately) eight to ten moderately inflated chambers in the last whorl increase gradually in size, except for the ultimate chamber, which may be nearly twice as high as the penultimate chamber. Sutures are gently curved and slightly depressed, but more distinctly incised near the umbilicus. Chamber walls are calcareous, thin, translucent, and finely perforate. The primary aperture is a low, arched slit at the base of the apertural face.

### Synonyms and Citations

*Nonionella labradorica* (Dawson). –Knudsen, 1971, p. 262, pl. 10, figs. 1–2.

*Nonion labradoricum* (Dawson). –Loeblich and Tappan, 1953, p. 86, pl. 17, figs. 1–2.

*Nonionellina labradorica* (Dawson). –Loeblich and Tappan, 1987, p. 617, pl. 689, figs. 8–17. –Narayan et al., 2005, p. 132, pl. 4, figs. 21–22.

### Biogeography

Arctic and Boreal areas (Alaska, Canada, Greenland, Spitzbergen, Norway, Denmark).

### Bathymetry

Neritic to bathyal.

### Chronostratigraphy

Pliocene? to Recent.

## References


Dawson, J.W., Notice of Tertiary fossils from Labrador, Maine, etc., and remarks on the climate of Canada in the newer Pliocene or Pleistocene period. Canadian Naturalist, vol. 5, pp. 188–200.

Knudsen, K.L., 1971, Late Quaternary Foraminifera from Vendsyssel, Denmark and Sandnes, Norway. Bulletin of the Geological Society of Denmark, vol. 21, pt 2–3, pp. 185–292.

Loeblich, A.R. Jr and Tappan, H., 1953, Studies of Arctic Foraminifera. Smithsonian Miscellaneous Collections, vol. 121, no. 7, pp. 1–150.

Loeblich, A.R. Jr and Tappan, H., 1987, Foraminiferal Genera and Their Classification. New York: Van Nostrand Reinhold, 970 pp.

Narayan, Y.R., Barnes, C.R., and Johns, M.J., 2005, Taxonomy and biostratigraphy of Cenozoic foraminifers from Shell Canada wells, Tofino Basin, offshore Vancouver Island. Micropaleontology, vol. 51, no. 2, pp. 101–167.

# *Nummoloculina contraria*

## (d'Orbigny), 1846

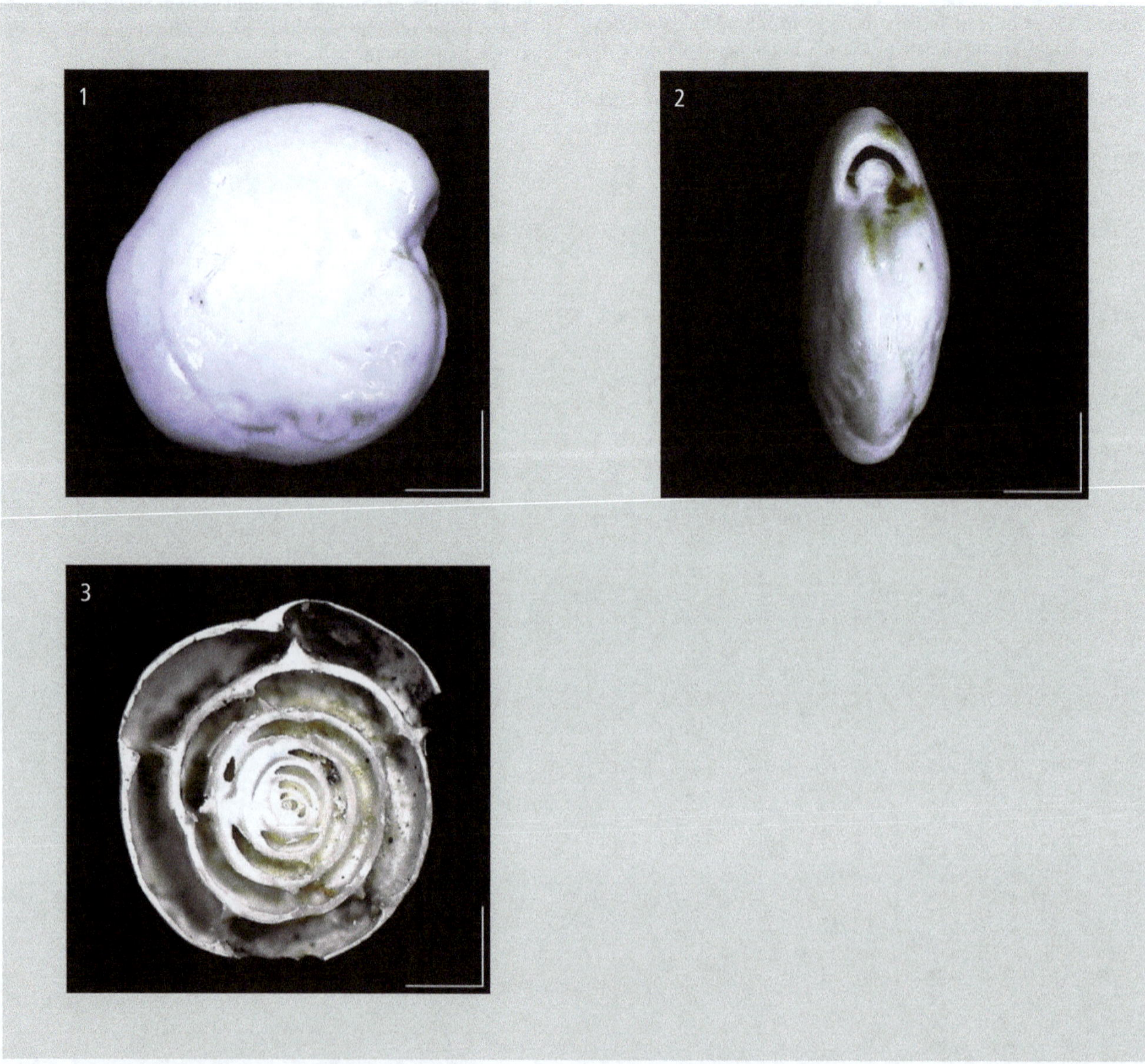


1. *Nummoloculina contraria.* Off Skye, North Atlantic Ocean; "Porcupine" Station AA, depth: 45–60 fathoms. Deposited in the micropaleontological collections of The Natural History Museum, London; NHMUK ZF2105. Lateral view, NHMUK PM DI 0475, scale bar = 246 μm.

2. *Nummoloculina contraria.* Off Skye, North Atlantic Ocean; "Porcupine" Station AA, depth: 45–60 fathoms. Deposited in the micropaleontological collections of The Natural History Museum, London; NHMUK ZF2105. Apertural view, NHMUK PM DI 0476, scale bar = 266 μm.

3. *Nummoloculina contraria.* Off Skye, North Atlantic Ocean; "Porcupine" Station AA, depth: 45–60 fathoms. Deposited in the micropaleontological collections of The Natural History Museum, London; NHMUK ZF2106. Cross-section view, NHMUK PM DI 0477, scale bar = 267 μm.

## Original Designation

*Biloculina contraria* d'Orbigny, 1846, p. 266, pl. 16, figs. 4–6.

## Type Specimen

Depository not given. Lectotype (No. 1981/03/431) designated by Papp and Schmid (1985, pl. 83, figs. 7–9), and deposited in the collections of the Austrian Geological Survey Museum, Vienna, Austria.

## Type Level

Tertiary (Middle Miocene, Badenian, Upper Lagenid Zone fide Papp and Schmid, 1985).

Very fine blue marl layer.

## Type Locality

Brick pits eight leagues south of Vienna, near Baden, Vienna Basin, Austria.

## Description

Test forms a trochospire; initially with two rapidly enlarging chambers per whorl, added in a plane less than 180° apart, later with up to five chambers per whorl, added in a single plane. The overall test shape is ovate in outline and biconvex in cross-section with a rounded periphery. The slightly inflated chambers increase gradually in size, and are separated by fine, slightly depressed sutures. Chamber walls are thick, calcareous, imperforate, smooth, and polished. The primary aperture is large and subtriangular with a broad flap at the end of the final chamber.

## Synonyms and Citations

*Planispirina contraria* (d'Orbigny). –Brady, 1884, p. 713, pl. 11, figs. 10–11.

*Nummoloculina contraria* (d'Orbigny). –Barker, 1960, pl. 11, figs. 10–11. –Papp and Schmid, 1985, p. 90, pl. 83, figs. 7–9. –Loeblich and Tappan, 1987, p. 347, pl. 355, figs. 17–23. –Jones, 1994, p. 27, pl. 11, figs. 10–11.

## Remarks

Type species of the genus *Nummoloculina* Steinmann, 1881.

## Biogeography

Worldwide.

## Bathymetry

Middle neritic to abyssal. Reported between 40 and 2160 fathoms at "Challenger" stations (Jones, 1994).

## Chronostratigraphy

Miocene to Recent.

# References


Barker, R.W., 1960, Taxonomic notes on the species figured by H.B. Brady in his report on the foraminifera dredged by H.M.S. Challenger during the years 1873–1876. Society of Economic Paleontologists and Mineralogists Special Publication, vol. 9, pp. 1–238.

Brady, H.B., 1884, Report on the foraminifera dredged by H.M.S. Challenger during the years 1873–1876. In Brady, H.B., ed., Report on the Scientific Results of the Voyage of the H.M.S. Challenger during the years 1873–1876. Zoology, vol. 9, pp. 1–814.

d'Orbigny, A.D., 1846, Foraminifères fossiles du Bassin Tertiaire de Vienne (Autriche). Paris: Gide et Compe, xxxvii + 312 pp.

Jones, R.W., 1994, The Challenger Foraminifera. Oxford: Oxford University Press, 149 pp.

Loeblich, A.R. Jr and Tappan, H., 1987, Foraminiferal Genera and Their Classification. New York: Van Nostrand Reinhold, 970 pp.

Papp, A. and Schmid, M.E., 1985, Die fossilen Foraminiferen des tertiaeren Beckens von Wien: Revision der Monographie Alcide d'Orbigny (1846). Abhandlungen der Geologischen Bundesanstalt, vol. 37, pp. 1–311.

# *Nuttallides truempyi*

## (Nuttall), 1930

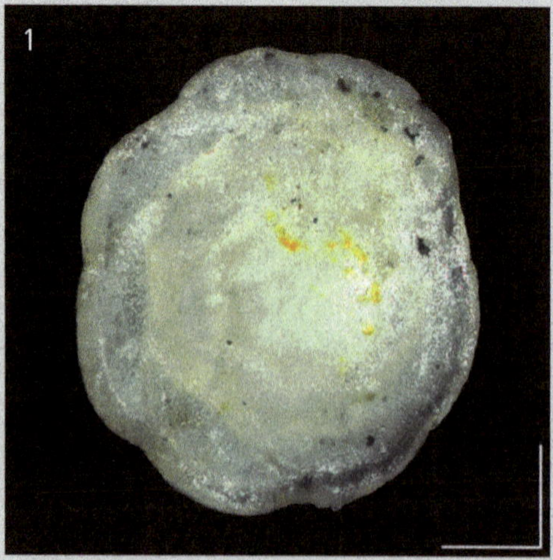

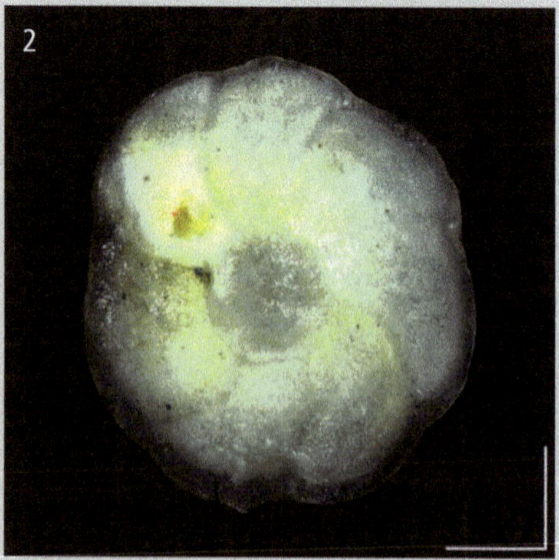

1. *Nuttallides truempyi.* Aragon, 1850 meters north and 85 meters west of El Tule, Mexico. Deposited in the collections of the United States National Museum, Washington, DC; CC 59492. Lectotype. Spiral view, NHMUK PM DI 0478, scale bar = 268 μm.

2. *Nuttallides truempyi.* Aragon, 1850 meters north and 85 meters west of El Tule, Mexico. Deposited in the collections of the United States National Museum, Washington, DC; CC 59492. Lectotype. Umbilical view, NHMUK PM DI 0479, scale bar = 271 μm.

### Original Designation

*Eponides truempyi* Nuttall, 1930, p. 287, pl. 24, figs. 9, 13–14.

### Type Specimen

Cotypes (CC 59491, CC 59492, and CC 59493) deposited in the collections of the United States National Museum, Washington, DC. Van Morkhoven et al. (1986) designated CC 59492 as lectotype (= Nuttall, 1930, pl. 24, fig. 13).

### Type Level

Not designated. Levels given: Upper Eocene, Chapapote Formation; Middle Eocene, Guayabal Formation; Lower Eocene Aragon and Chicontepec Formations.

### Type Locality

Not designated. Localities given: Río La Puerta, 200 meters upstream from La Antigua crossing; 1850 meters north and 85 meters west of El Tule; Alazan Well 33 from 2080 feet, Tampico region, Mexico.

### Description

Test forms a lenticular trochospiral, slightly lobulate in outline and planoconvex to unequally biconvex in cross-section, with an acute, keeled, imperforate periphery. The involute umbilical side exhibits a clear, imperforate, umbilical boss, and is more strongly convex than the evolute spiral side. The crescentic chambers, arranged in four to five whorls, increase slowly in size, and are all visible on the spiral side. The six to thirteen chambers in the last whorl are separated by narrow, flush, tangential sutures, curving into the peripheral keel on the spiral side, and by flush or slightly depressed, sinuous sutures on the umbilical side. Chamber walls are calcareous, finely perforate, and smooth. The primary aperture is an interiomarginal slit extending from the umbilical boss nearly to the periphery.

### Synonyms and Citations

*Asterigerina crassaformis* Cushman and Siegfus, 1935, p. 94, pl. 14, fig. 10.

*Nuttallides subtruempyi* Finlay, 1939, p. 521.

*Eponides bronnimanni* Cushman and Renz, 1946, p. 45, pl. 7, fig. 24.

*Nuttallides truempyi* (Nuttall). –Tjalsma and Lohmann, 1983, p. 17, pl. 6, fig. 4; pl. 17, figs. 4–5; pl. 21, figs. 1–4. –van Morkhoven et al., 1986, p. 288, pl. 96. –Loeblich and Tappan, 1987, p. 603, pl. 669, figs. 24–27. –Thomas, 1990, p. 590, pl. 3, figs. 1–2. –Bolli et al., 1994, p. 370, figs. 58.2–3, 88.13. –Nomura, 1995, p. 276, pl. 3, fig. 1. –Ortiz and Thomas, 2006, p. 123, pl. 9, figs. 7–8.

### Remarks

Type species of the genus *Nuttallides* Finlay, 1939.

*Nuttallides truempyi* (Nuttall) exhibits considerable variation in the shape of the test and the size of the umbilical boss. Beckmann (1954) distinguished two main morphotypes: a lenticular form with eight to thirteen chambers in the last whorl and a large, transparent umbilical mass, and a near planoconvex form with six to nine chambers in the last whorl and a smaller umbilical mass. Tjalsma and Lohmann (1983) did not detect any bathymetric trend in the distribution of these morphotypes. Mackensen and Berggren (1992) reported forms transitional between *Nuttallides truempyi* and *Nuttallides umbonifera* (Cushman) in Upper Eocene sediments from the southern Indian Ocean.

Van Morkhoven et al. (1986) tentatively synonymized *Asterigerina crassaformis* Cushman and Siegfus with *Nuttallides truempyi* (Nuttall). Ortiz and Thomas (2006) examined the holotype and paratypes of *Asterigerina crassaformis*, and considered them to be conspecific with the planoconvex morphotype of *Nuttallides truempyi*.

### Biogeography

Worldwide.

### Bathymetry

Middle bathyal to abyssal (van Morkhoven et al., 1986). Upper depth limit about 500–600 meters according to Berggren and Aubert (1983).

### Chronostratigraphy

Late Cretaceous (Campanian) to Late Eocene (Priabonian).

Berggren and Aubert (1983) considered the last occurrence of *Nuttallides truempyi* to indicate the Eocene–Oligocene boundary in deep-sea sediments. Tjalsma and Lohmann (1983) reported the last occurrence of this species as diachronous, ranging from late Middle Eocene at sites of less than 2000–2500 meters to latest Eocene at sites over 3000 meters. Berggren and Miller (1989) and Mackensen and Berggren (1992) put the last common occurrence of *Nuttallides truempyi* at the Middle–Late Eocene boundary and its last occurrence at the Eocene–Oligocene boundary.

## References


Beckmann, J.P., 1954, Die Foraminiferen der Oceanic Formation (Eocaen–Oligocaen) von Barbados, Kl. Antillen. Eclogae Geologicae Helvetiae, vol. 46, pp. 301–412.

Berggren, W.A. and Aubert, J., 1983, Paleogene benthonic foraminiferal biostratigraphy and bathymetry of the Central Coast Ranges of California. In Brabb, E.E., ed., Studies in Tertiary Stratigraphy of the California Coast Ranges. Professional Paper of the United States Geological Survey, vol. 1213, pp. 4–21.

Berggren, W.A., and Miller, K.G., 1989, Cenozoic bathyal and abyssal calcareous benthic foraminiferal zonations. Micropaleontology, vol. 35, pp. 308–320.

Bolli, H.M., Beckmann, J.P., and Saunders, J.B., 1994, Benthic Foraminiferal Biostratigraphy of the South Caribbean Region. Cambridge: Cambridge University Press, 408 pp.

Cushman, J.A. and Renz, H.H., 1946, The foraminiferal fauna of the Lizard Springs Formation of Trinidad, British West Indies. Special Publications of the Cushman Laboratory for Foraminiferal Research, vol. 18, pp. 1–48.

Cushman, J.A. and Siegfus, S.S., 1935, New species of foraminifera from the Kreyenhagen Shale of Fresno County, California. Contributions from the Cushman Laboratory for Foraminiferal Research, vol. 11, no. 4, pp. 90–95.

Finlay, H.J., 1939, New Zealand foraminifera: key species in stratigraphy – no. 1. Transactions of the Royal Society of New Zealand, vol. 68, no. 4, pp. 504–543.

Loeblich, A.R. Jr and Tappan, H., 1987, Foraminiferal Genera and Their Classification. New York: Van Nostrand Reinhold, 970 pp.

Mackensen, A. and Berggren, W.A. 1992. Paleogene benthic foraminifers from the southern Indian Ocean (Kerguelen Plateau): biostratigraphy and paleoecology. In Wise, S.W. Jr, Schlich, R., et al., ed., Scientific Results of the Proceedings of the Ocean Drilling Program, vol. 120, pp. 603–630.

Nomura, R., 1995, Paleogene to Neogene deep-sea paleoceanography in the eastern Indian Ocean: benthic foraminifera from ODP Sites 747, 757 and 758. Micropaleontology, vol. 41, pp. 251–290.

Nuttall, W.L.F., 1930, Eocene foraminifera from Mexico. Journal of Paleontology, vol. 4, pp. 271–293.

Ortiz, S. and Thomas, E., 2006, Lower–middle Eocene benthic foraminifera from the Fortuna Section (Betic Cordillera, southeastern Spain). Micropaleontology, vol. 52, pp. 97–150.

Thomas, E., 1990, Late Cretaceous through Neogene benthic foraminifers. In Barker, P.F., Kennett, J.P., et al., ed., Scientific Results of the Proceedings of the Ocean Drilling Program, vol. 113, pp. 571–594.

Tjalsma, R.C. and Lohmann, G.P., 1983, Paleocene–Eocene bathyal and abyssal benthic foraminifera from the Atlantic Ocean. Micropaleontology, Special Publication, no. 4, pp. 1–90.

van Morkhoven, F.P.C.M., Berggren, W.A., Edwards, A.S., et al., 1986, Cenozoic cosmopolitan deep-water benthic foraminifera. Bulletin des Centres de Recherches Exploration-Production Elf-Aquitaine, Mémoire, vol. 11, pp. 1–421.

# *Nuttallides umbonifera*

## (Cushman), 1933

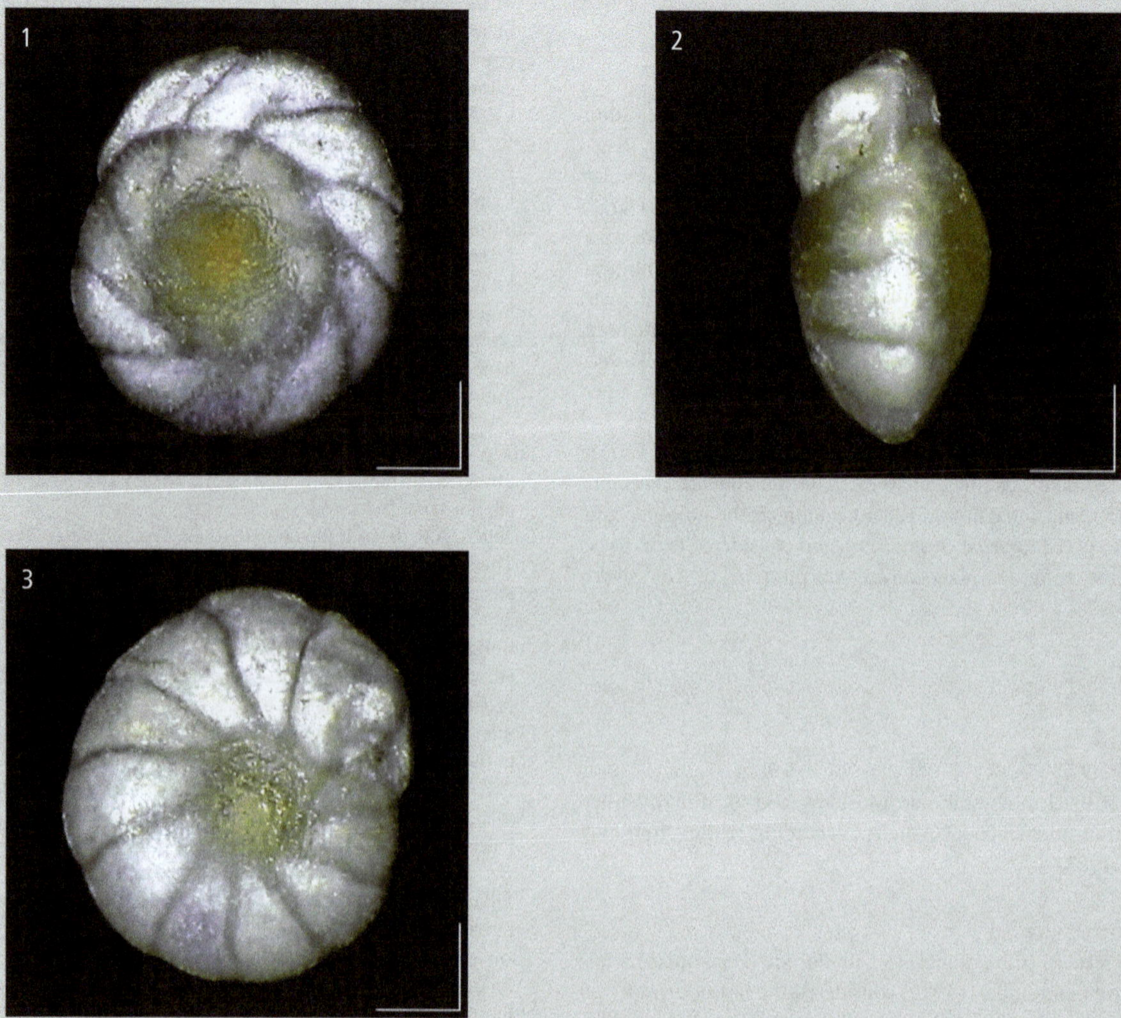

1. *Nuttallides umbonifera.* Walvis Ridge, South Atlantic; DSDP Leg 40, Site 363, Core 2, Section 4, 63–70 cm. Spiral view, NHMUK PM DI 0480, scale bar = 102 µm.

2. *Nuttallides umbonifera.* Walvis Ridge, South Atlantic; DSDP Leg 40, Site 363, Core 2, Section 4, 63–70 cm. Apertural view, NHMUK PM DI 0481, scale bar = 102 µm.

3. *Nuttallides umbonifera.* Walvis Ridge, South Atlantic; DSDP Leg 40, Site 363, Core 2, Section 4, 63–70 cm. Umbilical view, NHMUK PM DI 0482, scale bar = 88 µm.

## Original Designation

*Pulvinulinella umbonifera* Cushman, 1933, p. 90, pl. 9, fig. 9.

## Type Specimen

Holotype (CC 26162) deposited in the collections of the United States National Museum, Washington, DC.

## Type Level

Recent.

## Type Locality

"Albatross" Station H3833, off Paumoto Islands, latitude 15° 42′ S, longitude 148° 44′ W, 2243 fathoms.

## Description

Test forms a small, lenticular trochospire; slightly lobulate in outline, and unequally biconvex in cross-section, with a subacute, keeled, imperforate periphery. The involute umbilical side exhibits a clear, imperforate, umbilical boss, and is more strongly convex than the evolute spiral side. The six to twelve crescentic chambers in the final whorl increase slowly in size, and are separated by slightly depressed, slightly limbate, strongly oblique sutures curving into the peripheral keel on the spiral side, and by slightly depressed, sinuous sutures on the umbilical side. Chamber walls are calcareous, finely perforate, and smooth. The primary aperture is an interiomarginal slit extending from the umbilical boss nearly to the periphery.

## Synonyms and Citations

*Truncatulina pygmaea* Hantken. –Brady, 1884, p. 666, pl. 95, figs. 9–10.

*Eponides bradyi* Earland. –Barker, 1960, pl. 95, figs. 9–10.

*Nuttallides umbonifera* (Cushman). –Miller and Katz, 1987, p. 136, pl. 5, fig. 5. –Hermelin, 1989, p. 69, pl. 12, figs. 15–17. –Thomas, 1990, p. 590.

*Osangulariella umbonifera* (Cushman). –Jones, 1994, p. 99, pl. 95, figs. 9–10.

*Nuttallides umboniferus* (Cushman). –Nomura, 1995, p. 276, pl. 3, fig. 2.

## Remarks

*Nuttallides umbonifera* (Cushman) is regarded as an indicator of Antarctic Bottom Water in the Pacific and Indian Oceans (Corliss, 1979). In the Weddell Sea it is associated with carbonate corrosive bottom water (Mackensen et al., 1990).

## Biogeography

Worldwide. Recorded from the North Atlantic (Miller and Katz, 1987), the Weddell Sea, Antarctica (Thomas, 1990), the Gulf of Mexico (Katz and Miller, 1993), the Kerguelen Plateau (Mackensen and Berggren, 1992), the Indian Ocean (Nomura, 1995), and the Pacific Ocean (Hermelin, 1989).

## Bathymetry

Primarily lower bathyal to abyssal. Abyssal, greater than 3.5 kilometers (Miller and Katz, 1987), lower bathyal (Thomas, 1990), lower bathyal, rarely upper bathyal (Katz and Miller, 1993).

## Chronostratigraphy

Latest Paleocene or Early Eocene to Recent. Common to abundant in Oligocene to Miocene (Miller and Katz, 1987). Rare from topmost Paleocene to Early Eocene, common from Middle Eocene to Pleistocene (Thomas, 1990). First common occurrence at the Middle–Late Eocene boundary in ODP samples from the Kerguelen Plateau (Mackensen and Berggren, 1992).

# References


Barker, R.W., 1960, Taxonomic notes on the species figured by H.B. Brady in his report on the foraminifera dredged by H.M.S. Challenger during the years 1873–1876. Society of Economic Paleontologists and Mineralogists Special Publication, vol. 9, pp. 1–238.

Brady, H.B., 1884, Report on the foraminifera dredged by H.M.S. Challenger during the years 1873–1876. In Brady, H.B., ed., Report on the Scientific Results of the Voyage of the H.M.S. Challenger during the years 1873–1876. Zoology, vol. 9, pp. 1–814.

Corliss, B.H., 1979, Recent deep-sea benthonic foraminiferal distributions in the southeast Indian Ocean: inferred bottom water routes and ecological implications. Marine Geology, vol. 31, pp. 115–138.

Cushman, J.A., 1933, Some new Recent foraminifera from the tropical Pacific. Contributions from the Cushman Laboratory for Foraminiferal Research, vol. 9, pp. 77–95.

Hermelin, J.O.R., 1989, Pliocene benthic foraminifera from the Ontong-Java Plateau (Western Equatorial Pacific Ocean): faunal response to changing paleoenvironment. Special Publication of the Cushman Foundation for Foraminiferal Research, vol. 26, pp. 1–143.

Jones, R.W., 1994, The Challenger Foraminifera. Oxford: Oxford University Press, 149 pp.

Katz, M.E. and Miller, K.G., 1993, Latest Oligocene to Earliest Pliocene benthic foraminiferal biofacies of the northeastern Gulf of Mexico. Micropaleontology, vol. 39, no. 4, pp. 367–403.

Mackensen, A. and Berggren, W.A., 1992, Paleogene benthic foraminifers from the southern Indian Ocean (Kerguelen Plateau): biostratigraphy and paleoecology. In Wise, S.W. Jr, Schlich, R. et al., ed., Scientific Results of the Proceedings of the Ocean Drilling Program, vol. 120, pp. 603–630.

Mackensen, A., Grobe, H., Kuhn, G., and Fütterer, D.K., 1990, Benthic foraminiferal assemblages from the eastern Weddell Sea between 68° and 73° S: distribution, ecology and fossilization potential. Marine Micropaleontology, vol. 16, pp. 241–283.

Miller, K.G. and Katz, M.E., 1987, Oligocene to Miocene benthic foraminiferal and abyssal circulation changes in the North Atlantic. Micropaleontology, vol. 33, no. 2, pp. 97–149.

Nomura, R., 1995, Paleogene to Neogene deep-sea paleoceanography in the eastern Indian Ocean: benthic foraminifera from ODP Sites 747, 757 and 758. Micropaleontology, vol. 41, pp. 251–290.

Thomas, E., 1990, Late Cretaceous through Neogene benthic foraminifers. In Barker, P.F., Kennett, J.P., et al., ed., Scientific Results of the Proceedings of the Ocean Drilling Program, vol. 113, pp. 571–594.

# Oolina hexagona

## (Williamson), 1848

1. *Oolina hexagona*. Not localised, the slide contains syntypic material from Oban, Kyleakin, Lamlash Bay, Scarborough, Boston or March, British Isles. Deposited in the micropaleontological collections of The Natural History Museum, London; NHMUK 96.8.13.67. Lectotype. Lateral view, NHMUK PM DI 0483, scale bar = 42 µm.

### Original Designation

*Entosolenia squamosa* Montagu var. *hexagona* Williamson, 1848, p. 20, pl. 2, fig. 23.

### Type Specimen

Lectotype, designated by Jones (1984), is housed in the micropaleontological collections of The Natural History Museum, London, UK (Slide NHMUK 96.8.13.67).

### Type Level

Recent, surface sediments.

### Type Locality

Not designated. Localities given: Oban, Kyleakin, Lamlash Bay, Scarborough, Boston, and March, British Isles.

### Description

Test forms a unilocular ovate chamber, circular in cross-section. The chamber wall is calcareous, finely perforate, and ornamented with a pattern of raised polygonal ridges arranged in a honeycomb pattern. The primary aperture is round, terminal, and produced on a short neck, with a short entosolenian tube.

### Synonyms and Citations

*Lagena hexagona* (Williamson). –Brady, 1884, p. 472, pl. 58, fig. 33. –Cushman and Todd, 1945, p. 33, pl. 5, fig. 14.

*Oolina hexagona* (Williamson). –Barker, 1960, pl. 58, fig. 33. –Hermelin, 1989, p. 56, pl. 10, fig. 5. –Jones, 1994, p. 66, pl. 58, fig. 33. –Robertson, 1998, p. 100, pl. 37, fig. 2.

### Remarks

Differs from *Oolina melo* d'Orbigny in its reticulae, which are hexagonal rather than quadrate, and are arranged in slightly oblique rather than longitudinal rows (Robertson, 1998).

Jones (1994) referred the specimen illustrated in pl. 58, fig. 32 to *Oolina squamosa* (Montagu).

### Biogeography

Worldwide.

### Bathymetry

Bathyal to abyssal. Reported from 2425 fathoms at South Pacific "Challenger" Station (Jones, 1994).

### Chronostratigraphy

Miocene to Recent.

## References

Barker, R.W., 1960, Taxonomic notes on the species figured by H.B. Brady in his report on the foraminifera dredged by H.M.S. Challenger during the years 1873–1876. Society of Economic Paleontologists and Mineralogists Special Publication, vol. 9, pp. 1–238.

Brady, H.B., 1884, Report on the foraminifera dredged by H.M.S. Challenger during the years 1873–1876. In Brady, H.B., ed., Report on the Scientific Results of the Voyage of the H.M.S. Challenger during the years 1873–1876. Zoology, vol. 9, pp. 1–814.

Cushman, J.A. and Todd, R., 1945, Miocene foraminifera from Buff Bay, Jamaica. Special Publications of the Cushman Laboratory for Foraminiferal Research, vol. 15, pp. 1–73.

Hermelin, J.O.R., 1989, Pliocene benthic foraminifera from the Ontong-Java Plateau (Western Equatorial Pacific Ocean): faunal response to changing paleoenvironment. Special Publication of the Cushman Foundation for Foraminiferal Research, vol. 26, pp. 1–143.

Jones, R.W., 1984, On the designation of lectotypes for certain species and subspecies of unilocular hyaline foraminifera; Part 1 – Those housed in the British Museum (Natural History). Journal of Micropalaeontology, vol. 3, pp. 63–69.

Jones, R.W., 1994, The Challenger Foraminifera. Oxford: Oxford University Press, 149 pp.

Robertson, B.E., 1998, Systematics and paleoecology of the benthic foraminiferida from the Buff Bay section, Miocene of Jamaica. Micropaleontology, vol. 44, supplement 2, pp. 1–266.

Williamson, W.C., 1848, On the Recent British species of the genus Lagena. Annals and Magazine of Natural History, ser. 2, vol. 1, pp. 1–20.

# *Oridorsalis umbonatus*

## (Reuss), 1851

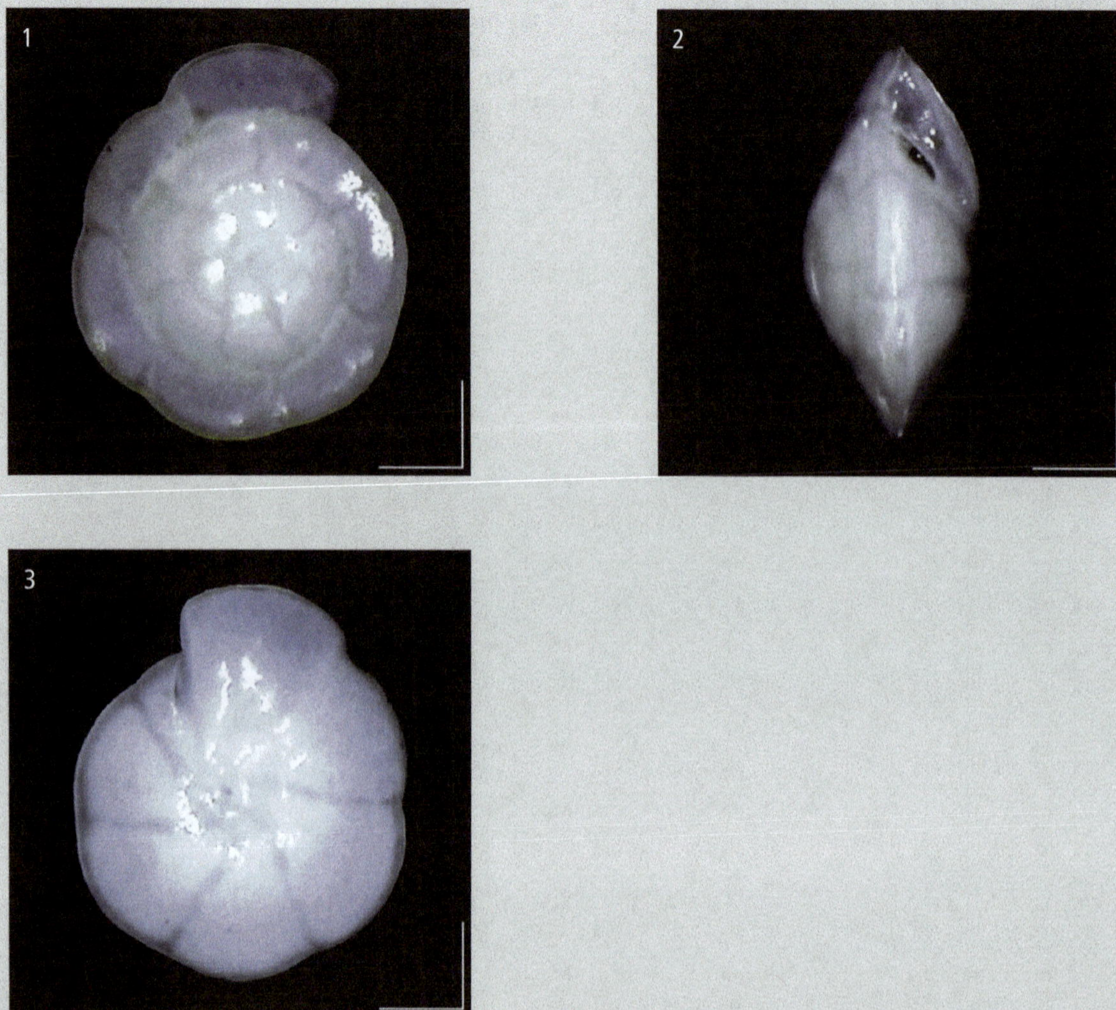


1. *Oridorsalis umbonatus.* West coast of Patagonia, Pacific Ocean; "Challenger" Station 308, depth: 175 fathoms. Deposited in the micropaleontological collections of The Natural History Museum, London; NHMUK ZF2250. Spiral view, NHMUK PM DI 0484, scale bar = 102 μm.

2. *Oridorsalis umbonatus.* West coast of Patagonia, Pacific Ocean; "Challenger" Station 308, depth: 175 fathoms. Deposited in the micropaleontological collections of The Natural History Museum, London; NHMUK ZF2250. Apertural view, NHMUK PM DI 0485, scale bar = 102 μm.

3. *Oridorsalis umbonatus.* West coast of Patagonia, Pacific Ocean; "Challenger" Station 308, depth: 175 fathoms. Deposited in the micropaleontological collections of The Natural History Museum, London; NHMUK ZF2250. Umbilical view, NHMUK PM DI 0486, scale bar = 88 μm.

## Original Designation
*Rotalina umbonata* Reuss, 1851, p. 75, pl. 5, fig. 35.

## Type Specimen
Depository not given.

## Type Level
Originally given as Eocene, Septarienthon.

## Type Locality
Not designated. Localities given: Hermsdorf and Freienwalde, near Berlin, Germany.

## Description
Test forms a lenticular, compressed, low trochospire; subcircular to slightly lobulate in outline, and unequally biconvex in cross-section, with an evolute, convex spiral side, an involute, less convex umbilical side, and a subacute, keeled periphery. The five to six moderately inflated chambers in the last whorl increase gradually in size, and are separated by slightly depressed sutures; radial on the spiral side and strongly sinuous on the umbilical side. Chamber walls are calcareous, finely perforate, and smooth. The primary aperture is an interiomarginal slit extending from the periphery almost to the closed umbilicus. Small supplementary apertures occur at the junction between spiral and intercameral sutures on the spiral side, and at the sinuous bend along the sutures on the umbilical side.

## Synonyms and Citations
*Pulvinulina umbonata* (Reuss). –Brady, 1884, p. 695, pl. 105, fig. 2.
*Truncatulina tenera* Brady, 1884, p. 665, pl. 95, fig. 11.
*Eponides umbonata* (Reuss). –Nuttall, 1932, p. 26, pl. 6, figs. 4–5.
*Oridorsalis umbonatus* (Reuss). –Tjalsma and Lohmann, 1983, p. 18, pl. 6, fig. 8. –Bolli et al., 1994, p. 372, figs. 58.10–13, 51. –Hermelin, 1989, p. 81, pl. 16, figs. 1–5. –van Marle, 1991, p. 138, pl. 11, figs. 13–15. –Kuhnt et al., 2002, p. 150, pl. 11. figs. 9–11. –Ortiz and Thomas, 2006, p. 124, pl. 9, fig. 9.
*Oridorsalis umbonata* (Reuss). –Whittaker, 1988, p. 137, pl. 19, figs. 1–3. –Jones, 1994, p. 99, pl. 95, fig. 11; p. 104, pl. 105, fig. 2.

## Remarks
*Oridorsalis umbonatus* (Reuss) displays variation in the degree of convexity of the test and the acuteness of the periphery (Tjalsma and Lohmann, 1983). Whittaker (1988) found that, in contrast to the specimens illustrated by Reuss (1851), the sutures on the umbilical side of the specimens he examined were curved, particularly at the inner end. Taxonomic affinities of this species are discussed in Whittaker (1988).

## Biogeography
Worldwide. Recorded from Europe, the Atlantic, Pacific, Gulf of Mexico, Indian Ocean, Mexico, eastern Indonesia (van Marle, 1991), Java, Guam, Papua New Guinea (see references in Hermelin, 1989), Weddell Sea, and Antarctica (Thomas, 1990).

## Bathymetry
Lower neritic to abyssal. Wide bathymetric distribution (Tjalsma and Lohmann, 1983). Reported between 166 and 3125 fathoms at "Challenger" stations (Jones, 1994), lower bathyal (Thomas, 1990).

## Chronostratigraphy
Middle Paleocene to Recent. Early Eocene to Middle Miocene in Trinidad (Bolli et al., 1994). Rare in the Paleocene and frequent to abundant during the latest Paleocene and Eocene (Tjalsma and Lohmann, 1983). Oligocene to Recent (Jones, 1994). Topmost Paleocene to Pleistocene, common from Middle Eocene onwards (Thomas, 1990).

# References


Barker, R.W., 1960, Taxonomic notes on the species figured by H.B. Brady in his report on the foraminifera dredged by H.M.S. Challenger during the years 1873–1876. Society of Economic Paleontologists and Mineralogists Special Publication, vol. 9, pp. 1–238.

Bolli, H.M., Beckmann, J.P., and Saunders, J.B., 1994, Benthic Foraminiferal Biostratigraphy of the South Caribbean Region. Cambridge: Cambridge University Press, 408 pp.

Brady, H.B., 1884, Report on the foraminifera dredged by H.M.S. Challenger during the years 1873–1876. In Brady, H.B., ed., Report on the Scientific Results of the Voyage of the H.M.S. Challenger during the years 1873–1876. Zoology, vol. 9, pp. 1–814.

Hermelin, J.O.R., 1989, Pliocene benthic foraminifera from the Ontong-Java Plateau (Western Equatorial Pacific Ocean): faunal response to changing paleoenvironment. Special Publication of the Cushman Foundation for Foraminiferal Research, vol. 26, pp. 1–143.

Jones, R.W., 1994, The Challenger Foraminifera. Oxford: Oxford University Press, 149 pp.

Kuhnt, W., Holbourn, A.E., and Zhao, Q., 2002, The early history of the South China Sea: Evolution of Oligocene–Miocene deep water environments. Revue de Micropaléontologie, vol. 45, pp. 99–159.

Nuttall, W.L.F., 1932, Lower Oligocene foraminifera from Mexico. Journal of Paleontology, vol. 6, pp. 3–35.

Ortiz, S. and Thomas, E., 2006, Lower–middle Eocene benthic foraminifera from the Fortuna Section (Betic Cordillera, southeastern Spain). Micropaleontology, vol. 52, pp. 97–150.

Reuss, A.E., 1851, Ueber die fossilen Foraminiferen und Entomostraceen der Septarienthone der Umgegend von Berlin. Zeitschrift der Deutschen Geologischen Gesellschaft, Berlin, vol. 3, pp. 49–91.

Thomas, E., 1990, Late Cretaceous through Neogene benthic foraminifers. In Barker, P.F., Kennett, J.P., et al., ed., Scientific Results of the Proceedings of the Ocean Drilling Program, vol. 113, pp. 571–594.

Tjalsma, R.C. and Lohmann, G.P., 1983, Paleocene–Eocene bathyal and abyssal benthic foraminifera from the Atlantic Ocean. Micropaleontology, Special Publication, no. 4, pp. 1–90.

van Marle, L.J., 1991, Eastern Indonesian Late Cenozoic smaller benthic foraminifera. Verhandelingen der Koninklijke Nederlanse Akademie van Wetenschappen, Afdeeling Natuurkunde, Eerste Reeks, vol. 34, pp. 1–328.

Whittaker, J.E., 1988, Benthic Foraminifera from Ecuador: Taxonomy and Distribution of Smaller Benthic Foraminifera from Coastal Ecuador (Late Oligocene–Late Pliocene). London: British Museum (Natural History), xi + 194 pp.

# Osangularia culter

## (Parker and Jones), 1865

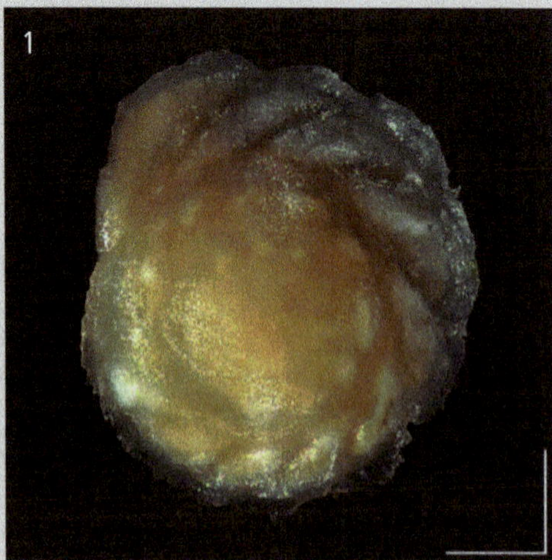

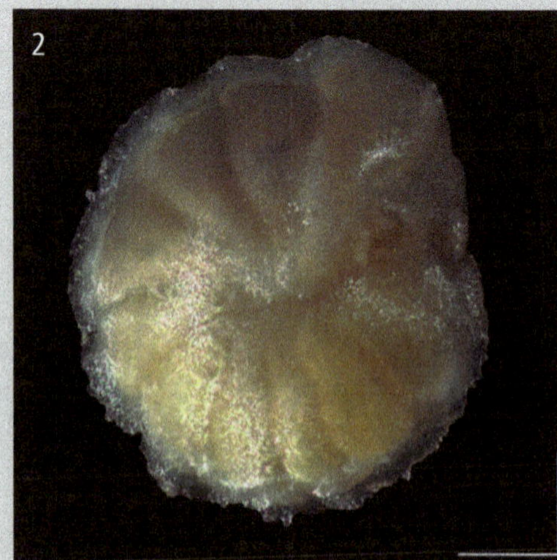

1. *Osangularia culter*. Deposited in the Lutze Collection (West Africa), no. 194. Institute of Geosciences, Kiel University, Germany. Spiral view, NHMUK PM DI 0487, scale bar = 130 μm.

2. *Osangularia culter*. Deposited in the Lutze Collection (West Africa), no. 194. Institute of Geosciences, Kiel University, Germany. Umbilical view, NHMUK PM DI 0743, scale bar = 135 μm.

### Original Designation

*Planorbulina farcta* var. *ungeriana* subvar. *culter* Parker and Jones, 1865, pp. 382, 421, pl. 19, fig. 1.

### Type Specimen

Depository not given.

### Type Level

Recent.

### Type Locality

Tropical Atlantic Ocean, latitude 2° 20′ N, longitude 28° 44′ W, 1080 fathoms.

### Description

Test forms a lenticular, low trochospire; subcircular in outline and unequally biconvex in cross-section with a more flattened, evolute spiral side, and a more convex, involute umbilical side. The periphery is acute with a broad, thin, serrate keel, approximately equal in breadth around the periphery. The eight or more slightly inflated chambers in the last whorl increase gradually in size, and are separated by oblique, slightly raised sutures on the spiral side, and by slightly depressed, straight or slightly curved sutures on the umbilical side. Chamber walls are calcareous, finely perforate, and smooth. The primary aperture is an interiomarginal slit, extending from the umbilicus to the periphery, splitting off perpendicularly close to the periphery, with an overall L-shape.

### Synonyms and Citations

*Anomalina bengalensis* Schwager, 1866, p. 259, pl. 7, fig. 111.
*Truncatulina culter* (Parker and Jones). –Brady, 1884, p. 668, pl. 96, fig. 3.
*Osangularia culter* (Parker and Jones). –Phleger et al., 1953, p. 42, pl. 9, figs. 11–16. –Belford, 1966, p. 173, pl. 35, figs. 1–5. –Kohl, 1985, p. 96, pl. 34, fig. 6. –Hermelin, 1989, p. 84, pl. 16, figs. 11–13. –van Marle, 1991, p. 140, pl. 12, figs. 2–4. –Robertson, 1998, p. 238, pl. 95, figs. 2–3. –Nomura, 1995, p. 278, pl. 4, fig. 8.
*Osangularia bengalensis* (Schwager). –Barker, 1960, pl. 96, fig. 3. –Jones, 1994, p. 100, pl. 96, fig. 3.

### Remarks

Hermelin (1989) reported that *Osangularia bengalensis* (Schwager) may represent a shallow-water form of *Osangularia culter*

(Parker and Jones), and regarded it to be a junior synonym of *Osangularia culter*.

### Biogeography
Worldwide.

### Bathymetry
Bathyal. Reported between 495 and 1760 meters in eastern Indonesia (van Marle, 1991). Middle to lower bathyal on the Sunda Shelf of the South China Sea (Szarek et al., 2009).

### Chronostratigraphy
Miocene to Recent.

## References


Barker, R.W., 1960, Taxonomic notes on the species figured by H.B. Brady in his report on the foraminifera dredged by H.M.S. Challenger during the years 1873–1876. Society of Economic Paleontologists and Mineralogists Special Publication, vol. 9, pp. 1–238.

Belford, D.J., 1966, Miocene and Pliocene smaller foraminifera from Papua and New Guinea. Australia Bureau of Mineral Resources Geology and Geophysics Bulletin, vol. 79, pp. 1–306.

Brady, H.B., 1884, Report on the foraminifera dredged by H.M.S. Challenger during the years 1873–1876. In Brady, H.B., ed., Report on the Scientific Results of the Voyage of the H.M.S. Challenger during the years 1873–1876. Zoology, vol. 9, pp. 1–814.

Hermelin, J.O.R., 1989, Pliocene benthic foraminifera from the Ontong-Java Plateau (Western Equatorial Pacific Ocean): faunal response to changing paleoenvironment. Special Publication of the Cushman Foundation for Foraminiferal Research, vol. 26, pp. 1–143.

Jones, R.W., 1994, The Challenger Foraminifera. Oxford: Oxford University Press, 149 pp.

Kohl, B., 1985. Early Pliocene benthic foraminifers from the Salina Basin, southeastern Mexico. Bulletin of American Paleontology, vol. 88, no. 322, 173 pp.

Nomura, R., 1995, Paleogene to Neogene deep-sea paleoceanography in the eastern Indian Ocean: benthic foraminifera from ODP Sites 747, 757 and 758. Micropaleontology, vol. 41, pp. 251–290.

Parker, W.K. and Jones, T.R., 1865, On some foraminifera from the North Atlantic and Arctic Oceans, including Davis Straits and Baffin's Bay. Philosophical Transactions of the Royal Society, vol. 155, pp. 325–441.

Phleger, F.B., Parker, F.L., and Pierson, J.F., 1953, North Atlantic foraminifera. Reports of the Swedish Deep-Sea Expedition, 1947–1948, vol. 7, pp. 3–122.

Robertson, B.E., 1998, Systematics and paleoecology of the benthic foraminiferida from the Buff Bay section, Miocene of Jamaica. Micropaleontology, vol. 44, supplement 2, pp. 1–266.

Szarek, R., Kuhnt, W., Kawamura, H., and Nishi, H., 2009, Distribution of recent benthic foraminifera along the continental slope of the Sunda Shelf (South China Sea). Marine Micropaleontology, vol. 71, pp. 41–59.

van Marle, L.J., 1991, Eastern Indonesian Late Cenozoic smaller benthic foraminifera. Verhandelingen der Koninklijke Nederlanse Akademie van Wetenschappen, Afdeeling Natuurkunde, Eerste Reeks, vol. 34, pp. 1–328.

# Osangularia velascoensis

## (Cushman), 1925

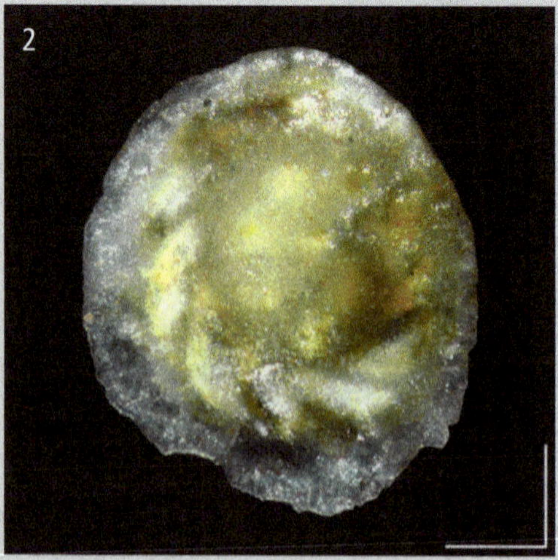


1. *Osangularia velascoensis.* Velasco Shale, 5 km southwest of La Bolsa, west side of Mactezuma River, Hacienda Santa Ines, State of San Luis Potosi, Mexico. Deposited in the collections of the United States National Museum, Washington, DC; CC 4344. Holotype. Umbilical view, NHMUK PM DI 0488, scale bar = 105 μm.

2. *Osangularia velascoensis.* Velasco Shale, 5 km southwest of La Bolsa, west side of Mactezuma River, Hacienda Santa Ines, State of San Luis Potosi, Mexico. Deposited in the collections of the United States National Museum, Washington, DC; CC 4344. Holotype. Spiral view, NHMUK PM DI 0489, scale bar = 105 μm.

### Original Designation

*Truncatulina velascoensis* Cushman, 1925, p. 20, pl. 3, fig. 2.

### Type Specimen

Holotype (CC 4344) deposited in the collections of the United States National Museum, Washington, DC.

### Type Level

Velasco Shale. Given as Upper Cretaceous by Cushman (1925).

### Type Locality

Five kilometers southwest of La Bolsa, on west side of the Moctezuma River, Hacienda Santa Ines, State of San Luis Potosi, Mexico.

### Description

Test forms a lenticular, low trochospire; subcircular in outline, and unequally biconvex in cross-section with a flatter, evolute spiral side, more convex, involute umbilical side, and acute, keeled periphery. The keel is thin, often appearing irregular and serrate, and decreasing in width in the last chambers. The ten or more slightly inflated chambers in the last whorl increase gradually in size, and are separated by oblique, raised sutures on the spiral side (spiral suture very distinct), and by slightly depressed, curved sutures on the umbilical side. Chamber walls are calcareous, finely perforate, and smooth. A wide umbilical boss covers the umbilicus. The primary aperture is an interiomarginal slit, extending from the umbilicus to the periphery, splitting off perpendicularly close to the periphery, with an overall L-shape.

### Synonyms and Citations

*Rotalia velascoensis* (Cushman). –White, 1928, p. 290, pl. 39, fig. 5.

*Pulvinulinella velascoensis* (Cushman). –Cushman and Jarvis, 1932, p. 48, pl. 14, fig. 6. –Cushman and Renz, 1946, p. 46, pl. 8, figs. 2–3.

*Osangularia velascoensis* (Cushman). –Tjalsma and Lohmann, 1983, p. 18, pl. 7, fig. 14. –Thomas, 1990, p. 590. –Bolli et al., 1994, p. 157, figs. 43.36–38. –Nomura, 1995, p. 278, pl. 4, fig. 7.

### Remarks

This is a distinctive species with its broad, thin keel and thickened, limbate sutures, confluent on the spiral side, leaving depressed areas over the chambers. Juvenile specimens have a relatively high convex umbilical side, with a large proloculus not yet covered by later enveloping chambers (Tjalsma and Lohmann, 1983).

### Biogeography

Worldwide. Recorded from Mexico (Cushman, 1925), the Atlantic (Tjalsma and Lohmann, 1983), the Weddell Sea, Antarctica (Thomas, 1990), Trinidad (Bolli et al., 1994), and the Indian Ocean (Nomura, 1995).

### Bathymetry

Wide bathymetric range with highest abundances at lower bathyal depths, in particular during the Late Paleocene (Tjalsma and Lohmann, 1983). Lower bathyal (Thomas, 1990).

### Chronostratigraphy

Paleocene (Danian to Thanetian). Paleocene, fairly common (Bolli et al., 1994). Rare to common in Biozones P1–P5 (Tjalsma and Lohmann, 1983).

## References


Bolli, H.M., Beckmann, J.P., and Saunders, J.B., 1994, Benthic Foraminiferal Biostratigraphy of the South Caribbean Region. Cambridge: Cambridge University Press, 408 pp.

Cushman, J.A., 1925, Some new foraminifera from the Velasco Shale of Mexico. Contributions from the Cushman Laboratory for Foraminiferal Research, vol. 1, no. 1, pp. 18–23.

Cushman, J.A. and Jarvis, P.W., 1932, Upper Cretaceous foraminifera from Trinidad. Proceedings of the United States National Museum, vol. 80, no. 14, pp. 1–60.

Cushman, J.A. and Renz, H.H., 1946, The foraminiferal fauna of the Lizard Springs Formation of Trinidad, British West Indies. Special Publications of the Cushman Laboratory for Foraminiferal Research, vol. 18, pp. 1–48.

Nomura, R., 1995, Paleogene to Neogene deep-sea paleoceanography in the eastern Indian Ocean: benthic foraminifera from ODP Sites 747, 757 and 758. Micropaleontology, vol. 41, pp. 251–290.

Thomas, E., 1990, Late Cretaceous through Neogene benthic foraminifers. In Barker, P.F., Kennett, J.P., et al., ed., Scientific Results of the Proceedings of the Ocean Drilling Program, vol. 113, pp. 571–594.

Tjalsma, R.C. and Lohmann, G.P., 1983, Paleocene–Eocene bathyal and abyssal benthic foraminifera from the Atlantic Ocean. Micropaleontology, Special Publication, no. 4, pp. 1–90.

White, M.P., 1928, Some index foraminifera of the Tampico Embayment area of Mexico (Part II). Journal of Paleontology, vol. 2, pp. 280–317.

# Parabrizalina porrecta

## (Brady), 1881

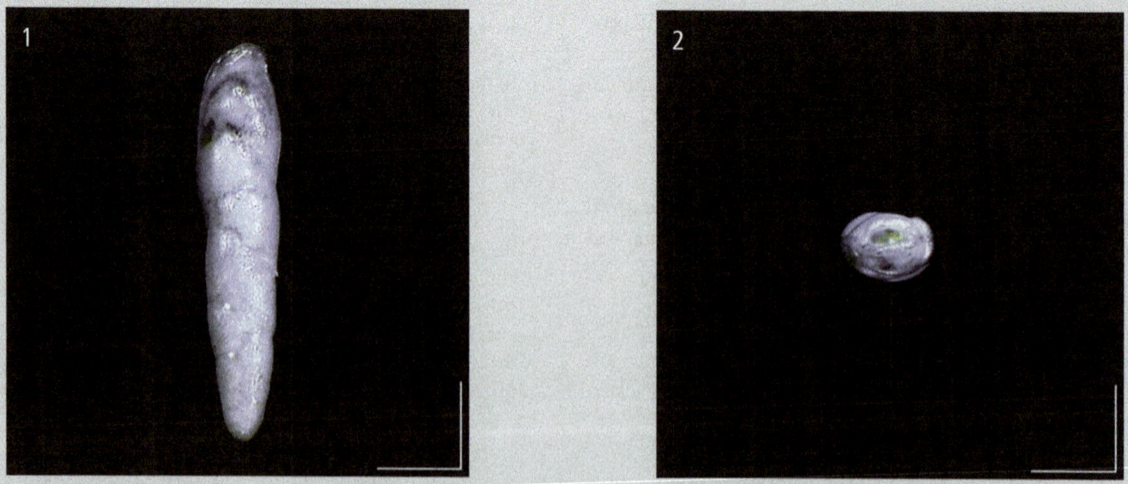

1. *Parabrizalina porrecta*. Off Culebra Island, West Indies; "Challenger" Station 24, depth: 390 fathoms. Deposited in the micropaleontological collections of The Natural History Museum, London; NHMUK ZF4970 (ex. NHMUK ZF1190). Lectotype. Lateral view, NHMUK PM DI 0490, scale bar = 209 μm.

2. *Parabrizalina porrecta*. Off Culebra Island, West Indies; "Challenger" Station 24, depth: 390 fathoms. Deposited in the micropaleontological collections of The Natural History Museum, London; NHMUK ZF4970 (ex. NHMUK ZF1190). Lectotype. Apertural view, NHMUK PM DI 0491, scale bar = 199 μm.

### Original Designation

*Bolivina porrecta* Brady, 1881, p. 57 (type description), illustrated by Brady, 1884, in pl. 52, fig. 22.

### Type Specimen

Lectotype (ZF4970) and paralectotype (ZF4971), designated by Revets (1996) from slide ZF1190, are housed in the micropaleontological collections of The Natural History Museum, London, UK.

### Type Level

Recent, surface sediments.

### Type Locality

Specimens figured by Brady (1884) are from "Challenger" Station 24, off Culebra Island, North of St Thomas's, West Indies, 390 fathoms.

### Description

Test forms an elongate, slightly tapered, slightly compressed, biserial series with a tendency toward rectilinearity in later chambers, elliptical in cross-section with slightly lobulate periphery. The moderately inflated chambers increase rapidly in height, and are separated by distinct, depressed, slightly curved sutures. The frontal overlap of chambers occupies almost the full width of the test. Chamber walls are calcareous, perforate, transparent, and smooth. The primary aperture is wide, terminal, elliptical, and surrounded by a thickened, slightly everted lip with a large internal toothplate.

### Synonyms and Citations

*Bolivina porrecta* Brady. –Brady, 1884, p. 418, pl. 52, fig. 22.

*Loxostoma porrecta* (Brady). –Cushman, 1937, p. 190, pl. 22, figs. 7–10.

*Loxostomum porrectum* (Brady). –Barker, 1960, pl. 52, fig. 22.

*Loxostomina porrecta* (Brady). –Loeblich and Tappan, 1994, p. 120, pl. 235, figs. 1–7.

*Parabrizalina porrecta* (Brady). –Loeblich and Tappan, 1987, p. 516, pl. 566, figs. 24–26. –Jones, 1994, p. 57, pl. 52, fig. 22. –Revets, 1996, p. 14, pl. 11, figs. 1–4.

### Remarks

Type species of the genus *Parabrizalina* Zweig-Strykowski and Reiss, 1976.

### Biogeography

Worldwide.

### Bathymetry

Middle neritic to middle bathyal. Reported between 37 and 420 fathoms at "Challenger" stations (Jones, 1994).

### Chronostratigraphy

Pliocene? to Recent.

## References


Barker, R.W., 1960, Taxonomic notes on the species figured by H.B. Brady in his report on the foraminifera dredged by H.M.S. Challenger during the years 1873–1876. Society of Economic Paleontologists and Mineralogists Special Publication, vol. 9, pp. 1–238.

Brady, H.B., 1881, Notes on some of the reticularian Rhizopoda of the "Challenger" Expedition, Part III. Quarterly Journal of Microscopical Science (New Series), vol. 21, pp. 31–71.

Brady, H.B., 1884, Report on the foraminifera dredged by H.M.S. Challenger during the years 1873–1876. In Brady, H.B., ed., Report on the Scientific Results of the Voyage of the H.M.S. Challenger during the years 1873–1876. Zoology, vol. 9, pp. 1–814.

Cushman, J.A., 1937, A monograph of the subfamily Virgulininae of the foraminiferal family Buliminidae. Special Publications of the Cushman Laboratory for Foraminiferal, vol. 9, pp. 1–228.

Jones, R.W., 1994, The Challenger Foraminifera. Oxford: Oxford University Press, 149 pp.

Loeblich, A.R. Jr and Tappan, H., 1987, Foraminiferal Genera and Their Classification. New York: Van Nostrand Reinhold, 970 pp.

Loeblich, A.R. Jr and Tappan, H., 1994, Foraminifera of the Sahul Shelf and Timor Sea. Special Publications of the Cushman Laboratory for Foraminiferal Research, vol. 31, pp. 1–661.

Revets, S.A., 1996, The generic revision of five families of Rotaliine Foraminifera. Special Publications of the Cushman Foundation for Foraminiferal Research, vol. 34, pp. 1–108.

# *Paratrochammina challengeri*

## Brönnimann and Whittaker, 1988

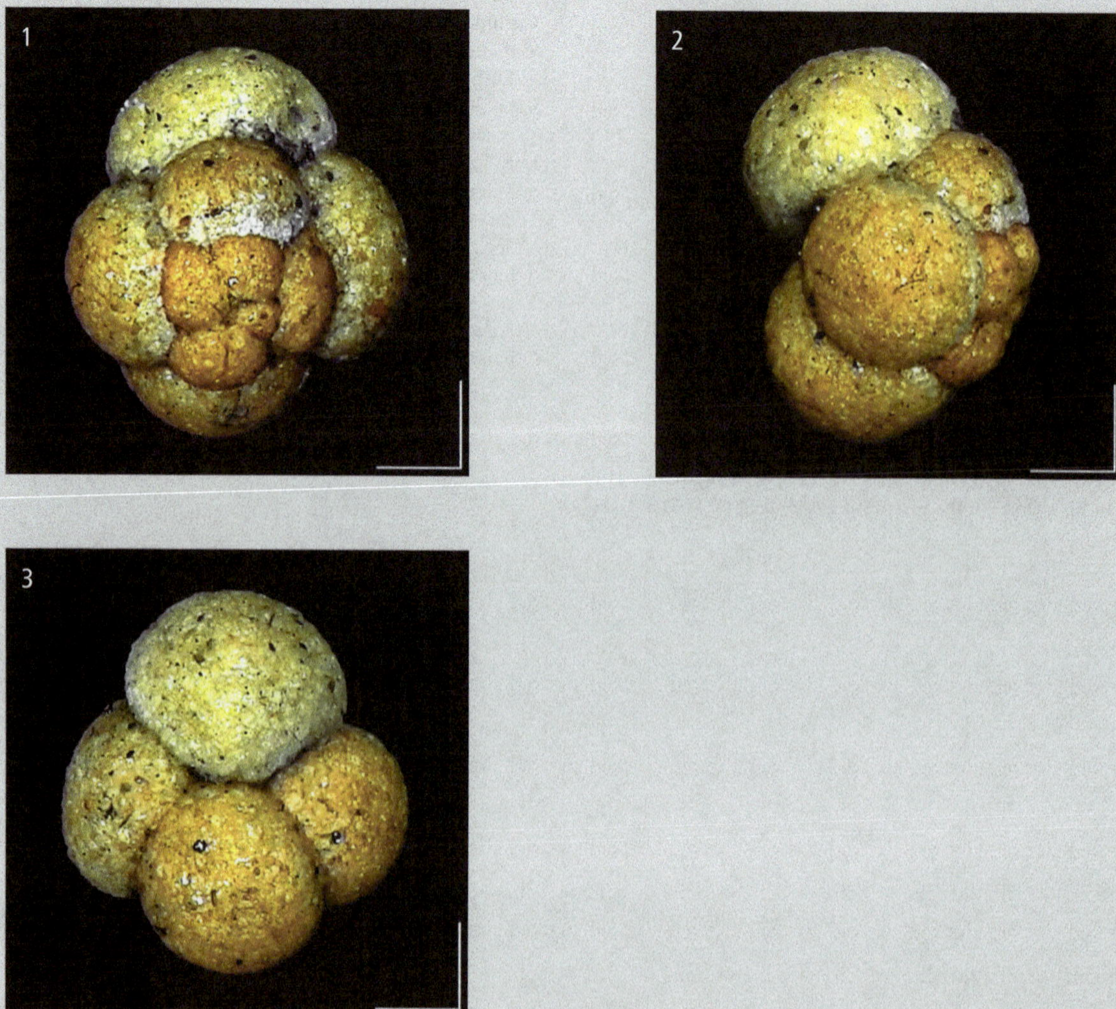


1. *Paratrochammina challengeri.* South Atlantic Ocean; "Challenger" Station 323, depth: 1900 fathoms. Deposited in the micropaleontological collections of The Natural History Museum, London; NHMUK ZF1538. Spiral view, NHMUK PM DI 0492, scale bar = 319 μm.

2. *Paratrochammina challengeri.* South Atlantic Ocean; "Challenger" Station 323, depth: 1900 fathoms. Deposited in the micropaleontological collections of The Natural History Museum, London; NHMUK ZF1538. Edge view, NHMUK PM DI 0493, scale bar = 337 μm.

3. *Paratrochammina challengeri.* South Atlantic Ocean; "Challenger" Station 323, depth: 1900 fathoms. Deposited in the micropaleontological collections of The Natural History Museum, London; NHMUK ZF1538. Umbilical view, NHMUK PM DI 0494, scale bar = 339 μm.

### Original Designation

*Paratrochammina challengeri* Brönnimann and Whittaker, 1988, p. 43, figs. 16H–K.

### Type Specimen

Holotype is housed in the micropaleontological collections of The Natural History Museum, London, UK, and is registered in slide ZF4135.

### Type Level

Recent.

### Type Locality

"Challenger" Station 323, 450 miles north of South Georgia, South Atlantic (50° 47′ W, 35° 39′ S), depth 1900 fathoms.

### Description

Test trochospiral, concavo-convex with evolute, convex spiral side, concave, involute umbilical side, and rounded periphery. Approximately four highly inflated chambers in the last whorl, increasing gradually in size, are separated by straight, markedly depressed sutures. The wall is finely agglutinated with abundant cement. The aperture is a low interiomarginal arch with an umbilical flap covering all previous flaps.

### Synonyms and Citations

*Haplophragmium globigeriniforme* (Parker and Jones). –Brady, 1884, p. 312, pl. 35, fig. 10 (not fig. 11).

*Ammoglobigerina globigeriniformis* (Parker and Jones). –Barker, 1960, pl. 35, fig. 10 (not fig. 11). –Milker and Schmiedl, 2012, p. 35, figs. 9.25–26.

*Trochammnina globigeriniformis* (Parker and Jones). –Zheng, 1988, p. 82, pl. 40, fig. 6.

*Trochamminopsis challengeri* (Brönnimann and Whittaker). –Charnock and Jones, 1990, p. 189, pl. 11, figs. 7–10; pl. 22, fig. 6.

*Paratrochammina challengeri* Brönnimann and Whittaker. –Jones, 1994, p. 41, pl. 35, fig. 10.

### Remarks

Brönnimann and Whittaker (1988) regarded Brady's specimen as being clearly distinct from *Lituola nautiloidea* Lamarck var. *globigeriformis* Parker and Jones, and proposed the new name *Paratrochammina challengeri* for this taxon.

### Biogeography

Worldwide.

### Bathymetry

Bathyal to abyssal, including sub-CCD. Common in lower bathyal environments on the Sunda Shelf in the South China Sea (Szarek et al., 2009). Charnock and Jones (1990) reported a more extended bathymetric range of 50 to 5029 meters in the North Atlantic.

### Chronostratigraphy

Late Cretaceous (Maastrichtian) to Recent (Jones, 1994).

## References


Barker, R.W., 1960, Taxonomic notes on the species figured by H.B. Brady in his report on the foraminifera dredged by H.M.S. Challenger during the years 1873–1876. Society of Economic Paleontologists and Mineralogists Special Publication, vol. 9, pp. 1–238.

Brady, H.B., 1884, Report on the foraminifera dredged by H.M.S. Challenger during the years 1873–1876. In Brady, H.B., ed., Report on the Scientific Results of the Voyage of the H.M.S. Challenger during the years 1873–1876. Zoology, vol. 9, pp. 1–814.

Brönnimann, P. and Whittaker, J.E., 1988, The Trochamminacea of the Discovery Reports. A review of the Trochamminacea (Protozoa: Foraminiferida) described from South Atlantic and Antarctic waters by Heron-Allen and Earland (1932) and Earland (1933; 1934; 1936). London: British Museum (Natural History), 152 pp.

Charnock, M.A. and Jones, R.W., 1990, Agglutinated foraminifera from the Palaeogene of the North Sea. In Hemleben, C., Kaminski, M.A., Kuhnt, W., and Scott, D.B., ed., Paleoecology, Biostratigraphy, Paleoceanography and Taxonomy of Agglutinated Foraminifera, NATO ASI Series, C: Mathematical and Physical Sciences. Dordrecht: Kluwer Academic, pp. 139–244.

Jones, R.W., 1994, The Challenger Foraminifera. Oxford: Oxford University Press, 149 pp.

Milker, Y. and Schmiedl, G., 2012, A taxonomic guide to modern benthic shelf foraminifera of the western Mediterranean Sea. Palaeontologia Electronica, vol. 15, Issue 2;16A, 134 pp.

Szarek, R., Kuhnt, W., Kawamura, H., and Nishi, H., 2009, Distribution of recent benthic foraminifera along the continental slope of the Sunda Shelf (South China Sea). Marine Micropaleontology, vol. 71, pp. 41–59.

Zheng, S.Y., 1988, The Agglutinated and Porcelaneous Foraminifera of the East China Sea. Beijing: China Ocean Press, 337 pp.

# *Patellina corrugata*

## Williamson, 1858

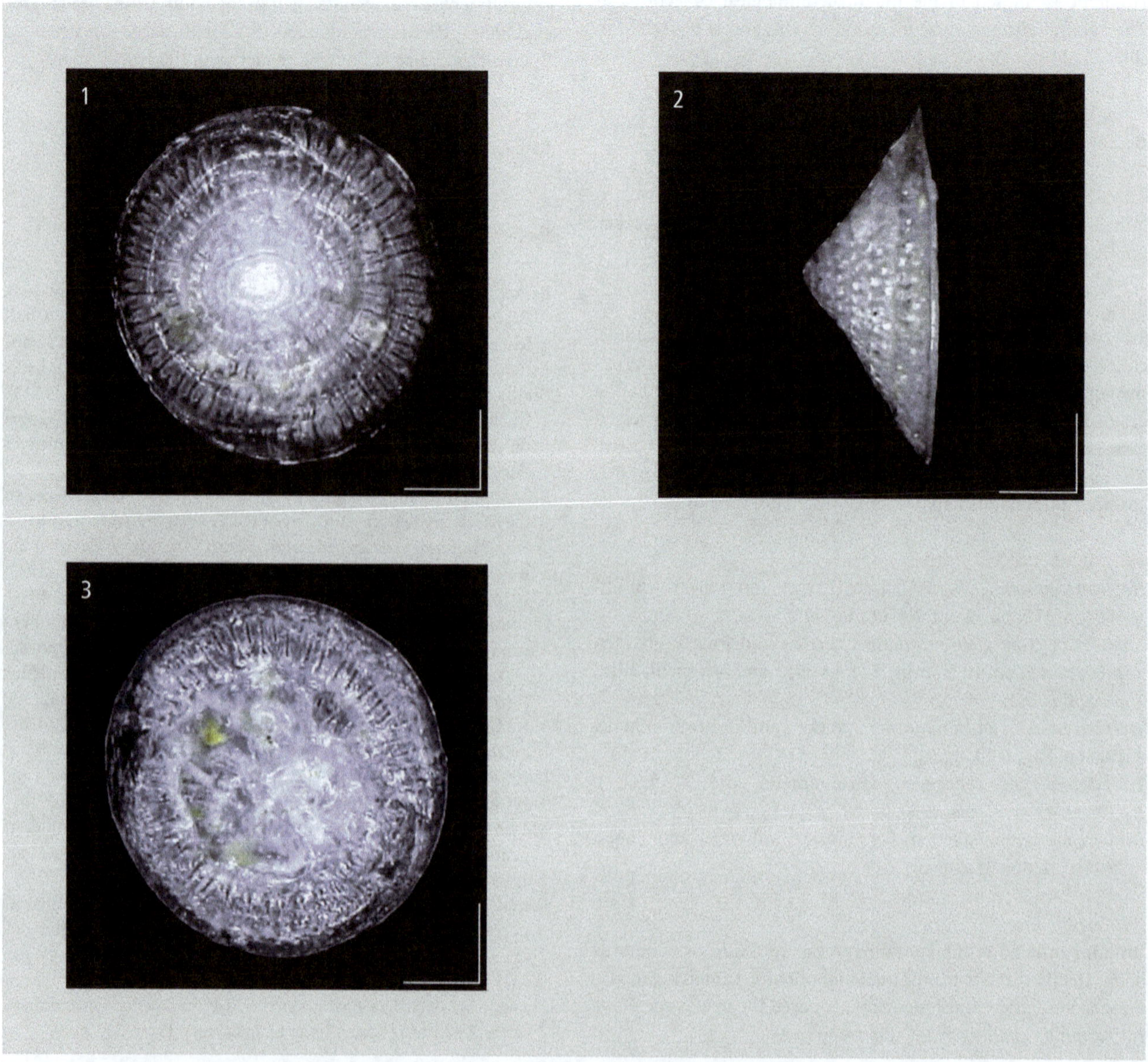

1. *Patellina corrugata.* Kerguelen Island, South Pacific Ocean; "Challenger" Station 149D, depth: 20–60 fathoms. Deposited in the micropaleontological collections of The Natural History Museum, London; NHMUK ZF2068. Spiral view, NHMUK PM DI 0495, scale bar = 58 μm.

2. *Patellina corrugata.* Kerguelen Island, South Pacific Ocean; "Challenger" Station 149D, depth: 20–60 fathoms. Deposited in the micropaleontological collections of The Natural History Museum, London; NHMUK ZF2068. Lateral view, NHMUK PM DI 0496, scale bar = 71 μm.

3. *Patellina corrugata.* Kerguelen Island, South Pacific Ocean; "Challenger" Station 149D, depth: 20–60 fathoms. Deposited in the micropaleontological collections of The Natural History Museum, London; NHMUK ZF2068. Umbilical view, NHMUK PM DI 0497, scale bar = 56 μm.

## Original Designation

*Patellina corrugata* Williamson, 1858, p. 46, pl. 3, figs. 86–89, 89a.

## Type Specimen

Specimens illustrated by Brady (1884) are deposited in the micropaleontological collections of The Natural History Museum, London, UK, and are registered in slides ZF2066–ZF2069.

## Type Level

Recent.

## Type Locality

Not designated. Localities given: Boston, dredged sands of Arran, Skye, Shetland, Brixham, and Fowey, British Isles.

## Description

Test forms a planoconvex trochospiral cone, with an evolute convex spiral side, an involute, flattened umbilical side, and a keeled periphery. The broad, low, crescentic chambers that comprise the whorls (two chambers per whorl) increase rapidly in breadth, and are separated by slightly raised, oblique sutures. Chambers are subdivided in their upper parts by short, radial septula extending from the spiral side, but not reaching the umbilical side, and form a scroll-like median septum on the umbilical side. Chamber walls are calcareous and smooth, with numerous large pores closed by sieve-plates. The primary aperture is a narrow, sinuous slit, covered by a broad apertural plate.

## Synonyms and Citations

*Patellina corrugata* Williamson. –Brady, 1884, p. 634, pl. 86, figs. 1–7. –Barker, 1960, pl. 86, figs. 1–7. –Loeblich and Tappan, 1987, p. 306, pl. 320, figs. 7–14. –Jones, 1994, p. 93, pl. 86, figs. 1–7. –Milker and Schmiedl, 2012, p. 43, figs. 11.21–23.

## Remarks

Type species of the genus *Patellina* Williamson, 1858.

Although the type locality was not designated, localities given include Boston, dredged sands of Arran, Skye, Shetland, Brixham, and Fowey, British Isles.

## Biogeography

Worldwide.

## Bathymetry

Neritic to middle bathyal. Reported between 7 and 620 fathoms at "Challenger" stations (Jones, 1994).

## Chronostratigraphy

Late Miocene to Recent (Jones, 1994).

# References


Barker, R.W., 1960, Taxonomic notes on the species figured by H.B. Brady in his report on the foraminifera dredged by H.M.S. Challenger during the years 1873–1876. Society of Economic Paleontologists and Mineralogists Special Publication, vol. 9, pp. 1–238.

Brady, H.B., 1884, Report on the foraminifera dredged by H.M.S. Challenger during the years 1873–1876. In Brady, H.B., ed., Report on the Scientific Results of the Voyage of the H.M.S. Challenger during the years 1873–1876. Zoology, vol. 9, pp. 1–814.

Jones, R.W., 1994, The Challenger Foraminifera. Oxford: Oxford University Press, 149 pp.

Loeblich, A.R. Jr and Tappan, H., 1987, Foraminiferal Genera and Their Classification. New York: Van Nostrand Reinhold, 970 pp.

Milker, Y. and Schmiedl, G., 2012, A taxonomic guide to modern benthic shelf foraminifera of the western Mediterranean Sea. Palaeontologia Electronica, vol. 15, Issue 2;16A, 134 pp.

Williamson, W.C., 1858, On the Recent Foraminifera of Great Britain. London: Ray Society, 107 pp.

# *Planularia australis*

## Chapman, 1941

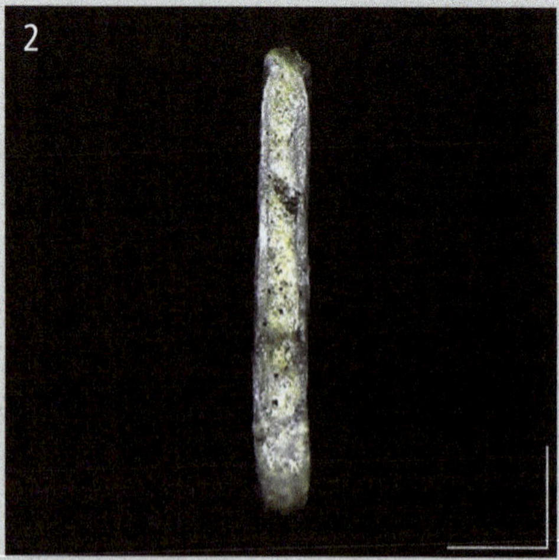


1. *Planularia australis*. Off the Philippines; "Challenger" Station 209, depth: 95 fathoms. Deposited in the micropaleontological collections of The Natural History Museum, London; NHMUK ZF1351. Lateral view, NHMUK PM DI 0498, scale bar = 209 μm.

2. *Planularia australis*. Off the Philippines; "Challenger" Station 209, depth: 95 fathoms. Deposited in the micropaleontological collections of The Natural History Museum, London; NHMUK ZF1351. Edge view, NHMUK PM DI 0499, scale bar = 171 μm.

### Original Designation

*Planularia australis* Chapman, 1941, p. 158, pl. 9, fig. 1.

### Type Specimen

Deposited in the collections of the National Museum of Victoria, Melbourne, Australia.

### Type Level

Recent.

### Type Locality

Sample E 3919 (37° 21′ 20″ S, 150° 24′ 25″ E) off the southeast coast of Australia near the boundary of Victoria and New South Wales, in 505 fathoms.

### Description

Test strongly compressed with sides flattened, flaring with an early partially planispiral stage, later becoming evolute with a keeled periphery. Additional keels are present on either side of the umbilical edge, as well as twin peripheral keels. The broad, low chambers increase rapidly in breadth initially, then more slowly, and are separated by thickened, slightly curved, raised sutures. Chamber walls are calcareous, perforate, and smooth. The primary aperture is radiate and terminal, and slightly produced.

### Synonyms and Citations

*Cristellaria tricarinella* Reuss. –Brady, 1884, p. 540, pl. 68, figs. 3–4.
*Planularia australis* Chapman. –Barker, 1960, pl. 68, figs. 3–4. –van Marle, 1991, p. 40. –Jones, 1994, p. 80, pl. 68, figs. 3–4.

### Biogeography

West Pacific Ocean and Indonesian archipelago. Brady's figured specimens are off the coast of New Zealand and the Philippines. Van Marle (1991) recorded *Planularia australis* Chapman in the Indonesian seas.

### Bathymetry

Lower neritic to upper bathyal. Reported between 50 and 536 meters in the Indonesian seas (van Marle, 1991), and between 95 and 155 fathoms at Pacific "Challenger" stations (Jones, 1994).

### Chronostratigraphy

Late Miocene to Recent (van Marle, 1991; Jones, 1994).

## References

Barker, R.W., 1960, Taxonomic notes on the species figured by H.B. Brady in his report on the foraminifera dredged by H.M.S. Challenger during the years 1873–1876. Society of Economic Paleontologists and Mineralogists Special Publication, vol. 9, pp. 1–238.

Brady, H.B., 1884, Report on the foraminifera dredged by H.M.S. Challenger during the years 1873–1876. In Brady, H.B., ed., Report on the Scientific Results of the Voyage of the H.M.S. Challenger during the years 1873–1876. Zoology, vol. 9, pp. 1–814.

Chapman, F., 1941, Report on the foraminiferal soundings and dredgings of the F.I.S. "Endeavour" along the continental shelf off the south-east coast of Australia. Transactions of the Royal Society of South Australia, vol. 65, no. 2, pp. 145–211.

Jones, R.W., 1994, The Challenger Foraminifera. Oxford: Oxford University Press, 149 pp.

van Marle, L.J., 1991, Eastern Indonesian Late Cenozoic smaller benthic foraminifera. Verhandelingen der Koninklijke Nederlanse Akademie van Wetenschappen, Afdeeling Natuurkunde, Eerste Reeks, vol. 34, pp. 1–328.

# *Planularia gemmata*

## (Brady), 1881

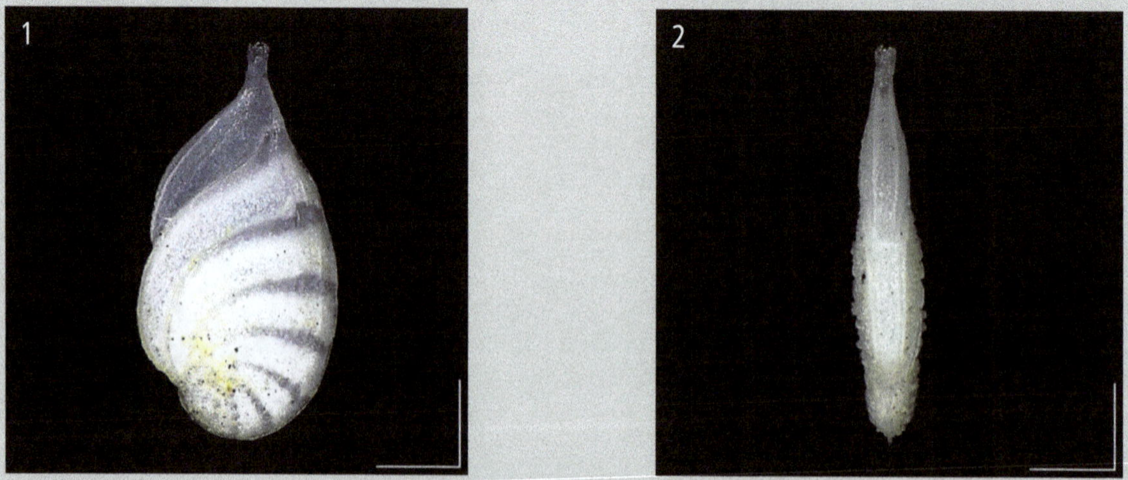


1. *Planularia gemmata.* Off Fiji, Pacific Ocean; "Challenger" Station 174C, depth: 210 fathoms. Deposited in the micropaleontological collections of The Natural History Museum, London; NHMUK ZF1326. Syntype. Lateral view, NHMUK PM DI 0500, scale bar = 305 μm.

2. *Planularia gemmata.* Off Fiji, Pacific Ocean; "Challenger" Station 174C, depth: 210 fathoms. Deposited in the micropaleontological collections of The Natural History Museum, London; NHMUK ZF1326. Syntype. Edge view, NHMUK PM DI 0501, scale bar = 326 μm.

### Original Designation

*Cristellaria gemmata* Brady, 1881, p. 64 (type description), illustrated by Brady, 1884, in pl. 38, figs. 4–6.

### Type Specimen

Figured specimens are housed in the micropaleontological collections of The Natural History Museum, London, UK, and are registered in slides ZF1326–ZF1328.

### Type Level

Recent, surface sediments.

### Type Locality

Figured specimens from "Challenger" Station 174C, latitude 19° 10′ S, longitude 178° 10′ E, off Kandavu, Fiji Islands in 210 fathoms (fig. 6), and from "Challenger" Station 209, latitude 10° 10′ N, longitude 123° 55′ E, off the Philippine Islands in 95–100 fathoms (fig. 7).

### Description

Test strongly compressed with sides only slightly raised, flaring with early partially planispiral stage, later becoming evolute with a keeled periphery. Faint additional keels may be present on either side of the periphery and umbilical edge. The broad, low chambers increase rapidly in breadth and are separated by thickened, curved sutures; the last chamber may be reduced. Chamber walls are calcareous, perforate, ornamented by short ribs or knobs, prominent along the sutures. The primary aperture is radiate and terminal.

### Synonyms and Citations

*Cristellaria gemmata* Brady, 1884, p. 554, pl. 71, figs. 6–7.

*Planularia gemmata* (Brady) –Barker, 1960, pl. 71, figs. 6–7. –van Marle, 1991, p. 41, pl. 1, figs. 4, 9. –Jones, 1994, p. 82, pl. 71, figs. 6–7.

### Biogeography

West Pacific Ocean and Indonesian archipelago (Brady, 1884; van Marle, 1991).

### Bathymetry

Lower neritic to upper bathyal. Reported between 50 and 300 meters in the Mahakam Delta, Indonesia (van Marle, 1991), and between 95 and 210 fathoms at "Challenger" stations (Jones, 1994).

## Chronostratigraphy

Miocene to Recent (Jones, 1994). Late Miocene to Recent (van Marle, 1991).

## References

Barker, R.W., 1960, Taxonomic notes on the species figured by H.B. Brady in his report on the foraminifera dredged by H.M.S. Challenger during the years 1873–1876. Society of Economic Paleontologists and Mineralogists Special Publication, vol. 9, pp. 1–238.

Brady, H.B., 1881, Notes on some of the reticularian Rhizopoda of the Challenger Expedition, Part III. Quarterly Journal of Microscopical Science (New Series), vol. 21, pp. 31–71.

Brady, H.B., 1884, Report on the foraminifera dredged by H.M.S. Challenger during the years 1873–1876. In Brady, H.B., ed., Report on the Scientific Results of the Voyage of the H.M.S. Challenger during the years 1873–1876. Zoology, vol. 9, pp. 1–814.

Jones, R.W., 1994, The Challenger Foraminifera. Oxford: Oxford University Press, 149 pp.

van Marle, L.J., 1991, Eastern Indonesian Late Cenozoic smaller benthic foraminifera. Verhandelingen der Koninklijke Nederlanse Akademie van Wetenschappen, Afdeeling Natuurkunde, Eerste Reeks, vol. 34, pp. 1–328

# *Planulina ambigua*

## (Franzenau), 1888

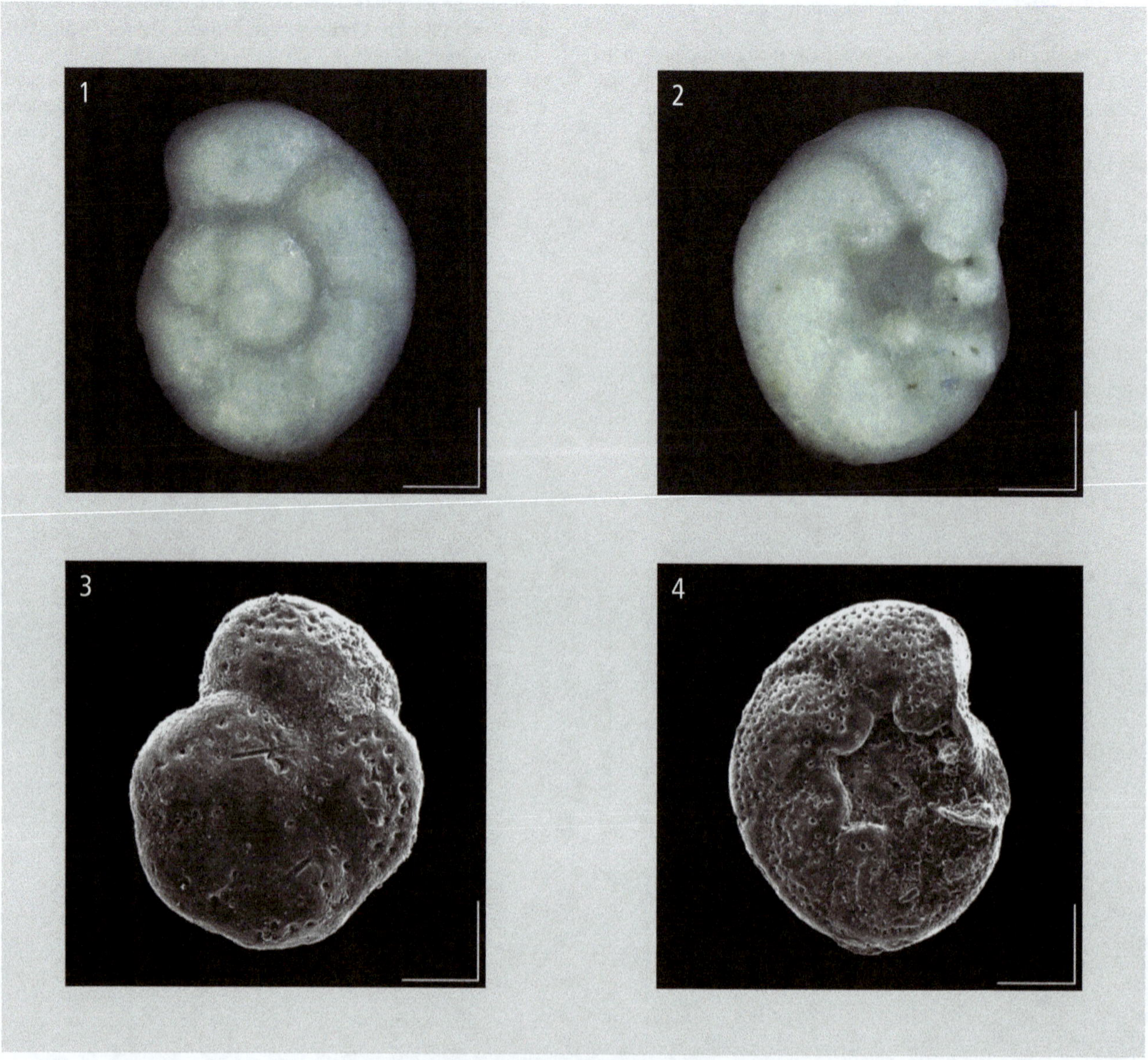


1. *Planulina ambigua*. Jamaica S6-32. Spiral view, NHMUK PM DI 0502, scale bar = 99 μm.
2. *Planulina ambigua*. Jamaica S6-32. Umbilical view, NHMUK PM DI 0503, scale bar = 99 μm.
3. *Planulina ambigua*. Jamaica S6-32. Spiral view, NHMUK PM DI 0504, scale bar = 67 μm.
4. *Planulina ambigua*. Jamaica S6-32. Umbilical view, NHMUK PM DI 0505, scale bar = 104 μm.

## Original Designation

*Rotalia ambigua* Franzenau, 1888, p. 106, pl. 2, figs. 9–11.

## Type Specimen

Depository not given.

## Type Level

Oligocene, sandy clay. Can be correlated with layer occurring at a depth of 245.63–302.81 meters in the boring of the artesian well in the city forest of Budapest.

## Type Locality

Boring number 13, 13.31 meters, drillled in connection with the building of the House of Parliament in Budapest, Hungary.

## Description

Test forms a compressed, discoidal, very low trochospire; subcircular or lobulate in outline, and planoconvex or concavo-convex in cross-section with a slightly convex, evolute spiral side, a flattened or slightly concave, partially evolute umbilical side, and a subacute, keeled periphery. The (approximately) six slightly inflated chambers in the final whorl increase rapidly in size, and are separated by curved, slightly depressed, limbate sutures. These sutures may become slightly raised, forming an elevated rim along the imperforate peripheral band. The umbilicus is broad and shallow, ornamented with raised ridges, and surrounded by periumbilical extensions of the sutures. Chamber walls are calcareous, and coarsely perforate on both sides of the test. The primary aperture is an equatorial slit with a narrow lip extending beneath the umbilical folium.

## Synonyms (following van Morkhoven, 1986) and Citations

*Planulina jabacoensis* Bermúdez, 1937, p. 246, pl. 21, figs. 10–11.
*Discorbis* (?) sp. Hagn, 1956, p. 158, pl. 14, fig. 15.
*Discorbis subvilardeboanus* (Rzehak). –Boltovskoy, 1978, p. 157, pl. 3, figs. 26–30.
*Planulina ambigua* (Franzenau). –van Morkhoven et al., 1986, p. 230, pl. 78A–B. –Katz and Miller, 1993, pl. 3, fig. 5.

## Remarks

*Planulina ambigua* (Franzenau) differs from *Discorbis subvilarbeboanus* (Rzehak) by having a larger test with six chambers in the final whorl (evolute on both sides), a wider umbilicus, and coarser perforations on chamber walls (see Remarks in van Morkhoven et al., 1986).

## Biogeography

Worldwide. Recorded from Eastern Europe, the Gulf of Mexico, the North and South Atlantic, and the Pacific (van Morkhoven et al., 1986).

## Bathymetry

Primarily upper to middle bathyal. May be found in the neritic zone (van Morkhoven et al., 1986). Also occurs at lower bathyal depths (M. Katz, pers. comm. 1998).

## Chronostratigraphy

Middle Eocene to Middle Miocene. *Planulina ambigua* occurred within Biozone N8 in the Spring Garden Member of Buff Bay, Jamaica (M. Katz, pers. comm. 1998), extending the range of this species into the Middle Miocene.

# References


Bermúdez, P.J., 1937, Nuevas especies de foraminiferos del Eoceno de las cercanias de Guanajay, Provincia Pinar del Rio, Memorias de la Sociedad Cubana de Historia Natural "Felipe Poey", vol. 11, pp. 237–247.

Boltovskoy, E., 1978, Late Cenozoic benthonic foraminifera of the Ninetyeast Ridge (Indian Ocean). Marine Geology, vol. 26, pp. 139–175.

Franzenau, A., 1888, Adat Budapest altalajának ismeretéhez. Földtani Közlöny, Budapest, vol. 18, no. 3–4, pp. 87–106 (157–174).

Hagn, H., 1956, Geologische und palaontologische Untersuchungen im Tertiar des Monte Brione und seiner Umgebung. Palaeontographica, vol. 107(A), pp. 67–210.

Katz, M.E. and Miller, K.G., 1993, Latest Oligocene to Earliest Pliocene benthic foraminiferal biofacies of the northeastern Gulf of Mexico. Micropaleontology, vol. 39, no. 4, pp. 367–403.

van Morkhoven, F.P.C.M., Berggren, W.A., Edwards, A.S., et al., 1986, Cenozoic cosmopolitan deep-water benthic foraminifera. Bulletin des Centres de Recherches Exploration-Production Elf-Aquitaine, Mémoire, vol. 11, pp. 1–421.

# *Planulina ariminensis*

## d'Orbigny, 1826

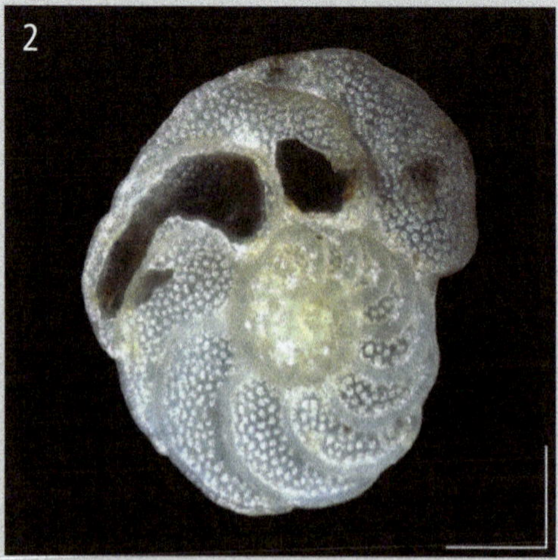

1. *Planulina ariminensis.* "Meteor" Cruise off West Africa; M25. Deposited in the Lutze Collection (West Africa), no. 169. Institute of Geosciences, Kiel University, Germany. Umbilical view, NHMUK PM DI 0506, scale bar = 171 µm.

2. *Planulina ariminensis.* "Meteor" Cruise off West Africa; M25. Deposited in the Lutze Collection (West Africa), no. 169. Institute of Geosciences, Kiel University, Germany. Spiral view, NHMUK PM DI 0507, scale bar = 170 µm.

### Original Designation

*Planulina ariminensis* d'Orbigny, 1826, p. 280, pl. 14, figs. 1–3.

### Type Specimen

Originally deposited in the collections of the Laboratoire de Paléontologie, Muséum d'Histoire Naturelle, Paris, France.

### Type Level

Recent.

### Type Locality

The Adriatic Sea near Rimini, Italy.

### Description

Test forms a strongly compressed, discoidal, very low trochospire; subcircular in outline and slightly lobulate in the final chambers, with a flattened, partially evolute, umbilical side, a flattened, evolute, spiral side, and an acute, keeled periphery. The ten to twelve narrow, curved chambers in the final whorl increase rapidly in size, and are separated by strongly curved, limbate sutures that are thick and slightly depressed in the final chambers. Chamber walls are calcareous, and generally more coarsely perforate on the spiral side. The primary aperture is an equatorial slit with a narrow lip, extending below the umbilical folium.

### Synonyms and Citations

*Hammonia subrotunda* Soldani, 1789, p. 50, pl. 40, fig. EE.
*Hammonia foliacea* Soldani, 1799, append., pl. 3, figs. O, P.
*Anomalina ariminensis* (d'Orbigny). –Brady, 1884, p. 674, pl. 93, figs. 10–11.
*Planulina ariminensis* (d'Orbigny). –van Morkhoven et al., 1986, p. 38, pl. 10. –Loeblich and Tappan, 1987, p. 580, pl. 633, figs. 1–4. –Robertson, 1998, p. 212, pl. 85, fig. 1. –Jones, 1994, p. 98, pl. 93, figs. 10–11. –Milker and Schmield, 2012, p. 106, figs. 24.3–4.

### Remarks

Type species of the genus *Planulina* d'Orbigny, 1826.

### Biogeography

Worldwide. Recorded from the North and South Atlantic, the Gulf of Mexico, the Caribbean, Europe, and off the Philippines (van Morkhoven et al., 1986).

### Bathymetry

Primarily lower neritic to upper bathyal (van Morkhoven et al., 1986).

### Chronostratigraphy

Late Miocene (Tortonian) to Recent.

## References


Brady, H.B., 1884, Report on the foraminifera dredged by H.M.S. Challenger during the years 1873–1876. In Brady, H.B., ed., Report on the Scientific Results of the Voyage of the H.M.S. Challenger during the years 1873–1876. Zoology, vol. 9, pp. 1–814.

d'Orbigny, A., 1826, Tableau méthodique de la classe des Céphalopodes. Annales des Sciences Naturelles, vol. 7, pp. 245–314.

Jones, R.W., 1994, The Challenger Foraminifera. Oxford: Oxford University Press, 149 pp.

Loeblich, A.R. Jr and Tappan, H., 1987, Foraminiferal Genera and their Classification. New York: Van Nostrand Reinhold, 970 pp.

Milker, Y. and Schmiedl, G., 2012, A taxonomic guide to modern benthic shelf foraminifera of the western Mediterranean Sea. Palaeontologia Electronica, vol. 15, Issue 2;16A, 134 pp.

Robertson, B.E., 1998, Systematics and paleoecology of the benthic foraminiferida from the Buff Bay section. Miocene of Jamaica. Micropaleontology, vol. 44, supplement 2, pp. 1–266.

Soldani, A., 1789, Testaceographiae ac Zoophytographiae parvae et microscopicae, tom. 1(1). Siena: F. Rossi.

Soldani, A., 1799, Testaceographiae ac Zoophytographiae parvae et microscopicae, tom. 1(4). Siena: F. Rossi.

van Morkhoven, F.P.C.M., Berggren, W.A., Edwards, A.S., et al., 1986, Cenozoic cosmopolitan deep-water benthic foraminifera. Bulletin des Centres de Recherches Exploration-Production Elf-Aquitaine, Mémoire, vol. 11, pp. 1–421.

# Planulina costata

## (Hantken), 1875

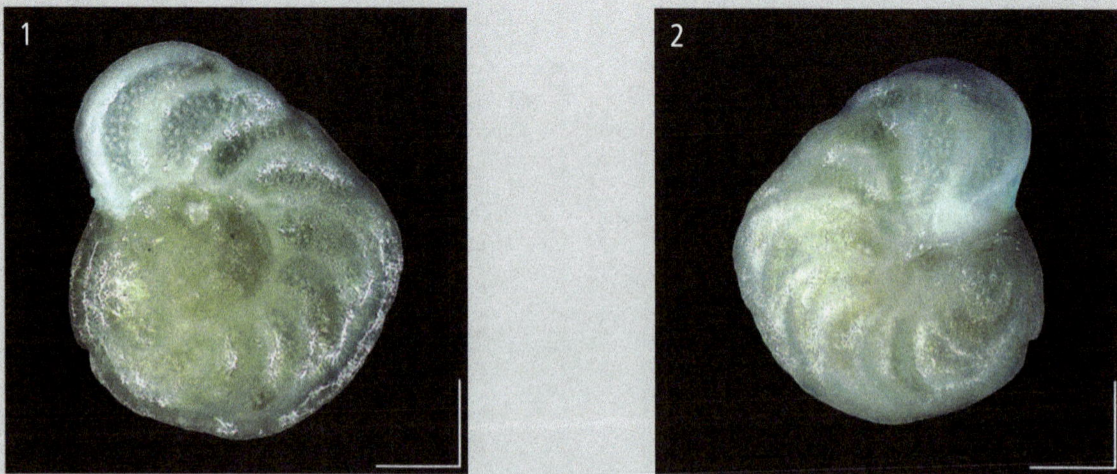

1. *Planulina costata.* North Atlantic; DSDP Leg 95, Site 612, Core 20, core catcher. Spiral view, NHMUK PM DI 0508, scale bar = 105 μm.
2. *Planulina costata.* North Atlantic; DSDP Leg 95, Site 612, Core 20, core catcher. Umbilical view, NHMUK PM DI 0509, scale bar = 111 μm.

### Original Designation

*Truncatulina costata* Hantken, 1875, p. 73, pl. 9, fig. 2a–c.

### Type Specimen

Holotype not designated.

### Type Level

Lower Oligocene, upper part of the *Clavulina szabói* Beds (Kleinzeller Tegel).

### Type Locality

Brick Quarry in Buda Hills, near Budapest, Hungary.

### Description

Test forms a compressed, discoidal, very low trochospire; subcircular and slightly lobulate in outline with a flattened and evolute spiral side, a slightly convex and partially evolute umbilical side (bearing a small umbonal depression), and an acute and keeled periphery. The ten to twelve narrow, curved, slightly inflated chambers in the final whorl increase rapidly in size, and are separated by strongly curved, limbate sutures that are thick and flush or slightly raised on the spiral side, and slightly depressed and thinning toward the umbilicus on the umbilical side. Chamber walls are calcareous, and densely and finely perforate on both sides of the test. The primary aperture is an equatorial slit with a narrow lip, extending below the umbilical folium.

### Synonyms (following van Morkhoven et al., 1986) and Citations

*Planulina wuellerstorfi* Schwager. –Nuttall, 1932, p. 31, pl. 4, figs. 14–15. –Sztràkos, 1979, p. 86, pl. 26, fig. 4.

*Planulina marialana* Hadley, 1934, p. 27, pl. 4, figs. 4–6.

*Planulina palmerae* van Bellen, 1941, in van Bellen et al., 1941, p. 1144, figs. 7–9.

*Planulina costata* (Hantken). –van Morkhoven et al., 1986, p. 212, pl. 72.

### Biogeography

Recorded from Europe, the North Caucasus, Crimea, Turkmenia, Mexico, the Gulf of Mexico, and off New Jersey (van Morkhoven et al., 1986).

### Bathymetry

Primarily upper to middle bathyal (van Morkhoven et al., 1986).

## Chronostratigraphy

Middle Eocene (Bartonian) to Early Miocene (Burdigalian). Doubtful in Early Miocene Biozones M3–M4 (van Morkhoven et al., 1986).

# References


Hadley, W.H., 1934, Some Tertiary foraminifera from the north coast of Cuba. Bulletin of American Paleontology, vol. 20, pp. 1–40.

Hantken, M., 1875, Die fauna der Clavulina szaboi-Schichten; Theil I – Foraminiferen. Könglich-Ungarische Geologische Anstalt, Mitteilugen Jahrbuch, vol. 4, pp. 1–93.

Nuttall, W.L.F., 1932, Lower Oligocene foraminifera from Mexico. Journal of Paleontology, vol. 6, pp. 3–35.

Sztràkos, L., 1979, La stratigraphie, paléoécologie et les Foraminifères de l'Oligocène supérieur du Nord-Est de la Hongrie. Cahiers de Micropaléontologie, vol. 3, pp. 3–95.

van Bellen, R.C., de Witt Puyt, J.F.C., Rutgers, A.C., and van Soest, J., 1941, Smaller foraminifera from the Lower Oligocene of Cuba. Proceedings of the Koninklijke Nederlandse Akademie van Wetenschappen, vol. 44, no. 9, pp. 1140–1146.

van Morkhoven, F.P.C.M., Berggren, W.A., Edwards, A.S., et al., 1986, Cenozoic cosmopolitan deep-water benthic foraminifera. Bulletin des Centres de Recherches Exploration-Production Elf-Aquitaine, Mémoire, vol. 11, pp. 1–421.

# Planulina foveolata

## (Brady), 1884

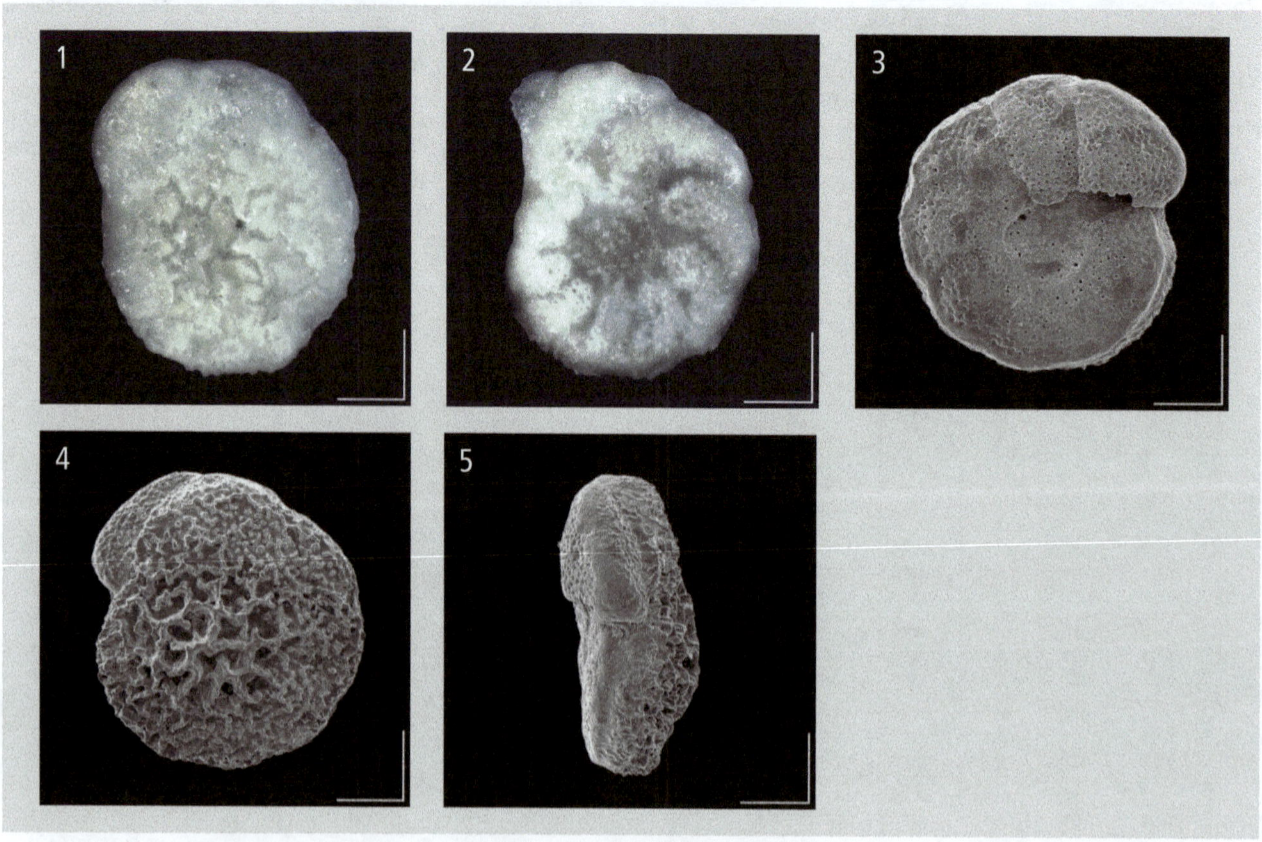

1. *Planulina foveolata.* Off Bermuda; "Challenger" Station 33, depth: 435 fathoms. Deposited in the micropaleontological collections of The Natural History Museum, London; NHMUK ZF1082. Umbilical view, NHMUK PM DI 0510, scale bar = 121 μm.

2. *Planulina foveolata.* Off Bermuda; "Challenger" Station 33, depth: 435 fathoms. Deposited in the micropaleontological collections of The Natural History Museum, London; NHMUK ZF1082. Spiral view, NHMUK PM DI 0511, scale bar = 105 μm.

3. *Planulina foveolata.* Off Bermuda; "Challenger" Station 33, depth: 435 fathoms. Deposited in the micropaleontological collections of The Natural History Museum, London; NHMUK ZF1082. Umbilical view, NHMUK PM DI 0512, scale bar = 145 μm.

4. *Planulina foveolata.* Off Bermuda; "Challenger" Station 33, depth: 435 fathoms. Deposited in the micropaleontological collections of The Natural History Museum, London; NHMUK ZF1082. Spiral view, NHMUK PM DI 0513, scale bar = 149 μm.

5. *Planulina foveolata.* Off Bermuda; "Challenger" Station 33, depth: 435 fathoms. Deposited in the micropaleontological collections of The Natural History Museum, London; NHMUK ZF1082. Edge view, NHMUK PM DI 0514, scale bar = 142 μm.

### Original Designation

*Anomalina foveolata* Brady, 1884, p. 674, pl. 94, fig. 1a–c.

### Type Specimen

Figured specimen (ZF1082) housed in the micropaleontological collections of The Natural History Museum, London, UK.

### Type Level

Recent, shelly sand.

### Type Locality

Dredged off Bermuda in 435 fathoms.

### Description

Test forms a compressed, discoidal, and very low trochospiral with nearly flat, parallel sides, and a subquadrate cross-section, with a broad, angled, keeled periphery. The umbilical side is partially evolute, flattened or slightly concave, and the spiral side is strongly evolute, and flattened or slightly convex. The nine to eleven narrow, curved, slightly inflated chambers in the final whorl increase gradually in size, and are separated by oblique sutures, becoming raised over the early portion on the spiral side, and by slightly depressed, strongly curved, limbate sutures on the umbilical side. Chamber walls are calcareous, coarsely perforate, and fairly smooth on the umbilical side, and sculptured and deeply pitted on the spiral side. The primary aperture is an equatorial slit with a narrow lip extending beneath the umbilical folium.

### Synonyms and Citations

*Discorbina biconcava* Parker and Jones. –Flint, 1899, p. 327, pl. 72, fig. 5.

*Planulina foveolata* (Brady). –Barker, 1960, pl. 94, fig. 1. –van Morkhoven et al., 1986, p. 26, pl. 5, figs. 1–3. –Jones, 1994, p. 98, pl. 94, fig. 1.

### Remarks

Useful indicator for Pliocene–Pleistocene outer neritic to upper bathyal environments (van Morkhoven et al., 1986).

### Biogeography

Low latitudes (America, Caribbean, Gulf of Mexico). Bahamas, Florida, southern Massachusetts, Maryland, Caribbean, Trinidad, Venezuela, and Gulf of Mexico (van Morkhoven et al., 1986); Bermuda (Jones, 1994).

### Bathymetry

Lower neritic to upper bathyal. Predominantly shelf, rare at greater depths (van Morkhoven et al., 1986).

### Chronostratigraphy

Latest Miocene to Recent.

## References

Barker, R.W., 1960, Taxonomic notes on the species figured by H.B. Brady in his report on the foraminifera dredged by H.M.S. Challenger during the years 1873–1876. Society of Economic Paleontologists and Mineralogists Special Publication, vol. 9, pp. 1–238.

Brady, H.B., 1884, Report on the foraminifera dredged by H.M.S. Challenger during the years 1873–1876. In Brady, H.B., ed., Report on the Scientific Results of the Voyage of the H.M.S. Challenger during the years 1873–1876. Zoology, vol. 9, pp. 1–814.

Flint, J.M., 1899, Recent foraminifera. A descriptive catalogue of specimens dredged by the US Fish Commission Steamer Albatross. Report of the United States National Museum for 1897, pt 1, pp. 249–349.

Jones, R.W., 1994, The Challenger Foraminifera. Oxford: Oxford University Press, 149 pp.

van Morkhoven, F.P.C.M., Berggren, W.A., Edwards, A.S., et al., 1986, Cenozoic cosmopolitan deep-water benthic foraminifera. Bulletin des Centres de Recherches Exploration-Production Elf-Aquitaine, Mémoire, vol. 11, pp. 1–421.

# *Planulina mexicana*

## Cushman, 1927

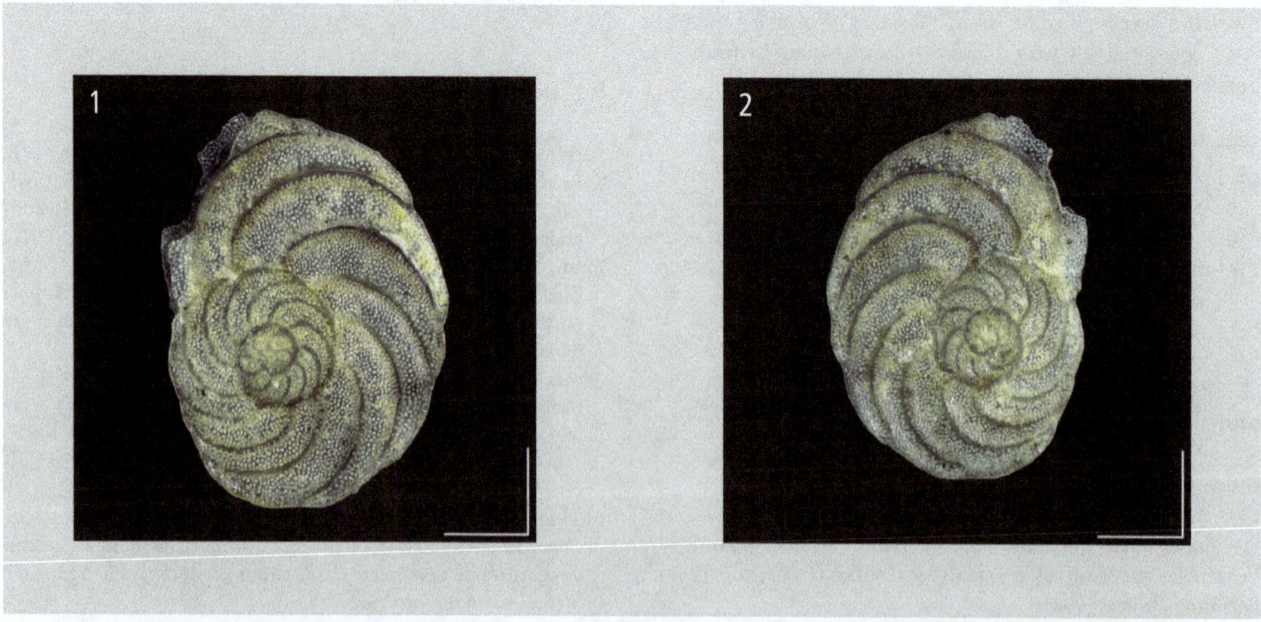

1. *Planulina mexicana.* Alazan Clay, Vera Cruz, Mexico; M.-51-V. Deposited in the collections of the United States National Museum, Washington, DC; USNM 369303. Holotype. Lateral view, NHMUK PM DI 0515, scale bar = 222 μm.

2. *Planulina mexicana.* Alazan Clay, Vera Cruz, Mexico; M.-51-V. Deposited in the collections of the United States National Museum, Washington, DC; USNM 369303. Holotype. Lateral view, NHMUK PM DI 0516, scale bar = 228 μm.

### Original Designation
*Planulina mexicana* Cushman, 1927, p. 113, pl. 23, fig. 5.

### Type Specimen
Holotype (USNM 369303) deposited in the collections of the United States National Museum, Washington, DC.

### Type Level
Probably Lower Oligocene, Alazan Clay (van Morkhoven et al., 1986). Originally given as Upper Eocene by Cushman (1927).

### Type Locality
Rio Buena Vista, 0.5 kilometer south and 25° east from Tumbadero Hacienda House, Vera Cruz, Mexico.

### Description
Test forms a strongly compressed, discoidal, very low trochospire, with flattened, evolute spiral and umbilical sides, and a truncate periphery. The ten to twelve narrow, curved chambers in the final whorl increase rapidly in size, and are separated by limbate, slightly depressed, strongly curved sutures. Chamber walls are calcareous and finely perforate on both sides of the test. The primary aperture is an equatorial slit with a narrow lip extending beneath the umbilical folium.

### Synonyms and Citations
*Planulina mexicana* Cushman. –van Morkhoven et al., 1986, p. 161, pl. 53.

### Biogeography
Recorded in Ecuador (Cushman and Stainforth, 1951), Florida, Mexico, Cuba, Puerto Rico, the Dominican Republic, Venezuela, and Trinidad (van Morkhoven et al., 1986).

### Bathymetry
Lower neritic to upper bathyal (van Morkhoven et al., 1986).

### Chronostratigraphy
Late Eocene (Priabonian) to Middle Miocene (Serravallian).

## References


Cushman, J.A., 1927, New and interesting foraminifera from Mexico and Texas. Contributions from the Cushman Laboratory for Foraminiferal Research, vol. 3, pp. 111–117.

Cushman, J.A. and Stainforth, R.M., 1951, Tertiary foraminifera of coastal Ecuador – Part 1, Eocene. Journal of Paleontology, vol. 25, no. 2, pp. 129–164.

van Morkhoven, F.P.C.M., Berggren, W.A., Edwards, A.S., et al., 1986, Cenozoic cosmopolitan deep-water benthic foraminifera. Bulletin des Centres de Recherches Exploration-Production Elf-Aquitaine, Mémoire, vol. 11, pp. 1–421.

# Planulina renzi

## Cushman and Stainforth, 1945

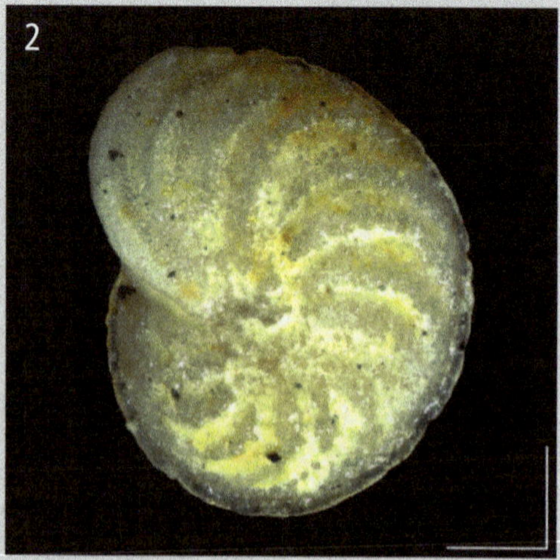


1. *Planulina renzi*. Cipero zone VII (upper), Trinidad; Sample Rz.425. Deposited in the collections of the United States National Museum, Washington, DC; USNM 44002. Holotype. Spiral view, NHMUK PM DI 0517, scale bar = 126 μm.

2. *Planulina renzi*. Cipero zone VII (upper), Trinidad; Sample Rz.425. Deposited in the collections of the United States National Museum, Washington, DC; USNM 44002. Holotype. Umbilical view, NHMUK PM DI 0518, scale bar = 130 μm.

### Original Designation

*Planulina renzi* Cushman and Stainforth, 1945, p. 72, pl. 15, fig. 1.

### Type Specimen

Holotype (USNM 44002) deposited in the collections of the United States National Museum, Washington, DC.

### Type Level

Middle Miocene, Cipero Formation, *Globorotalia fohsi* Zone, light colored marls. Originally given as Upper Oligocene by Cushman and Stainforth (1945).

### Type Locality

Sample Rz.425, Trinidad Leaseholds Catalog No. 61418, Trinidad Government Cadastral coordinates: N 228,300 links, E 353,600 links, situated at the base of the cliff in the southern (north-dipping) limb of the syncline, 120 feet northeastward from the boundary between the upper and middle zones, 880 feet in a straight line southwestward from Rz.108 ("Cipero Nose"), along the coast between the mouth of the Cipero River, at the point where the Trinidad Government railway turns inland, south of San Fernando, Trinidad.

### Description

Test forms a compressed, discoidal, very low trochospire; unequally biconvex in cross-section, with a strongly evolute, nearly flat or slightly convex spiral side, a partially evolute, slightly more convex umbilical side, and an acute, keeled periphery. The twelve to fourteen narrow, curved, slightly inflated chambers in the final whorl increase gradually in size, and are separated by broad, limbate, raised, strongly curved sutures. Chamber walls are calcareous, coarsely perforate, and granular in appearance. The central area of the spiral side is ornamented by several bosses that coalesce with the raised sutures. The primary aperture is an equatorial slit with a narrow lip extending beneath the umbilical folium.

### Synonyms (following van Morkhoven et al., 1986) and Citations

*Truncatulina wuellerstorfi* (Schwager). –Nuttall, 1928, p. 98, pl. 7, fig. 12.

*Planulina marialana* Hadley var. *gigas* Keijzer, 1945, p. 206, pl. 5, fig. 77.

*Planulina renzi* Cushman and Stainforth. –van Morkhoven et al., 1986, p. 133, pl. 43. –Miller and Katz, 1987, p. 136, pl. 6, fig. 1.

–Katz and Miller, 1993, pl. 3, fig. 6. –Bolli et al., 1994, p. 369, figs. 57.3, 88.12. –Robertson, 1998, p. 214, pl. 85, fig. 2.

### Remarks

*Planulina renzi* Cushman and Stainforth differs from *Planulina wuellerstorfi* (Schwager) by its less flattened, more robust, generally larger test with granular wall texture. Intergrades between *Planulina renzi* and *Planulina wuellerstorfi* occur in the Miocene, when the stratigraphic range of the two species overlaps.

### Biogeography

Worldwide.

### Bathymetry

Primarily bathyal. Occurs at abyssal depths from the Oligocene to early Middle Miocene. Became restricted to bathyal depths in the later part of its range (van Morkhoven et al., 1986).

### Chronostratigraphy

Early Oligocene (Rupelian) to Late Miocene (Messinian). Van Morkhoven et al. (1986) noted that a small and probably related form occurs in the Late Eocene (Biozones P16 and P17).

## References


Bolli, H.M., Beckmann, J.P., and Saunders, J.B., 1994, Benthic Foraminiferal Biostratigraphy of the South Caribbean Region. Cambridge: Cambridge University Press, 408 pp.

Cushman, J.A. and Stainforth, R.M., 1945, The foraminifera of the Cipero Marl Formation of Trinidad, British West Indies. Contributions from the Cushman Laboratory for Foraminiferal Research, vol. 14, pp. 3–75.

Katz, M.E. and Miller, K.G., 1993, Latest Oligocene to Earliest Pliocene benthic foraminiferal biofacies of the northeastern Gulf of Mexico. Micropaleontology, vol. 39, no. 4, pp. 367–403.

Keijzer, F.G., 1945, Outline of the geology of the eastern part of the province of Oriente, Cuba (E of 76 degrees WL), with notes on the geology of other parts of the island. Utrecht University, Geographische en Geologische Mededelingen, Physiographisch-Geologische Reeks, ser. 2, no. 6, pp. 1–238.

Miller, K.G. and Katz, M.E., 1987, Oligocene to Miocene benthic foraminiferal and abyssal circulation changes in the North Atlantic. Micropaleontology, vol. 33, no. 2, pp. 97–149.

Nuttall, W.L.F., 1928, Tertiary foraminifera from the Naparima region of Trinidad (British West Indies). Quarterly Journal of the Geological Society of London, vol. 84, pp. 57–117.

Robertson, B.E., 1998, Systematics and paleoecology of the benthic foraminiferida from the Buff Bay section, Miocene of Jamaica. Micropaleontology, vol. 44, supplement 2, pp. 1–266.

van Morkhoven, F.P.C.M., Berggren, W.A., Edwards, A.S., et al., 1986, Cenozoic cosmopolitan deep-water benthic foraminifera. Bulletin des Centres de Recherches Exploration-Production Elf-Aquitaine, Mémoire, vol. 11, pp. 1–421.

# *Planulina rugosa*

## (Phleger and Parker), 1951

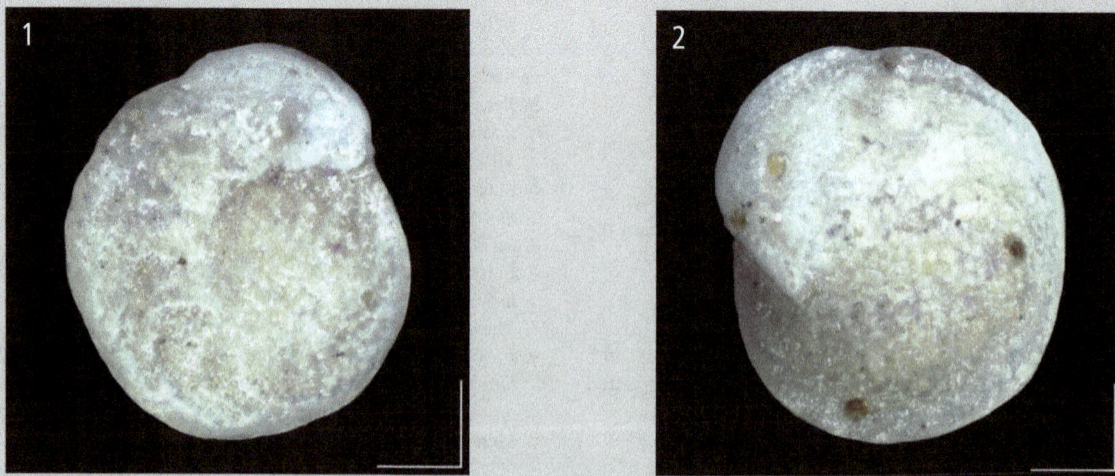


1. *Planulina rugosa*. Gulf of Mexico; Station 451, 27° 27′ N, depth 914 meters. Deposited in the collections of the United States National Museum, Washington, DC. Holotype. Spiral view, NHMUK PM DI 0519, scale bar = 180 μm.

2. *Planulina rugosa*. Gulf of Mexico; Station 451, 27° 27′ N, depth 914 meters. Deposited in the collections of the United States National Museum, Washington, DC. Holotype. Umbilical view, NHMUK PM DI 0520, scale bar = 183 μm.

### Original Designation

*Cibicides rugosa* Phleger and Parker, 1951, p. 31, pl. 17, figs. 5–6.

### Type Specimen

Deposited in the collections of the United States National Museum, Washington, DC.

### Type Level

Recent.

### Type Locality

Bottom sample at Station 451, latitude 27° 27′ N, longitude 93° 04′ W, Northwest Gulf of Mexico; 914 meters.

### Description

Test forms a compressed, discoidal, very low trochospire; subcircular in outline and planoconvex in cross-section, with a flattened or slightly convex, evolute, spiral side, a convex, partially evolute, umbilical side, and a subacute periphery. The ten to twelve narrow, curved, slightly inflated chambers in the final whorl increase rapidly in size, and are separated by broad, limbate, raised sutures on the dorsal side, and by flush, limbate sutures on the ventral side. Sutures curve back sharply along the periphery on both sides, and may be slightly depressed in the last three chambers. Chamber walls are calcareous and coarsely perforate, with a coarsely papillate ornamentation on the umbilical surface forming an almost reticulate pattern. The primary aperture is an equatorial slit with a narrow lip extending beneath the umbilical folium.

### Synonyms and Citations

*Planulina rugosa* (Phleger and Parker). –van Morkhoven et al., 1986, p. 45, pl. 13. –Katz and Miller, 1993, p. 394, pl. 2, fig. 3. –Robertson, 1998, p. 214, pl. 85, fig. 3.

### Remarks

*Planulina rugosa* (Phleger and Parker) exhibits some variation in the degree of ornamentation.

### Biogeography

Worldwide.

### Bathymetry

Bathyal to abyssal.

### Chronostratigraphy

Middle Miocene (Serravallian) to Recent.

## References


Katz, M.E. and Miller, K.G., 1993, Latest Oligocene to Earliest Pliocene benthic foraminiferal biofacies of the northeastern Gulf of Mexico. Micropaleontology, vol. 39, no. 4, pp. 367–403.

Phleger, F.B. and Parker, F.L., 1951, Ecology of foraminifera, northwest Gulf of Mexico; Part II – Foraminifera species. Memoirs of the Geological Society of America, vol. 46, pp. 1–64.

Robertson, B.E., 1998, Systematics and paleoecology of the benthic foraminiferida from the Buff Bay section, Miocene of Jamaica. Micropaleontology, vol. 44, supplement 2, pp. 1–266.

van Morkhoven, F.P.C.M., Berggren, W.A., Edwards, A.S., et al., 1986, Cenozoic cosmopolitan deep-water benthic foraminifera. Bulletin des Centres de Recherches Exploration-Production Elf-Aquitaine, Mémoire, vol. 11, pp. 1–421.

# Planulina subtenuissima

## (Nuttall), 1928

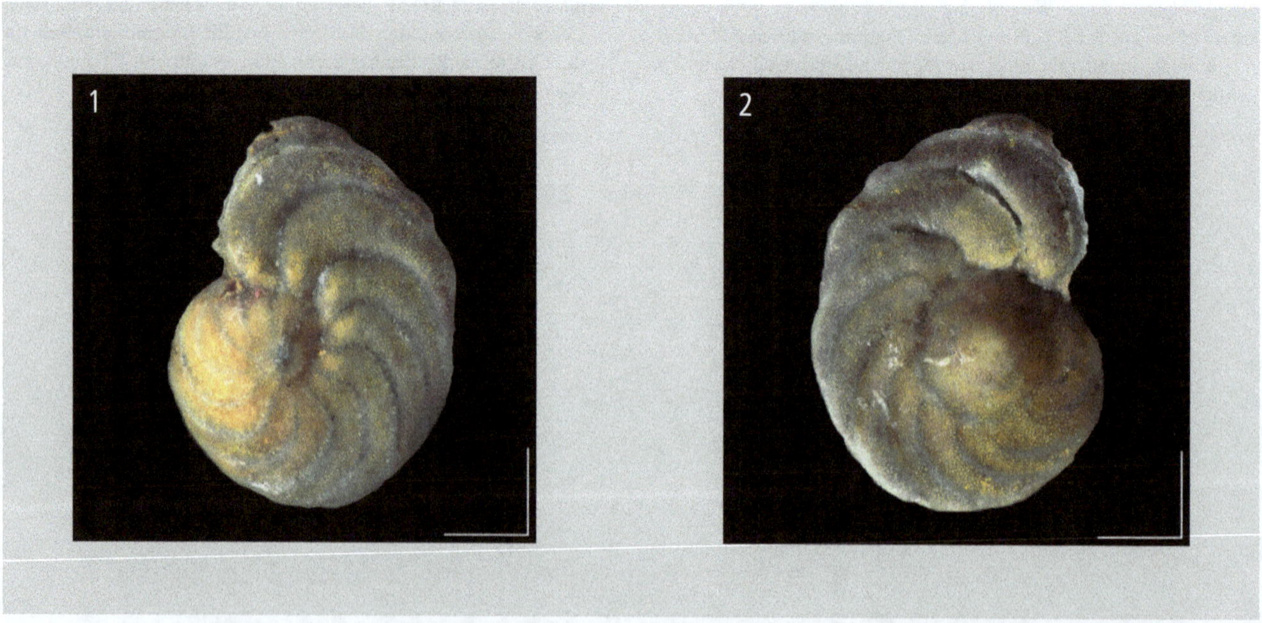

1. *Planulina subtenuissima.* Williamsonville Clay, Williamsonville Estate, Naparima Region of Trinidad. Deposited in Slide 221, specimen no. 2 WE 1317, Nuttall Collection, Sedgwick Museum, Cambridge, England. Cotype. Umbilical view, NHMUK PM DI 0521, scale bar = 127 μm.

2. *Planulina subtenuissima.* Williamsonville Clay, Williamsonville Estate, Naparima Region of Trinidad. Deposited in Slide 221, specimen no. 2 WE 1317, Nuttall Collection, Sedgwick Museum, Cambridge, England. Cotype. Spiral view, NHMUK PM DI 0522, scale bar = 127 μm.

### Original Designation

*Anomalina subtenuissima* Nuttall, 1928, p. 100, pl. 7, figs. 13, 15; textfig. 6.

### Type Specimen

Cotypes deposited in the Sedgwick Museum, Cambridge, UK. Van Morkhoven et al. (1986) designated as lectotype specimen no. 9 on the type slide (=Nuttall, 1928, pl. 7, fig. 15).

### Type Level

Lower Miocene (approximately Zone N7), Williamsville Clay (van Morkhoven et al., 1986). Originally given as Lower Miocene or Upper Oligocene by Nuttall (1928).

### Type Locality

Williamsville Estate, Naparima region of Trinidad, West Indies.

### Description

Test forms a strongly compressed, discoidal, very low trochospire; unequally biconvex, with a slightly convex, evolute, spiral side, a slightly more convex, partially evolute, umbilical side, and an acute periphery. The ten to thirteen narrow, curved, slightly inflated chambers in the final whorl increase rapidly in size, and are separated by narrow, strongly curved, slightly depressed sutures. Early whorls are indistinct, below a smooth, glossy boss on both sides of the test. Chamber walls are calcareous and finely perforate on both sides of the test. The primary aperture is an equatorial slit with a narrow lip extending beneath the umbilical folium.

### Synonyms and Citations

*Planulina subtenuissima* (Nuttall). –van Morkhoven et al., 1986, p. 198, pl. 67A–B. –Robertson, 1998, p. 214, pl. 86, fig. 1. –Ortiz and Thomas, 2006, p. 128, pl. 10, figs. 7–8.

### Remarks

*Planulina subtenuissima* (Nuttall) differs from *Eponides vortex* Galloway and Heminway, which has a thicker test, narrower final whorl, and lacks depressed sutures (van Morkhoven et al., 1986).

### Biogeography

Worldwide. Recorded from Cuba, the Dominican Republic, Barbados, Trinidad, Venezuela, the Gulf of Mexico, the North and South Atlantic, the eastern Pacific, and Morocco (van Morkhoven et al., 1986).

## Bathymetry

Primarily upper bathyal. Rare in lower bathyal sediments (Katz and Miller, 1993).

## Chronostratigraphy

Early Oligocene (Rupelian) to Middle Miocene (Langhian).

# References


Katz, M.E. and Miller, K.G., 1993, Latest Oligocene to Earliest Pliocene benthic foraminiferal biofacies of the northeastern Gulf of Mexico. Micropaleontology, vol. 39, no. 4, pp. 367–403.

Nuttall, W.L.F., 1928, Tertiary foraminifera from the Naparima region of Trinidad (British West Indies). Quarterly Journal of the Geological Society of London, vol. 84, pp. 57–117.

Ortiz, S. and Thomas, E., 2006, Lower-middle Eocene benthic foraminifera from the Fortuna Section (Betic Cordillera, southeastern Spain). Micropaleontology, vol. 52, pp. 97–150.

Robertson, B.E., 1998, Systematics and paleoecology of the benthic foraminiferida from the Buff Bay section, Miocene of Jamaica. Micropaleontology, vol. 44, supplement 2, pp. 1–266.

van Morkhoven, F.P.C.M., Berggren, W.A., Edwards, A.S., et al., 1986, Cenozoic cosmopolitan deep-water benthic foraminifera. Bulletin des Centres de Recherches Exploration-Production Elf-Aquitaine, Mémoire, vol. 11, pp. 1–421.

# *Planulina wuellerstorfi*

## (Schwager), 1866

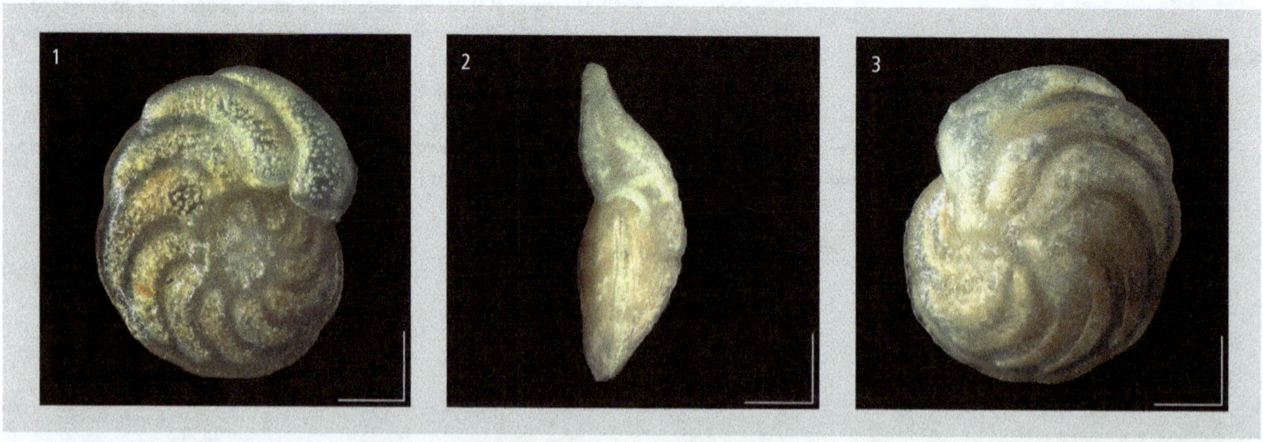


1. *Planulina wuellerstorfi*. "Meteor" Cruise, off West Africa; M25. Deposited in the Lutze Collection (West Africa), no. 200. Institute of Geosciences, Kiel University, Germany. Spiral view, NHMUK PM DI 0523, scale bar = 186 µm.

2. *Planulina wuellerstorfi*. "Meteor" Cruise, off West Africa; M25. Deposited in the Lutze Collection (West Africa), no. 200. Institute of Geosciences, Kiel University, Germany. Apertural view, NHMUK PM DI 0524, scale bar = 186 µm.

3. *Planulina wuellerstorfi*. "Meteor" Cruise, off West Africa; M25. Deposited in the Lutze Collection (West Africa), no. 200. Institute of Geosciences, Kiel University, Germany. Umbilical view, NHMUK PM DI 0525, scale bar = 187 µm.

### Original Designation

*Anomalina wuellerstorfi* Schwager, 1866, p. 258, pl. 7, figs. 105, 107.

### Type Specimen

Depository not given. Neotype (Mf 18) designated by Srinivasan and Sharma (1980) deposited in the Micropaleontology Section, Department of Geology, Banaras Hindu University, Banaras, India.

### Type Level

Upper Tertiary, lower and upper clays.

### Type Locality

Car Nicobar, Andaman Sea.

### Description

Test forms a compressed, discoidal, very low trochospire; plano-convex in cross-section, with a flattened, evolute, spiral side, a slightly convex, partially evolute, umbilical side, and a truncate, keeled periphery. The (approximately) ten narrow, curved, slightly inflated chambers in the final whorl increase rapidly in size, and are separated by thickened, strongly curved sutures, slightly depressed in final chambers on the spiral side, and sinuoid or hooked in final chambers on the umbilical side. Chamber walls are calcareous, coarsely perforate on the spiral side, and finely perforate on the umbilical side. The primary aperture is an equatorial slit with a narrow lip extending beneath the umbilical folium.

### Synonyms and Citations

*Truncatulina wuellerstorfi* (Schwager). –Brady, 1884, pp. 627, 662, pl. 93, figs. 8–9.

*Planulina bradii* Tolmachoff. –Barker, 1960, pl. 93, fig. 8.

*Planulina wuellerstorfi* (Schwager). –Barker, 1960 pl. 93, fig. 9. –van Morkhoven et al., 1986, p. 48, pl. 14. –Miller and Katz, 1987, p. 136, pl. 6, fig. 2. –Robertson, 1998, p. 216, pl. 86, fig. 2. –Kuhnt et al., 2002, p. 151, pl. 12. figs. 18–20.

*Cibicides wuellerstorfi* (Schwager). –Pflum and Frerichs, 1976, p. 116, pl. 4, figs. 2–4.

*Cibicidoides wuellerstorfi* (Schwager). –Nomura, 1995, p. 272, fig. 15. –Jones 1994, p. 98, pl. 93, figs. 8–9.

### Remarks

*Planulina wuellerstorfi* (Schwager) has been previously referred to several other genera including *Anomalina*, *Cibicides*, *Cibicidoides*, and *Fontbotia*. We follow van Morkhoven et al. (1986) in placing the species into the genus *Planulina*, as subsequent reassignment to *Cibicidoides* (Whittaker, 1988), *Fontbotia* (González-Donoso and Linares, 1970), or *Cibicides* (Sen Gupta, 1989) did not fully resolve the generic assignation of this species.

*Planulina renzi* Cushman and Stainforth differs from *Planulina wuellerstorfi* (Schwager) by its less flattened, more robust, generally larger test and its granular wall texture. Intergrades between *Planulina renzi* and *Planulina wuellerstorfi* occur in the Miocene, when the stratigraphic range of the two species overlaps.

*Planulina wuellerstorfi* generally has an epifaunal mode of life, and is one of the most commonly used benthic foraminifers to reconstruct bottom water $\delta^{18}O$ and $\delta^{13}C$. However, *Planulina wuellerstorfi* may exhibit negative shifts in $\delta^{13}C$ in areas of strong seasonal productivity, where fluffy layers develop at the sea floor (see Mackensen and Bickert, 1999).

### Biogeography

Worldwide. Widely recorded from the North and South Atlantic, the Norwegian–Greenland Sea, the Mediterranean, the Caribbean, the Gulf of Mexico, the Pacific, the Sea of Japan, Japan, Okinawa, Saipan, the Philippines, Fiji, the Netherlands East Indies, New Guinea, New Zealand, the Tasman Sea, the Coral Sea, the Bay of Bengal, the Indian Ocean, and India (see Observed Occurrences in van Morkhoven et al., 1986).

### Bathymetry

Primarily bathyal. *Planulina wuellerstorfi* is a low–intermediate organic flux indicator (Altenbach et al., 1999) that may be used as a bathymetric indicator for water depths over about 800 meters.

### Chronostratigraphy

Middle Miocene (Langhian) to Recent.

## References


Altenbach, A.V., Pflaumann, U., Schiebel, R., Thies, A., Timm, S., and Trauth, M., 1999, Scaling percentages and distributional patterns of benthic foraminifera with flux rates of organic carbon. Journal of Foraminiferal Research, vol. 29, pp. 173–185.

Barker, R.W., 1960, Taxonomic notes on the species figured by H.B. Brady in his report on the foraminifera dredged by H.M.S. Challenger during the years 1873–1876. Society of Economic Paleontologists and Mineralogists Special Publication, vol. 9, pp. 1–238.

González-Donoso, J.M. and Linares, D., 1970, Datos sobre los foraminiferos del Tortonense de Alcalá la Real (Jaén). Revista Española de Micropaleontologia, vol. 2, pp. 235–242.

Jones, R.W., 1994, The Challenger Foraminifera. Oxford: Oxford University Press, 149 pp.

Kuhnt, W., Holbourn, A.E., and Zhao, Q., 2002, The early history of the South China Sea: Evolution of Oligocene–Miocene deep water environments. Revue de Micropaléontologie, vol. 45, pp. 99–159.

Mackensen, A. and Bickert, T., 1999, Stable carbon isotopes in benthic foraminifera: proxies of deep and bottom water circulation and new production. In Fischer, G. and Wefer, G., ed., Use of Proxies in Paleoceanography: Examples from the South Atlantic. Berlin: Springer Verlag, pp. 229–254.

Miller, K.G. and Katz, M.E., 1987, Oligocene to Miocene benthic foraminiferal and abyssal circulation changes in the north Atlantic. Micropaleontology, vol. 33, no. 2, pp. 97–149.

Nomura, R., 1995, Paleogene to Neogene deep-sea paleoceanography in the eastern Indian Ocean: benthic foraminifera from ODP Sites 747, 757 and 758. Micropaleontology, vol. 41, pp. 251–290.

Pflum, C.E. and Frerichs, W.E., 1976, Gulf of Mexico deep-water foraminifers. Cushman Foundation, Special Publication, vol. 14, pp. 1–125.

Robertson, B.E., 1998, Systematics and paleoecology of the benthic foraminiferida from the Buff Bay section, Miocene of Jamaica. Micropaleontology, vol. 44, supplement 2, pp. 1–266.

Schwager, C., 1866, Fossile Foraminiferen von Kar Nikobar, Reise der Österreichischen Fregatte Novara um Erde in den Jahren 1857, 1858, 1859 unten den Befehlen des Commodore B. Von Wüllerstorf-Urbair. Geologischer Theil, vol. 2, no. 1, Geologische Beobachtung, no. 2, Paläontologische Mittheilung, pp. 187–268.

Sen Gupta, B.K., 1989, Morphology and generic placement of the foraminifer "Anomalina" wuellerstorfi Schwager. Journal of Paleontology, vol. 63, no. 5, pp. 706–713.

Srinivasan, M.S. and Sharma, V., 1980, Schwager's Car Nicobar Foraminifera in the Reports of the Novara Expedition – A Revision. New Delhi: Today & Tomorrow's Printers and Publishers, 83 pp.

van Morkhoven, F.P.C.M., Berggren, W.A., Edwards, A.S., et al., 1986, Cenozoic cosmopolitan deep-water benthic foraminifera. Bulletin des Centres de Recherches Exploration-Production Elf-Aquitaine, Mémoire, vol. 11, pp. 1–421.

Whittaker, J.E., 1988, Benthic Foraminifera from Ecuador: Taxonomy and Distribution of Smaller Benthic Foraminifera from Coastal Ecuador (Late Oligocene–Late Pliocene). London: British Museum (Natural History), xi + 194 pp.

# Plectofrondicularia parri

## Finlay, 1939

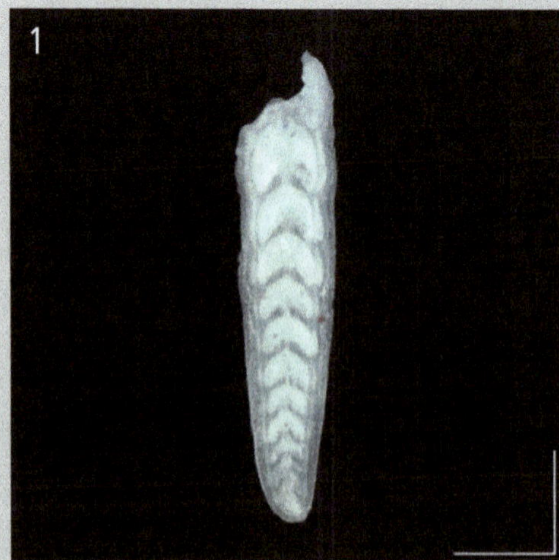


1. *Plectofrondicularia parri.* Locality BB-16, Buff Bay, Jamaica. Broken specimen. Lateral view, NHMUK PM DI 0526, scale bar = 172 μm.

### Original Designation

*Plectofrondicularia parri* Finlay, 1939, p. 516, pl. 68, fig. 4.

### Type Specimen

Deposited in the New Zealand Geological Survey Collections, Wellington, New Zealand.

### Type Level

Lower Miocene (N6–N7), Awamoan Stage, All Day Bay Beds. Originally given as Middle Miocene by Finlay (1939).

### Type Locality

Locality 5273, 60 chains at 150° from Trig V, Kakanui Beach, New Zealand. Other locality given: Pukeuri, 4 miles north of Oamaru, New Zealand (fig. 4).

### Description

Test initially biserial becoming uniserial; lanceolate in outline with a subacute initial portion. Periphery is truncate with keels at each angle of the test and additional lateral keels. Chambers are slightly inflated, and arched centrally. In the uniserial portion chambers are separated by thick, slightly depressed, chevron-shaped sutures. Chamber walls are calcareous, finely perforate, smooth in the upper part of the test, and ornamented by two or three longitudinal costae in the lower half of the test. The primary aperture is terminal, radial or with a denticulate rim.

### Synonyms and Citations

*Plectofrondicularia pohana* Finlay, 1939, p. 516, pl. 68, fig. 3.

*Plectofrondicularia parri* Finlay. –van Morkhoven et al., 1986, p. 128, pl. 41. –Robertson, 1998, p. 56, pl. 19, fig. 1.

### Remarks

Van Morkhoven et al. (1986) compared *Plectofrondicularia parri* Finlay with type material deposited in the United States National Museum. They noted that *Plectofrondicularia floridana* Cushman is broader, flatter, and has weakly developed peripheral ridges and a single central rib; *Plectofrondicularia mexicana* (Cushman) is smaller with subparallel sides, and is non-costate; *Plectofrondicularia trinitatensis* Cushman and Jarvis is more slender and rectilinear.

The upper stratigraphic range (Late Miocene, Biozones M13b/M14), given by van Morkhoven et al. (1986), was extended by the occurrence of *Plectofrondicularia parri* in the Bowden Formation Jamaica (87-J-252) of Pliocene (Biozone PL3) age (M. Katz, pers. comm. 1998).

### Biogeography

Worldwide. Recorded from the North and South Atlantic, the Pacific, the Gulf of Mexico, and Europe (van Morkhoven et al., 1986).

### Bathymetry

Lower neritic to middle bathyal (van Morkhoven et al., 1986).

### Chronostratigraphy

Early Miocene (Aquitanian) to Pliocene (Piacenzian). Doubtful in Oligocene Biozones P21–P22 (van Morkhoven et al., 1986).

## References


Finlay, H.J., 1939, New Zealand foraminifera: Key species in stratigraphy – No. 1. Transactions of the Royal Society of New Zealand, vol. 68, no. 4, pp. 504–543.

Robertson, B.E., 1998, Systematics and paleoecology of the benthic foraminiferida from the Buff Bay section, Miocene of Jamaica. Micropaleontology, vol. 44, supplement 2, pp. 1–266.

van Morkhoven, F.P.C.M., Berggren, W.A., Edwards, A.S., et al., 1986, Cenozoic cosmopolitan deep-water benthic foraminifera. Bulletin des Centres de Recherches Exploration-Production Elf-Aquitaine, Mémoire, vol. 11, pp. 1–421.

# *Plectofrondicularia paucicostata*

## Cushman and Jarvis, 1929

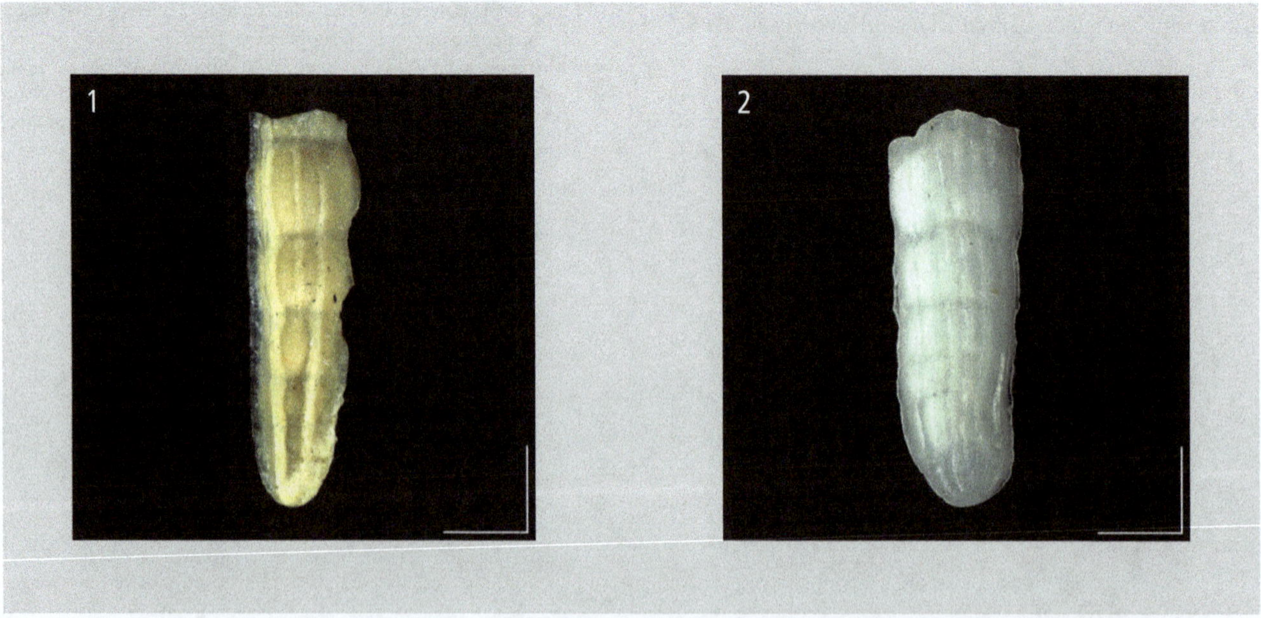


1. *Plectofrondicularia paucicostata.* "Bamboo Clay", Cipero Section Station 11, Trinidad. Deposited in the collections of the United States National Museum, Washington, DC; CC 10073. Holotype. Lateral view, NHMUK PM DI 0527, scale bar = 335 μm.

2. *Plectofrondicularia paucicostata.* Gulf of Mexico; Eureka Core E67-128. Deposited in the collections of the United States National Museum, Washington, DC. Specimen illustrated in van Morkhoven et al. 1986, Ref. no. 91. Lateral view, NHMUK PM DI 0528, scale bar = 111 μm.

### Original Designation

*Plectofrondicularia paucicostata* Cushman and Jarvis, 1929, p. 10, pl. 2, figs. 11–13.

### Type Specimen

Holotype (CC 10073) and paratypes (CC 10072) deposited in the collections of the United States National Museum, Washington, DC.

### Type Level

Eocene, "Bamboo Clay".

### Type Locality

Cipero Section, Trinidad.

### Description

Test large, forming an initial compressed biserial series, but becoming uniserial and lanceolate in outline, with a rounded initial portion and a truncate periphery. The initial biserial stage may be loosely twisted, and at a slight angle to the uniserial portion. Chambers are slightly inflated, arched centrally in the uniserial portion, and separated by slightly depressed sutures. Chamber walls are calcareous, finely perforate, and ornamented by prominent longitudinal costae that generally extend from the base of the test to the aperture (although the last chamber may be partly smooth). The primary aperture is an elongate, terminal slit.

### Synonyms and Citations

*Plectofrondicularia lirata* Bermúdez, 1937, p. 240, pl. 20, fig. 6.

*Plectofrondicularia kerni* Cook. –Mallory, 1959, p. 212, pl. 18, fig. 2; pl. 33, fig. 10.

*Plectofrondicularia paucicostata* Cushman and Jarvis. –van Morkhoven et al., 1986, p. 273, pl. 91.

### Biogeography

Worldwide.

### Bathymetry

Middle bathyal to abyssal (van Morkhoven et al., 1986).

### Chronostratigraphy

Middle Eocene (Lutetian) to Early Oligocene (Rupelian). Doubtful occurrence in the Early Eocene (van Morkhoven et al., 1986).

## References

Bermúdez, P.J., 1937, Nuevas especies de foraminiferos del Eoceno de las cercanias de Guanajay, Provincia Pinar del Rio. Memorias de la Sociedad Cubana de Historia Natural "Felipe Poey", vol. 11, pp. 237–247.

Cushman, J.A. and Jarvis, P.W., 1929, New foraminifera from Trinidad. Contributions from the Cushman Laboratory for Foraminiferal Research, vol. 5, pp. 6–17.

Mallory, V.S., 1959, Lower Tertiary Biostratigraphy of the California Coast Ranges. Tulsa, OK: American Association of Petroleum Geologists, 416 pp.

van Morkhoven, F.P.C.M., Berggren, W.A., Edwards, A.S., et al., 1986, Cenozoic cosmopolitan deep-water benthic foraminifera. Bulletin des Centres de Recherches Exploration-Production Elf-Aquitaine, Mémoire, vol. 11, pp. 1–421.

# Plectofrondicularia vaughani

## Cushman, 1927

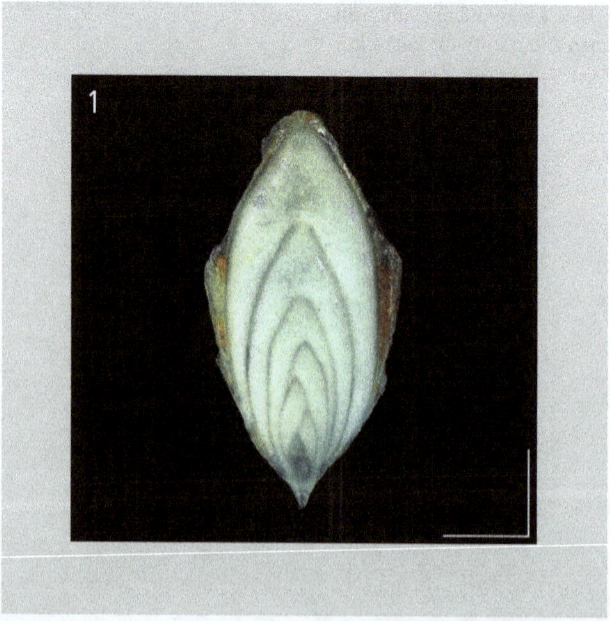

1. *Plectofrondicularia vaughani*. Bowden Formation, Jamaica, 87-J-251. Lateral view, NHMUK PM DI 0529, scale bar = 158 μm.

### Original Designation

*Plectofrondicularia vaughani* Cushman, 1927, p. 112, pl. 23, fig. 3.

### Type Specimen

Holotype (No. 369300) deposited in the collections of the United States National Museum, Washington, DC.

### Type Level

Oligocene, Alazan Clay. Originally given as Upper Eocene by Cushman (1927).

### Type Locality

Rio Buena Vista, 9.8 kilometers in a straight line upstream from Tumbadero, Vera Cruz, Mexico.

### Description

Test elongate, forming an initial strongly compressed, fragile, biserial series, but becoming uniserial and ovate in outline with an elliptical cross-section, a distinctly acute initial portion, and a truncate, keeled periphery. Chambers are very slightly inflated, arched centrally in the uniserial portion, and separated by thick, slightly depressed, strongly curved sutures. Chamber walls are calcareous, finely perforate, and smooth. The primary aperture is terminal, radial or with a denticulate rim.

### Synonyms and Citations

*Plectofrondicularia deyaensis* Colom, 1943, p. 321, pl. 22, figs. 33–38.

*Plectofrondicularia jarvisi* Cushman and Todd, 1945, p. 38, pl. 6, fig. 4.

*Plectofrondicularia ruthvenmurrayi* Cushman and Stainforth, 1945, p. 37, pl. 5, fig. 14.

*Plectofrondicularia vaughani* Cushman. –van Morkhoven et al., 1986, p. 130, pl. 42. –Robertson, 1998, p. 58, pl. 19, fig. 2. –Kuhnt et al., 2002, p. 151, pl. 12. figs. 1–3.

### Remarks

Specimens mostly found broken, as test is fragile. The upper stratigraphic range of *Plectofrondicularia vaughani*, reported as Late Miocene (Biozones M13b/M14) by van Morkhoven et al. (1986), was extended to the Early Pliocene (Biozone PL2) by its occurrence in the Bowden Formation, Jamaica (M. Katz, pers. comm. 1998).

### Biogeography

Worldwide. Recorded from the Caribbean, Venezuela, California, Mexico, the Gulf of Mexico, the North and South Atlantic, and Europe (van Morkhoven et al., 1986).

### Bathymetry

Primarily lower neritic to middle bathyal (van Morkhoven et al., 1986). Rarely lower bathyal (>2%) (Katz and Miller, 1993).

### Chronostratigraphy

Early Oligocene to Early Pliocene.

## References


Colom, G., 1943, Nuevas especies y subespecies de foraminiferos fosiles de Mallorca. Boletín de la Real Sociedad Española de Historia Natural, vol. 41, pp. 317–335.

Cushman, J.A., 1927, New and interesting foraminifera from Mexico and Texas. Contributions from the Cushman Laboratory for Foraminiferal Research, vol. 3, pp. 111–117.

Cushman, J.A. and Stainforth, R.M., 1945, The foraminifera of the Cipero Marl Formation of Trinidad, British West Indies. Contributions from the Cushman Laboratory for Foraminiferal Research, vol. 14, pp. 3–75.

Cushman, J.A. and Todd, R., 1945, Miocene foraminifera from Buff Bay, Jamaica. Special Publications of the Cushman Laboratory for Foraminiferal Research, vol. 15, pp. 1–73.

Katz, M.E. and Miller, K.G., 1993, Latest Oligocene to Earliest Pliocene benthic foraminiferal biofacies of the northeastern Gulf of Mexico. Micropaleontology, vol. 39, no. 4, pp. 367–403.

Kuhnt, W., Holbourn, A.E., and Zhao, Q., 2002, The early history of the South China Sea: Evolution of Oligocene–Miocene deep water environments. Revue de Micropaléontologie, vol. 45, pp. 99–159.

Robertson, B.E., 1998, Systematics and paleoecology of the benthic foraminiferida from the Buff Bay section, Miocene of Jamaica. Micropaleontology, vol. 44, supplement 2, pp. 1–266.

van Morkhoven, F.P.C.M., Berggren, W.A., Edwards, A.S., et al., 1986, Cenozoic cosmopolitan deep-water benthic foraminifera. Bulletin des Centres de Recherches Exploration-Production Elf-Aquitaine, Mémoire, vol. 11, pp. 1–421.

# *Pleurostomella brevis*

## Schwager, 1866

1. *Pleurostomella brevis.* Ki Islands, Central Pacific Ocean. "Challenger" Station 192, depth: 129 fathoms. Deposited in the micropaleontological collections of The Natural History Museum, London; NHMUK ZF2121. Lateral view, NHMUK PM DI 0530, scale bar = 170 μm.

2. *Pleurostomella brevis.* Ki Islands, Central Pacific Ocean. "Challenger" Station 192, depth: 129 fathoms. Deposited in the micropaleontological collections of The Natural History Museum, London; NHMUK ZF2121. Apertural view, NHMUK PM DI 0531, scale bar = 171 μm.

### Original Designation

*Pleurostomella brevis* Schwager, 1866, p. 239, pl. 6, fig. 81.

### Type Specimen

Depository not given. Neotype (Mf 18) designated by Srinivasan and Sharma (1980) deposited in the Micropaleontology Section, Department of Geology, Banaras Hindu University, Banaras, India.

### Type Level

Neogene, lower and upper clays.

### Type Locality

Car Nicobar, Andaman Sea.

### Description

Test forms a robust, elongate, ovate, initially biserial series, later lax-uniserial with a circular cross-section. The maximum width is located in the middle of the test. The highly inflated chambers increase rapidly in size (final chamber occupying more than half the length of the test), and are separated by slightly depressed, curved sutures. Chamber walls are calcareous, translucent, smooth, and finely perforate. The primary aperture is terminal with a hood and two teeth.

### Synonyms and Citations

*Pleurostomella brevis* Schwager. –Brady, 1884, p. 411, pl. 51, fig. 20a–b. –Barker, 1960, pl. 51, fig. 20. –Srinivasan and Sharma, 1980, p. 58, pl. 8, figs. 1–2. –Jones, 1994, p. 56, pl. 51, fig. 20. –Hayward, 2002, p. 302, pl. 1, figs. 25–26.

### Remarks

Reported by Hayward (2002) as one of the species that became globally extinct during the mid–Pleistocene Climatic Transition. Hayward and Kawagata (2005) suggested that Brady's figured specimens from "Challenger" Station 192 in the Banda Sea, Indonesia, include relict or reworked forms, possibly mixed with Holocene taxa.

### Biogeography

Worldwide.

### Bathymetry

Lower neritic to abyssal.

## Chronostratigraphy

Miocene to Recent?

## References


Barker, R.W., 1960, Taxonomic notes on the species figured by H.B. Brady in his report on the foraminifera dredged by H.M.S. Challenger during the years 1873–1876. Society of Economic Paleontologists and Mineralogists Special Publication, vol. 9, pp. 1–238.

Brady, H.B., 1884, Report on the foraminifera dredged by H.M.S. Challenger during the years 1873–1876. In Brady, H.B., ed., Report on the Scientific Results of the Voyage of the H.M.S. Challenger during the years 1873–1876. Zoology, vol. 9, pp. 1–814.

Hayward, B., 2002, Late Pliocene to Middle Pleistocene extinctions of deep-sea benthic foraminifera ("Stilostomella extinction") in the southwest Pacific. Journal of Foraminiferal Research, vol. 32, pp. 274–307.

Hayward, B. and Kawagata, S., 2005, Extinct foraminifera in Brady's Challenger Report. Journal of Micropalaeontology, vol. 24, pp. 171–175.

Jones, R.W., 1994, The Challenger Foraminifera. Oxford: Oxford University Press, 149 pp.

Schwager, C., 1866, Fossile Foraminiferen von Kar Nikobar, Reise der Österreichischen Fregatte Novara um Erde in den Jahren 1857, 1858, 1859 unten den Befehlen des Commodore B. Von Wüllerstorf-Urbair. Geologischer Theil, vol. 2, no. 1, Geologische Beobachtung, no. 2, Paläontologische Mittheilung, pp. 187–268.

Srinivasan, M.S. and Sharma, V., 1980, Schwager's Car Nicobar Foraminifera in the Reports of the Novara Expedition – A Revision. New Delhi: Today & Tomorrow's Printers and Publishers, 83 pp.

# Praecystammina globigerinaeformis

## Krasheninnikov, 1973

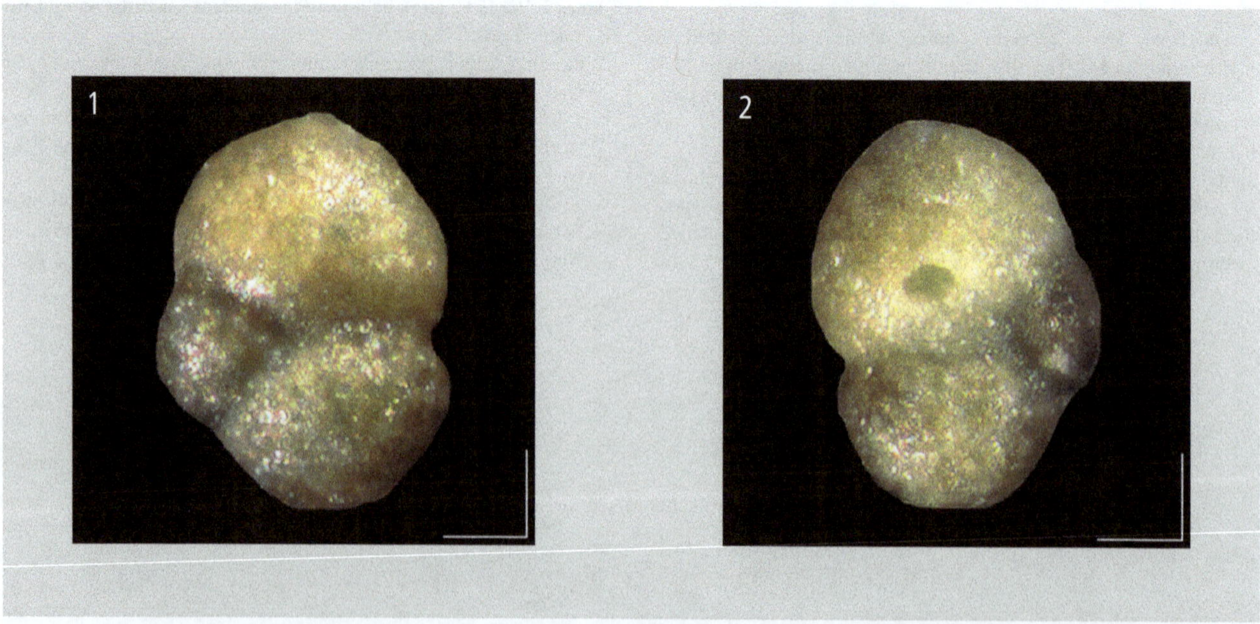


1. *Praecystammina globigerinaeformis.* East of Izu-Bonin Trench, North Pacific; DSDP Leg 20, Site 196, Core 2, 1–4 cm. Lateral view, NHMUK PM DI 0532, scale bar = 82 μm.

2. *Praecystammina globigerinaeformis.* East of Izu-Bonin Trench, North Pacific; DSDP Leg 20, Site 196, Core 2, 1–4 cm. Lateral view, NHMUK PM DI 0533, scale bar = 82 μm.

### Original Designation

*Praecystammina globigerinaeformis* Krasheninnikov, 1973, p. 211, pl. 2, figs. 1–2.

### Type Specimen

Housed at the Geological Institute of the Academy of Sciences, Moscow. The holotype is registered as N 4011/20; the paratype is registered as N 4011/21.

### Type Level

Upper Cretaceous (Santonian–Campanian).

### Type Locality

Holotype from DSDP Leg 20, Hole 198A, Core Catcher sample. Paratype is from Section 1, Core 4, 138–140 cm interval. Hole 198A was drilled in a water depth of 5858 meters in the abyssal plain north of Marcus Island in the Northwest Pacific Ocean (25° 49.54′ N, 154° 35.05′ E).

### Description

Small, *Globigerina*-like, irregularly evolute trochospire with initial streptospiral coiling, a rounded periphery, and a trilobate outline. The three inflated, reniform chambers in the last whorl are separated by slightly curved and depressed sutures. Chamber walls are finely agglutinated, well silicified, and with a smooth, glassy finish. The primary aperture is areal, oval, and bordered by a thin lip.

### Synonyms and Citations

*Praecystammina globigerinaeformis* Krasheninnikov. –Hemleben and Tröster, 1984, p. 521, pl. 4, fig. 8. –Moullade et al., 1988, pp. 365–366, pl. 7, fig. 6.

### Remarks

*Cystammina sveni* Gradstein and Kaminski differs from *Praecystammina globigerinaeformis* by its more tightly coiled, less evolute test with fewer chambers in the last whorl and by its slit-like aperture (Kaminski and Gradstein, 2005). The two species also have different stratigraphic ranges: *Cystammina sveni* is typical of Paleocene bathyal or deeper marine sediments extending into the Middle Eocene, occurring rarely in the Campanian to Maastrichtian.

### Biogeography

Worldwide. Recorded from the Pacific Ocean (Krasheninnikov, 1973), the Indian Ocean (Krasheninnikov, 1974), and the North Atlantic (Moullade et al., 1988).

### Bathymetry

Abyssal (typically sub-CCD).

### Chronostratigraphy

Late Cretaceous (Turonian to Maastrichtian).

## References

Hemleben, C. and Tröster, J., 1984, Campanian–Maestrichtian deep-water foraminifers from Hole 543A, Deep Sea Drilling Project. In Bij-Duval, B., Moore, J.C., et al., ed., Initial Reports of the Deep Sea Drilling Project, Washington (U.S. Government Printing Office), vol. 78A, pp. 509–532.

Kaminski, M.A. and Gradstein, F.M., 2005, Atlas of Palaeogene Cosmopolitan Deep-Water Agglutinated Foraminifera. Grzybowski Foundation Special Publication, no. 10, 547 + vii pp.

Krasheninnikov, V.A., 1973, Cretaceous benthonic foraminifera, Leg 20, Deep Sea Drilling Project. In Heezen, B.C., MacGregor, I.D., et al., ed., Initial Reports of the Deep Sea Drilling Project, Washington (U.S. Government Printing Office), vol. 20, pp. 205–221.

Krasheninnikov, V.A., 1974, Upper Cretaceous benthonic agglutinated foraminifera, Leg 27 of the Deep Sea Drilling Project. In Veevers, J.J., Heirtzler, J.R., et al., ed., Initial Reports of the Deep Sea Drilling Project, Washington (U.S. Government Printing Office), vol. 27, pp. 531–662.

Moullade, M., Kuhnt, W., and Thurow, J., 1988, Agglutinated benthic foraminifers from Upper Cretaceous variegated clays of the North Atlantic Ocean (DSDP Leg 93 and ODP Leg 103). In Boillot, G., Winterer, E.L., et al., ed. Scientific Results of the Proceedings of the Ocean Drilling Program, vol. 103, pp. 349–377.

# Psammosphaera fusca

## Schulze, 1875

1. *Psammosphaera fusca*. West of Ireland, Atlantic Ocean; "Porcupine" Station 16, depth not given. Deposited in the micropaleontological collections of The Natural History Museum, London; NHMUK ZF2190. Lateral view, NHMUK PM DI 0534, scale bar = 549 μm.

2. *Psammosphaera fusca*. Off Sombrero Island, West Indies; "Challenger" Station 23, depth: 450 fathoms. Deposited in the micropaleontological collections of The Natural History Museum, London; NHMUK ZF2193. Cross-section view, NHMUK PM DI 0535, scale bar = 195 μm.

### Original Designation

*Psammosphaera fusca* Schulze, 1875, p. 113, pl. 2, fig. 8a–f.

### Type Specimen

Depository not given.

### Type Level

Recent, silt and small stones.

### Type Locality

Localities given: a depth of 106 fathoms in the entrance of the Bommel Fjord and a depth of 137 to 217 fathoms in the mouth of the Korsfjord.

### Description

Test is unilocular, spherical or subspherical with a subcircular cross-section. Chamber walls are agglutinated, consisting of a single layer of sand grains cemented over an inner organic lining. Agglutinated grains may be of very uneven size, and may include calcareous foraminiferal tests. The primary aperture is single, round, large, and without a lip.

### Synonyms and Citations

*Psammosphaera fusca* Schulze. –Brady, 1884, p. 250, pl. 18, figs. 1–8. –Loeblich and Tappan, 1987, p. 28, pl. 19, figs. 2–3. –Kaminski et al., 1992, p. 255, pl. 1, figs. 16–17. –Jones, 1994, p. 31, pl. 18, figs. 1–8.

*Psammosphaera fusca* Schulze, emend. Heron-Allen and Earland, 1913. –Kaminski and Gradstein, 2005, pp. 125–128, fig. 8, pl. 8.

### Remarks

Type species of the genus *Psammosphaera* Schulze, 1875. Jones (1994) considered *Psammosphaera parva* Flint, originally distinguished on the basis of its spicular construction, to be a junior synonym of *Psammosphaera fusca*.

Specimens may reach several millimeters in size. Specimens up to 4 mm were reported by Schulze (1875).

### Biogeography

Worldwide. See Observed Occurrences in Kaminski and Gradstein (2005).

### Bathymetry

Lower neritic to abyssal. Reported between 45 and 2750 fathoms at "Challenger" stations (Jones, 1994). Also common in Arctic fjords (Kaminski and Gradstein, 2005).

### Chronostratigraphy

Early Cretaceous (Hauterivian?) to Recent.

## References


Brady, H.B., 1884, Report on the foraminifera dredged by H.M.S. Challenger during the years 1873–1876. In Brady, H.B., ed., Report on the Scientific Results of the Voyage of the H.M.S. Challenger during the years 1873–1876. Zoology, vol. 9, pp. 1–814.

Heron-Allen, E. and Earland, A., 1913, On some foraminifera from the North Sea, etc., dredged by the Fisheries Cruiser "Goldseeker" (International North Sea Investigations – Scotland). II. On the distribution of Saccammina sphaerica (M. Sars) and Psammosphaera fusca (Schulze) in the North Sea: particularly with reference to the suggested identity of the two species. Journal of the Royal Microscopical Society, vol. 1913, pp. 1–26.

Jones, R.W., 1994, The Challenger Foraminifera. Oxford: Oxford University Press, 149 pp.

Kaminski, M.A. and Gradstein, F.M., 2005, Atlas of Palaeogene Cosmopolitan Deep-Water Agglutinated Foraminifera. Grzybowski Foundation Special Publication, no. 10, 547 + vii pp.

Kaminski, M.A., Gradstein, F., and Geroch, S., 1992, Uppermost Jurassic to Lower Cretaceous deep-water benthic foraminiferal assemblages from Site 765 on the Argo Abyssal Plain. In Gradstein, F., Ludden, J.N., et al., ed., Scientific Results of the Proceedings of the Ocean Drilling Program, vol. 123, pp. 239–269.

Loeblich, A.R. Jr and Tappan, H., 1987, Foraminiferal Genera and Their Classification. New York: Van Nostrand Reinhold, 970 pp.

Schulze, F.E., 1875, Zoologische Ergebnisse der Nordseefart, vom 21 Juli bis 9 September 1872: I, Rhizopoden. II. Jahresberichte Kommission zur Untersuchung der Deutschen Meer in Kiel für die Jahr 1872, 1873, pp. 99–114.

# Pseudoclavulina amorpha

## (Cushman), 1926

1. *Pseudoclavulina amorpha.* Velasco Shale, Hacienda El Limon, Vera Cruz, Mexico; Franco Espanola Well no. 7, 385 feet. Deposited in the collections of the United States National Museum, Washington, DC; No. 5153. Holotype. Lateral view, NHMUK PM DI 0536, scale bar = 467 μm.

### Original Designation

*Clavulina amorpha* Cushman, 1926, p. 589, pl. 17, fig. 5.

### Type Specimen

Holotype (No. 5153) deposited in the collections of the United States National Museum, Washington, DC.

### Type Level

Paleocene, Velasco Shale. Originally Given as Upper Cretaceous by Cushman (1926).

### Type Locality

Eastern part of the Hacienda El Limón, State of San Luis Potosi, Mexico.

### Description

Test forms a large, elongate, initially triserial series with a triangular cross-section, becoming uniserial and circular in cross-section section during ontogeny. Chambers are inflated, broad, and low in the uniserial portion, separated by slightly depressed sutures. In the triserial portion sutures are flush. Chamber walls are coarsely agglutinated, canaliculate, and smoothly finished with abundant cement. The primary aperture is circular and terminal without a distinct tooth.

### Synonyms and Citations

*Clavulina amorpha* Cushman. –Cushman and Jarvis, 1928, p. 93, pl. 13, fig. 9.

*Pseudoclavulina amorpha* (Cushman). –Cushman and Renz, 1946, p. 21, pl. 2, fig. 23. –Bolli et al., 1994, p. 96, fig. 25.22.

*Clavulinoides amorpha* (Cushman). –Kaminski et al., 1988, p. 194, pl. 8, fig. 13.

### Biogeography

Worldwide.

### Bathymetry

Bathyal to abyssal.

### Chronostratigraphy

Late Cretaceous to Eocene?

## References

Bolli, H.M., Beckmann, J.P., and Saunders, J.B., 1994, Benthic Foraminiferal Biostratigraphy of the South Caribbean Region. Cambridge: Cambridge University Press, 408 pp.

Cushman, J.A., 1926, The foraminifera of the Velasco Shale of the Tampico Embayment. Bulletin of the American Association of Petroleum Geologists, vol. 10, no. 6, pp. 581–612.

Cushman, J.A. and Jarvis, P.W., 1928, Cretaceous foraminifera from Trinidad. Contributions from the Cushman Laboratory for Foraminiferal Research, vol. 4, pp. 85–103.

Cushman, J.A. and Renz, H.H., 1946, The foraminiferal fauna of the Lizard Springs Formation of Trinidad, British West Indies. Special Publications of the Cushman Laboratory for Foraminiferal Research, vol. 18, pp. 1–48.

Kaminski, M.A., Gradstein, F.M., Berggren, W.A., Geroch, S., and Beckmann, J.P., 1988, Agglutinated foraminiferal assemblages from Trinidad: taxonomy, stratigraphy and paleobathymetry. In Gradstein, F.M. and Rögl, F., ed., Second International Workshop on Agglutinated Foraminifera, Vienna 1986, Abhandlungen der Geologischen Bundesanstalt, vol. 41, pp. 155–228.

# *Pseudonodosinella elongata*

## (Grzybowski), 1898

1. *Pseudonodosinella elongata.* Silesian Unit; Karaś stream 75, Biecz near Gorlice, Poland. Deposited in the Geroch Collection, Jagiellonian University, Kraków, Poland. Lateral view, NHMUK PM DI 0537, scale bar = 514 μm.

### Original Designation

*Reophax elongata* Grzybowski, 1898, p. 279, pl. 10, figs. 19–20.

### Type Specimen

Reference material (UJ-132-P, 1/26) and lectotype designated by Kaminski and Geroch (1993) are deposited in the Grzybowski Collection at the Jagiellonian University, Kraków, Poland. Lectotype is illustrated by Kaminski and Geroch (1993) in pl. 3, fig. 2 (=syntype illustrated by Grzybowski in pl. 10, fig. 19).

### Type Level

Upper Eocene, red and gray clay. Silesian Unit, Polish Carpathians.

### Type Locality

Not originally designated. Reported from exploration wells near Krosno, Poland, by Grzybowski (1898). Lectotype is from Potok H-33 well, 25 meters.

### Description

Test forms an elongate, uniserial series with a subcircular cross-section, and large, elongate, pyriform chambers that gradually increase in length. Each new chamber overlaps the preceding chamber. Chamber walls are agglutinated, composed of medium- to coarse-grained particles held in organic cement, and smoothly finished. The primary aperture is round and terminal at the end of a short neck.

### Synonyms and Citations

*Reophax elongatus* Grzybowski. –Morgiel and Olszewska, 1981, p. 9, pl. 2, fig. 11. –Geroch and Nowak, 1984, pl. 1, figs. 5–6; pl. 5, fig. 6. –Kaminski and Geroch, 1993, p. 250, pl. 3, figs. 1–5.

*Pseudonodosinella elongata* (Grzybowski). –Kaminski and Gradstein, 2005, pp. 256–258, fig. 48, pl. 48.

### Remarks

Often found as broken fragments representing single chambers.

Kaminski and Gradstein (2005) remarked that the specimens illustrated by Brady (1884, pl. 31, figs. 3–4) as *Reophax nodulosa* Brady are probably conspecific, although they are larger and possess more chambers. Hofker (1972) tentatively placed these specimens in the synonymy of his new species *Hormosina mortenseni*, and Jones (1994) subsequently reported Brady's specimens as *Hormosina mortenseni.*

### Biogeography

Worldwide. Common in Carpathian "flysch" sediments. See Observed Occurrences in Kaminski and Gradstein (2005).

### Bathymetry

Bathyal to abyssal.

### Chronostratigraphy

Eocene to Middle Miocene (Recent, if *Hormosina mortenseni* Hofker is regarded as synonymous, see Kaminski and Gradstein, 2005).

## References


Brady, H.B., 1884, Report on the foraminifera dredged by H.M.S. Challenger during the years 1873–1876. In Brady, H.B., ed., Report on the Scientific Results of the Voyage of the H.M.S. Challenger during the years 1873–1876. Zoology, vol. 9, pp. 1–814.

Geroch, S. and Nowak, W., 1984, Proposal of zonation for the late Tithonian–late Eocene, based upon arenaceous foraminifera from the outer Carpathians, Poland. In Oertli, H., ed., Benthos '83, 2nd International Symposium on Benthic Foraminifera, Pau (France), April 11–15, 1983. Elf Aquitaine, ESSO REP and TOTAL CFP, Pau and Bordeaux, pp. 225–239.

Grzybowski, J., 1898, Otwornice pokładów naftonosnych okolicy Krosna. Rozprawy Wydziału Matematyczno-Przyrodniczego, Akademia Umiejętności w Krakowie, serya 2, vol. 33, pp. 257–305.

Hofker, J., 1972, Primitive Agglutinated Foraminifera. Leiden: E.J. Brill, 95 pp.

Kaminski, M. A. and Geroch, S., 1993, A revision of foraminiferal species in the Grzybowski Collection. In Kaminski, M.A., Geroch, S., and Kaminski, D.G., ed., The Origins of Applied Micropalaeontology: The School of Jósef Grzybowski (with a taxonomic revision of the Grzyboswski Collection). Grzybowski Foundation Special Publication, no. 1, pp. 239–330.

Kaminski, M.A. and Gradstein, F.M., 2005, Atlas of Palaeogene Cosmopolitan Deep-Water Agglutinated Foraminifera. Grzybowski Foundation Special Publication, no. 10, 547 + vii pp.

Jones, R.W., 1994, The Challenger Foraminifera. Oxford: Oxford University Press, 149 pp.

Morgiel, J. and Olszewska, B., 1981, Biostratigraphy of the Polish External Carpathians based on agglutinated foraminifera. Micropaleontology, vol. 27, no. 1, pp. 1–30.

# *Pseudonodosinella nodulosa*

## (Brady, 1879) emend. Loeblich and Tappan, 1987

1. *Pseudonodosinella nodulosa.* South Atlantic Ocean; "Challenger" Station 323, depth: 1900 fathoms. Deposited in the micropaleontological collections of The Natural History Museum, London; NHMUK ZF2281. Syntype. Lateral view, NHMUK PM DI 0538, scale bar = 275 μm.

### Original Designation

*Reophax nodulosa* Brady, 1879, p. 52, pl. 4, figs. 7–8.

### Type Specimen

Brady's figured specimen (pl. 4, fig. 8) and specimens figured by Brady (1884) are preserved in the micropaleontological collections of The Natural History Museum, London, UK (slide ZF2282). Kaminski and Gradstein (2005) reported that the lectotype, designated by Loeblich and Tappan (1987) and figured in pl. 4, fig. 7 by Brady (1879), is missing. Kaminski and Cetean (2011) designated the seven-chambered specimen in the lower part of slide ZF2282 as the neolectotype.

### Type Level

Recent.

### Type Locality

Not originally designated. Localities given: "Challenger stations" in the South Atlantic and in the North and South Pacific Oceans, in water depth of 1400 to 2000 fathoms. Lectotype designated by Loeblich and Tappan (1987) is from "Challenger" Station 323 in the western South Atlantic (1900 fathoms, east of Buenos Aires).

### Description

Test forms an elongate, uniserial series, rectilinear or slightly arched in overall shape, and tapered with a subcircular cross-section. The ovate, overlapping chambers (wider at the base) increase gradually in size, and are separated by horizontal, depressed sutures. Chamber interiors are pyriform; chamber walls are thin, finely agglutinated, smoothly finished with a non-calcareous cement, and thickened around the aperture. The primary aperture is round, terminal, without a neck, and slightly produced.

### Synonyms and Citations

*Reophax subnodulosa* Grzybowski, 1898, p. 279, pl. 10, figs. 17–18.

*Reophax nodulosa* Brady. –Brady, 1884, p. 294, pl. 31, figs. 6–9 (not figs. 1–5). –Barker, 1960, pl. 31, figs. 6–9 (not figs. 1–5).

*Reophax nodulosus* Brady. –Kaminski and Geroch, 1993, p. 250, pl. 3, figs. 6–10. –Jones, 1994, p. 38, pl. 31, figs. 6–9.

*Pseudonodosinella nodulosa* (Brady) emend. Loeblich and Tappan, 1987, p. 61, pl. 46, figs. 5–6. –Kaminski and Gradstein, 2005, pp. 259–262, fig. 49, pl. 49.

### Remarks

Type species of the genus *Pseudonodosinella* Saidova, 1970.

Loeblich and Tappan (1987) emended the definition of *Pseudonodosinella nodulosa* (Brady) to include only the smaller specimens figured by Brady (1884). These authors referred the large, coarse specimens figured in pl. 31, figs. 1–2 and possibly 5 to *Nodosinum gaussicum* (Rhumbler).

Kaminski and Gradstein (2005) suggested that the specimens illustrated as *Reophax nodulosa* by Brady (1884) in pl. 31, figs. 3–4 are probably conspecific with *Pseudonodosinella elongata* (Grzybowski), although they are larger and possess more chambers. Hofker (1972) tentatively placed these specimens in the synonymy of his new species *Hormosina mortenseni*, and Jones (1994) subsequently reported them as *Hormosina mortenseni.*

### Biogeography

Worldwide. See Observed Occurrences in Kaminski and Gradstein (2005).

### Bathymetry

Bathyal to abyssal.

### Chronostratigraphy

Late Cretaceous? to Recent. Predominantly Eocene to Recent (Kaminski and Gradstein, 2005).

## References


Barker, R.W., 1960, Taxonomic notes on the species figured by H.B. Brady in his report on the foraminifera dredged by H.M.S. Challenger during the years 1873–1876. Society of Economic Paleontologists and Mineralogists Special Publication, vol. 9, pp. 1–238.

Brady, H.B., 1879, Notes on some of the reticularian Rhizopoda of the Challenger Expedition, Part I; On new or little known Arenaceous types. Quarterly Journal of Microscopical Science (New Series), vol. 19, pp. 20–63.

Brady, H.B., 1884, Report on the foraminifera dredged by H.M.S. Challenger during the years 1873–1876. In Brady, H.B., ed., Report on the Scientific Results of the Voyage of the H.M.S. Challenger during the years 1873–1876. Zoology, vol. 9, pp. 1–814.

Grzybowski, J., 1898, Otwornice pokładów naftonosnych okolicy Krosna. Rozprawy Wydziału Matematyczno-Przyrodniczego, Akademia Umiejętności w Krakowie, serya 2, vol. 33, pp. 257–305.

Jones, R.W., 1994, The Challenger Foraminifera. Oxford: Oxford University Press, 149 pp.

Kaminski, M.A. and Cetean, C.G., 2011. Lectotypes of type species of Agglutinated Foraminiferal Genera in the Collections of the Natural History Museum, London. Part 2. Hippocrepinina and Hormosinina. In Kaminski, M.A. and Filipescu, S., ed., Proceedings of the Eighth International Workshop on Agglutinated Foraminifera, Grzybowski Foundation Special Publication, no. 16, pp. 61–69.

Kaminski, M.A. and Geroch, S., 1993, A revision of foraminiferal species in the Grzybowski Collection. In Kaminski, M.A., Geroch, S., and Kaminski, D.G., ed., The Origins of Applied Micropalaeontology: The School of Jósef Grzybowski (with a taxonomic revision of the Grzyboswski Collection). Grzybowski Foundation Special Publication, no. 1, pp. 239–330.

Kaminski, M.A. and Gradstein, F.M., 2005, Atlas of Palaeogene Cosmopolitan Deep-Water Agglutinated Foraminifera. Grzybowski Foundation Special Publication, no. 10, 547 + vii pp.

Loeblich, A.R. Jr and Tappan, H., 1987, Foraminiferal Genera and Their Classification. New York: Van Nostrand Reinhold, 970 pp.

# *Pseudonodosinella troyeri*

## (Tappan), 1960

1. *Pseudonodosinella troyeri.* Ubionka 241, Verovice shales, Poland. Deposited in the Geroch Collection (as *Reophax minutus*), Jagiellonian University, Kraków, Poland. Lateral view, NHMUK PM DI 0539, scale bar = 143 μm.

### Original Designation

*Reophax troyeri* Tappan, 1960, p. 291, pl. 1, figs. 10–12.

### Type Specimen

Deposited in the collections of the United States National Museum, Washington, DC.

### Type Level

Cretaceous (Albian), Topagoruk Formation.

### Type Locality

Holotype from well cuttings at a depth of 1870–1880 feet in South Barrow test well no. 1, latitude 71° 19′ 12.07″ N, longitude 156° 42′ 15.34″ W, southeast of Point Barrow, northern Alaska. Figured paratypes from well cuttings at a depth 1319–1329 feet in Arcon Point, Barrow test well no. 1, latitude 71° 19′ 30″ N, longitude 156° 40′ W, southwest of Point Barrow, and from Umiat test well no. 2, latitude 69° 23′ 04″ N, longitude 152° 05′ 01″ W, south of Umiat, northern Alaska.

### Description

Test forms a small, elongate, flattened, uniserial series; rectilinear or slightly arched in overall shape, and lobulate and slightly tapered in outline. The three to six globular, overlapping chambers increase rapidly in size, and are separated by horizontal, depressed sutures. Chamber walls are finely agglutinated, multilayered, non-calcareous, smoothly finished, and thickened around the aperture. The primary aperture is simple, terminal, and produced on a low, broad neck.

### Synonyms and Citations

*Reophax minutus* Tappan. –Geroch and Nowak, 1984, p. 226, pl. 1, fig. 9. –King et al., 1989, pl. 8.1, fig. 19.

*Reophax* cf. *troyeri* Tappan. –Weidich, 1990, p. 86, pl. 34, fig. 18.

*Pseudonodosinella troyeri* (Tappan). –Geroch and Kaminski, 1995, p. 118, pl. 1, figs. 1–2, 4–17.

### Remarks

Microspheric forms of *Pseudonodosinella troyeri* (Tappan) may have up to six chambers, whereas megalospheric forms usually have three chambers (Geroch and Kaminski, 1995). Early Cretaceous specimens tapered at both ends and with a produced aperture, which have been previously assigned to *Reophax minutus* Tappan, closely resemble the type specimen of *Pseudonodosinella troyeri* (see Geroch and Kaminski, 1995).

### Biogeography

Worldwide.

### Bathymetry

Bathyal to abyssal. Common in flysch-type assemblages.

### Chronostratigraphy

Cretaceous (Barremian to Turonian) (Geroch and Nowak, 1984).

## References


Geroch, S. and Kaminski, M.A., 1995, An emendation of some Cretaceous species of "Reophax" (Foraminiferida) from northwest Europe and Poland. In Kaminski, M.A., Geroch, S., and Gasinski, M.A., ed., Proceedings of the Fourth International Workshop on Agglutinated Foraminifera. Grzybowski Foundation Special Publication, no. 3, pp. 117–122.

Geroch, S. and Nowak, W., 1984, Proposal of zonation for the late Tithonian–late Eocene, based upon arenaceous foraminifera from the outer Carpathians, Poland. In Oertli, H., ed., Benthos '83, 2nd International Symposium on Benthic Foraminifera, Pau (France), April 11–15, 1983. Elf Aquitaine, ESSO REP and TOTAL CFP, Pau and Bordeaux, pp. 225–239.

King, C., Bailey, H.W., Burton, C.A., and King, A.D., 1989, Cretaceous of the North Sea. In Jenkins, D.G. and Murray, J.W., ed., Stratigraphical Atlas of Fossil Foraminifera, 2nd edn. Chichester: Ellis Horwood Ltd, pp. 312–417.

Tappan, H., 1960, Cretaceous biostratigraphy of Northern Alaska. Bulletin of the American Association of Petroleum Geologists, vol. 44, no. 3, pp. 273–297.

Weidich, K., 1990, Die kalkalpine Unterkreide und ihre Foraminiferenfauna. Zitteliana, vol. 17, pp. 1–312.

# Pseudoreophax cisownicensis

## Geroch, 1961

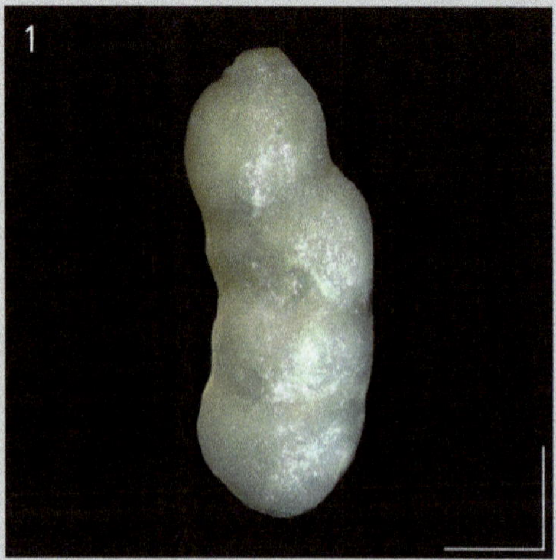

1. *Pseudoreophax cisownicensis.* Lipnik 11/59, Grodziszcze Shale, Silesian Unit, Poland. Deposited in the Geroch Collection, Jagiellonian University, Kraków, Poland. Lateral view, NHMUK PM DI 0540, scale bar = 97 μm.

### Original Designation

*Pseudoreophax cisownicensis* Geroch, 1961, p. 160, pl. 17, figs. 1–20, textfigs. 1.1–5.

### Type Specimen

Deposited in the Geroch Collection at the Jagiellonian University, Kraków, Poland.

### Type Level

Lower Cretaceous (Valanginian–Hauterivian), Upper Cieszyn Shales. Also occurs in the Grodziszcze Beds and the lowermost Verovice Beds.

### Type Locality

Holotype (fig. 16, textfig. 1.2), and specimens in figs. 17–20 from Cisownica, Cieszyn sheet; other figured specimens (figs. 1–15, textfigs. 1.1, 1.3–5) from Lipnik, Bielsko sheet; also found at Sulkowice-Ubionka, Wadowice sheet; all in the Carpathians of Poland.

### Description

Test forms an elongate, rectilinear or arched, narrow uniserial series with a circular cross-section. The microspheric generation has an early stage with a single trochospiral whorl, before becoming uniserial. The megalospheric generation, composed of three to seven chambers, remains uniserial throughout. The inflated chambers have asymmetric connections (visible in well-silicified topotype material), and are rotated with respect to one another, giving the test a "pseudobiserial" appearance. Chamber walls are finely agglutinated, and solid with a smooth finish. The primary aperture is terminal, and round.

### Synonyms and Citations

*Ammobaculites irregularis* (Gümbel). –Sliter, 1980, pl. 2, fig. 14.
*Bigenerina jurassica* (Haeusler). –Gradstein, 1983, pl. 1, fig. 1.
*Pseudoreophax cisownicensis.* –Geroch and Nowak, 1984, p. 230, pl. 3, figs. 14–18. –Riegraf and Luterbacher, 1989, p. 1093, pl. 2, fig. 22. –Kaminski et al., 1992, p. 255, pl. 7, figs. 5–6. –Kuhnt, 1995, pl. 2, figs. 1–9; pl. 3, fig. 12. –Holbourn and Kaminski, 1997, p. 52, pl. 22, fig. 4.

### Remarks

The partial range of this species characterizes the *Pseudoreophax cisownicensis* Biozone (Late Tithonian to Berriasian) of Geroch and Nowak (1984).

### Biogeography

Worldwide. Recorded from the Polish Carpathians (Geroch and Nowak, 1984), North Atlantic (Riegraf and Luterbacher, 1989), the Betic Cordillera of Spain (Kuhnt, 1995), and the Indian Ocean (Kaminski et al., 1992).

### Bathymetry

Bathyal to abyssal. Generally sub-CCD.

### Chronostratigraphy

Late Jurassic (Tithonian) to Early Cretaceous (Barremian) (Geroch and Nowak, 1984).

### References


Geroch, S., 1961, Pseudoreophax a new genus of foraminifer from the Neocomian in the Flysch Carpathians. Rocznik Polskiego Towarzystwa Geologicznego, vol. 31, pp. 159–165.

Geroch, S. and Nowak, W., 1984, Proposal of zonation for the late Tithonian–late Eocene, based upon arenaceous foraminifera from the outer Carpathians, Poland. In Oertli, H., ed., Benthos '83, 2nd International Symposium on Benthic Foraminifera, Pau (France), April 11–15, 1983. Elf Aquitaine, ESSO REP and TOTAL CFP, Pau and Bordeaux, pp. 225–239.

Gradstein, F.M., 1983, Paleoecology and stratigraphy of Jurassic abyssal foraminifera in the Blake Bahama Basin, Deep Sea Drilling Project Site 534. In Sheridan, R.E., Gradstein, F.M., et al., ed., Initial Reports of the Deep Sea Drilling Project, Washington (U.S. Government Printing Office), vol. 76, pp. 537–559.

Holbourn, A.E.L. and Kaminski, M.A., 1997, Cretaceous deep-water benthic foraminifera of the Indian Ocean. Grzybowski Foundation Special Publication, no. 4, pp. 1–175.

Kaminski, M.A., Gradstein, F., and Geroch, S., 1992, Uppermost Jurassic to Lower Cretaceous deep-water benthic foraminiferal assemblages from Site 765 on the Argo Abyssal Plain. In Gradstein, F., Ludden, J.N., et al., ed., Scientific Results of the Proceedings of the Ocean Drilling Program, vol. 123, pp. 239–269.

Kuhnt, W., 1995, Deep-water agglutinated foraminifera from the Lower Cretaceous (Neocomian) "complex à Aptychus" Formation (Corridor de Boyar, Betic Cordillera, southern Spain). Journal of Micropalaeontology, vol. 14, no. 1, pp. 37–52.

Riegraf, W. and Luterbacher, H., 1989, Benthonische Foraminiferen aus de Unterkreide des "Deep Sea Drilling Project" (Leg 1–79). Geologische Rundschau, vol. 78, no. 3, pp. 1063–1120.

Sliter, W.V., 1980, Mesozoic foraminifers and deep sea benthic environments from Deep Sea Drilling Project Sites 415 and 416, eastern North Atlantic. In Lancelot, Y., Winterer, E.L., et al., ed., Initial Reports of the Deep Sea Drilling Project, Washington (U.S. Government Printing Office), vol. 50, pp. 353–427.

# *Pseudouvigerina plummerae*

## Cushman, 1927

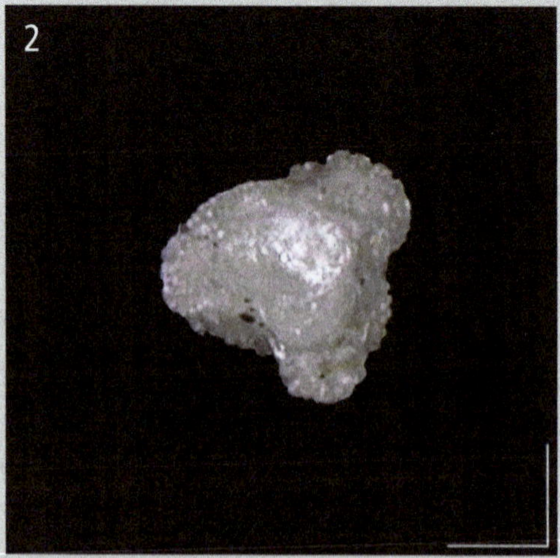

1. *Pseudouvigerina plummerae.* Pecan Gap, Texas. Heron-Allen and Earland Collection. Deposited in the micropaleontological collections of The Natural History Museum, London; NHMUK P47556. Paratype. Lateral view, NHMUK PM DI 0541, scale bar = 53 μm.

2. *Pseudouvigerina plummerae.* Pecan Gap, Texas. Heron-Allen and Earland Collection. Deposited in the micropaleontological collections of The Natural History Museum, London; NHMUK P47556. Paratype. Apertural view, NHMUK PM DI 0542, scale bar = 29 μm.

### Original Designation

*Pseudouvigerina plummerae* Cushman, 1927, p. 115, pl. 23, fig. 8a–b.

### Type Specimen

Holotype (CC 6700) deposited in the collections of the United States National Museum, Washington, DC.

### Type Level

Upper Cretaceous, Pecan Gap Chalk.

### Type Locality

Roadside ditch along pike, 1.8 mile northeast of Rockall, Texas, USA.

### Description

Test forms a small, subpyramidal, high trochospire, reducing through ontogeny to a biserial series, with a triangular cross-section, and a crenulate, truncate periphery. Chambers are slightly inflated, increase rapidly in height, and are separated by slightly depressed sutures. Chamber walls are calcareous, finely perforate, smooth or granular, with crenulate margins. The primary aperture is round, terminal, and bordered by a low rim.

### Synonyms and Citations

*?Pseudouvigerina* sp. Nyong and Olsson, 1984, p. 477, pl. 1, figs. 14–15.

*Pseudouvigerina plummerae* Cushman. –Cushman, 1946, p. 116, pl. 49, figs. 14–16. –Bolli et al., 1994, p. 133, fig. 35.34.

### Biogeography

Recorded from Texas (Cushman, 1927), the Gulf coastal region (Cushman, 1946), the western North Atlantic (Nyong and Olsson, 1984), and Trinidad (Bolli et al., 1994).

### Bathymetry

Neritic to upper bathyal.

### Chronostratigraphy

Late Cretaceous (Campanian to Maastrichtian).

## References

Bolli, H.M., Beckmann, J.P., and Saunders, J.B., 1994, Benthic Foraminiferal Biostratigraphy of the South Caribbean Region. Cambridge: Cambridge University Press, 408 pp.

Cushman, J.A., 1927, New and interesting foraminifera from Mexico and Texas. Contributions from the Cushman Laboratory for Foraminiferal Research, vol. 3, pp. 111–117.

Cushman, J.A., 1946, Upper Cretaceous foraminifera of the Gulf coastal region of the United States and adjacent areas. Professional Paper of the United States Geological Survey, vol. 206, 241 pp.

Nyong, E.E. and Olsson, R.K., 1984, A paleoslope model for Campanian to lower Maestrichtian Foraminifera of the North American Basin and adjacent continental margin. Marine Micropaleontology, vol. 8, pp. 437–477.

# *Pullenia bulloides*

## (d'Orbigny), 1846

1. *Pullenia bulloides.* "Meteor" Cruise, West Africa; M25. Deposited in the Lutze Collection (West Africa), no. 184. Institute of Geosciences, Kiel University, Germany. Umbilical view, NHMUK PM DI 0543, scale bar = 115 μm.

2. *Pullenia bulloides.* "Meteor" Cruise, West Africa; M25. Deposited in the Lutze Collection (West Africa), no. 184. Institute of Geosciences, Kiel University, Germany. Apertural view, NHMUK PM DI 0544, scale bar = 115 μm.

3. *Pullenia bulloides.* "Meteor" Cruise, West Africa; M25. Deposited in the Lutze Collection (West Africa), no. 184. Institute of Geosciences, Kiel University, Germany. Umbilical view, NHMUK PM DI 0545, scale bar = 119 μm.

### Original Designation
*Nonionina bulloides* d'Orbigny, 1846, p. 107, pl. 5, figs. 9–10.

### Type Specimen
Depository not given. A lectotype (C748) was designated by Loeblich and Tappan (1964) from the d'Orbigny Collection, and is deposited in the Laboratoire de Paléontologie, Muséum d'Histoire Naturelle, Paris, France.

### Type Level
Tertiary (Middle Miocene), Tegel.

### Type Locality
Near the village of Nussdorf, north of Vienna, Austria. Original material from d'Orbigny, 1826, from Coroncina, near Siena, Tuscany, Italy.

### Description
Test forms a subglobular, involute planispire; subcircular in outline with a broadly rounded periphery. The (approximately) four to five strongly inflated chambers in the final whorl increase gradually in size, and are separated by flush or slightly depressed, radial or slightly curved sutures. Chamber walls are calcareous, finely perforate, and smooth. The primary aperture is a narrow, arcuate, interiomarginal, crescentic slit, extending across the periphery to the umbilici.

### Synonyms and Citations
*Nonionina bulloides* d'Orbigny, 1826, p. 293.

*Pullenia sphaeroides* (d'Orbigny). –Brady, 1884, p. 615, pl. 84, figs. 12–13.

*Pullenia bulloides* (d'Orbigny). –Cushman and Todd, 1943, p. 13, pl. 2, figs. 15–18. –Barker, 1960, pl. 84, figs. 12–13. –Papp and Schmid, 1985, p. 45, pl. 34, figs. 7–9. –Loeblich and Tappan, 1987, p. 621, pl. 696, figs. 3–4. –Miller and Katz, 1987, p. 136, pl. 4, figs. 4a–b. –Hermelin, 1989, p. 78, pl. 15, figs. 4–5. –Thomas, 1990, p. 590. –van Marle, 1991, p. 192, pl. 20, figs. 13–15. –Jones, 1994, p. 92, pl. 84, figs. 12–13. –Robertson, 1998, p. 230, pl. 92, fig. 1. –Kuhnt et al., 2002, p. 152, pl. 12. figs. 14–17.

### Remarks
Type species of the genus *Pullenia* Parker and Jones, 1862.

D'Orbigny named his species in 1826 from Pliocene material from the Sienna area (Coroncina, Tuscany) in Italy. No type figure or description were provided, and therefore *Nonionina bulloides* d'Orbigny, 1826 is a nomen nudum. D'Orbigny (1846) later described and illustrated this species from the Middle Miocene of the Vienna Basin. Papp and Schmid illustrated a specimen from d'Orbigny's material in the Hauer-d'Orbigny Collection, housed at the Geologische Bundesanstalt in Vienna, Austria (pl. 34, figs. 7–9). These authors did not designate a lectotype, as reported by Jones (1994). A lectotype from the Miocene of Nussdorf, Austria, was designated by Loeblich and Tappan (1964) from the d'Orbigny

Collection in the Laboratoire de Paléontologie, Muséum d'Histoire Naturelle, Paris, France.

*Pullenia bulloides* (d'Orbigny) shows close affinity to *Pullenia eocenica* Cushman and Siegfus and to *Pullenia subsphaerica* Parr. These forms were described as slightly compressed with five (*Pullenia eocenica*) or four (*Pullenia subsphaerica*) chambers in the last whorl. Comparison of type material may determine whether they are conspecific.

### Biogeography

Worldwide.

### Bathymetry

Bathyal to abyssal. Reported between 300 and 2750 fathoms at "Challenger" stations (Jones, 1994) and between 545 and 2119 meters with higher frequencies between 1654 and 2119 meters in eastern Indonesia (van Marle, 1991); lower bathyal (Thomas, 1990; Katz and Miller, 1993); abyssal, greater than 3.5 kilometers in the Gulf of Mexico (Miller and Katz, 1987).

### Chronostratigraphy

Latest Paleocene (Thanetian) to Recent. The species occurs rarely in the topmost Paleocene. Abundant in the Oligocene and Miocene (Miller and Katz, 1987).

## References


Barker, R.W., 1960, Taxonomic notes on the species figured by H.B. Brady in his report on the foraminifera dredged by H.M.S. Challenger during the years 1873–1876, Society of Economic Paleontologists and Mineralogists Special Publication, vol. 9, pp. 1–238.

Brady, H.B., 1884, Report on the foraminifera dredged by H.M.S. Challenger during the years 1873–1876. In Brady, H.B., ed., Report on the Scientific Results of the Voyage of the H.M.S. Challenger during the years 1873–1876. Zoology, vol. 9, pp. 1–814.

Cushman, J.A. and Todd, R., 1943, The genus Pullenia and its species. Contributions from the Cushman Laboratory for Foraminiferal Research, vol. 19, pp. 1–24.

d'Orbigny, A., 1826, Tableau méthodique de la classe des Céphalopodes. Annales des Sciences Naturelles, vol. 7, pp. 245–314.

d'Orbigny, A.D., 1846, Foraminifères fossiles du Bassin Tertiaire de Vienne (Autriche). Paris: Gide et Compe, xxxvii + 312 pp.

Hermelin, J.O.R., 1989, Pliocene benthic foraminifera from the Ontong-Java Plateau (Western Equatorial Pacific Ocean): faunal response to changing paleoenvironment. Special Publication of the Cushman Foundation for Foraminiferal Research, vol. 26, pp. 1–143.

Jones, R.W., 1994, The Challenger Foraminifera. Oxford: Oxford University Press, 149 pp.

Katz, M.E. and Miller, K.G., 1993, Latest Oligocene to Earliest Pliocene benthic foraminiferal biofacies of the northeastern Gulf of Mexico. Micropaleontology, vol. 39, no. 4, pp. 367–403.

Kuhnt, W., Holbourn, A.E., and Zhao, Q., 2002, The early history of the South China Sea: Evolution of Oligocene–Miocene deep water environments. Revue de Micropaléontologie, vol. 45, pp. 99–159.

Loeblich, A.R. Jr and Tappan, H., 1964, Sarcodina, chiefly "Thecamoebians" and Foraminiferida. In Moore, R.C., ed., Treatise on Invertebrate Paleontology, Part C, Protista 2 (2 vols). Lawrence, KS: Geological Society of America and University of Kansas Press, pp. 1–900.

Loeblich, A.R. Jr and Tappan, H., 1987, Foraminiferal Genera and Their Classification. New York: Van Nostrand Reinhold, 970 pp.

Miller, K.G. and Katz, M.E., 1987, Oligocene to Miocene benthic foraminiferal and abyssal circulation changes in the north Atlantic. Micropaleontology, vol. 33, no. 2, pp. 97–149.

Papp, A. and Schmid, M.E., 1985, Die fossilen Foraminiferen des tertiaeren Beckens von Wien: Revision der Monographie Alcide d'Orbigny (1846). Abhandlungen der Geologischen Bundesanstalt, vol. 37, pp. 1–311.

Robertson, B.E., 1998, Systematics and paleoecology of the benthic foraminiferida from the Buff Bay section, Miocene of Jamaica. Micropaleontology, vol. 44, supplement 2, pp. 1–266.

Thomas, E., 1990, Late Cretaceous through Neogene benthic foraminifers. In Barker, P.F., Kennett, J.P., et al., ed., Scientific Results of the Proceedings of the Ocean Drilling Program, vol. 113, pp. 571–594.

van Marle, L.J., 1991, Eastern Indonesian Late Cenozoic smaller benthic foraminifera. Verhandelingen der Koninklijke Nederlanse Akademie van Wetenschappen, Afdeeling Natuurkunde, Eerste Reeks, vol. 34, pp. 1–328.

# *Pullenia coryelli*

## White, 1929

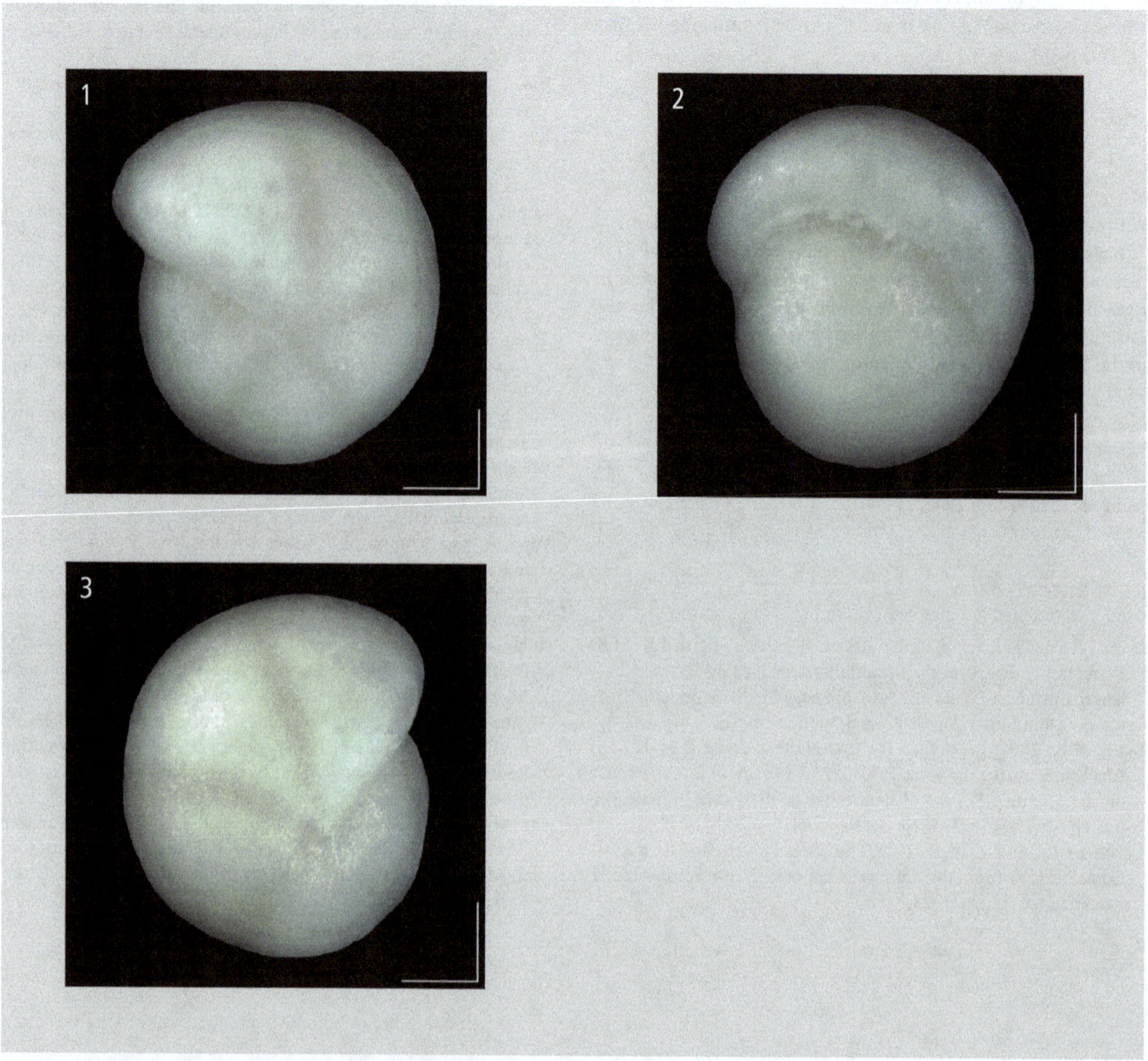


1. *Pullenia coryelli.* Northeast Georgia Rise, South Atlantic; DSDP Leg 114, Hole 700B, Core 32, Section 1, 15–19 cm. Umbilical view, NHMUK PM DI 0546, scale bar = 170 μm.

2. *Pullenia coryelli.* Northeast Georgia Rise, South Atlantic; DSDP Leg 114, Hole 700B, Core 32, Section 1, 15–19 cm. Oblique apertural view, NHMUK PM DI 0547, scale bar = 174 μm.

3. *Pullenia coryelli.* Northeast Georgia Rise, South Atlantic; DSDP Leg 114, Hole 700B, Core 32, Section 1, 15–19 cm. Umbilical view, NHMUK PM DI 0548, scale bar = 170 μm.

## Original Designation

*Pullenia coryelli* White, 1929, p. 56, pl. 5, fig. 22.

## Type Specimen

Holotype deposited in Columbia University, New York, Paleontology Collection, No. 19928.

## Type Level

Given as Upper Cretaceous, Velasco Formation, shale.

## Type Locality

Locality No. 23, Columbus Station on the Tampico Monterey railroad line, Mexico.

## Description

Test forms a subglobular, involute planispire; subcircular in outline, and with a broadly rounded periphery. The characteristically five to six strongly inflated chambers in the final whorl increase gradually in size, and are separated by flush or slightly depressed, radial or slightly curved sutures. Chamber walls are calcareous, finely perforate, and smooth. The primary aperture is a narrow, arcuate, interiomarginal, crescentic slit at the base of the final chamber, extending from the periphery to the umbilici.

## Synonyms and Citations

*Pullenia sphaeroides* (d'Orbigny). –Cushman, 1926, p. 605, pl. 21, fig. 2.

*Pullenia quaternaria* (Reuss). –Cushman, 1931, p. 313, pl. 36, fig. 4.

*Pullenia coryelli* White. –Cushman and Todd, 1943, p. 6, pl. 1, fig. 18. –Cushman, 1946, p. 147, pl. 60, figs. 10–11. –Tjalsma and Lohmann, 1983, p. 18, pl. 5, figs. 5a–b. –Bolli et al., 1994, p. 151, figs. 41.23–24.

## Remarks

*Pullenia cretacea* Cushman differs from *Pullenia coryelli* White in having a slightly more compressed test, fewer chambers in the final whorl, and a higher apertural face (Cushman and Todd, 1943).

## Biogeography

Worldwide. Texas, Trinidad, Arkansas (Cushman and Todd, 1943); Gulf coastal region (Cushman, 1946); Atlantic Ocean (Tjalsma and Lohmann, 1983); Weddell Sea, Antarctic (Thomas, 1990).

## Bathymetry

Bathyal to abyssal (Tjalsma and Lohmann, 1983).

## Chronostratigraphy

Late Cretaceous (Maastrichtian) to Late Paleocene (Bolli et al., 1994). Rare to abundant in Biozones P1–P5 (Tjalsma and Lohmann, 1983). Late Paleocene (Thomas, 1990).

# References


Bolli, H.M., Beckmann, J.P., and Saunders, J.B., 1994, Benthic Foraminiferal Biostratigraphy of the South Caribbean Region. Cambridge: Cambridge University Press, 408 pp.

Cushman, J.A., 1926, The foraminifera of the Velasco Shale of the Tampico Embayment. Bulletin of the American Association of Petroleum Geologists, vol. 10, no. 6, pp. 581–612.

Cushman, J.A., 1931, The Foraminifera of the Saratoga Chalk. Journal of Paleontology, vol. 5, pp. 297–315.

Cushman, J.A., 1946, Upper Cretaceous foraminifera of the Gulf coastal region of the United States and adjacent areas. Professional Paper of the United States Geological Survey, vol. 206, 241 pp.

Cushman, J.A. and Todd, R., 1943, The genus Pullenia and its species. Contributions from the Cushman Laboratory for Foraminiferal Research, vol. 19, pp. 1–24.

Thomas, E., 1990, Late Cretaceous through Neogene benthic foraminifers. In Barker, P.F., Kennett, J.P., et al., ed., Scientific Results of the Proceedings of the Ocean Drilling Program, vol. 113, pp. 571–594.

Tjalsma, R.C. and Lohmann, G.P., 1983, Paleocene–Eocene bathyal and abyssal benthic foraminifera from the Atlantic Ocean. Micropaleontology, Special Publication, no. 4, pp. 1–90.

White, M.P., 1929, Some index foraminifera of the Tampico Embayment area of Mexico (Part III). Journal of Paleontology, vol. 3, no. 1, pp. 30–58.

# *Pullenia cretacea*

## Cushman, 1936

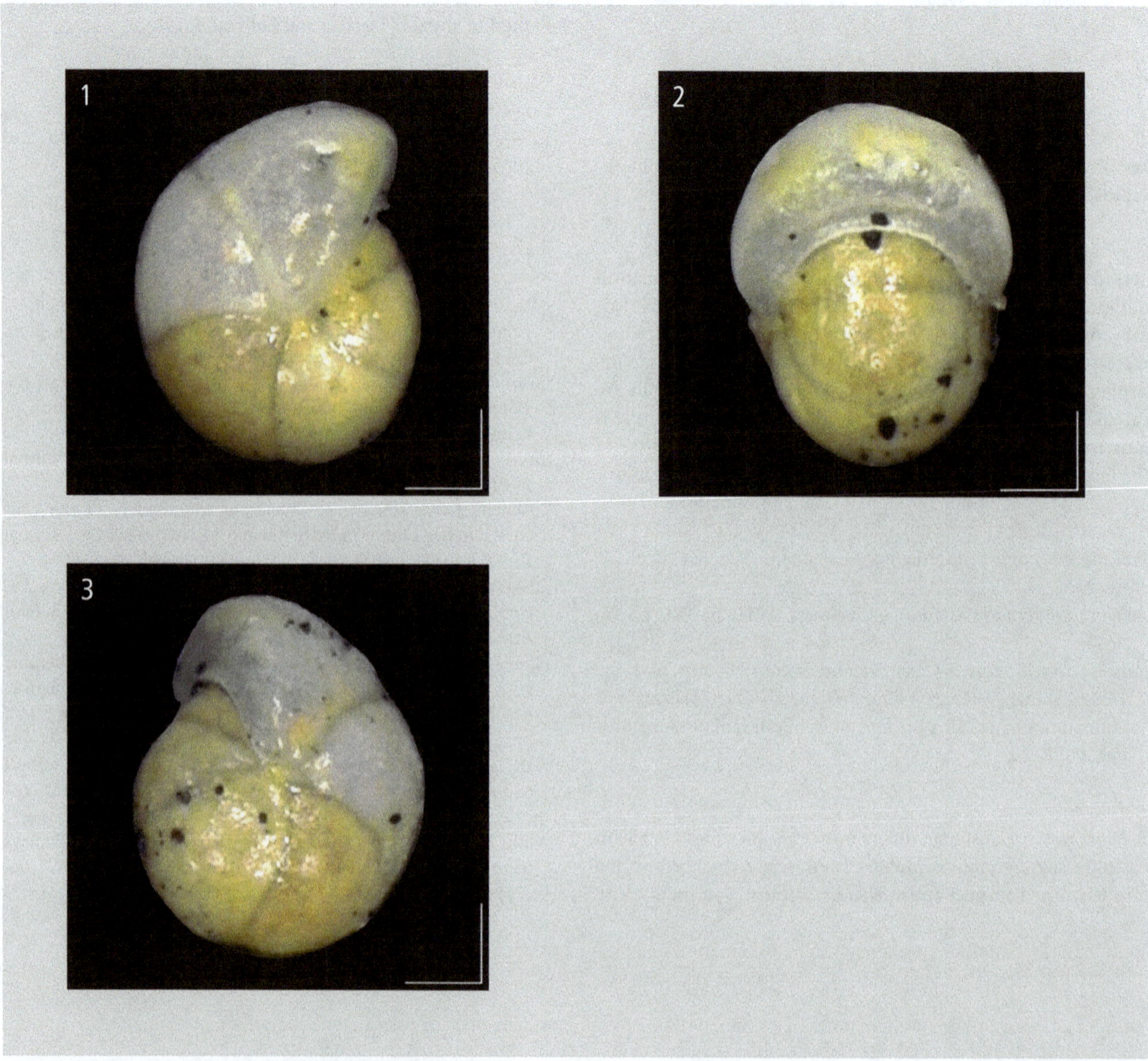

1. *Pullenia cretacea.* Selma Chalk, Tennessee, USA. Deposited in the collections of the United States National Museum, Washington, DC; CC 23527. Holotype. Umbilical view, NHMUK PM DI 0549, scale bar = 50 μm.

2. *Pullenia cretacea.* Selma Chalk, Tennessee, USA. Deposited in the collections of the United States National Museum, Washington, DC; CC 23527. Holotype. Apertural view, NHMUK PM DI 0550, scale bar = 46 μm.

3. *Pullenia cretacea.* Selma Chalk, Tennessee, USA. Deposited in the collections of the United States National Museum, Washington, DC; CC 23527. Holotype. Umbilical view, NHMUK PM DI 0551, scale bar = 49 μm.

## Original Designation

*Pullenia cretacea* Cushman, 1936, p. 75, pl. 13, fig. 8.

## Type Specimen

Holotype (CC 23527) deposited in the collections of the United States National Museum, Washington, DC.

## Type Level

Upper Cretaceous, Selma Chalk.

## Type Locality

One and a half miles west of Sardis, on Sardis-Henderson Road, Henderson County, Tennessee, USA.

## Description

Test forms a close-coiled, involute planispire; often somewhat compressed, with a subglobular outline, and a broadly rounded periphery. The (approximately) five slightly inflated chambers in the last whorl increase gradually in size, and are separated by distinct, slightly depressed, radial, or slightly curved sutures. Chamber walls are calcareous, distinctly perforate, and smooth. The primary aperture is a low, narrow, arcuate opening at the base of the final chamber, extending from one umbilicus to the other, with a slight overhanging lip.

## Synonyms and Citations

*Pullenia cretacea* Cushman. –Cushman and Todd, 1943, p. 7, pl. 1, fig. 14. –Bolli et al., 1994, p. 152, figs. 41.27–28.

## Remarks

*Pullenia cretacea* Cushman differs from *Pullenia coryelli* White in having a slightly more compressed test, fewer chambers in the final whorl, and a higher apertural face (Cushman and Todd, 1943).

## Biogeography

Localities given by Cushman and Todd (1943): Texas (lower Navarro to upper Taylor), Tennessee, Mississippi, Alabama (Selma Chalk), Nebraska (Cretaceous), Columbia (Colon Fathoms). Recorded in Trinidad by Bolli et al. (1994).

## Bathymetry

Bathyal.

## Chronostratigraphy

Late Cretaceous (Santonian to Maastrichtian) in Trinidad (Bolli et al., 1994).

## References


Bolli, H.M., Beckmann, J.P., and Saunders, J.B., 1994, Benthic Foraminiferal Biostratigraphy of the South Caribbean Region. Cambridge: Cambridge University Press, 408 pp.

Cushman, J.A., 1936, Cretaceous foraminifera of the family Chilostomellidae. Contributions from the Cushman Laboratory for Foraminiferal Research, vol. 12, no. 4, pp. 1–77.

Cushman, J.A. and Todd, R., 1943, The genus Pullenia and its species. Contributions from the Cushman Laboratory for Foraminiferal Research, vol. 19, pp. 1–24.

# *Pullenia quinqueloba*

## (Reuss), 1851

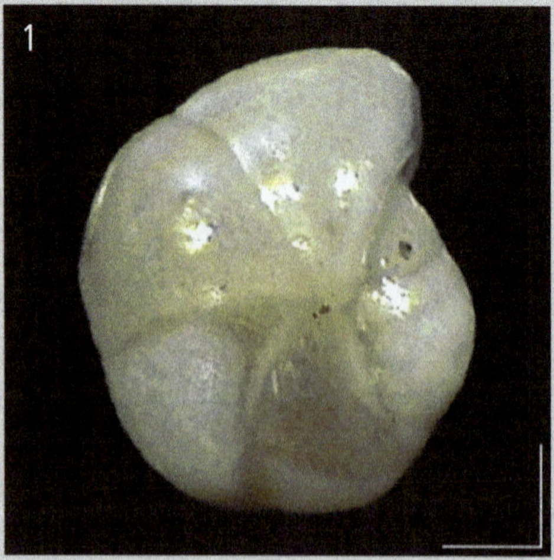

1. *Pullenia quinqueloba.* West Africa; Beleg-Zelle, Art 151-200. Deposited in the Lutze Collection (West Africa). Institute of Geosciences, Kiel University, Germany. Umbilical view, NHMUK PM DI 0552, scale bar = 81 μm.

2. *Pullenia quinqueloba.* West Africa; Beleg-Zelle, Art 151-200. Deposited in the Lutze Collection (West Africa). Institute of Geosciences, Kiel University, Germany. Apertural view, NHMUK PM DI 0553, scale bar = 83 μm.

### Original Designation

*Nonionina quinqueloba* Reuss, 1851, p. 71, pl. 5, fig. 31.

### Type Specimen

Depository not given.

### Type Level

Originally given as Eocene, Septarienthon.

### Type Locality

Hermsdorf near Berlin, Germany.

### Description

Test forms a subglobular, involute planispire; somewhat compressed, and slightly lobulate in outline, with a rounded periphery. The (typically) five triangular, inflated chambers in the final whorl increase gradually in size, and are separated by depressed, slightly curved, nearly radial sutures. Chamber walls are calcareous, finely perforate, and smooth. The primary aperture is a narrow, arcuate, interiomarginal, crescentic slit, extending across the periphery to the umbilici.

### Synonyms and Citations

*Pullenia subcarinata* (d'Orbigny). –Barker, 1960, pl. 84, figs. 14–15.

*Pullenia quinqueloba* (Reuss). –Brady, 1884, p. 617, pl. 84, figs. 14–15. –Cushman and Todd, 1943, p. 10, pl. 2, fig. 5; pl. 3, fig. 8. –Bermúdez, 1949, p. 276, pl. 21, figs. 32–33. –Tjalsma and Lohmann, 1983, p. 36, pl. 16, fig. 2. –Miller and Katz, 1987, p. 138, pl. 4, fig. 2. –Thomas, 1990, p. 590. –van Marle, 1991, p. 194, pl. 20, figs. 11–12. –Bolli et al., 1994, p. 152, figs. 41.31–32. –Jones, 1994, p. 92, pl. 84, figs. 14–15. –Kuhnt et al., 2002, p. 152, pl. 12, figs. 18–21. –Ortiz and Thomas, 2006, p. 128, pl. 10, fig. 10.

### Remarks

Cushman and Todd (1943) distinguished the varietal form *Pullenia quinqueloba* var. *angusta* (p. 10, pl. 2, figs. 3–4) on the basis of its smaller, narrower test. Bermúdez (1949), Tjalsma and Lohmann (1983), and Miller and Katz (1987) considered four-chambered forms to fall with the range of variability of *Pullenia quinqueloba* (Reuss).

### Biogeography

Worldwide.

### Bathymetry

Neritic to abyssal. Reported between 20 and 2750 fathoms at "Challenger" stations (Jones, 1994) and between 711 and 2119 meters in eastern Indonesia (van Marle, 1991); lower bathyal (Thomas, 1990); abyssal, greater than 3.5 kilometers (Miller and Katz, 1987).

### Chronostratigraphy

Early Paleocene (Danian) to Recent. Middle Paleocene to Early Eocene in Trinidad (Bolli et al., 1994). Rare to common throughout the Eocene (Tjalsma and Lohmann, 1983).

## References


Barker, R.W., 1960, Taxonomic notes on the species figured by H.B. Brady in his report on the foraminifera dredged by H.M.S. Challenger during the years 1873–1876. Society of Economic Paleontologists and Mineralogists Special Publication, vol. 9, pp. 1–238.

Bermúdez, P.J., 1949, Tertiary smaller foraminifera of the Dominican Republic. Special Publications of the Cushman Laboratory for Foraminiferal Research, no. 25, pp. 1–322.

Bolli, H.M., Beckmann, J.P., and Saunders, J.B., 1994, Benthic Foraminiferal Biostratigraphy of the South Caribbean Region. Cambridge: Cambridge University Press, 408 pp.

Brady, H.B., 1884, Report on the foraminifera dredged by H.M.S. Challenger during the years 1873–1876. In Brady, H.B., ed., Report on the Scientific Results of the Voyage of the H.M.S. Challenger during the years 1873–1876. Zoology, vol. 9, pp. 1–814.

Cushman, J.A. and Todd, R., 1943, The genus Pullenia and its species. Contributions from the Cushman Laboratory for Foraminiferal Research, vol. 19, pp. 1–24.

Jones, R.W., 1994, The Challenger Foraminifera. Oxford, Oxford University Press, 149 pp.

Kuhnt, W., Holbourn, A.E., and Zhao, Q., 2002, The early history of the South China Sea: Evolution of Oligocene–Miocene deep water environments. Revue de Micropaléontologie, vol. 45, pp. 99–159.

Miller, K.G. and Katz, M.E., 1987, Oligocene to Miocene benthic foraminiferal and abyssal circulation changes in the North Atlantic. Micropaleontology, vol. 33, no. 2, pp. 97–149.

Ortiz, S. and Thomas, E., 2006, Lower-middle Eocene benthic foraminifera from the Fortuna Section (Betic Cordillera, southeastern Spain). Micropaleontology, vol. 52, pp. 97–150.

Reuss, A.E., 1851, Ueber die fossilen Foraminiferen und Entomostraceen der Septarienthone der Umgegend von Berlin. Zeitschrift der Deutschen Geologischen Gesellschaft, Berlin, vol. 3, pp. 49–91.

Thomas, E., 1990, Late Cretaceous through Neogene benthic foraminifers. In Barker, P.F., Kennett, J.P., et al., ed., Scientific Results of the Proceedings of the Ocean Drilling Program, vol. 113, pp. 571–594.

Tjalsma, R.C. and Lohmann, G.P., 1983, Paleocene–Eocene bathyal and abyssal benthic foraminifera from the Atlantic Ocean. Micropaleontology, Special Publication, no. 4, pp. 1–90.

van Marle, L.J., 1991, Eastern Indonesian Late Cenozoic smaller benthic foraminifera. Verhandelingen der Koninklijke Nederlanse Akademie van Wetenschappen, Afdeeling Natuurkunde, Eerste Reeks, vol. 34, pp. 1–328.

# *Pyramidulina sceptrum*

## (Reuss), 1863

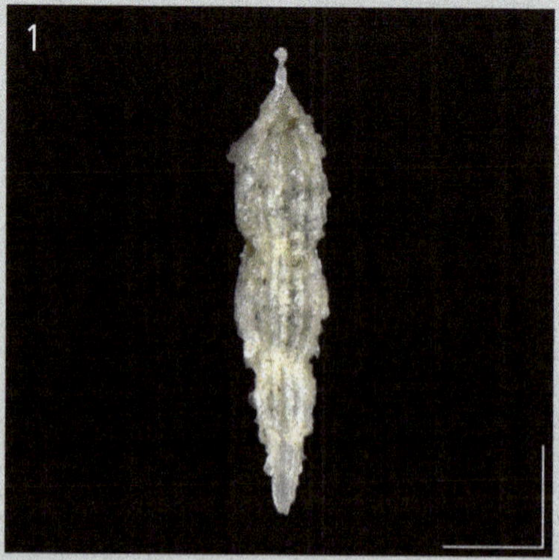


1. *Pyramidulina sceptrum*. Copt Point, Folkestone. F.W. Millett Collection, 1894. Deposited in the micropaleontological collections of The Natural History Museum, London; NHMUK PF70838(1). Lateral view, NHMUK PM DI 0554, scale bar = 145 μm.

### Original Designation

*Nodosaria (Nodosaria) sceptrum* Reuss, 1863, p. 37, pl. 2, fig. 3.

### Type Specimen

Depository not given.

### Type Level

Lower Cretaceous, Upper Hils, clay bed with *Crioceras emerici*.

### Type Locality

Bohnenkamp near Braunschweig, North Germany.

### Description

Test forms an elongate, uniserial series; rectilinear in overall shape, with a subcircular cross-section. The globular or subspherical chambers are broader at their base, increase gradually in size, and are separated by straight, horizontal, depressed sutures. Chamber walls are calcareous, perforate, and ornamented by fine, longitudinal costae. The primary aperture is terminal at the end of a short neck.

### Synonyms and Citations

*Nodosaria tubifera* Reuss, 1863, p. 37, pl. 2, fig. 4.

*Nodosaria sceptrum* Reuss. –Berthelin, 1880, p. 32. –Neagu, 1975, pl. 70, figs. 3–5, 7.

*Pyramidulina sceptrum* (Reuss). –Meyn and Vespermann, 1994, p. 100, pl. 12, figs. 1–14; pl. 13, figs. 1–16; pl. 14, figs. 1–17.

### Biogeography

Worldwide.

### Bathymetry

Neritic to bathyal.

### Chronostratigraphy

Late Jurassic (Oxfordian) to Early Cretaceous (Albian)

## References


Berthelin, G., 1880, Mémoire sur les foraminifères fossiles de l'étage albien de Montcley (Doubs). Mémoires de la Société géologique de France, ser. 31, no. 5, pp. 1–84.

Meyn, H. and Vespermann, J., 1994, Taxonomic revision of Lower Cretaceous Foraminifera described by Roemer (1839, 1841, 1842), Koch (1851), and Reuss (1863) from Lower Cretaceous Saxony. Senckenbergiana Lethaea, vol. 74, no. 1/2, pp. 49–272.

Neagu, T., 1975, Monographie de la faune des Foraminifères Éocrétacés du couloir de Dîmbovicioara, de Codlea et des Monts Persani (Couches de Carhaga). Memorii Institutul de Geologie si Geofizica, Bucaresti, vol. 35, pp. 1–141.

Reuss, A.E., 1863, Beiträge zur Kenntniss der tertiären Foraminiferen-Fauna (Zweite Folge). Sitzungsberichte der Kaiserlichen Akademie der Wissenschaften in Wien, Mathematisch-Naturwissenschaftliche Classe, vol. 48, no. 1, pp. 36–71.

# Pyrgo comata

## (Brady), 1881

1. *Pyrgo comata.* Culebra Islands, West Indies; "Challenger" Station 24, depth: 390 fathoms. Deposited in the micropaleontological collections of The Natural History Museum, London; NHMUK ZF1143. Syntype. Lateral view, NHMUK PM DI 0555, scale bar = 171 μm.

2. *Pyrgo comata.* Culebra Islands, West Indies; "Challenger" Station 24, depth: 390 fathoms. Deposited in the micropaleontological collections of The Natural History Museum, London; NHMUK ZF1143. Syntype. Apertural view, NHMUK PM DI 0556, scale bar = 191 μm.

### Original Designation

*Biloculina comata* Brady, 1881, p. 45 (type description), illustrated in Brady, 1884, pl. 3, fig. 9.

### Type Specimen

Figured specimen is housed in the micropaleontological collections of The Natural History Museum, London, UK, and is registered in slide ZF1143.

### Type Level

Recent, surface sediments.

### Type Locality

Not designated. Figured specimen from "Challenger" Station 24, Culebra Island, north of St Thomas, West Indies, 390 fathoms.

### Description

Test forms a slightly compressed, biloculine series in the adult form; nearly circular in outline, with a subcircular cross-section, and a subrounded, keeled periphery. Chamber walls are calcareous, porcelaneous, and imperforate, ornamented by numerous longitudinal striae. The primary aperture is oval and terminal, with a bifid tooth.

### Synonyms and Citations

*Biloculina comata* Brady. –Brady, 1884, p. 144, pl. 3, fig. 9.

*Pyrgo comata* (Brady). –Barker, 1960, pl. 3, fig. 9. –Jones, 1994, p. 19, pl. 3, fig. 9.

### Biogeography

Worldwide.

### Bathymetry

Neritic to abyssal. Reported between 15 and 2900 fathoms at "Challenger" stations (Jones, 1994).

### Chronostratigraphy

Miocene to Recent.

## References

Barker, R.W., 1960, Taxonomic notes on the species figured by H.B. Brady in his report on the foraminifera dredged by H.M.S. Challenger during the years 1873–1876, Society of Economic Paleontologists and Mineralogists Special Publication, vol. 9, pp. 1–238.

Brady, H.B., 1881, Notes on some of the reticularian Rhizopoda of the Challenger Expedition, Part III: Quarterly Journal of Microscopical Science (New Series), vol. 21, pp. 31–71.

Brady, H.B., 1884, Report on the foraminifera dredged by H.M.S. Challenger during the years 1873–1876. In Brady, H.B., ed., Report on the Scientific Results of the Voyage of the H.M.S. Challenger during the years 1873–1876. Zoology, vol. 9, pp. 1–814.

Jones, R.W., 1994, The Challenger Foraminifera. Oxford: Oxford University Press, 149 pp.

# *Pyrgo fornasinii*

## Chapman and Parr, 1935

1. *Pyrgo fornasinii.* "Meteor" Cruise, West Africa; M25. Deposited in the Lutze Collection (West Africa), no. 94. Institute of Geosciences, Kiel University, Germany. Lateral view, NHMUK PM DI 0557, scale bar = 440 μm.

### Original Designation

*Pyrgo fornasinii* Chapman and Parr, 1935, p. 5, illustrated as *Biloculina ringens* in Brady, 1884, pl. 2, fig. 7.

### Type Specimen

Specimen figured by Brady (1884) as *Biloculina ringens* Lamarck is housed in the micropaleontological collections of The Natural History Museum, London, and is registered in slide ZF1159.

### Type Level

Recent.

### Type Locality

Not designated. Brady's specimen is from "Challenger" Station 24, West Indies, 390 fathoms. Localities given in Schlumberger (1891): Dredging no. 1 of the "Travailleur," 1880, in the Gulf of Gascony, 1850 meters; Moroccan coast and Atlantic Ocean.

### Description

Test forms a biloculine series in the adult form; subcircular or ovate in outline, and biconvex in cross-section, with an acute, extended, keeled periphery. Chamber walls are calcareous, porcelaneous, smooth, and imperforate. The primary aperture is an elongate, terminal slit, with a short bifid tooth.

### Synonyms and Citations

*Biloculina ringens* Lamarck (non *Miliolites ringens* Lamarck). –Brady, 1884, p. 142, pl. 2, fig. 7.

*Biloculina bradyi* Schlumberger (non Fornasini), 1891, p. 557, textfigs. 15–19; pl. 10, figs. 63–71.

*Pyrgo sarsi* (Schlumberger). –Jones, 1994, p. 18, pl. 2, fig. 7. –Loeblich and Tappan, 1994, p. 54, pl. 94, figs. 1–9.

*Pyrgo fornasinii* Chapman and Parr. –Barker, 1960, pl. 2, fig. 7.

### Remarks

The specimen figured by Brady (1884) as *Biloculina ringens* Lamarck was subsequently renamed *Biloculina bradyi* by Schlumberger (1891), who overlooked that this name had already been used by Fornasini for another species. Chapman and Parr (1935) subsequently gave it the new name *Pyrgo fornasinii*. Jones (1994) considered *Pyrgo fornasinii* to be a junior synonym of *Biloculina sarsi* Schlumberger.

### Biogeography

Worldwide.

### Bathymetry

Primarily bathyal.

### Chronostratigraphy

Early Miocene (Aquitanian?) to Recent.

## References

Barker, R.W., 1960, Taxonomic notes on the species figured by H.B. Brady in his report on the foraminifera dredged by H.M.S. Challenger during the years 1873–1876, Society of Economic Paleontologists and Mineralogists Special Publication, vol. 9, pp. 1–238.

Brady, H.B., 1884, Report on the foraminifera dredged by H.M.S. Challenger during the years 1873–1876. In Brady, H.B., ed., Report on the Scientific Results of the Voyage of the H.M.S. Challenger during the years 1873–1876. Zoology, vol. 9, pp. 1–814.

Chapman, F. and Parr, W.J., 1935, Foraminifera and Ostracoda from soundings made by the trawler "Bonthorpe" in the Great Australian Bight. Journal of the Royal Society of Western Australia, vol. 21, pp. 1–7.

Jones, R.W., 1994, The Challenger Foraminifera. Oxford: Oxford University Press, 149 pp.

Loeblich, A.R. Jr and Tappan, H., 1994, Foraminifera of the Sahul Shelf and Timor Sea. Special Publications of the Cushman Laboratory for Foraminiferal Research, vol. 31, pp. 1–661.

Schlumberger, C., 1891, Révision des Biloculines des grands fonds. Mémoires de la Société Zoologique de France, vol. 4, pp. 542–579.

# *Pyrgo lucernula*

## (Schwager), 1866

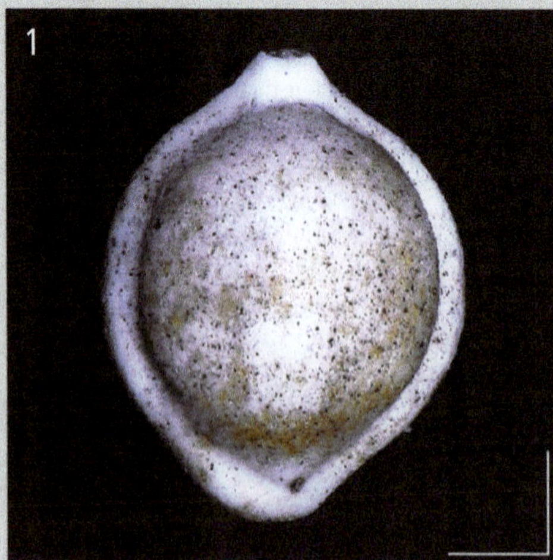


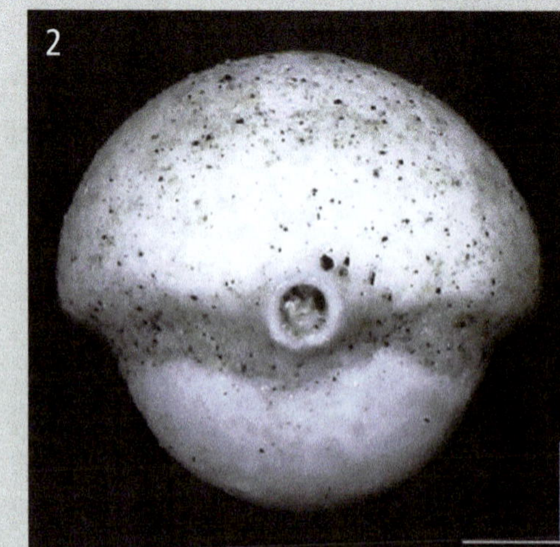


1. *Pyrgo lucernula.* Off the Azores, Atlantic Ocean; "Challenger" Station 76, depth: 900 fathoms. Deposited in the micropaleontological collections of The Natural History Museum, London; NHMUK ZF1139. Lateral view, NHMUK PM DI 0558, scale bar = 391 μm.

2. *Pyrgo lucernula.* Off the Azores, Atlantic Ocean; "Challenger" Station 76, depth: 900 fathoms. Deposited in the micropaleontological collections of The Natural History Museum, London; NHMUK ZF1139. Apertural view, NHMUK PM DI 0559, scale bar = 219 μm.

### Original Designation

*Biloculina lucernula* Schwager, 1866, p. 202, pl. 4, figs. 14, 17a–b.

### Type Specimen

Depository not given. Neotype (Mf 13) designated by Srinivasan and Sharma (1980) deposited in the Micropaleontology Section, Department of Geology, Banaras Hindu University, Banaras, India.

### Type Level

Upper Tertiary, upper and lower clays.

### Type Locality

Car Nicobar, Andaman Sea.

### Description

Test forms a biloculine series in the adult form; ovate in outline and biconvex in cross-section, with a rounded periphery. Embracing chambers are moderately inflated; only the last two chambers are visible. Chamber walls are calcareous, imperforate, appearing finely agglutinated. The primary aperture is oval and terminal with a distinct, bifid tooth.

### Synonyms and Citations

*Biloculina bulloides* d'Orbigny. –Brady, 1884, p. 142, pl. 2, figs. 5–6.
*Miliolina trigonula* (Lamarck). –Brady, 1884, p. 142, pl. 3, fig. 14.
*Pyrgo lucernula* (Schwager). –Barker, 1960, pl. 2, figs. 5–6; pl. 3, fig. 14 (not fig. 6). –Srinivasan and Sharma, 1980, p. 21, pl. 3, figs. 10–11. –van Morkhoven et al., 1986, p. 43, pl. 12. –Jones, 1994, p. 18, pl. 2, figs. 5–6; p. 19, pl. 3, fig. 14 (not fig. 6).

### Biogeography

Worldwide.

### Bathymetry

Middle bathyal to abyssal (van Morkhoven et al., 1986). Reported between 300 and 1000 fathoms at "Challenger" stations (Jones, 1994).

### Chronostratigraphy

Middle Miocene to Recent (Jones, 1994).

## References


Barker, R.W., 1960, Taxonomic notes on the species figured by H.B. Brady in his report on the foraminifera dredged by H.M.S. Challenger during the years 1873–1876, Society of Economic Paleontologists and Mineralogists Special Publication, vol. 9, pp. 1–238.

Brady, H.B., 1884, Report on the foraminifera dredged by H.M.S. Challenger during the years 1873–1876. In Brady, H.B., ed., Report on the Scientific Results of the Voyage of the H.M.S. Challenger during the years 1873–1876. Zoology, vol. 9, pp. 1–814.

Jones, R.W., 1994, The Challenger Foraminifera. Oxford: Oxford University Press, 149 pp.

Schwager, C., 1866, Fossile Foraminiferen von Kar Nikobar, Reise der Österreichischen Fregatte Novara um Erde in den Jahren 1857, 1858, 1859 unten den Befehlen des Commodore B. Von Wüllerstorf-Urbair. Geologischer Theil, vol. 2, no. 1, Geologische Beobachtung, no. 2, Paläontologische Mittheilung, pp. 187–268.

Srinivasan, M.S. and Sharma, V., 1980, Schwager's Car Nicobar Foraminifera in the Reports of the Novara Expedition – A Revision. New Delhi: Today & Tomorrow's Printers and Publishers, 83 pp.

van Morkhoven, F.P.C.M., Berggren, W.A., Edwards, A.S., et al., 1986, Cenozoic cosmopolitan deep-water benthic foraminifera. Bulletin des Centres de Recherches Exploration-Production Elf-Aquitaine, Mémoire, vol. 11, pp. 1–421.

# Pyrgo murrhina

## (Schwager), 1866

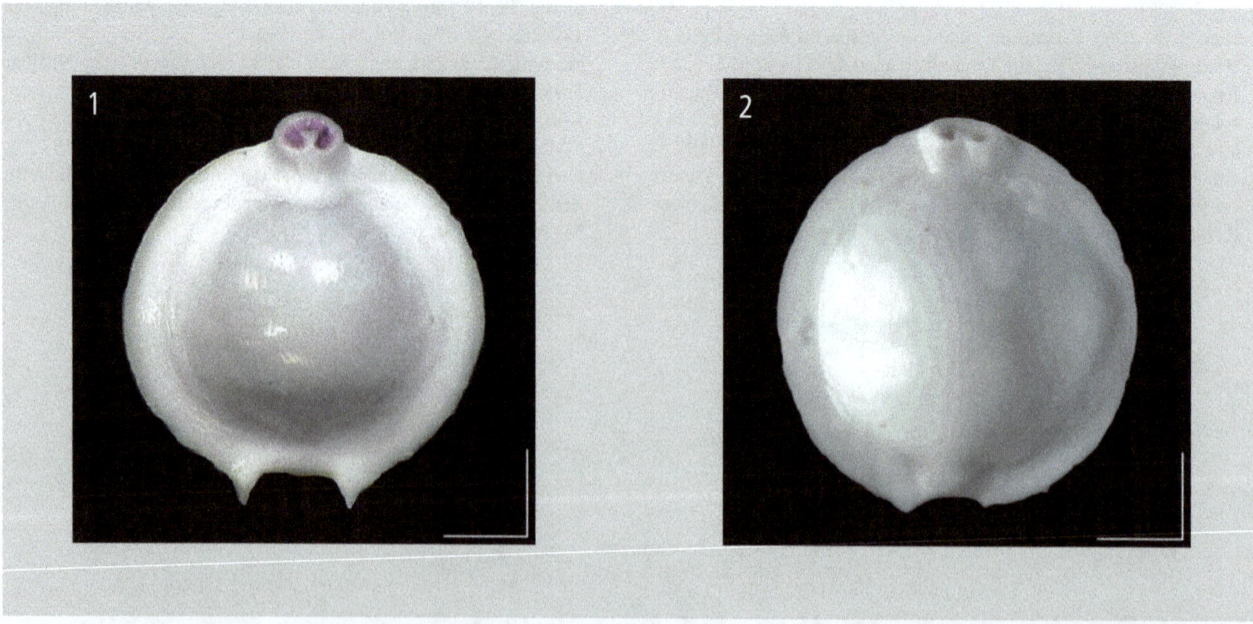


1. *Pyrgo murrhina.* South China Sea; Station 17965, latitude 06° 09.40′ N, longitude 112° 33.30′ E, water depth: 889 meters, surface sample. Deposited in the South China Sea Collection, Institute of Geosciences, Kiel University, Germany. Lateral view, NHMUK PM DI 0560, scale bar = 102 μm.

2. *Pyrgo murrhina.* "Meteor" Cruise, West Africa; M25. Deposited in the Lutze Collection (West Africa), no. 94. Institute of Geosciences, Kiel University, Germany. Lateral view, NHMUK PM DI 0561, scale bar = 123 μm.

### Original Designation

*Biloculina murrhina* Schwager, 1866, p. 203, pl. 4, fig. 15.

### Type Specimen

Depository not given. Neotype (Mf 41) designated by Srinivasan and Sharma (1980) deposited in the Micropaleontology Section, Department of Geology, Banaras Hindu University, Banaras, India.

### Type Level

Upper Tertiary, upper and lower clays.

### Type Locality

Car Nicobar, Andaman Sea.

### Description

Test forms a slightly compressed, biloculine series in the adult form; subcircular in outline, and elliptical in cross-section, with an acute, extended, keeled periphery. Keel may be smooth or dentate, and is interrupted at a point opposite to the aperture to form a sinus. Chamber walls are calcareous, porcelaneous, smooth, and imperforate. The primary aperture is oval, and terminal with a short bifid tooth.

### Synonyms and Citations

*Pyrgo murrhina* (Schwager). –Cushman and Todd, 1945, p. 12, pl. 2, fig. 5. –van Morkhoven et al., 1986, p. 50, pl. 15. –Jones, 1994, p. 18, pl. 2, figs. 10–11, 15. –Robertson, 1998, p. 34, pl. 10, fig. 2.

### Remarks

*Pyrgo murrhina* (Schwager) varies considerably in size and shape. The characteristic sinus in the keel is often lacking in adult specimens (Phleger et al., 1953).

### Biogeography

Worldwide. Recorded from the Caribbean, the Gulf of Mexico, the Mediterranean, the Indo-Pacific region, Okinawa, Japan, the central North Pacific, and the Atlantic (van Morkhoven et al., 1986).

### Bathymetry

Bathyal to abyssal. Reported between 1180 and 1900 fathoms at "Challenger" stations (Jones, 1994).

### Chronostratigraphy

Middle Miocene to Recent.

## References

Cushman, J.A. and Todd, R., 1945, Miocene foraminifera from Buff Bay, Jamaica. Special Publications of the Cushman Laboratory for Foraminiferal Research, vol. 15, pp. 1–73.

Jones, R.W., 1994, The Challenger Foraminifera. Oxford: Oxford University Press, 149pp.

Phleger, F.B., Parker, F.L., and Pierson, J.F., 1953, North Atlantic foraminifera, Reports of the Swedish Deep-Sea Expedition, 1947–1948, vol. 7, pp. 3–122.

Robertson, B.E., 1998, Systematics and paleoecology of the benthic foraminiferida from the Buff Bay section, Miocene of Jamaica. Micropaleontology, vol. 44, supplement 2, pp. 1–266.

Schwager, C., 1866, Fossile Foraminiferen von Kar Nikobar, Reise der Österreichischen Fregatte Novara um Erde in den Jahren 1857, 1858, 1859 unten den Befehlen des Commodore B. Von Wüllerstorf-Urbair. Geologischer Theil, vol. 2, no. 1, Geologische Beobachtung, no. 2, Paläontologische Mittheilung, pp. 187–268.

Srinivasan, M.S. and Sharma, V., 1980, Schwager's Car Nicobar Foraminifera in the Reports of the Novara Expedition – A Revision. New Delhi: Today & Tomorrow's Printers and Publishers, 83 pp.

van Morkhoven, F.P.C.M., Berggren, W.A., Edwards, A.S., et al., 1986, Cenozoic cosmopolitan deep-water benthic foraminifera. Bulletin des Centres de Recherches Exploration-Production Elf-Aquitaine, Mémoire, vol. 11, pp. 1–421.

# *Pyrgo serrata*

## (Bailey), 1861

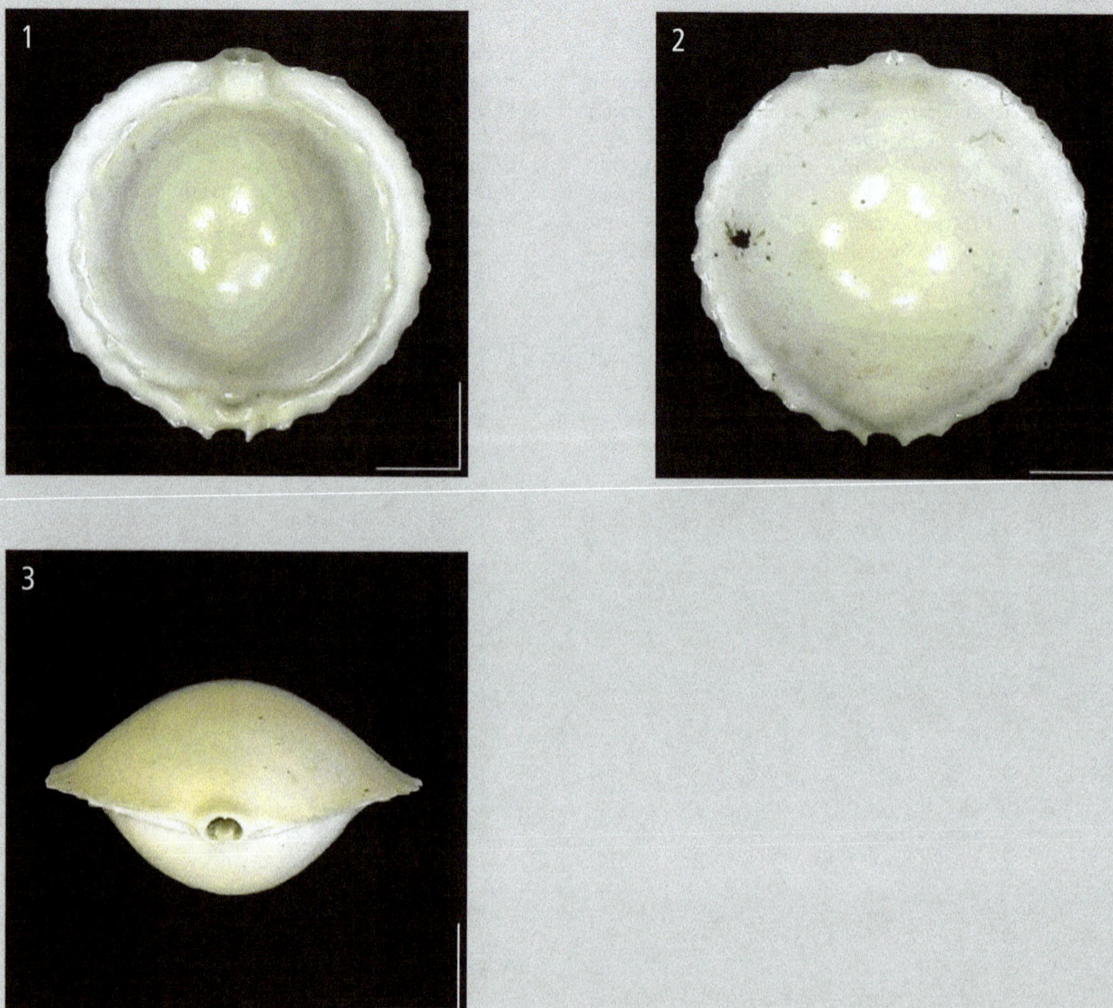


1. *Pyrgo serrata.* Northeast coast of New Zealand, South Pacific Ocean; "Challenger" Station 168, depth: 1100 fathoms. Deposited in the micropaleontological collections of The Natural History Museum, London; NHMUK ZF1148. Lateral view, NHMUK PM DI 0562, scale bar = 390 μm.

2. *Pyrgo serrata.* Northeast coast of New Zealand, South Pacific Ocean; "Challenger" Station 168, depth: 1100 fathoms. Deposited in the micropaleontological collections of The Natural History Museum, London; NHMUK ZF1148. Lateral view, NHMUK PM DI 0563, scale bar = 383 μm.

3. *Pyrgo serrata.* Northeast coast of New Zealand, South Pacific Ocean; "Challenger" Station 168, depth: 1100 fathoms. Deposited in the micropaleontological collections of The Natural History Museum, London; NHMUK ZF1148. Apertural view, NHMUK PM DI 0564, scale bar = 357 μm.

## Original Designation

*Biloculina serrata* Bailey, 1861, p. 350, pl. 8, fig. E.

## Type Specimen

Figured specimen is housed in the micropaleontological collections of The Natural History Museum, London, UK, and is registered in slide ZF1143.

## Type Level

Recent.

## Type Locality

United States Coast Survey, Position 14, 150 fathoms, Gulf Stream.

## Description

Test forms a slightly compressed, biloculine series in the adult form; nearly circular in outline, elliptical in cross-section, with a broad, serrate keel. Chamber walls are calcareous, porcelaneous, and imperforate. The primary aperture is oval and terminal, with a bifid tooth.

## Synonyms and Citations

*Biloculina depressa* d'Orbigny var. *serrata* Brady, 1884, p. 146, pl. 3, fig. 3.

*Pyrgo serrata* (Bailey). –Barker, 1960, pl. 3, fig. 3. –Jones, 1994, p. 19, pl. 3, fig. 3.

## Biogeography

Worldwide.

## Bathymetry

Neritic to abyssal. Reported between 580 and 1750 fathoms at "Challenger" stations (Jones, 1994).

## Chronostratigraphy

Miocene to Recent (Jones, 1994).

# References


Bailey, J.W., 1861, Notes on new species of microscopical organisms, chiefly from the Para River, South America. Boston Journal of Natural Histroy, vol. 7, pp. 329–352.

Barker, R.W., 1960, Taxonomic notes on the species figured by H.B. Brady in his report on the foraminifera dredged by H.M.S. Challenger during the years 1873–1876, Society of Economic Paleontologists and Mineralogists Special Publication, vol. 9, pp. 1–238.

Brady, H.B., 1884, Report on the foraminifera dredged by H.M.S. Challenger during the years 1873–1876. In Brady, H.B., ed., Report on the Scientific Results of the Voyage of the H.M.S. Challenger during the years 1873–1876. Zoology, vol. 9, pp. 1–814.

Jones, R.W., 1994, The Challenger Foraminifera. Oxford: Oxford University Press, 149 pp.

# *Pyrulina cylindroides*

## (Roemer), 1838

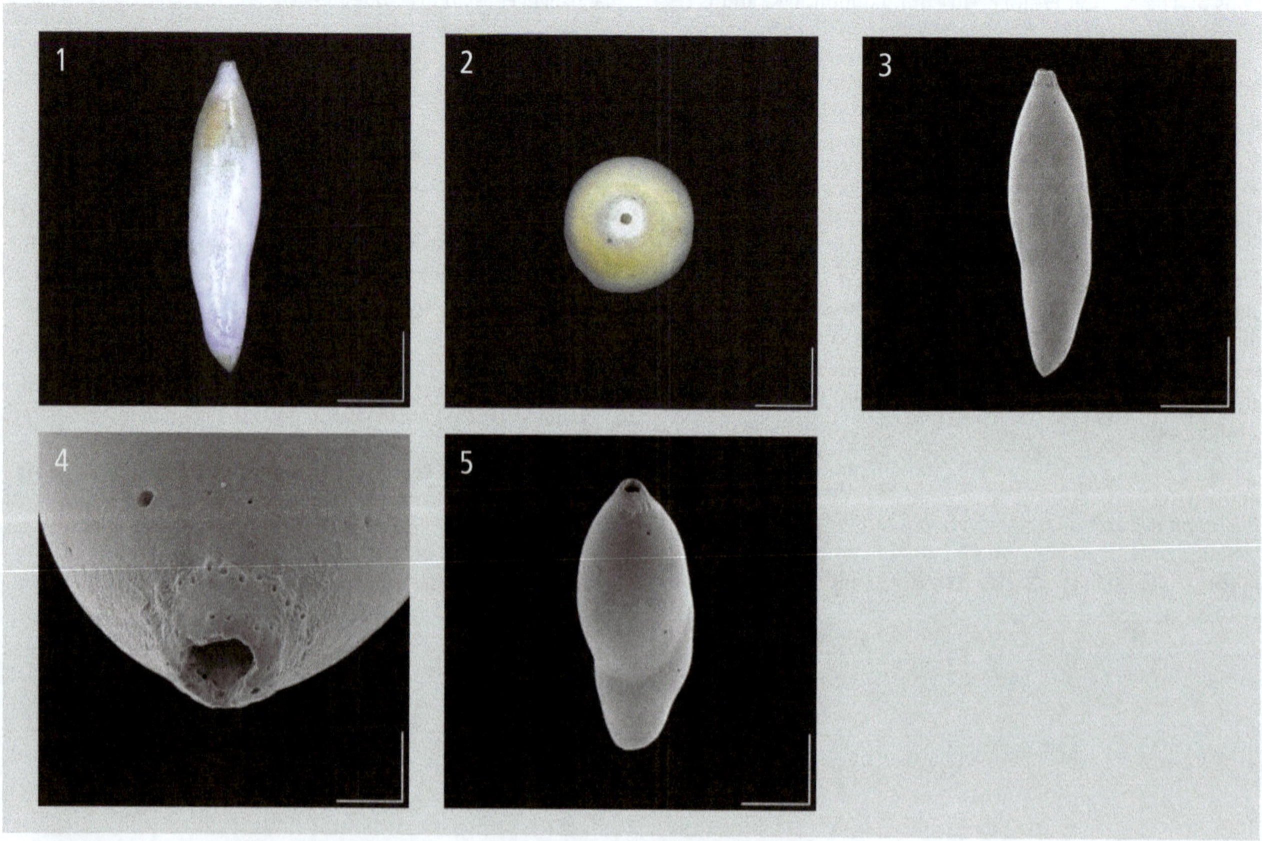

1. *Pyrulina cylindroides.* South Pacific Ocean; "Challenger" Station 296, depth: 1825 fathoms. Deposited in the micropaleontological collections of The Natural History Museum, London; NHMUK ZF2140. Lateral view, NHMUK PM DI 0565, scale bar = 213 μm.

2. *Pyrulina cylindroides.* South Pacific Ocean; "Challenger" Station 296, depth: 1825 fathoms. Deposited in the micropaleontological collections of The Natural History Museum, London; NHMUK ZF2140. Apertural view, NHMUK PM DI 0566, scale bar = 56 μm.

3. *Pyrulina cylindroides.* South Pacific Ocean; "Challenger" Station 296, depth: 1825 fathoms. Deposited in the micropaleontological collections of The Natural History Museum, London; NHMUK ZF2140. Lateral view, NHMUK PM DI 0567, scale bar = 233 μm.

4. *Pyrulina cylindroides.* South Pacific Ocean; "Challenger" Station 296, depth: 1825 fathoms. Deposited in the micropaleontological collections of The Natural History Museum, London; NHMUK ZF2140. Detail view, NHMUK PM DI 0568, scale bar = 37 μm.

5. *Pyrulina cylindroides.* South Pacific Ocean; "Challenger" Station 296, depth: 1825 fathoms. Deposited in the micropaleontological collections of The Natural History Museum, London; NHMUK ZF2140. Oblique apertural view, NHMUK PM DI 0569, scale bar = 158 μm.

## Original Designation

*Polymorphina (Polymorphinen) cylindroides* Roemer, 1838, p. 385, pl. 3, fig. 26a–b.

## Type Specimen

Depository not given.

## Type Level

Lower Tertiary, marine sands.

## Type Locality

North Germany.

## Description

Test forms an elongate, fusiform series; slightly tapered at both extremities, and circular in cross-section. The inflated chambers, initially added in plane approximately 120° apart and later biserial, increase rapidly in height, and are separated by flush sutures. Chamber walls are calcareous, finely perforate, and smooth. The primary aperture is terminal and radiate.

## Synonyms and Citations

*Polymorphina lanceolata* Reuss. –Brady, 1884, p. 564, pl. 72, fig. 5 (not fig. 6).

*Pyrulina cylindroides* (Roemer). –Barker, 1960, pl. 72, fig. 5 (not fig. 6). –McNeil and Caldwell, 1981, pl. 17, fig. 16. –Riegraf and Luterbacher, 1989, pl. 4, figs. 23–24. –Jones, 1994, p. 83, pl. 72, fig. 5. –Holbourn and Kaminski, 1997, p. 73, pl. 43, fig. 6.

## Biogeography

Worldwide.

## Bathymetry

Bathyal.

## Chronostratigraphy

Late Jurassic (Tithonian?) to Recent.

# References


Barker, R.W., 1960, Taxonomic notes on the species figured by H.B. Brady in his report on the foraminifera dredged by H.M.S. Challenger during the years 1873–1876, Society of Economic Paleontologists and Mineralogists Special Publication, vol. 9, pp. 1–238.

Brady, H.B., 1884, Report on the foraminifera dredged by H.M.S. Challenger during the years 1873–1876. In Brady, H.B., ed., Report on the Scientific Results of the Voyage of the H.M.S. Challenger during the years 1873–1876. Zoology, vol. 9, pp. 1–814.

Holbourn, A.E.L. and Kaminski, M.A., 1997, Cretaceous deep-water benthic foraminifera of the Indian Ocean. Grzybowski Foundation Special Publication, no. 4, pp. 1–175.

Jones, R.W., 1994, The Challenger Foraminifera. Oxford: Oxford University Press, 149 pp.

McNeil, D.H. and Caldwell, W.G.E., 1981, Cretaceous rocks and their foraminifera in the Manitoba Escarpment. Special Papers of the Geological Association of Canada, vol. 21, pp. 1–439.

Riegraf, W. and Luterbacher, H., 1989, Benthonische Foraminiferen aus de Unterkreide des "Deep Sea Drilling Project" (Leg 1–79). Geologische Rundschau, vol. 78, no. 3, pp. 1063–1120.

Roemer, F.A., 1838, Die Cephalopoden des Nord-Deutschen tertiären Meersandes. Neues Jahrbuch für Mineralogie, Geognosie, Geologie und Petrefakten-Kunde, vol. 1838, pp. 381–394.

# *Quadratobuliminella pyramidalis*

## de Klasz, 1953

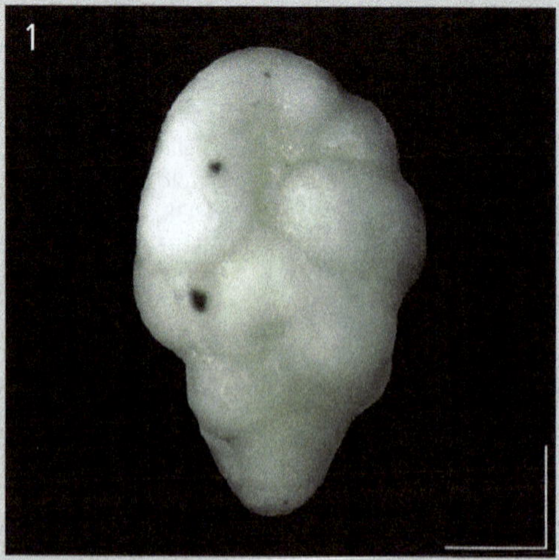


1. *Quadratobuliminella pyramidalis.* Gulf of Mexico; Eureka Core E68-139. Deposited in the collections of the United States National Museum, Washington, DC. Specimen illustrated in van Morkhoven et al., 1986, Ref. no.104. Lateral view, NHMUK PM DI 0570, scale bar = 168 μm.

### Original Designation

*Quadratobuliminella pyramidalis* de Klasz, 1953, p. 435, textfigs. 1–2.

### Type Specimen

Holotype (Prot. 321) and paratype (Prot. 322) deposited in the Bavarian State collections for Paleontology and Historical Geology, Munich, Germany.

### Type Level

Upper Cretaceous or Lower Paleocene, Bucheck Beds (grayish-brown marl). Paratype from the Danian.

### Type Locality

Holotype from a trench along the footpath west of the railroad bridge at Wiesen, north of Eisenärzt, near Traunstein, Upper Bavaria, southeastern Germany. Paratype from Bidart, south of Biarritz, Basses Pyrénées, southwestern France.

### Description

Test forms an elongate, flaring trochospire with four chambers per whorl; tapered in outline, and subquadrate in cross-section, with a subrounded initial portion, and rounded apertural end. The inflated chambers (initially globular, becoming cylindrical) increase rapidly in size, and are separated by distinct, depressed sutures. Chamber walls are calcareous, finely perforate, and smooth, except for the final four chambers, which are ornamented by thin grooves radiating from the aperture. The primary aperture is situated at the junction of the final four chambers, and is loop-shaped, bordered by a lip and with an internal toothplate.

### Synonyms and Citations

*Quadratobuliminella pyramidalis* de Klasz. –Tjalsma and Lohmann, 1983, p. 19, pl. 4, fig. 1; pl. 12, fig. 6. –van Morkhoven et al., 1986, p. 320, pl. 104. –Loeblich and Tappan, 1987, p. 522, pl. 572, figs. 19–20. –Nomura, 1995, p. 274, pl. 2, fig. 12.

### Remarks

Type species of the genus *Quadratobuliminella* de Klasz, 1953.

### Biogeography

Worldwide. Recorded from the Pacific, the North and South Atlantic, the Gulf of Mexico, the Caribbean and Europe (van Morkhoven et al., 1986).

### Bathymetry

Bathyal to abyssal.

### Chronostratigraphy

Primarily Paleocene. Occurs more rarely in the Early to Middle Eocene (Tjalsma and Lohmann, 1983).

## References


de Klasz, I., 1953, Quadratobuliminella n. gen., eine neue Foraminiferengattung von der Wende Kreide-Tertiaer. Neues Jahrbuch für Geologie und Paläontologie, Monatshefte, vol. 10, pp. 434–436.

Loeblich, A.R. Jr and Tappan, H., 1987, Foraminiferal Genera and Their Classification. New York: Van Nostrand Reinhold, 970 pp.

Nomura, R., 1995, Paleogene to Neogene deep-sea paleoceanography in the eastern Indian Ocean: benthic foraminifera from ODP Sites 747, 757 and 758. Micropaleontology, vol. 41, pp. 251–290.

Tjalsma, R.C. and Lohmann, G.P., 1983, Paleocene–Eocene bathyal and abyssal benthic foraminifera from the Atlantic Ocean. Micropaleontology, Special Publication, no. 4, pp. 1–90.

van Morkhoven, F.P.C.M., Berggren, W.A., Edwards, A.S., et al., 1986, Cenozoic cosmopolitan deep-water benthic foraminifera. Bulletin des Centres de Recherches Exploration-Production Elf-Aquitaine, Mémoire, vol. 11, pp. 1–421.

# *Ramulina globulifera*

## Brady, 1879

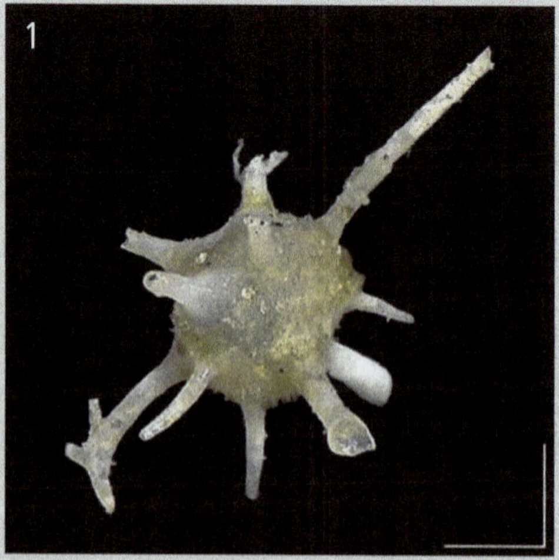


1. *Ramulina globulifera.* Off Fiji, Pacific Ocean; "Challenger" Station 174C, depth: 210 fathoms. Deposited in the micropaleontological collections of The Natural History Museum, London; specimen from slide NHMUK ZF2253. Syntype. Lateral view, NHMUK PM DI 0571, scale bar = 346 μm.

### Original Designation

*Ramulina globulifera* Brady, 1879, p. 272, pl. 8, figs. 22–23.

### Type Specimen

Figured specimens (ZF2253–ZF2255) deposited in the micropaleontological collections of The Natural History Museum, London, UK. Hypotype (USNM 470478) designated by Loeblich and Tappan (1994), and deposited in the collections of the United States National Museum, Washington, DC.

### Type Level

Recent.

### Type Locality

Not designated. Localities given: "Challenger" dredgings near island groups in the North Atlantic or South Pacific. Recent, 145–300 fathoms. Hypotype is from the northeastern Timor Sea.

### Description

Test forms a branching series, composed of globular or irregular chambers of different sizes, loosely connected by narrow stolons extending from different portions of the chambers. Chamber walls are calcareous, finely perforate, and hispid. The primary aperture is round at the end of the open stolons.

### Synonyms and Citations

*Ramulina globulifera* Brady. –Brady, 1884, p. 587, pl. 76, figs. 22–28. –Barker, 1960, pl. 76, figs. 22–28. –Loeblich and Tappan, 1994, p. 84, pl. 149, fig. 17. –Jones, 1994, p. 88, pl. 76, figs. 22–28.

### Remarks

*Ramulina globulifera* Brady closely resembles the Cretaceous species *Ramulina globotubulosa* described by Cushman (1938) from the Selma Chalk in Tennessee. Specimens closely resembling *Ramulina globulifera* Brady were reported under the name *Ramulina tetrahadrelis* Ludbrook in Lower Cretaceous cores from the Indian Ocean (Holbourn and Kaminski, 1997).

### Biogeography

Worldwide.

### Bathymetry

Neritic to bathyal. Reported between 95 and 435 fathoms at "Challenger" stations (Jones, 1994).

### Chronostratigraphy

Cretaceous? to Recent.

## References

Barker, R.W., 1960, Taxonomic notes on the species figured by H.B. Brady in his report on the foraminifera dredged by H.M.S. Challenger during the years 1873–1876, Society of Economic Paleontologists and Mineralogists Special Publication, vol. 9, pp. 1–238.

Brady, H.B., 1879, Notes on some of the reticularian Rhizopoda of the Challenger Expedition, Part I; On new or little known Arenaceous types. Quarterly Journal of Microscopical Science (New Series), vol. 19, pp. 20–63.

Brady, H.B., 1884, Report on the foraminifera dredged by H.M.S. Challenger during the years 1873–1876. In Brady, H.B., ed., Report on the Scientific Results of the Voyage of the H.M.S. Challenger during the years 1873–1876. Zoology, vol. 9, pp. 1–814.

Cushman, J.A., 1938, Additional new species of American Cretaceous foraminifera. Contributions from the Cushman Laboratory for Foraminiferal Research, vol. 14, pp. 31–50.

Holbourn, A.E.L. and Kaminski, M.A., 1997, Cretaceous deep-water benthic foraminifera of the Indian Ocean. Grzybowski Foundation Special Publication, no. 4, pp. 1–175.

Jones, R.W., 1994, The Challenger Foraminifera. Oxford: Oxford University Press, 149 pp.

Loeblich, A.R. Jr and Tappan, H., 1994, Foraminifera of the Sahul Shelf and Timor Sea. Special Publications of the Cushman Laboratory for Foraminiferal Research, vol. 31, pp. 1–661.

# Rectuvigerina mexicana

## (Cushman), 1926

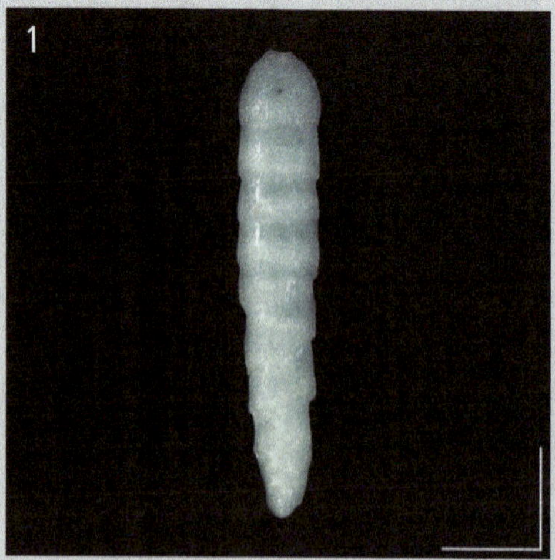

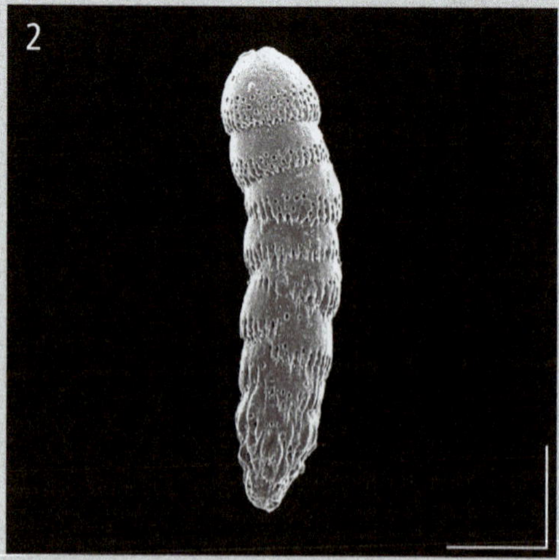

1. *Rectuvigerina mexicana.* Gulf of Mexico; Eureka Core E68-151A, depth: 4628 feet. Lateral view, NHMUK PM DI 0572, scale bar = 132 μm.
2. *Rectuvigerina mexicana.* Gulf of Mexico; Eureka Core E68-151A, depth: 4628 feet. Lateral view, NHMUK PM DI 0573, scale bar = 131 μm.

### Original Designation

*Siphogenerina mexicana* Cushman, 1926, p. 15, pl. 5, fig. 4.

### Type Specimen

Holotype (No. 353174) deposited in the collections of the United States National Museum, Washington, DC. Holotype not found in February 1984, presumed lost.

### Type Level

Oligocene (van Morkhoven et al., 1986), Alazan Clay. Originally given as Upper Eocene by Cushman (1926).

### Type Locality

Rio Buena Vista, 0.5 kilometer and 25° East from Tumbadero Hacienda House, Vera Cruz, Mexico.

### Description

Test forms an initial trochospire, later becoming uniserial and cylindrical, with circular cross-section. The chambers are inflated (up to eight in the uniserial portion), increase gradually in size, and are separated by distinct, depressed sutures. Chamber walls are calcareous, coarsely perforate, and ornamented by fine, short, longitudinal striae in the lower portion of the test. Upper chambers are smooth, and translucent. The primary aperture is terminal and round, flush or at the end of a slight neck, with a phialine lip, and with an internal toothplate.

### Synonyms and Citations

*Rectuvigerina mexicana* (Cushman).–van Morkhoven et al., 1986, p. 150, pl. 49.–Katz and Miller, 1993, pl. 1, fig. 3.–Robertson, 1998, p. 142, pl. 55, fig. 5.

### Biogeography

Worldwide. Recorded from Puerto Rico, Trinidad, the Gulf of Mexico, the North Atlantic, and Mexico (van Morkhoven et al., 1986).

### Bathymetry

Bathyal.

### Chronostratigraphy

Middle Eocene (Lutetian) to Late Miocene (Tortonian).

## References


Cushman, J.A., 1926. Foraminifera of the genera Siphogenerina and Pavonina. Proceedings of the United States National Museum, vol. 67, no. 25, pp. 1–24.

Katz, M.E. and Miller, K.G., 1993, Latest Oligocene to Earliest Pliocene benthic foraminiferal biofacies of the northeastern Gulf of Mexico. Micropaleontology, vol. 39, no. 4, pp. 367–403.

Robertson, B.E., 1998, Systematics and paleoecology of the benthic foraminiferida from the Buff Bay section, Miocene of Jamaica. Micropaleontology, vol. 44, supplement 2, pp. 1–266.

van Morkhoven, F.P.C.M., Berggren, W.A., Edwards, A.S., et al., 1986, Cenozoic cosmopolitan deep-water benthic foraminifera. Bulletin des Centres de Recherches Exploration-Production Elf-Aquitaine, Mémoire, vol. 11, pp. 1–421.

# *Rectuvigerina multicostata*

## (Cushman and Jarvis), 1929

1. *Rectuvigerina multicostata*. Gulf of Mexico; Eureka Core E68-151A, depth: 4574 feet. Lateral view, NHMUK PM DI 0574, scale bar = 186 μm.

### Original Designation

*Siphogenerina multicostata* Cushman and Jarvis, 1929, p. 14, pl. 3, fig. 6.

### Type Specimen

Holotype (CC 10056) deposited in the collections of the United States National Museum, Washington, DC.

### Type Level

Probably Miocene (van Morkhoven et al., 1986). Originally given as Tertiary (Eocene or Miocene) "Green clay" by Cushman and Jarvis (1929).

### Type Locality

Cipero Section, Trinidad.

### Description

Test forms an initially triserial portion, later becoming an elongate uniserial series with a circular cross-section. The inflated chambers (approximately three in uniserial portion) increase gradually in size, and are separated by distinct, depressed sutures. Chamber walls are calcareous, finely perforate, and ornamented by numerous longitudinal costae. Costae are restricted to individual chambers, and may form short spines at the base of the early chambers. A basal spine is present in well-preserved specimens. The primary aperture is a terminal, round opening at the end of a cylindrical neck, with a phialine lip and internal toothplate.

### Synonyms (following van Morkhoven et al., 1986) and Citations

*Siphogenerina multicostata* Cushman and Jarvis var. *optima*. –Cushman, 1943, p. 91, pl. 16, figs. 9–10.

*Siphogenerina spinea* Bermúdez, 1949, p. 223, pl. 13, figs. 78–79.

*Rectuvigerina royoi* Bermúdez and Fuenmayor. –Boltovskoy, 1978, p. 167, pl. 6, figs. 38–39.

*Rectuvigerina multicostata* (Cushman and Jarvis). –van Morkhoven et al., 1986, p. 115, pl. 36. –Loeblich and Tappan, 1987, p. 519, pl. 569, figs. 28–29. –Katz and Miller, 1993, pl. 1, fig. 2. –Robertson, 1998, p. 144, pl. 55, figs. 6–7.

### Remarks

Type species of the genus *Rectuvigerina* Mathews, 1945.

According to van Morkhoven et al. (1986), megalospheric forms have their greatest breadth in the middle part of the test, in contrast to microspheric forms, which are broadest in the final chamber. Megalospheric forms are also generally smaller and have more-depressed sutures.

## Biogeography

Worldwide. Recorded from Cuba, Puerto Rico, the Dominican Republic, Barbados, Trinidad, Venezuela, California, the Gulf of Mexico, the Indian Ocean, the North Atlantic, the Tasman Sea, the Coral Sea, and Chatham Rise (van Morkhoven et al., 1986).

## Bathymetry

Bathyal. Abundant in areas receiving high organic export flux from enhanced surface productivity.

## Chronostratigraphy

Late Oligocene to Pliocene.

# References


Bermúdez, P.J., 1949, Tertiary smaller foraminifera of the Dominican Republic. Special Publications of the Cushman Laboratory for Foraminiferal Research, no. 25, pp. 1–322.

Boltovskoy, E., 1978, Late Cenozoic benthonic foraminifera of the Ninetyeast Ridge (Indian Ocean). Marine Geology, vol. 26, pp. 139–175.

Cushman, J.A., 1943, Some new foraminifera from the Tertiary of the Island of St. Croix. Contributions from the Cushman Laboratory for Foraminiferal Research, vol. 19, pp. 90–93.

Cushman, J.A. and Jarvis, P.W., 1929, New foraminifera from Trinidad. Contributions from the Cushman Laboratory for Foraminiferal Research, vol. 5, pp. 6–17.

Katz, M.E. and Miller, K.G., 1993, Latest Oligocene to Earliest Pliocene benthic foraminiferal biofacies of the northeastern Gulf of Mexico. Micropaleontology, vol. 39, no. 4, pp. 367–403.

Loeblich, A.R. Jr and Tappan, H., 1987, Foraminiferal Genera and Their Classification. New York: Van Nostrand Reinhold, 970 pp.

Robertson, B.E., 1998, Systematics and paleoecology of the benthic foraminiferida from the Buff Bay section, Miocene of Jamaica. Micropaleontology, vol. 44, supplement 2, pp. 1–266.

van Morkhoven, F.P.C.M., Berggren, W.A., Edwards, A.S., et al., 1986, Cenozoic cosmopolitan deep-water benthic foraminifera. Bulletin des Centres de Recherches Exploration-Production Elf-Aquitaine, Mémoire, vol. 11, pp. 1–421.

# Rectuvigerina nodifera

## (Cushman and Kleinpell), 1934

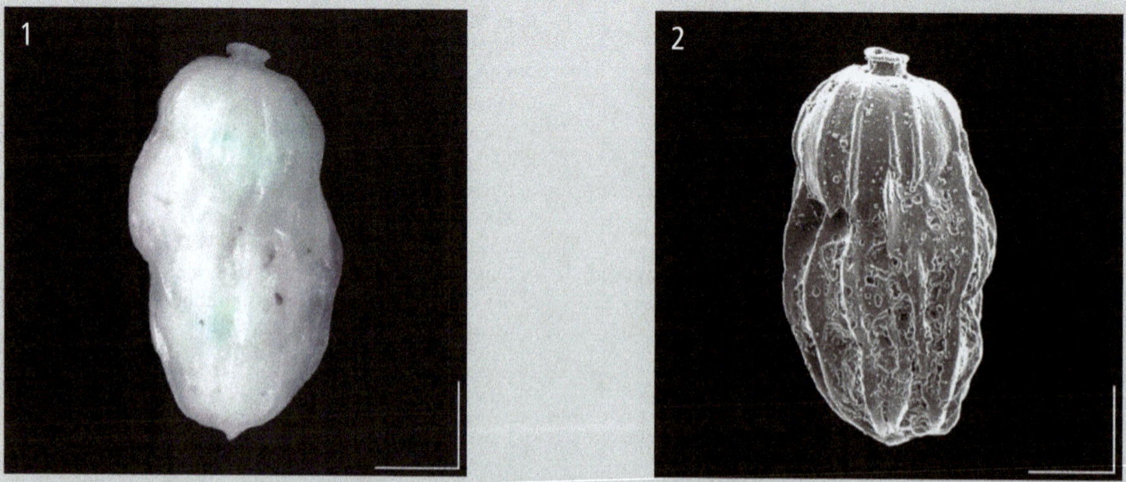

1. *Rectuvigerina nodifera.* Gulf of Mexico; Eureka Core E68-136, depth: 2824 feet. Lateral view, NHMUK PM DI 0575, scale bar = 157 μm.
2. *Rectuvigerina nodifera.* Gulf of Mexico; Eureka Core E68-136, depth: 2824 feet. Lateral view, NHMUK PM DI 0576, scale bar = 163 μm.

### Original Designation

*Siphogenerina nodifera* Cushman and Kleinpell, 1934, p. 13, pl. 2, figs. 15–16.

### Type Specimen

Holotype (No. 20141) deposited in the collections of the United States National Museum, Washington, DC.

### Type Level

Miocene, more than 160 feet above base of Temblor Formation.

### Type Locality

Carneros, Creek, California.

### Description

Test forms an initial triserial series, approximately two and a half times as long as broad, later becoming an elongate uniserial series with a circular cross-section. The inflated chambers increase gradually in size, and are separated by distinct, depressed sutures. Chamber walls are calcareous, finely perforate, and ornamented by ten to twelve prominent, thick, flange-like, longitudinal costae that extend from the base of the test to about halfway up the ultimate chamber. The primary aperture is a terminal, round opening, with a phialine lip and internal toothplate, situated at the end of a short, cylindrical neck.

### Synonyms and Citations

*Rectuvigerina nodifera* (Cushman and Kleinpell).–van Morkhoven et al., 1986, p. 240, pl. 81.–Katz and Miller, 1993, pl. 4, fig. 1.

### Remarks

The flange-like, longitudinal costae are very characteristic of *Rectuvigerina nodifera* (Cushman and Kleinpell).

### Biogeography

Worldwide. Recorded from the Gulf Coast, California, Venezuela, Ecuador, the North Atlantic, Europe, the eastern Pacific, and the Gulf of Mexico (van Morkhoven et al., 1986).

### Bathymetry

Primarily upper bathyal (van Morkhoven et al., 1986).

### Chronostratigraphy

Early Oligocene to Early Miocene.

## References


Cushman, J.A. and Kleinpell, R.M., 1934, New and unrecorded foraminifera from the California Miocene. Contributions from the Cushman Laboratory for Foraminiferal Research, vol. 10, pp. 1–23.

Katz, M.E. and Miller, K.G., 1993, Latest Oligocene to Earliest Pliocene benthic foraminiferal biofacies of the northeastern Gulf of Mexico. Micropaleontology, vol. 39, no. 4, pp. 367–403.

van Morkhoven, F.P.C.M., Berggren, W.A., Edwards, A.S., et al., 1986, Cenozoic cosmopolitan deep-water benthic foraminifera. Bulletin des Centres de Recherches Exploration-Production Elf-Aquitaine, Mémoire, vol. 11, pp. 1–421.

# *Rectuvigerina stonei*

## (Bermúdez), 1949

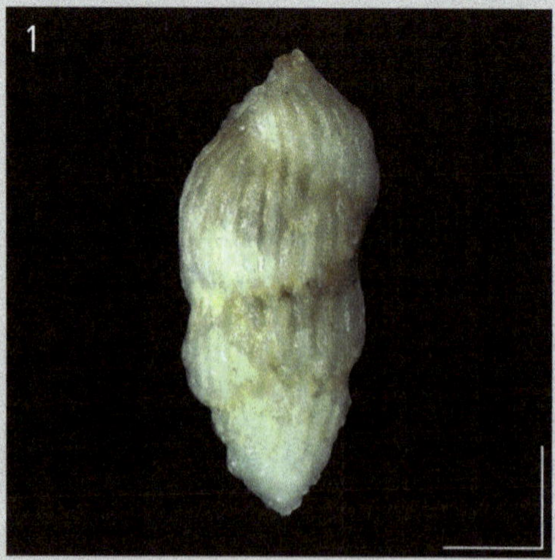


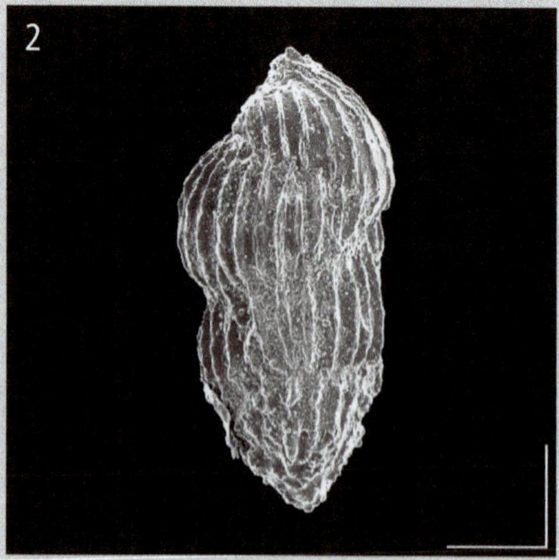


1. *Rectuvigerina stonei.* Chevron Main Pass Block 254, offshore Alabama, Gulf of Mexico; Well OCS-G 1655 no. 1, water depth: 92 metres, sample depth: 2185 meters below sea floor. Sidewall core. Lateral view, NHMUK PM DI 0577, scale bar = 182 μm.

2. *Rectuvigerina stonei.* Holmes Reef, Queensland Plateau, Northeast Australian Margin; ODP Leg 133, Hole 811A, Core 12, Section 5, 50–54 cm. Lateral view, NHMUK PM DI 0578, scale bar = 202 μm.

### Original Designation
*Siphogenerina stonei* Bermúdez, 1949, p. 223, pl. 14, figs. 19–20.

### Type Specimen
Holotype (CC 62575) and paratype (CC 62576) deposited in the collections of the United States National Museum, Washington, DC.

### Type Level
Miocene (van Morkhoven et al., 1986), Sombrerito Formation; tan, massive, calcareous, silty sandy, fossiliferous clay shale. Originally given as Middle Oligocene by Bermúdez (1949).

### Type Locality
Sample H-1077, from the Cañada La Horqueta, a small arroyo confluent of the Rio del Medio, approximately 4 kilometers northeast of Padre de las Casas village, Azua Province, Dominican Republic.

### Description
Test forms an initial triserial series, followed by a uniserial series, with a circular cross-section. The inflated chambers (approximately three or four in uniserial stage) increase gradually in size, and are separated by depressed sutures. Chamber walls are calcareous, finely perforate, and ornamented by longitudinal, sharp costae, extending as short, stout spines in the initial chambers, and becoming twisted in a single direction just below the apertural neck. The primary aperture is a round opening, with a phialine lip, and an internal toothplate at the end of a short, cylindrical neck.

### Synonyms and Citations
?*Uvigerina longa* Cushman and Bermúdez, 1937, p. 109, pl. 16, figs. 5–6.

*Rectuvigerina stonei* (Bermúdez). –van Morkhoven et al., 1986, p. 215, pl. 73.

### Biogeography
Worldwide. Recorded from the North and South Atlantic and the East Pacific (as *Rectuvigerina* sp. 1) by van Morkhoven (1981) and as *Rectuvigerina stonei* from the Gulf of Mexico (van Morkhoven et al., 1986). Also recorded from the North Atlantic by Katz et al. (2003) and along the Wilkes Land Continental Margin (Escutia et al., 2011).

### Bathymetry

Lower neritic to bathyal.

### Chronostratigraphy

Late Oligocene to Early Miocene.

## References

Bermúdez, P.J., 1949, Tertiary smaller foraminifera of the Dominican Republic. Special Publications of the Cushman Laboratory for Foraminiferal Research, no. 25, pp. 1–322.

Cushman, J.A. and Bermúdez, P.J., 1937, Further new species of foraminifera from the Eocene of Cuba. Contributions from the Cushman Laboratory for Foraminiferal Research, vol. 13, pp. 1–29.

Escutia, C., Brinkhuis, H., Klaus, A., and the Expedition 318 Scientists, 2011, Proc. IODP, 318: Tokyo (Integrated Ocean Drilling Program Management International, Inc.). doi:10.2204/iodp.proc.318.104.2011.

Katz, M.E., Tjalsma, R.C., and Miller, K.G., 2003, Oligocene bathyal to abyssal benthic foraminifera of the Atlantic Ocean. Micropaleontology, vol. 49, supplement 2, pp. 1–45.

van Morkhoven, F.P.C.M., 1981, Cosmopolitan Tertiary bathyal benthic foraminifera. Transactions of the Gulf Coast Association of Geological Societies, Supplement 31, 445 abstracts with accompanying range chart.

van Morkhoven, F.P.C.M., Berggren, W.A., Edwards, A.S., et al., 1986, Cenozoic cosmopolitan deep-water benthic foraminifera. Bulletin des Centres de Recherches Exploration-Production Elf-Aquitaine, Mémoire, vol. 11, pp. 1–421.

# Rectuvigerina striata

## (Schwager), 1866

1. *Rectuvigerina striata.* Holmes Reef, Queensland Plateau, Northeast Australian Margin; ODP Leg 133, Hole 811A, Core 12, Section 5, 50–54 cm. Lateral view, NHMUK PM DI 0579, scale bar = 137 μm.

2. *Rectuvigerina striata.* Holmes Reef, Queensland Plateau, Northeast Australian Margin; ODP Leg 133, Hole 811A, Core 12, Section 5, 50–54 cm. Lateral view, NHMUK PM DI 0580, scale bar = 145 μm.

### Original Designation

*Dimorphina striata* Schwager, 1866, p. 251, pl. 7, figs. 2, 99.

### Type Specimen

Depository not given. Neotype (Mf 41) designated by Srinivasan and Sharma (1980) deposited in the Micropaleontology Section, Department of Geology, Banaras Hindu University, Banaras, India.

### Type Level

Pliocene (van Morkhoven et al., 1986). Originally given as Late Tertiary by Schwager (1866).

### Type Locality

Car Nicobar, Andaman Sea.

### Description

Test forms an initially triserial series, followed by a uniserial series; subcylindrical or tapered in outline, with a circular cross-section. The inflated chambers increase gradually in size, and are separated by distinct, depressed sutures that are crenulate in the final part of the test. A basal spine is present in rare specimens. Chamber walls are calcareous, finely perforate, and ornamented by fine, discontinuous, longitudinal costae. The primary aperture is terminal, round, and situated at the end of a cylindrical neck, with a phialine lip and an internal toothplate.

### Synonyms and Citations

*Siphogenerina rerensis* Finlay, 1939, p. 108, pl. 11, fig. 8.

*Siphogenerina kugleri* Cushman and Renz, 1941, p. 2, pl. 3, fig. 23.

*Sagrina nodosa* Parker and Jones. –Brady, 1884, p. 583, pl. 114, fig. 18.

*Rectuvigerina nicoli* Mathews, 1945, p. 593. –Barker, 1960, pl. 114, fig. 18. –Jones, 1994, p. 113, pl. 114, fig. 18.

*Rectuvigerina striata* (Schwager). –Srinivasan and Sharma, 1980, p. 51, pl. 7, fig. 13. –van Morkhoven et al., 1986, p. 110, pl. 34. –van Marle, 1991, p. 98, pl. 7, figs. 4–5. –Robertson, 1998, p. 146, pl. 56, figs. 1–2. –Hayward, 2002, p. 300, pl. 1, figs. 12–13.

### Remarks

Microspheric forms have distinctly triangular chambers in the triserial part, in contrast to megalospheric forms, which have a broadly rounded initial portion.

Revised stratigraphy at the Eureka boreholes in the Gulf of Mexico (Katz and Miller, 1993) extended the first occurrence of this species to the base of Biozone P22, in contrast to the Biozone M6 occurrence reported by van Morkhoven et al. (1986).

### Biogeography

Worldwide. Recorded from Madagascar, the Pacific, the North and South Atlantic, the Bismarck Sea, Papua New Guinea, Java, Borneo, the Tasman Sea, the Coral Sea, Chatham Rise, Okinawa, the Gulf of Mexico, and the Caribbean (van Morkhoven et al., 1986).

### Bathymetry

Primarily upper to middle bathyal (van Morkhoven et al., 1986).

### Chronostratigraphy

Late Oligocene (Biozone P22) to Recent.

## References


Barker, R.W., 1960, Taxonomic notes on the species figured by H.B. Brady in his report on the foraminifera dredged by H.M.S. Challenger during the years 1873–1876, Society of Economic Paleontologists and Mineralogists Special Publication, vol. 9, pp. 1–238.

Brady, H.B., 1884, Report on the foraminifera dredged by H.M.S. Challenger during the years 1873–1876. In Brady, H.B., ed., Report on the Scientific Results of the Voyage of the H.M.S. Challenger during the years 1873–1876. Zoology, vol. 9, pp. 1–814.

Cushman, J.A. and Renz, H.H., 1941, New Oligocene–Miocene foraminifera from Venezuela. Contributions from the Cushman Laboratory for Foraminiferal Research, vol. 17, pp. 1–27.

Finlay, H.J., 1939, New Zealand foraminifera: key species in stratigraphy – No. 1. Transactions of the Royal Society of New Zealand, vol. 68, no. 4, pp. 504–543.

Hayward, B., 2002, Late Pliocene to Middle Pleistocene extinctions of deep-sea benthic foraminifera ("Stilostomella extinction") in the southwest Pacific. Journal of Foraminiferal Research, vol. 32, pp. 274–307.

Jones, R.W., 1994, The Challenger Foraminifera. Oxford: Oxford University Press, 149 pp.

Katz, M.E. and Miller, K.G., 1993, Latest Oligocene to Earliest Pliocene benthic foraminiferal biofacies of the northeastern Gulf of Mexico. Micropaleontology, vol. 39, no. 4, pp. 367–403.

Mathews, R.D., 1945, Rectuvigerina, a new genus of foraminifera from a restudy of Siphogenerina. Journal of Paleontology, vol. 19, no. 6, pp. 588–606.

Robertson, B.E., 1998, Systematics and paleoecology of the benthic foraminiferida from the Buff Bay section, Miocene of Jamaica. Micropaleontology, vol. 44, supplement 2, pp. 1–266.

Schwager, C., 1866, Fossile Foraminiferen von Kar Nikobar, Reise der Österreichischen Fregatte Novara um Erde in den Jahren 1857, 1858, 1859 unten den Befehlen des Commodore B. Von Wüllerstorf-Urbair. Geologischer Theil, vol. 2, no. 1, Geologische Beobachtung, no. 2, Paläontologische Mittheilung, pp. 187–268.

Srinivasan, M.S. and Sharma, V., 1980, Schwager's Car Nicobar Foraminifera in the Reports of the Novara Expedition – A Revision. New Delhi: Today & Tomorrow's Printers and Publishers, 83 pp.

van Marle, L.J., 1991, Eastern Indonesian Late Cenozoic smaller benthic foraminifera. Verhandelingen der Koninklijke Nederlanse Akademie van Wetenschappen, Afdeeling Natuurkunde, Eerste Reeks, vol. 34, pp. 1–328.

van Morkhoven, F.P.C.M., Berggren, W.A., Edwards, A.S., et al., 1986, Cenozoic cosmopolitan deep-water benthic foraminifera. Bulletin des Centres de Recherches Exploration-Production Elf-Aquitaine, Mémoire, vol. 11, pp. 1–421.

# *Recurvoides contortus*

## Earland, 1934

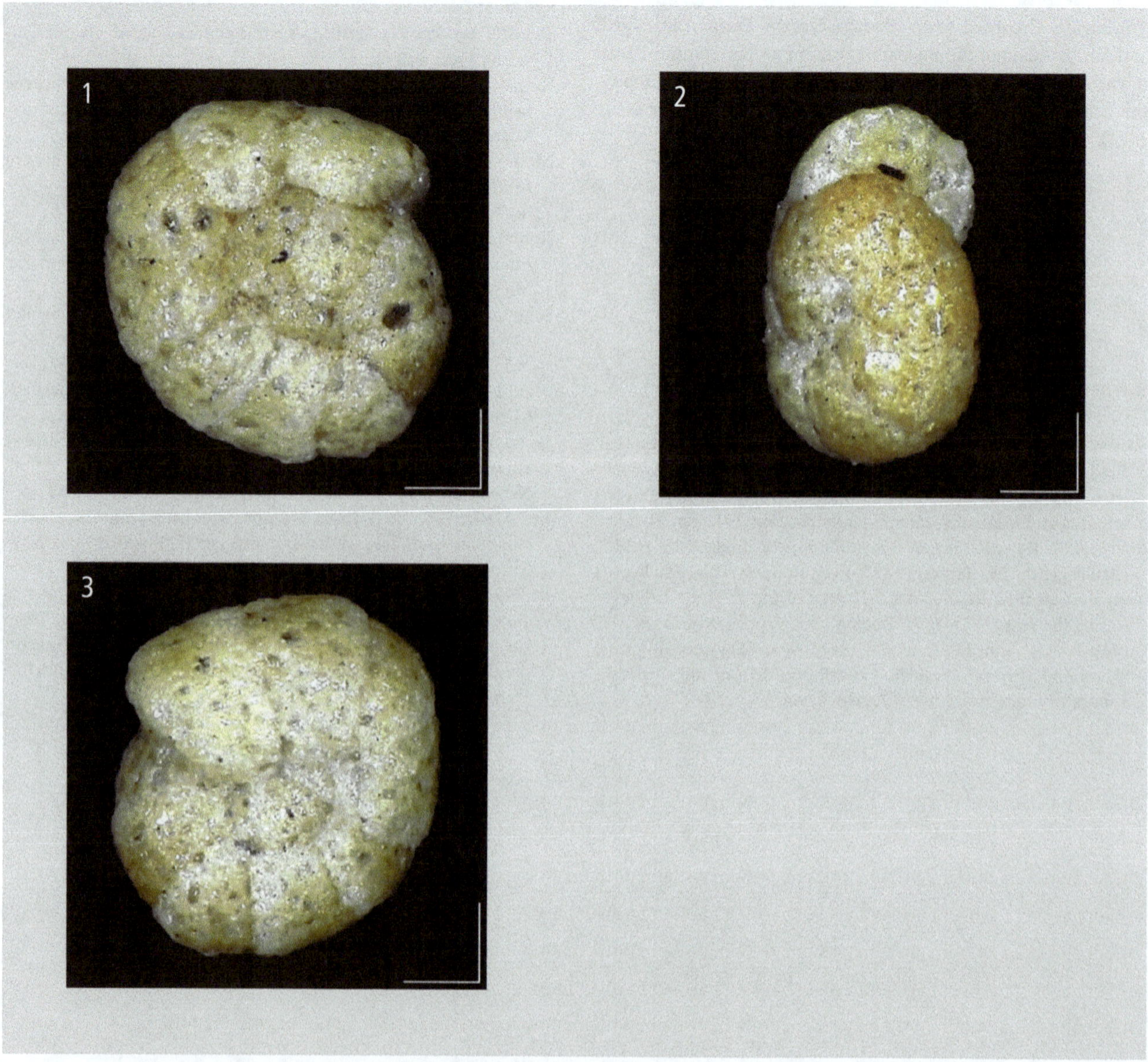


1. *Recurvoides contortus.* Off Cape Bowles, Clarence Island, Antarctic; "Discovery" Station 170. Deposited in the micropaleontological collections of The Natural History Museum, London; NHMUK ZF3470. Lateral view, NHMUK PM DI 0581, scale bar = 224 μm.

2. *Recurvoides contortus.* Off Cape Bowles, Clarence Island, Antarctic; "Discovery" Station 170. Deposited in the micropaleontological collections of The Natural History Museum, London; NHMUK ZF3470. Apertural view, NHMUK PM DI 0582, scale bar = 170 μm.

3. *Recurvoides contortus.* Off Cape Bowles, Clarence Island, Antarctic; "Discovery" Station 170. Deposited in the micropaleontological collections of The Natural History Museum, London; NHMUK ZF3470. Lateral view, NHMUK PM DI 0583, scale bar = 195 μm.

## Original Designation

*Recurvoides contortus* Earland, 1934, p. 91, pl. 10, figs. 7–19.

## Type Specimen

Figured specimens are housed in the "type slide" collection (ZF3470) of Heron-Allen and Earland at The Natural History Museum, London, UK.

## Type Level

Recent.

## Type Locality

Not designated. Localities given: 65 R.R.S. "Discovery" and "William Scoresby" stations from latitude 55° 53′ S to 70° 34′ S and from longitude 19° 36′ W to 100° 39′ W, Antarctic region, between 50 and 4224 meters water depth.

## Description

Test exhibits a "streptospiral recurvoidiform" type of coiling (Kaminski and Gradstein, 2005), with a tendency to become trochospiral to planispiral in the last whorl; test has a subcircular outline with rounded periphery. The umbilical area is depressed on one side of the test, and flattened or slightly convex on the other side. Chambers are moderately inflated (approximately ten in the last whorl), and are separated by slightly depressed sutures. Chamber walls are agglutinated, thin, and often roughly finished. The primary aperture is a small areal opening, with a distinct lip.

## Synonyms (following Kaminski and Gradstein, 2005) and Citations

*Haplophragmoides scitula* (Brady). –Wiesner, 1931, p. 96, pl. 12, fig. 141.

*Haplophragmoides scitulum* (Brady) "oval variety". –Earland, 1933, pl. 3, figs. 11–12.

*Recurvoides laevigatum* Höglund, 1947, p. 150, textfigs. 117–118, pl. 11, fig. 6.

*Recurvoides contortus gurgitis* Saidova, 1961, p. 27, pl. 8, fig. 38.

*Recurvoides contortus* Earland. –Loeblich and Tappan, 1987, p. 83, pl. 68, figs. 7–9. Kaminski and Gradstein, 2005, pp. 405–408, fig. 96, pl. 96.

## Remarks

Type species of the genus *Recurvoides* Earland, 1934.

*Recurvoides contortus* Earland differs from *Labrospira scitula* (Brady), which is planispirally coiled, and has two depressed umbilici. Kaminski and Gradstein (2005) considered the Northwest Pacific abyssal form *Recurvoides contortus gurgitis* Saidova to be identical to *Recurvoides contortus*, whereas the shallower form *Recurvoides contortus sublittoralis* Saidova differed in its mode of coiling, number of chambers in the last whorl, and wall texture.

## Biogeography

Worldwide. See Observed Occurrences in Kaminski and Gradstein (2005).

## Bathymetry

Bathyal to abyssal; neritic at high latitudes (Kaminski and Gradstein, 2005).

## Chronostratigraphy

Recent (Kaminski and Gradstein, 2005).

## References


Earland, A., 1933, Foraminifera, Part II: South Georgia. Discovery Reports, vol. 7, pp. 27–138.

Earland, A., 1934, Foraminifera, Part III: the Falklands sector of the Antarctic (excluding South Georgia). Discovery Reports, vol. 10, pp. 1–208.

Höglund, H., 1947, Foraminifera in the Gullmar Fjord and the Skagerak. Zoologiska Bidrag Från Uppsala, vol. 26, pp. 1–328.

Loeblich, A.R. Jr and Tappan, H., 1987, Foraminiferal Genera and Their Classification. New York: Van Nostrand Reinhold, 970 pp.

Kaminski, M.A. and Gradstein, F.M., 2005, Atlas of Palaeogene Cosmopolitan Deep-Water Agglutinated Foraminifera. Grzybowski Foundation Special Publication, no. 10, 547 + vii pp.

Saidova, Kh.M., 1961, Ekologiya foraminifer i paleogeografiya dal'nevostochnykh morey SSSR, I severo-zapadnoy chasti Tikhogo Okeans. Moscow: Instituta Okeanologii, Akademiya Nauk SSSR, 232 pp.

Wiesner, H., 1931, Die Foraminiferen der Deutsche Südpolar Expedition 1901–1903. Deutsche Südpolar Expedition, vol. 20, 165 pp.

# *Reophanus berggreni*

## Gradstein and Kaminski, 1997

1. *Reophanus berggreni.* North Sea; Phillips (UK) Well 22.14-1x, 7090 feet. Deposited in the micropaleontological collections of The Natural History Museum, London; NHMUK PF62943. Holotype. Lateral view, NHMUK PM DI 0584, scale bar = 688 μm.

### Original Designation

*Reophanus berggreni* Gradstein and Kaminski, 1997, p. 217, textfig. 4, pl. 5, figs. 1–6.

### Type Specimen

Holotype (PF62943) is housed in the micropaleontological collections of The Natural History Museum, London, UK.

### Type Level

Lower Oligocene.

### Type Locality

Holotype is from well Phillips (UK) 22/14–1x, 7090 feet, North Sea. Paratypes are from well Phillips (UK) 16/29–2x, 7020 feet, and from Amoco Norway 2/8–1 at 7200–7500 feet.

### Description

Test is comprised of massive, spherical to ovate chambers, arranged in a rectilinear series. Chambers are connected by short, thick stolons. Chamber walls are thick, finely agglutinated, of several grains in thickness. The primary aperture is round, and at the open end of the stolon.

### Synonyms and Citations

*Hormosinella carpenteri* Brady. –Charnock and Jones, 1990, p. 163, pl. 4, figs. 8–9; pl. 15, fig. 10.

*Reophanus berggreni* Gradstein and Kaminski. –Kaminski and Gradstein, 2005, pp. 263–265, fig. 50, pl. 50.

### Remarks

Specimens mostly observed as individual chambers with stolons at either end of the test. This species occurs consistently in the coarse fraction (over 500 μm) in the North Sea and offshore Norway (Gradstein and Kaminski, 1997).

### Biogeography

North Sea and offshore Norway. Also reported from the Beaufort-Mackenzie Basin and Celebes Sea (ODP Hole 767C), but not from the Labrador Sea (Kaminski and Gradstein, 2005).

### Bathymetry

Primarily bathyal. Not observed in abyssal ODP sites in the Norwegian Sea and North Atlantic (Kaminski and Gradstein, 2005).

### Chronostratigraphy

Middle Eocene to Middle Miocene. Middle Eocene (Biozone NSR6) to mid-Oligocene (Biozone NSR7B) in North Sea; ranges to Middle Miocene in Arctic (Kaminski and Gradstein, 2005).

## References


Charnock, M.A. and Jones, R.W., 1990, Agglutinated foraminifera from the Palaeogene of the North Sea. In Hemleben, C., Kaminski, M.A., Kuhnt, W., and Scott, D.B., ed., Paleoecology, Biostratigraphy, Paleoceanography and Taxonomy of Agglutinated Foraminifera, NATO ASI Series, C: Mathematical and Physical Sciences. Dordrecht: Kluwer Academic, pp. 139–244.

Gradstein, F.M. and Kaminski, M.A., 1997, New species of Paleogene deep-water agglutinated foraminifera from the North Sea and Norwegian Sea. Annales Societatis Geologorum Poloniae, vol. 67, pp. 217–229.

Kaminski, M.A. and Gradstein, F.M., 2005, Atlas of Palaeogene Cosmopolitan Deep-Water Agglutinated Foraminifera. Grzybowski Foundation Special Publication, no. 10, 547 + vii pp.

# Reophax pilulifer

## Brady, 1884

1. *Reophax pilulifer.* North Atlantic Ocean; "Porcupine" Station 31, depth: 1360 fathoms. Deposited in the micropaleontological collections of The Natural History Museum, London; NHMUK ZF4878 from slide ZF2284. Lectotype. Lateral view, NHMUK PM DI 0585, scale bar = 335 μm.

### Original Designation

*Reophax pilulifera* Brady, 1884, p. 292, p. 30, figs. 18–20.

### Type Specimen

Figured specimens are housed in the micropaleontological collections of The Natural History Museum, London, UK. Lectotype (ZF4878) was designated by Kaminski and Gradstein (2005) from slide ZF2284.

### Type Level

Recent.

### Type Locality

"Porcupine" Station 31 in the North Atlantic, 1360 fathoms.

### Description

Test forms an elongate, uniserial series; rectilinear or slightly arcuate in shape, and composed of three to five large, globular to pyriform chambers that increase rapidly in size. Later chambers are slightly overlapping. Chamber walls are thin, agglutinated of a single layer of coarse grains, and smoothly finished with abundant organic cement. The primary aperture is a terminal, round opening.

### Synonyms and Citations

*Hormosina pilulifera* (Brady). –Charnock and Jones, 1990, p. 162, pl. 4, fig. 5; pl. 15, fig. 7. –Jones, 1994, p. 37, pl. 30, figs. 18–20.

*Reophax pilulifer* Brady. –Barker, 1960, pl. 30, figs. 18–20. –Kaminski and Gradstein, 2005, pp. 272–274, fig. 53, pl. 53.

### Remarks

Displays considerable variation in size and in the composition of agglutinated particles.

### Biogeography

Worldwide. See detailed occurrences in Kaminski and Gradstein (2005).

### Bathymetry

Bathyal to abyssal. Reported between 400 and 2435 fathoms at "Challenger" stations (Jones, 1994).

### Chronostratigraphy

Late Cretaceous to Recent (Kaminski and Gradstein, 2005).

## References

Barker, R.W., 1960, Taxonomic notes on the species figured by H.B. Brady in his report on the foraminifera dredged by H.M.S. Challenger during the years 1873–1876. Society of Economic Paleontologists and Mineralogists Special Publication, vol. 9, pp. 1–238.

Brady, H.B., 1884, Report on the foraminifera dredged by H.M.S. Challenger during the years 1873–1876. In Brady, H.B., ed., Report on the Scientific Results of the Voyage of the H.M.S. Challenger during the years 1873–1876. Zoology, vol. 9, pp. 1–814.

Charnock, M.A. and Jones, R.W., 1990, Agglutinated foraminifera from the Palaeogene of the North Sea. In Hemleben, C., Kaminski, M.A., Kuhnt, W., and Scott, D.B., ed., Paleoecology, Biostratigraphy, Paleoceanography and Taxonomy of Agglutinated Foraminifera, NATO ASI Series, C: Mathematical and Physical Sciences. Dordrecht: Kluwer Academic, pp. 139–244.

Jones, R.W., 1994, The Challenger Foraminifera. Oxford: Oxford University Press, 149 pp.

Kaminski, M.A. and Gradstein, F.M., 2005, Atlas of Palaeogene Cosmopolitan Deep-Water Agglutinated Foraminifera. Grzybowski Foundation Special Publication, no. 10, 547 + vii pp.

# Reophax subfusiformis

## Earland 1933 emend. Höglund, 1947

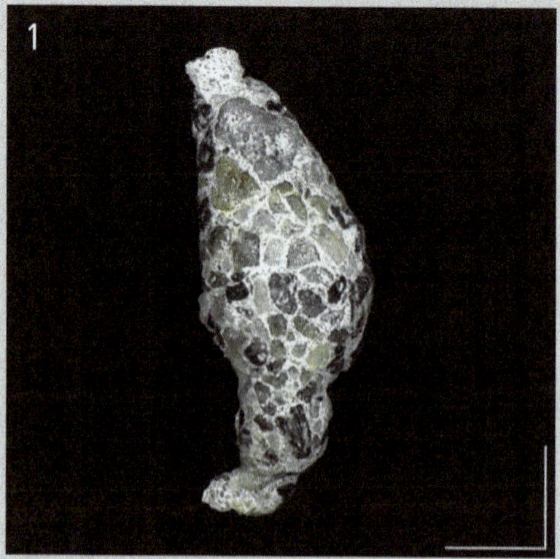


1. *Reophax subfusiformis.* South Georgia; "Discovery" Station D144, depth: 155–178 meters. Deposited in the micropaleontological collections of The Natural History Museum, London; NHMUK ZF3487. Lateral view, NHMUK PM DI 0586, scale bar = 444 μm.

### Original Designation

*Reophax subfusiformis* Earland, 1933, p. 74, pl. 2, figs. 16–19.

### Type Specimen

Figured specimens are housed in the Heron-Allen and Earland Collection at The Natural History Museum, London, UK (slide ZF3487).

### Type Level

Recent.

### Type Locality

Originally described by Earland (1933) from "Discovery" Station D144, 155–178 meters, near South Georgia. Höglund (1947) redescribed this species from the Gullmar Fjord, Skagerak and North Sea. Höglund's stations are between 30 and 220 meters water depth.

### Description

Test forms an elongate, straight or slightly arched, uniserial series, composed of two to four typically flattened chambers. The last chamber is markedly larger, and has an elongate, pyriform shape. Chambers increase rapidly in size, and are separated by depressed sutures. Chamber walls are agglutinated, and thin, with poorly sorted grains held in organic cement. The primary aperture is round, and terminal, opening at the end of a wide, extended neck.

### Synonyms and Citations

*Reophax subfusiformis* Earland emend. Höglund, 1947, p. 82, pl. 9, figs. 1–4, pl. 26, figs. 1–36, pl. 27, figs. 1–19, textfigs. 43–50. –Kaminski et al., 1988, p. 187, pl. 2, figs. 18–19. –Kaminski and Gradstein, 2005, pp. 275–277, fig. 54, pl. 54.

### Remarks

Earland's specimens from Georgia are up to 2.2 millimeters in length. Fossil specimens are generally smaller (<1 millimeter). Höglund regarded *Reophax subfusiformis* Earland as extremely variable, and included some of the specimens described by Goës (1894) as *Reophax scorpiurus* de Montfort and *Reophax pilulifer* Brady in the concept of the species.

### Biogeography

Primarily high latitudes (Southern Ocean, North Atlantic, North Sea, Norwegian–Greenland Sea, Labrador Sea, Labrador Margin); rarely recorded in the Tethyan region (Kaminski and Gradstein, 2005).

### Bathymetry

Neritic to abyssal.

### Chronostratigraphy

Late Cretaceous (Campanian) to Recent (Kaminski and Gradstein, 2005).

## References

Earland, A., 1933, Foraminifera. Part II. South Georgia Discovery Reports, vol. 7, pp. 27–138.

Goës, A., 1894, A synopsis of the Arctic and Scandinavian Recent marine Foraminifera hitherto discovered. Kongl. Svenska Vetenskaps Akademiens Handlingar, vol. 25, no. 9, pp. 3–127.

Höglund, H., 1947, Foraminifera in the Gullmar Fjord and the Skagerak. Zoologiska Bidrag Från Uppsala, vol. 26, pp. 1–328.

Kaminski, M.A. and Gradstein, F.M., 2005, Atlas of Palaeogene Cosmopolitan Deep-Water Agglutinated Foraminifera. Grzybowski Foundation Special Publication, no. 10, 547 + vii pp.

Kaminski, M.A., Gradstein, F.M., Berggren, W.A., Geroch, S., and Beckmann, J.P., 1988, Agglutinated foraminiferal assemblages from Trinidad: taxonomy, stratigraphy and paleobathymetry. In Gradstein, F.M. and Rögl, F., ed., Second International Workshop on Agglutinated Foraminifera, Vienna 1986. Abhandlungen der Geologischen Bundesanstalt, vol. 41, pp. 155–228.

# Reticulophragmium amplectens

## (Grzybowski), 1898

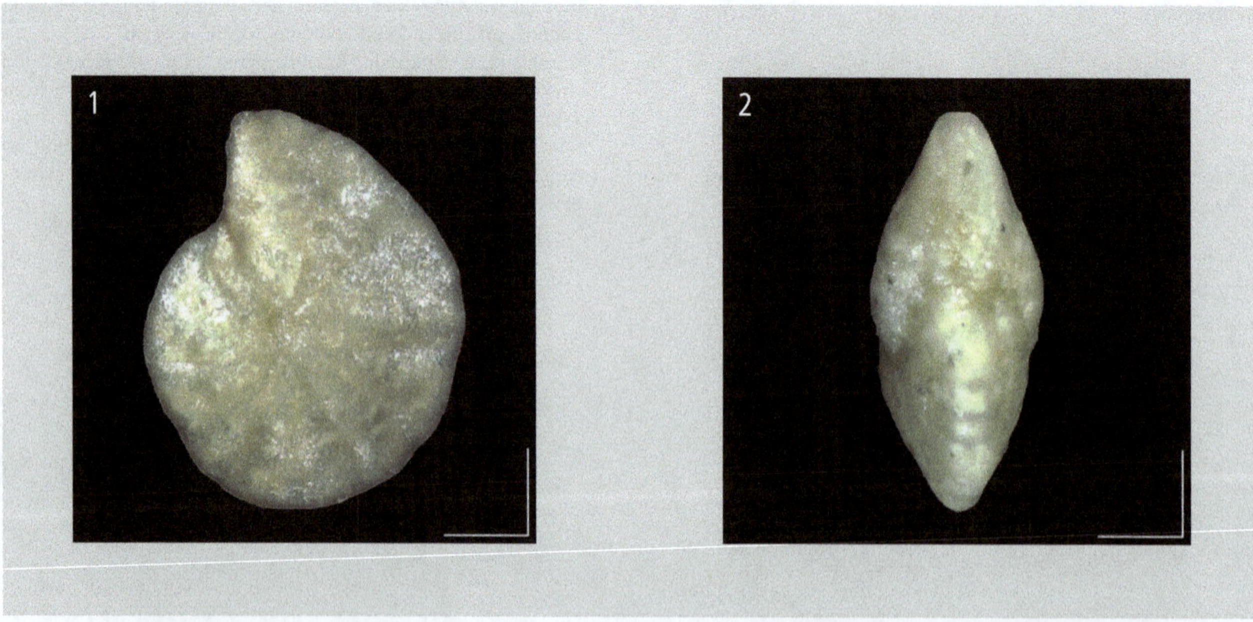

1. *Reticulophragmium amplectens.* Red shales, Biecz, Poland. Deposited in the Geroch Collection, Jagiellonian University, Kraków, Poland. Lateral view, NHMUK PM DI 0587, scale bar = 200 μm.

2. *Reticulophragmium amplectens.* Red shales, Biecz, Poland. Deposited in the Geroch Collection, Jagiellonian University, Kraków, Poland. Apertural view, NHMUK PM DI 0588, scale bar = 187 μm.

### Original Designation

*Cyclammina amplectens* Grzybowski, 1898, p. 292, pl. 12, figs. 1–3.

### Type Specimen

Lectotype (UJ-132-P, 1/77) designated by Kaminski and Geroch (1993), and deposited in the Grzybowski Collection at the Jagiellonian University, Kraków, Poland.

### Type Level

Middle Eocene, red and gray clay, Silesian unit of the Polish Carpathians.

### Type Locality

Wells drilled at Potok, Krosno district, Poland. Lectotype is from 190 meters in the Old Klobassa well no. 5.

### Description

Test forms an involute, biumbilicate planispire, with a circular or lobate outline, a subacute periphery, and a shallow, depressed umbilicus. Early chambers are usually without alveoles; subsequent chambers develop unbranched alveoles, often elongated in shape near the sutures and along the periphery. Sutures are straight or slightly curved, and flush in the early chambers, becoming slightly depressed in later whorls. Chamber walls are usually finely agglutinated with much cement, although larger agglutinated grains may be found in well-preserved specimens. The primary aperture is an interiomarginal slit, with a thin lip.

### Synonyms and Citations

*Cyclammina grangei* Finlay, 1940, p. 448, pl. 62, figs. 3–4.

*Cyclammina subcarpatica* Majzon, 1943, p. 156, pl. 2, fig. 8a–b.

*Cyclammina amplectens* (Grzybowski). –Geroch and Nowak, 1984, pl. 3, figs. 5–7, 9–10.

*Reticulophragmium amplectens* (Grzybowski). –King, 1989, p. 458, pl. 9.2, figs. 16–18.

*Cyclammina (Reticulophragmium) amplectens* Grzybowski. –Charnock and Jones, 1990, p. 176, pl. 8, figs. 1–5; pl. 18, fig. 3.

*Reticulophragmium amplectens* (Grzybowski). –Kaminski and Geroch, 1993, p. 266, pl. 11, figs. 5–7c. –Gradstein et al., 1994, pl. 4, figs. 1–5, pl. 5, figs. 1–10. –Kaminski and Gradstein, 2005, pp. 491–496, fig. 123, pl. 123.

### Remarks

Megalospheric forms typically possess ten to thirteen chambers in the last whorl, whereas microspheric forms have twelve to seventeen chambers; specimens display much variation in the position and amount of alveoles (Kaminski and Geroch, 1993).

### Biogeography

Worldwide. Recorded from the North Sea (King, 1989; Charnock and Jones, 1990; Gradstein et al., 1994), Poland, Czechoslovakia, Ukraine, Romania (Kaminski and Geroch, 1993), Tunisia (Charnock and Jones, 1990), and New Zealand (Hornibrook et al., 1989). See also Observed Occurrences in Kaminski and Gradstein (2005).

### Bathymetry

Bathyal to abyssal.

### Chronostratigraphy

Late Early Eocene to Early Oligocene in the Atlantic and Tethys, extending to Late Miocene in the Arctic (Kaminski and Gradstein, 2005). The partial range and optimum occurrence of *Reticulophragmium amplectens* (Grzybowski) characterises the Middle Eocene *Cyclammina amplectens* Biozone of Geroch and Nowak (1984).

## References


Charnock, M.A. and Jones, R.W., 1990, Agglutinated foraminifera from the Palaeogene of the North Sea. In Hemleben, C., Kaminski, M.A., Kuhnt, W., and Scott, D.B., ed., Paleoecology, Biostratigraphy, Paleoceanography and Taxonomy of Agglutinated Foraminifera, NATO ASI Series, C: Mathematical and Physical Sciences, Dordrecht: Kluwer Academic, pp. 139–244.

Finlay, H.J., 1940, New Zealand Foraminifera; key species in stratigraphy – No. 4. Transactions of the Royal Society of New Zealand, vol. 69, no. 4, pp. 448–472.

Geroch, S. and Nowak, W., 1984, Proposal of zonation for the late Tithonian–late Eocene, based upon arenaceous foraminifera from the outer Carpathians, Poland. In Oertli, H., ed., Benthos '83, 2nd International Symposium on Benthic Foraminifera, Pau (France), April 11–15, 1983. Elf Aquitaine, ESSO REP and TOTAL CFP, Pau and Bordeaux, pp. 225–239.

Gradstein, F.M., Kaminski, M.A., Berggren, W.A., and D'Iorio, M.A., 1994, Cenozoic biostratigraphy of the Central North Sea and Labrador Shelf. Micropaleontology, vol. 40 (Supplement), pp. 1–152.

Grzybowski, J., 1898, Otwornice pokładów naftonosnych okolicy Krosna. Rozprawy Wydziału Matematyczno-Przyrodniczego, Akademia Umiejętności w Krakowie, serya 2, vol. 33, pp. 257–305.

Hornibrook, N. de, Brazier, R.C., and Strong, C.P., 1989, Manual of New Zealand Permian to Pleistocene foraminiferal biostratigraphy. New Zealand Geological Survey, Paleontological Bulletin 1989, pp. 1–175.

Kaminski, M.A. and Geroch, S., 1993, A revision of foraminiferal species in the Grzybowski Collection. In Kaminski, M.A., Geroch, S., and Kaminski, D.G., ed., The Origins of Applied Micropalaeontology: The School of Jósef Grzybowski (with a taxonomic revision of the Grzyboswski Collection). Grzybowski Foundation Special Publication, no. 1, pp. 239–330.

Kaminski, M.A. and Gradstein, F.M., 2005, Atlas of Palaeogene Cosmopolitan Deep-Water Agglutinated Foraminifera. Grzybowski Foundation Special Publication, no. 10, 547 + vii pp.

King, C., 1989, Cenozoic of the North Sea. In Jenkins, D.G. and Murray, J.W., ed., Stratigraphical Atlas of Fossil Foraminifera, 2nd edn. Chichester: Ellis Horwood, pp. 418–489.

Majzon, L., 1943, Adatok egyes Kárpátaljai Flisrétegekhez, tekintettel a Globotruncanákra: Beiträge zur Kenntniss einiger Flysch-Schichten des Karpaten-Vorlandes mit Rücksicht auf die Globotruncanen. A magyar Királyi Földtani Intézet, Évkönnyve, vol. 37, pp. 1–170.

# *Reussella szajnochae*

## (Grzybowski), 1896

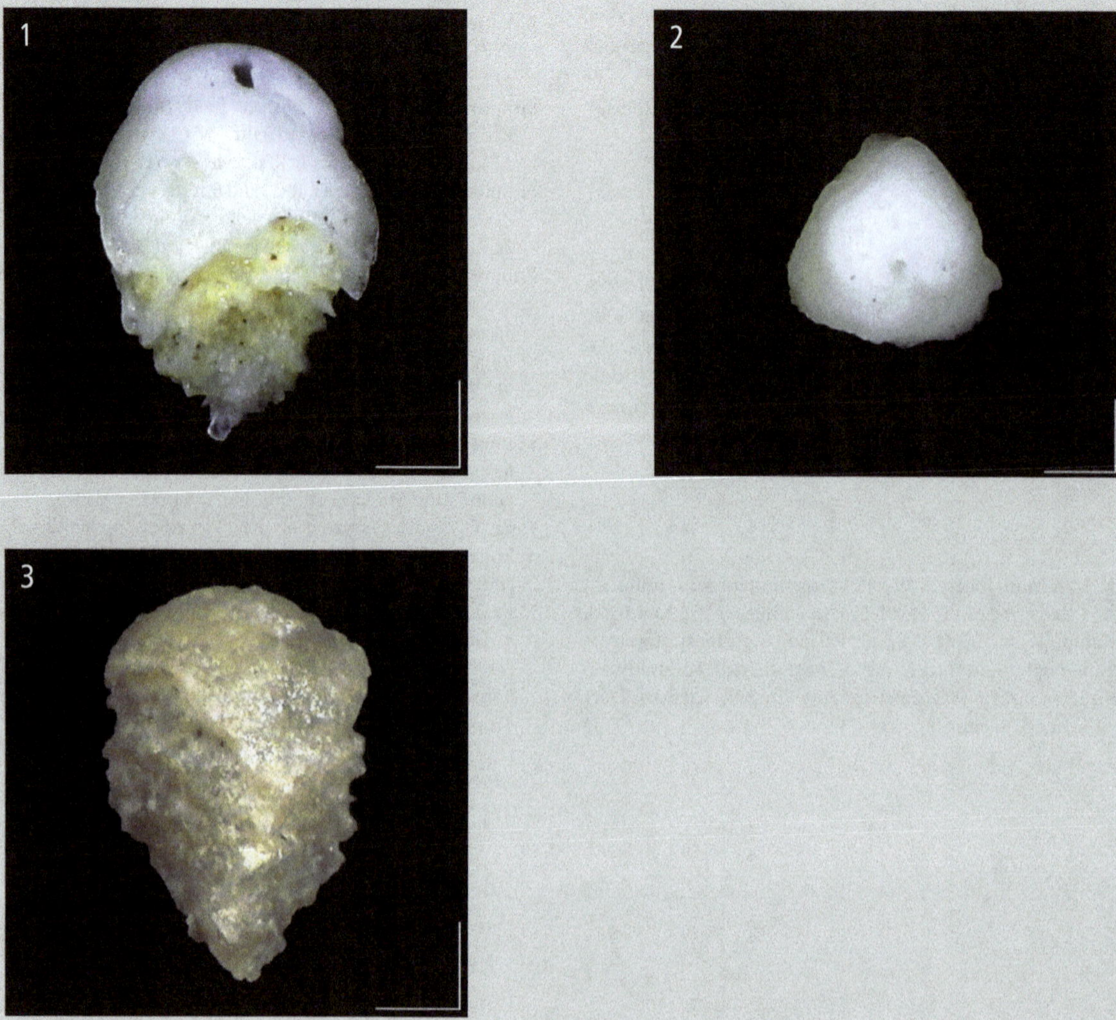

1. *Reussella szajnochae.* Vigo Seamount, North Atlantic; DSDP Hole 398D, Core 48, Section 5, 112–115 cm. Lateral view, NHMUK PM DI 0589, scale bar = 169 μm.

2. *Reussella szajnochae.* Vigo Seamount, North Atlantic; DSDP Hole 398D, Core 48, Section 5, 112–115 cm. Apertural view, NHMUK PM DI 0590, scale bar = 190 μm.

3. *Reussella szajnochae.* Subsilesian Unit. Variegated marls; Węglówka borehole 5, depth: 304–310 meters. Deposited in the Geroch Collection, Jagiellonian University, Kraków, Poland. Lateral view, NHMUK PM DI 0591, scale bar = 59 μm.

## Original Designation

*Verneuilina szajnochae* Grzybowski, 1896, p. 287, pl. 9, fig. 19.

## Type Specimen

Deposited in the Grzybowski Collection at the Jagiellonian University, Kraków, Poland.

## Type Level

Campanian red clay.

## Type Locality

Exploratory coal shaft (64 to 70 meters below top), about 500 meters southwest of the courthouse in Wadowice, near Kraków, Poland.

## Description

Test forms a triserial coil that is triangular with sharp angles, and slightly concave sides. Chambers low and oblique, increasing gradually in size, and separated by arcuate, limbate, raised sutures that meet at the center of each side. Chamber walls are calcareous, perforate, and ornamented by sharp, projecting spines at the edges of adjacent sides. The proloculus may have a basal spine. The ultimate chamber is highly convex on the upper side. The primary aperture is a slit at the base of the final chamber, with an internal toothplate.

## Synonyms and Citations

*Bulimina limbata* White, 1929, p. 48, pl. 5, fig. 9.

*Reussella californica* Cushman and Goudkoff, 1944, p. 59, pl. 10, figs. 3–5.

*Reussella szajnochae* (Grzybowski). –Nyong and Olsson, 1984, p. 474, pl. 3, fig. 14.

*Angulogerina szajnochae* (Grzybowski). –Thomas, 1990, p. 589, pl. 2, fig. 9.

*Pyramidina szajnochae* (Grzybowski). –Bolli et al., 1994, p. 134, figs. 36.1–2.

## Biogeography

Worldwide. Recorded from Poland (Grzybowski, 1896), Weddell Sea, Antarctica (Thomas, 1990), North Atlantic (Nyong and Olsson, 1984), Trinidad (Bolli et al., 1994).

## Bathymetry

Bathyal. Predominantly lower bathyal according to Nyong and Olsson (1984).

## Chronostratigraphy

Late Cretaceous (Campanian) to Early Paleocene (Danian). First occurrence just below the Cretaceous–Paleogene boundary in the Weddell Sea, Antarctica (Thomas, 1990), Campanian to Early Maastrichtian in Trinidad (Bolli et al., 1994).

# References


Bolli, H.M., Beckmann, J.P., and Saunders, J.B., 1994, Benthic Foraminiferal Biostratigraphy of the South Caribbean Region. Cambridge: Cambridge University Press, 408 pp.

Cushman, J.A. and Goudkoff, P.P., 1944, Some foraminifera from the upper Cretaceous of California. Contributions from the Cushman Laboratory for Foraminiferal Research, vol. 20, pp. 53–64.

Grzybowski, J., 1896, Otwornice czerwonych iłów z Wadowic. Rozprawy Wydziału Matematyczno-Przyrodniczego, Akademia Umiejętności w Krakowie, serya 2, vol. 30, pp. 261–308.

Nyong, E.E. and Olsson, R.K., 1984, A Paleoslope model for Campanian to lower Maestrichtian foraminifera of the North American Basin and adjacent continental margin. Marine Micropaleontology, vol. 8, pp. 437–477.

Thomas, E., 1990, Late Cretaceous through Neogene benthic foraminifers. In Barker, P.F., Kennett, J.P., et al., ed., Scientific Results of the Proceedings of the Ocean Drilling Program, vol. 113, pp. 571–594.

White, M.P., 1929, Some index foraminifera of the Tampico Embayment area of Mexico (Part III). Journal of Paleontology, vol. 3, no. 1, pp. 30–58.

# *Rhabdammina discreta*

## Brady, 1881

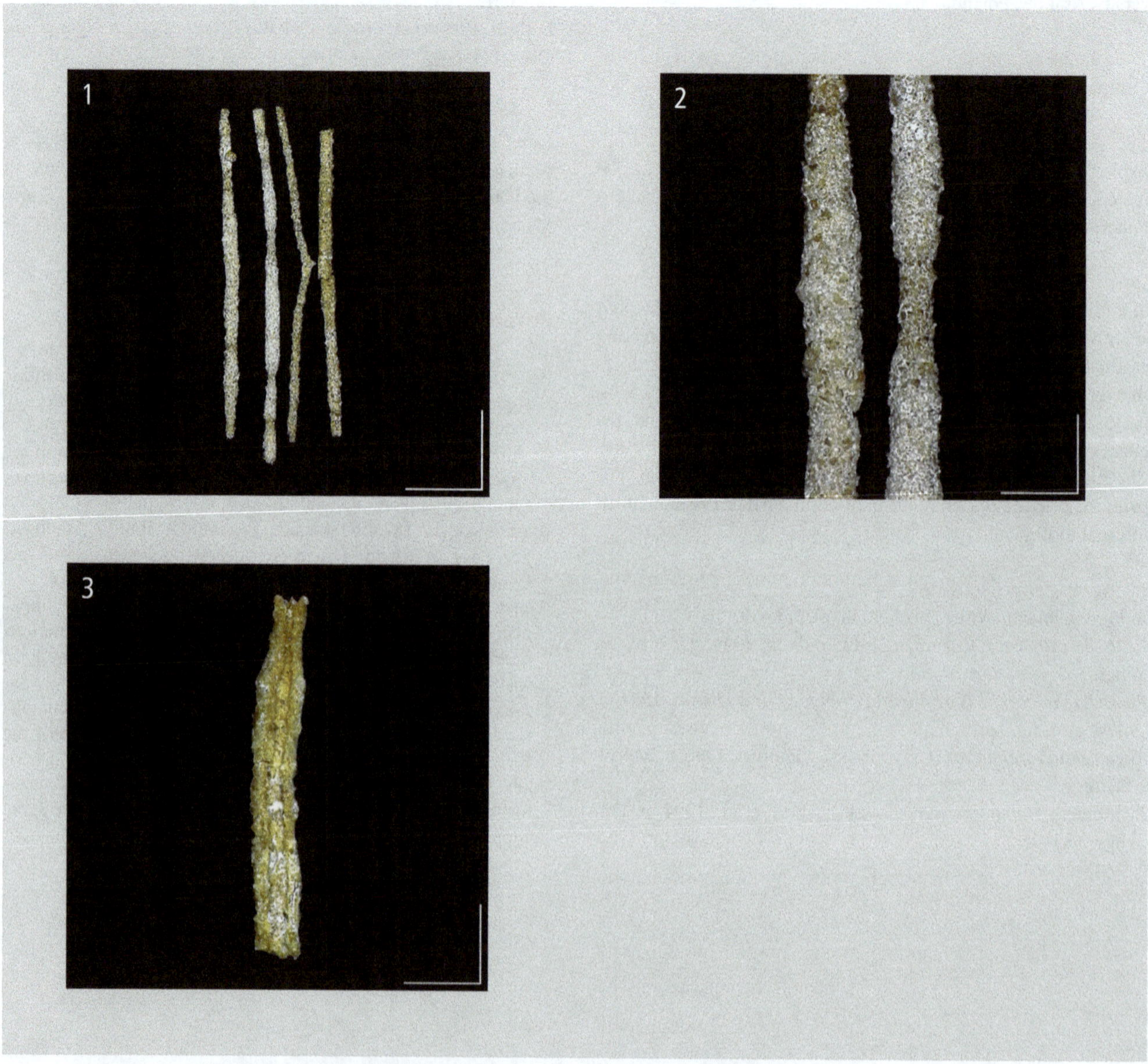


1. *Rhabdammina discreta.* Faroe Channel. Deposited in the micropaleontological collections of The Natural History Museum, London; NHMUK ZF2299. Syntypes. Lateral view, NHMUK PM DI 0592, scale bar = 3.94 mm.

2. *Rhabdammina discreta.* Faroe Channel. Deposited in the micropaleontological collections of The Natural History Museum, London; NHMUK ZF2299. Syntypes. Detailed view, showing constrictions. NHMUK PM DI 0592, scale bar = 1.2 mm.

3. *Rhabdammina discreta.* Off Northwest Ireland. North Atlantic Ocean; "Porcupine" Station 23, depth: 630 fathoms. Deposited in the micropaleontological collections of The Natural History Museum, London; NHMUK ZF2301. Cross-section view, NHMUK PM DI 0594, scale bar = 1.00 mm.

## Original Designation

*Rhabdammina discreta* Brady, 1881, p. 48 (type description), illustrated in Brady, 1884, pl. 22, figs. 7–10.

## Type Specimen

Specimens illustrated by Brady (slides ZF2299–ZF2302) housed in the micropaleontological collections of The Natural History Museum, London, UK. Lectotype (slide ZF4862, ex slide ZF2300) designated as *Psammosiphonella discreta* by Kaminski and Gradstein (2005).

## Type Level

Recent.

## Type Locality

Not designated. Specimens illustrated by Brady (1884) from the Faroe Channel (fig. 7, slide ZF2299), from "Porcupine" Station 4, in the North Atlantic (fig. 8, slide ZF2300), from "Porcupine" Station 23, off Northwest Ireland, 630 fathoms (fig. 9, slide ZF2301) and from "Valorous" Station 5, off Sydney, Australia, 410 fathoms (fig. 10, slide ZF2302).

## Description

Test forms an elongate cylinder, open at both ends, straight or nearly straight, with external constrictions at irregular intervals, giving a somewhat nodose appearance. The test wall is coarsely agglutinated, non-calcareous, and well cemented, with a smooth interior. The primary aperture is a simple, round opening at the end of the open tube.

## Synonyms and Citations

*Rhabdammina discreta* Brady. –Brady, 1884, p. 268, pl. 22, figs. 7–10.
*Rhabdammina subdiscreta* Grzybowski, 1896, p. 275, pl. 8, figs. 5–6.
*Psammosiphonella discreta* (Brady). –Barker, 1960, pl. 22, figs. 7–10. –Kaminski and Gradstein, 2005, pp. 117–119, fig. 5, pl. 5/6.
*Psammosiphonella* ex gr. *discreta* (Brady). –King et al., 1989, p. 458, pl. 9.2, figs. 13–14.
*Rhabdammina discreta* Brady. –Charnock and Jones, 1990, p. 152, pl. 1, fig. 25; pl. 13, fig. 23. –Jones, 1994, p. 32, pl. 22, figs. 7–10.

## Remarks

*Rhabdammina discreta* Brady varies considerably in size; Brady (1881) reported specimens up to 1 inch in size. *Rhabdammina discreta* differs from *Rhabdammina abyssorum* Sars by the presence of external constrictions at irregular intervals along the test.

## Biogeography

Worldwide. See Observed Occurrences in Kaminski and Gradstein (2005).

## Bathymetry

Bathyal to abyssal. Recorded between 20 and 2475 fathoms at "Challenger" stations, shallower at high latitudes (Jones, 1994).

## Chronostratigraphy

Late Cretaceous to Recent. Turonian to Middle Miocene in the North Sea (Charnock and Jones, 1990).

# References


Barker, R.W., 1960, Taxonomic notes on the species figured by H.B. Brady in his report on the foraminifera dredged by H.M.S. Challenger during the years 1873–1876. Society of Economic Paleontologists and Mineralogists Special Publication, vol. 9, pp. 1–238.

Brady, H.B., 1881, Notes on some of the reticularian Rhizopoda of the Challenger Expedition, Part III. Quarterly Journal of Microscopical Science (New Series), vol. 21, pp. 31–71.

Brady, H.B., 1884, Report on the foraminifera dredged by H.M.S. Challenger during the years 1873–1876. In Brady, H.B., ed., Report on the Scientific Results of the Voyage of the H.M.S. Challenger during the years 1873–1876. Zoology, vol. 9, pp. 1–814.

Charnock, M.A. and Jones, R.W., 1990, Agglutinated foraminifera from the Palaeogene of the North Sea. In Hemleben, C., Kaminski, M.A., Kuhnt, W., and Scott, D.B., ed., Paleoecology, Biostratigraphy, Paleoceanography and Taxonomy of Agglutinated Foraminifera, NATO ASI Series, C: Mathematical and Physical Sciences. Dordrecht, Kluwer Academic, pp. 139–244.

Grzybowski, J., 1896, Otwornice czerwonych iłów z Wadowic. Rozprawy Wydziału Matematyczno-Przyrodniczego, Akademia Umiejętności w Krakowie, serya 2, vol. 30, pp. 261–308.

Jones, R.W., 1994, The Challenger Foraminifera. Oxford: Oxford University Press, 149 pp.

Kaminski, M.A. and Gradstein, F.M., 2005, Atlas of Paleogene Cosmopolitan Deep-Water Agglutinated Foraminifera. Grzybowski Foundation Special Publication, no. 10, 547 + vii pp.

King, C., 1989, Cenozoic of the North Sea. In Jenkins, D.G. and Murray, J.W., ed., Stratigraphical Atlas of Fossil Foraminifera, 2nd edn. Chichester: Ellis Horwood, pp. 418–489.

# *Rhabdammina linearis*

## Brady, 1879

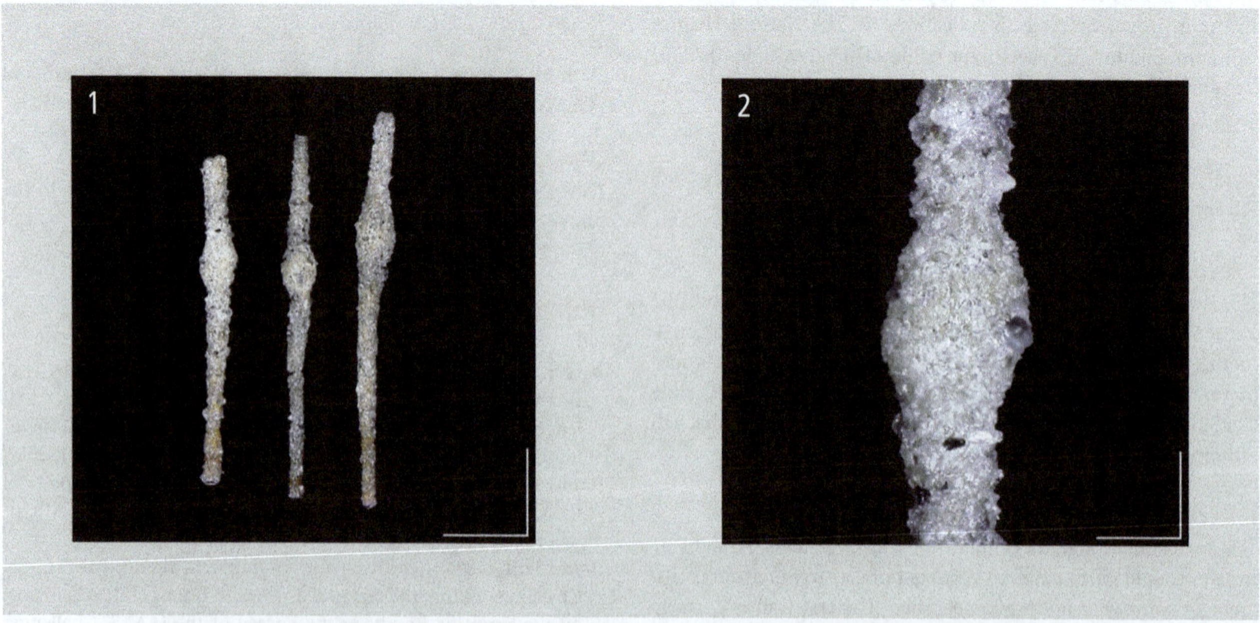


1. *Rhabdammina linearis.* South Atlantic Ocean; "Challenger" Station 323, depth: 1900 fathoms. Deposited in the micropaleontological collections of The Natural History Museum, London; NHMUK ZF2304. From left to right: syntype, lectotype (ZF4867), paralectotype (ZF4868). Lateral view, NHMUK PM DI 0595, scale bar = 1107 μm.

2. *Rhabdammina linearis.* South Atlantic Ocean; "Challenger" Station 323, depth: 1900 fathoms. Deposited in the micropaleontological collections of The Natural History Museum, London; NHMUK ZF2304. Detail of syntype showing swollen central area. NHMUK PM DI 0596, scale bar = 376 μm.

### Original Designation

*Rhabdammina linearis* Brady, 1879, p. 37, pl. 3, figs. 10–11.

### Type Specimen

Figured specimens are housed in the micropaleontological collections of The Natural History Museum, London, UK, and are registered in slides ZF2304–ZF2307. Lectotype (slide ZF4867, originally 3rd specimen from left in top row of slide ZF2304) was designated by Kaminski and Gradstein (2005).

### Type Level

Recent.

### Type Locality

Not designated. Localities given: "Challenger" dredgings, off Sombrero Island and off Culebra Island. West Indies (450 and 390 fathoms, respectively). Lectotype is from a deeper sounding (1900 fathoms), off the coast of South America, at latitude 35° 59′ S.

### Description

Test forms an elongate cylinder, straight or slightly curved, with narrow tubular portions extending in opposite directions from an inflated central area. The test wall is relatively thick, and medium to fine grained. Cement is non-calcareous. The primary aperture is the opening at the end of the tube.

### Synonyms and Citations

*Rhabdammina linearis* Brady.–Brady, 1884, p. 269, pl. 22, figs. 1–3, 5–6 (not fig. 4).–Jones, 1994, p. 32, pl. 22, figs. 1–3, 5–6. Kaminski and Gradstein, 2005, pp. 122–124, fig. 7, pl. 7.

*Oculosiphon linearis* (Brady). –Barker, 1960, pl. 22, figs. 1–3, 5–6 (not fig. 4). –Loeblich and Tappan, 1987, p. 23, pl. 15, fig. 1.

### Remarks

This is the type species by original designation of the genus *Oculosiphon* Avnimelech, 1952, regarded by Jones (1994) as a junior synonym of *Rhabdammina* Sars, 1869.

Gooday (1986) and Jones (1994) referred the specimen figured in pl. 22, fig. 4 by Brady (1884) to *Rhabdammina neglecta* Gooday. The latter species is characterized by irregularly curved, branched tubes, whereas *Rhabdammina linearis* Brady has straight tubes arising from the proloculus.

### Biogeography

Worldwide. See Observed Occurrences in Kaminski and Gradstein (2005).

### Bathymetry

Bathyal to abyssal.

### Chronostratigraphy

Cretaceous to Recent (Kaminski and Gradstein, 2005).

## References


Barker, R.W., 1960, Taxonomic notes on the species figured by H.B. Brady in his report on the foraminifera dredged by H.M.S. Challenger during the years 1873–1876. Society of Economic Paleontologists and Mineralogists Special Publication, vol. 9, pp. 1–238.

Brady, H.B., 1879, Notes on some of the reticularian Rhizopoda of the Challenger Expedition, Part I; On new or little known Arenaceous types. Quarterly Journal of Microscopical Science (New Series), vol. 19, pp. 20–63.

Brady, H.B., 1884, Report on the foraminifera dredged by H.M.S. Challenger during the years 1873–1876. In Brady, H.B., ed., Report on the Scientific Results of the Voyage of the H.M.S. Challenger during the years 1873–1876. Zoology, vol. 9, pp. 1–814.

Gooday, A.J., 1986, The genus Rhabdammina in the northeast Atlantic: A new species, a redescription of R. major de Folin, 1887, and some speculations on species relationships. Journal of Foraminiferal research, vol. 16, pp. 150–160.

Jones, R.W., 1994, The Challenger Foraminifera. Oxford: Oxford University Press, 149 pp.

Kaminski, M.A. and Gradstein, F.M., 2005, Atlas of Paleogene Cosmopolitan Deep-Water Agglutinated Foraminifera. Grzybowski Foundation Special Publication, no. 10, 547 + vii pp.

Loeblich, A.R. Jr and Tappan, H., 1987, Foraminiferal Genera and Their Classification. New York: Van Nostrand Reinhold, 970 pp.

# *Rhizammina indivisa*

## Brady, 1884

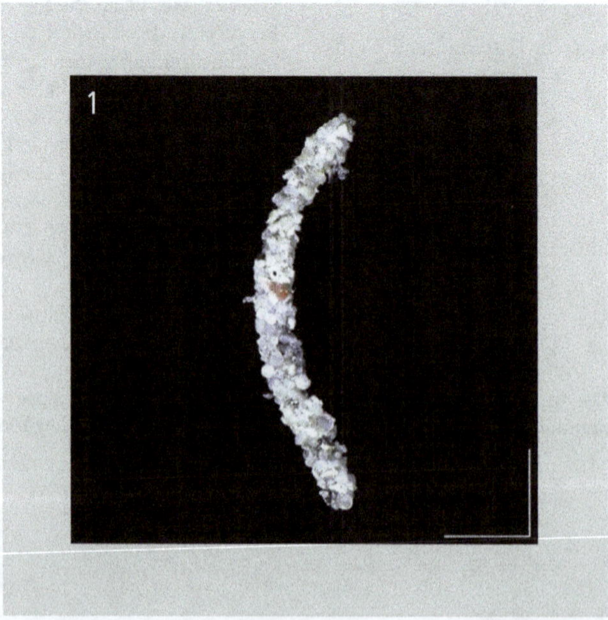


1. *Rhizammina indivisa.* "Challenger" Station 174, depth: 210 fathoms. Deposited in the micropaleontological collections of The Natural History Museum, London; from slide NHMUK1958.9.15.111-127. Lateral view, NHMUK PM DI 0597, scale bar = 356 μm.

### Original Designation
*Rhizammina indivisa* Brady, 1884, p. 277, pl. 29, figs. 5–7.

### Type Specimen
Figured specimens are housed in the micropaleontological collections of The Natural History Museum, London, UK.

### Type Level
Recent, surface sediments.

### Type Locality
Not designated. Locality given: Faröe Channel, 540 fathoms.

### Description
Test forms an elongate, unbranched, tube, rarely straight, and often tapered toward the ends, with a circular cross-section. The test wall is thin, flexible, and coarsely agglutinated, often appearing transversely wrinkled. The primary aperture is the terminal opening at the end of the tube.

### Synonyms and Citations
*Testulosiphon indivisus* Brady. –Barker, 1960, pl. 29, figs. 5–7. –Jones, 1994, p. 36, pl. 29, figs. 5–7.

### Remarks
Type species of the genus *Testulosiphon* Avnimelech, 1952, considered by Loeblich and Tappan (1987) to be a junior synonym of *Rhizammina* Brady, 1879.

### Biogeography
Worldwide.

### Bathymetry
Bathyal to abyssal.

### Chronostratigraphy
Early Cretaceous? to Recent. Reports from Cretaceous and Cenozoic deposits are questionable, as fossilized tests do not preserve the orginal flexible wall.

## References

Barker, R.W., 1960, Taxonomic notes on the species figured by H.B. Brady in his report on the foraminifera dredged by H.M.S. Challenger during the years 1873–1876. Society of Economic Paleontologists and Mineralogists Special Publication, vol. 9, pp. 1–238.

Brady, H.B., 1884, Report on the foraminifera dredged by H.M.S. Challenger during the years 1873–1876. In Brady, H.B., ed., Report on the Scientific Results of the Voyage of the H.M.S. Challenger during the years 1873–1876. Zoology, vol. 9, pp. 1–814.

Jones, R.W., 1994, The Challenger Foraminifera. Oxford: Oxford University Press, 149 pp.

Loeblich, A.R. Jr and Tappan, H., 1987, Foraminiferal Genera and Their Classification. New York: Van Nostrand Reinhold, 970 pp.

# Rzehakina epigona

## (Rzehak), 1895

1. *Rzehakina epigona.* Magura Unit, 5.7. 91-12, Variegated shales; Siary near Gorlice, Poland. Deposited in the Geroch Collection, Jagiellonian University, Kraków, Poland. Lateral view, NHMUK PM DI 0598, scale bar = 93 μm.

### Original Designation

*Silicina epigona* Rzehak, 1895, p. 214, pl. 6, fig. 1.

### Type Specimen

Type specimens from the Carpathian flysch are not preserved in the Rzehak collections and are presumed lost (Bubik and Kaminski, 2000). Topotypes from Zdaunek, Mähren, are housed in the Czech Geological Survey, Brno Branch.

### Type Level

Upper Maastrichtian Zdounky Unit (marginal facies of Silesian Unit) of the Outer Carpathians in Moravia (Kaminski and Gradstein, 2005). Originally given as Lower Tertiary by Rzehak (1895).

### Type Locality

Zdounky, Moravia. Bubik and Kaminski (2000) designated a "neotype locality" in a small abandoned quarry, southeast of Zdounky village.

### Description

Test forms a compressed, semi-involute, planispire; ovate in outline, slightly extended at both ends, and elliptical in cross-section, with a concave central umbilicus, and a subacute, carinate periphery. Later chambers are narrow (half a coil in length), and arranged planispirally, partially overlapping earlier whorls. Chamber walls are finely agglutinated with much organic cement, and smoothly finished. The primary aperture is a small, terminal slit or oval opening, without a tooth.

### Synonyms and Citations

*Spiroloculina occulta* Grzybowski, 1901, p. 261, pl. 7, fig. 25.

*Rzehakina epigona* (Rzehak) var. *lata* Cushman and Jarvis, 1928, p. 93, pl. 13, fig. 11.

*Rzehakina epigona epigona* (Rzehak). –Krasheninnikov and Pflaumann, 1978, p. 569, pl. 3, figs. 5a–b, 11. –Geroch and Nowak, 1984, pl. 3, fig. 12.

*Rzehakina epigona* (Rzehak). –van Morkhoven et al., 1986, p. 317, pl. 103, figs. 1–3. –Loeblich and Tappan, 1987, p. 54, pl. 41, figs. 7–11. –Kaminski and Geroch, 1993, p. 274, pl. 15, fig. 3a–c. –Bolli et al., 1994, p. 71, figs. 19.7–8. –Gradstein et al., 1994, pl. 9, figs. 14–17. –Kaminski and Gradstein, 2005, pp. 201–205, fig. 31, pl. 31.

### Remarks

Type species of the genus *Rzehakina* Cushman, 1927.

*Rzehakina epigona* (Rzehak) differs from *Rzehakina inclusa* (Grzybowski) by its more evolute, flatter test and the lack of an umbilical boss obscuring all but the last chambers of the test. It differs from *Rzehakina fissistomata* (Grzybowski), which has a more evolute, flatter test with a distinct neck.

### Biogeography

Worldwide. See Observed Occurrences in Kaminski and Gradstein (2005).

### Bathymetry

Primarily bathyal, rare in upper abyssal flysh-type assemblages (Kaminski and Gradstein, 2005).

### Chronostratigraphy

Late Cretaceous to Early Eocene (Biozone P9); rare in Turonian to Santonian (Kaminski and Gradstein, 2005). Doubtful in Middle Eocene (Biozones P10 to P13) according to van Morkhoven et al. (1986).

## References


Bolli, H.M., Beckmann, J.P., and Saunders, J.B., 1994, Benthic Foraminiferal Biostratigraphy of the South Caribbean Region. Cambridge: Cambridge University Press, 408 pp.

Bubik, M. and Kaminski, M.A., 2000, Remarks on the type locality and current status of the foraminiferal species Rzehakina epigona (Rzehak, 1895). In Hart, M.B., Kaminski, M.A., and Smart, C.W., ed., Proceedings of the Fifth International Workshop on Agglutinated Foraminifera. Grzybowski Special Publication, vol. 7, pp. 71–80.

Cushman, J.A. and Jarvis, P.W., 1928, Cretaceous foraminifera from Trinidad. Contributions from the Cushman Laboratory for Foraminiferal Research, vol. 4, pp. 85–103.

Geroch, S. and Nowak, W., 1984, Proposal of zonation for the late Tithonian–late Eocene, based upon arenaceous foraminifera from the outer Carpathians, Poland. In Oertli, H., ed., Benthos '83, 2nd International Symposium on Benthic Foraminifera, Pau (France), April 11–15, 1983. Elf Aquitaine, ESSO REP and TOTAL CFP, Pau and Bordeaux, pp. 225–239.

Gradstein, F.M., Kaminski, M.A., Berggren, W.A., and D'Iorio, M.A., 1994, Cenozoic biostratigraphy of the Central North Sea and Labrador Shelf. Micropaleontology, vol. 40 (Supplement), pp. 1–152.

Grzybowski, J., 1901, Otwornice warstw inoceramowych okolicy Gorlic. Rozprawy Wydziału Matematyczno-Przyrodniczego, Akademia Umiejętności w Krakowie, serya 2, vol. 41, pp. 219–286.

Kaminski, M.A. and Geroch, S., 1993, A revision of foraminiferal species in the Grzybowski Collection. In Kaminski, M.A., Geroch, S., and Kaminski, D.G., ed., The Origins of Applied Micropalaeontology: The School of Jósef Grzybowski (with a taxonomic revision of the Grzyboswski Collection). Grzybowski Foundation Special Publication, no. 1, pp. 239–330.

Kaminski, M.A. and Gradstein, F.M., 2005, Atlas of Palaeogene Cosmopolitan Deep-Water Agglutinated Foraminifera. Grzybowski Foundation Special Publication, no. 10, 547 + vii pp.

Krasheninnikov, V.A. and Pflaumann, U., 1978, Cretaceous agglutinated foraminifera of the Atlantic Ocean off West Africa (Leg 41, Deep Sea Drilling Project). In Lancelot, Y., Seibold, E., et al., ed., Initial Reports of the Deep Sea Drilling Project, Washington (U.S. Government Printing Office), vol. 41, pp. 565–580.

Loeblich, A.R. Jr and Tappan, H., 1987, Foraminiferal Genera and Their Classification. New York: Van Nostrand Reinhold, 970 pp.

Rzehak, A., 1895, Ueber einige merkwürdige Foraminiferen aus österreichischen Tertiär. Annalen Naturhistorisches Hofmuseum, Wien, vol. 10, pp. 213–330.

van Morkhoven, F.P.C.M., Berggren, W.A., Edwards, A.S., et al., 1986, Cenozoic cosmopolitan deep-water benthic foraminifera. Bulletin des Centres de Recherches Exploration-Production Elf-Aquitaine, Mémoire, vol. 11, pp. 1–421.

# Rzehakina fissistomata

## (Grzybowski), 1901

1. *Rzehakina fissistomata.* Magura Unit, 5.7. 91-12, Variegated shales; Siary near Gorlice, Poland. Deposited in the Geroch Collection, Jagiellonian University, Kraków, Poland. Lateral view, NHMUK PM DI 0599, scale bar = 141 μm.

### Original Designation

*Spiroloculina fissistomata* Grzybowski, 1901, p. 261, pl. 7, figs. 22–24.

### Type Specimen

Not originally designated. No syntypes are preserved in the Grzybowski Collection. Kaminski and Geroch (1993, p. 273, pl. 15, fig. 2) designated a neotype from Siary 131/1DW. The neotype is deposited in the Grzybowski Collection at the Jagiellonian University, Kraków, Poland.

### Type Level

Probably Late Paleocene, Magura Unit of the Polish Carpathians.

### Type Locality

Grzybowski (1901) reported specimens from the Kalinka-1 well in Ropica Polska, and an outcrop in a stream in the village of Sekowa. The neotype is from an outcrop in a stream in the village of Siary.

### Description

Test forms an elliptical, flattened, semi-involute planispire; slightly extended at both ends, with an ovate outline and a carinate periphery. Earlier whorls are arranged in an irregular spiral, later chambers are narrow (half a coil in length), and arranged planispirally, partially overlapping earlier whorls. Four to six chambers are visible between the broad, embracing final chambers. Chamber walls are finely agglutinated with organic cement, and smoothly finished. The primary aperture is a small, terminal slit or oval opening without a tooth at the end of a distinct neck.

### Synonyms and Citations

*Rzehakina epigona fissistomata* (Grzybowski). –Krasheninnikov and Pflaumann, 1978, p. 569, pl. 3, figs. 1–3. –Geroch and Nowak, 1984, pl. 3, figs. 8, 13.

*Rzehakina fissistomata* (Grzybowski). –Kaminski and Geroch, 1993, p. 272, pl. 15, figs. 1–2. –Kaminski and Gradstein, 2005, pp. 206–208, fig. 32, pl. 32.

### Remarks

*Rzehakina fissistomata* (Grzybowski) differs from *Rzehakina epigona* (Rzehak) and *Rzehakina inclusa* (Grzybowski) by its flatter, more evolute test and distinct apertural neck. The outline of *Rzehakina fissistomata* (Grzybowski) is also usually more asymmetrical.

### Biogeography

Worldwide.

### Bathymetry

Primarily bathyal.

### Chronostratigraphy

Late Cretaceous (Maastrichtian) to Paleocene. An Early Paleocene *Rzehakina epigona fissistomata* Biozone was defined by Geroch and Nowak (1984) for the Polish Carpathians.

## References


Geroch, S. and Nowak, W., 1984, Proposal of zonation for the late Tithonian–late Eocene, based upon arenaceous foraminifera from the outer Carpathians, Poland. In Oertli, H., ed., Benthos '83, 2nd International Symposium on Benthic Foraminifera, Pau (France), April 11–15, 1983. Elf Aquitaine, ESSO REP and TOTAL CFP, Pau and Bordeaux, pp. 225–239.

Grzybowski, J., 1901, Otwornice warstw inoceramowych okolicy Gorlic. Rozprawy Wydziału Matematyczno-Przyrodniczego, Akademia Umiejętności w Krakowie, serya 2, vol. 41, pp. 219–286.

Kaminski, M.A. and Geroch, S., 1993, A revision of foraminiferal species in the Grzybowski Collection. In Kaminski, M.A., Geroch, S. and Kaminski, D.G., ed., The Origins of Applied Micropalaeontology: The School of Jósef Grzybowski (with a taxonomic revision of the Grzyboswski Collection). Grzybowski Foundation Special Publication, no. 1, pp. 239–330.

Kaminski, M.A. and Gradstein, F.M., 2005, Atlas of Palaeogene Cosmopolitan Deep-Water Agglutinated Foraminifera. Grzybowski Foundation Special Publication, no. 10, 547 + vii pp.

Krasheninnikov, V.A. and Pflaumann, U., 1978, Cretaceous agglutinated foraminifera of the Atlantic Ocean off West Africa (Leg 41, Deep Sea Drilling Project). In Lancelot, Y., Seibold, E., et al., ed., Initial Reports of the Deep Sea Drilling Project, Washington (U.S. Government Printing Office), vol. 41, pp. 565–580.

# Rzehakina inclusa

## (Grzybowski), 1901

1. *Rzehakina inclusa.* Lower Istebna Beds; Czarna, Wisełka 118. Deposited in the Geroch Collection, Jagiellonian University, Kraków, Poland. Lateral view, NHMUK PM DI 0600, scale bar = 90 μm.

### Original Designation

*Spiroloculina inclusa* Grzybowski, 1901, p. 260, pl. 7, fig. 20.

### Type Specimen

Not originally designated. No specimens labeled *Spiroloculina inclusa* are preserved in the Grzybowski Collection. Kaminski and Geroch (1993) designated a lectotype (UJ-133-P 2/1) from one of Grzybowski's samples. The lectotype is housed in the Grzybowski Collection at the Jagiellonian University, Kraków, Poland.

### Type Level

Probably Upper Cretaceous, *Inoceramus* beds, Magura Unit of Polish Carpathians.

### Type Locality

An outcrop containing *Inoceramus* fragments in the village of Siary, near Gorlice, Poland.

### Description

Test forms an elliptical, biconvex, involute planispire; slightly extended at both ends, with an ovate outline and a carinate periphery. Earlier whorls are arranged in an irregular spiral; later chambers are narrow (half a coil in length), and arranged planispirally, partially overlapping earlier whorls. The last two chambers envelop earlier whorls, and the umbilical region is filled by a thick central boss, obscuring previous whorls. Chamber walls are finely agglutinated with organic cement and smoothly finished. The primary aperture is a small, terminal slit or oval opening without a tooth.

### Synonyms and Citations

*Rzehakina epigona inclusa* (Grzybowski). –Krasheninnikov and Pflaumann, 1978, p. 569, pl. 3, figs. 6a–b. –Geroch and Nowak, 1984, pl. 3, fig. 11.

*Rzehakina inclusa* (Grzybowski). –Kaminski and Geroch, 1993, p. 272, pl. 15, figs. 4a–c. –Kaminski and Gradstein, 2005, pp. 209–211, fig. 33, pl. 33.

### Remarks

*Rzehakina epigona* (Rzehak) differs from *Rzehakina inclusa* (Grzybowski) by its more evolute, flatter test and the lack of an umbilical boss obscuring all but the last chambers of the test.

### Biogeography

Worldwide. See Observed Occurrences in Kaminski and Gradstein (2005).

### Bathymetry

Primarily bathyal.

### Chronostratigraphy

Late Cretaceous (Santonian) to Early Paleocene.

## References


Geroch, S. and Nowak, W., 1984, Proposal of zonation for the late Tithonian–late Eocene, based upon arenaceous foraminifera from the outer Carpathians, Poland. In Oertli, H., ed., Benthos '83, 2nd International Symposium on Benthic Foraminifera, Pau (France), April 11–15, 1983. Elf Aquitaine, ESSO REP and TOTAL CFP, Pau and Bordeaux, pp. 225–239.

Grzybowski, J., 1901, Otwornice warstw inoceramowych okolicy Gorlic. Rozprawy Wydziału Matematyczno-Przyrodniczego, Akademia Umiejętności w Krakowie, serya 2, vol. 41, pp. 219–286.

Kaminski, M.A. and Geroch, S., 1993, A revision of foraminiferal species in the Grzybowski Collection. In Kaminski, M.A., Geroch, S. and Kaminski, D.G., ed., The Origins of Applied Micropalaeontology: The School of Jósef Grzybowski (with a taxonomic revision of the Grzyboswski Collection). Grzybowski Foundation Special Publication, no. 1, pp. 239–330.

Kaminski, M.A. and Gradstein, F.M., 2005, Atlas of Palaeogene Cosmopolitan Deep-Water Agglutinated Foraminifera. Grzybowski Foundation Special Publication, no. 10, 547 + vii pp.

Krasheninnikov, V.A. and Pflaumann, U., 1978, Cretaceous agglutinated foraminifera of the Atlantic Ocean off West Africa (Leg 41, Deep Sea Drilling Project). In Lancelot, Y., Seibold, E., et al., ed., Initial Reports of the Deep Sea Drilling Project, Washington (U.S. Government Printing Office), vol. 41, pp. 565–580.

# Saccorhiza ramosa

## (Brady), 1879

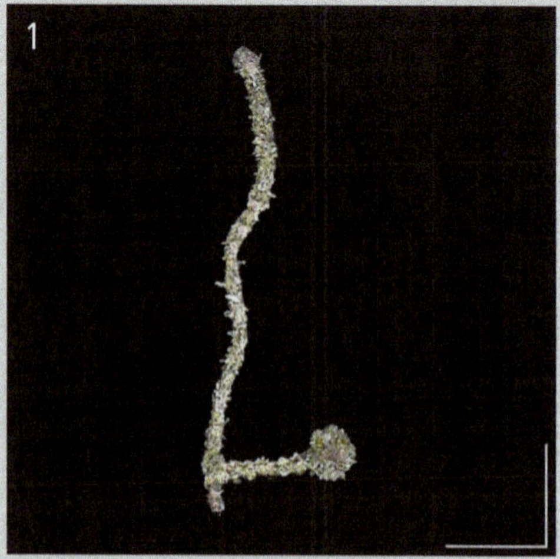


1. *Saccorhiza ramosa.* South China Sea; Station 17956, latitude 13° 50.90′ N, longitude 112° 35.30′ E, water depth: 3387 meters, surface sample. Deposited in the South China Sea Collection, Institute of Geosciences, Kiel University, Germany. Lateral view, NHMUK PM DI 0601, scale bar = 274 μm.

### Original Designation

*Hyperammina ramosa* Brady, 1879, p. 33, pl. 3, figs. 14–15.

### Type Specimen

Specimens illustrated by Brady (1879) are housed in the micropaleontological collections of The Natural History Museum, London, UK, and are registered in slides ZF1595 and ZF3612. Loeblich and Tappan (1964) designated as lectotype the specimen in slide ZF3612 (ex slide ZF1596), illustrated in fig. 15 of Brady (1879).

### Type Level

Recent.

### Type Locality

Not designated. Localities given: North Polar Expedition Dredgings, "Porcupine" dredgings in North Atlantic, "Challenger" stations at various locations in North and South Atlantic and North and South Pacific. Lectotype is from North Pacific "Challenger" Station 246 (2050 fathoms).

### Description

Test forms a large, elongate, tubular mass, with a globular proloculus followed by irregular or branching tubular chambers with terminal stolons. Chamber walls are agglutinated with non-calcareous cement, often exhibiting abundant sponge spicules, giving the test a hirsute appearance. The aperture is a round opening at the open end of stolons.

### Synonyms and Citations

*Hyperammina ramosa* Brady. –Brady, 1884, p. 261, pl. 23, figs. 15–19.

*Saccorhiza ramosa* (Brady). –Cushman, 1910, p. 65, pl. 30, figs. 3–4, textfig. 81. –Cushman, 1921, p. 54, pl. 4, fig. 5. –Barker, 1960, pl. 23, figs. 15–19. –Burke, 1981, p. 19, pl. 1, fig. 3. –Schröder, 1986, p. 35, pl. 7, fig. 1. –Zheng, 1988, p. 31, pl. 3, figs. 6–9, pl. 7, figs. 1–2. –Inoue, 1989, pl. 18, fig. 1. –Loeblich and Tappan, 1987, p. 43, pl. 32, figs. 13–15. –Loeblich and Tappan, 1994, p. 14, pl. 1, figs. 4–5. –Jones, 1994, p. 33, pl. 23, figs. 15–19.

## Remarks

Type species of the genus *Saccorhiza* Eimer and Fickert, 1899.

## Biogeography

Worldwide.

## Bathymetry

Neritic to abyssal. Reported between 60 and 3000 fathoms at "Challenger" stations (Jones 1994).

## Chronostratigraphy

Cretaceous to Recent.

# References


Barker, R.W., 1960, Taxonomic notes on the species figured by H.B. Brady in his report on the foraminifera dredged by H.M.S. Challenger during the years 1873–1876. Society of Economic Paleontologists and Mineralogists Special Publication, vol. 9, pp. 1–238.

Brady, H.B., 1879, Notes on some of the reticularian Rhizopoda of the Challenger Expedition, Part I; On new or little known Arenaceous types. Quarterly Journal of Microscopical Science (New Series), vol. 19, pp. 20–63.

Brady, H.B., 1884, Report on the foraminifera dredged by H.M.S. Challenger during the years 1873–1876. In Brady, H.B., ed., Report on the Scientific Results of the Voyage of the H.M.S. Challenger during the years 1873–1876. Zoology, vol. 9, pp. 1–814.

Burke, S.C., 1981, Recent benthic foraminifera of the Ontong-Java Plateau. Journal of Foraminiferal Research, vol. 11, pp. 1–19.

Cushman, J.A., 1910, A monograph of the Foraminifera of the North Pacific Ocean. Part 1. Astrorhizidae and Lituolidae. Bulletin of the United States National Museum, vol. 71, no. 1, pp. 1–134.

Cushman, J.A., 1921, Foraminifera of the Philippine and adjacent seas. Bulletin of the United States National Museum, vol. 100, pp. 1–608.

Inoue, Y., 1989, Northwest Pacific Foraminifera as paleoenvironmental indicators. Science Reports of the Institute of Geosciences, University of Tsukuba (Section B, Geological Sciences), vol. 10, pp. 57–162.

Jones, R.W., 1994, The Challenger Foraminifera. Oxford: Oxford University Press, 149 pp.

Loeblich, A.R. Jr and Tappan, H., 1964, Sarcodina, chiefly "Thecamoebians" and Foraminiferida. In Moore, R.C., ed., Treatise on Invertebrate Paleontology, Part C, Protista 2 (2 vols). Lawrence, KS: Geological Society of America and University of Kansas Press, pp. 1–900.

Loeblich, A.R. Jr and Tappan, H., 1987, Foraminiferal Genera and Their Classification. New York: Van Nostrand Reinhold, 970 pp.

Loeblich, A.R. Jr and Tappan, H., 1994, Foraminifera of the Sahul Shelf and Timor Sea. Special Publications of the Cushman Laboratory for Foraminiferal Research, vol. 31, pp. 1–661.

Schröder, C.J., 1986, Deep-Water Arenaceous Foraminifera in the Northwest Atlantic Ocean: Canadian Technical Report of Hydrography and Ocean Sciences No. 71. Atlantic Geoscience Centre, Bedford Institute of Oceanography, 191 pp.

Zheng, S.Y., 1988, The Agglutinated and Porcelaneous Foraminifera of the East China Sea. Beijing: China Ocean Press, 337 pp.

# *Sigmavirgulina tortuosa*

## (Brady), 1881

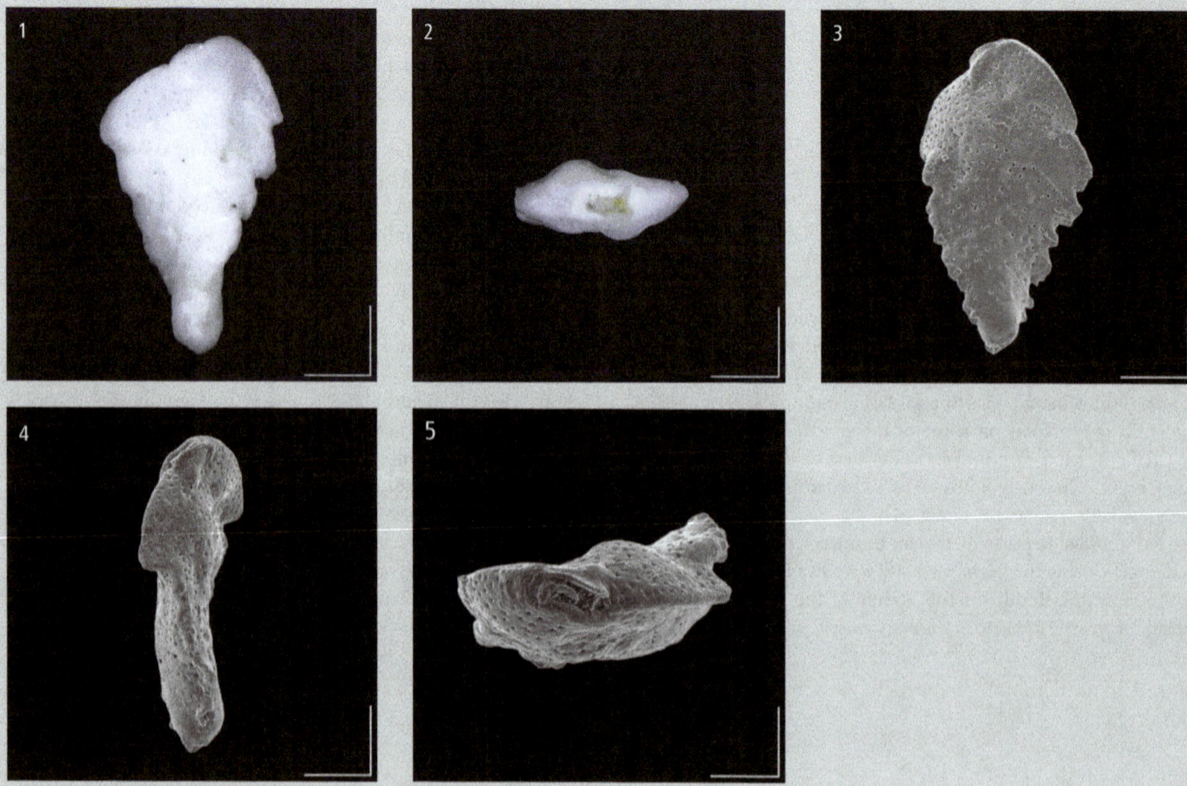

1. *Sigmavirgulina tortuosa*. Cape Verde Islands, Atlantic Ocean; "Challenger" Station 352, depth: 11 fathoms. Deposited in the micropaleontological collections of The Natural History Museum, London; NHMUK ZF1202. Syntype. Lateral view, NHMUK PM DI 0602, scale bar = 80 μm.

2. *Sigmavirgulina tortuosa*. Cape Verde Islands, Atlantic Ocean; "Challenger" Station 352, depth: 11 fathoms. Deposited in the micropaleontological collections of The Natural History Museum, London; NHMUK ZF1202. Syntype. Apertural view, NHMUK PM DI 0603, scale bar = 46 μm.

3. *Sigmavirgulina tortuosa*. Cape Verde Islands, Atlantic Ocean; "Challenger" Station 352, depth: 11 fathoms. Deposited in the micropaleontological collections of The Natural History Museum, London; NHMUK ZF1202. Syntype. Lateral view, NHMUK PM DI 0604, scale bar = 95 μm.

4. *Sigmavirgulina tortuosa*. Cape Verde Islands, Atlantic Ocean; "Challenger" Station 352, depth: 11 fathoms. Deposited in the micropaleontological collections of The Natural History Museum, London; NHMUK ZF1202. Syntype. Edge view, NHMUK PM DI 0605, scale bar = 95 μm.

5. *Sigmavirgulina tortuosa*. Cape Verde Islands, Atlantic Ocean; "Challenger" Station 352, depth: 11 fathoms. Deposited in the micropaleontological collections of The Natural History Museum, London; NHMUK ZF1202. Syntype. Apertural view, NHMUK PM DI 0606, scale bar = 59 μm.

## Original Designation

*Bulimina (Bolivina) tortuosa* Brady 1881, p. 57 (type description), illustrated as *Bolivina tortuosa* in Brady, 1884, pl. 52, figs. 31–34.

## Type Specimen

Figured specimens are housed in the micropaleontological collections of The Natural History Museum, London, UK, and are registered in slides ZF1201 and ZF1202.

## Type Level

Recent.

## Type Locality

Not designated. Figured specimens are from "Challenger" Station 219A, Nares Harbour, Admiralty Islands (17 fathoms) and "Challenger" Station 325A, off Saint Vincent, Cape Verde Islands (11 fathoms).

## Description

Test forms a compressed, flaring, biserial series; slightly greater in length than width, with an acute or keeled, lobulate periphery. Early chambers are added more than 180° apart, giving the appearance of coiling along the longitudinal axis. The slightly inflated chambers increase rapidly in width, slightly overhanging the margins of earlier chambers, and are separated by thin, depressed, curved sutures. Chamber walls are calcareous, coarsely perforate, and smooth. The primary aperture is an elongate, oval, subterminal slit with a lip and an internal toothplate.

## Synonyms and Citations

*Bolivina tortuosa* Brady. –Brady 1884, p. 420, pl. 52, figs. 31–34. –Barker, 1960, pl. 52, figs. 31–32.

*Sigmavirgulina (?) tortuosa* (Brady) var. *atlantica* (Cushman). –Barker 1960, p. 108. pl. 52, figs. 33–34.

*Sigmavirgulina tortuosa* (Brady). –Loeblich and Tappan, 1987, p. 531, pl. 579, figs. 1–5. –Jones, 1994, p. 58, pl. 52, figs. 31–34. –Robertson, 1998, p. 166, pl. 62, fig. 6. –Milker and Schmiedl, 2012, p. 93, fig. 21.12.

## Remarks

Type species of the genus *Sigmavirgulina* Loeblich and Tappan, 1957.

## Biogeography

Worldwide.

## Bathymetry

Primarily neritic to upper bathyal. Reported between 2 and 420 fathoms at "Challenger" stations (Jones 1994).

## Chronostratigraphy

Miocene to Recent (Jones 1994).

## References


Barker, R.W., 1960, Taxonomic notes on the species figured by H.B. Brady in his report on the foraminifera dredged by H.M.S. Challenger during the years 1873–1876. Society of Economic Paleontologists and Mineralogists Special Publication, vol. 9, pp. 1–238.

Brady, H.B., 1881, Notes on some of the reticularian Rhizopoda of the Challenger Expedition, Part III. Quarterly Journal of Microscopical Science (New Series), vol. 21, pp. 31–71.

Brady, H.B., 1884, Report on the foraminifera dredged by H.M.S. Challenger during the years 1873–1876. In Brady, H.B., ed., Report on the Scientific Results of the Voyage of the H.M.S. Challenger during the years 1873–1876. Zoology, vol. 9, pp. 1–814.

Jones, R.W., 1994, The Challenger Foraminifera. Oxford: Oxford University Press, 149 pp.

Loeblich, A.R. Jr and Tappan, H., 1987, Foraminiferal Genera and Their Classification. New York: Van Nostrand Reinhold, 970 pp.

Milker, Y. and Schmiedl, G., 2012, A taxonomic guide to modern benthic shelf foraminifera of the western Mediterranean Sea. Palaeontologia Electronica, vol. 15, Issue 2;16A, 134 pp.

Robertson, B.E., 1998, Systematics and paleoecology of the benthic foraminiferida from the Buff Bay section, Miocene of Jamaica. Micropaleontology, vol. 44, supplement 2, pp. 1–266.

# *Sigmoilopsis schlumbergeri*

## (Silvestri), 1904

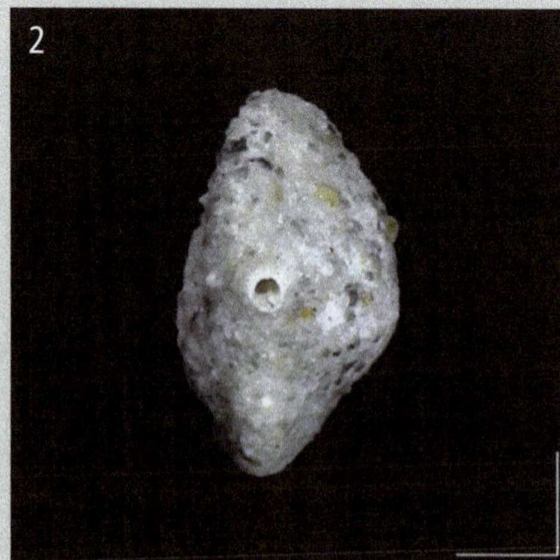

1. *Sigmoilopsis schlumbergeri*. Porcupine Basin, west of Ireland. R.W. Jones Collection. Lateral view, NHMUK PM DI 0607, scale bar = 175 µm.
2. *Sigmoilopsis schlumbergeri*. Porcupine Basin, west of Ireland. R.W. Jones Collection. Apretural view, NHMUK PM DI 0608, scale bar = 91 µm.

### Original Designation

*Sigmoilina schlumbergeri* Silvestri, 1904, pp. 267, 269 (type description), illustrated in Schlumberger, 1887, pl. 7, figs. 12–14, textfigs. 6–7.

### Type Specimen

Depository not given.

### Type Level

Recent.

### Type Locality

Gulf of Gascony, France (600 to 1200 meters).

### Description

Test forms an ovate, initially quinqueloculine series, then rapidly changes the plane of coiling, initially appearing sigmoid in section, then gradually becoming biserial in adult specimens. The moderately inflated chambers are half a coil in length, with a floor and a narrow lumen. Chamber walls are thick, and agglutinated with calcareous cement including sand, calcareous shell fragments, and sponge spicules. Aperture is a terminal, round opening with a small tooth.

### Synonyms and Citations

*Planispira celata* (Costa). –Brady, 1884, p. 197, pl. 8, figs. 1–4.
*Sigmoilina schlumbergeri* Silvestri. –Schlumberger, 1887, p. 481, pl. 7, figs. 12–14, textfigs. 6–7.
*Sigmoilopsis schlumbergeri* (Silvestri). –Barker, 1960, pl. 8, figs. 1–4. –van Morkhoven et. al. 1986, p. 57, pl. 18, figs. 1a–d. –Loeblich and Tappan, 1987, p. 350, pl. 356, figs. 8–13. –Loeblich and Tappan, 1994, p. 59, pl. 103, figs. 9–12. –Jones, 1994, p. 23, pl. 8, figs. 1–4. –Robertson, 1998, p. 35, pl. 11, fig. 5. –Milker and Schmiedl, 2012, p. 69, figs. 18.7–8.

### Remarks

*Sigmoilopsis schlumbergeri* (Silvestri) closely resembles *Sigmoilopsis celata* (Costa). However, it has an acutely angled periphery and clearly delineated sutures, whereas *Sigmoilopsis celata* has a more rounded periphery and less conspicuous sutures.

### Biogeography

Worldwide. Reported from the Indian Ocean, Antarctic, Coral Sea, Tasman Sea, Chatham Rise, New Zealand, Indonesia, Taiwan, Okinawa, Japan, Atlantic, North Sea, Mediterranean, Caribbean, and Gulf of Mexico (see Observed Occurrences in van Morkhoven et al., 1986).

### Bathymetry

Neritic to abyssal. Reported between 28 and 1630 fathoms at "Challenger" stations (Jones 1994).

### Chronostratigraphy

Early Miocene to Recent.

## References


Barker, R.W., 1960, Taxonomic notes on the species figured by H.B. Brady in his report on the foraminifera dredged by H.M.S. Challenger during the years 1873–1876. Society of Economic Paleontologists and Mineralogists Special Publication, vol. 9, pp. 1–238.

Brady, H.B., 1884, Report on the foraminifera dredged by H.M.S. Challenger during the years 1873–1876. In Brady, H.B., ed., Report on the Scientific Results of the Voyage of the H.M.S. Challenger during the years 1873–1876. Zoology, vol. 9, pp. 1–814.

Jones, R.W., 1994, The Challenger Foraminifera. Oxford: Oxford University Press, 149 pp.

Katz, M.E. and Miller, K.G., 1993, Latest Oligocene to Earliest Pliocene benthic foraminiferal biofacies of the northeastern Gulf of Mexico. Micropaleontology, vol. 39, no. 4, pp. 367–403.

Loeblich, A.R. Jr and Tappan, H., 1987, Foraminiferal Genera and Their Classification. New York: Van Nostrand Reinhold, 970 pp.

Loeblich, A.R. Jr and Tappan, H., 1994, Foraminifera of the Sahul Shelf and Timor Sea. Special Publications of the Cushman Laboratory for Foraminiferal Research, vol. 31, pp. 1–661.

Milker, Y. and Schmiedl, G., 2012, A taxonomic guide to modern benthic shelf foraminifera of the western Mediterranean Sea. Palaeontologia Electronica, vol. 15, Issue 2;16A, 134 pp.

Robertson, B.E., 1998, Systematics and paleoecology of the benthic foraminiferida from the Buff Bay section, Miocene of Jamaica. Micropaleontology, vol. 44, supplement 2, pp. 1–266.

Schlumberger, C., 1887, Note sur le genre Planispira. Bulletin de la Société Zoologique de France, vol. 12, pp. 105–118.

Silvestri, A., 1904, Richerche strutturali su alcune forme dei Trube di Bonfornello (Palermo). Memorie della Pontificia Accademia Romana dei Nuovi Lincei, vol. 22, pp. 235–276.

van Morkhoven, F.P.C.M., Berggren, W.A., Edwards, A.S., et al., 1986, Cenozoic cosmopolitan deep-water benthic foraminifera. Bulletin des Centres de Recherches Exploration-Production Elf-Aquitaine, Mémoire, vol. 11, pp. 1–421.

# *Siphogenerinoides brevispinosa*

## Cushman, 1939

1. *Siphogenerinoides brevispinosa.* Arkadelphia Clay, 5.5 miles northeast of Hope, Hemsptead Co., Arkansas, USA. Deposited in the collections of the United States National Museum, Washington, DC; CC 24664. Holotype Lateral view, NHMUK PM DI 0609, scale bar = 82 μm.

2. *Siphogenerinoides brevispinosa.* Maud Rise, Weddell Sea, Antarctica; ODP Leg 113, Hole 690B, Core 18, Section 6, 42–44 cm. Lateral view, NHMUK PM DI 0610, scale bar = 46 μm.

3. *Siphogenerinoides brevispinosa.* Maud Rise, Weddell Sea, Antarctica; ODP Leg 113, Hole 690B, Core 18, Section 6, 42–44 cm. Lateral view, NHMUK PM DI 0611, scale bar = 51 μm.

## Original Designation

*Siphogenerinoides brevispinosa* Cushman, 1939, p. 92, pl. 16, fig. 7.

## Type Specimen

Holotype (CC 24664) deposited in the collections of the United States National Museum, Washington, DC.

## Type Level

Originally given as ?Uppermost Cretaceous or Lower Midway, Arkadelphia marl (?) by Cushman (1939).

## Type Locality

Five and a half miles northeast of Hope, Hempstead County, Arkansas, USA.

## Description

Test forms an initially small, elongate, slender, slightly compressed, biserial series, becoming uniserial and rectilinear in the later stage with nearly parallel sides and a subcircular cross-section. The numerous broad and low chambers, which increase gradually in size and become more inflated, are separated by depressed sutures (horizontal in the uniserial portion). Chamber walls are calcareous, finely perforate, and ornamented by short spines. The primary aperture is an elliptical, terminal opening, bordered by an everted lip, with an internal toothplate.

## Synonyms and Citations

*Siphogenerinoides brevispinosa* Cushman. –Thomas, 1990, p. 591, pl. 1, fig. 7.

## Remarks

*Siphogenerinoides brevispinosa* Cushman is a small species that is frequently overlooked, as it is mainly found in the fine fraction. It is one of the most common species in Late Paleocene to Early Eocene assemblages from the Weddell Sea (Thomas 1990).

## Biogeography

Worldwide.

## Bathymetry

Lower bathyal (Thomas 1990).

## Chronostratigraphy

Early Paleocene to Early Eocene.

# References


Cushman, J.A. 1939, New American Cretaceous Foraminifera. Contributions from the Cushman Laboratory for Foraminiferal Research, vol. 15, no. 4, pp. 89–93.

Thomas, E., 1990, Late Cretaceous through Neogene benthic foraminifers. In Barker, P.F., Kennett, J.P., et al., ed., Scientific Results of the Proceedings of the Ocean Drilling Program, vol. 113, pp. 571–594.

# *Siphogenerinoides eleganta*

## (Plummer), 1927

1. *Siphogenerinoides eleganta*. Maud Rise, Weddell Sea, Antarctica; ODP Leg 113, Hole 689B, Core 21, Section 2, 40–44 cm. Lateral view, NHMUK PM DI 0612, scale bar = 64 μm.

2. *Siphogenerinoides eleganta*. Northeast Georgia Rise, South Atlantic; ODP Leg 114, Hole 698B, Core 21, Section 2, 40–44 cm. Lateral view, NHMUK PM DI 0613, scale bar = 52 μm.

### Original Designation

*Siphogenerina eleganta* Plummer, 1927, p. 126, pl. 8, fig. 1.

### Type Specimen

Cotypes (No. 33067) deposited in the Walker Museum, University of Chicago, Chicago, Illinois, USA. Duplicate set deposited at the Bureau of Economic Geology, University of Texas, Austin, Texas, USA.

### Type Level

Eocene, lower part of the upper Midway Formation, clay above the Tehuacana Limestone horizon.

### Type Locality

Station 46, clay pit, Mexia Brick Works, about 1 mile west of the town of Mexia, Limestone County, Texas, USA.

### Description

Test forms an initially small, elongate, slightly compressed, biserial series, becoming uniserial with alternating chambers, and subcircular in cross-section. The oblique chambers increase gradually in size, and are separated by depressed, oblique sutures. Chamber walls are calcareous, perforate, and ornamented by irregular, longitudinal striae that disappear upward. The primary aperture is an elliptical, terminal opening, bordered by an everted lip with an internal toothplate.

### Synonyms and Citations

*Siphogenerinoides eleganta* (Plummer). –Thomas, 1990, p. 591, pl. 1, fig. 6. –Ortiz and Thomas, 2006, p. 132, pl. 11, figs. 3–4.

### Remarks

*Siphogenerinoides eleganta* (Plummer) is a small species that is frequently overlooked, as it is mainly found in the fine fraction. Specimens from the Weddell Sea, figured by Thomas (1990) and herein, do not possess a well-developed uniserial portion, possibly representing juvenile forms. In these specimens, the ornamentation extends over the whole test, whereas Plummer's specimens possess smooth upper chambers in the uniserial portion.

### Biogeography

Worldwide.

### Bathymetry
Lower bathyal (Thomas 1990).

### Chronostratigraphy
First occurrence at the Early–Middle Eocene boundary (Biozones P9/P10), last occurrence in the Middle Eocene (Biozones P11/P12) in the Weddell Sea (Thomas 1990).

## References


Ortiz, S. and Thomas, E., 2006, Lower-middle Eocene benthic foraminifera from the Fortuna Section (Betic Cordillera, southeastern Spain). Micropaleontology, vol. 52, pp. 97–150.

Plummer, H.J., 1927, Foraminifera of the Midway Formation in Texas. Bulletin of the University of Texas Bureau of Economic Geology and Technology, vol. 2644, pp. 1–206.

Thomas, E., 1990, Late Cretaceous through Neogene benthic foraminifers. In Barker, P.F., Kennett, J.P., et al., ed., Scientific Results of the Proceedings of the Ocean Drilling Program, vol. 113, pp. 571–594.

# *Siphonina pozonensis*

## Cushman and Renz, 1941

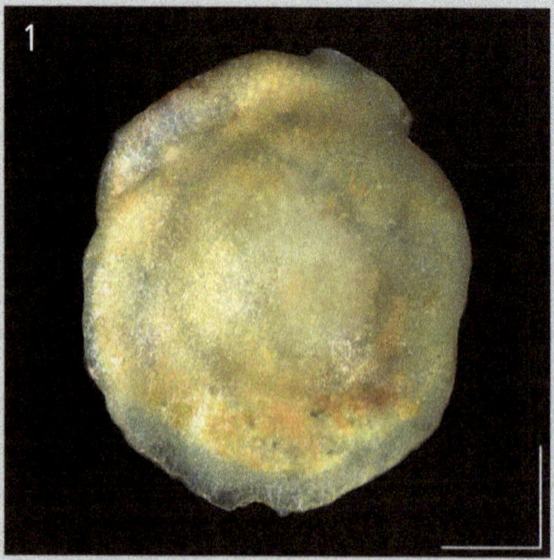

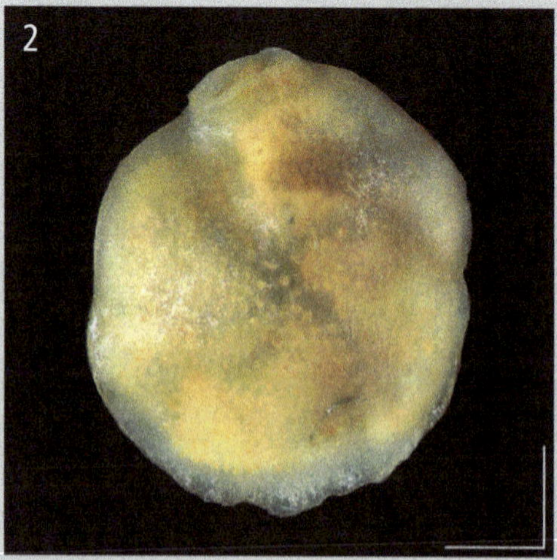

1. *Siphonina pozonensis.* Agua Salada Formation, Isidro, 33.2 km east (79° 25′) of Pueblo Pirita, District of Zamora, State of Falcón, Venezuela. Deposited in the collections of the United States National Museum, Washington, DC; CC 35913. Holotype. Spiral view, NHMUK PM DI 0614, scale bar = 112 μm.

2. *Siphonina pozonensis.* Agua Salada Formation, Isidro, 33.2 km east (79° 25′) of Pueblo Pirita, District of Zamora, State of Falcón, Venezuela. Deposited in the collections of the United States National Museum, Washington, DC; CC 35913. Holotype. Umbilical view, NHMUK PM DI 0615, scale bar = 109 μm.

### Original Designation

*Siphonina pozonensis* Cushman and Renz, 1941, p. 24, pl. 4, fig. 3.

### Type Specimen

Holotype (CC 35913) deposited in the collections of the United States National Museum, Washington, DC.

### Type Level

Miocene, Upper Agua Salada Formation, Zone 4.

### Type Locality

Sample L.P. 167, from Isidro, 33.2 kilometers east (79° 25′) of Pueblo Piritu, District of Zamora, State of Falcón, Venezuela.

### Description

Test forms a trochospire; subcircular and slightly lobulate in outline, and unequally biconvex in cross-section, with a slightly more convex, involute umbilical side, an evolute spiral side, and an acute peripheral keel (except for the final chambers). The (approximately) five slightly inflated, crescentic chambers in the last whorl are separated by flush, slightly limbate sutures; strongly oblique on the spiral side, and curved on the umbilical side. Chamber walls are calcareous, and coarsely perforate on both sides of the test. The primary aperture is areal, elliptical, nearly equatorial in position, and bordered by a phialine lip.

### Synonyms and Citations

*Siphonina pozonensis* Cushman and Renz. –van Morkhoven et al., 1986, p. 145, pl. 47. –Bolli et al., 1994, p. 364, figs. 80.8, 73.

### Remarks

*Siphonina pozonensis* Cushman and Renz differs from *Siphonina tenuicarinata* Cushman in having a more convex test and a broader keel that disappears by the final chamber.

### Biogeography

Recorded from Trinidad, Mexico, Venezuela, and Cuba (van Morkhoven et al. 1986).

### Bathymetry

Upper bathyal (van Morkhoven et al. 1986).

### Chronostratigraphy

Early to Middle Miocene.

## References

Bolli, H.M., Beckmann, J.P., and Saunders, J.B., 1994, Benthic Foraminiferal Biostratigraphy of the South Caribbean Region. Cambridge: Cambridge University Press, 408 pp.

Cushman, J.A. and Renz, H.H., 1941, New Oligocene–Miocene foraminifera from Venezuela. Contributions from the Cushman Laboratory for Foraminiferal Research, vol. 17, pp. 1–27.

van Morkhoven, F.P.C.M., Berggren, W.A., Edwards, A.S., et al., 1986, Cenozoic cosmopolitan deep-water benthic foraminifera. Bulletin des Centres de Recherches Exploration-Production Elf-Aquitaine, Mémoire, vol. 11, pp. 1–421.

# *Siphonina tenuicarinata*

## Cushman, 1927

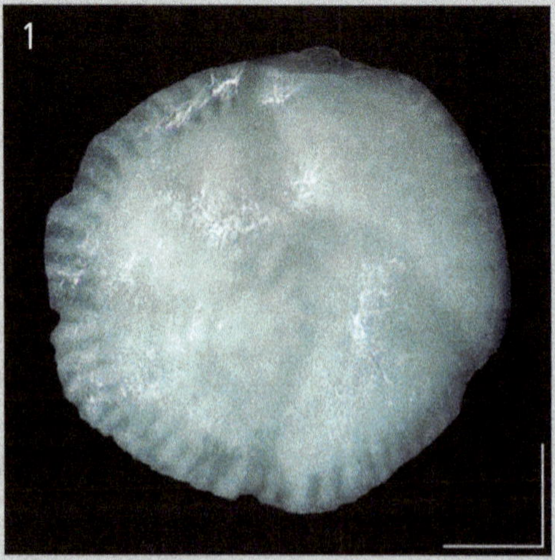

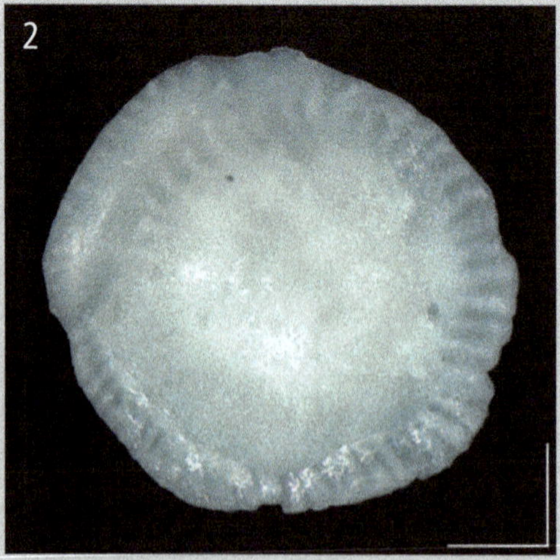

1. *Siphonina tenuicarinata.* Huasteca Formation North Bank, Rio Tuxpan 500–600 m downstream from Tumbadera landing, Vera Cruz, Mexico. Neotype. Umbilical view, NHMUK PM DI 0616, scale bar = 89 μm.

2. *Siphonina tenuicarinata.* Huasteca Formation North Bank, Rio Tuxpan 500–600 m downstream from Tumbadera landing, Vera Cruz, Mexico. Neotype. Spiral view, NHMUK PM DI 0617, scale bar = 90 μm.

### Original Designation

*Siphonina tenuicarinata* Cushman 1927, p. 166, pl. 26, figs. 11–12.

### Type Specimen

Holotype was never received by the United States National Museum. Van Morkhoven et al. (1986) designated a neotype (USNM 406406), originally deposited in the collections of the Paleontological Research Institute, Ithaca, NY, USA.

### Type Level

Neotype from Huasteca Formation (Upper Palma Real).

### Type Locality

Neotype from north bank of Rio Tuxpan, 500–600 meters downstream from Tumbadero Landing, Vera Cruz, Mexico.

### Description

Test forms a trochospire; subcircular and slightly lobulate in outline, nearly biconvex in cross-section with a slightly more convex, involute umbilical side, an evolute spiral side, and an acute, keeled periphery. The (approximately) five slightly inflated, crescentic chambers increase gradually in size, and are separated by flush, oblique sutures (forming an imperforate pinwheel pattern on the spiral side), and by curved, slightly depressed sutures on the umbilical side. Chamber walls are calcareous, and coarsely perforate on both sides of the test. Pores along the periphery of the last whorl indent the peripheral keel. The primary aperture is areal, elliptical, nearly equatorial in position, and bordered by a phialine lip.

### Synonyms and Citations

*Siphonina tenuicarinata* Cushman. –Tjalsma, 1983, p. 743, pl. 5, fig. 4. –van Morkhoven et al., 1986, p. 206, pl. 70. –Katz and Miller, 1993, pl. 5, fig. 1. –Robertson, 1998, p. 198, pl. 76, fig. 2.

### Remarks

*Siphonina tenuicarinata* Cushman differs from *Siphonina pozonensis* Cushman in having a flatter test and a continuous keel around the entire periphery.

Revised stratigraphy at the Eureka boreholes in the Gulf of Mexico (Katz and Miller 1993) has extended the last occurrence of this species to Biozone M7, in contrast to the M6 occurrence reported by van Morkhoven et al. (1986).

## Biogeography

Worldwide. Recorded from Trinidad, Venezuela, Mexico, Puerto Rico, Haiti, the Dominican Republic, and the North and South Atlantic (van Morkhoven et al. 1986).

## Bathymetry

Lower neritic to middle bathyal (van Morkhoven et al. 1986).

## Chronostratigraphy

Late Eocene to Middle Miocene.

# References

Cushman, J.A., 1927, Some characteristic Mexican fossil foraminifera. Journal of Paleontology, vol. 1, pp. 147–172.

Katz, M.E. and Miller, K.G., 1993, Latest Oligocene to Earliest Pliocene benthic foraminiferal biofacies of the northeastern Gulf of Mexico. Micropaleontology, vol. 39, no. 4, pp. 367–403.

Robertson, B.E., 1998, Systematics and paleoecology of the benthic foraminiferida from the Buff Bay section, Miocene of Jamaica. Micropaleontology, vol. 44, supplement 2, pp. 1–266.

Tjalsma, R.C., 1983, Eocene to Miocene benthic foraminifera from DSDP site 516, Rio Grande Rise, South Atlantic. In Barker, P.F., Carlson, R.L., Johnson, D.A., et al., ed., Initial Reports of the Deep Sea Drilling Project, Washington (U.S. Government Printing Office), vol. 72, pp. 731–755.

van Morkhoven, F.P.C.M., Berggren, W.A., Edwards, A.S., et al., 1986, Cenozoic cosmopolitan deep-water benthic foraminifera. Bulletin des Centres de Recherches Exploration-Production Elf-Aquitaine, Mémoire, vol. 11, pp. 1–421.

# *Siphotextularia concava*

## (Karrer), 1868

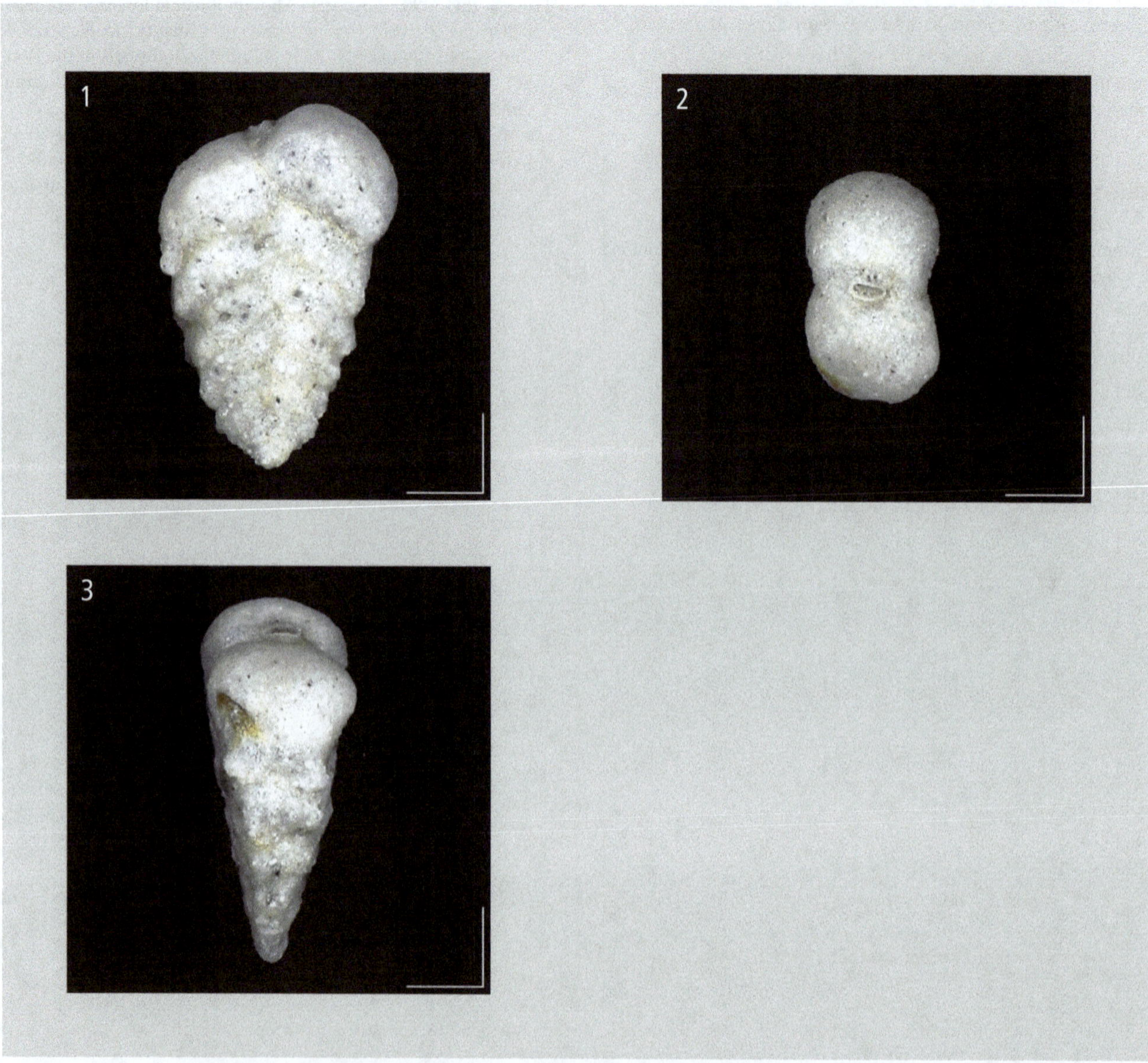


1. *Siphotextularia concava.* West of Patagonia, East Pacific Ocean; "Challenger" Station 308, depth: 175 fathoms. Deposited in the micropaleontological collections of The Natural History Museum, London; NHMUK ZF2443. Lateral view, NHMUK PM DI 0618, scale bar = 200 μm.

2. *Siphotextularia concava.* West of Patagonia, East Pacific Ocean; "Challenger" Station 308, depth: 175 fathoms. Deposited in the micropaleontological collections of The Natural History Museum, London; NHMUK ZF2443. Apertural view, NHMUK PM DI 0619, scale bar = 204 μm.

3. *Siphotextularia concava.* West of Patagonia, East Pacific Ocean; "Challenger" Station 308, depth: 175 fathoms. Deposited in the micropaleontological collections of The Natural History Museum, London; NHMUK ZF2443. Edge view, NHMUK PM DI 0620, scale bar = 221 μm.

## Original Designation
*Plecanium concavum* Karrer, 1868, p. 129, pl. 1, fig. 3.

## Type Specimen
Depository not given.

## Type Level
Miocene, Gainfahrn Marl, very sandy calcareous clay.

## Type Locality
Kostej im Banat, Romania.

## Description
Test forms an elongate, tapered, slightly lobulate, slightly compressed, biserial series, subquadrate in cross-section. The moderately inflated chambers increase gradually in size, and are separated by oblique, slightly depressed sutures. Chamber walls are finely agglutinated with calcareous cement, and canaliculate. The primary aperture is an areal slit, bordered by a lip, produced on a short neck.

## Synonyms and Citations
*Textularia concava* (Karrer). –Brady, 1884, p. 360, pl. 42, figs. 13–14 (not pl. 43, fig. 11).

*Siphotextularia concava* (Karrer). –Barker, 1960, pl. 42, figs. 13–14. –Jones, 1994, p. 47, pl. 42, figs. 13–14.

## Biogeography
Worldwide.

## Bathymetry
Neritic to upper bathyal. Reported from water depths of 17 and 175 fathoms at two Pacific "Challenger" stations (Jones 1994).

## Chronostratigraphy
Late Miocene to Recent (Jones 1994).

# References


Barker, R.W., 1960, Taxonomic notes on the species figured by H.B. Brady in his report on the foraminifera dredged by H.M.S. Challenger during the years 1873–1876. Society of Economic Paleontologists and Mineralogists Special Publication, vol. 9, pp. 1–238.

Brady, H.B., 1884, Report on the foraminifera dredged by H.M.S. Challenger during the years 1873–1876. In Brady, H.B., ed., Report on the Scientific Results of the Voyage of the H.M.S. Challenger during the years 1873–1876. Zoology, vol. 9, pp. 1–814.

Jones, R.W., 1994, The Challenger Foraminifera. Oxford: Oxford University Press, 149 pp.

Karrer, F., 1868, Die Miocene Foraminiferenfauna von Kostej im Banat. Sitzungsberichte der Kaiserlichen Akademie der Wissenschaften Wien, Mathematisch-Naturwissenschaftliche Klasse, vol. 58, no. 1, pp. 121–193.

# *Siphouvigerina porrecta*

## (Brady), 1879

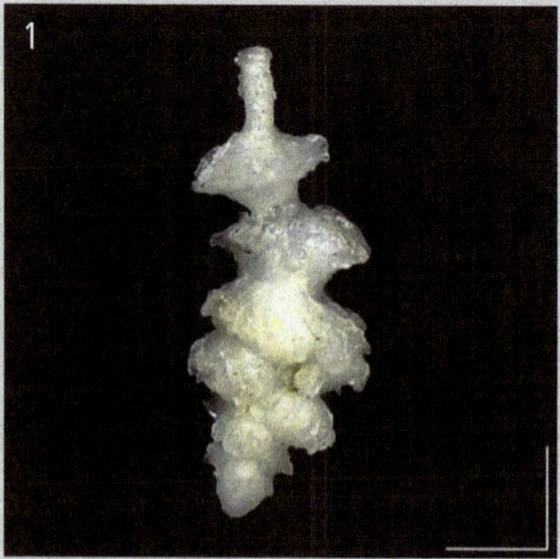


1. *Siphouvigerina porrecta.* Torres Strait, Pacific Ocean; "Challenger" Station 185, depth: 155 fathoms. Deposited in the micropaleontological collections of The Natural History Museum, London; NHMUK ZF2576. Syntype. Lateral view, NHMUK PM DI 0621, scale bar = 119 μm.

### Original Designation

*Uvigerina porrecta* Brady, 1879, p. 274, pl. 8, figs. 15–16.

### Type Specimen

Figured specimens are housed in the micropaleontological collections of The Natural History Museum, London, UK, and are registered in slide ZF2576.

### Type Level

Recent.

### Type Locality

Not designated. Localities given: off Bermuda, 435 fathoms; off Papua, 155 fathoms; and a point about 10° north of the Equator, in nearly the same longitude as the latter, 1850 fathoms.

### Description

Test forms an initially triserial series, later becoming loosely triserial, then biserial with sharp edges around equatorial margins of chambers. The umbrella-shaped chambers increase rapidly in size, are attached to the preceding apertural neck, and are separated by constricted sutures in the juvenile stage. Chamber walls are calcareous, perforate, and ornamented by fine striae. The primary aperture is a terminal, round opening at the end of a tubular neck, bordered by a phialine lip.

### Synonyms and Citations

*Uvigerina porrecta* Brady. –Brady, 1884, p. 577, pl. 74, figs. 21–23. –Barker, 1960, pl. 74, figs. 21–23.

*Siphouvigerina porrecta* (Brady). –Jones, 1994, p. 86, pl. 74, figs. 21–23.

### Biogeography

Worldwide.

### Bathymetry

Neritic to bathyal. Reported between 12 and 1850 fathoms at "Challenger" stations (Jones 1994).

### Chronostratigraphy

Miocene to Recent (Jones 1994).

## References


Barker, R.W., 1960, Taxonomic notes on the species figured by H.B. Brady in his report on the foraminifera dredged by H.M.S. Challenger during the years 1873–1876. Society of Economic Paleontologists and Mineralogists Special Publication, vol. 9, pp. 1–238.

Brady, H.B., 1879, Notes on some of the reticularian Rhizopoda of the Challenger Expedition, Part I; On new or little known Arenaceous types. Quarterly Journal of Microscopical Science (New Series), vol. 19, pp. 20–63.

Brady, H.B., 1884, Report on the foraminifera dredged by H.M.S. Challenger during the years 1873–1876. In Brady, H.B., ed., Report on the Scientific Results of the Voyage of the H.M.S. Challenger during the years 1873–1876. Zoology, vol. 9, pp. 1–814.

Jones, R.W., 1994, The Challenger Foraminifera. Oxford: Oxford University Press, 149 pp.

# *Sphaeroidina bulloides*

**d'Orbigny, 1826**

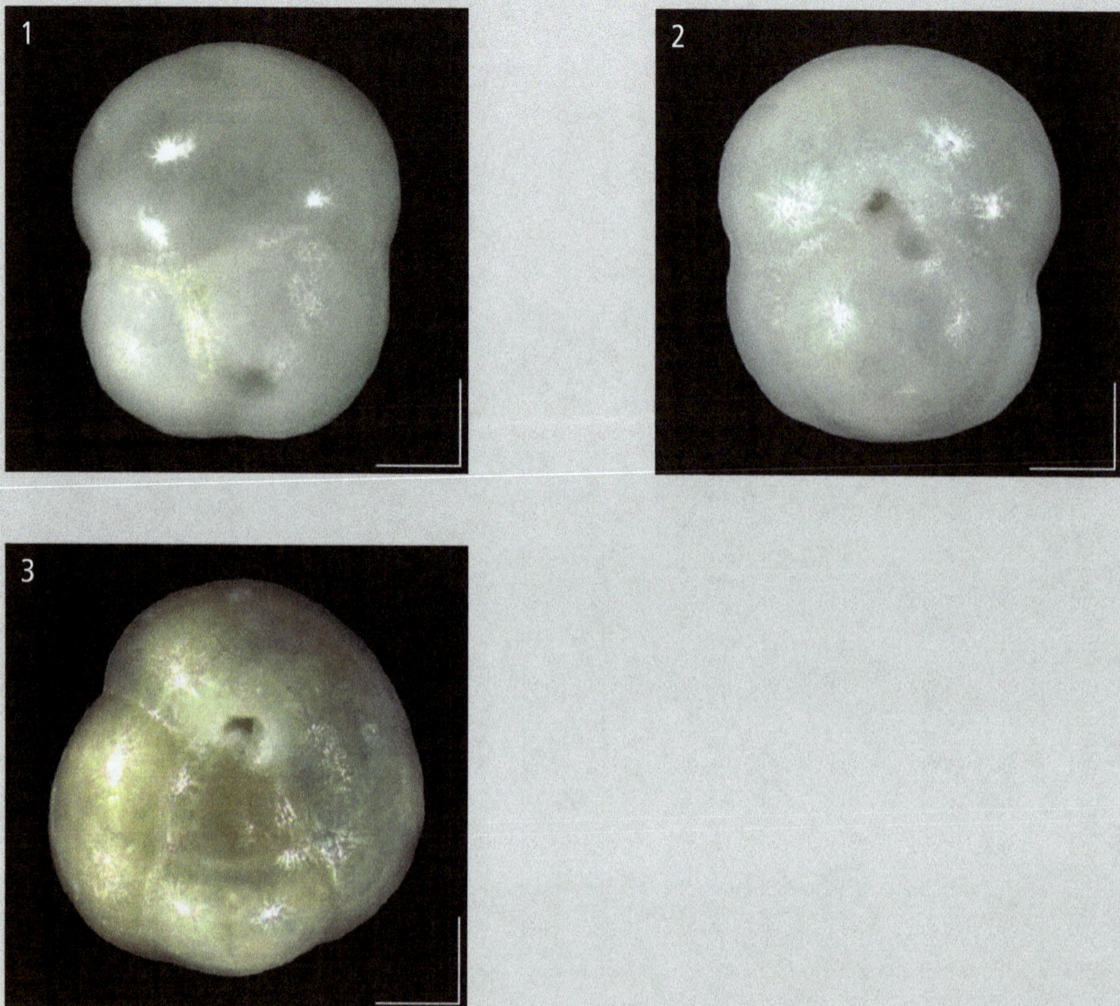


1. *Sphaeroidina bulloides.* Gulf of Mexico. R.W. Jones Collection. Spiral view, NHMUK PM DI 0622, scale bar = 88 μm.
2. *Sphaeroidina bulloides.* Gulf of Mexico. R.W. Jones Collection. Umbilical view, NHMUK PM DI 0623, scale bar = 88 μm.
3. *Sphaeroidina bulloides.* Gulf of Mexico; Eureka Core E68-136, depth: 2026 feet. Umbilical view, NHMUK PM DI 0624, scale bar = 147 μm.

## Original Designation

*Sphaeroidina bulloides* d'Orbigny 1826, p. 267, Model No. 65, illustrated in Parker et al., 1865, pl. 2, fig. 58.

## Type Specimen

Deposited in the collections of the Laboratoire de Paléontologie, Muséum d'Histoire Naturelle, Paris, France.

## Type Level

Not designated.

## Type Locality

Not designated. Localities given: L'île de France, la mer Adriatique near Rimini, Italy (Recent) near Sienna, Italy (Fossil). Papp and Schmid (1985) illustrated specimens from material collected in Nussdorf by Hauer, and previously described by d'Orbigny (1846).

## Description

Test forms a subglobular, variable coil. The strongly embracing, hemispherical, globular chambers are centered over the preceding aperture, but diverge in direction from previous apertural planes by angles of 90–180°. Chambers are separated by slightly depressed sutures; walls are calcareous, imperforate, and smooth. The primary aperture is a crescentic opening, bordered by a lip, and with a simple or bifid tooth; the aperture is commonly located close to the junction of three chambers.

## Synonyms and Citations

*Sphaeroidina austriaca* d'Orbigny, 1846, p. 284, pl. 20, figs. 19–21.

*Sexloculina haueri* Czjzek, 1848, p. 149, pl. 13, figs. 35–38.

*Sphaeroidina ciperana* Cushman and Todd, 1949, p. 67, pl. 12, fig. 14.

*Sphaeroidina compacta* Cushman and Todd, 1949, p. 19, pl. 4, fig. 14.

*Sphaeroidina compressa* Cushman and Todd, 1949, p. 20, pl. 4, figs. 15–16.

*Sphaeroidina nitida* Cushman and Todd, 1949, p. 20, pl. 4, fig. 13.

*Sphaeroidina japonica* Asano, 1953, p. 19, pl. 2, figs. 43–44.

*Sphaeroidina bulloides* d'Orbigny. –Papp and Schmid, 1985, p. 96, pl. 90, figs. 7–12. –van Morkhoven et al., 1986, p. 80, pl. 24. –Jones, 1994, p. 91, pl. 84, figs. 1–2 (not figs. 3–7). –Robertson, 1998, p. 196, pl. 74, fig. 4. –Milker and Schmiedl, 2012, p. 100, figs. 23.3–4.

## Biogeography

Worldwide.

## Bathymetry

Neritic to abyssal. Reported between 85 and 2600 fathoms at "Challenger" stations (Jones, 1994).

## Chronostratigraphy

Early Oligocene to Recent.

# References


Asano, K., 1953, Miocene foraminifera from the Noto Peninsula, Ishikawa Prefecture. Short Papers, Institute of Geology and Paleontology, Tohoku University, vol. 5, pp. 1–21.

Cushman, J.A. and Todd, R., 1949, The genus Sphaeroidina and its species. Special Publications of the Cushman Laboratory for Foraminiferal Research, vol. 25, pp. 11–21.

Czjzek, J., 1848, Beitrag zur Kenntniss der fossilen Foraminiferen des Wiener Beckens. Naturwissenschaftliche Abhadlungen, Wien, vol. 2, no. 1, pp. 137–150.

d'Orbigny, A., 1826, Tableau méthodique de la classe des Céphalopodes. Annales des Sciences Naturelles, vol. 7, pp. 245–314.

d'Orbigny, A.D., 1846, Foraminifères fossiles du Bassin Tertiaire de Vienne (Autriche). Paris: Gide et Compe, xxxvii + 312 pp.

Jones, R.W., 1994, The Challenger Foraminifera. Oxford: Oxford University Press, 149 pp.

Milker, Y. and Schmiedl, G., 2012, A taxonomic guide to modern benthic shelf foraminifera of the western Mediterranean Sea. Palaeontologia Electronica, vol. 15, Issue 2; 16A, 134 pp.

Papp, A. and Schmid, M.E., 1985, Die fossilen Foraminiferen des tertiaeren Beckens von Wien: Revision der Monographie Alcide d'Orbigny (1846). Abhandlungen der Geologischen Bundesanstalt, vol. 37, pp. 1–311.

Parker, W.K., Jones, T.R., and Brady, H.B., 1865, On the nomenclature of the foraminifera, Pt. 12: the species enumerated by d'Orbigny in the "Annales des Sciences Naturelles", vol. 7, 1826. Annals and Magazine of Natural History, series no. 3, vol. 16, pp. 15–41.

Robertson, B.E., 1998, Systematics and paleoecology of the benthic foraminiferida from the Buff Bay section, Miocene of Jamaica. Micropaleontology, vol. 44, supplement 2, pp. 1–266.

van Morkhoven, F.P.C.M., Berggren, W.A., Edwards, A.S., et al., 1986, Cenozoic cosmopolitan deep-water benthic foraminifera. Bulletin des Centres de Recherches Exploration-Production Elf-Aquitaine, Mémoire, vol. 11, pp. 1–421.

# Spiroplectammina spectabilis

## (Grzybowski), 1898 emend. Kaminski, 1984

1. *Spiroplectammina spectabilis.* Karaś Potok, Poland. Deposited in the Geroch Collection, Jagiellonian University, Kraków, Poland. Megalospheric form. Lateral view, NHMUK PM DI 0625, scale bar = 514 μm.

2. *Spiroplectammina spectabilis.* Karaś Potok, Poland. Deposited in the Geroch Collection, Jagiellonian University, Kraków, Poland. Megalospheric form. Lateral view, NHMUK PM DI 0626, scale bar = 503 μm.

3. *Spiroplectammina spectabilis.* Karaś Potok, Poland. Deposited in the Geroch Collection, Jagiellonian University, Kraków, Poland. Microspheric form. Lateral view, NHMUK PM DI 0627, scale bar = 509 μm.

### Original Designation
*Spiroplecta spectabilis* Grzybowski, 1898, p. 293, pl. 12, fig. 12.

### Type Specimen
Not originally designated. Kaminski and Geroch (1993) designated a lectotype from one of Grzybowski's slides (UJ-132-P. 1/82a) housed in the Grzybowski Collection at the Jagiellonian University, Poland.

### Type Level
Paleogene, red and gray clay, Silesian Unit of the Polish Carpathians.

### Type Locality
Not originally designated. Lectotype is from Hanover-Galicia well no. 3. 170 meters at Potok, near Krosno, Poland.

### Description
Test forms an initial planispire, later becoming a biserial series. In megalospheric forms, the initial planispiral portion, consisting of four to seven chambers, may be wider than the biserial portion. In microspheric forms, the planispiral portion is much smaller. In both generations, the biserial part is rhomboidal in cross-section with nearly parallel sides. However, in microspheric forms, it may increase in breadth initially, then decrease distally. Chambers in the biserial part are low, and numerous, with flush or slightly depressed sutures, inclined approximately 60° to the long axis of the test. The peripheral margin is acute, and may be weakly keeled. Chamber walls are imperforate, finely agglutinated with organic cement, and smoothly finished. The primary aperture is a narrow interiomarginal slit.

### Synonyms (following Kaminski and Gradstein 2005) and Citations
*Spiroplecta brevis* Grzybowski, 1898, p. 293, pl. 12, fig. 13.
*Spiroplecta costidorsata* Grzybowski, 1898, p. 294, pl. 12, fig. 11.
*Spiroplecta foliacea* Grzybowski, 1898, p. 293, pl. 12, fig. 14–15.
*Spiroplecta clotho* Grzybowski, 1901, p. 283, pl. 7, fig. 18.
*Spiroplectoides californica* Cushman and Campbell, 1934, p. 70, pl. 9, figs. 13–14.
*Spiroplectammina mexiaensis* Lalicker, 1935, p. 43, pl. 6, figs. 5–6.
*Spiroplectoides directa* Cushman and Siegfus, 1939, p. 26, pl. 6, figs. 7–8.
*Spiroplectammina grzybowskii* Frizzell, 1943, p. 339, pl. 55, figs. 12a–13.
*Spiroplectammina perplexa* Israelsky, 1951, p. 12, pl. 3, figs. 9–14.
*Spiroplectammina brunswickensis* Todd and Kniker, 1952, p. 6, pl. 1, fig. 16.
*Spiroplectammina (Spiroplectinella) spectabilis* (Grzybowski). –Charnock and Jones, 1990, p. 182, pl. 9, figs. 15–18; pl. 21, fig. 5.
*Spiroplectammina spectabilis* (Grzybowski) emended Kaminski, 1984, p. 29, pl. 12, figs. 1–9, pl. 13, figs. 1–8. –Kaminski et al., 1990, p. 369, pl. 6, figs. 5–6. –Kaminski and Geroch, 1993, p. 267, pl. 12, figs. 4a–5c. –Gradstein et al., 1994, pl. 2, figs. 14–18. –Kaminski and Gradstein, 2005, pp. 429–435, fig. 104, pl. 104.

## Remarks

This species exhibits considerable variability, which has contributed to the multiplicity of junior synonyms (Kaminski, 1984).

## Biogeography

Worldwide. *Spiroplectammina spectabilis* (Grzybowski) was originally described from the Paleogene of the Carpathian flysch (Grzybowski 1898), and has been subsequently found in Europe, the North Sea, the North and South Pacific, and the North and South Atlantic (see Observed Occurrences in Kaminski and Gradstein 2005).

## Bathymetry

Bathyal to abyssal. During the early portion of its range, *Spiroplectammina spectabilis* appears to have been restricted to flysch-type assemblages, since it does not occur in Maastrichtian assemblages from the abyssal plains (Kuhnt and Kaminski 1990).

## Chronostratigraphy

Late Cretaceous (Maastrichtian) to Late Eocene. In its type area, the first frequent occurrence of this species marks the base of the Late Paleocene *Spiroplectammina spectabilis* Biozone of Geroch and Nowak (1984). However, *Spiroplectammina spectabilis* has a diachronous first occurrence, which may reflect the immigration of the species into various ocean basins. For instance, it occurs in the Maastrichtian in Trinidad (Kaminski et al. 1988), on Sakhalin Island (Serova 1987), and in the lower Paleocene of other areas of the Atlantic. At many localities, *Spiroplectammina spectabilis* displays a distinctive abundance maximum immediately after the Cretaceous–Paleogene boundary in Biozone P0. Its last common occurrence occurs near the base of the uppermost Paleocene to lowermost Eocene tuff horizon in the North Sea and Barents Sea, and near the Paleocene–Eocene boundary along the Labrador Margin (Gradstein et al. 1994). In these regions, its final occurrence is generally within the Middle Eocene. In paleobathymetrically deeper localities, *Spiroplectammina spectabilis* ranges into the Upper Eocene. In the Polish Carpathians, its last occurrence coincides with the base of the uppermost Eocene "*Globigerina* Marls". At ODP Site 647 in the Labrador Sea, its last occurrence coincides precisely with the location of the Eocene–Oligocene boundary, as determined by nannofossil- and magnetostratigraphy (Kaminski et al. 1989).

## References


Charnock, M.A. and Jones, R.W., 1990, Agglutinated foraminifera from the Palaeogene of the North Sea. In Hemleben, C., Kaminski, M.A., Kuhnt, W., and Scott, D.B., ed., Paleoecology, Biostratigraphy, Paleoceanography and Taxonomy of Agglutinated Foraminifera, NATO ASI Series, C: Mathematical and Physical Sciences. Dordrecht: Kluwer Academic, pp. 139–244.

Cushman, J.A. and Campbell, A.S., 1934, A new Spiroplectoides from the Cretaceous of California. Contributions from the Cushman Laboratory for Foraminiferal Research, vol. 10, no. 3, p. 70.

Cushman, J.A. and Siegfus, S.S., 1939, Some new and interesting foraminifera from the Kreyenhagen Shale of California. Contributions from the Cushman Laboratory for Foraminiferal Research, vol. 15, no. 2, pp. 1–26.

Frizzell, D.L., 1943, Upper Cretaceous foraminifera from northwestern Peru. Journal of Paleontology, vol. 17, pp. 337.

Geroch, S. and Nowak, W., 1984, Proposal of zonation for the late Tithonian–late Eocene, based upon arenaceous foraminifera from the outer Carpathians, Poland. In Oertli, H., ed., Benthos '83, 2nd International Symposium on Benthic Foraminifera, Pau (France), April 11–15, 1983. Elf Aquitaine, ESSO REP and TOTAL CFP, Pau and Bordeaux, pp. 225–239.

Gradstein, F.M., Kaminski, M.A., Berggren, W.A., and D'Iorio, M.A., 1994, Cenozoic biostratigraphy of the Central North Sea and Labrador Shelf. Micropaleontology, vol. 40 (Supplement), pp. 1–152.

Grzybowski, J., 1898, Otwornice pokładów naftonosnych okolicy Krosna. Rozprawy Wydziału Matematyczno-Przyrodniczego, Akademia Umiejętności w Krakowie, serya 2, vol. 33, pp. 257–305.

Grzybowski, J., 1901, Otwornice warstw inoceramowych okolicy Gorlic. Rozprawy Wydziału Matematyczno-Przyrodniczego, Akademia Umiejętności w Krakowie, serya 2, vol. 41, pp. 219–286.

Israelsky, M.C., 1951, Foraminifera of the Lodo Formation, central California; General introduction and Part 1, arenaceous foraminifera. Professional Paper of the United States Geological Survey, no. 240-A, pp. 1–29.

Kaminski, M.A., 1984, Shape variation in Spiroplectammina spectabilis (Grzybowski). Acta Palaeontologica Polonica, vol. 29, pp. 29–46.

Kaminski, M.A. and Geroch, S., 1993, A revision of foraminiferal species in the Grzybowski Collection. In Kaminski, M.A., Geroch, S., and Kaminski, D.G., ed., The Origins of Applied Micropalaeontology: The School of Jósef Grzybowski (with a taxonomic revision of the Grzyboswski Collection). Grzybowski Foundation Special Publication, no. 1, pp. 239–330.

Kaminski, M.A. and Gradstein, F.M., 2005, Atlas of Palaeogene Cosmopolitan Deep-Water Agglutinated Foraminifera. Grzybowski Foundation Special Publication, no. 10, 547 + vii pp.

Kaminski, M.A., Gradstein, F.M., and Berggren, W.A., 1989, Paeogene benthic foraminifer biostratigraphy and paleoecology at site 647, southern Labrador Sea. In Srivastava, S.P., Arthur, M.A., Clement, B., et al., ed., Scientific Results of the Proceedings of the Ocean Drilling Program, vol. 105, pp. 705–730.

Kaminski, M.A., Gradstein, F.M., Berggren, W.A., Geroch, S., and Beckmann, J.P., 1988, Agglutinated foraminiferal assemblages from Trinidad: taxonomy, stratigraphy and paleobathymetry. In Gradstein, F.M. and Rögl, F., ed., Second International Workshop on Agglutinated Foraminifera, Vienna 1986, Abhandlungen der Geologischen Bundesanstalt, vol. 41, pp. 155–228.

Kaminski, M.A., Gradstein, F.M., Goll, R.M., and Greig, D., 1990, Biostratigraphy and paleoecology of deep-water aggutinated foraminifera at ODP Site 643, Norwegian–Greenland Sea. In Hemleben, C., Kaminski, M.A., Kuhnt, W., and Scott, D.B., ed., Paleoecology, Biostratigraphy, Paleoceanography and Taxonomy of Agglutinated Foraminifera. NATO ASI Series, Series C: Mathematical and Physical Sciences, vol. 327. Dordrecht: Kluwer Academic, pp. 345–386.

Kuhnt, W. and Kaminski, M.A., 1990, Paleoecology of Late Cretaceous to Paleocene deep-water agglutinated foraminifera from the North Atlantic and western Tethys. In Hemleben, C., Kaminski, M.A., Kuhnt, W., and Scott, D.B., ed., Paleoecology, Biostratigraphy, Paleoceanography and Taxonomy of Agglutinated Foraminifera. NATO ASI Series, Series C: Mathematical and Physical Sciences. Dordrecht: Kluwer Academic, vol. 327, pp. 433–505.

Lalicker, C.G., 1935, New Cretaceous Textulariidae. Contributions from the Cushman Laboratory for Foraminiferal Research, vol. 11, pp. 39–51.

Serova, M. Ya., 1987, Foraminifery i biostratigrafia Severnoy Patsifiki na rubiezhie Mela i Paleogena. Izdatelstvo "Nauka," Moscow, pp. 1–144.

Todd, R. and Kniker, H.T., 1952, An Eocene foraminiferal fauna from the Agua Fresca shale of Magallanes Province, southernmost Chile. Special Publications of the Cushman Foundation for Foraminiferal Research, no. 1, pp. 1–28.

# *Spiroplectinella subhaeringensis*

## (Grzybowski), 1896

1. *Spiroplectinella subhaeringensis.* Żywiec, Subsilesian Unit of Polish Carpathians. Deposited in the Geroch Collection, Jagiellonian University, Kraków, Poland. Lateral view, NHMUK PM DI 0628, scale bar = 341 μm.

### Original Designation

*Textularia subhaeringensis* var. α Grzybowski, 1896, p. 285, pl. 9, Fig. 16.

*Textularia subhaeringensis* var. β Grzybowski, 1896, p. 285, pl. 9, Fig. 13.

### Type Specimen

Not originally designated. Specimens corresponding to Grzybowski's illustrations, found by Liszka and Liszkowa (1981), are deposited in the Grzybowski Collection at the Jagiellonian University, Kraków, Poland (slide nos UJ-115P, 44, 47).

### Type Level

Campanian, variegated shales and marls, Subsilesian Unit of Polish Carpathians.

### Type Locality

Exploratory coal shaft about 500 meters southwest of the courthouse in Wadowice, near Kraków, Poland. Between 64 and 70 meters below the top.

### Description

Test forms an initial planispire, later becoming an elongate, flaring biserial series, increasing markedly in thickness, and developing a serrate outline and a lozenge-shaped cross-section. The microspheric generation differs from the macrospheric generation by its more compressed, elongate shape, and its lanceolate outline, with shorter spines at the periphery. The low and broad chambers are separated by arched, depressed sutures. Chamber walls are imperforate, medium to finely agglutinated, and smoothly cemented with calcareous cement. The primary aperture is a low arch at the base of the last chamber.

### Synonyms and Citations

*Textularia excolata* Cushman, 1926, p. 585, pl. 15, fig. 9.

*Spiroplectammina subhaeringensis* (Grzybowski). –Liszka and Liszkowa, 1981, p. 178, pl. 3, figs. 6–7. –Hemleben and Tröster, 1984, p. 522, pl. 3, figs. 21–23. –Geroch and Nowak, 1984, pl. 4, figs. 1–4. –Kaminski and Gradstein, 2005, pp. 444–448, fig. 108, pl. 108.

*Spiroplectammina (Spiroplectinella) subhaeringensis* (Grzybowski). –Charnock and Jones, 1990, p. 183, pl. 21, fig. 6.

### Remarks

Grzybowski (1896) suggested that *Spiroplectammina subhaeringensis* was ancestral to *Vulvulina pennatula* (Batsch).

### Biogeography

Worldwide.

### Bathymetry

Bathyal to abyssal (Kaminski and Gradstein 2005).

### Chronostratigraphy

Late Cretaceous (Campanian) to Paleocene (Kaminski and Gradstein 2005).

## References


Charnock, M.A. and Jones, R.W., 1990, Agglutinated foraminifera from the Palaeogene of the North Sea. In Hemleben, C., Kaminski, M.A., Kuhnt, W., and Scott, D.B., ed., Paleoecology, Biostratigraphy, Paleoceanography and Taxonomy of Agglutinated Foraminifera, NATO ASI Series, C: Mathematical and Physical Sciences. Dordrecht: Kluwer Academic, pp. 139–244.

Cushman, J.A., 1926, The foraminifera of the Velasco Shale of the Tampico Embayment. Bulletin of the American Association of Petroleum Geologists, vol. 10, no. 6, pp. 581–612.

Geroch, S. and Nowak, W., 1984, Proposal of zonation for the late Tithonian–late Eocene, based upon arenaceous foraminifera from the outer Carpathians, Poland. In Oertli, H., ed., Benthos '83, 2nd International Symposium on Benthic Foraminifera, Pau (France), April 11–15, 1983. Elf Aquitaine, ESSO REP and TOTAL CFP, Pau and Bordeaux, pp. 225–239.

Grzybowski, J., 1896, Otwornice czerwonych iłów z Wadowic. Rozprawy Wydziału Matematyczno-Przyrodniczego, Akademia Umiejętności w Krakowie, serya 2, vol. 30, pp. 261–308.

Hemleben, C. and Tröster, J., 1984, Campanian–Maestrichtian deep-water foraminifers from Hole 543A, Deep Sea Drilling Project. In Bij-Duval, B., Moore, J.C., et al., ed., Initial Reports of the Deep Sea Drilling Project, Washington (U.S. Government Printing Office), vol. 78A, pp. 509–532.

Kaminski, M.A. and Gradstein, F.M., 2005, Atlas of Palaeogene Cosmopolitan Deep-Water Agglutinated Foraminifera. Grzybowski Foundation Special Publication, no. 10, 547 + vii pp.

Liszka, S. and Liszkowa, J., 1981, Revision of J. Grzybowski's paper (1896) "Foraminifera of the red clays from Wadowice". Rocznik Polskiego Towarzystwa Geologicznego, vol. 51, pp. 153–208.

# *Stensioeina beccariiformis*

## (White), 1928

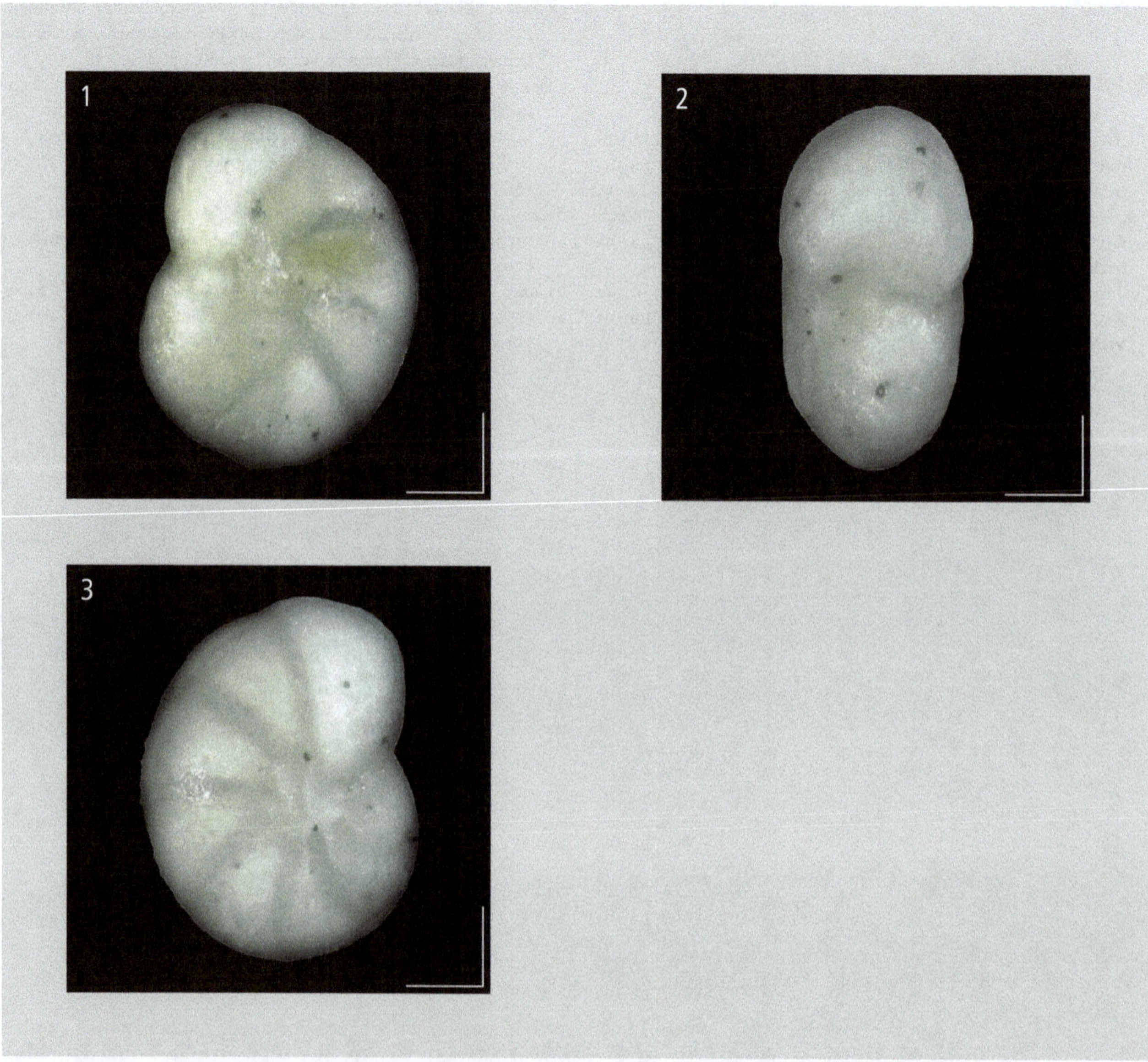


1. *Stensioeina beccariiformis.* Gulf of Mexico; Eureka Core E68-139. Deposited in the collections of the United States National Museum, Washington, DC. Specimen illustrated in van Morkhoven et al. 1986, Ref. no. 113. Spiral view, NHMUK PM DI 0629, scale bar = 136 μm.

2. *Stensioeina beccariiformis.* Gulf of Mexico; Eureka Core E68-139. Deposited in the collections of the United States National Museum, Washington, DC. Specimen illustrated in van Morkhoven et al. 1986, Ref. no. 113. Edge view, NHMUK PM DI 0630, scale bar = 131 μm.

3. *Stensioeina beccariiformis.* Gulf of Mexico; Eureka Core E68-139. Deposited in the collections of the United States National Museum, Washington, DC. Specimen illustrated in van Morkhoven et al. 1986, Ref. no. 113. Umbilical view, NHMUK PM DI 0631, scale bar = 138 μm.

## Original Designation

*Rotalia beccariiformis* White, 1928, p. 287, pl. 39, fig. 2.

## Type Specimen

Holotype (No. 19892) deposited in the collections of the Department of Invertebrate Paleontology at the American Museum of Natural History, New York, USA.

## Type Level

Upper Cretaceous, Lower Velasco Formation, shale. Rare in lower portion, common at top of section.

## Type Locality

One half kilometer west of Velasco Station, Tampico Embayment, Mexico.

## Description

Test forms a low trochospire; planoconvex or unequally biconvex in cross-section, with a flattened or slightly convex, evolute, spiral side, a moderately convex, involute umbilical side, and a rounded periphery. The (approximately) eight to ten moderately inflated chambers in the last whorl increase gradually in size, and are separated by curved sutures, flush on the spiral side, and limbate on the umbilical side. Chamber walls are calcareous, coarsely perforate, and smooth on the spiral side; more coarsely perforate on the umbilical side that is also ornamented by lines radiating irregularly from the central area. These range from thin, depressed, thread-like lines occasionally divided into beads near the umbilicus, to thick, depressed lines that subdivide toward the periphery, and may be interspersed with beads (van Morkhoven et al., 1986). The primary aperture is an umbilical–extraumbilical, elliptical opening bordered by a thick lip, and continuing into the umbilicus under an umbilical flap. Successive umbilicate flaps are partially imbricate, covering the umbilicus.

## Synonyms (following van Morkhoven et al. 1986) and Citations

*Rotalia beccarii* (Linné). –Wright,1881, p. 132, pl. 27, fig. 15.
*Gyroidina infrafossa* Finlay, 1940, p. 462, pl. 66, figs. 181–183.
*Rotalia parvula* ten Dam, 1944, p. 121, pl. 4, fig. 1.
*Pseudovalvulineria vombensis* Brotzen, 1945, p. 50, pl. 1, figs. 12–13.
*Gyroidina caucasica* Subbotina, 1947, p. 100, pl. 3, figs. 23–25.
*Stensioeina praecaucasica* Vasilenko, 1961, p. 71, pl. 12, figs. 2, 5.
*Anomalina whitei* Martin, 1964, p. 106, pl. 16, fig. 4.
*Pseudovalvulineria* sp. McGowran, 1965, p. 49, pl. 3, fig. 4.
*Anomalina beccariiformis* (White). –Cushman and Renz, 1946, p. 48, pl. 8, figs. 21–22.
*Stensioeina beccariiformis* (White). –van Morkhoven et al., 1986, p. 346, pl. 113A–D.
*Gavelinella beccariiformis* (White). –Thomas, 1990, p. 590, pl. 3, fig. 5. –Bolli et al., 1994, p. 161, figs. 46.4–6.

## Remarks

*Rotalia parvula* ten Dam is a shallow-water morphotype, commonly recorded in post-Danian neritic deposits of northern Europe.

## Biogeography

Worldwide. See Observed Occurrences in van Morkhoven et al. (1986). Weddell Sea, Antarctica (Thomas, 1990).

## Bathymetry

Bathyal to abyssal. Lower bathyal in Weddell Sea (Thomas, 1990).

## Chronostratigraphy

Late Cretaceous (Santonian) to Late Paleocene.

# References


Bolli, H.M., Beckmann, J.P., and Saunders, J.B., 1994, Benthic Foraminiferal Biostratigraphy of the South Caribbean Region. Cambridge: Cambridge University Press, 408 pp.

Brotzen, F., 1945, De geologiska resultaten från borrningarna vid Hoellviken: Preliminar rapport – Del 1: Kritan. Årsbok Sveriges Geologiska Undersökning, vol. 38 (1944), no. 7, pp. 3–64.

Cushman, J.A. and Renz, H.H., 1946, The foraminiferal fauna of the Lizard Springs Formation of Trinidad, British West Indies. Special Publications of the Cushman Laboratory for Foraminiferal Research, vol. 18, pp. 1–48.

Finlay, H.J., 1940, New Zealand Foraminifera; key species in stratigraphy – No. 4. Transactions of the Royal Society of New Zealand, vol. 69, no. 4, pp. 448–472.

Linnaeus, C., 1758, Systema Naturae, 10th edn. Lipsiae (Leipzig): G. Engelman, vol. 1, 824 pp.

Martin, L., 1964, Upper Cretaceous and Lower Tertiary foraminifera from Fresno County, California. Jahrbuch der Geologischen Bundesanst, Austria, vol. 9, pp. 1–128.

McGowran, B., 1965, Two Paleocene foraminiferal faunas from the Wangerrip Group, Pebble Point Coastal Section, Western Victoria. Proceedings of the Royal Society of Victoria, vol. 79, no. 1, pp. 9–74.

Subbotina, N.N., 1947, Microfauna of the oil fields of the Caucasus, Emba, and Central Asia. In Foraminifera of the Danian and Paleogene deposits of the North Caucasus. Leningrad: Vsesoyuznogo Neftyanogo Nauchno-Issledovatel'skogo, Geologo-razvedochnogo Instituta (VNIGRI), pp. 39–160.

ten Dam, A., 1944, Die Stratigraphische Gliederung des niederlaendlichen Paleozaens und Eozaens nach Foraminiferen (mit Ausnahme von Sued-Limburg). Mededeelingen van de Geologische Stichting, ser. C, vol. 5, no. 3, pp. 1–142.

Thomas, E., 1990, Late Cretaceous through Neogene benthic foraminifers. In Barker, P.F., Kennett, J.P., et al., ed., Scientific Results of the Proceedings of the Ocean Drilling Program, vol. 113, pp. 571–594.

van Morkhoven, F.P.C.M., Berggren, W.A., Edwards, A.S., et al., 1986, Cenozoic cosmopolitan deep-water benthic foraminifera. Bulletin des Centres de Recherches Exploration-Production Elf-Aquitaine, Mémoire, vol. 11, pp. 1–421.

Vasilenko, V.P., 1961, Foraminifera from the Upper Cretaceous of the Mangyshlak Peninsula. Trudy Vsesoyuznogo Neftyanogo Nauchno-Issledovatel'skogo, Geologo-razvedochnogo Instituta (VNIGRI) (New Series), vol. 171, pp. 1–487.

White, M.P., 1928, Some index foraminifera of the Tampico Embayment area of Mexico (Part II). Journal of Paleontology, vol. 2, pp. 280–317.

Wright, J., 1881, A list of the post-Tertiary foraminifera of the north-east of Ireland. Proceedings of the Belfast Naturalists' Field Club (New Series), vol. 1 (1873–80), pp. 149–163.

# *Stensioeina excolata*

## (Cushman), 1926

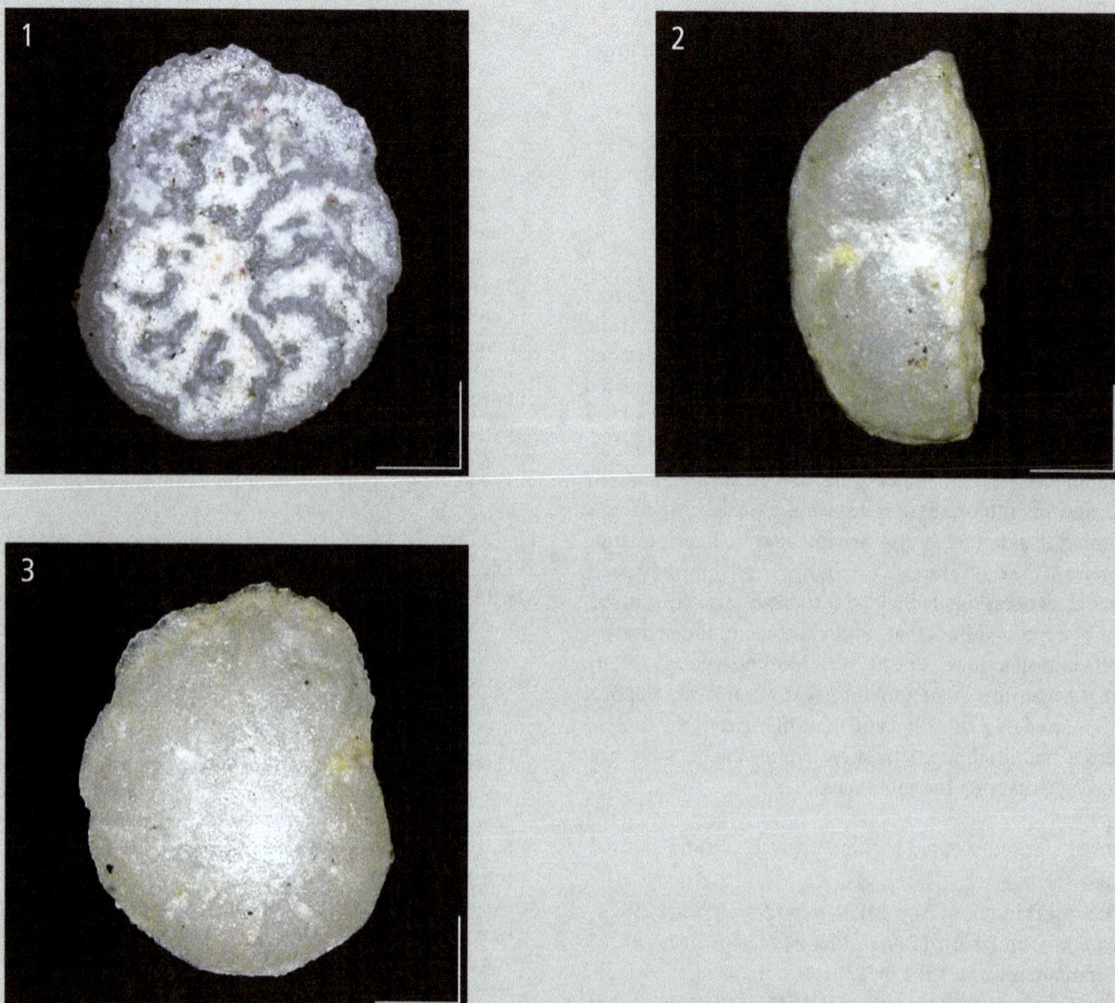

1. *Stensioeina excolata.* Mendez Formation, San Luis Potosi, Mexico. Heron-Allen and Earland Collection; deposited in the micropaleontological collections of The Natural History Museum, London; NHMUK P47558. Metatype. Spiral view, NHMUK PM DI 0632, scale bar = 95 μm.

2. *Stensioeina excolata.* Mendez Formation, San Luis Potosi, Mexico. Heron-Allen and Earland Collection; deposited in the micropaleontological collections of The Natural History Museum, London; NHMUK P47558. Metatype. Edge view, NHMUK PM DI 0633, scale bar = 91 μm.

3. *Stensioeina excolata.* Mendez Formation, San Luis Potosi, Mexico. Heron-Allen and Earland Collection; deposited in the micropaleontological collections of The Natural History Museum, London; NHMUK P47558. Metatype. Umbilical view, NHMUK PM DI 0634, scale bar = 87 μm.

## Original Designation

*Truncatulina excolata* Cushman, 1926, p. 22, pl. 3, fig. 2.

## Type Specimen

Holotype (No. 5104) deposited in the collections of the United States National Museum, Washington, DC.

## Type Level

Upper Cretaceous, Upper portion of the Papagallos Shales (Mendez Shale).

## Type Locality

Shale along the railroad near Coco, Hacienda El Limón, San Luis Potosi, Mexico.

## Description

Test forms a low trochospire; planoconvex in cross-section, with a flattened, evolute, spiral side, a convex, involute umbilical side, and a sharply angled periphery. The (approximately) ten crescentic chambers in the last whorl increase gradually in size, and are separated by curved, strongly raised, limbate sutures on the spiral side, and by gently curved, flush or slightly raised, limbate sutures on the umbilical side. Chamber walls are calcareous, coarsely perforate, and ornamented by raised, irregular costae, mainly following the spiral suture lines and giving the test a roughly sculptured spiral side. The primary aperture is an umbilical–extraumbilical, elliptical opening, bordered by a thick lip, and continuing into the umbilicus under an umbilical flap. Successive umbilicate flaps are partially imbricate, and provide a covering over the umbilicus.

## Synonyms and Citations

*Stensioeina labyrinthica* Cushman and Dorsey, 1940, p. 3, pl. 1, fig. 5.
*Stensioeina excolata* (Cushman). –van Morkhoven et al., 1986, p. 380, pl. 125. –Bolli et al., 1994, p. 161, figs. 45.31–33.

## Biogeography

Worldwide.

## Bathymetry

Lower neritic to bathyal.

## Chronostratigraphy

Late Cretaceous (Campanian to Maastrichtian).

## References


Bolli, H.M., Beckmann, J.P., and Saunders, J.B., 1994, Benthic Foraminiferal Biostratigraphy of the South Caribbean Region. Cambridge: Cambridge University Press, 408 pp.

Cushman, J.A., 1926, Some foraminifera from the Mendez Shale of eastern Mexico. Contributions from the Cushman Laboratory for Foraminiferal Research, vol. 2, no. 1, pp. 16–26.

Cushman, J.A. and Dorsey, A.L., 1940, The genus Stensiöiona and its species. Contributions from the Cushman Laboratory for Foraminiferal Research, vol. 16, pp. 1–6.

van Morkhoven, F.P.C.M., Berggren, W.A., Edwards, A.S., et al., 1986, Cenozoic cosmopolitan deep-water benthic foraminifera. Bulletin des Centres de Recherches Exploration-Production Elf-Aquitaine, Mémoire, vol. 11, pp. 1–421.

# *Stilostomella abyssorum*

## (Brady), 1881

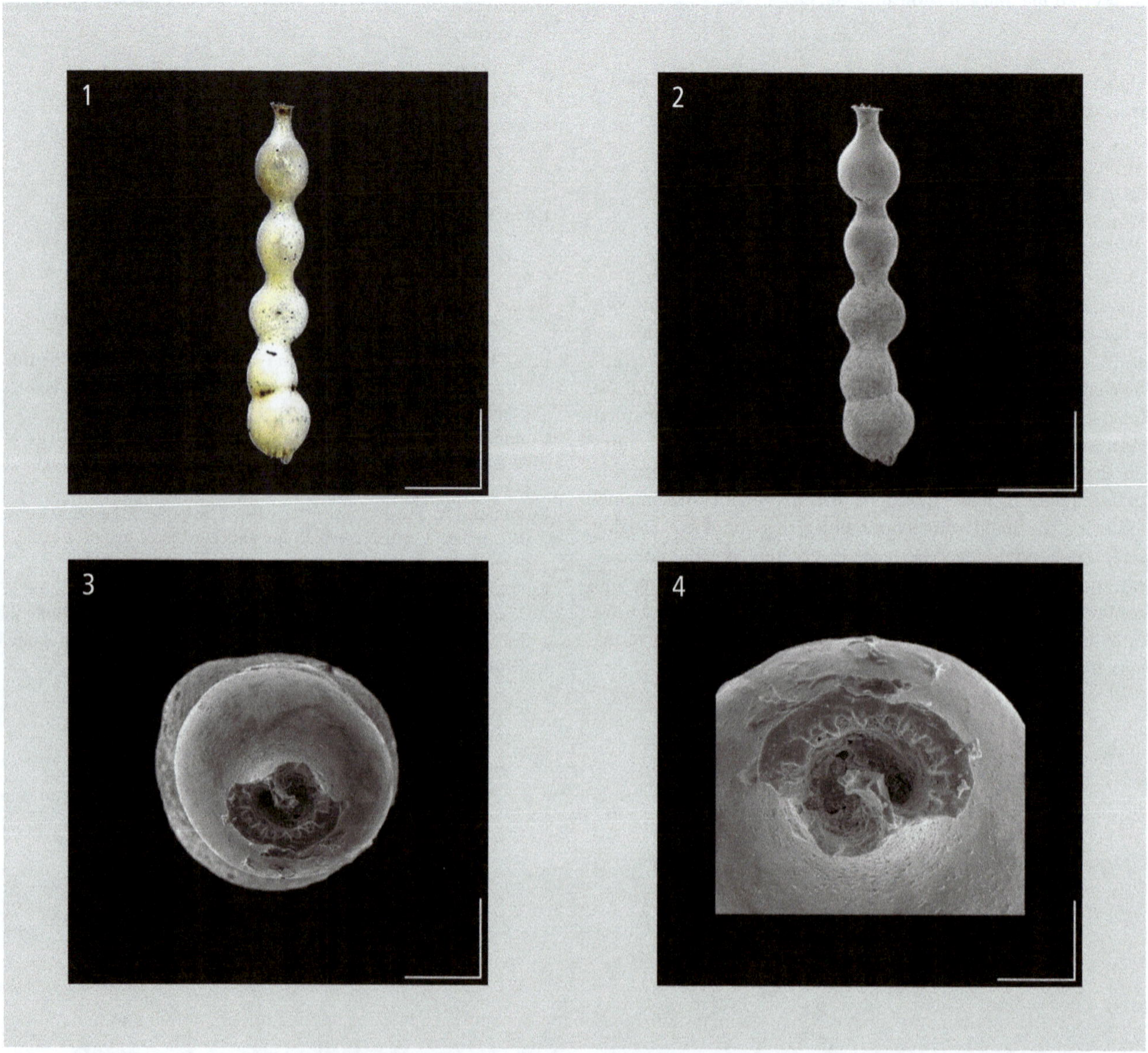

1. *Stilostomella abyssorum.* South Pacific Ocean; "Challenger" Station 296, depth: 1825 fathoms. Deposited in the micropaleontological collections of The Natural History Museum, London; NHMUK ZF1926. Paratype. Lateral view, NHMUK PM DI 0635, scale bar = 566 μm.

2. *Stilostomella abyssorum.* South Pacific Ocean; "Challenger" Station 296, depth: 1825 fathoms. Deposited in the micropaleontological collections of The Natural History Museum, London; NHMUK ZF1926. Paratype. Lateral view, NHMUK PM DI 0636, scale bar = 562 μm.

3. *Stilostomella abyssorum.* South Pacific Ocean; "Challenger" Station 296, depth: 1825 fathoms. Deposited in the micropaleontological collections of The Natural History Museum, London; NHMUK ZF1926. Paratype. Apertural view, NHMUK PM DI 0637, scale bar = 155 μm.

4. *Stilostomella abyssorum.* South Pacific Ocean; "Challenger" Station 296, depth: 1825 fathoms. Deposited in the micropaleontological collections of The Natural History Museum, London; NHMUK ZF1926. Paratype. Detail, NHMUK PM DI 0638, scale bar = 61 μm.

## Original Designation

*Nodosaria (?) abyssorum* Brady, 1881, p. 63 (type description), illustrated in Brady, 1884, pl. 63, figs. 8–9.

## Type Specimen

Specimens figured in Brady (1884) are housed in the micropaleontological collections of The Natural History Museum, London, UK, and registered in slides ZF1926 and ZF3649. Lectotype (slide ZF3649) designated by Loeblich and Tappan (1964).

## Type Level

Recent.

## Type Locality

Not designated. Found only at "Challenger" Station 296, latitude 38° 06′ S, longitude 88° 02′ W in the South Pacific Ocean, southwest of Juan Fernandez Island, 1825 fathoms.

## Description

Test forms an elongate, large, robust, uniserial series; rectilinear or slightly arcuate in shape, and circular in cross-section. The (approximately) five subspherical, inflated chambers are separated by constricted, slightly limbate sutures. Chambers may vary in shape and size (first chamber is often the largest). Chamber walls are calcareous, finely perforate, and smooth, with a ring of short, stout spines on the initial chamber. The primary aperture is an arcuate or V-shaped terminal opening on a short, broad neck, with a phialine lip and small, tooth-like projection (in mature specimens).

## Synonyms and Citations

*Nodosaria (?) abyssorum* Brady. –Brady, 1884, p. 504, pl. 63, figs. 8–9.

*Siphonodosaria abyssorum* (Brady). –Loeblich and Tappan, 1987, p. 540, pl. 585, figs. 5–7. –Hermelin, 1989, p. 61, pl. 11, figs. 2–5. –Robertson, 1998, p. 178, pl. 66, fig. 7.

*Stilostomella abyssorum* (Brady). –Barker, 1960, pl. 63, figs. 8–9. –Jones, 1994, p. 74, pl. 63, figs. 8–9; supplementary pl. 2, figs. 8–9.

## Remarks

Type species of the genus *Siphonodosaria* Silvestri, 1924, regarded by Jones (1994) as a junior synonym of *Stilostomella* Guppy, 1894.

Hermelin (1989) considered the species *Ellipsonodosaria nuttalli* described by Cushman and Jarvis (1934, p. 72, pl. 10, fig. 6) to be a junior synonym of *Stilostomella abyssorum* (Brady).

Listed by Hayward and Kawagata (2005) as one of the species that became globally extinct during the mid-Pleistocene Climatic Transition. These authors suggested that Brady's figured specimens from "Challenger" Station 296 in the southeastern Pacific Ocean include relict or reworked forms, possibly mixed with Holocene taxa.

## Biogeography

Worldwide.

## Bathymetry

Bathyal to abyssal.

## Chronostratigraphy

Late Eocene to Recent?

## References


Barker, R.W., 1960, Taxonomic notes on the species figured by H.B. Brady in his report on the foraminifera dredged by H.M.S. Challenger during the years 1873–1876. Society of Economic Paleontologists and Mineralogists Special Publication, vol. 9, pp. 1–238.

Brady, H.B., 1881, Notes on some of the reticularian Rhizopoda of the Challenger Expedition, Part III. Quarterly Journal of Microscopical Science (New Series), vol. 21, pp. 31–71.

Brady, H.B., 1884, Report on the foraminifera dredged by H.M.S. Challenger during the years 1873–1876. In Brady, H.B., ed., Report on the Scientific Results of the Voyage of the H.M.S. Challenger during the years 1873–1876. Zoology, vol. 9, pp. 1–814.

Cushman, J.A. and Jarvis, P.W., 1934, Some interesting new uniserial foraminifera from Trinidad. Contributions from the Cushman Laboratory for Foraminiferal Research, vol. 10, pp. 71–75.

Hayward, B. and Kawagata, S., 2005, Extinct foraminifera in Brady's Challenger Report. Journal of Micropalaeontology, vol. 24, pp. 171–175.

Hermelin, J.O.R., 1989, Pliocene benthic foraminifera from the Ontong-Java Plateau (Western Equatorial Pacific Ocean): faunal response to changing paleoenvironment. Special Publication of the Cushman Foundation for Foraminiferal Research, vol. 26, pp. 1–143.

Jones, R.W., 1994, The Challenger Foraminifera. Oxford: Oxford University Press, 149 pp.

Loeblich, A.R. Jr and Tappan, H., 1964, Sarcodina, chiefly "Thecamoebians" and Foraminiferida. In Moore, R.C., ed., Treatise on Invertebrate Paleontology, Part C, Protista 2 (2 vols). Lawrence, KS: Geological Society of America and University of Kansas Press, pp. 1–900.

Loeblich, A.R. Jr and Tappan, H., 1987, Foraminiferal Genera and Their Classification. New York: Van Nostrand Reinhold, 970 pp.

Robertson, B.E., 1998, Systematics and paleoecology of the benthic foraminiferida from the Buff Bay section, Miocene of Jamaica. Micropaleontology, vol. 44, supplement 2, pp. 1–266.

# *Stilostomella aculeata*

## (Cushman and Renz), 1948

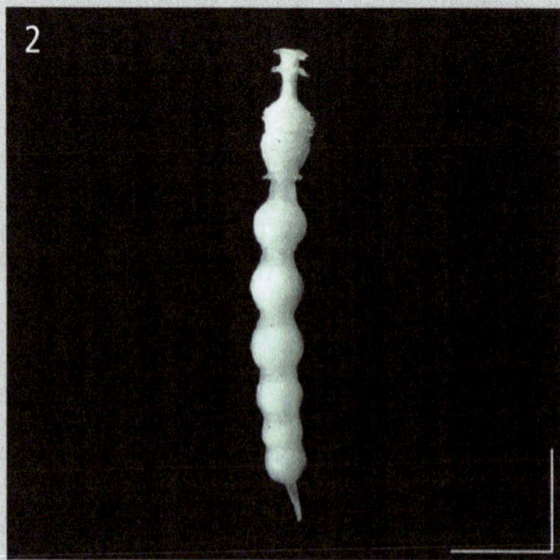


1. *Stilostomella aculeata.* Hospital Hill Marl, west slope of Hospital Hill, San Fernando, Trinidad. Deposited in the collections of the United States National Museum, Washington, DC; CC 57461. Holotype. Lateral view, NHMUK PM DI 0639, scale bar = 304 μm.

2. *Stilostomella aculeata.* Allison Guyot, Mid-Pacific Mountains, Northwest Pacific; ODP Leg 142, Hole 865B, Core 4, Section 5, 120 cm. Lateral view, NHMUK PM DI 0640, scale bar = 324 μm.

### Original Designation

*Ellipsonodosaria nuttalli* Cushman and Jarvis var. *aculeata* Cushman and Renz, 1948, p. 32, pl. 6, fig. 10.

### Type Specimen

Holotype (CC 57461) deposited in the collections of the United States National Museum, Washington, DC.

### Type Level

Upper Eocene, San Fernando Group, Hospital Hill Formation, light greenish-gray, indurated, nodular marl and marly clay, poorly bedded.

### Type Locality

Northwest slope of Hospital Hill, the site of the San Fernando Hospital, on the east side of the road coming from San Fernando (King's Wharf) and leading to Point Bontour (Sawmill), Trinidad. Trinidad Government Cadastral Coordinates: North 234,850 links; East 355,850 links.

### Description

Test forms an elongate, large, robust, uniserial series; rectilinear or slightly arcuate in overall shape, and circular in cross-section. The (approximately) five subspherical, inflated chambers are separated by constricted, slightly limbate sutures. Chamber walls are calcareous, finely perforate, and spinose on the lower part of chambers, with a single basal spine on the proloculus. The primary aperture is an arcuate or V-shaped terminal opening on a short, broad neck, with a phialine lip and small, tooth-like projection (in mature specimens).

### Synonyms and Citations

*Stilostomella abyssorum* (Brady) var. *aculeata* (Cushman and Renz). –Beckmann, 1953, p. 369, pl. 21, fig. 34.

*Siphonodosaria abyssorum* (Brady) var. *aculeata* (Cushman and Renz). –Resig, 1976, pl. 2, fig. 12.

*Stilostomella aculeata* (Cushman and Renz). –Tjalsma and Lohmann, 1983, p. 36, pl. 14, fig. 12. –Miller and Katz, 1987, p. 138, pl. 1, fig. 11. –Thomas, 1990, p. 590.

### Remarks

There is considerable variation in the degree of constriction of the chambers and spinose ornamentation of *Stilostomella aculeata* (Cushman and Renz). Smooth specimens closely resemble *Stilostomella abyssorum* (Brady). Hermelin (1989) considered the species *Ellipsonodosaria nuttalli* Cushman and Jarvis to be a junior synonym of *Stilostomella abyssorum.*

### Biogeography

Worldwide. Reported from Trinidad (Cushman and Renz, 1948); the Weddell Sea, Antarctic (Thomas, 1990); North Atlantic (Miller and Katz, 1987).

### Bathymetry

Predominantly lower bathyal and abyssal. Wide bathymetric range, but relatively more abundant at deeper sites (Tjalsma and Lohmann, 1983), lower bathyal (Thomas, 1990), abyssal, greater than 3.5 kilometers (Miller and Katz, 1987).

### Chronostratigraphy

Early Eocene (Ypresian) to Recent. First occurrence in the Early Eocene, following the benthic foraminiferal extinction (E. Thomas, pers. comm. 1998). Early Eocene (Biozone P7) upward, particularly abundant in Biozones P11–P14 (Tjalsma and Lohmann, 1983).

## References


Beckmann, J.P., 1953, Die Foraminiferen der Oceanic Formation (Eocaen–Oligocaen) von Barbados, Kl. Antillen. Eclogae Geologicae Helvetiae, vol. 46, pp. 301–412.

Cushman, J.A. and Renz, H.H., 1948, Eocene foraminifera of the Navet and Hospital Hill formations of Trinidad, B.W.I. Special Publications of the Cushman Laboratory for Foraminiferal Research, vol. 24, pp. 1–42.

Hermelin, J.O.R., 1989, Pliocene benthic foraminifera from the Ontong-Java Plateau (Western Equatorial Pacific Ocean): faunal response to changing paleoenvironment. Special Publication of the Cushman Foundation for Foraminiferal Research, vol. 26, pp. 1–143.

Miller, K.G. and Katz, M.E., 1987, Oligocene to Miocene benthic foraminiferal and abyssal circulation changes in the North Atlantic. Micropaleontology, vol. 33, no. 2, pp. 97–149.

Resig, J.M., 1976, Benthic foraminiferal stratigraphy, eastern margin, Nazca Plate. In Yeats, R.S., Hart, S.R., et al., ed. Initial Reports of the Deep Sea Drilling Project, Washington (U.S. Government Printing Office), vol. 34, pp. 743–759.

Thomas, E., 1990, Late Cretaceous through Neogene benthic foraminifers. In Barker, P.F., Kennett, J.P., et al., ed., Scientific Results of the Proceedings of the Ocean Drilling Program, vol. 113, pp. 571–594.

Tjalsma, R.C. and Lohmann, G.P., 1983, Paleocene–Eocene bathyal and abyssal benthic foraminifera from the Atlantic Ocean. Micropaleontology, Special Publication, no. 4, pp. 1–90.

# *Stilostomella alexanderi*

## (Cushman), 1936

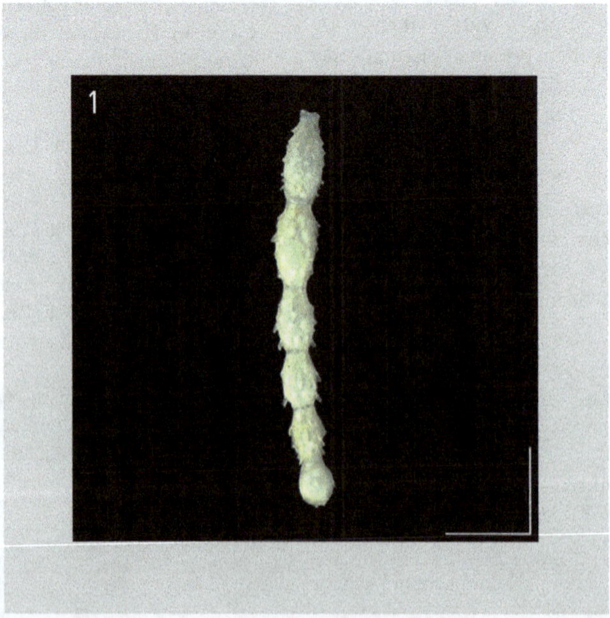

1. *Stilostomella alexanderi.* Taylor Marl, road cut 14.4 miles south of Paris, 0.9 mile north of Lake City, Delta Co., Texas. Deposited in the collections of the United States National Museum, Washington, DC; CC 23254. Holotype. Lateral view, NHMUK PM DI 0641, scale bar = 72 μm.

### Original Designation

*Ellipsonodosaria alexanderi* Cushman, 1936, p. 52, pl. 9, figs. 6–9.

### Type Specimen

Holotype (CC 23254) deposited in the collections of the United States National Museum, Washington, DC.

### Type Level

Upper Cretaceous, Taylor Marl.

### Type Locality

Roadcut 14.4 miles south of Paris, 0.9 mile north of Lake City, Delta County, Texas, USA.

### Description

Test forms an elongate, uniserial series; rectilinear or slightly arcuate in overall shape, and circular in cross-section. The subspherical, inflated chambers, which increase gradually in length, and become ultimately twice as long as broad, are separated by constricted sutures. Chamber walls are calcareous, finely perforate, and ornamented with short, backward-pointing spines. Early stages of microspheric forms may only have a single ring of spines in the lower part of the chambers. Adult forms have numerous spines, rather irregularly scattered over the surface. The primary aperture is arcuate or V-shaped terminal opening on a short, broad neck, with a phialine lip and small, tooth-like projection (in mature specimens).

### Synonyms and Citations

*Ellipsonodosaria alexanderi* Cushman var. *impensia* Cushman, 1938, p. 48, pl. 8, figs. 4–5. –Cushman and Renz, 1947, p. 48, pl. 12, figs. 9–10.

*Stilostomella alexanderi* (Cushman). –Bolli et al., 1994, p. 145, figs. 38.27–28.

### Biogeography

Originally described from the Upper Cretaceous in Texas. Reported from the southern USA, Peru, Venezuela, and Trinidad.

### Bathymetry

Bathyal.

### Chronostratigraphy

Late Cretaceous (Santonian to Maastrichtian) (Bolli et al., 1994).

## References

Bolli, H.M., Beckmann, J.P., and Saunders, J.B., 1994, Benthic Foraminiferal Biostratigraphy of the South Caribbean Region. Cambridge: Cambridge University Press, 408 pp.

Cushman, J.A., 1936, Some American Cretaceous species of Ellipsonodosaria and Chrysalogonium. Contributions from the Cushman Laboratory for Foraminiferal Research, vol. 12, no. 3, pp. 51–55.

Cushman, J.A., 1938, Additional new species of American Cretaceous foraminifera. Contributions from the Cushman Laboratory for Foraminiferal Research, vol. 14, pp. 31–50.

Cushman, J.A. and Renz, H.H., 1947, Further notes on the Cretaceous foraminifera of Trinidad. Contributions from the Cushman Laboratory for Foraminiferal Research, vol. 23, pp. 31–51.

# *Stilostomella annulifera*

## (Cushman and Bermúdez), 1936

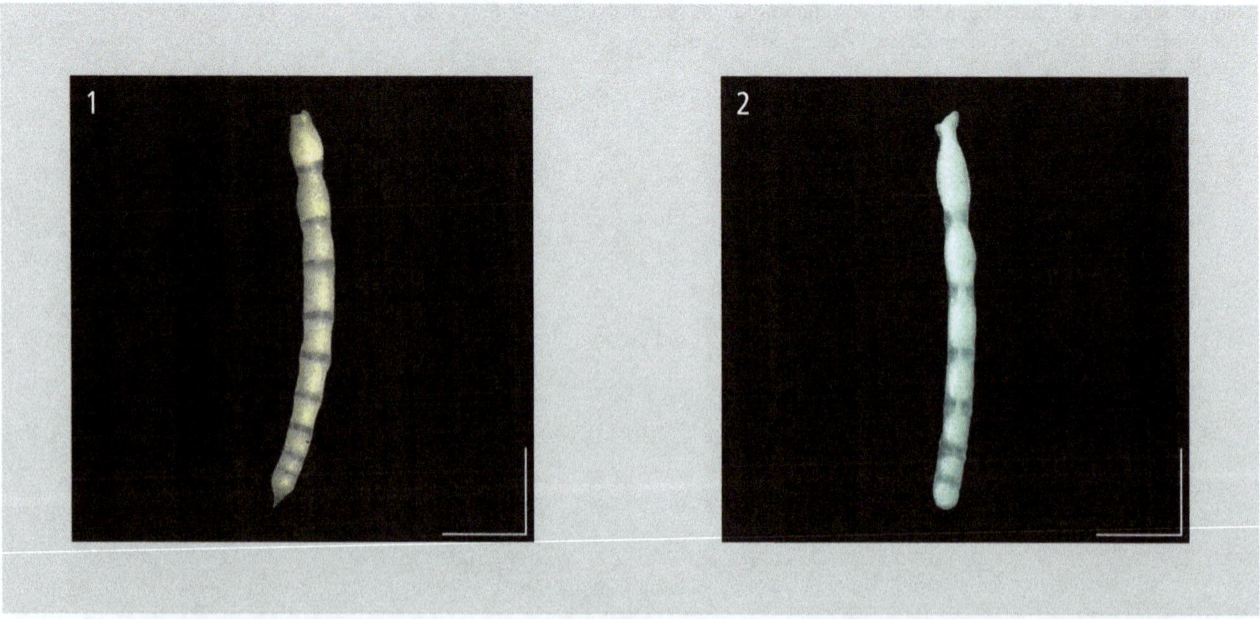

1. *Stilostomella annulifera.* Bermúdez Station 257 (under the library of Havana University), Cuba. Deposited in the collections of the United States National Museum, Washington, DC; CC 23100. Paratype. Lateral view, NHMUK PM DI 0642, scale bar = 274 μm.

2. *Stilostomella annulifera.* Eastern Equatorial Pacific; DSDP Leg 85, Hole 573B, Core 26, core catcher. Lateral view, NHMUK PM DI 0643, scale bar = 222 μm.

### Original Designation

*Ellipsonodosaria annulifera* Cushman and Bermúdez, 1936, p. 28, pl. 5, figs. 8–9.

### Type Specimen

Holotype (CC 23099) deposited in the collections of the United States National Museum, Washington, DC.

### Type Level

Probably Middle Eocene.

### Type Locality

Bermúdez Station 257, under library of Havana University, Havana, Cuba.

### Description

Test forms an elongate, large, slender, uniserial series; rectilinear or slightly arcuate in overall shape, and circular in cross-section. The subspherical, and moderately inflated chambers increase gradually in length, and are separated by limbate sutures forming broad glassy bands between the chambers. Final chambers may be more elongated, and separated by more distinct constrictions. Chamber walls are calcareous, finely perforate, and smooth, with a single basal spine on the proloculus. The primary aperture is an arcuate or V-shaped terminal opening on a short, broad neck, with a phialine lip and small, tooth-like projection (in mature specimens).

### Synonyms and Citations

*Stilostomella annulifera* (Cushman and Bermúdez). –Thomas, 1990, p. 590, pl. 1, fig. 4.

*Siphonodosaria annulifera* (Cushman and Bermúdez). –Bolli et al., 1994, p. 358, figs. 87.40, 40a. –Robertson, 1998, p. 178, pl. 66, figs. 1–2. –Ortiz and Thomas, 2006, p. 132, pl. 111, fig. 6.

### Biogeography

Worldwide. Reported from Cuba (Cushman and Bermúdez, 1936), Barbados (Bolli et al., 1994), Jamaica (Robertson, 1998), Spain (Ortiz and Thomas, 2006), and Weddell Sea, Antarctica (Thomas, 1990).

### Bathymetry

Bathyal. Lower bathyal (Thomas, 1990).

### Chronostratigraphy

Early Eocene to Middle Miocene.

## References


Bolli, H.M., Beckmann, J.P., and Saunders, J.B., 1994, Benthic Foraminiferal Biostratigraphy of the South Caribbean Region. Cambridge: Cambridge University Press, 408 pp.

Cushman, J.A. and Bermúdez, P.J., 1936, New genera and species of foraminifera from the Eocene of Cuba. Contributions from the Cushman Laboratory for Foraminiferal Research, vol. 12, no. 2, pp. 27–38.

Ortiz, S. and Thomas, E., 2006, Lower–middle Eocene benthic foraminifera from the Fortuna Section (Betic Cordillera, southeastern Spain). Micropaleontology, vol. 52, pp. 97–150.

Robertson, B.E., 1998, Systematics and paleoecology of the benthic foraminiferida from the Buff Bay section, Miocene of Jamaica. Micropaleontology, vol. 44, supplement 2, pp. 1–266.

Thomas, E., 1990, Late Cretaceous through Neogene benthic foraminifers. In Barker, P.F., Kennett, J.P., et al., ed., Scientific Results of the Proceedings of the Ocean Drilling Program, vol. 113, pp. 571–594.

# *Stilostomella consobrina*

## (d'Orbigny), 1846

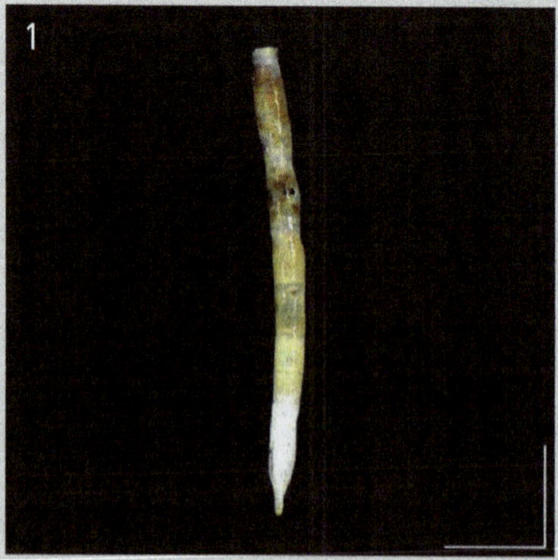

1. *Stilostomella consobrina*. North of Juan Fernandez, East Pacific Ocean; "Challenger" Station 300, depth: 1370 fathoms. Deposited in the micropaleontological collections of The Natural History Museum, London; NHMUK ZF1937. Lateral view, NHMUK PM DI 0644, scale bar = 275 μm.

### Original Designation

*Dentalina consobrina* d'Orbigny, 1846, p. 46, pl. 2, figs. 1–3.

### Type Specimen

Lectotype designated and illustrated by Papp and Schmid (1985, pl. 11, fig. 3); original material deposited in the Hauer-d'Orbigny Collection, Geologische, Bundesanstalt, Vienna, Nos 1977/01/23, 1981/03/42 (lectotype).

### Type Level

Middle Miocene, very fine blue marl.

### Type Locality

Brick pits eight leagues south of Vienna, near Baden, Vienna Basin, Austria.

### Description

Test forms an elongate, slender, uniserial series; rectilinear or slightly arcuate in overall shape, and circular in cross-section, with a pointed initial portion. The moderately inflated chambers vary in height, and are separated by flush sutures. Chamber walls are calcareous, finely perforate, and smooth. The primary aperture is an arcuate or V-shaped, terminal opening on a short, broad neck, with a phialine lip and small, tooth-like projection (in mature specimens).

### Synonyms and Citations

*Nodosaria consobrina* (d'Orbigny). –Brady, 1884, p. 501, pl. 62, figs. 23–24.

*Siphonodosaria consobrina* (d'Orbigny). –Hermelin, 1989, p. 61, pl. 11, figs. 6–7. –Bolli et al., 1994, p. 359, figs. 87.41, 86.

*Stilostomella consobrina* (d'Orbigny). –Barker, 1960, pl. 62, figs. 23–24. –Papp and Schmid, 1985, p. 29, pl. 11, figs. 1–5. –Jones, 1994, p. 74, pl. 62, figs. 23–24.

### Remarks

Listed by Hayward and Kawagata (2005) as one of the species that became globally extinct during the mid-Pleistocene Climatic Transition. These authors suggested that Brady's figured specimens from "Challenger" Station 191A in the Banda Sea, Indonesia, and Station 300 in the southeastern Pacific Ocean include relict or reworked forms, possibly mixed with Holocene taxa.

### Biogeography

Worldwide.

### Bathymetry

Primarily bathyal. Reported between 129 and 1375 fathoms at "Challenger" stations (Jones, 1994).

### Chronostratigraphy

Late Eocene to Recent?

## References


Barker, R.W., 1960, Taxonomic notes on the species figured by H.B. Brady in his report on the foraminifera dredged by H.M.S. Challenger during the years 1873–1876. Society of Economic Paleontologists and Mineralogists Special Publication, vol. 9, pp. 1–238.

Bolli, H.M., Beckmann, J.P., and Saunders, J.B., 1994, Benthic Foraminiferal Biostratigraphy of the South Caribbean Region. Cambridge: Cambridge University Press, 408 pp.

Brady, H.B., 1884, Report on the foraminifera dredged by H.M.S. Challenger during the years 1873–1876. In Brady, H.B., ed., Report on the Scientific Results of the Voyage of the H.M.S. Challenger during the years 1873–1876. Zoology, vol. 9, pp. 1–814.

d'Orbigny, A.D., 1846, Foraminifères fossiles du Bassin Tertiaire de Vienne (Autriche). Paris: Gide et Compe, xxxvii + 312 pp.

Hayward, B. and Kawagata, S., 2005, Extinct foraminifera in Brady's Challenger Report. Journal of Micropalaeontology, vol. 24, pp. 171–175.

Hermelin, J.O.R., 1989, Pliocene benthic foraminifera from the Ontong-Java Plateau (Western Equatorial Pacific Ocean): faunal response to changing paleoenvironment. Special Publication of the Cushman Foundation for Foraminiferal Research, vol. 26, pp. 1–143.

Jones, R.W., 1994, The Challenger Foraminifera. Oxford: Oxford University Press, 149 pp.

Papp, A. and Schmid, M.E., 1985, Die fossilen Foraminiferen des tertiaeren Beckens von Wien: Revision der Monographie Alcide d'Orbigny (1846). Abhandlungen der Geologischen Bundesanstalt, vol. 37, pp. 1–311.

# Stilostomella lepidula

## (Schwager), 1866

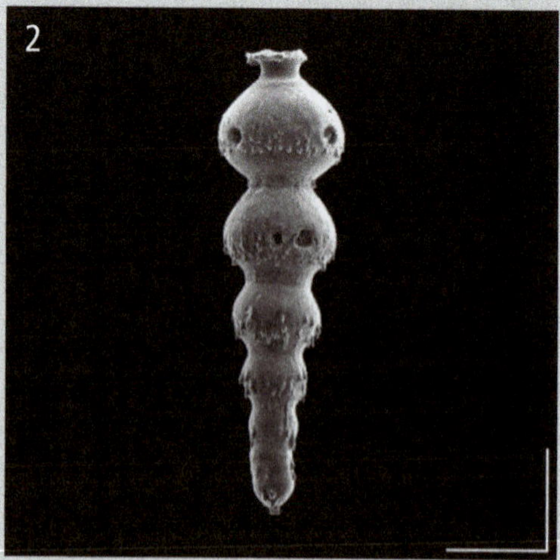



1. *Stilostomella lepidula.* East African Continental Rise, Northwest Indian Ocean; DSDP Leg 25, Site 241, Core 11, Section 3, 66–71 cm. Lateral view, NHMUK PM DI 0645, scale bar = 78 μm.

2. *Stilostomella lepidula.* Laccadive-Chagos Ridge, Arabian Sea; DSDP Leg 23, Site 219, Core 2, Section 5, 35–37 cm. Lateral view, NHMUK PM DI 0646, scale bar = 194 μm.

### Original Designation

*Nodosaria lepidula* Schwager, 1866, p. 210, pl. 5, figs. 27–28.

### Type Specimen

Neotype (Mf 18) designated by Srinivasan and Sharma (1980) deposited in the Micropaleontology Section, Department of Geology, Banaras Hindu University, Banaras, India.

### Type Level

Upper Tertiary, lower and upper clays.

### Type Locality

Car Nicobar, Andaman Sea.

### Description

Test forms an elongate, slender, uniserial series; rectilinear or slightly arcuate in overall shape, slightly tapered in outline, and circular in cross-section. The eight to sixteen subspherical, inflated chambers increase gradually in size, and are separated by strongly constricted sutures. Chamber walls are calcareous, and finely perforate, ornamented by a ring of short, fine spines at the base of chambers, and occasionally in the middle of chambers. A single, short, basal spine may be present on the proloculus. The primary aperture is an arcuate or V-shaped, terminal opening on a short, broad neck, with a phialine lip and small, tooth-like projection (in mature specimens).

### Synonyms and Citations

*Stilostomella lepidula* (Schwager). –Srinivasan and Sharma, 1980, p. 46, pl. 7, figs. 1–6. –Thomas, 1990, p. 590. –Boersma, 1990, pl. 2, figs. 5–6. –Nomura, 1995, p. 274, pl. 2, fig. 7.

*Nodogenerina lepidula* (Schwager). –Loeblich and Tappan, 1994, p. 133, pl. 261, figs. 18–20.

*Siphonodosaria lepidula* f. *lepidula* (Schwager). –Hayward, 2002, p. 305, pl. 3, figs. 25–32.

### Remarks

Hayward (2002) reported that specimens figured as *Sagrina virgula* Brady by Brady (1884) in pl. 76, figs. 8–10, and refigured by Jones (1994) as *Nodogenerina antillea* (Cushman) in pl. 76, figs. 9–10, suppl. pl. 2, fig. 1, and as *Nodogenerina virgula* Brady in pl. 76, fig. 8, suppl. pl. 2, figs. 2–3, 15, fall with the range of variability of *Stilostomella lepidula* (Schwager).

*Stilostomella lepidula* was listed by Hayward and Kawagata (2005) as one of the species that became globally extinct during the mid-Pleistocene Climatic Transition. These authors suggested that Brady's figured specimens from "Challenger" Station 120 in the western equatorial Atlantic and Station 192 in the Banda Sea, Indonesia, include relict or reworked forms, possibly mixed with Holocene taxa.

### Biogeography

Worldwide. Andaman Sea (Schwager, 1866), Western Timor Sea (Loeblich and Tappan, 1994). Weddell Sea, Antarctica (Thomas, 1990), Indian Ocean (Boersma, 1990; Nomura, 1995), Southwest Pacific (Hayward, 2002).

### Bathymetry

Primarily bathyal.

### Chronostratigraphy

Eocene to Recent?

## References


Boersma, A. 1990, Late Oligocene to late Pliocene benthic foraminifers from depth traverses in the central Indian Ocean. In Duncan, R.A., Backman, J., Peterson, L.C. et al., ed., Scientific Results of the Proceedings of the Ocean Drilling Program, vol. 115, pp. 315–379.

Brady, H.B., 1884, Report on the foraminifera dredged by H.M.S. Challenger during the years 1873–1876. In Brady, H.B., ed., Report on the Scientific Results of the Voyage of the H.M.S. Challenger during the years 1873–1876. Zoology, vol. 9, pp. 1–814.

Hayward, B., 2002, Late Pliocene to Middle Pleistocene extinctions of deep-sea benthic foraminifera ("Stilostomella extinction") in the southwest Pacific. Journal of Foraminiferal Research, vol. 32, pp. 274–307.

Hayward, B. and Kawagata, S., 2005, Extinct foraminifera in Brady's Challenger Report. Journal of Micropalaeontology, vol. 24, pp. 171–175.

Loeblich, A.R. Jr and Tappan, H., 1994, Foraminifera of the Sahul Shelf and Timor Sea. Special Publications of the Cushman Laboratory for Foraminiferal Research, vol. 31, pp. 1–661.

Nomura, R., 1995, Paleogene to Neogene deep-sea paleoceanography in the eastern Indian Ocean: benthic foraminifera from ODP Sites 747, 757 and 758. Micropaleontology, vol. 41, pp. 251–290.

Schwager, C., 1866, Fossile Foraminiferen von Kar Nikobar, Reise der Österreichischen Fregatte Novara um Erde in den Jahren 1857, 1858, 1859 unten den Befehlen des Commodore B. Von Wüllerstorf-Urbair. Geologischer Theil, vol. 2, no. 1, Geologische Beobachtung, no. 2, Paläontologische Mittheilung, pp. 187–268.

Srinivasan, M.S. and Sharma, V., 1980, Schwager's Car Nicobar Foraminifera in the Reports of the Novara Expedition – A Revision. New Delhi: Today & Tomorrow's Printers and Publishers, 83 pp.

Thomas, E., 1990, Late Cretaceous through Neogene benthic foraminifers. In Barker, P.F., Kennett, J.P., et al., ed., Scientific Results of the Proceedings of the Ocean Drilling Program, vol. 113, pp. 571–594.

# *Stilostomella modesta*

## (Bermúdez), 1937

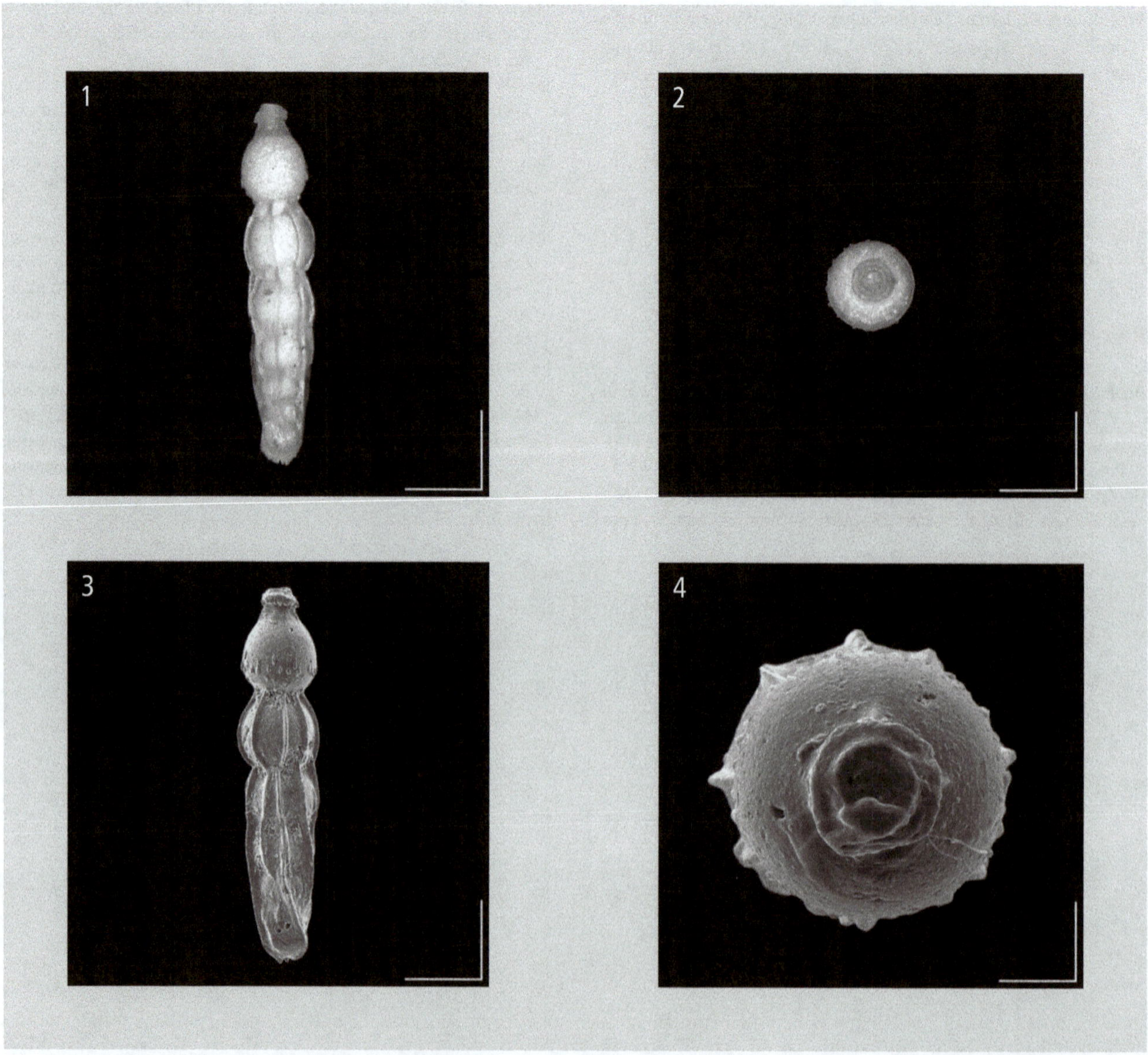


1. *Stilostomella modesta.* North Biscay, North Atlantic; DSDP Leg 48, Site 401, Core 6, Section 3, 76–80 cm. Lateral view, NHMUK PM DI 0647, scale bar = 112 μm.

2. *Stilostomella modesta.* North Biscay, North Atlantic; DSDP Leg 48, Site 401, Core 6, Section 3, 76–80 cm. Apertural view, NHMUK PM DI 0648, scale bar = 110 μm.

3. *Stilostomella modesta.* North Biscay, North Atlantic; DSDP Leg 48, Site 401, Core 6, Section 3, 76–80 cm. Lateral view, NHMUK PM DI 0649, scale bar = 116 μm.

4. *Stilostomella modesta.* North Biscay, North Atlantic; DSDP Leg 48, Site 401, Core 6, Section 3, 76–80 cm. Apertural view, NHMUK PM DI 0650, scale bar = 32 μm.

## Original Designation

*Ellipsonodosaria modesta* Bermúdez, 1937, p. 238, pl. 20, fig. 3.

## Type Specimen

Originally deposited in the Bermúdez Collection (No. 1989).

## Type Level

Upper Eocene, Principe Formation.

## Type Locality

Bermúdez 275: Batey de Noroña, 3.7 kilometers north of Guanajay on the railtrack of Central Toledo, Province Pinar del Rio, Cuba.

## Description

Test forms an elongate, small, uniserial series; rectilinear or slightly arcuate in overall shape, and circular in cross-section, consisting of up to ten chambers. The last three chambers are markedly inflated, and separated by strongly depressed sutures. Chamber walls are calcareous, finely perforate, and ornamented by numerous fine costae, except for the last chamber that is finely spinose. The primary aperture is an arcuate or V-shaped, terminal opening on a short, broad neck with a phialine lip and small, tooth-like projection (in mature specimens).

## Synonyms and Citations

*Pyramidulina modesta* (Bermúdez). –Bolli et al., 1994, p. 331, fig. 63.14.

*Stilostomella modesta* (Bermúdez). –Tjalsma, 1983, p. 743, pl. 1, fig. 10. –Bolli et al., 1994, p. 359, figs. 87.38, 38a.

*Strictocostella modesta* (Bermúdez). –Hayward, 2002, p. 306, pl. 3, figs. 51–52.

## Biogeography

Worldwide. Reported from Cuba (Bermúdez, 1937), Barbados (Bolli et al., 1994), and the Southwest Pacific (Hayward, 2002).

## Bathymetry

Primarily bathyal.

## Chronostratigraphy

Eocene to Late Pliocene.

# References

Bermúdez, P.J., 1937, Nuevas especies de foraminiferos del Eoceno de las cercanias de Guanajay, Provincia Pinar del Rio. Memorias de la Sociedad Cubana de Historia Natural "Felipe Poey", vol. 11, pp. 237–247.

Bolli, H.M., Beckmann, J.P., and Saunders, J.B., 1994, Benthic Foraminiferal Biostratigraphy of the South Caribbean Region. Cambridge: Cambridge University Press, 408 pp.

Hayward, B., 2002, Late Pliocene to Middle Pleistocene extinctions of deep-sea benthic foraminifera ("Stilostomella extinction") in the southwest Pacific. Journal of Foraminiferal Research, vol. 32, pp. 274–307.

Tjalsma, R.C., 1983, Eocene to Miocene benthic foraminifera from DSDP site 516, Rio Grande Rise, South Atlantic. In Barker, P.F., Carlson, R.L., Johnson, D.A., et al., ed., Initial Reports of the Deep Sea Drilling Project, Washington (U.S. Government Printing Office), vol. 72, pp. 731–755.

# *Stilostomella paleocenica*

## (Cushman and Todd), 1946

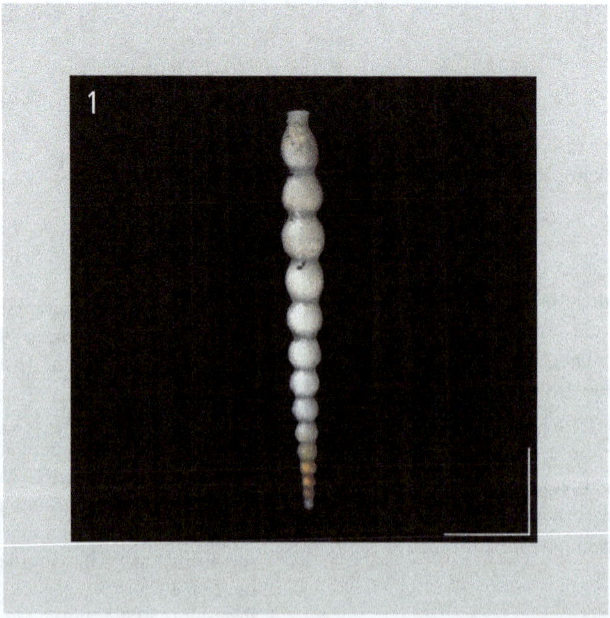

1. *Stilostomella paleocenica.* Loc 1 NE 1/4 NW1/4 sec. 16, T.1N.R.12W., Little Rock, Arkansas. Deposited in the collections of the United States National Museum, Washington, DC; CC 46415. Holotype. Lateral view, NHMUK PM DI 0651, scale bar = 658 μm.

### Original Designation

*Ellipsonodosaria paleocenica* Cushman and Todd, 1946, p. 61, pl. 10, fig. 26.

### Type Specimen

Holotype (CC 46415) deposited in the collections of the United States National Museum, Washington, DC.

### Type Level

Paleocene.

### Type Locality

About 1000 feet south of Roosevelt Rd viaduct, 200 feet east of railroad tracks, in small gully heading east, just upstream from small abandoned bridge, NE 1/4, NW 1/4, sec. 16, T. 1 N, R. 12 W, Little Rock, Pulaski County, Arkansas, USA.

### Description

Test forms an elongate, large, slender, uniserial series; rectilinear or slightly arcuate in overall shape, circular in cross-section, and slightly tapered in outline from the very narrow proloculus to the broadest final chambers. The inflated, subspherical, and numerous (up to fourteen) chambers increase very gradually in size, and are separated by strongly depressed sutures, forming narrow inter-chamber bands. Chamber walls are calcareous, finely perforate, and smooth or slightly hispid, with a single basal spine on the proloculus. The last chamber is more elongated than previous chambers, and is also more ornamented. The primary aperture is an arcuate or V-shaped terminal opening on a short, broad neck, with a phialine lip and small, tooth-like projection (in mature specimens).

### Synonyms and Citations

*Stilostomella paleocenica* (Cushman and Todd). –Bolli et al., 1994, p. 145, figs. 38.29–30.

### Remarks

In Trinidad specimens are usually smaller than the holotype (Bolli et al., 1994). Differs from *Ellipsonodosaria plummerae* Cushman in the shorter, more inflated chambers and smoother surface.

### Biogeography

Worldwide. Reported from Arkansas and Alabama (Cushman and Todd, 1946), from Trinidad (Bolli et al., 1994), and from ODP sites in the equatorial central Pacific Ocean (Nomura and Takata, 2005).

### Bathymetry

Primarily bathyal.

### Chronostratigraphy

Late Cretaceous (Santonian) to Eocene.

## References

Bolli, H.M., Beckmann, J.P., and Saunders, J.B., 1994, Benthic Foraminiferal Biostratigraphy of the South Caribbean Region. Cambridge: Cambridge University Press, 408 pp.

Cushman, J.A. and Todd, R., 1946, A foraminiferal fauna from the Paleocene of Arkansas. Contributions from the Cushman Laboratory for Foraminiferal Research, vol. 22, no. 2, pp. 45–65.

Nomura, R. and Takata, H., 2005, Data report: Paleocene/Eocene benthic foraminifers, ODP Leg 199, Sites 1215, 1220, and 1221, equatorial central Pacific Ocean. In Wilson, P.A., Lyle, M., and Firth, J.V., ed., Scientific Results of the Proceedings of the Ocean Drilling Program, vol. 199, pp. 1–34 [Online].

# *Stilostomella spinescens*

## (Reuss), 1851

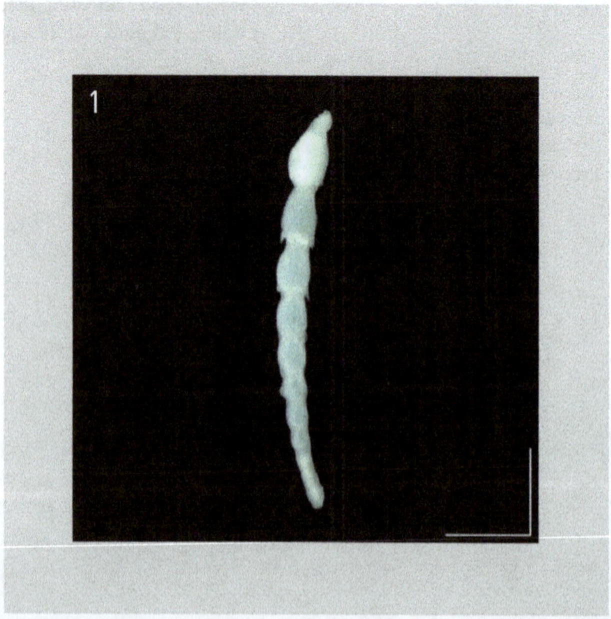

1. *Stilostomella spinescens.* Allison Guyot, Mid-Pacific Mountains, Northwest Pacific; ODP Leg 142, Hole 865B, Core 4, Section 1, 120 cm. Lateral view, NHMUK PM DI 0652, scale bar = 145 μm.

### Original Designation
*Dentalina spinescens* Reuss, 1851, p. 62, pl. 3, fig. 10.

### Type Specimen
Depository not given.

### Type Level
Eocene, Septarienthon.

### Type Locality
Hermsdorf, near Berlin, Germany.

### Description
Test forms an elongate, slender, uniserial series; slightly arcuate in overall shape, slightly tapered in outline, and circular in cross-section. The (approximately) ten to twelve chambers increase gradually in size, and are separated by horizontal sutures, strongly depressed in the upper part of the test. Later chambers are sub-spherical, and markedly inflated, earlier chambers are cylindrical. Chamber walls are calcareous, finely perforate, and ornamented by a few short, downward-pointing spines at the base of the upper chambers; the initial chamber bears a short basal spine. The primary aperture is an arcuate or V-shaped terminal opening on a short, broad neck, with a phialine lip and small, tooth-like projection (in mature specimens).

### Synonyms and Citations
*Laevidentalina spinescens* (Reuss). –Bolli et al., 1994, 102, figs. 38.19–20.

### Remarks
*Stilostomella spinescens* is common in high latitude Eocene sites, and is a high productivity indicator showing strong fluctuations in abundance (E. Thomas, pers. comm. 1998).

### Biogeography
Worldwide.

### Bathymetry
Bathyal?

### Chronostratigraphy
Paleocene to Miocene?

## References

Bolli, H.M., Beckmann, J.P., and Saunders, J.B., 1994, Benthic Foraminiferal Biostratigraphy of the South Caribbean Region. Cambridge: Cambridge University Press, 408 pp.

Reuss, A.E., 1851, Ueber die fossilen Foraminiferen und Entomostraceen der Septarienthone der Umgegend von Berlin. Zeitschrift der Deutschen Geologischen Gesellschaft, Berlin, vol. 3, pp. 49–91.

# *Stilostomella stephensoni*

## (Cushman), 1936

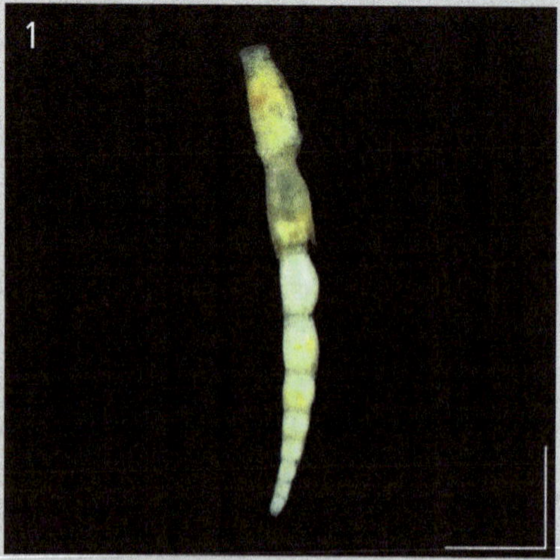


1. *Stilostomella stephensoni.* Taylor Marl, 7.5 miles from Terrell on road to Crandall, Kaufman Co., Texas. Deposited in the collections of the United States National Museum, Washington, DC; CC 23256. Holotype. Lateral view, NHMUK PM DI 0653, scale bar = 170 µm.

### Original Designation

*Ellipsonodosaria stephensoni* Cushman, 1936, p. 52, pl. 9, figs. 10–15.

### Type Specimen

Holotype (CC 23256) deposited in the collections of the United States National Museum, Washington, DC.

### Type Level

Upper Cretaceous, Taylor Marl.

### Type Locality

Seven and a half miles from Terrell on road to Crandall, Kaufman County, Texas, USA.

### Description

Test forms an elongate, slender, slightly arcuate, uniserial series; slightly tapered in outline, and subcircular in cross-section. The inflated, subcylindrical chambers exhibit their greatest breadth near the base, increase rapidly in height, and are separated by strongly depressed sutures. Chamber walls are calcareous, finely perforate, and ornamented by a ring of short, downward-pointing spines at the base of later chambers, occasionally with a few scattered spines above. The primary aperture is an arcuate or V-shaped terminal opening on a short, broad neck, with a phialine lip and small, tooth-like projection (in mature specimens).

### Synonyms and Citations

*Ellipsonodosaria stephensoni* Cushman. –Cushman, 1946, p. 134, pl. 56, figs. 2–7.

### Biogeography

Recorded in Upper Cretaceous deposits of the Gulf coastal region in the USA (Cushman, 1946).

### Bathymetry

Bathyal.

### Chronostratigraphy

Late Cretaceous.

## References

Cushman, J.A., 1936, Some American Cretaceous species of Ellipsonodosaria and Chrysalogonium. Contributions from the Cushman Laboratory for Foraminiferal Research, vol. 12, no. 3, pp. 51–55.

Cushman, J.A., 1946, Upper Cretaceous foraminifera of the Gulf coastal region of the United States and adjacent areas. Professional Paper of the United States Geological Survey, vol. 206, 241 pp.

# *Stilostomella subspinosa*

## (Cushman), 1943

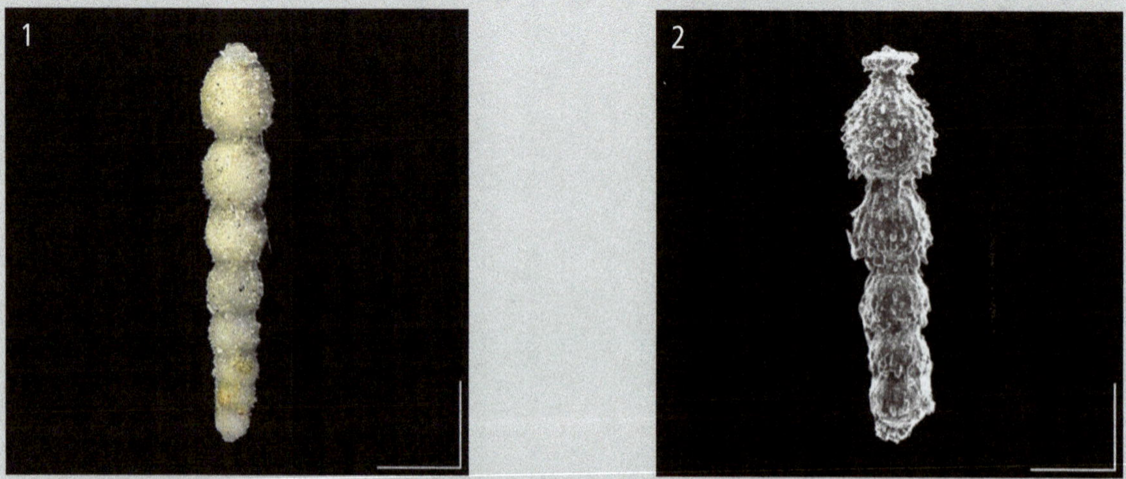

1. *Stilostomella subspinosa*. Cipero Section no. 14 (green clay), San Fernando, Trinidad. Deposited in the collections of the United States National Museum, Washington, DC; CC 21484. Holotype. Lateral view, NHMUK PM DI 0654, scale bar = 512 μm.

2. *Stilostomella subspinosa*. Allison Guyot, Mid-Pacific Mountains, Northwest Pacific; ODP Leg 143, Hole 865B, Core 15, Section 3, 52–57 cm. Lateral view, NHMUK PM DI 0655, scale bar = 88 μm.

### Original Designation

*Ellipsonodosaria subspinosa* Cushman, 1943, p. 92, pl. 16, figs. 6–7b.

### Type Specimen

Holotype (CC 21484) deposited in the collections of the United States National Museum, Washington, DC.

### Type Level

Lower Middle Miocene, green clay.

### Type Locality

Cipero section, San Fernando, Trinidad.

### Description

Test forms an elongate, robust uniserial series; rectilinear or slightly arcuate in overall shape, subcylindrical and slightly tapered in outline, and circular in cross-section with the greatest breadth near the apertural end. The subspherical, strongly inflated chambers increase gradually in size, and are separated by strongly depressed sutures (not forming narrow bands between chambers as in *Stilostomella paleocenica*). Chamber walls are calcareous, finely perforate, and ornamented by short, stout spines, either entirely covering the chambers or, in early stages, confined to the lower portion of the chamber wall. The primary aperture is an arcuate or V-shaped terminal opening on a short, broad neck, with a phialine lip and small, tooth-like projection (in mature specimens).

### Synonyms and Citations

*Ellipsonodosaria* sp. Cushman and Jarvis, 1934, pl. 10, figs. 4–5.

*Stilostomella subspinosa* (Cushman). –Tjalsma and Lohmann, 1983, p. 36, pl. 14, figs. 16–17. –Miller and Katz, 1987, p. 138, pl. 1, fig. 12. –Thomas, 1990, p. 590. –Bolli et al., 1994, p. 360, figs. 63.30, 30a. –Nomura, 1995, p. 274, pl. 2, fig. 8.

*Siphonodosaria subspinosa* (Cushman). –Robertson, 1998, p. 180, pl. 67, fig. 3.

### Remarks

*Stilostomella subspinosa* (Cushman) displays considerable variability in size and ornamentation. Spines may be more delicate than on the original type material and may tend to coalesce into fine costae in the lower part of the test (Tjalsma and Lohmann, 1983).

### Biogeography

Worldwide. Reported from the Indian Ocean (Nomura, 1995), Weddell Sea, Antarctica (Thomas, 1990), North Atlantic (Miller and Katz, 1987), Trinidad (Cushman, 1943), and Jamaica (Robertson, 1998).

### Bathymetry

Bathyal to abyssal. Lower bathyal (Thomas, 1990), abyssal, greater than 3.5 kilometers (Miller and Katz, 1987).

### Chronostratigraphy

Middle Eocene to Late Miocene. Common from Middle Eocene onwards (Thomas, 1990). Abundant in Oligocene and Miocene (Miller and Katz, 1987).

## References


Bolli, H.M., Beckmann, J.P., and Saunders, J.B., 1994, Benthic Foraminiferal Biostratigraphy of the South Caribbean Region. Cambridge: Cambridge University Press, 408 pp.

Cushman, J.A., 1943, Some new foraminifera from the Tertiary of the Island of St. Croix. Contributions from the Cushman Laboratory for Foraminiferal Research, vol. 19, pp. 90–93.

Cushman, J.A. and Jarvis, P.W., 1934, Some interesting new uniserial foraminifera from Trinidad. Contributions from the Cushman Laboratory for Foraminiferal Research, vol. 10, pp. 71–75.

Miller, K.G. and Katz, M.E., 1987, Oligocene to Miocene benthic foraminiferal and abyssal circulation changes in the North Atlantic. Micropaleontology, vol. 33, no. 2, pp. 97–149.

Nomura, R., 1995, Paleogene to Neogene deep-sea paleoceanography in the eastern Indian Ocean: benthic foraminifera from ODP Sites 747, 757 and 758. Micropaleontology, vol. 41, pp. 251–290.

Robertson, B.E., 1998, Systematics and paleoecology of the benthic foraminiferida from the Buff Bay section, Miocene of Jamaica. Micropaleontology, vol. 44, supplement 2, pp. 1–266.

Thomas, E., 1990, Late Cretaceous through Neogene benthic foraminifers. In Barker, P.F., Kennett, J.P., et al., ed., Scientific Results of the Proceedings of the Ocean Drilling Program, vol. 113, pp. 571–594.

Tjalsma, R.C. and Lohmann, G.P., 1983, Paleocene–Eocene bathyal and abyssal benthic foraminifera from the Atlantic Ocean. Micropaleontology, Special Publication, no. 4, pp. 1–90.

# *Tappanina selmensis*

## (Cushman), 1933 emend. Brotzen, 1948

1. *Tappanina selmensis.* Walvis Ridge, South Atlantic; DSDP Leg 74, Site 527, Core 24, Section 2, 20–22 cm. Lateral view, NHMUK PM DI 0656, scale bar = 42 μm.

2. *Tappanina selmensis.* Walvis Ridge, South Atlantic; DSDP Leg 74, Site 527, Core 24, Section 2, 20–22 cm. Lateral view, NHMUK PM DI 0657, scale bar = 42 μm.

3. *Tappanina selmensis.* Walvis Ridge, South Atlantic; DSDP Leg 74, Site 527, Core 24, Section 2, 20–22 cm. Edge view, NHMUK PM DI 0658, scale bar = 39 μm.

4. *Tappanina selmensis.* Allison Guyot, Mid-Pacific Mountains, Northwest Pacific; ODP Leg 143, Hole 865C, Core 12, Section 4, 2–3 cm. Lateral view, NHMUK PM DI 0659, scale bar = 49 μm.

## Original Designation

*Bolivinita selmensis* Cushman, 1933, p. 58, pl. 7, figs. 3–4.

## Type Specimen

Holotype (CC 19043) and paratypes (CC 19046) of Cushman (1933) deposited in the collections of the United States National Museum, Washington, DC. Specimens figured by Brotzen (1948) deposited in the collections of the Swedish Geological Survey, Stockholm.

## Type Level

Upper Cretaceous, Selma Chalk.

## Type Locality

New Corinth Highway, 13.5 miles south of Selmer, McNairy County, Tennessee, USA.

Specimens figured by Brotzen (1948) from "Gjuteriet" (foundry) well at Ystad, southern Sweden; Middle Paleocene, Selandian.

## Description

Test forms a small, elongate, biserial series; tapered and crenulated in outline, rhomboidal or rectangular in cross-section, with weakly concave, broad sides, and convex or truncated narrow sides. The moderately inflated, angled chambers increase gradually in size, are sharply edged near the sutures, and carinate distally from the sutures giving the test a distinctive zigzag peripheral outline. Sutures become increasingly depressed, slightly limbate, and curved. Chamber walls are calcareous, finely perforate, and smooth, except for the lateral carinae on the outer margins of the chambers. The primary aperture is a narrow, elongate slit, bordered by a raised lip on one side, and with an internal toothplate on the other.

## Synonyms and Citations

*Bolivinita crawfordensis* Jennings, 1936, p. 28, pl. 3, fig. 14.
*Bolivinita costifera* Cushman, 1937, p. 105, pl. 15, fig. 15.
*Bolivinita exigua* Glaessner, 1937, p. 369, pl. 2, fig. 17.
*Eouvigerina excavata* Cushman, 1940, p. 66, pl. 11, fig. 18.
*Bolivinita glaessneri* Finlay, 1947, p. 277, pl. 4, figs. 55–58.
*Bolivinita bandyi* Said and Kenawy, 1956, p. 141, pl. 3, fig. 49.
*Bolivinita selmensis* Cushman, emend. Brotzen, 1948, p. 56, pl. 9, fig. 7, textfig. 16.
*Tappanina selmensis* (Cushman). –Tjalsma and Lohmann, 1983, p. 37, pl. 11, fg. 4. –Loeblich and Tappan, 1987, p. 499, pl. 550, figs. 5–8. –Revets, 1996, p. 17, pl. 15, figs. 4–7.
*Tappanina selmensis* (Cushman) emend. Brotzen.–van Morkhoven et al., 1986, p. 332, pl. 108.

## Remarks

Type species of the genus *Tappanina* Montanaro Gallitelli, 1955.

Brotzen (1948) considered *Bolivinita selmensis* Cushman to be highly variable, and emended the definition of this species to include *Bolivinita crawfordensis* Jennings, *Bolivinita costifera* Cushman, and *Bolivinita exigua* Glaessner.

## Biogeography

Worldwide. See Observed Occurrences in van Morkhoven et al. (1986).

## Bathymetry

Lower neritic to bathyal.

## Chronostratigraphy

Late Cretaceous (Santonian) to Early Eocene. Maastrichtian to Early Eocene (Biozone P6b) according (van Morkhoven et al., 1986), Santonian at DSDP Site 357 according (Sliter, 1977), common until middle Early Eocene in the Weddell Sea (Thomas, 1990) and on the Kerguelen Plateau (Mackensen and Berggren, 1992).

## References


Brotzen, F., 1948, The Swedish Paleocene and its Foraminiferal Fauna. Årsbok Sveriges Geologiska Undersökning, vol. 42, no. 2, pp. 1–140.

Cushman, J.A., 1933, New American Cretaceous Foraminifera. Contributions from the Cushman Laboratory for Foraminiferal Research, vol. 9, pp. 49–64.

Cushman, J.A., 1937, A few new species of American Cretaceous foraminifera. Contributions from the Cushman Laboratory for Foraminiferal Research, vol. 13, pp. 100–105.

Cushman, J.A., 1940, Midway foraminifera from Alabama. Contributions from the Cushman Laboratory for Foraminiferal Research, vol. 16, no. 3, pp. 51–73.

Finlay, H.J., 1947, New Zealand Foraminifera; Key species in stratigraphy – No. 5. New Zealand Journal of Science and Technology, vol. 28, no. 5 (sec. B), pp. 259–292.

Glaessner, M.F., 1937, Studien über Foraminiferen aus der Kreide und dem Tertiär des Kaukasus; 1. Die Foraminiferen der ältesten Tertiärschichten des Nordwest-Kaukas. Problemy Paleontologii, Paleontologichesjaya Laboratoriya Moskovskogo Gosudarstvennogo Universiteta, vol. 2–3, pp. 349–408.

Jennings, P.H., 1936, A microfauna from the Monmouth of basal Rancocas groups of New Jersey. Bulletin of American Paleontology, vol. 23, no. 78, pp. 3–76.

Loeblich, A.R. Jr and Tappan, H., 1987, Foraminiferal Genera and Their Classification. New York: Van Nostrand Reinhold, 970 pp.

Mackensen, A. and Berggren, W.A. 1992. Paleogene benthic foraminifers from the southern Indian Ocean (Kerguelen Plateau): biostratigraphy and paleoecology. In Wise, S.W. Jr, Schlich, R., et al., ed., Scientific Results of the Proceedings of the Ocean Drilling Program, vol. 120, pp. 603–630.

Revets, S.A., 1996, The generic revision of five families of Rotaliine Foraminifera. Special Publications of the Cushman Foundation for Foraminiferal Research, vol. 34, pp. 1–108.

Said, R. and Kenawy, A., 1956, Upper Cretaceous and Lower Tertiary foraminifera from northern Sinai, Egypt. Micropaleontology, vol. 2, no. 2, pp. 105–173.

Sliter, W.V., 1977, Cretaceous benthic foraminifers from the western South Atlantic Leg 39, Deep Sea Drilling Project. In Supko, P.R., Perch-Nielsen, K., et al., ed., Initial Reports of the Deep Sea Drilling Project, Washington (U.S. Government Printing Office), vol. 39, pp. 657–697.

Thomas, E., 1990, Late Cretaceous through Neogene benthic foraminifers. In Barker, P.F., Kennett, J.P., et al., ed., Scientific Results of the Proceedings of the Ocean Drilling Program, vol. 113, pp. 571–594.

Tjalsma, R.C. and Lohmann, G.P., 1983, Paleocene–Eocene bathyal and abyssal benthic foraminifera from the Atlantic Ocean. Micropaleontology, Special Publication, no. 4, pp. 1–90.

van Morkhoven, F.P.C.M., Berggren, W.A., Edwards, A.S., et al., 1986, Cenozoic cosmopolitan deep-water benthic foraminifera. Bulletin des Centres de Recherches Exploration-Production Elf-Aquitaine, Mémoire, vol. 11, pp. 1–421.

# Transversigerina senni

## (Cushman and Renz), 1941

1. *Transversigerina senni.* Agua Salada Formation, Isidro, 34.9 km east (81°) of Pueblo Pirita, District Zamura, State Falcon, Venezuela. Deposited in the collections of the United States National Museum, Washington, DC; No. 35912. Holotype. Lateral view, NHMUK PM DI 0660, scale bar = 404 µm.

### Original Designation

*Siphogenerina senni* Cushman and Renz, 1941, p. 22, pl. 3, figs. 21–22.

### Type Specimen

Holotype (No. 35912) deposited in the collections of the United States National Museum, Washington, DC.

### Type Level

Middle Miocene, originally given as Oligocene to Lower Miocene (van Morkhoven et al., 1986).

### Type Locality

Sample L.P. 149 from Isidro, 34.9 kilometers east (81°) of Pueblo Piritu, District of Zamora, State of Falcón, Venezuela.

### Description

Test forms an initial triserial series, becoming an elongate uniserial series, which is up to approximately three and a half to four times as long as broad, with a subcylindrical outline in megalospheric forms, a tapered outline in microspheric forms, and a circular cross-section in both. The slightly inflated chambers increase gradually in size, and are separated by depressed sutures, arched between the adjacent costae of uniserial chambers. Chamber walls are calcareous, finely perforate, and ornamented by eight to ten prominent longitudinal costae, continuous over the sutures, and ending at the base of the last chamber, which is smooth. The primary aperture is a terminal, round opening at the end of a short, cylindrical neck, with a phialine lip and an internal toothplate.

### Synonyms and Citations

*Siphogenerina cumingsi* Galloway and Hemingway, 1941, p. 433, pl. 34, figs. 6–7.

*Siphogenerina yumuriana* Palmer, 1941, p. 185, pl. 15, figs. 3–4.

*Rectuvigerina senni* (Cushman and Renz). –van Morkhoven et al., 1986, p. 174, pl. 58. –Katz and Miller, 1993, p. 392, pl. 1, fig. 6.

### Remarks

*Transversigerina senni* (Cushman and Renz) differs from *Transversigerina transversa* (Cushman) by its smaller size, fewer costae, and nearly parallel sides in the megalospheric generation, and by its longer, more tapering test in the microspheric generation.

### Biogeography

Worldwide. Reported from the Dominican Republic, Trinidad, Barbados, Venezuela, Europe, the East Pacific, the North Atlantic, and the Gulf of Mexico (van Morkhoven et al., 1986).

### Bathymetry

Bathyal.

### Chronostratigraphy

Late Oligocene to Middle Miocene. Doubtful occurrences in the Early Oligocene.

## References


Cushman, J.A. and Renz, H.H., 1941, New Oligocene–Miocene foraminifera from Venezuela. Contributions from the Cushman Laboratory for Foraminiferal Research, vol. 17, pp. 1–27.

Galloway, J.J. and Hemingway, C.E., 1941, The Tertiary Foraminifera of Porto Rico. New York Academy of Science, Scientific Survey of Porto Rico and the Virgin Islands, vol. 3, no. 4, pp. 275–491.

Katz, M.E. and Miller, K.G., 1993, Latest Oligocene to Earliest Pliocene benthic foraminiferal biofacies of the northeastern Gulf of Mexico. Micropaleontology, vol. 39, no. 4, pp. 367–403.

Palmer, D.K., 1941, Foraminifera of the upper Oligocene Cojimar Formation of Cuba Part 4. Memorias de la Sociedad Cubana de Historia Natural "Filipe Poey", vol. 15, no. 2, pp. 181–200.

van Morkhoven, F.P.C.M., Berggren, W.A., Edwards, A.S., et al., 1986, Cenozoic cosmopolitan deep-water benthic foraminifera. Bulletin des Centres de Recherches Exploration-Production Elf-Aquitaine, Mémoire, vol. 11, pp. 1–421.

# Transversigerina transversa

## (Cushman), 1918

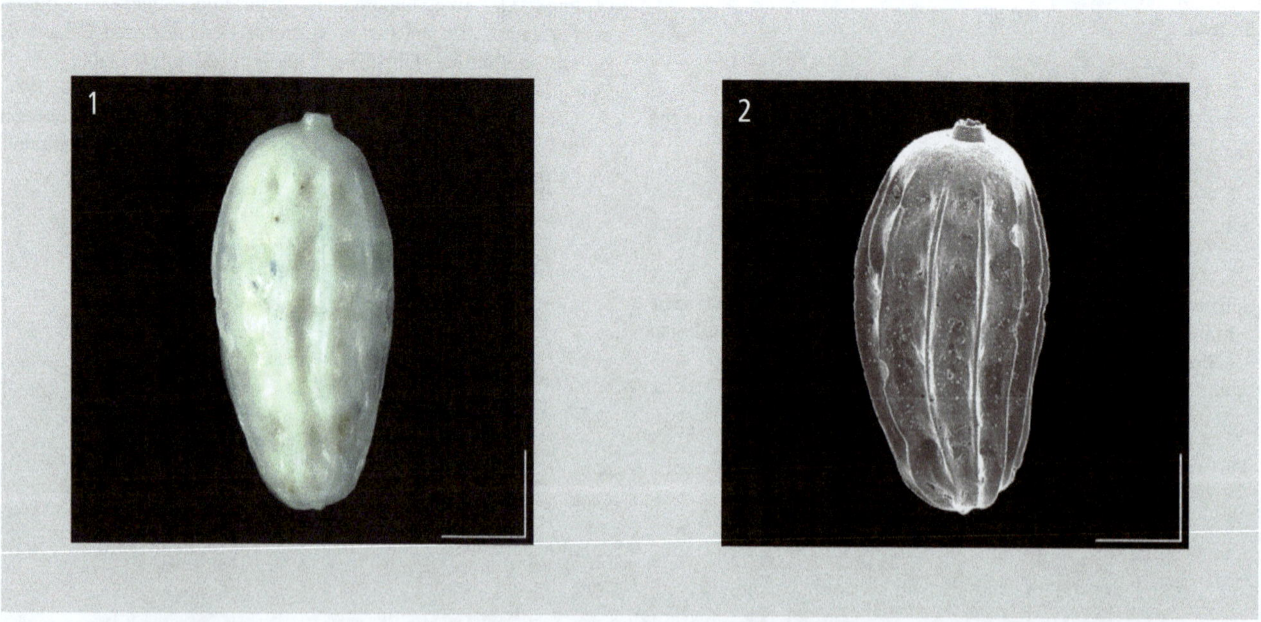


1. *Transversigerina transversa*. Gulf of Mexico; Eureka Core E68-136, depth: 2606 feet. Lateral view, NHMUK PM DI 0661, scale bar = 256 μm.
2. *Transversigerina transversa*. Gulf of Mexico; Eureka Core E68-136, depth: 2606 feet. Lateral view, NHMUK PM DI 0662, scale bar = 273 μm.

### Original Designation

*Siphogenerina raphanus* (Parker and Jones) var. *transversus* Cushman, 1918, p. 64, pl. 22, fig. 8.

### Type Specimen

Holotype (No. 324646) deposited in the collections of the United States National Museum, Washington, DC.

### Type Level

Miocene, Lower part of the Culebra Formation, dark, well laminated, very soft, carbonaceous clay rocks. Originally given as Oligocene by Cushman (1918).

### Type Locality

US Geological Survey Station No. 6010, north of Pedro Miguel Locks to Paraiso Bridge, Panama Canal Zone.

### Description

Test forms an initially triserial series, followed by an elongate, stout, robust, uniserial stage, with a subcylindrical outline and a circular cross-section. The inflated chambers increase gradually in size, and are separated by strongly depressed sutures, arched between adjacent costae of uniserial chambers. Chamber walls are calcareous, finely perforate, and ornamented by approximately twelve prominent longitudinal costae that are continuous over the sutures, ending at the base of the last chamber. The primary aperture is a terminal, round opening at the end of a short, cylindrical neck, with a phialine lip and internal toothplate.

### Synonyms and Citations

*Rectuvigerina transversa* (Cushman).–van Morkhoven et al., 1986, p. 171, pl. 57.–Loeblich and Tappan, 1987, p. 520, pl. 570, figs. 15–16.–Katz and Miller, 1993, p. 392, pl. 1, fig. 1.

### Remarks

Type species of the genus *Transversigerina* Mathews, 1945.

*Transversigerina senni* (Cushman and Renz) differs from *Transversigerina transversa* (Cushman) by its smaller size, fewer costae, and nearly parallel sides in the megalospheric generation, and by its longer, more tapering test in the microspheric generation.

Revised stratigraphy at the Eureka boreholes in the Gulf of Mexico (Katz and Miller, 1993) has extended the last occurrence of this species to Biozone M9b, in contrast to the Biozone M7 occurrence reported by van Morkhoven et al. (1986).

### Biogeography

Worldwide. Reported from California, Colombia, Cuba, the Dominican Republic, Trinidad, Mexico, Venezuela, Europe, the East Pacific, the North and South Atlantic, and the Gulf of Mexico (van Morkhoven et al., 1986).

### Bathymetry

Primarily upper bathyal. Very rare (<1%), throughout bathyal (Katz and Miller 1993).

### Chronostratigraphy

Latest Oligocene to Middle Miocene (Serravallian).

## References

Cushman, J.A., 1918, The smaller fossil foraminifera of the Panama Canal Biozone. Bulletin of the United States National Museum, vol. 103, pp. 45–87.

Katz, M.E. and Miller, K.G., 1993, Latest Oligocene to Earliest Pliocene benthic foraminiferal biofacies of the northeastern Gulf of Mexico. Micropaleontology, vol. 39, no. 4, pp. 367–403.

Loeblich, A.R. Jr and Tappan, H., 1987, Foraminiferal Genera and Their Classification. New York: Van Nostrand Reinhold, 970 pp.

van Morkhoven, F.P.C.M., Berggren, W.A., Edwards, A.S., et al., 1986, Cenozoic cosmopolitan deep-water benthic foraminifera. Bulletin des Centres de Recherches Exploration-Production Elf-Aquitaine, Mémoire, vol. 11, pp. 1–421.

# *Trifarina angulosa*

## (Williamson), 1858

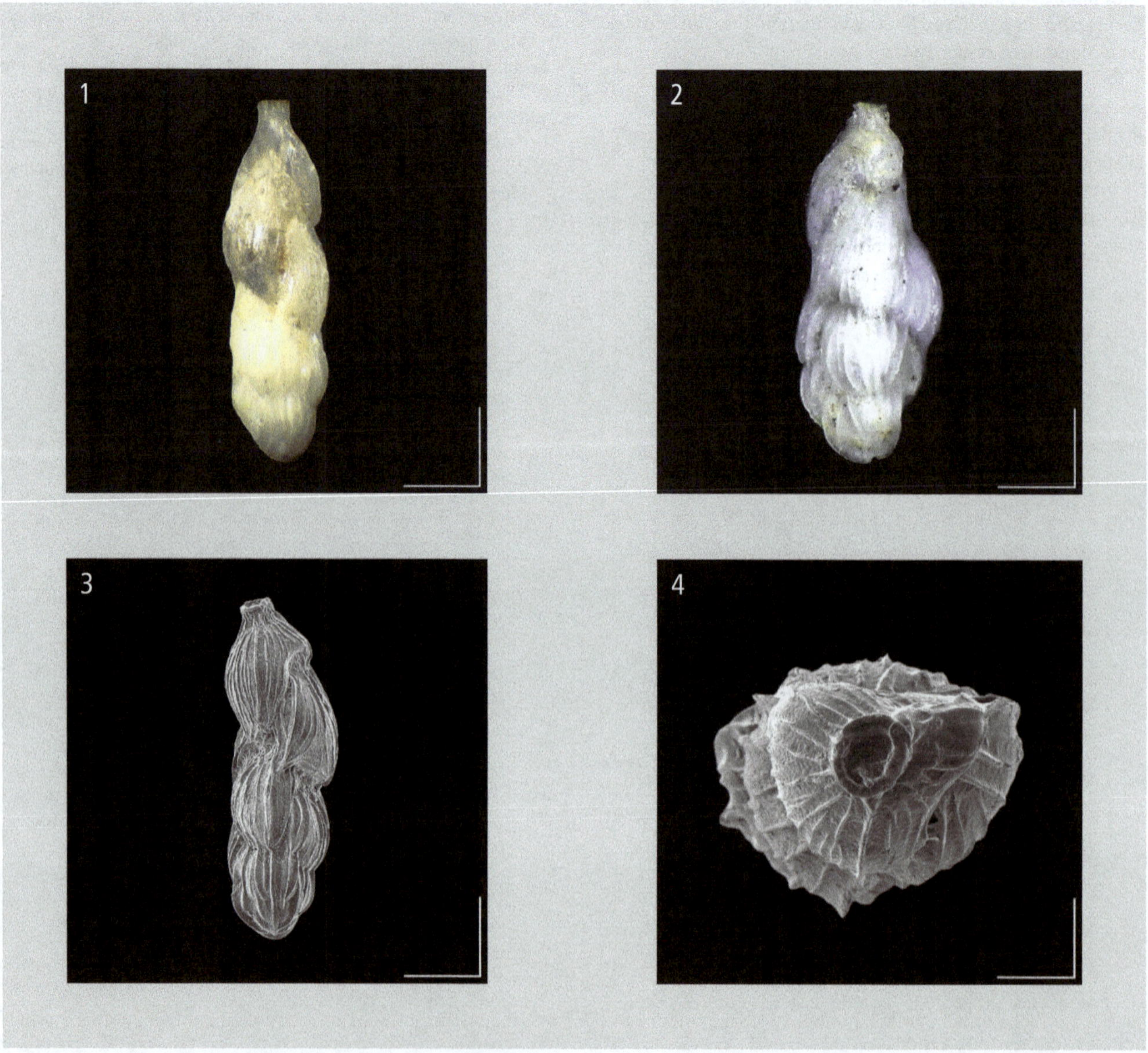


1. *Trifarina angulosa*. Prince Edward Island; "Challenger" Station 145. Deposited in the micropaleontological collections of The Natural History Museum, London; NHMUK ZF2563. Syntype. Lateral view, NHMUK PM DI 0663, scale bar = 143 μm.

2. *Trifarina angulosa*. Prince Edward Island; "Challenger" Station 145. Deposited in the micropaleontological collections of The Natural History Museum, London; NHMUK ZF2563. Syntype. Lateral view, NHMUK PM DI 0664, scale bar = 159 μm.

3. *Trifarina angulosa*. Prince Edward Island; "Challenger" Station 145. Deposited in the micropaleontological collections of The Natural History Museum, London; NHMUK ZF2563. Syntype. Lateral view, NHMUK PM DI 0665, scale bar = 184 μm.

4. *Trifarina angulosa*. Prince Edward Island; "Challenger" Station 145. Deposited in the micropaleontological collections of The Natural History Museum, London; NHMUK ZF2563. Syntype. Apertural view, NHMUK PM DI 0666, scale bar = 79 μm.

## Original Designation

*Uvigerina angulosa* Williamson, 1858, p. 67, pl. 5, fig. 140.

## Type Specimen

Specimens figured by Brady (1884) are housed in the micropaleontological collections of The Natural History Museum, London, UK, and are registered in slide ZF2563.

## Type Level

Recent.

## Type Locality

Not designated. Localities given: Boston, dredged sands of Torquay, Arran, Shetland, and Scilly, British Isles. Brady's specimens are from "Challenger" Station 145, Prince Edward Island.

## Description

Test forms an initially triserial series, later becoming an elongate, rectilinear, uniserial series, with a triangular cross-section, and an acute periphery. The moderately inflated chambers increase rapidly in height, and are separated by curved, depressed sutures. Chamber walls are calcareous, finely perforate, and ornamented by discontinuous, longitudinal costae. The primary aperture is a terminal, round opening at the end of a short neck, with a lip and internal toothplate.

## Synonyms and Citations

*Uvigerina angulosa* Williamson. –Brady, 1884, p. 576, pl. 74, figs. 15–16 (not figs. 17–18).

*Angulogerina angulosa* (Williamson). –Barker, 1960, pl. 74, figs. 15–16. –van Marle, 1991, p. 109, pl. 7, figs. 6–7. –Milker and Schmiedl, 2012, p. 90, figs. 21.2–4.

*Trifarina angulosa* (Williamson). –Jones, 1994, p. 86, pl. 74, figs. 15–16.

## Remarks

*Trifarina bradyi* Cushman differs from *Trifarina angulosa* (Williamson) by its more regularly arranged, symmetrically triangular, smooth test.

## Biogeography

Worldwide.

## Bathymetry

Lower neritic to upper bathyal.

## Chronostratigraphy

Middle Miocene to Recent (Jones, 1994).

## References


Barker, R.W., 1960, Taxonomic notes on the species figured by H.B. Brady in his report on the foraminifera dredged by H.M.S. Challenger during the years 1873–1876. Society of Economic Paleontologists and Mineralogists Special Publication, vol. 9, pp. 1–238.

Brady, H.B., 1884, Report on the foraminifera dredged by H.M.S. Challenger during the years 1873–1876. In Brady, H.B., ed., Report on the Scientific Results of the Voyage of the H.M.S. Challenger during the years 1873–1876. Zoology, vol. 9, pp. 1–814.

Jones, R.W., 1994, The Challenger Foraminifera. Oxford: Oxford University Press, 149 pp.

Milker, Y. and Schmiedl, G., 2012, A taxonomic guide to modern benthic shelf foraminifera of the western Mediterranean Sea. Palaeontologia Electronica, vol. 15, Issue 2;16A, 134 pp.

van Marle, L.J., 1991, Eastern Indonesian Late Cenozoic smaller benthic foraminifera. Verhandelingen der Koninklijke Nederlanse Akademie van Wetenschappen, Afdeeling Natuurkunde, Eerste Reeks, vol. 34, pp. 1–328.

Williamson, W.C., 1858, On the Recent Foraminifera of Great Britain. London: Ray Society, 107 pp.

# *Trifarina bradyi*

## Cushman, 1923

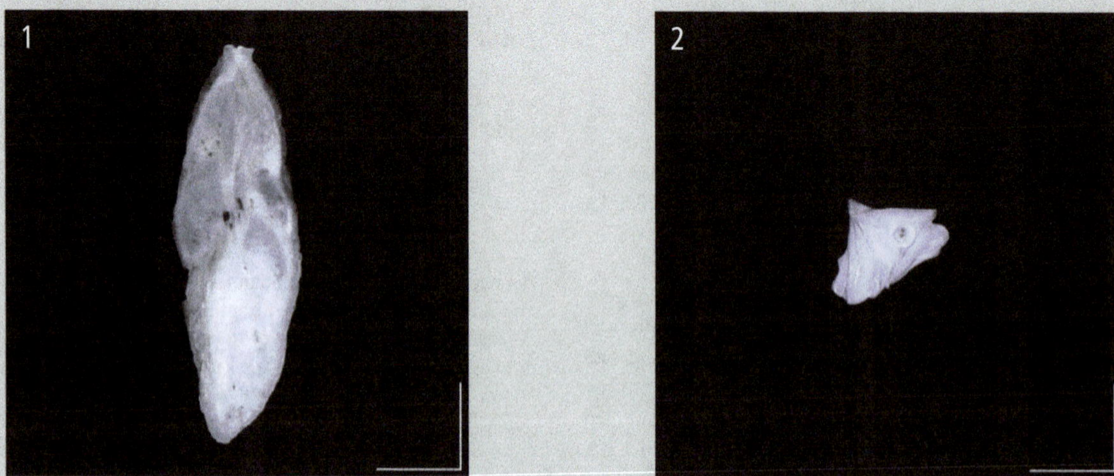

1. *Trifarina bradyi.* Off Bermuda; "Challenger" Station 33, depth: 435 fathoms. Deposited in the micropaleontological collections of The Natural History Museum, London; NHMUK ZF2309. Lateral view, NHMUK PM DI 0667, scale bar = 94 μm.

2. *Trifarina bradyi.* Off Bermuda; "Challenger" Station 33, depth: 435 fathoms. Deposited in the micropaleontological collections of The Natural History Museum, London; NHMUK ZF2309. Apertural view, NHMUK PM DI 0668, scale bar = 90 μm.

### Original Designation
*Trifarina bradyi* Cushman, 1923, p. 99, pl. 22, figs. 3–9.

### Type Specimen
Deposited in United States National Museum, Washington, DC. (Nos 17442, 17443 (type), 17444–17452).

### Type Level
Recent, surface sediments.

### Type Locality
"Albatross" Station D2150 (13° 34′ 45″ N, 80° 21′ 10″ W) in 382 fathoms, Caribbean Sea. Other localities mentioned in original description include "Albatross" Station D2677 (32° 39′ 00″ N, 76° 50′ 30″ W) in 478 fathoms, off the eastern coast of the USA.

### Description
Test forms an initially triserial series, later becoming an elongate, rectilinear, uniserial series, with slightly tapered ends, a triangular cross-section, and an acute, keeled periphery. The moderately inflated chambers increase gradually in size, and are separated by distinct, thick, curved sutures. Chamber walls are calcareous, finely perforate, and smooth. The primary aperture is a terminal, round opening at the end of a short neck, with a lip and an internal toothplate,

### Synonyms and Citations
*Rhabdogonium tricarinatum* d'Orbigny. –Brady, 1884, p. 525, pl. 67, figs. 1–3.

*Trifarina bradyi* Cushman. –Barker, 1960, pl. 67, figs. 1–3. –Loeblich and Tappan, 1987, p. 526, pl. 574, figs. 10–13. –van Marle, 1991, p. 110, pl. 7, figs. 8–9. –Jones, 1994, p. 78, pl. 67, figs. 1–3.

### Remarks
Type species of the genus *Trifarina* Cushman, 1923.

*Trifarina bradyi* Cushman differs from *Trifarina angulosa* (Williamson) by its more regularly arranged, symmetrically triangular, smooth test. It differs from *Trifarina carinata* (Cushman) by its less inflated chambers, lacking longitudinal costae.

### Biogeography

Worldwide.

### Bathymetry

Lower neritic to bathyal. Reported between 12 and 1360 fathoms at "Challenger" stations (Jones, 1994), and between 60 and 1560 meters in the eastern Indian Ocean (van Marle, 1991).

### Chronostratigraphy

Miocene to Recent (Jones, 1994).

## References


Barker, R.W., 1960, Taxonomic notes on the species figured by H.B. Brady in his report on the foraminifera dredged by H.M.S. Challenger during the years 1873–1876. Society of Economic Paleontologists and Mineralogists Special Publication, vol. 9, pp. 1–238.

Brady, H.B., 1884, Report on the foraminifera dredged by H.M.S. Challenger during the years 1873–1876. In Brady, H.B., ed., Report on the Scientific Results of the Voyage of the H.M.S. Challenger during the years 1873–1876. Zoology, vol. 9, pp. 1–814.

Cushman, J.A., 1923, The foraminifera of the Vicksburg Group. Professional Paper of the United States Geological Survey, vol. 133, pp. 11–71.

Jones, R.W., 1994, The Challenger Foraminifera. Oxford: Oxford University Press, 149 pp.

Loeblich, A.R. Jr and Tappan, H., 1987, Foraminiferal Genera and Their Classification. New York: Van Nostrand Reinhold, 970 pp.

van Marle, L.J., 1991, Eastern Indonesian Late Cenozoic smaller benthic foraminifera. Verhandelingen der Koninklijke Nederlanse Akademie van Wetenschappen, Afdeeling Natuurkunde, Eerste Reeks, vol. 34, pp. 1–328.

# Trifarina carinata

## (Cushman), 1927

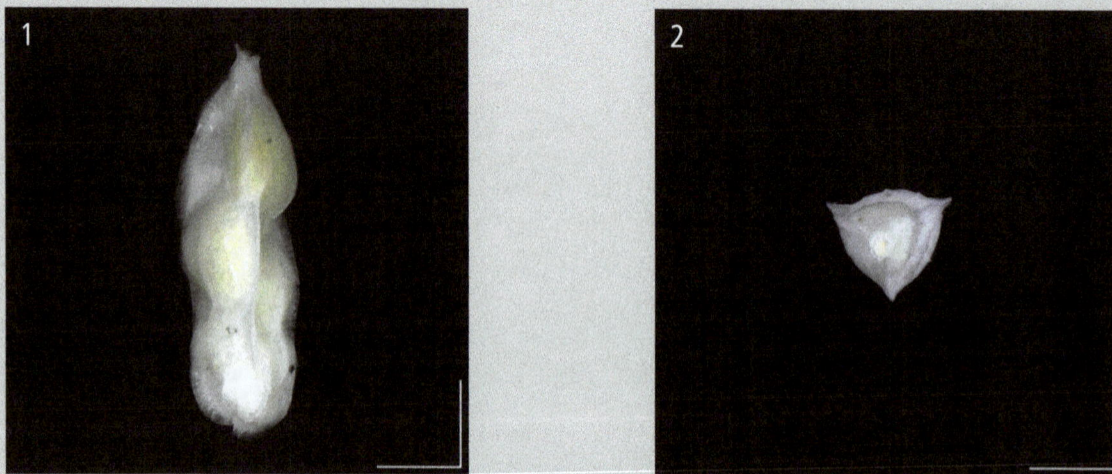


1. *Trifarina carinata.* North of Juan Fernandez, East Pacific Ocean; "Challenger" Station 300, depth: 1375 fathoms. Deposited in the micropaleontological collections of The Natural History Museum, London; NHMUK ZF2564. Lateral view, NHMUK PM DI 0669, scale bar = 264 μm.

2. *Trifarina carinata.* North of Juan Fernandez, East Pacific Ocean; "Challenger" Station 300, depth: 1375 fathoms. Deposited in the micropaleontological collections of The Natural History Museum, London; NHMUK ZF2564. Apertural view, NHMUK PM DI 0670, scale bar = 250 μm.

### Original Designation
*Angulogerina carinata* Cushman, 1927, p. 159, pl. 4, fig. 3.

### Type Specimen
Holotype deposited in the United States National Museum, Washington, DC (No. 20299).

### Type Level
Recent.

### Type Locality
US Coast and Geodetic Survey ship "Guide" Station 12 (25), latitude 32° 28′ N, longitude 118° 20′ W, 429 fathoms, Pacific Ocean.

### Description
Test forms an initially triserial series, later becoming an elongate, rectilinear, uniserial series, with a triangular cross-section, and an acute, keeled periphery. The sharp carinated edges often bear fine, radial tubules. The moderately inflated chambers increase rapidly in height, and are separated by curved, depressed sutures. Chamber walls are calcareous, finely perforate, and smooth or with a few fine, discontinuous, longitudinal costae. The primary aperture is a terminal, round opening at the end of a short neck, with a lip and an internal toothplate.

### Synonyms and Citations
*Uvigerina angulosa* Williamson. –Brady, 1884, pl. 74, figs. 17–18.
*Angulogerina carinata* Cushman var. *bradyana* Cushman. –Barker, 1960, pl. 74, fig. 18.
*Angulogerina carinata* Cushman. –Barker, 1960, pl. 74, fig. 17. –Resig, 1981, pl. 2, fig. 1.
*Trifarina carinata* (Cushman). –Jones, 1994, p. 86, pl. 74, figs. 17–18.

### Remarks
*Trifarina carinata* (Cushman) differs from *Trifarina bradyi* (Cushman) by its more inflated chambers, distinct longitudinal costae, and sharp carinated edges, often with fine, radial tubules.

Jones (1994) considered the variety *bradyana*, which differs in the degree of ornamentation, to be superfluous.

### Biogeography
Worldwide.

### Bathymetry

Primarily bathyal. Reported from "Challenger" Station 300, north of Juan Fernandez in the East Pacific, 1375 fathoms water depth (Jones, 1994). Upper to middle bathyal (350–625 meters) off Peru (Resig, 1981).

### Chronostratigraphy

Miocene to Recent.

## References


Barker, R.W., 1960, Taxonomic notes on the species figured by H.B. Brady in his report on the foraminifera dredged by H.M.S. Challenger during the years 1873–1876. Society of Economic Paleontologists and Mineralogists Special Publication, vol. 9, pp. 1–238.

Brady, H.B., 1884, Report on the foraminifera dredged by H.M.S. Challenger during the years 1873–1876. In Brady, H.B., ed., Report on the Scientific Results of the Voyage of the H.M.S. Challenger during the years 1873–1876. Zoology, vol. 9, pp. 1–814.

Cushman, J.A., 1927, Recent foraminifera from off the West coast of America. Bulletin of the Scripps Institute of Oceanography, Technical Series, vol. 1, pp. 119–188.

Jones, R.W., 1994, The Challenger Foraminifera. Oxford: Oxford University Press, 149 pp.

Resig., J., 1981, Biogeography of benthic foraminifera of the northern Nazca Plate and adjacent continental margin. Geological Society of America, Memoir 154, pp. 619–666.

# *Triloculina tricarinata*

## d'Orbigny, 1826

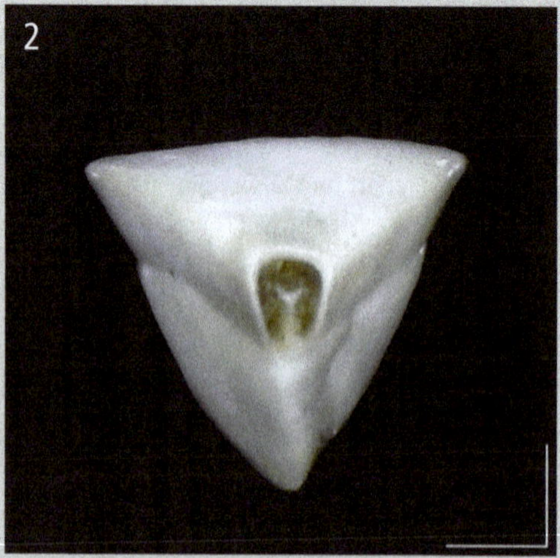


1. *Triloculina tricarinata.* Shell Cove, Port Jackson, Australia. Deposited in the micropaleontological collections of The Natural History Museum, London; NHMUK ZF1906. Lateral view, NHMUK PM DI 0671, scale bar = 140 μm.

2. *Triloculina tricarinata.* Shell Cove, Port Jackson, Australia. Deposited in the micropaleontological collections of The Natural History Museum, London; NHMUK ZF1906. Apertural view, NHMUK PM DI 0672, scale bar = 121 μm.

### Original Designation

*Triloculina tricarinata* d'Orbigny, 1826, p. 299, Model No. 94, illustrated in Parker et al., 1865, pl. 1, fig. 8.

### Type Specimen

Deposited in the collections of the Laboratoire de Paléontologie, Muséum d'Histoire Naturelle, Paris, France.

### Type Level

Recent.

### Type Locality

Red Sea.

### Description

Test forms an ovate, triloculine series (except for the early stage) that is triangular in cross-section, with straight or slightly concave sides. Three chambers are visible externally. Chamber walls are calcareous, porcelaneous, imperforate, and smooth. The primary aperture is a terminal, round opening, with a small bifid tooth.

### Synonyms and Citations

*Miliolina tricarinata* (d'Orbigny). –Brady, 1884, p. 165, pl. 3, fig. 17a–b.

*Triloculina tricarinata* d'Orbigny. –Phleger et al., 1953, p. 28, pl. 5, fig. 21. –Barker, 1960, p. 6, pl. 3, fig. 17a–b. –Hermelin, 1989, p. 38, pl. 3, figs. 6–7. –Milker and Schmiedl, 2012, p. 68, figs. 17.23–24.

*Triloculina tricarinata* sensu Parker et al., 1865. –Jones, 1994, p. 20, pl. 3, fig. 17.

### Remarks

See Jones (1994) for detailed taxonomic history.

### Biogeography

Worldwide.

### Bathymetry

Neritic to upper abyssal. Reported between 6 and 2350 fathoms at "Challenger" stations (Jones, 1994).

## Chronostratigraphy

Miocene to Recent (Jones, 1994).

## References


Barker, R.W., 1960, Taxonomic notes on the species figured by H.B. Brady in his report on the foraminifera dredged by H.M.S. Challenger during the years 1873–1876. Society of Economic Paleontologists and Mineralogists Special Publication, vol. 9, pp. 1–238.

Brady, H.B., 1884, Report on the foraminifera dredged by H.M.S. Challenger during the years 1873–1876. In Brady, H.B., ed., Report on the Scientific Results of the Voyage of the H.M.S. Challenger during the years 1873–1876. Zoology, vol. 9, pp. 1–814.

d'Orbigny, A., 1826, Tableau méthodique de la classe des Céphalopodes. Annales des Sciences Naturelles, vol. 7, pp. 245–314.

Hermelin, J.O.R., 1989, Pliocene benthic foraminifera from the Ontong-Java Plateau (Western Equatorial Pacific Ocean): faunal response to changing paleoenvironment. Special Publication of the Cushman Foundation for Foraminiferal Research, vol. 26, pp. 1–143.

Jones, R.W., 1994, The Challenger Foraminifera. Oxford: Oxford University Press, 149 pp.

Milker, Y. and Schmiedl, G., 2012, A taxonomic guide to modern benthic shelf foraminifera of the western Mediterranean Sea. Palaeontologia Electronica, vol. 15, Issue 2;16A, 134 pp.

Parker, W.K., Jones, T.R., and Brady, H.B., 1865, On the nomenclature of the foraminifera, Pt. 12: the species enumerated by d'Orbigny in the "Annales des Sciences Naturelles", vol. 7, 1826. Annals and Magazine of Natural History, series no. 3, vol. 16, pp. 15–41.

Phleger, F.B., Parker, F.L., and Pierson, J.F., 1953, North Atlantic foraminifera. Reports of the Swedish Deep-Sea Expedition, 1947–1948, vol. 7, pp. 3–122.

# *Triloculina trigonula*

## (Lamarck), 1804

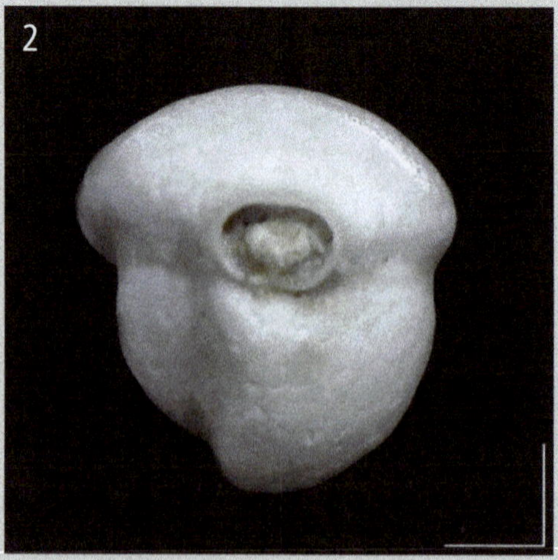


1. *Triloculina trigonula*. Off the Isle of Skye, North Atlantic Ocean; "Porcupine" Station AA, depth: 45–60 fathoms. Deposited in the micropaleontological collections of The Natural History Museum, London; NHMUK ZF1910. Lateral view, NHMUK PM DI 0673, scale bar = 201 μm.

2. *Triloculina trigonula*. Off the Isle of Skye, North Atlantic Ocean; "Porcupine" Station AA, depth: 45–60 fathoms. Deposited in the micropaleontological collections of The Natural History Museum, London; NHMUK ZF1910. Apertural view, NHMUK PM DI 0674, scale bar = 165 μm.

### Original Designation

*Miliolites trigonula* Lamarck, 1804, p. 351, pl. 17 (15), fig. 4.

### Type Specimen

Originally deposited in the author's and M. Defrance's offices, Paris, France.

### Type Level

Fossil.

### Type Locality

Grignon, near Paris, France.

### Description

Test forms an ovate, triloculine series (except for the early stage), triangular in cross-section, with convex sides and a subrounded periphery. Three chambers are visible externally. Chamber walls are calcareous, porcelaneous, imperforate, and smooth. The primary aperture is a terminal, round opening with a small bifid tooth.

### Synonyms and Citations

*Miliolina trigonula* (Lamarck). –Brady, 1884, p. 164, pl. 3, figs. 15–16 (not fig. 14). –Williamson, 1858, p. 83, pl. 7, figs. 180–182.

*Triloculina trigonula* (Lamarck). –d'Orbigny, 1826, p. 299, pl. 16, figs. 5–9, Mod. No. 93. –Barker, 1960, pl. 3, figs. 15–16. –Hermelin, 1989, p. 39, pl. 3, fig. 8. –Jones, 1994, p. 20, pl. 3, figs. 15–16.

### Biogeography

Worldwide.

### Bathymetry

Lower neritic to bathyal. Shoreline to 100 fathoms (Jones, 1994). Reported from lower bathyal zone in eastern Pacific Ocean (see Hermelin, 1989).

### Chronostratigraphy

Early Eocene to Recent (Jones, 1994).

## References

Barker, R.W., 1960, Taxonomic notes on the species figured by H.B. Brady in his report on the foraminifera dredged by H.M.S. Challenger during the years 1873–1876. Society of Economic Paleontologists and Mineralogists Special Publication, vol. 9, pp. 1–238.

Brady, H.B., 1884, Report on the foraminifera dredged by H.M.S. Challenger during the years 1873–1876. In Brady, H.B., ed., Report on the Scientific Results of the Voyage of the H.M.S. Challenger during the years 1873–1876. Zoology, vol. 9, pp. 1–814.

d'Orbigny, A., 1826, Tableau méthodique de la classe des Céphalopodes. Annales des Sciences Naturelles, vol. 7, pp. 245–314.

Hermelin, J.O.R., 1989, Pliocene benthic foraminifera from the Ontong-Java Plateau (Western Equatorial Pacific Ocean): faunal response to changing paleoenvironment. Special Publication of the Cushman Foundation for Foraminiferal Research, vol. 26, pp. 1–143.

Jones, R.W., 1994, The Challenger Foraminifera. Oxford: Oxford University Press, 149 pp.

Lamarck, J.B., 1804, Suite des memoires sur les fossiles des environs de Paris. Annales Muséum National d'Histoire Naturelle, vol. 5, pp. 349–357.

Williamson, W.C., 1858, On the Recent Foraminifera of Great Britain. London: Ray Society, 107 pp.

# *Trochammina abrupta*

## Geroch, 1966

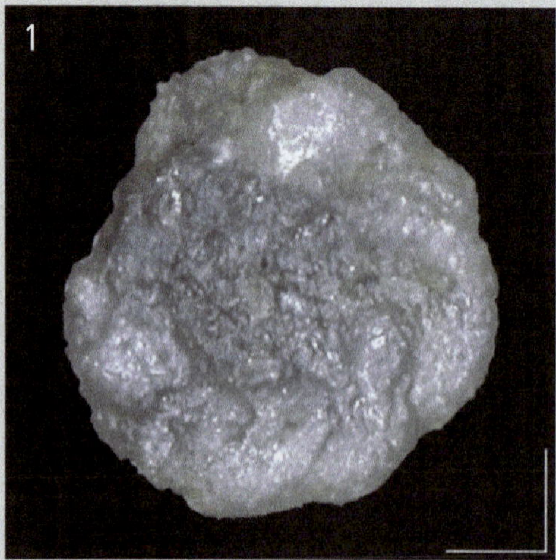


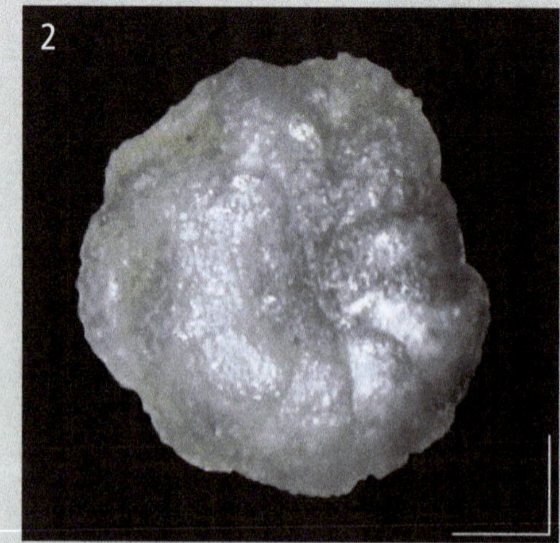


1. *Trochammina abrupta*. Straconka, near Bielsko, Poland; Verovice Beds, Silesian Unit, Sample 39/57. Deposited in the collections of the United States National Museum, Washington, DC; USNM 460968. Paratype. Spiral view, NHMUK PM DI 0675, scale bar = 64 μm.

2. *Trochammina abrupta*. Straconka, near Bielsko, Poland; Verovice Beds, Silesian Unit, Sample 39/57. Deposited in the collections of the United States National Museum, Washington, DC; USNM 460968. Paratype. Umbilical view, NHMUK PM DI 0676, scale bar = 72 μm.

### Original Designation
*Trochammina abrupta* Geroch, 1966, p. 468, pl. 12, figs. 13–17.

### Type Specimen
Original material deposited in the micropaleontological collections of the Jagiellonian University, Kraków, Poland.

### Type Level
Barremian–Aptian, Verovice Shales.

### Type Locality
Straconka, near Bielsko, Silesia, Poland.

### Description
Test forms a small, trochospiral series; slightly lobulate in outline, and planoconvex in cross-section, with an acute periphery, a flattened, evolute spiral side, and a convex, involute umbilical side, with a deep, wide umbilicus. The seven to eight triangular, moderately inflated chambers in the final whorl increase rapidly in size, and are separated by flush or slightly depressed, radial sutures. Chamber walls are finely agglutinated, non-calcareous (with considerable siliceous cement), and smoothly finished. The primary aperture is an interiomarginal, umbilical–extraumbilical arch, bordered by a narrow lip.

### Synonyms and Citations
*Trochammina* sp. Geroch, 1959, p. 120, pl. 12, fig. 4.

*Trochammina abrupta* Geroch. –Geroch and Nowak, 1984, pl. 2, figs. 11–12; pl. 7, figs. 1–4. –Riegraf and Luterbacher, 1989, p. 1092, pl. 2, figs. 4–5. –Kaminski et al., 1992, p. 256, pl. 5, figs. 14–15, pl. 6, figs. 1–2. –Holbourn and Kaminski, 1997, p. 49, pl. 18, figs. 11–13; pl. 19, fig. 1.

### Remarks
Genus used in the sense of *Trochammina* sensu lato. Two morphotypes were recognized in the DSDP and ODP material from the Indian Ocean: one morphotype with a high, convex umbilical side with a small, slightly depressed umbilicus; a second broader morphotype with more open coiling, a lower umbilical side, and a more pronounced umbilicus (Kaminski et al., 1992). The latter type corresponds to the topotypes from Poland and to the specimen illustrated by Riegraf and Luterbacher (1989).

### Biogeography

Worldwide.

### Bathymetry

Bathyal to abyssal. Common in flysch-type assemblages.

### Chronostratigraphy

Cretaceous (Berriasian to Cenomanian). Geroch and Nowak (1984) reported a stratigraphic range from the Hauterivian to early Cenomanian for this species. Occurrences in DSDP and ODP material from the Indian Ocean extended the lower range to the Tithonian or Berriasian (Holbourn and Kaminski, 1997).

## References


Geroch, S., 1959, Stratigraphic significance of arenaceous foraminifera in the Carpathian flysch. Paläontologische Zeitung, vol. 33, pp. 113–122.

Geroch, S., 1966, Lower Cretaceous small foraminifera of the Silesian series, Polish Carpathians. Rocznik Polskiego Towarzystwa Geologicznego, vol. 36, pp. 413–480.

Geroch, S. and Nowak, W., 1984, Proposal of zonation for the late Tithonian–late Eocene, based upon arenaceous foraminifera from the outer Carpathians, Poland. In Oertli, H., ed., Benthos '83, 2nd International Symposium on Benthic Foraminifera, Pau (France), April 11–15, 1983. Elf Aquitaine, ESSO REP and TOTAL CFP, Pau and Bordeaux, pp. 225–239.

Holbourn, A.E.L. and Kaminski, M.A., 1997, Cretaceous deep-water benthic foraminifera of the Indian Ocean. Grzybowski Foundation Special Publication, no. 4, pp. 1–175.

Kaminski, M.A., Gradstein, F., and Geroch, S., 1992, Uppermost Jurassic to Lower Cretaceous deep-water benthic foraminiferal assemblages from Site 765 on the Argo Abyssal Plain. In Gradstein, F., Ludden, J.N., et al., ed., Scientific Results of the Proceedings of the Ocean Drilling Program, vol. 123, pp. 239–269.

Riegraf, W. and Luterbacher, H., 1989, Benthonische Foraminiferen aus de Unterkreide des "Deep Sea Drilling Project" (Leg 1–79). Geologische Rundschau, vol. 78, no. 3, pp. 1063–1120.

# *Trochammina inflata*

## (Montagu), 1808

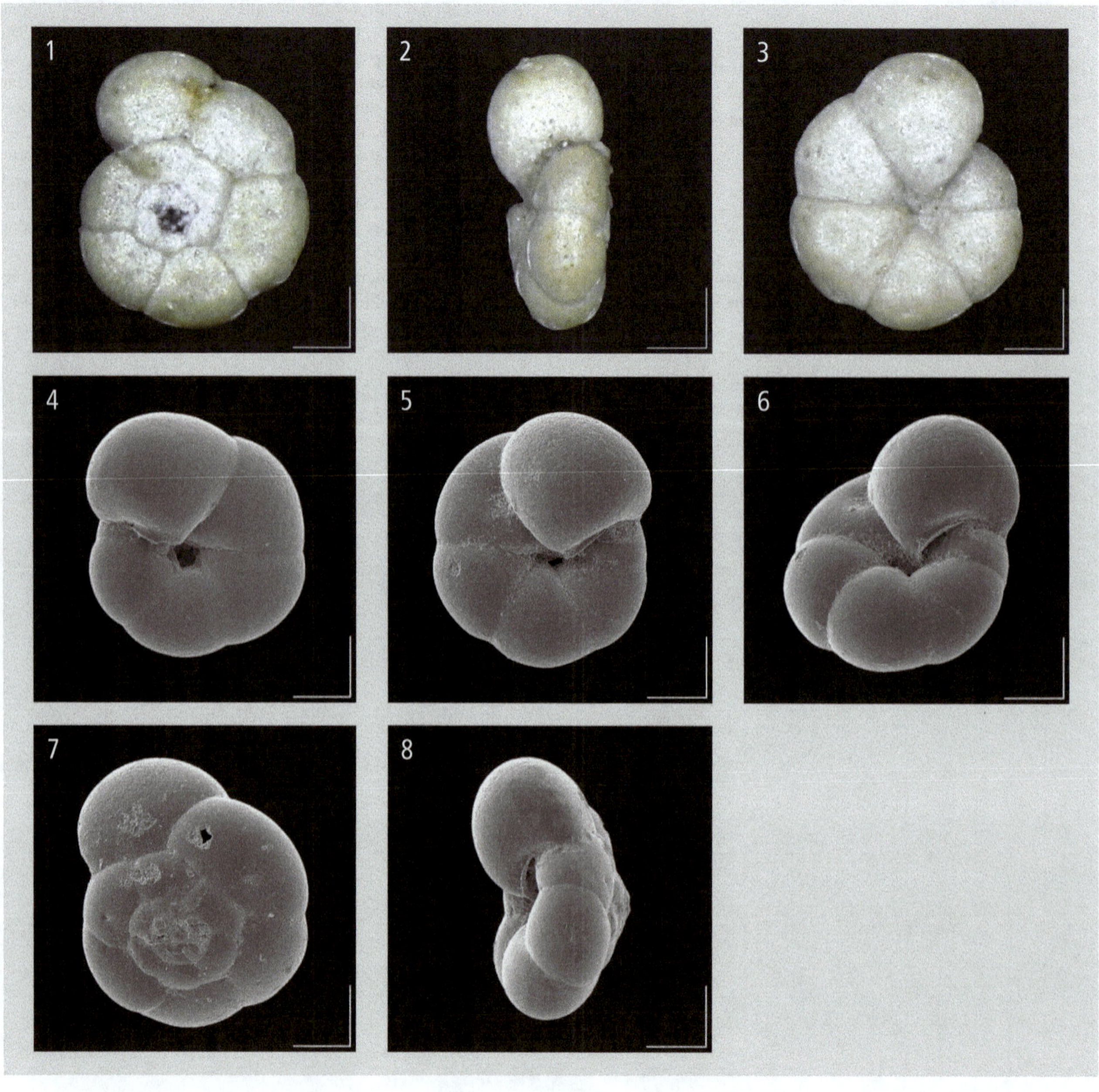


*Please see opposite for captions*

## Original Designation

*Nautilus inflatus* Montagu, 1808, p. 81, pl. 18, fig. 3.

## Type Specimen

Neotype designated by Brönnimann and Whittaker (1984) from the brackish waters of Broadsands, Torquay, Devon, England. Neotype deposited in the micropaleontological collections of The Natural History Museum, London, UK, registered in slide ZF4267.

## Type Level

Recent.

## Type Locality

Devon, England.

## Description

Test forms a lobulate trochospire; concavo-convex in cross-section, with a rounded periphery. The (approximately) five moderately inflated chambers in the last whorl increase gradually in size, and are separated by radial, depressed sutures. Chamber walls are agglutinated, consisting of three main layers: an organic inner layer, a thick, imperforate, middle agglutinated layer with grains of variable size, and a thin, veneer-like, organic outer layer. The primary aperture is an interiomarginal umbilical–extraumbilical arch, with a narrow lip.

## Synonyms and Citations

*Trochammina inflata* (Montagu). –Brady, 1884, p. 338, pl. 41, fig. 4. –Barker, 1960, pl. 41, fig. 4.
–Brönnimann and Whittaker, 1984, p. 311, figs. 1–11. –Loeblich and Tappan, 1987, p. 122, pl. 129, figs. 20–23. –Brönnimann and Whittaker, 1988, p. 26, pl. 1, fig. 1. –Jones, 1994, p. 46, pl. 41, fig. 4.

## Remarks

Type species of the genus *Trochammina* Parker and Jones, 1859.

See Brönnimann and Whittaker (1988) for detailed description of wall structure.

## Biogeography

Coasts of Northwest Europe (Jones, 1994).

## Bathymetry

Neritic.

## Chronostratigraphy

Late Pliocene to Recent.

## References


Barker, R.W., 1960, Taxonomic notes on the species figured by H.B. Brady in his report on the foraminifera dredged by H.M.S. Challenger during the years 1873–1876. Society of Economic Paleontologists and Mineralogists Special Publication, vol. 9, pp. 1–238.

Brady, H.B., 1884, Report on the foraminifera dredged by H.M.S. Challenger during the years 1873–1876. In Brady, H.B., ed., Report on the Scientific Results of the Voyage of the H.M.S. Challenger during the years 1873–1876. Zoology, vol. 9, pp. 1–814.

Brönnimann, P. and Whittaker, J.E., 1984, A neotype for Trochammina inflata (Montagu) (Protozoa, Foraminifera) with notes on the wall structure. Bulletin of the British Museum (Natural History), (Zoology Series), vol. 46, pp. 311–315.

Brönnimann, P. and Whittaker, J.E., 1988, The trochamminaceous test and the taxonomic criteria used in the classification of the Superfamily Trochamminacea. Abhandlungen der Geologischen Bundesanstalt, Wien, vol. 41, pp. 23–39.

Jones, R.W., 1994, The Challenger Foraminifera. Oxford: Oxford University Press, 149 pp.

Loeblich, A.R. Jr and Tappan, H., 1987, Foraminiferal Genera and Their Classification. New York: Van Nostrand Reinhold, 970 pp.

Montagu, G., 1808, Supplement to Testacea Britannica. Exeter: S. Woolmer, v + 183 pp.


1. *Trochammina inflata.* Thames/Medway Estuary, Southeast England; Borehole 95 for Gillingham Northern Link Road, (2.75/2.80) meters. J.E. Whittaker Collection. Spiral view, NHMUK PM DI 0677, scale bar = 98 μm.

2. *Trochammina inflata.* Thames/Medway Estuary, Southeast England; Borehole 95 for Gillingham Northern Link Road, (2.75/2.80) meters. J.E. Whittaker Collection. Apertural view, NHMUK PM DI 0678, scale bar = 96 μm.

3. *Trochammina inflata.* Thames/Medway Estuary, Southeast England; Borehole 95 for Gillingham Northern Link Road, (2.75/2.80) meters. J.E. Whittaker Collection. Umbilical view, NHMUK PM DI 0679, scale bar = 101 μm.

4. *Trochammina inflata.* Arne Peninsula, Poole Harbour, Southern England. J.E. Whittaker Collection. Umbilical view, NHMUK PM DI 0680, scale bar = 104 μm.

5. *Trochammina inflata.* Arne Peninsula, Poole Harbour, Southern England. J.E. Whittaker Collection. Umbilical view, NHMUK PM DI 0681, scale bar = 133 μm.

6. *Trochammina inflata.* Arne Peninsula, Poole Harbour, Southern England. J.E. Whittaker Collection. Slightly oblique view to show aperture detail, NHMUK PM DI 0682, scale bar = 127 μm.

7. *Trochammina inflata.* Arne Peninsula, Poole Harbour, Southern England. J.E. Whittaker Collection. Spiral view, NHMUK PM DI 0683, scale bar = 126 μm.

8. *Trochammina inflata.* Arne Peninsula, Poole Harbour, Southern England. J.E. Whittaker Collection. Edge view, NHMUK PM DI 0684, scale bar = 134 μm.

# *Trochammina quinqueloba*

## Geroch, 1959

1. *Trochammina quinqueloba*. Black shales, Verovice, Stepina, near Jasto, Silesian Poland. Deposited in the collections of the United States National Museum, Washington, DC; USNM 460975. Paratype. Spiral view, NHMUK PM DI 0685, scale bar = 51 μm.

2. *Trochammina quinqueloba*. Black shales, Verovice, Stepina, near Jasto, Silesian Poland. Deposited in the collections of the United States National Museum, Washington, DC; USNM 460975. Paratype. Edge view, NHMUK PM DI 0686, scale bar = 51 μm.

3. *Trochammina quinqueloba*. Black shales, Verovice, Stepina, near Jasto, Silesian Poland. Deposited in the collections of the United States National Museum, Washington, DC; USNM 460975. Paratype. Umbilical view, NHMUK PM DI 0687, scale bar = 51 μm.

## Original Designation

*Trochammina quinqueloba* Geroch, 1959, p. 118, pl. 12, figs. 1–3.

## Type Specimen

Original material deposited in the micropaleontological collections of the Jagiellonian University, Kraków, Poland.

## Type Level

Lower Cretaceous (Hauterivian to Barremian), Lower Verovice (Wernsdorf beds).

## Type Locality

Straconka, near Bielsko, Silesia, Poland.

## Description

Test forms a small, trochospiral series; slightly lobulate in outline, and planoconvex in cross-section, with an acute periphery, a flattened or slightly convex, evolute spiral side, and a convex, conical, involute umbilical side with a small, closed umbilicus. The five triangular, inflated chambers in the final whorl increase gradually in size, and are separated by flush or slightly depressed, straight, radial sutures. Chamber walls are finely agglutinated, non-calcareous (with considerable siliceous cement), and smoothly finished. The primary aperture is an interiomarginal, umbilical–extraumbilical arch, bordered by a narrow lip.

## Synonyms and Citations

*Trochammina quinqueloba* Geroch. –Kuznetsova, 1974, pl. 1, fig. 5. –Geroch and Nowak, 1984, pl. 2, figs. 16–18, pl. 7, figs. 5–8. –Nagy et al., 1995, pl. 1, figs. 14–16. –Kuhnt, 1995, pl. 3, figs. 1–3. –Holbourn and Kaminski, 1997, p. 49, pl. 19, figs. 2–6.

## Biogeography

Worldwide. Reported from Nepal (Nagy et al., 1995), Polish Carpathians (Geroch and Nowak, 1984), Spain (Kuhnt, 1995), Indian Ocean (Kuznetsova, 1974; Kaminski et al., 1992; Holbourn and Kaminski, 1997).

## Bathymetry

Bathyal to abyssal. Common in flysch-type assemblages.

## Chronostratigraphy

Jurassic (Tithonian) to Cretaceous (Albian) according to Geroch and Nowak (1984).

# References


Geroch, S., 1959, Stratigraphic significance of arenaceous foraminifera in the Carpathian flysch. Paläontologische Zeitung, vol. 33, pp. 113–122.

Geroch, S. and Nowak, W., 1984, Proposal of zonation for the late Tithonian–late Eocene, based upon arenaceous foraminifera from the outer Carpathians, Poland. In Oertli, H., ed., Benthos '83, 2nd International Symposium on Benthic Foraminifera, Pau (France), April 11–15, 1983. Elf Aquitaine, ESSO REP and TOTAL CFP, Pau and Bordeaux, pp. 225–239.

Holbourn, A.E.L. and Kaminski, M.A., 1997, Cretaceous deep-water benthic foraminifera of the Indian Ocean. Grzybowski Foundation Special Publication, no. 4, pp. 1–175.

Kaminski, M.A., Gradstein, F. and Geroch, S., 1992, Uppermost Jurassic to Lower Cretaceous deep-water benthic foraminiferal assemblages from Site 765 on the Argo Abyssal Plain. In Gradstein, F., Ludden, J.N., et al., ed., Scientific Results of the Proceedings of the Ocean Drilling Program, vol. 123, pp. 239–269.

Kuhnt, W., 1995, Deep-water agglutinated foraminifera from the Lower Cretaceous (Neocomian) "complex à Aptychus" Formation (Corridor de Boyar, Betic Cordillera, southern Spain). Journal of Micropalaeontology, vol. 14, no. 1, pp. 37–52.

Kuznetsova, K.I., 1974, Distribution of Benthonic Foraminifera in Upper Jurassic and Lower Cretaceous Deposits at Site 261, DSDP Leg 27, in the Eastern Indian Ocean. In Veevers, J.J., Heirtzler, J.R., et al., ed., Initial Reports of the Deep Sea Drilling Project, Washington (U.S. Government Printing Office), vol. 27, pp. 673–681.

Nagy, J., Gradstein, F.M., Kaminski, M.A., and Holbourn, A.E.L., 1995, Foraminiferal morphogroups, paleoenvironments and new taxa from Jurassic and Cretaceous stata of Thakkhola, Nepal. In Kaminski, M.A., Geroch, S., and Gasinski, M.A., ed., Proceedings of the Fourth International Workshop on Agglutinated Foraminifera, Grzybowski Foundation Special Publication, no. 3, pp. 181–209.

# *Trochamminoides coronatum*

## (Brady), 1879

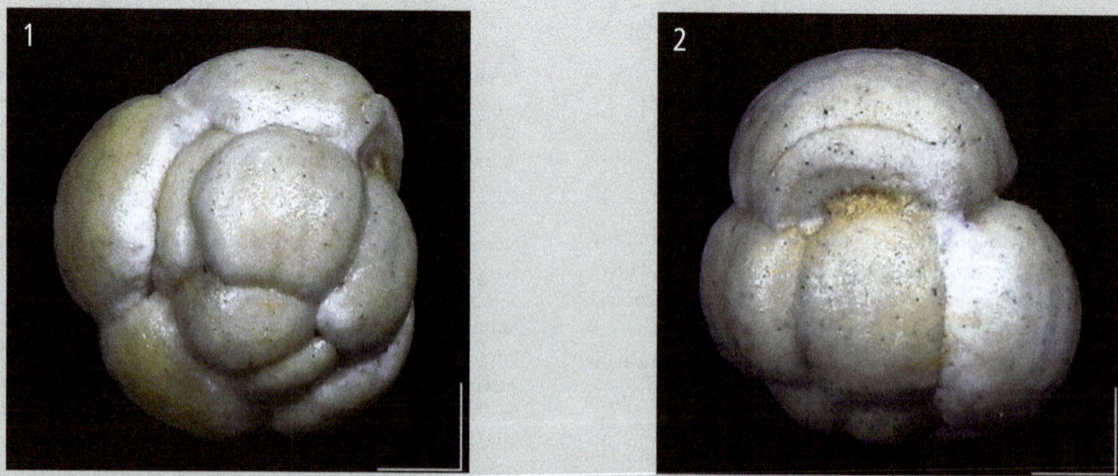


1. *Trochamminoides coronatum*. Off Pernambuco, West Atlantic; "Challenger" Station 120, depth: 675 fathoms. Deposited in the micropaleontological collections of The Natural History Museum, London; NHMUK ZF2498. Lateral view, NHMUK PM DI 0688, scale bar = 265 μm.

2. *Trochamminoides coronatum*. Off Pernambuco, West Atlantic; "Challenger" Station 120, depth: 675 fathoms. Deposited in the micropaleontological collections of The Natural History Museum, London; NHMUK ZF2498. Apertural view, NHMUK PM DI 0689, scale bar = 253 μm.

### Original Designation

*Trochammina coronata* Brady, 1879, p. 58, pl. 5, fig. 15.

### Type Specimen

Not designated.

### Type Level

Recent.

### Type Locality

Not designated. Localities given: western side of the North and South Atlantic, at two stations near the West Indies (390 and 450 fathoms), and two stations off the coast of South America (675 and 1900 fathoms).

### Description

Test is initially streptospiral, becoming planispiral, with a rounded periphery. The highly inflated chambers increase gradually in size, and are separated by distinct, fine, depressed sutures. Chamber walls are finely agglutinated, and well cemented. The primary aperture is an arch at the base of the apertural face.

### Synonyms and Citations

*Trochammina coronata* Brady. –Brady, 1884, p. 340, pl. 40, figs. 10–12.

*Trochammina conglobata* Brady. –Brady, 1884, p. 340, pl. 40, figs. 8–9. –Barker, 1960, pl. 40, figs. 8–9.

*Haplophragmoides coronatus* (Brady). –Barker, 1960, pl. 40, figs. 10–12.

*Trochamminoides coronatus* (Brady). –Morgiel and Olszewska, 1981, p. 12, pl. 3, fig. 2.

*Conglophragmium coronatum* (Brady). –Charnock and Jones, 1990, p. 167, pl. 5, figs. 2–3, pl. 16, fig. 2. –Jones, 1994, p. 45, pl. 40, figs. 8–12.

### Remarks

Several authors including Liszka and Liszkowa (1981) and Charnock and Jones (1990) included *Trochamminoides subcoronatus* Grzybowski in the synonymy of *Trochamminoides coronatum* (Brady). Kaminski and Gradstein (2005) retained *Trochamminoides subcoronatus* as a distinct species on account of its smaller size with fewer whorls, lack of a deep umbilical depression, and older stratigraphic range (Campanian to Eocene).

## Biogeography

Worldwide.

## Bathymetry

Bathyal to abyssal. Reported between 390 and 3950 fathoms at "Challenger" stations (Jones, 1994).

## Chronostratigraphy

Late Cretaceous (Santonian) to Recent, if *Trochamminoides subcoronatus* is treated as a synonym.

# References


Barker, R.W., 1960, Taxonomic notes on the species figured by H.B. Brady in his report on the foraminifera dredged by H.M.S. Challenger during the years 1873–1876. Society of Economic Paleontologists and Mineralogists Special Publication, vol. 9, pp. 1–238.

Brady, H.B., 1879, Notes on some of the reticularian Rhizopoda of the Challenger Expedition, Part I: on new or little known Arenaceous types. Quarterly Journal of Microscopical Science (New Series), vol. 19, pp. 20–63.

Brady, H.B., 1884, Report on the foraminifera dredged by H.M.S. Challenger during the years 1873–1876. In Brady, H.B., ed., Report on the Scientific Results of the Voyage of the H.M.S. Challenger during the years 1873–1876. Zoology, vol. 9, pp. 1–814.

Charnock, M.A. and Jones, R.W., 1990, Agglutinated foraminifera from the Palaeogene of the North Sea. In Hemleben, C., Kaminski, M.A., Kuhnt, W., and Scott, D.B., ed., Paleoecology, Biostratigraphy, Paleoceanography and Taxonomy of Agglutinated Foraminifera, NATO ASI Series, C: Mathematical and Physical Sciences. Dordrecht: Kluwer Academic, pp. 139–244.

Jones, R.W., 1994, The Challenger Foraminifera. Oxford: Oxford University Press, 149 pp.

Kaminski, M.A. and Gradstein, F.M., 2005, Atlas of Palaeogene Cosmopolitan Deep-Water Agglutinated Foraminifera. Grzybowski Foundation Special Publication, no. 10, 547 + vii pp.

Liszka, S. and Liszkowa, J., 1981, Revision of J. Grzybowski's paper (1896) "Foraminifera of the red clays from Wadowice". Rocznik Polskiego Towarzystwa Geologicznego, vol. 51, pp. 153–208.

Morgiel, J. and Olszewska, B., 1981, Biostratigraphy of the Polish External Carpathians based on agglutinated foraminifera. Micropaleontology, vol. 27, no. 1, pp. 1–30.

# *Trochamminoides grzybowskii*

## Kaminski and Geroch, 1992

1. *Trochamminoides grzybowskii.* Off Culebra Island, West Indies; "Challenger" Station 24, depth: 390 fathoms. Deposited in the micropaleontological collections of The Natural History Museum, London; NHMUK ZF2511. Spiral view, NHMUK PM DI 0690, scale bar = 237 μm.

### Original Designation

*Trochamminoides grzybowskii* Kaminski and Geroch, 1992, p. 64, fig. 1.

### Type Specimen

The specimen of *Trochammina elegans* figured by Grzybowski is not preserved; an unfigured syntype, registered as 1/58, was designated as lectotype by Kaminski and Geroch (1992), and is housed in the Grzybowski Collection of the Jagiellonian University, Kraków, Poland.

### Type Level

Eocene, Silesian Unit of the Polish Carpathians.

### Type Locality

Not originally designated. Lectotype is from gray clay collected along the bank of the Wislok River in Kroscienko Nizne, near Krosno, Poland.

### Description

Test forms a planispiral to slightly glomospiral coil of approximately three whorls, oval in outline, and with many chambers (approximately fifteen in the last whorl). These chambers are spherical, tightly appressed, and increase slowly in size. In immersion the spherical chamber interiors can be seen to be connected by a thin tube (Kaminski and Gradstein, 2005). Chamber walls are thick and medium to finely agglutinated. The primary aperture is areal, situated just above the base of the last chamber.

### Synonyms and Citations

*Trochammina proteus* Karrer. –Brady, 1884, p. 341, pl. 40, fig. 3 (not figs. 1–2).

*Trochamminoides proteus* Karrer. –Barker, 1960, pl. 40, fig. 3 (not figs. 1–2).

*Trochamminoides grzybowskii* Kaminski and Geroch. –Kaminski and Geroch, 1993, p. 260, pl. 8, fig. 7. –Jones, 1994, p. 44, pl. 40, fig. 3. –Kaminski and Gradstein, 2005, pp. 311–314, fig. 66, pl. 66.

### Remarks

*Trochamminoides grzybowskii* Kaminski and Geroch was originally described as *Trochammina elegans* by Grzybowski (1898, p. 287, pl. 11, fig. 10). Kaminski and Geroch (1992) renamed the species *Trochamminoides grzybowskii*, as the name *Trochammina elegans* Grzybowski was a junior synonym of *Trochammina elegans* Egger.

*Trochamminoides proteus* Karrer is coiled more irregularly, and has chambers generally longer than wide in contrast to *Trocham-*

*minoides grzybowskii*, which is predominantly planispiral and has more regular chambers of nearly equal size.

### Biogeography

Worldwide. See Observed Occurrences in Kaminski and Gradstein (2005).

### Bathymetry

Bathyal to abyssal.

### Chronostratigraphy

Late Cretaceous (Coniacian) to Eocene (Kaminski and Gradstein, 2005).

## References


Barker, R.W., 1960, Taxonomic notes on the species figured by H.B. Brady in his report on the foraminifera dredged by H.M.S. Challenger during the years 1873–1876. Society of Economic Paleontologists and Mineralogists Special Publication, vol. 9, pp. 1–238.

Brady, H.B., 1884, Report on the foraminifera dredged by H.M.S. Challenger during the years 1873–1876. In Brady, H.B., ed., Report on the Scientific Results of the Voyage of the H.M.S. Challenger during the years 1873–1876. Zoology, vol. 9, pp. 1–814.

Grzybowski, J., 1898, Otwornice pokładów naftonosnych okolicy Krosna. Rozprawy Wydziału Matematyczno-Przyrodniczego, Akademia Umiejętności w Krakowie, serya 2, vol. 33, pp. 257–305.

Jones, R.W., 1994, The Challenger Foraminifera. Oxford: Oxford University Press, 149 pp.

Kaminski, M.A. and Geroch, S., 1992, Trochamminoides grzybowskii, nom. nov., a new name for Trochammina elegens Grzybowski, 1898 (Foraminiferida). Journal of Micropalaeontology, vol. 11, p. 64.

Kaminski, M.A. and Geroch, S., 1993, A revision of foraminiferal species in the Grzybowski Collection. In Kaminski, M.A., Geroch, S., and Kaminski, D.G., ed., The Origins of Applied Micropalaeontology: The School of Jósef Grzybowski (with a taxonomic revision of the Grzyboswski Collection). Grzybowski Foundation Special Publication, no. 1, pp. 239–330.

Kaminski, M.A. and Gradstein, F.M., 2005, Atlas of Palaeogene Cosmopolitan Deep-Water Agglutinated Foraminifera. Grzybowski Foundation Special Publication, no. 10, 547 + vii pp.

# *Turrilina brevispira*

## ten Dam, 1944

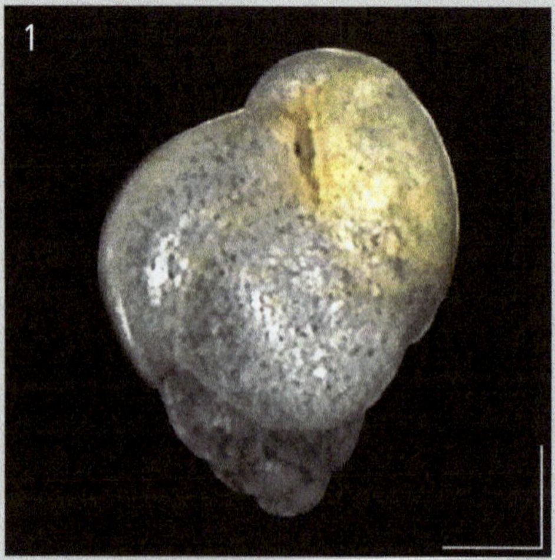


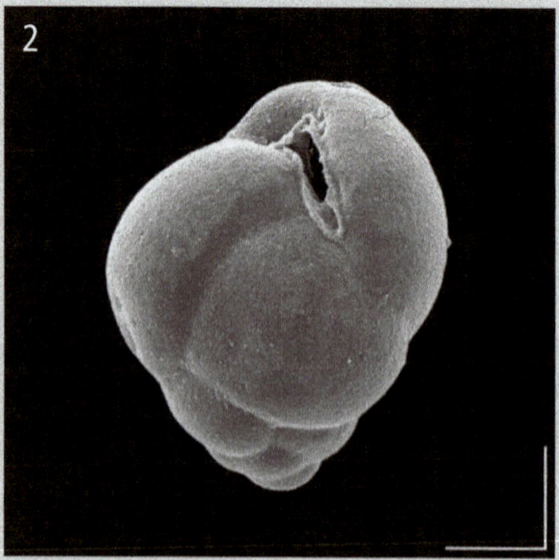


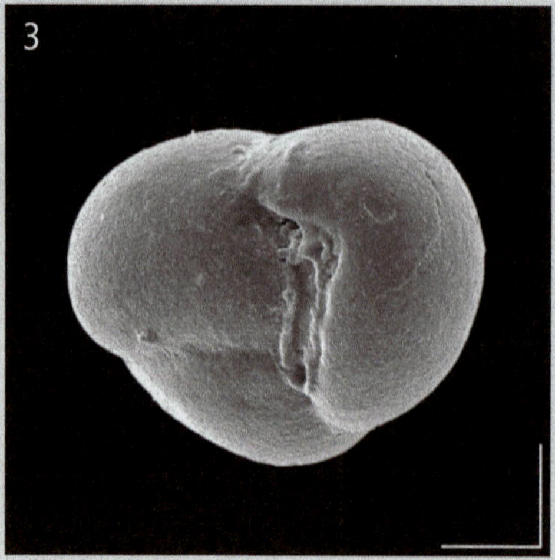


1. *Turrilina brevispira*. London Clay Formation, Sheppey, Kent, Southeast England. Deposited in The micropaleontological collections of The Natural History Museum, London; NHMUK P51113. Oblique lateral view, NHMUK PM DI 0691, scale bar = 72 μm.

2. *Turrilina brevispira*. London Clay Formation, Sheppey, Kent, Southeast England. Deposited in the micropaleontological collections of The Natural History Museum, London; NHMUK P51113. Oblique lateral view, NHMUK PM DI 0692, scale bar = 43 μm.

3. *Turrilina brevispira*. London Clay Formation, Sheppey, Kent, Southeast England. Deposited in the micropaleontological collections of The Natural History Museum, London; NHMUK P51113. Apertural view, NHMUK PM DI 0693, scale bar = 37 μm.

## Original Designation

*Turrilina brevispira* ten Dam, 1944, p. 110, pl. 3, fig. 14.

## Type Specimen

Holotype (No. F 553) deposited in the collections of the Netherlands Geological Survey, Haarlem.

## Type Level

Eocene.

## Type Locality

Boring Ijhorst (deep boring N.N.5, of Bataafsche Petroleum Maatschappij), in the eastern provinces of the Netherlands, 579–580 meters.

## Description

Test forms an initial trochospiral series, later becoming triserial with a subcircular periphery. The highly inflated chambers increase very rapidly in size (last whorl equivalent to about two-thirds of the test), and are separated by distinct, slightly depressed sutures. Chamber walls are calcareous, finely perforate, and smooth. The primary aperture is an oval opening parallel to the basal suture, bordered by a lip, and with an internal toothplate.

## Synonyms and Citations

*Turrilina brevispira* ten Dam. –van Morkhoven et al., 1986, p. 305, pl. 100, figs. 1–4. –Ernst et al., 2006, p. 102, pl. 2, fig. v. –Stassen et al., 2012, p. 158, fig. 5.15.

## Remarks

*Turrilina brevispira* became a common component of benthic assemblages in the aftermath of the Paleocene–Eocene Thermal Maximum (Stassen et al., 2012).

## Biogeography

Worldwide.

## Bathymetry

Lower neritic to abyssal.

## Chronostratigraphy

Early Eocene (Ypresian) to Late Eocene (Priabonian).

## References


Ernst, S.R., Guasti, E., Dupuis, C., and Speijer, R.P., 2006, Environmental perturbation in the southern Tethys across the Paleocene/Eocene boundary (Dababiya, Egypt): Foraminiferal and clay mineral records. Marine Micropaleontology, vol. 60, pp. 89–111.

Stassen, P., Steurbaut, E., Morsi, A.-M.M., and Speijer, R.P., 2012, Biotic impact of Eocene thermal maximum 2 in a shelf setting (Dababiya, Egypt). Austrian Journal of Earth Sciences, vol. 105, pp. 154–160.

ten Dam, A., 1944, Die Stratigraphische Gliederung des niederlaendlichen Paleozaens und Eozaens nach Foraminiferen (mit Ausnahme von Sued-Limburg). Mededeelingen van de Geologische Stichting, ser. C, vol. 5, no. 3, pp. 1–142.

van Morkhoven, F.P.C.M., Berggren, W.A., Edwards, A.S., et al., 1986, Cenozoic cosmopolitan deep-water benthic foraminifera. Bulletin des Centres de Recherches Exploration-Production Elf-Aquitaine, Mémoire, vol. 11, pp. 1–421.

# *Uvigerina aculeata*

## d'Orbigny, 1846

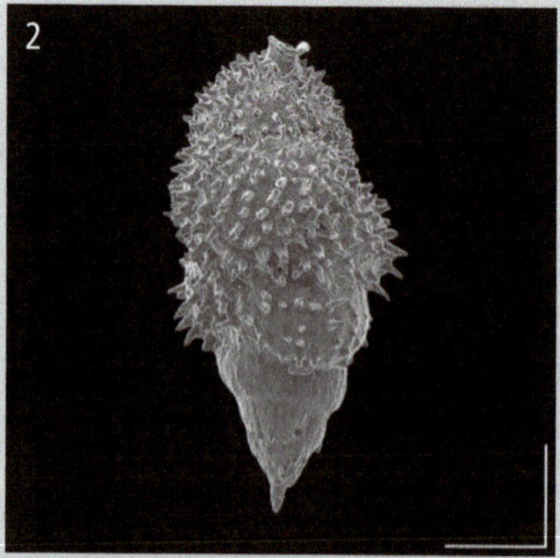

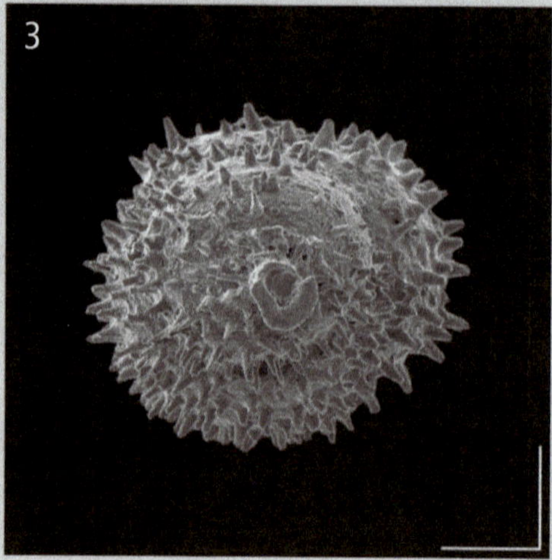

1. *Uvigerina aculeata.* Off Ki Islands, Pacific Ocean; "Challenger" Station 191A, depth: 580 fathoms. Deposited in the micropaleontological collections of The Natural History Museum, London; NHMUK ZF2562. Lateral view, NHMUK PM DI 0694, scale bar = 340 μm.

2. *Uvigerina aculeata.* Off Ki Islands, Pacific Ocean; "Challenger" Station 191A, depth: 580 fathoms. Deposited in the micropaleontological collections of The Natural History Museum, London; NHMUK ZF2562. Lateral view, NHMUK PM DI 0695, scale bar = 327 μm.

3. *Uvigerina aculeata.* Off Ki Islands, Pacific Ocean; "Challenger" Station 191A, depth: 580 fathoms. Deposited in the micropaleontological collections of The Natural History Museum, London; NHMUK ZF2562. Apertural view, NHMUK PM DI 0696, scale bar = 206 μm.

## Original Designation

*Uvigerina aculeata* d'Orbigny, 1846, p. 191, pl. 11, figs. 27–28.

## Type Specimen

Not designated. Original material deposited in the Hauer-d'Orbigny Collection, Geologische Bundesanstalt, Vienna.

## Type Level

Middle Miocene, "Tegel".

## Type Locality

Vicinity of the village of Nussdorf, north of Vienna, Vienna Basin, Austria.

## Description

Test forms an elongate, fusiform, triserial series, with a subcircular cross-section. The moderately inflated chambers increase gradually in size, and are separated by curved, depressed sutures. Chamber walls are calcareous, perforate, and ornamented by numerous fine spines and fine striations (early chambers). The primary aperture is a terminal, round opening at the end of a tubular neck, bordered by a phialine lip.

## Synonyms and Citations

*Euuvigerina aculeata* (d'Orbigny). –Barker, 1960, pl. 75, figs. 1–3. –Loeblich and Tappan, 1987, p. 523, pl. 572, figs. 23–25.

*Uvigerina aculeata* d'Orbigny. –Brady, 1884, p. 578, pl. 75, figs. 1–3. –Jones, 1994, p. 86, pl. 75, figs. 1–3.

## Remarks

Type species of the genus *Euuvigerina* Thalmann, 1952, regarded by Revets as a junior synonym of *Uvigerina* d'Orbigny (Jones, 1994).

Papp and Schmid (1985) placed *Uvigerina aculeata* d'Orbigny in the synonymy of *Uvigerina grilli* Schmid, which is not followed here, as *Uvigerina grilli* appears to have very different ornamentation.

## Biogeography

Worldwide.

## Bathymetry

Primarily bathyal. Reported between 580 and 1900 fathoms at "Challenger" stations (Jones, 1994).

## Chronostratigraphy

Miocene to Recent.

# References


Barker, R.W., 1960, Taxonomic notes on the species figured by H.B. Brady in his report on the foraminifera dredged by H.M.S. Challenger during the years 1873–1876. Society of Economic Paleontologists and Mineralogists Special Publication, vol. 9, pp. 1–238.

Brady, H.B., 1884, Report on the foraminifera dredged by H.M.S. Challenger during the years 1873–1876. In Brady, H.B., ed., Report on the Scientific Results of the Voyage of the H.M.S. Challenger during the years 1873–1876. Zoology, vol. 9, pp. 1–814.

d'Orbigny, A.D., 1846, Foraminifères fossiles du Bassin Tertiaire de Vienne (Autriche). Paris: Gide et Compe, xxxvii + 312 pp.

Jones, R.W., 1994, The Challenger Foraminifera. Oxford: Oxford University Press, 149 pp.

Loeblich, A.R. Jr and Tappan, H., 1987, Foraminiferal Genera and Their Classification. New York: Van Nostrand Reinhold, 970 pp.

Papp, A. and Schmid, M.E., 1985, Die fossilen Foraminiferen des tertiaeren Beckens von Wien: Revision der Monographie Alcide d'Orbigny (1846). Abhandlungen der Geologischen Bundesanstalt, vol. 37, pp. 1–311.

# *Uvigerina adelinensis*

## Palmer and Bermúdez, 1949

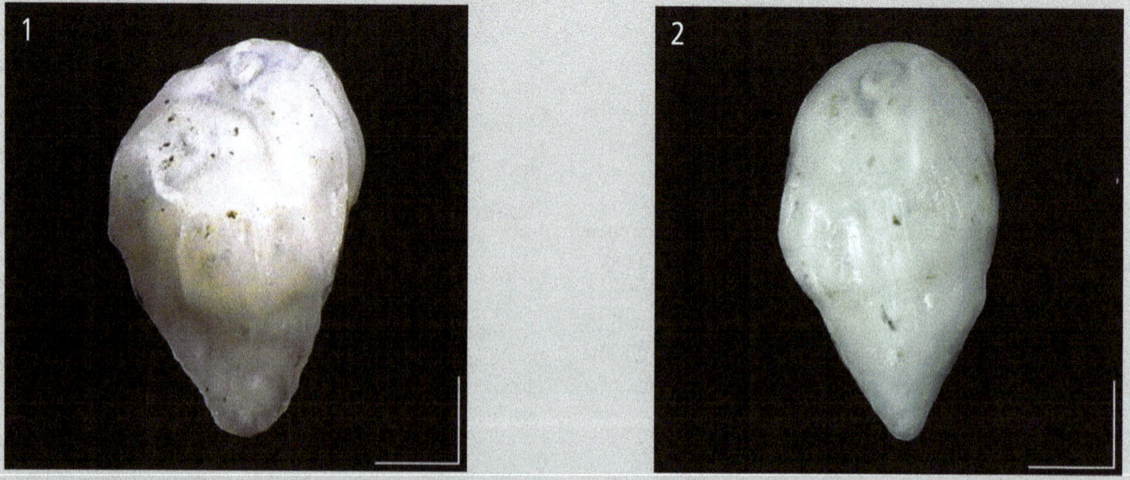

1. *Uvigerina adelinensis.* Cuba; Sullivan Borehole M. 23, 304 feet. Lateral view, NHMUK PM DI 0697, scale bar = 229 μm.
2. *Uvigerina adelinensis.* Cuba; Sullivan Borehole M. 23, 304 feet. Lateral view, NHMUK PM DI 0698, scale bar = 152 μm.

### Original Designation

*Uvigerina adelinensis* Palmer and Bermúdez, 1949, described in Bermúdez, 1949, and illustrated as *Uvigerina cubana* in Palmer and Bermúdez, 1936a, pl. 17, figs. 5–6.

### Type Specimen

Cotypes (No. C-36) deposited in the Dorothy K. Palmer Collection at the Paleontological Research Institution, Ithaca, NY, USA.

### Type Level

Upper Oligocene (van Morkhoven et al., 1986), soft, buff marl. Originally given as Early Oligocene by Palmer and Bermúdez (1936a).

### Type Locality

Shallow-water well on Finca Adelina, located approximately 9.5 kilometers east of Colón, Matamzas Province, and 0.5 kilometer north of the Central Highway, Cuba.

### Description

Test forms a large, robust, triserial series; about twice as long as broad, with greater breadth in the middle portion of the test, tapered in outline, and triangular in cross-section. The inflated, angular chambers increase gradually in size, and are separated by sutures obscured by ornamentation in the early part of the test, but slightly depressed in the final part. Chamber walls are calcareous, finely perforate, and ornamented by three wide flanges, and by irregular, longitudinal costae, occasionally continuous over several chambers. This ornamentation is reduced on the final chamber, which is often smooth. The primary aperture is a terminal opening at the end of a broad, cylindrical neck, bordered by a phialine lip, and with an internal toothplate.

### Synonyms and Citations

*Uvigerina cubana* Palmer and Bermúdez, 1936a–b, p. 292, pl. 17, figs. 5–6.

*Uvigerina adelinensis* Palmer and Bermúdez, 1949, p. 203 in Bermúdez, 1949. New name for *Uvigerina cubana* Palmer and Bermúdez.

*Uvigerina adelinensis* Palmer and Bermúdez. –Boersma, 1984, p. 1, pl. 1, figs. 1–4. –van Morkhoven et al., 1986, p. 180, pl. 61A–B.

### Remarks

Specimens may lack small costae between the three broad flanges. *Uvigerina adelinensis* Palmer and Bermúdez differs from *Uvigerina gallowayi* Cushman by its more tapering shape and the lack of continuous costae over the entire length of the test.

*Uvigerina cubana* Palmer and Bermúdez was renamed *Uvigerina adelinensis* Palmer and Bermúdez by Bermúdez (1949) because the name was preoccupied, once *Uvigerina gardnerae* Cushman and Applin var. *cubana* Hadley became elevated to specific rank in 1934.

### Biogeography

Worldwide. Recorded from the Pacific, North Atlantic, Gulf of Mexico, and Caribbean (van Morkhoven et al., 1986).

### Bathymetry

Upper to middle bathyal.

### Chronostratigraphy

Late Eocene (Priabonian) to Middle Miocene (Serravallian). Doubtful reports of occurrences in Biozones P15–P17 (van Morkhoven et al., 1986).

## References


Bermúdez, P.J., 1949, Tertiary smaller foraminifera of the Dominican Republic. Special Publications of the Cushman Laboratory for Foraminiferal Research, no. 25, pp. 1–322.

Boersma, A., 1984, Handbook of Common Tertiary Uvigerina. Stony Point, NY: Microclimates Press, 207 pp.

Hadley, W.H., 1934, Some Tertiary foraminifera from the north coast of Cuba. Bulletin of American Paleontology, vol. 20, pp. 1–40.

Palmer, D.K. and Bermúdez, P.J., 1936a, An Oligocene foraminiferal fauna from Cuba; Part I. Memorias de la Sociedad Cubana de Historia Natural "Filipe Poey", vol. 10, no. 4, pp. 227–271.

Palmer, D.K. and Bermúdez, P.J., 1936b, An Oligocene foraminiferal fauna from Cuba; Part II. Memorias de la Sociedad Cubana de Historia Natural "Filipe Poey", vol. 10, no. 5, pp. 273–317.

van Morkhoven, F.P.C.M., Berggren, W.A., Edwards, A.S., et al., 1986, Cenozoic cosmopolitan deep-water benthic foraminifera. Bulletin des Centres de Recherches Exploration-Production Elf-Aquitaine, Mémoire, vol. 11, pp. 1–421.

# *Uvigerina auberiana*

## d'Orbigny, 1839

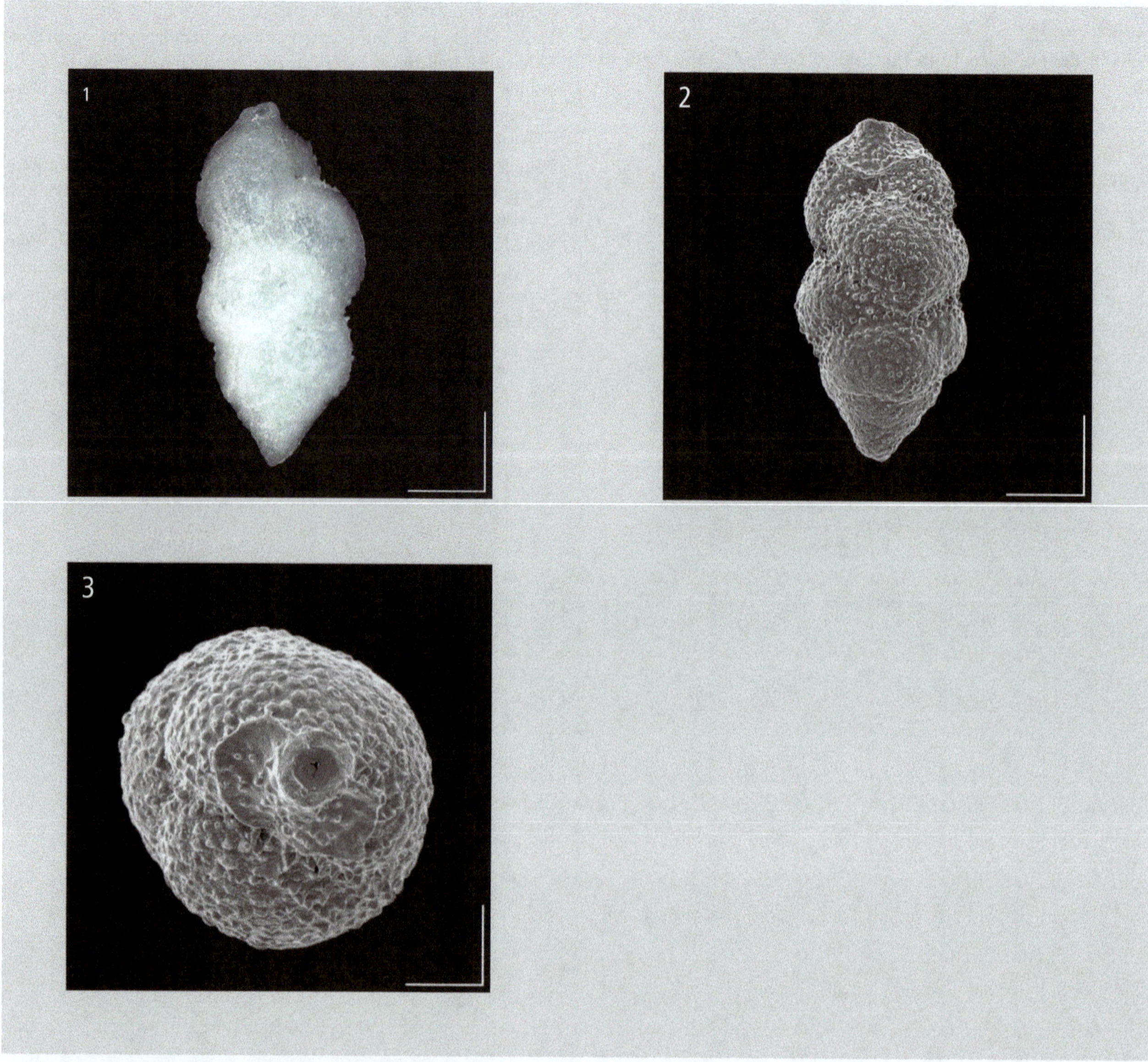


1. *Uvigerina auberiana*. East Java. R.W. Jones Collection. Lateral view, NHMUK PM DI 0699, scale bar = 130 μm.
2. *Uvigerina auberiana*. East Java. R.W. Jones Collection. Lateral view, NHMUK PM DI 0700, scale bar = 130 μm.
3. *Uvigerina auberiana*. East Java. R.W. Jones Collection. Apertural view, NHMUK PM DI 0701, scale bar = 71 μm.

## Original Designation

*Uvigerina auberiana* d'Orbigny, 1839, p. 106, pl. 2, figs. 23–24.

## Type Specimen

Depository not given.

## Type Level

Recent, marine sands.

## Type Locality

Not designated. Localities given: Cuba, Jamaica, and Martinique.

## Description

Test forms an elongate, triserial series, showing a tendency to become loosely triserial. Overall test shape is fusiform, about twice as long as broad (greatest breadth in the middle of the test), with a subrounded initial portion, a rounded apertural end, and a circular cross-section. The inflated chambers increase gradually in size, and are separated by depressed sutures. Chamber walls are calcareous, finely perforate, and ornamented by small spines randomly distributed over the whole test. The primary aperture is a terminal, round opening at the end of a broad, cylindrical neck, bordered by a phialine lip, and with an internal toothplate.

## Synonyms and Citations

*Uvigerina asperula* var. *auberiana*. –Brady, 1884, p. 579, pl. 75, fig. 9.
*Uvigerina asperula* Czjzek. –Brady, 1884, p. 578, pl. 75, figs. 6–8. –Barker, 1960, pl. 75, figs. 6–9.
*Uvigerina auberiana* d'Orbigny 1839. –Boersma, 1984, p. 4, pl. 1, figs. 1–5. –Hermelin, 1989, p. 64, pl. 12, figs. 4–5. –Jones, 1994, p. 86, pl. 75, figs. 6–9.

## Remarks

Hermelin (1989) illustrated two specimens from the Pliocene of the Western Pacific with multiple apertures.

## Biogeography

Worldwide. Reported from the Tasman Sea (Miocene), Barbados, Italy (Tortonian), Algeria, Caribbean, Ecuador, the West African upwelling, Pacific Ocean, Indian Ocean, Atlantic Ocean, Guam, and Mexico (Boersma, 1984).

## Bathymetry

Neritic to abyssal. Reported between 37 and 2335 fathoms at "Challenger" stations (Jones, 1994). Common in the upper to middle bathyal zone in areas of high surface production (Boersma, 1984).

## Chronostratigraphy

Early Oligocene to Recent (Boersma, 1984).

# References


Barker, R.W., 1960, Taxonomic notes on the species figured by H.B. Brady in his report on the foraminifera dredged by H.M.S. Challenger during the years 1873–1876. Society of Economic Paleontologists and Mineralogists Special Publication, vol. 9, pp. 1–238.

Boersma, A., 1984, Handbook of common Tertiary Uvigerina. Stony Point, NY: Microclimates Press, 207 pp.

Brady, H.B., 1884, Report on the foraminifera dredged by H.M.S. Challenger during the years 1873–1876. In Brady, H.B., ed., Report on the Scientific Results of the Voyage of the H.M.S. Challenger during the years 1873–1876. Zoology, vol. 9, pp. 1–814.

d'Orbigny, A.D., 1839, Foraminifères. In Sagra, R. de la, ed., Histoire physique, politique et naturelle de l'île de Cuba. Paris: A. Bertrand, pp. 1–224, 12 pls.

Hermelin, J.O.R., 1989, Pliocene benthic foraminifera from the Ontong-Java Plateau (Western Equatorial Pacific Ocean): faunal response to changing paleoenvironment. Special Publication of the Cushman Foundation for Foraminiferal Research, vol. 26, pp. 1–143.

Jones, R.W., 1994, The Challenger Foraminifera. Oxford: Oxford University Press, 149 pp.

# Uvigerina basicordata

## Cushman and Renz, 1941

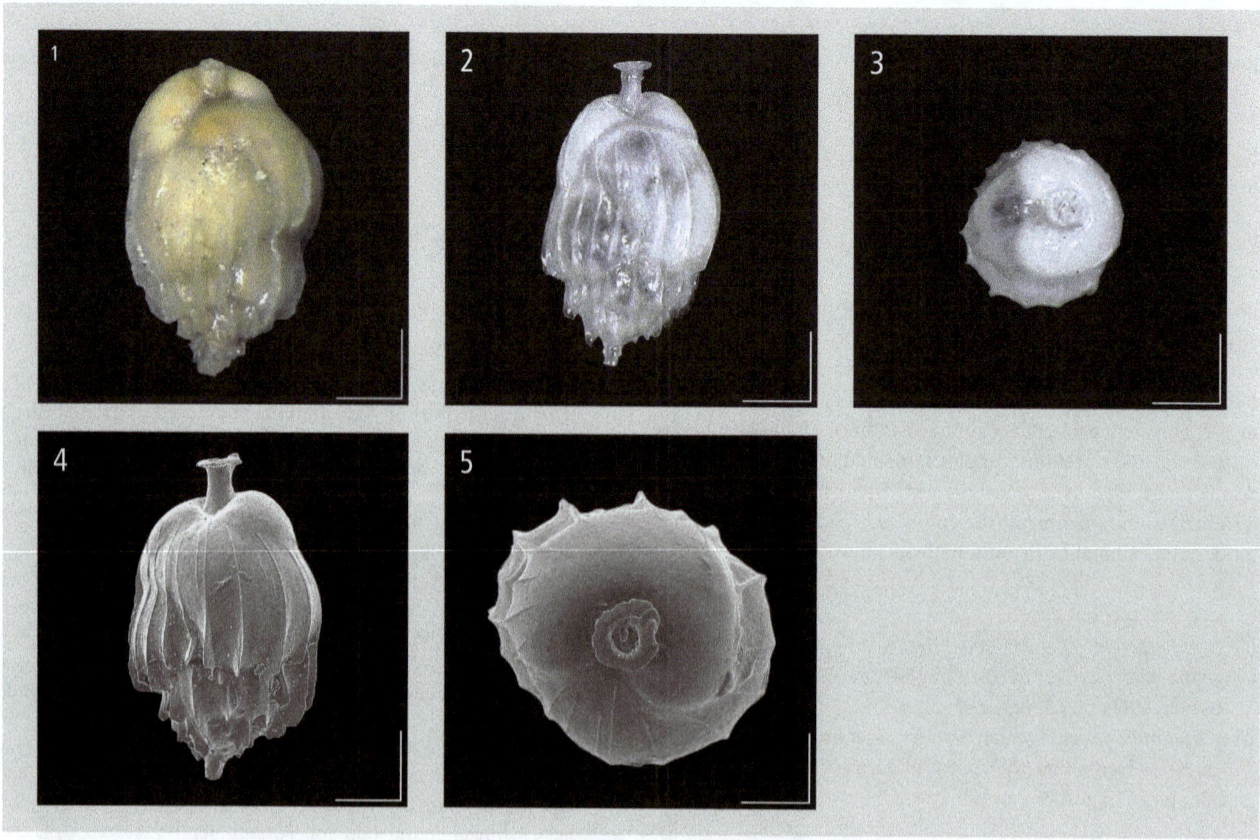

1. *Uvigerina basicordata*. Lower Agua Salada Formation, Tocuyo, 18.7 km south (202° 45′) of San Juan de los Cayos, District of Acosta, State Falcón, Venezuela. Deposited in the collections of the United States National Museum, Washington, DC; CC 35924. Holotype. Lateral view, NHMUK PM DI 0702, scale bar = 158 μm.
2. *Uvigerina basicordata*. Gulf of Mexico; Eureka Core E68-151A, depth: 4588 feet. Lateral view, NHMUK PM DI 0703, scale bar = 141 μm.
3. *Uvigerina basicordata*. Gulf of Mexico; Eureka Core E68-151A, depth: 4588 feet. Apertural view, NHMUK PM DI 0704, scale bar = 135 μm.
4. *Uvigerina basicordata*. Gulf of Mexico; Eureka Core E68-151A, depth: 4588 feet. Lateral view, NHMUK PM DI 0705, scale bar = 133 μm.
5. *Uvigerina basicordata*. Gulf of Mexico; Eureka Core E68-151A, depth: 4588 feet. Apertural view, NHMUK PM DI 0706, scale bar = 88 μm.

### Original Designation

*Uvigerina gallowayi* Cushman var. *basicordata* Cushman and Renz, 1941, p. 21, pl. 2, fig. 18.

### Type Specimen

Holotype (CC 35924) deposited in the collections of the United States National Museum, Washington, DC.

### Type Level

Lower Miocene, Lower Agua Salada Formation. Originally given as Upper Oligocene by Cushman and Renz (1941).

### Type Locality

Sample Su. A. 637, from Tocuyo, 18.7 kilometers south (20° 45′) of San Juan de Los Cayos, District of Acosta, State of Falcón, Venezuela.

### Description

Test forms an almost rectangular triserial series; about twice as long as broad, but with greater breadth in the mid-portion of the test, with a prominent basal spine, an almost rectangular outline, and a circular cross-section. The inflated chambers increase rapidly in size, and are separated by depressed sutures. Chamber walls are calcareous, finely perforate, and ornamented with four to five continuous or discontinuous, longitudinal costae extending as spines at the base of the chambers. This ornamentation is reduced on the final chamber, which is often smooth. The primary aperture is a terminal opening at the end of an elongate, thin, cylindrical neck, bordered by a phialine lip, and with an internal toothplate.

### Synonyms and Citations

*Uvigerina basicordata* Cushman and Renz. –van Morkhoven et al., 1986, p. 193, pl. 65. –Katz and Miller, 1993, pl. 4, fig. 2.

### Remarks

*Uvigerina basicordata* Cushman and Renz differs from *Uvigerina gallowayi* Cushman by its shorter, broader shape and by the presence of discontinuous costae (van Morkhoven et al., 1986) and basal spinose projections (Cushman and Renz, 1941).

### Biogeography

Worldwide. Recorded from Trinidad, Puerto Rico, California, Venezuela, Ecuador, eastern Pacific, North Atlantic, and Gulf of Mexico (van Morkhoven et al., 1986).

### Bathymetry

Bathyal.

### Chronostratigraphy

Early Miocene to Middle Miocene.

## References


Cushman, J.A. and Renz, H.H., 1941, New Oligocene–Miocene foraminifera from Venezuela. Contributions from the Cushman Laboratory for Foraminiferal Research, vol. 17, pp. 1–27.

Katz, M.E. and Miller, K.G., 1993, Latest Oligocene to Earliest Pliocene benthic foraminiferal biofacies of the northeastern Gulf of Mexico. Micropaleontology, vol. 39, no. 4, pp. 367–403.

van Morkhoven, F.P.C.M., Berggren, W.A., Edwards, A.S., et al., 1986, Cenozoic cosmopolitan deep-water benthic foraminifera. Bulletin des Centres de Recherches Exploration-Production Elf-Aquitaine, Mémoire, vol. 11, pp. 1–421.

# Uvigerina carapitana

## Hedberg, 1937

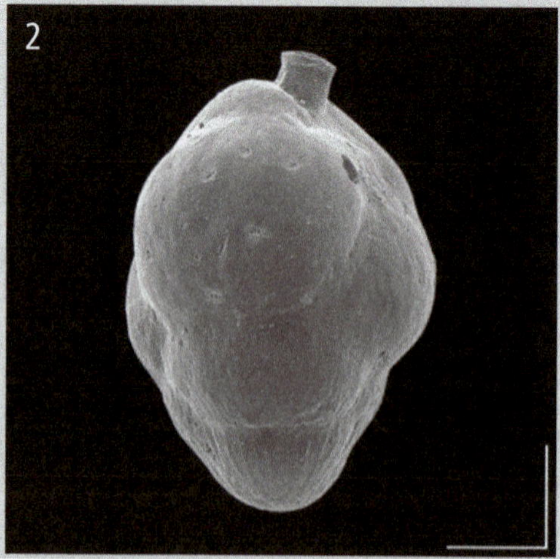


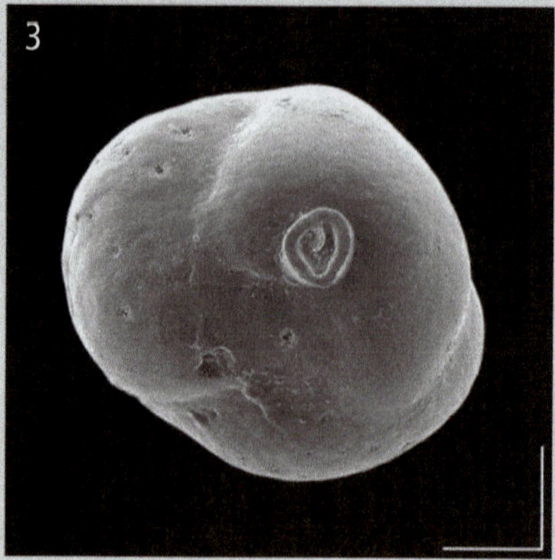


1. *Uvigerina carapitana.* Gulf of Mexico; Eureka Core E68-136, depth: 2394 feet. Lateral view, NHMUK PM DI 0707, scale bar = 142 μm.
2. *Uvigerina carapitana.* Gulf of Mexico; Eureka Core E68-136, depth: 2394 feet. Lateral view, NHMUK PM DI 0708, scale bar = 146 μm.
3. *Uvigerina carapitana.* Gulf of Mexico; Eureka Core E68-136, depth: 2394 feet. Apertural view, NHMUK PM DI 0709, scale bar = 109 μm.

## Original Designation

*Uvigerina carapitana* Hedberg, 1937, p. 677, pl. 91, fig. 20.

## Type Specimen

Holotype (CC 23602) deposited in the collections of the United States National Museum, Washington, DC.

## Type Level

Lower Miocene (van Morkhoven et al., 1986), Carapita Formation. Originally given as middle Tertiary by Hedberg (1937).

## Type Locality

North of the town of Santa Inés on Quebrada Carapita, a small tributary of the Rio Querecal, District of Libertad, State of Anzoategui, northeastern Venezuela.

## Description

Test forms a robust, fusiform, triserial series; about one and a half to two times as long as broad, with greater breadth in the middle portion of the test, a tapered outline, and a triangular cross-section. Chambers are slightly inflated, angular, increase rapidly in size, and are separated by curved, depressed sutures. The ultimate chamber is markedly square. Chamber walls are calcareous, finely perforate, and smooth or ornamented by small segments of faint, longitudinal striae, more evident in the early part of the test. The primary aperture is a terminal opening at the end of a broad, short neck set into a depression, bordered by a phialine lip, with an internal toothplate.

## Synonyms and Citations

*Uvigerina canariensis* d'Orbigny var. Cushman, 1918, p. 63, pl. 22, fig. 6.

*Uvigerina quadrata* Coryell and Rivero, 1940, p. 343, pl. 42, fig. 27.

*Uvigerina carapitana* Hedberg. –Bermúdez, 1949, p. 202, pl. 13, fig. 1. –Boersma, 1984, p. 28, pl. 1, figs. 1–5. –van Morkhoven et al., 1986, p. 124, pl. 40A–B. –Katz and Miller, 1993, pl. 4, fig. 6.

## Remarks

*Uvigerina carapitana* Hedberg resembles *Uvigerina laviculata* Coryell and Rivero, differing mainly in the degree of ornamentation: *Uvigerina carapitana* has faintly striated early chambers, whereas *Uvigerina laviculata* is finely striated throughout, except for the last chamber.

## Biogeography

Worldwide? Recorded from the Gulf of Mexico, the Caribbean islands, South American marginal basins, and Spain (Boersma, 1984).

## Bathymetry

Bathyal.

## Chronostratigraphy

Early Miocene (Burdigalian) to Late Miocene (Messinian).

# References


Bermúdez, P.J., 1949, Tertiary smaller foraminifera of the Dominican Republic. Special Publications of the Cushman Laboratory for Foraminiferal Research, no. 25, pp. 1–322.

Boersma, A., 1984, Handbook of Common Tertiary Uvigerina. Stony Point, NY: Microclimates Press, 207 pp.

Coryell, H.N. and Rivero, F.C., 1940, A Miocene microfauna of Haiti. Journal of Paleontology, vol. 14, pp. 324–344.

Cushman, J.A., 1918, The smaller fossil foraminifera of the Panama Canal Biozone. Bulletin of the United States National Museum, vol. 103, pp. 45–87.

Hedberg, H.D., 1937, Foraminifera of the middle Tertiary Carapitana formation of northeastern Venezuela. Journal of Paleontology, vol. 11, pp. 661–697.

Katz, M.E. and Miller, K.G., 1993, Latest Oligocene to Earliest Pliocene benthic foraminiferal biofacies of the northeastern Gulf of Mexico. Micropaleontology, vol. 39, no. 4, pp. 367–403.

van Morkhoven, F.P.C.M., Berggren, W.A., Edwards, A.S., et al., 1986, Cenozoic cosmopolitan deep-water benthic foraminifera. Bulletin des Centres de Recherches Exploration-Production Elf-Aquitaine, Mémoire, vol. 11, pp. 1–421.

# *Uvigerina havanensis*

## Cushman and Bermúdez, 1936

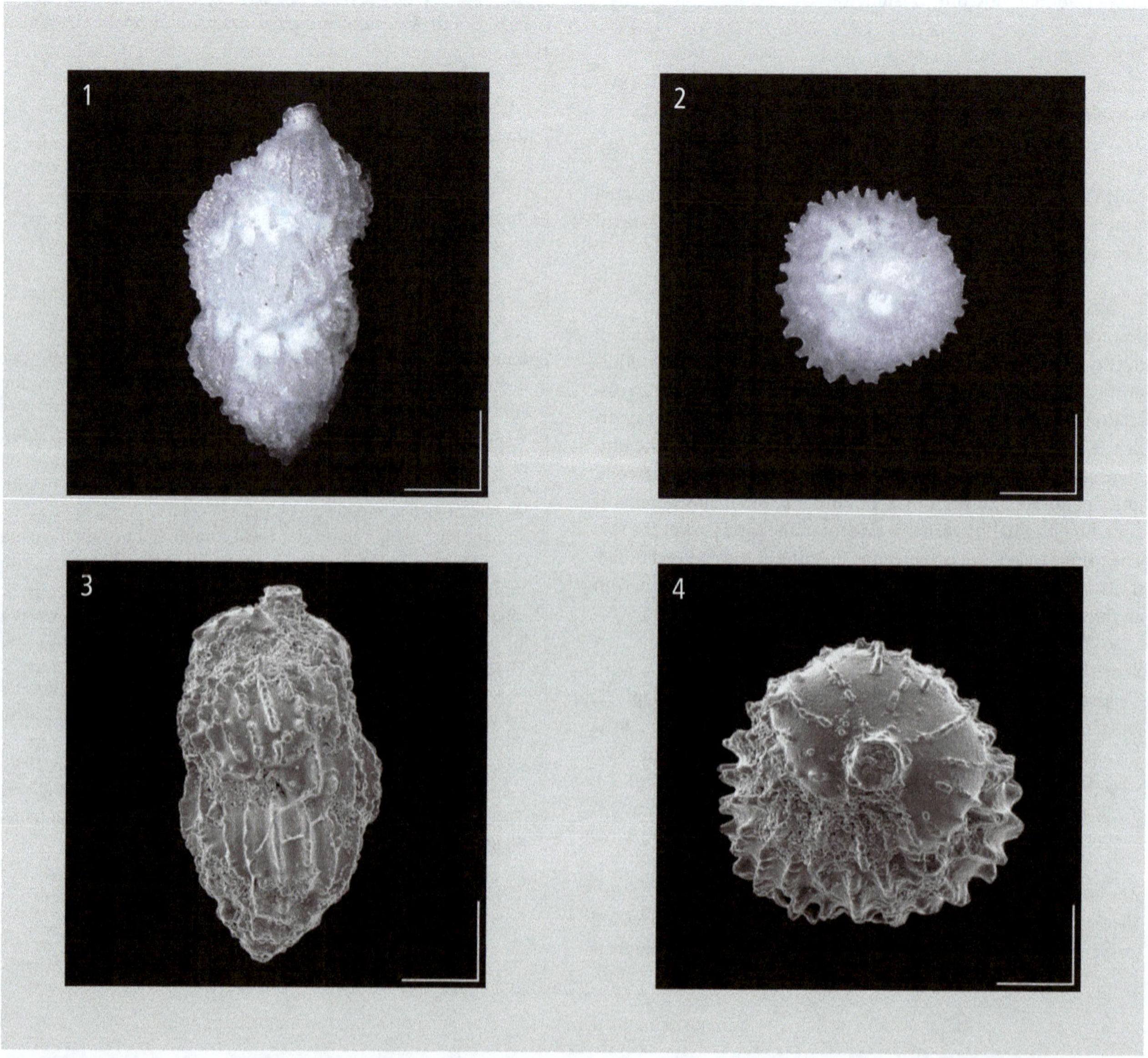

1. *Uvigerina havanensis*. Goban Spur, North Atlantic; DSDP Leg 80, Hole 549A, Core 15, Section 1, 116–121 cm. Lateral view, NHMUK PM DI 0710, scale bar = 118 μm.

2. *Uvigerina havanensis*. Goban Spur, North Atlantic; DSDP Leg 80, Hole 549A, Core 15, Section 1, 116–121 cm. Apertural view, NHMUK PM DI 0711, scale bar = 115 μm.

3. *Uvigerina havanensis*. Goban Spur, North Atlantic; DSDP Leg 80, Hole 549A, Core 15, Section 1, 116–121 cm. Lateral view, NHMUK PM DI 0712, scale bar = 123 μm.

4. *Uvigerina havanensis*. Goban Spur, North Atlantic; DSDP Leg 80, Hole 549A, Core 15, Section 1, 116–121 cm. Apertural view, NHMUK PM DI 0713, scale bar = 86 μm.

## Original Designation

*Uvigerina havanensis* Cushman and Bermúdez, 1936, p. 59, pl. 10, figs. 19–21.

## Type Specimen

Holotype (CC 23282) and paratypes (CC 23283 and CC 23284) deposited in the collections of the United States National Museum, Washington, DC.

## Type Level

Eocene.

## Type Locality

Bermúdez Station 31, 1 kilometer North of Arroyo Arenas, on the road to Jaimanitas (water well), Havana Province, Cuba.

## Description

Test forms an elongate, triserial series, approximately two and a half times as long as broad, with a tendency to become loosely triserial. The test has its greatest breadth at the last-formed chambers, is tapered in outline, and is circular in cross-section. The inflated chambers increase markedly in height, and are separated by depressed sutures. Chamber walls are calcareous, perforate, and ornamented by numerous prominent, raised, sharp longitudinal costae (discontinuous in adjacent chambers), veering in a single direction around the apertural neck. Costae of initial chambers extend as short, blunt spines. The primary aperture is a terminal, round opening at the end of a short neck, with an internal toothplate, and bordered by a phialine lip.

## Synonyms and Citations

*Uvigerina havanensis* Cushman and Bermúdez. –Boersma, 1984, pp. 70–73, pl. 1, figs. 1–5. –van Morkhoven et al., 1986, p. 238, pl. 80. –Miller and Katz, 1987, p. 140, pl. 2, fig. 1.

## Biogeography

Worldwide. Recorded from Europe, the North and South Atlantic, the East Pacific, and the Gulf of Mexico (van Morkhoven et al., 1986).

## Bathymetry

Primarily bathyal. Also found in lower bathyal to abyssal deposits (M. Katz, pers. comm. 1998).

## Chronostratigraphy

Middle Eocene (Lutetian) to Late Oligocene (Chattian). Doubtful occurrences in Early Miocene Biozones M1 and M2.

# References


Boersma, A., 1984, Handbook of Common Tertiary Uvigerina. Stony Point, NY: Microclimates Press, 207 pp.

Cushman, J.A. and Bermúdez, P.J., 1936, Additional new species of foraminifera and a new genus from the Eocene of Cuba. Contributions from the Cushman Laboratory for Foraminiferal Research, vol. 12, no. 3, pp. 55–63.

Miller, K.G. and Katz, M.E., 1987, Oligocene to Miocene benthic foraminiferal and abyssal circulation changes in the North Atlantic. Micropaleontology, vol. 33, no. 2, pp. 97–149.

van Morkhoven, F.P.C.M., Berggren, W.A., Edwards, A.S., et al., 1986, Cenozoic cosmopolitan deep-water benthic foraminifera. Bulletin des Centres de Recherches Exploration-Production Elf-Aquitaine, Mémoire, vol. 11, pp. 1–421.

# *Uvigerina hispida*

## Schwager, 1866

1. *Uvigerina hispida*. Gulf of Mexico; Eureka Core E68-151A. Deposited in the collections of the United States National Museum, Washington, DC. Specimen illustrated in van Morkhoven et al., 1986, Ref. no. 20. Lateral view, NHMUK PM DI 0714, scale bar = 335 μm.

### Original Designation

*Uvigerina hispida* Schwager, 1866, p. 249, pl. 7, fig. 95.

### Type Specimen

Depository not given. Neotype (Mf 9) and topotype (Mf 9, megalospheric form), designated as *Euuvigerina hispida* (Schwager) by Srinivasan and Sharma (1980), deposited in the Micropaleontology Section, Department of Geology, Banaras Hindu University, Banaras, India.

### Type Level

Upper Tertiary, upper and lower clays.

### Type Locality

Car Nicobar, Andaman Sea.

### Description

Test forms an elongate, fusiform, triserial series (showing a marked tendency to become loosely triserial), approximately two and a half times as long as broad, widest in the middle, tapered in outline in microspheric forms, subcylindrical in outline in megalospheric forms, and with a circular cross-section in both. A basal spine is present in some well-preserved specimens. The inflated chambers increase markedly in height, and are separated by distinctly depressed sutures. Chamber walls are calcareous, perforate, and ornamented by dense, acicular spines (microspheric forms) or blunt spines to coarse papillae (megalospheric forms). The primary aperture is a terminal, round opening at the end of a short neck, bordered by a phialine lip, and with an internal toothplate.

### Synonyms and Citations

*Uvigerina rustica* Cushman and Edwards, 1938, p. 83, pl. 14, fig. 6.
*Euuvigerina hispida* (Schwager). –Srinivasan and Sharma, 1980, pl. 7, figs. 14–15.
*Uvigerina hispida* Schwager. –Boersma, 1984, p. 76, pl. 1, figs. 1–4. –van Morkhoven et al., 1986, p. 62, pl. 20, figs. 1–4. –Miller and Katz, 1987, p. 140, pl. 2, fig. 2. –Katz and Miller, 1993, pl. 4, fig. 7. –Robertson, 1998, p. 154, pl. 58, fig. 3. –Kuhnt et al., 2002, p. 158, pl. 14. figs. 5–7. –Ortiz and Thomas, 2006, p. 134, pl. 11, Fig. 8.
*Neouvigerina hispida* (Schwager). –Hayward, 2002, p. 301, pl. 1, fig. 14.

### Remarks

Revised stratigraphy at the Eureka boreholes in the Gulf of Mexico (Katz and Miller, 1993) has extended the first occurrence of this species to Biozone M1, rather than the Biozone M2 occurrence reported by van Morkhoven et al. (1986).

### Biogeography

Worldwide. Recorded from Venezuela, Ecuador, Trinidad, the Gulf of Mexico, the Atlantic, the Mediterranean, the Pacific, California, Europe, East Indies, New Guinea, the Tasman and Coral Seas, the Indian Ocean, and India (van Morkhoven et al., 1986).

### Bathymetry

Bathyal.

### Chronostratigraphy

Early Miocene (Aquitanian) to Recent.

## References


Boersma, A., 1984, Handbook of Common Tertiary Uvigerina. Stony Point, NY: Microclimates Press, 207 pp.

Cushman, J.A. and Edwards, P.G., 1938, Notes on the Oligocene species of Uvigerina and Angulogerina. Contributions from the Cushman Laboratory for Foraminiferal Research, vol. 14, pp. 74–89.

Hayward, B., 2002, Late Pliocene to Middle Pleistocene extinctions of deep-sea benthic foraminifera ("Stilostomella extinction") in the southwest Pacific. Journal of Foraminiferal Research, vol. 32, pp. 274–307.

Katz, M.E. and Miller, K.G., 1993, Latest Oligocene to Earliest Pliocene benthic foraminiferal biofacies of the northeastern Gulf of Mexico. Micropaleontology, vol. 39, no. 4, pp. 367–403.

Kuhnt, W., Holbourn, A.E., and Zhao, Q., 2002, The early history of the South China Sea: Evolution of Oligocene–Miocene deep water environments. Revue de Micropaléontologie, vol. 45, pp. 99–159.

Miller, K.G. and Katz, M.E., 1987, Oligocene to Miocene benthic foraminiferal and abyssal circulation changes in the North Atlantic. Micropaleontology, vol. 33, no. 2, pp. 97–149.

Ortiz, S. and Thomas, E., 2006, Lower-middle Eocene benthic foraminifera from the Fortuna Section (Betic Cordillera, southeastern Spain). Micropaleontology, vol. 52, pp. 97–150.

Robertson, B.E., 1998, Systematics and paleoecology of the benthic foraminiferida from the Buff Bay section, Miocene of Jamaica. Micropaleontology, vol. 44, supplement 2, pp. 1–266.

Schwager, C., 1866, Fossile Foraminiferen von Kar Nikobar, Reise der Österreichischen Fregatte Novara um Erde in den Jahren 1857, 1858, 1859 unten den Befehlen des Commodore B. Von Wüllerstorf-Urbair. Geologischer Theil, vol. 2, no. 1, Geologische Beobachtung, no. 2, Paläontologische Mittheilung, pp. 187–268.

Srinivasan, M.S. and Sharma, V., 1980, Schwager's Car Nicobar Foraminifera in the Reports of the Novara Expedition – A Revision. New Delhi: Today & Tomorrow's Printers and Publishers, 83 pp.

van Morkhoven, F.P.C.M., Berggren, W.A., Edwards, A.S., et al., 1986, Cenozoic cosmopolitan deep-water benthic foraminifera. Bulletin des Centres de Recherches Exploration-Production Elf-Aquitaine, Mémoire, vol. 11, pp. 1–421.

# Uvigerina laviculata

## Coryell and Rivero, 1940

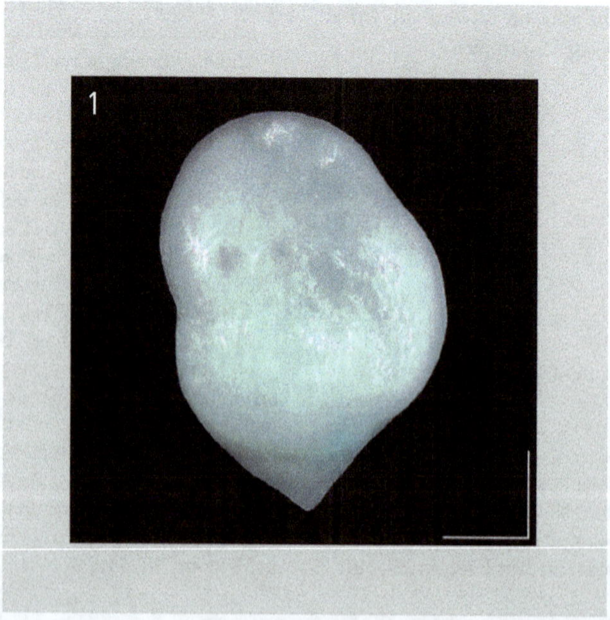


1. *Uvigerina laviculata.* Gulf of Mexico; Eureka Core E68-136, depth: 2506 feet. Lateral view, NHMUK PM DI 0715, scale bar = 176 μm.

### Original Designation

*Uvigerina laviculata* Coryell and Rivero, 1940, p. 343, pl. 44, fig. 24.

### Type Specimen

Depository not given.

### Type Level

Middle Miocene, Port-au-Prince beds; rather unconsolidated bluish-gray marl and fine-grained sands interbedded with layers of yellowish limestones.

### Type Locality

Near Port-au-Prince, Haiti.

### Description

Test forms a large, fusiform, triserial series; tapered in outline, with a subacute initial portion, a rounded apertural end (greater breadth in the upper half of the test), and a circular cross-section. The inflated chambers increase rapidly in size (except in the more elongate morphotypes where they remain more uniform), and are separated by slightly depressed sutures. Chamber walls are calcareous, finely perforate, and ornamented by sixteen to twenty fine, longitudinal striae, except in the last whorl, which is nearly smooth. Striae are generally restricted to individual chambers, and may be fragmented into segments intercalated between more continuous ones. The primary aperture is a terminal, round opening at the end of a broad, short neck set in a depression, bordered by a phialine lip, and with an internal toothplate.

### Synonyms and Citations

*Uvigerina striatissima* Perconig, 1955, p. 187, pl. 3, figs. 1–4.

*Uvigerina laviculata* Coryell and Rivero. –Cushman and Todd, 1945, p. 50, pl. 7, fig. 25. –Bermúdez, 1949, p. 207, pl. 13, fig. 34. –Boersma, 1984, pl. 1, pp. 91–94, figs. 1–4.

### Remarks

Boersma (1984) compared specimens of *Uvigerina laviculata* Coryell and Rivero from Haiti and specimens of *Uvigerina striatissima* Perconig from the Upper Miocene and Lower Pliocene of Italy, and concluded that they were identical.

*Uvigerina longistriata* Perconig differs from *Uvigerina laviculata* Coryell and Rivero by having a smaller number (about thirteen) of coarser striations per chamber (Boersma, 1984).

*Uvigerina carapitana* Hedberg resembles *Uvigerina laviculata* Coryell and Rivero, differing mainly in the degree of ornamentation: *Uvigerina carapitana* has faintly striated early chambers, whereas *Uvigerina laviculata* is finely striated throughout, except for the last chamber.

### Biogeography

Worldwide. Recorded from the Caribbean, South America, the circum-Mediterranean, and the Para-Tethys (Boersma, 1984).

### Bathymetry

Lower neritic to upper bathyal. Not found in close proximity to upwelling (Boersma, 1984).

### Chronostratigraphy

Early Miocene (Burdigalian) to Middle Pliocene.

## References


Bermúdez, P.J., 1949, Tertiary smaller foraminifera of the Dominican Republic. Special Publications of the Cushman Laboratory for Foraminiferal Research, no. 25, pp. 1–322.

Boersma, A., 1984, Handbook of common Tertiary Uvigerina. Stony Point, NY: Microclimates Press, 207 pp.

Coryell, H.N. and Rivero, F.C., 1940, A Miocene microfauna of Haiti. Journal of Paleontology, vol. 14, pp. 324–344.

Cushman, J.A. and Todd, R., 1945, Miocene foraminifera from Buff Bay, Jamaica. Special Publications of the Cushman Laboratory for Foraminiferal Research, vol. 15, pp. 1–73.

Perconig, E., 1955, Due nuove specie di Uvigerina del Neogene della Planura Padana. Bollettino del Servizio Geologico d'Italia, vol. 77, p. 187.

# Uvigerina mediterranea

## Hofker, 1932

1. *Uvigerina mediterranea.* Gulf of Mexico; Eureka Core E68-136. Deposited in the collections of the United States National Museum, Washington, DC. Illustrated in van Morkhoven et al., 1986, Ref. no. 1. Lateral view, NHMUK PM DI 0716, scale bar = 167 μm.

### Original Designation

*Uvigerina mediterranea* Hofker, 1932, pp. 118, 119, pls. 32a–g.

### Type Specimen

Not designated.

### Type Level

Recent.

### Type Locality

Ammontatura, Gulf of Naples, south of Posillipo, Italy; 200–300 meters.

### Description

Test forms an elongate, robust, fusiform, triserial series, showing a tendency to become loosely triserial, about two and a half times as long as broad, and with a circular cross-section. The inflated chambers increase markedly in height, and are separated by strongly depressed sutures. Chamber walls are calcareous, perforate, and ornamented by twelve to fourteen prominent, longitudinal costae per chamber, either restricted to individual chambers or crossing over suture lines, and extending to the final chamber. The primary aperture is a terminal, round opening at the end of a short, broad neck, with an internal toothplate, bordered by a phialine lip.

### Synonyms and Citations

*Uvigerina finisterrensis* Colom, 1952, p. 27, pl. 4, figs. 1–2, 13–14, textfigs. 4.1–8.

*Uvigerina pygmaea* d'Orbigny. –Brady, 1884, p. 575, pl. 74, figs. 11–12.

*Euuvigerina peregrina* (Cushman). –Barker, 1960, pl. 74, figs. 11–12.

*Uvigerina mediterranea* Hofker. –van Morkhoven et al., 1986, p. 16, pl. 1. –Lutze, 1986, p. 40, pl. 5, figs. 1–7. –Jones, 1994, p. 86, pl. 74, figs. 11–12.

### Remarks

*Uvigerina mediterranea* Hofker differs from *Uvigerina peregrina* Cushman by its larger and more robust test with fewer costae (Boersma, 1984).

### Biogeography

Worldwide. Common along the margins of the Atlantic, Mediterranean, North Sea, East Pacific, and in the Gulf of Mexico (van Morkhoven et al., 1986).

## Bathymetry

Lower neritic to upper bathyal.

## Chronostratigraphy

Late Pliocene to Recent.

# References


Barker, R.W., 1960, Taxonomic notes on the species figured by H.B. Brady in his report on the foraminifera dredged by H.M.S. Challenger during the years 1873–1876. Society of Economic Paleontologists and Mineralogists Special Publication, vol. 9, pp. 1–238.

Boersma, A., 1984, Handbook of Common Tertiary Uvigerina. Stony Point, NY: Microclimates Press, 207 pp.

Brady, H.B., 1884, Report on the foraminifera dredged by H.M.S. Challenger during the years 1873–1876. In Brady, H.B., ed., Report on the Scientific Results of the Voyage of the H.M.S. Challenger during the years 1873–1876. Zoology, vol. 9, pp. 1–814.

Colom, G., 1952, Foraminiferos de las costas de Galicia (campañas del "Xauen" en 1949 y 1950). Boletín del Instituto Español de Oceanografía, vol. 51, pp. 3–59.

Hofker, J., 1932, Notizen ueber die Foraminiferen des Golfes von Neapel; III – Die Foraminiferenfauna der Ammontatura. Pubblicazioni della Stazione Zoologica, Napoli, vol. 12, no. 1, pp. 61–145.

Jones, R.W., 1994, The Challenger Foraminifera. Oxford: Oxford University Press, 149 pp.

Lutze, G.F., 1986, Uvigerina species of the eastern North Atlantic. Utrecht Micropaleontological Bulletins, vol. 35, pp. 21–46.

van Morkhoven, F.P.C.M., Berggren, W.A., Edwards, A.S., et al., 1986, Cenozoic cosmopolitan deep-water benthic foraminifera. Bulletin des Centres de Recherches Exploration-Production Elf-Aquitaine, Mémoire, vol. 11, pp. 1–421.

# *Uvigerina mexicana*

## Nuttall, 1932

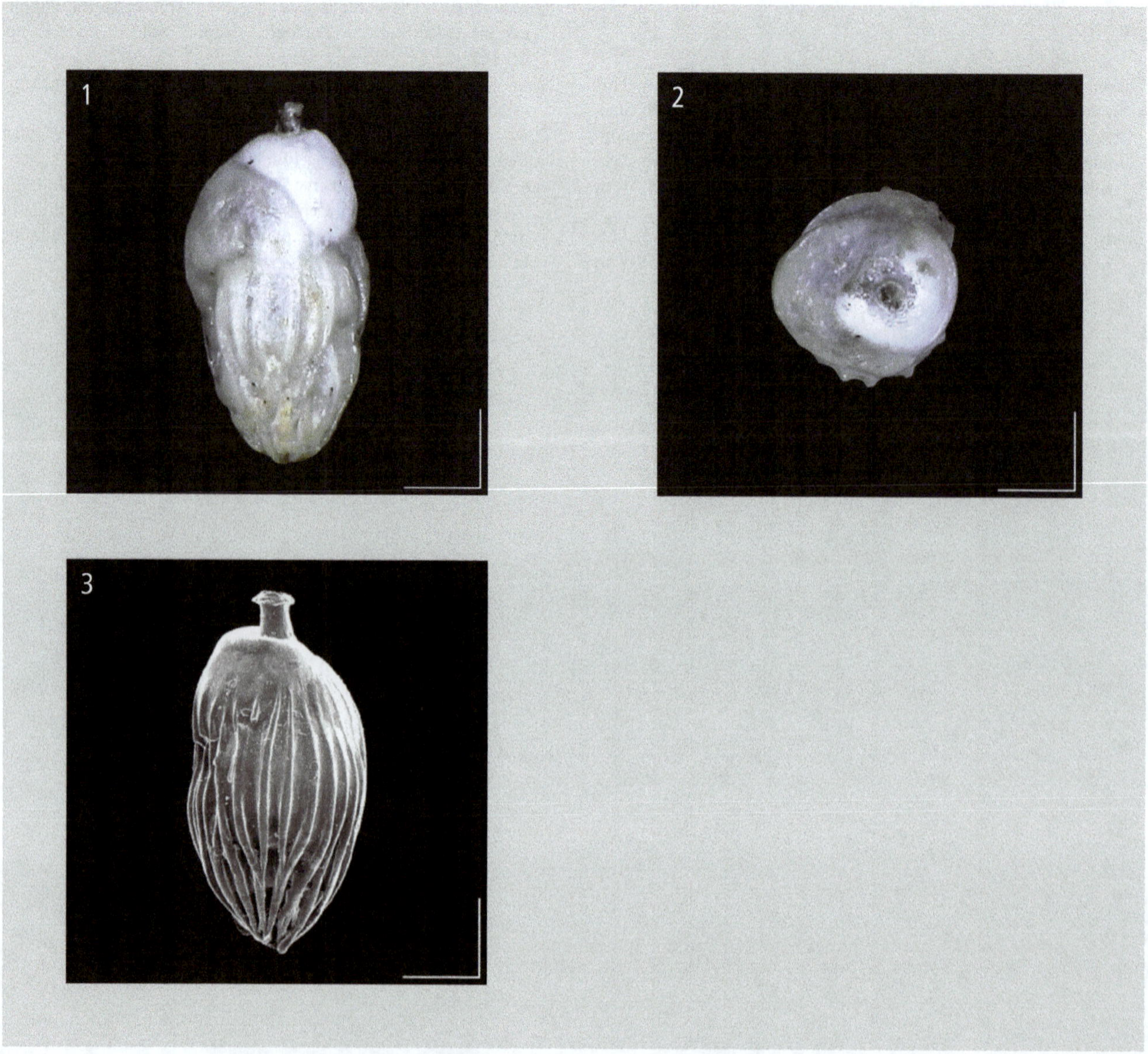

1. *Uvigerina mexicana.* Base of Alazan Formation, near Potrero del Llano, Vera Cruz State, East Mexico. Deposited in the micropaleontological collections of The Natural History Museum, London; NHMUK P39573-93. Lateral view, NHMUK PM DI 0717, scale bar = 143 μm.

2. *Uvigerina mexicana.* Base of Alazan Formation, near Potrero del Llano, Vera Cruz State, East Mexico. Deposited in the micropaleontological collections of The Natural History Museum, London; NHMUK P39573-93. Apertural view, NHMUK PM DI 0718, scale bar = 137 μm.

3. *Uvigerina mexicana.* Gulf of Mexico; Eureka Core E68-151A, depth: 4628 feet. Lateral view, NHMUK PM DI 0719, scale bar = 183 μm.

## Original Designation

*Uvigerina mexicana* Nuttall, 1932, p. 22, pl. 5, figs. 12–13.

## Type Specimen

Illustrated syntypes (CC 16450) and unillustrated syntypes (CC 16439 and CC 16451) deposited in the collections of the United States National Museum, Washington, DC. Van Morkhoven designated as lectotype an ovoid specimen on the right side of slide CC 16450 (=Nuttall, 1932, pl. 5, fig. 12).

## Type Level

Lower Oligocene, Lower Alazan Shale.

## Type Locality

San Miguel village, situated just south of the boundary of Palma Real and Llano Grande, Tampico Embayment, Vera Cruz, Mexico.

## Description

Test forms an elongate, short, stout, fusiform, triserial series, showing a tendency to become loosely triserial, about two and a half times as long as broad (the broadest part of which is situated approximately one-third of the way down from the aperture), and with a circular cross-section. The inflated chambers increase markedly in height and are separated by sutures, indistinct in the initial two-thirds of the test, then depressed. Chamber walls are calcareous, finely perforate, and ornamented by numerous prominent, longitudinal costae, continuous over adjacent chambers and more developed in the lower half of the test. Costae of initial chambers extend as short, blunt spines at the initial portion of the test. The primary aperture is a terminal, round opening at the end of a short neck set into a depression, with an internal toothplate, and bordered by a phialine lip.

## Synonyms (following van Morkhoven et al., 1986) and Citations

?*Uvigerina semivestita* Bermúdez, 1937, p. 143, pl. 17, figs. 8–9.

*Uvigerina mexicana* Nuttall var. *bulbosa* Galloway and Hemingway, 1941, p. 431, pl. 33, fig. 10.

*Uvigerina mexicana* Nuttall subsp. *nuda* Gohrbandt, 1962, p. 85, pl. 3, figs. 11–12; pl. 4, fig. 1.

*Uvigerina mexicana* Nuttall. –Tjalsma, 1983, p. 743, pl. 2, figs. 4–5. –Boersma, 1984, pp. 110–113, pl. 1, figs. 1–5. –van Morkhoven et al., 1986, p. 228, pl. 77, figs. 1–4. –Katz and Miller, 1993, pl. 4, fig. 4.

## Remarks

*Uvigerina mexicana* Nuttall differs from *Uvigerina basicordata* Cushman and Renz by its denser ornamentation and more continuous costae over the length of the test (M. Katz, pers. comm. 1998).

## Biogeography

Worldwide. Recorded from Cuba, Puerto Rico, the Dominican Republic, Trinidad, Venezuela, Ecuador, California, Europe, the North and South Atlantic, the East Pacific, and the Gulf of Mexico (van Morkhoven et al., 1986).

## Bathymetry

Lower neritic to bathyal (van Morkhoven et al., 1986).

## Chronostratigraphy

Late Eocene (Priabonian) to Early Miocene (Burdigalian).

# References


Bermúdez, P.J., 1937, Nuevas especies de foraminiferos del Eoceno de Cuba. Memorias de la Sociedad Cubana de Historia Natural "Felipe Poey", vol. 11, pp. 137–150.

Boersma, A., 1984, Handbook of Common Tertiary Uvigerina. Stony Point, NY: Microclimates Press, 207 pp.

Galloway, J.J. and Hemingway, C.E., 1941, The Tertiary Foraminifera of Porto Rico. New York Academy of Science, Scientific Survey of Porto Rico and the Virgin Islands, vol. 3, no. 4, pp. 275–491.

Gohrbandt, K., 1962, Die Kleinforaminiferenfauna des obereozaenen Anteils der Reingruber Serie bei Brunendorf (Bezirk Korneuburg, Niedersterreich). Mitteilungen der Geologischen Gesellschaft in Wien, vol. 54 (1961), pp. 55–145.

Katz, M.E. and Miller, K.G., 1993, Latest Oligocene to Earliest Pliocene benthic foraminiferal biofacies of the northeastern Gulf of Mexico. Micropaleontology, vol. 39, no. 4, pp. 367–403.

Nuttall, W.L.F., 1932, Lower Oligocene foraminifera from Mexico. Journal of Paleontology, vol. 6, pp. 3–35.

Tjalsma, R.C., 1983, Eocene to Miocene benthic foraminifera from DSDP site 516, Rio Grande Rise, South Atlantic. In Barker, P.F., Carlson, R.L., Johnson, D.A., et al., ed., Initial Reports of the Deep Sea Drilling Project, Washington (U.S. Government Printing Office), vol. 72, pp. 731–755.

van Morkhoven, F.P.C.M., Berggren, W.A., Edwards, A.S., et al., 1986, Cenozoic cosmopolitan deep-water benthic foraminifera. Bulletin des Centres de Recherches Exploration-Production Elf-Aquitaine, Mémoire, vol. 11, pp. 1–421.

# *Uvigerina pigmaea*

## d'Orbigny, 1826

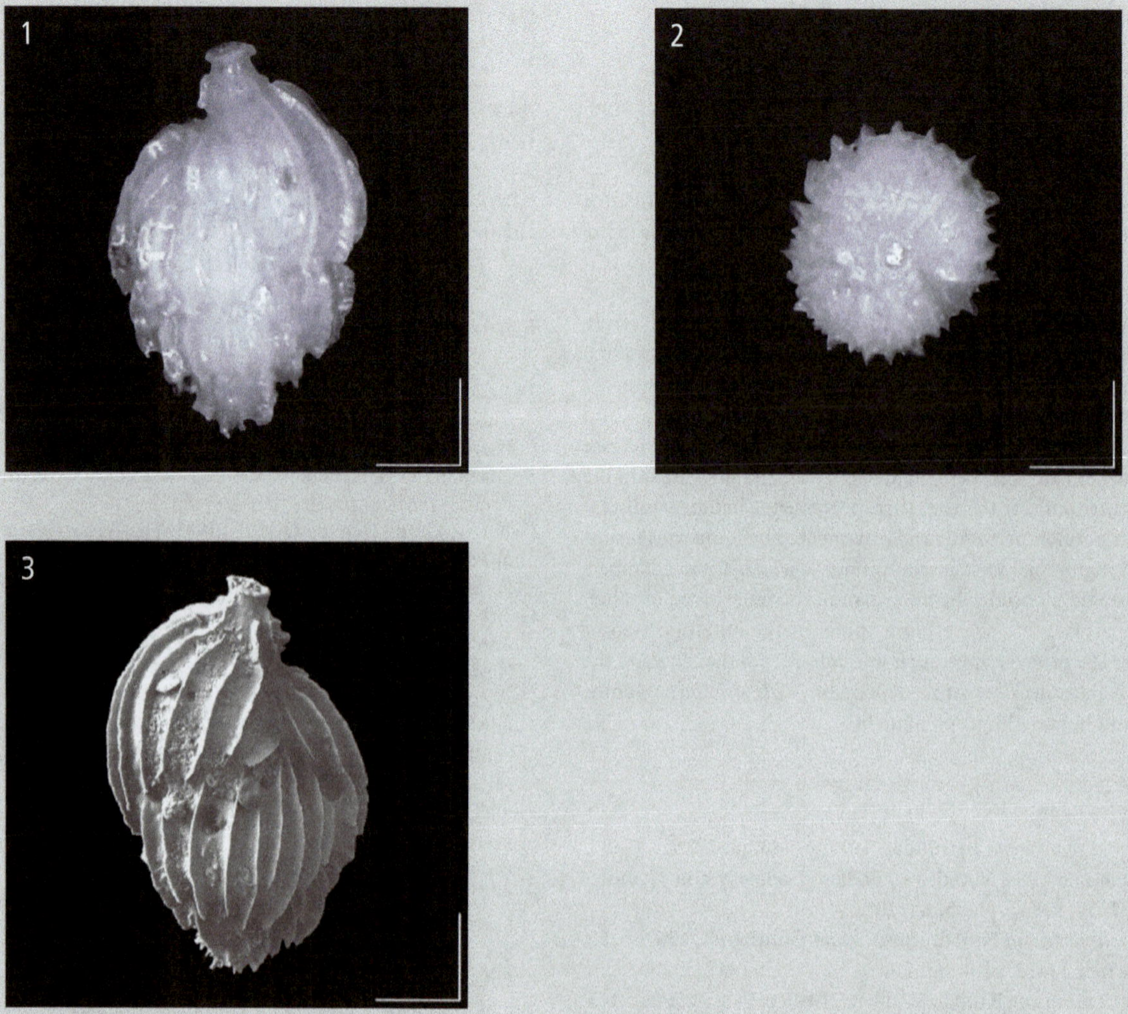

1. *Uvigerina pigmaea.* Gulf of Mexico; Eureka Core E68-136, depth: 2506 feet. Lateral view, NHMUK PM DI 0720, scale bar = 78 μm.
2. *Uvigerina pigmaea.* Gulf of Mexico; Eureka Core E68-136, depth: 2506 feet. Apertural view, NHMUK PM DI 0721, scale bar = 53 μm.
3. *Uvigerina pigmaea.* Gulf of Mexico; Eureka Core E68-136, depth: 2506 feet. Lateral view, NHMUK PM DI 0722, scale bar = 62 μm.

## Original Designation

*Uvigerina pigmea* d'Orbigny, 1826, p. 269, pl. 12, figs. 8–9.

## Type Specimen

Deposited in the collections of the Laboratoire de Paléontologie, Muséum d'Histoire Naturelle, Paris, France.

## Type Level

Fossil.

## Type Locality

Near Sienna, Italy.

## Description

Test forms a small, elongate, fusiform, triserial series, showing a tendency to become loosely triserial, about one and a half times as long as broad, slightly tapered with a subrounded initial portion (greatest breadth at or above the middle of the test), and circular in cross-section. The inflated chambers increase gradually in size and are separated by distinct, depressed sutures. Chamber walls are calcareous, finely perforate, and ornamented by thickened, low, longitudinal costae. Costae may be irregular in length and orientation or may be fragmented into spines or nodes. The primary aperture, often bearing spines, is a terminal, round opening at the end of a short neck, bordered by a phialine lip, and with an internal toothplate.

## Synonyms and Citations

*Uvigerina pigmea* d'Orbigny. –Bermúdez, 1949, p. 209, pl. 13, fig. 44. –Lutze, 1986, p. 36, pl. 3, figs. 6–8.

*Uvigerina pigmaea* d'Orbigny. –Boersma, 1984, pp. 127–130, pl. 1, figs. 1–6; pl. 2, figs. 1–5. –Katz and Miller, 1993, pl. 4, fig. 3.

## Remarks

The spelling of *Uvigerina pigmaea* d'Orbigny was amended by Boersma (1984).

*Uvigerina pigmaea* d'Orbigny differs from *Uvigerina capayana* Hedburg by its longer and less inflated test and from *Uvigerina azuana* Bermúdez by its more fusiform and markedly less inflated test. Both *Uvigerina capayana* and *Uvigerina azuana* are only found in shallower sediments, whereas *Uvigerina pigmaea* occurs in upper and middle bathyal sediments (Boersma, 1984). *Uvigerina pigmaea* also resembles *Uvigerina mediterranea* Hofker, *Uvigerina peregrina* Cushman, and *Uvigerina parvula* Cushman and Thomas. However, the stratigraphic ranges of *Uvigerina mediterranea* and *Uvigerina peregrina* do not extend below the Middle Pliocene and *Uvigerina pigmaea* has a deeper paleodepth range than *Uvigerina parvula* (M. Katz, pers. comm. 1998).

## Biogeography

Worldwide. Reported from the circum-Caribbean, circum-Mediterranean, and coastal South America (Boersma, 1984).

## Bathymetry

Upper to middle bathyal.

## Chronostratigraphy

Late Eocene to Early Pliocene.

# References


Bermúdez, P.J., 1949, Tertiary smaller foraminifera of the Dominican Republic. Special Publications of the Cushman Laboratory for Foraminiferal Research, no. 25, pp. 1–322.

Boersma, A., 1984, Handbook of Common Tertiary Uvigerina. Stony Point, NY: Microclimates Press, 207 pp.

d'Orbigny, A., 1826, Tableau méthodique de la classe des Céphalopodes. Annales des Sciences Naturelles, vol. 7, pp. 245–314.

Katz, M.E. and Miller, K.G., 1993, Latest Oligocene to Earliest Pliocene benthic foraminiferal biofacies of the northeastern Gulf of Mexico. Micropaleontology, vol. 39, no. 4, pp. 367–403.

Lutze, G.F., 1986, Uvigerina species of the eastern North Atlantic. Utrecht Micropaleontological Bulletins, vol. 35, pp. 21–46.

# *Uvigerina proboscidea*

## Schwager, 1866

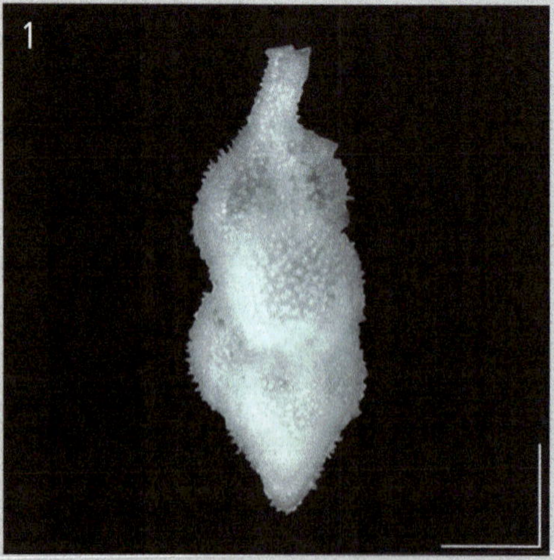


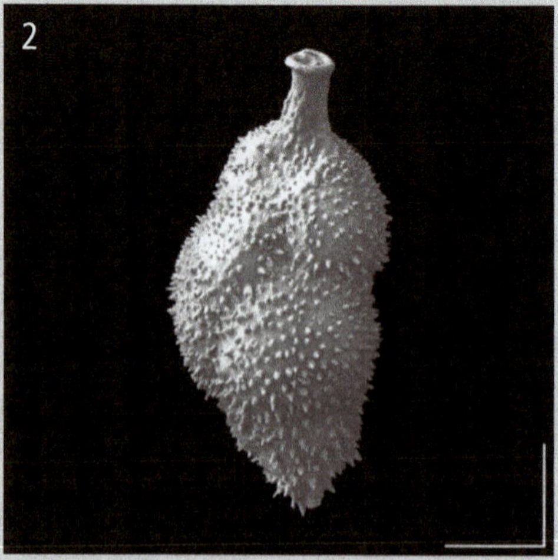


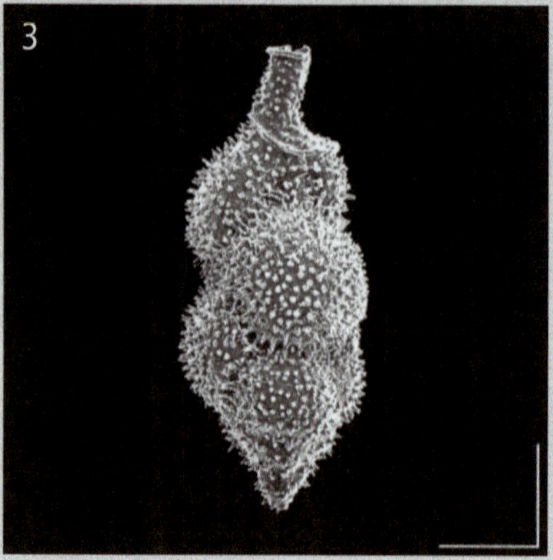


1. *Uvigerina proboscidea*. Gulf of Mexico; Eureka Core E66-73, depth: 2898 feet. Lateral view, NHMUK PM DI 0723, scale bar = 111 μm.
2. *Uvigerina proboscidea*. Gulf of Mexico; Eureka Core E66-73, depth: 2898 feet. Lateral view, NHMUK PM DI 0724, scale bar = 63 μm.
3. *Uvigerina proboscidea*. Gulf of Mexico; Eureka Core E66-73, depth: 2898 feet. Lateral view, NHMUK PM DI 0725, scale bar = 110 μm.

## Original Designation

*Uvigerina proboscidea* Schwager, 1866, p. 250, pl. 7, fig. 96.

## Type Specimen

Depository not given. Neotype (Mf 11) designated as *Neouvigerina proboscidea* (Schwager) by Srinivasan and Sharma (1980) deposited in the Micropaleontology Section, Department of Geology, Banaras Hindu University, Banaras, India.

## Type Level

Upper Tertiary, upper and lower clays.

## Type Locality

Car Nicobar, Andaman Sea.

## Description

Test elongate, fusiform, triserial, showing marked tendency to become loosely triserial, about two and a half times as long as broad, widest in the middle of the test, circular in cross-section, and with a distinct basal spine. Inflated chambers, increasing markedly in height, are separated by distinctly depressed sutures. Chamber walls are calcareous, perforate, and finely hispid. The primary aperture is terminal, round with an internal toothplate, bordered by a phialine lip, and at the end of a short neck.

## Synonyms and Citations

*Uvigerina proboscidea* Schwager. –Cushman and Todd, 1945, p. 50, pl. 7, figs. 28–29. –Bermúdez, 1949, p. 209, pl. 13, fig. 45. –Boersma, 1984, pp. 131–134, pl. 1, figs. 1–5. –van Morkhoven et al., 1986, p. 28, pl. 6. –Katz and Miller, 1993, pl. 4, fig. 8. –Nomura, 1995, p. 274, pl. 2, fig. 5.

*Neouvigerina proboscidea* Schwager. –Srinivasan and Sharma, 1980, p. 52, pl. 7, fig. 21.

## Remarks

Revised stratigraphy at the Eureka boreholes in the Gulf of Mexico (Katz and Miller, 1993) has extended the first occurrence of this species to Biozone P22, in contrast to the Biozone M14 occurrence reported by van Morkhoven et al. (1986).

## Biogeography

Worldwide. Recorded from the Andaman Sea, Indian Ocean, the Pacific, Atlantic, Caribbean, Europe, and the Gulf of Mexico (van Morkhoven et al., 1986).

## Bathymetry

Bathyal.

## Chronostratigraphy

Late Oligocene to Recent.

# References


Bermúdez, P.J., 1949, Tertiary smaller foraminifera of the Dominican Republic. Special Publications of the Cushman Laboratory for Foraminiferal Research, no. 25, pp. 1-322.

Boersma, A., 1984, Handbook of Common Tertiary Uvigerina. Stony Point, NY: Microclimates Press, 207 pp.

Cushman, J.A. and Todd, R., 1945, Miocene foraminifera from Buff Bay, Jamaica. Special Publications of the Cushman Laboratory for Foraminiferal Research, vol. 15, pp. 1–73.

Katz, M.E. and Miller, K.G., 1993, Latest Oligocene to Earliest Pliocene benthic foraminiferal biofacies of the northeastern Gulf of Mexico. Micropaleontology, vol. 39, no. 4, pp. 367–403.

Nomura, R., 1995, Paleogene to Neogene deep-sea paleoceanography in the eastern Indian Ocean: benthic foraminifera from ODP Sites 747, 757 and 758. Micropaleontology, vol. 41, pp. 251–290.

Schwager, C., 1866, Fossile Foraminiferen von Kar Nikobar, Reise der Österreichischen Fregatte Novara um Erde in den Jahren 1857, 1858, 1859 unten den Befehlen des Commodore B. Von Wüllerstorf-Urbair. Geologischer Theil, vol. 2, no. 1, Geologische Beobachtung, no. 2, Paläontologische Mittheilung, pp. 187–268.

Srinivasan, M.S. and Sharma, V., 1980, Schwager's Car Nicobar Foraminifera in the Reports of the Novara Expedition – A Revision. New Delhi: Today & Tomorrow's Printers and Publishers, 83 pp.

van Morkhoven, F.P.C.M., Berggren, W.A., Edwards, A.S., et al., 1986, Cenozoic cosmopolitan deep-water benthic foraminifera. Bulletin des Centres de Recherches Exploration-Production Elf-Aquitaine, Mémoire, vol. 11, pp. 1–421.

# *Uvigerina spinulosa*

## Hadley, 1934

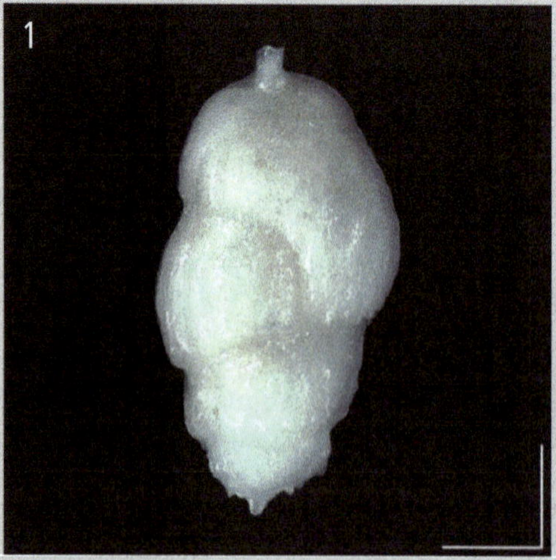

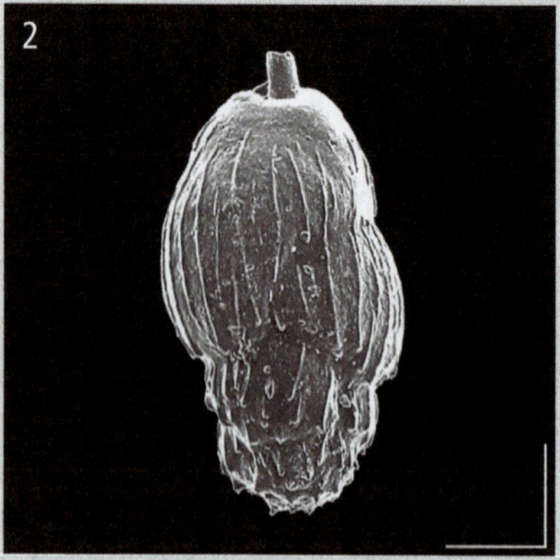

1. *Uvigerina spinulosa.* Rio Grande Rise, South Atlantic; DSDP Leg 39, Site 357, Core 18, Section 3, 49–57 cm. Lateral view, NHMUK PM DI 0726, scale bar = 108 μm.

2. *Uvigerina spinulosa.* Rio Grande Rise, South Atlantic; DSDP Leg 39, Site 357, Core 18, Section 3, 49–57 cm. Lateral view, NHMUK PM DI 0727, scale bar = 112 μm.

### Original Designation

*Uvigerina canariensis* d'Orbigny var. *spinulosa* Hadley, 1934, p. 18, pl. 2, fig. 17.

### Type Specimen

Deposited in the collections of the Paleontological Research Institution, Ithaca, NY, USA.

### Type Level

Tertiary, fine, white marl.

### Type Locality

Station 0–1, in bank 50 meters southeast of the entrance gate to the Cuban Naval Academy, Marel, at intersection with the road going up the hill toward La Merced, Pinar del Río Province, Cuba.

### Description

Test forms an elongate, triserial series, showing a tendency to become loosely triserial, about two and a half times as long as broad (the broadest part is approximately one-third of the way down from the aperture), and with a circular cross-section. The initial portion is rounded or pointed with a short, prominent, basal spine in well-preserved specimens. The inflated chambers increase markedly in height, and are separated by depressed sutures. Chamber walls are calcareous, finely perforate, and ornamented by faint, longitudinal striae (about ten to fifteen per chamber), more or less continuous over adjacent chambers, but fading away in final chambers. Initial chambers often bear short, blunt spines. The primary aperture is a terminal, round opening at the end of a short neck set into a depression, bordered by a thin, flaring phialine lip, and with an internal toothplate.

### Synonyms and Citations

*Uvigerina spinulosa.* –Tjalsma, 1983, p. 743, pl. 2, figs. 1–2. –Boersma, 1984, p. 163, pl. 1, figs. 1–6. –van Morkhoven et al., 1986, p. 219, pl. 75. –Thomas, 1985, p. 678, pl. 2, fig. 2.

### Remarks

Striae vary from coarser and fewer in number to faintly developed striae (Boersma, 1984).

### Biogeography

Worldwide. Recorded from the Dominican Republic, Mexico, Barbados, the Gulf of Mexico, the North and South Atlantic, and the Tasman Sea (van Morkhoven et al., 1986).

### Bathymetry

Lower neritic to bathyal.

### Chronostratigraphy

Late Eocene to Early Miocene (Burdigalian).

## References

Boersma, A., 1984, Handbook of Common Tertiary Uvigerina. Stony Point, NY: Microclimates Press, 207 pp.

Hadley, W.H., 1934, Some Tertiary foraminifera from the north coast of Cuba. Bulletin of American Paleontology, vol. 20, pp. 1–40.

Thomas, E., 1985, Late Eocene to Recent deep-sea benthic foraminifers from the central equatorial Pacific Ocean. In Mayer, L., Theyer, F., et al., ed., Initial Reports of the Deep Sea Drilling Project, Washington (U.S. Government Printing Office), vol. 85, pp. 655–694.

Tjalsma, R.C., 1983, Eocene to Miocene benthic foraminifera from DSDP site 516, Rio Grande Rise, South Atlantic. In Barker, P.F., Carlson, R.L., Johnson, D.A., et al., ed., Initial Reports of the Deep Sea Drilling Project, Washington (U.S. Government Printing Office), vol. 72, pp. 731–755.

van Morkhoven, F.P.C.M., Berggren, W.A., Edwards, A.S., et al., 1986, Cenozoic cosmopolitan deep-water benthic foraminifera. Bulletin des Centres de Recherches Exploration-Production Elf-Aquitaine, Mémoire, vol. 11, pp. 1–421.

# *Uvigerinammina jankoi*

## Majzon, 1943

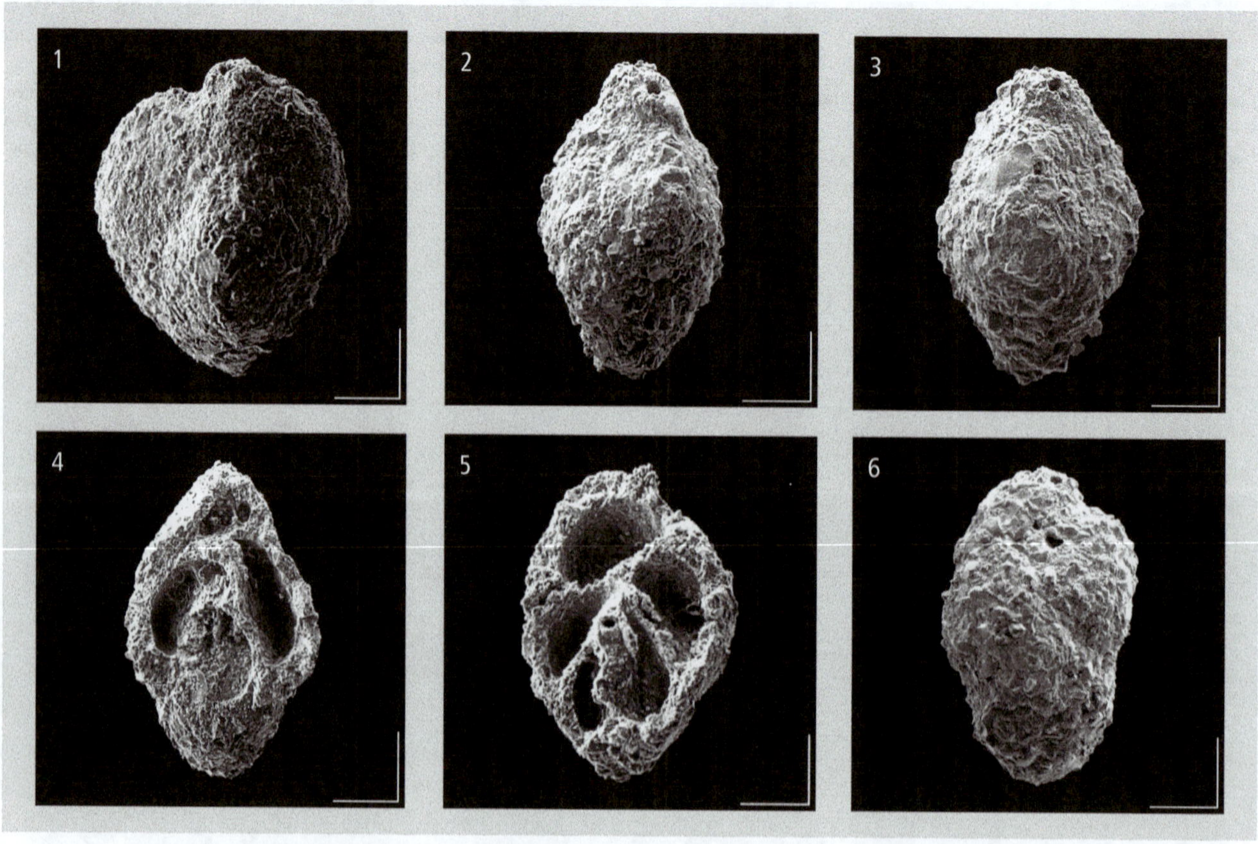


1. *Uvigerinammina jankoi*. Continental Rise, North American Basin; DSDP Leg 93, Hole 603B, Core 25, Section 1, 90–94 cm. Lateral view, NHMUK PM DI 0728, scale bar = 72 μm.

2. *Uvigerinammina jankoi*. Galicia Margin, northwest of the Iberian Peninsula, North Atlantic; ODP Leg 103, Hole 641A, Core 5, Section 1, 5–9 cm. Lateral view, NHMUK PM DI 0729, scale bar = 64 μm.

3. *Uvigerinammina jankoi*. Galicia Margin, northwest of the Iberian Peninsula, North Atlantic; ODP Leg 103, Hole 641A, Core 5, Section 1, 5–9 cm. Lateral view, NHMUK PM DI 0730, scale bar = 72 μm.

4. *Uvigerinammina jankoi*. Continental Rise, North American Basin; DSDP Leg 93, Hole 603B, Core 25, Section 1, 90–94 cm. Cross-section view, NHMUK PM DI 0731, scale bar = 61 μm.

5. *Uvigerinammina jankoi*. Continental Rise, North American Basin; DSDP Leg 93, Hole 603B, Core 25, Section 1, 90–94 cm. Cross-section view, NHMUK PM DI 0732, scale bar = 82 μm.

6. *Uvigerinammina jankoi*. Galicia Margin, northwest of the Iberian Peninsula, North Atlantic; ODP Leg 103, Hole 641A, Core 5, Section 1, 5–9 cm. Lateral view, NHMUK PM DI 0733, scale bar = 84 μm.

### Original Designation

*Uvigerinammina jankoi* Majzon, 1943, p. 158, pl. 2, fig. 15.

### Type Specimen

Depository not given.

### Type Level

Upper Cretaceous, Carpathian Sandstone (Carpathian Flysch) in tuffaceous *Trochamminoides* beds.

### Type Locality

Gernyes, Hungary.

### Description

Test forms a small, triserial series, with a subcircular or tricarinate cross-section. The inflated chambers increase rapidly in size, and are separated by distinct, depressed sutures. Chamber walls are finely agglutinated, and organically cemented. The primary aperture is a rounded, terminal opening, without a distinct neck.

### Synonyms and Citations

*Uvigerinammina jankoi* Majzon. –Krasheninnikov, 1974, p. 642, pl. 6, figs. 9–10. –Hemleben and Tröster, 1984, p. 522, pl. 4, fig. 15. –Geroch and Nowak, 1984, pl. 2, fig. 10. –Moullade et al., 1988, p. 367, pl. 10, figs. 1–6. –Kuhnt, 1990, p. 327, pl. 6, figs. 12–13.

### Biogeography

Worldwide.

### Bathymetry

Bathyal to abyssal. Common in sub-CCD deposits.

### Chronostratigraphy

Late Cretaceous (Turonian to Campanian). Geroch and Nowak (1984) defined the *Uvigerinammina jankoi* Biozone, based on the Turonian to Campanian occurrence of this species in the Polish Carpathians.

## References


Geroch, S. and Nowak, W., 1984, Proposal of zonation for the late Tithonian–late Eocene, based upon arenaceous foraminifera from the outer Carpathians, Poland. In Oertli, H., ed., Benthos '83, 2nd International Symposium on Benthic Foraminifera, Pau (France), April 11–15, 1983. Elf Aquitaine, ESSO REP and TOTAL CFP, Pau and Bordeaux, pp. 225–239.

Hemleben, C. and Tröster, J., 1984, Campanian–Maestrichtian deep-water foraminifers from Hole 543A, Deep Sea Drilling Project. In Bij-Duval, B., Moore, J.C., et al., ed., Initial Reports of the Deep Sea Drilling Project, Washington (U.S. Government Printing Office), vol. 78A, pp. 509–532.

Krasheninnikov, V.A., 1974, Upper Cretaceous benthonic agglutinated foraminifera, Leg 27 of the Deep Sea Drilling Project. In Veevers, J.J., Heirtzler, J.R., et al., ed., Initial Reports of the Deep Sea Drilling Project, Washington (U.S. Government Printing Office), vol. 27, pp. 531–662.

Kuhnt, W., 1990, Agglutinated foraminifera of western Mediterranean Upper Cretaceous pelagic limestones (Umbrian Apennines, Italy, and Betic Cordillera, Southern Spain). Micropaleontology, vol. 36, no. 4, pp. 297–330.

Majzon, L., 1943, Adatok egyes Kárpátaljai Flisrétegekhez, tekintettel a Globotruncanákra: Beiträge zur Kenntniss einiger Flysch-Schichten des Karpaten-Vorlandes mit Rücksicht auf die Globotruncanen. A magyar Királyi Földtani Intézet, Évkönnyve, vol. 37, pp. 1–170.

Moullade, M., Kuhnt, W., and Thurow, J., 1988, Agglutinated benthic foraminifers from Upper Cretaceous variegated clays of the North Atlantic Ocean (DSDP Leg 93 and ODP Leg 103). In Boillot, G., Winterer, E.L., et al., ed., Scientific Results of the Proceedings of the Ocean Drilling Program, vol. 103, pp. 349–377.

# *Veleroninoides scitulus*

## (Brady), 1881

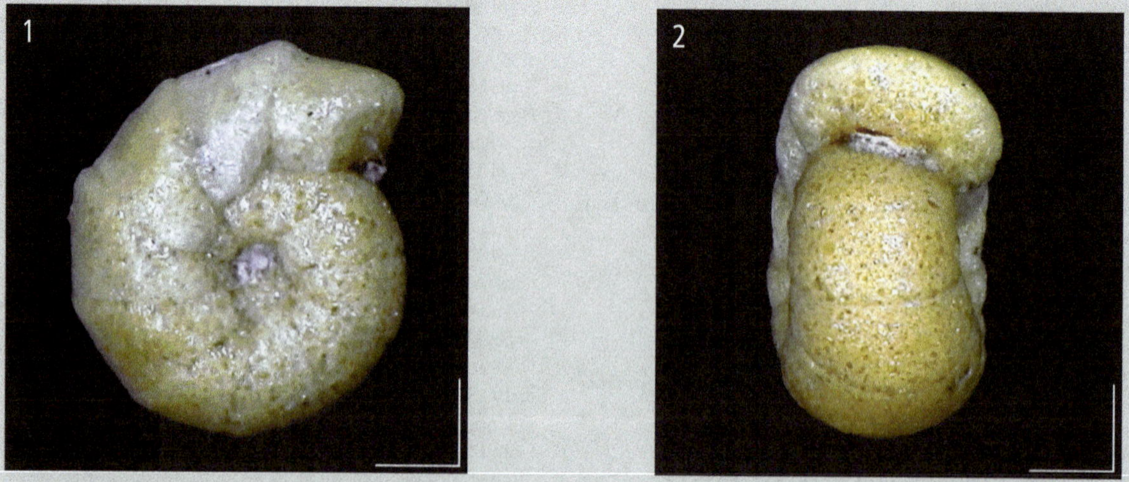

1. *Veleroninoides scitulus*. Faroe Channel; "Knight Errant" Station 7, depth: 530 fathoms. Deposited in the micropaleontological collections of The Natural History Museum, London; NHMUK ZF1551. Syntype. Umbilical view, NHMUK PM DI 0734, scale bar = 178 μm.

2. *Veleroninoides scitulus*. Faroe Channel; "Knight Errant" Station 7, depth: 530 fathoms. Deposited in the micropaleontological collections of The Natural History Museum, London; NHMUK ZF1551. Syntype. Apertural view, NHMUK PM DI 0735, scale bar = 183 μm.

### Original Designation

*Lituola (Haplophragmium) scitulum* Brady, 1881, p. 50 (type description), illustrated as *Haplophragmium scitulum* in Brady, 1884, pl. 34, figs. 11–13.

### Type Specimen

Figured specimens are housed in the micropaleontological collections of The Natural History Museum, London, UK, and are registered in slides ZF1551–ZF1553.

### Type Level

Recent.

### Type Locality

Not given. Figured specimens from "Porcupine" Station 47, latitude 59° 34′ N, longitude 7° 18′ W, Faröe Channel, North Atlantic, 542 fathoms; and "Knight Errant" Station 7, latitude 59° 37′ N, longitude 7° 19′ W, Faröe Channel, North Atlantic, 530 fathoms.

### Description

Test forms an evolute, biumbilicate, slightly lobulate planispire, with a rounded periphery. The numerous, inflated, broad and low chambers increase slowly in size, and are separated by fine, straight, slightly depressed sutures. Chamber walls are agglutinated, and smoothly finished with non-calcareous cement. The primary aperture is an areal, elliptical slit bordered by a lip, near the base of the apertural face.

### Synonyms and Citations

*Haplophragmium scitulum* Brady. –Brady, 1884, p. 308, pl. 34, figs. 11–13.

*Alveolophragmium scitulum* (Brady). –Barker, 1960, pl. 34, figs. 11–13.

*Labrospira scitula* (Brady). –Charnock and Jones, 1990, p. 172, pl. 6, figs. 9–10; pl. 17, fig. 6.

*Cribrostomoides scitulus* (Brady). –van Marle, 1991, p. 240.

*Veleroninoides scitulus* (Brady). –Jones, 1994, p. 41, pl. 34, figs. 11–13.

### Remarks

Charnock and Jones (1990) used the species concept in a broad sense, including in the synonymy forms originally described from Cretaceous and Paleogene sediments under various generic affiliations. The stratigraphic range given follows this broad concept.

### Biogeography
Worldwide. Reported from the Atlantic, North Sea, Tethys, and Indonesian seas.

### Bathymetry
Bathyal to upper abyssal. Reported between 400 and 2900 fathoms at "Challenger" stations (Jones, 1994).

### Chronostratigraphy
Late Cretaceous (Santonian?) to Recent (Jones, 1994).

## References


Barker, R.W., 1960, Taxonomic notes on the species figured by H.B. Brady in his report on the foraminifera dredged by H.M.S. Challenger during the years 1873–1876. Society of Economic Paleontologists and Mineralogists Special Publication, vol. 9, pp. 1–238.

Brady, H.B., 1881, Notes on some of the reticularian Rhizopoda of the Challenger Expedition, Part III. Quarterly Journal of Microscopical Science (New Series), vol. 21, pp. 31–71.

Brady, H.B., 1884, Report on the foraminifera dredged by H.M.S. Challenger during the years 1873–1876. In Brady, H.B., ed., Report on the Scientific Results of the Voyage of the H.M.S. Challenger during the years 1873–1876. Zoology, vol. 9, pp. 1–814.

Charnock, M.A. and Jones, R.W., 1990, Agglutinated foraminifera from the Palaeogene of the North Sea. In Hemleben, C., Kaminski, M.A., Kuhnt, W., and Scott, D.B., ed., Paleoecology, Biostratigraphy, Paleoceanography and Taxonomy of Agglutinated Foraminifera, NATO ASI Series, C: Mathematical and Physical Sciences. Dordrecht: Kluwer Academic, pp. 139–244.

Jones, R.W., 1994, The Challenger Foraminifera. Oxford: Oxford University Press, 149 pp.

van Marle, L.J., 1991, Eastern Indonesian Late Cenozoic smaller benthic foraminifera. Verhandelingen der Koninklijke Nederlanse Akademie van Wetenschappen, Afdeeling Natuurkunde, Eerste Reeks, vol. 34, pp. 1–328.

# *Verneuilinoides neocomiensis*

## (Mjatliuk), 1939

1. *Verneuilinoides neocomiensis*. Barrow Group. Northern Exmouth Plateau, Carnarvon Basin, Western Australia; ODP Leg 122. R.W. Jones Collection. Lateral view, NHMUK PM DI 0736, scale bar = 91 μm.

### Original Designation

*Verneuilina neocomiensis* Mjatliuk, 1939, p. 50, pl. 1, figs. 12–13.

### Type Specimen

Holotype (No. 944) deposited in the collections of the Oil Geological Institute, St Petersburg.

### Type Level

Lower Cretaceous, horizon with abundant *Haplophragmoides nonioninoides* (Reuss), gray, compact sandy clay with impressions of pelecypods (fide Ellis and Messina).

### Type Locality

Vostokneft Well no. 1, at a depth of 172–195 meters (holotype from 192–195 meters), near the station Ozinki on the Ryazan-Uralsk railroad, southern part of the Obshchi Syrt, Saratov District, USSR (fide Ellis and Messina).

### Description

Test forms an elongate, triserial series, gently flaring, and with a subrounded cross-section. The seven or more sets of moderately inflated chambers are separated by distinct, depressed sutures. A tendency to become biserial in the last whorl is apparent in some specimens. Chamber walls are usually finely agglutinated, and well cemented. The primary aperture is an interiomarginal arch.

### Synonyms and Citations

*Verneuilinoides asperulus* Crespin, 1963, pl. 15, figs. 8–12.

*Verneuilinoides neocomiensis* (Mjatliuk). –Kuznetsova, 1974, pl. 1, fig. 6a–c. –Geroch and Nowak, 1984, pl. 2, fig. 4. –Kuhnt, 1995, p. 50, pl. 4, figs. 12–16. –Holbourn and Kaminski, 1997, p. 53, pl. 23, figs. 7–9.

### Remarks

*Verneuilinoides asperulus* appears to be a coarse-grained morphotype of *Verneuilinoides neocomiensis* (Mjatliuk).

Geroch (1966) noted that the variability of *Verneuilinoides neocomiensis* (Mjatliuk) and *Verneuilinoides subfiliformis* Bartenstein is wide, and that transitional forms exist between these two species.

### Biogeography

Worldwide.

### Bathymetry

Bathyal to abyssal.

### Chronostratigraphy

Early Cretaceous (Berriasian to Albian). The partial range of this species characterizes the *Verneuilinoides neocomiensis* Biozone (Late Tithonian to Valanginian) of Geroch and Nowak (1984).

## References


Crespin, I., 1963, Lower Cretaceous Arenaceous Foraminifera of Australia. Bulletin of the Bureau of Mineral Resources, Geology and Geophysics, Australia, vol. 66, pp. 1–110.

Geroch, S., 1966, Lower Cretaceous small foraminifera of the Silesian series, Polish Carpathians. Rocznik Polskiego Towarzystwa Geologicznego, vol. 36, pp. 413–480.

Geroch, S. and Nowak, W., 1984, Proposal of zonation for the late Tithonian–late Eocene, based upon arenaceous foraminifera from the outer Carpathians, Poland. In Oertli, H., ed., Benthos '83, 2nd International Symposium on Benthic Foraminifera, Pau (France), April 11–15, 1983. Elf Aquitaine, ESSO REP and TOTAL CFP, Pau and Bordeaux, pp. 225–239.

Holbourn, A.E.L. and Kaminski, M.A., 1997, Cretaceous deep-water benthic foraminifera of the Indian Ocean. Grzybowski Foundation Special Publication, no. 4, pp. 1–175.

Kuhnt, W., 1995, Deep-water agglutinated foraminifera from the Lower Cretaceous (Neocomian) "complex à Aptychus" Formation (Corridor de Boyar, Betic Cordillera, southern Spain). Journal of Micropalaeontology, vol. 14, no. 1, pp. 37–52.

Kuznetsova, K.I., 1974, Distribution of Benthonic Foraminifera in Upper Jurassic and Lower Cretaceous Deposits at Site 261, DSDP Leg 27, in the Eastern Indian Ocean. In Veevers, J.J., Heirtzler, J.R., et al., ed., Initial Reports of the Deep Sea Drilling Project, Washington (U.S. Government Printing Office), vol. 27, pp. 673–681.

Mjatliuk, E.V., 1939, Foraminifera of the Upper Jurassic and Lower Cretaceous of the region of the Middle Volga River and the Great Syrte (in Russian). Trudy Vsesoyuznogo Nauchno-Issledovatel'skogo Geologorazvedochnogo Instituta VNIGRI, vol. 120, pp. 1–76.

# *Vulvulina pennatula*

## (Batsch), 1791

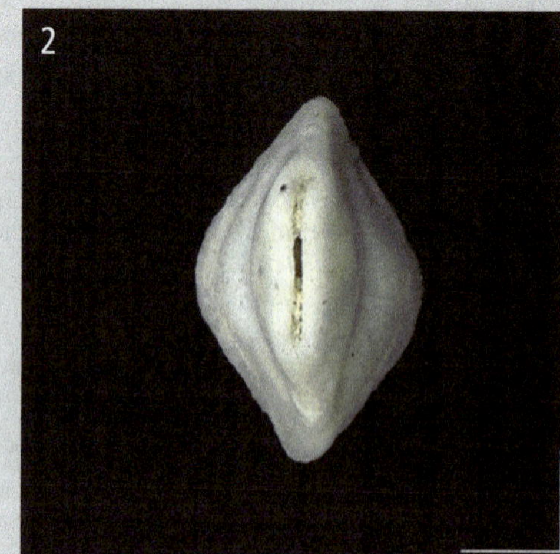

1. *Vulvulina pennatula.* Off Culebra Island, West Indies; "Challenger" Station 24, depth: 390 fathoms. Deposited in the micropaleontological collections of The Natural History Museum, London; NHMUK ZF1126. Lateral view, NHMUK PM DI 0737, scale bar = 500 μm.

2. *Vulvulina pennatula.* Off Culebra Island, West Indies; "Challenger" Station 24, depth: 390 fathoms. Deposited in the micropaleontological collections of The Natural History Museum, London; NHMUK ZF1126. Apertural view, NHMUK PM DI 0738, scale bar = 403 μm.

### Original Designation

*Nautilus (Orthoceras) pennatula* Batsch, 1791, pp. 3, 5, pl. 4, fig. 13a–e.

### Type Specimen

Depository not given.

### Type Level

Not designated, probably Recent.

### Type Locality

Shores of Italy [Probably Rimini, on the Adriatic Coast (Cushman, 1931)].

### Description

Test is elongate, flattened, and flaring, with an acute periphery. The microspheric generation with initial planispire, later becomes biserial, then uniserial in well-developed specimens. The numerous, broad, low chambers are slightly inflated, increase rapidly in size, and curve strongly downwards. Sutures are curved and depressed. Chamber walls are finely agglutinated with calcareous cement, and smoothly finished. The primary aperture is a broad, low interiomarginal arch in the early stage, later becoming a narrow, elongate, terminal slit in the adult stage.

### Synonyms and Citations

*Vulvulina capreolus* d'Orbigny, 1826, p. 264.

*Bigenerina capreolus* (d'Orbigny). –Brady, 1884, p. 372, pl. 45, figs. 1–4.

*Bigenerina pennatula* (Batsch). –Brady, 1884, p. 373, pl. 45, figs. 5–8.

*Vulvulina pennatula* (Batsch). –Cushman, 1931, p. 69. –Barker, 1960, pl. 45, figs. 1–8. –Loeblich and Tappan, 1987, p. 113, pl. 120, figs. 19–21. –van Marle, 1991, p. 244, pl. 25, figs. 15–16. –Jones, 1994, p. 49, pl. 45, figs. 1–8.

### Remarks

Type species of the genus *Vulvulina* d'Orbigny, 1826.

### Biogeography

Worldwide.

### Bathymetry

Bathyal. Reported between 350 and 675 fathoms at "Challenger" stations (Jones, 1994) and between 344 and 1760 meters in Indonesian seas (van Marle, 1991).

### Chronostratigraphy

Early Eocene to Recent.

## References

Barker, R.W., 1960, Taxonomic notes on the species figured by H.B. Brady in his report on the foraminifera dredged by H.M.S. Challenger during the years 1873–1876. Society of Economic Paleontologists and Mineralogists Special Publication, vol. 9, pp. 1–238.

Batsch, A.I.G.C., 1791, Sechs Kupfertafeln mit Conchylien des Seesandes, gezeichnet und gestochen. Jena, von A.J.G.K. Batsch, 6 pls.

Brady, H.B., 1884, Report on the foraminifera dredged by H.M.S. Challenger during the years 1873–1876. In Brady, H.B., ed., Report on the Scientific Results of the Voyage of the H.M.S. Challenger during the years 1873–1876. Zoology, vol. 9, pp. 1–814.

Cushman, J.A., 1931, Notes on the Foraminifera described by Batsch in 1791. Contributions from the Cushman Laboratory for Foraminiferal Research, vol. 7, pp. 62–72.

d'Orbigny, A., 1826, Tableau méthodique de la classe des Céphalopodes. Annales des Sciences Naturelles, vol. 7, pp. 245–314.

Jones, R.W., 1994, The Challenger Foraminifera. Oxford: Oxford University Press, 149 pp.

Loeblich, A.R. Jr and Tappan, H., 1987, Foraminiferal Genera and Their Classification. New York: Van Nostrand Reinhold, 970 pp.

van Marle, L.J., 1991, Eastern Indonesian Late Cenozoic smaller benthic foraminifera. Verhandelingen der Koninklijke Nederlanse Akademie van Wetenschappen, Afdeeling Natuurkunde, Eerste Reeks, vol. 34, pp. 1–328.

# *Vulvulina spinosa*

## Cushman, 1927

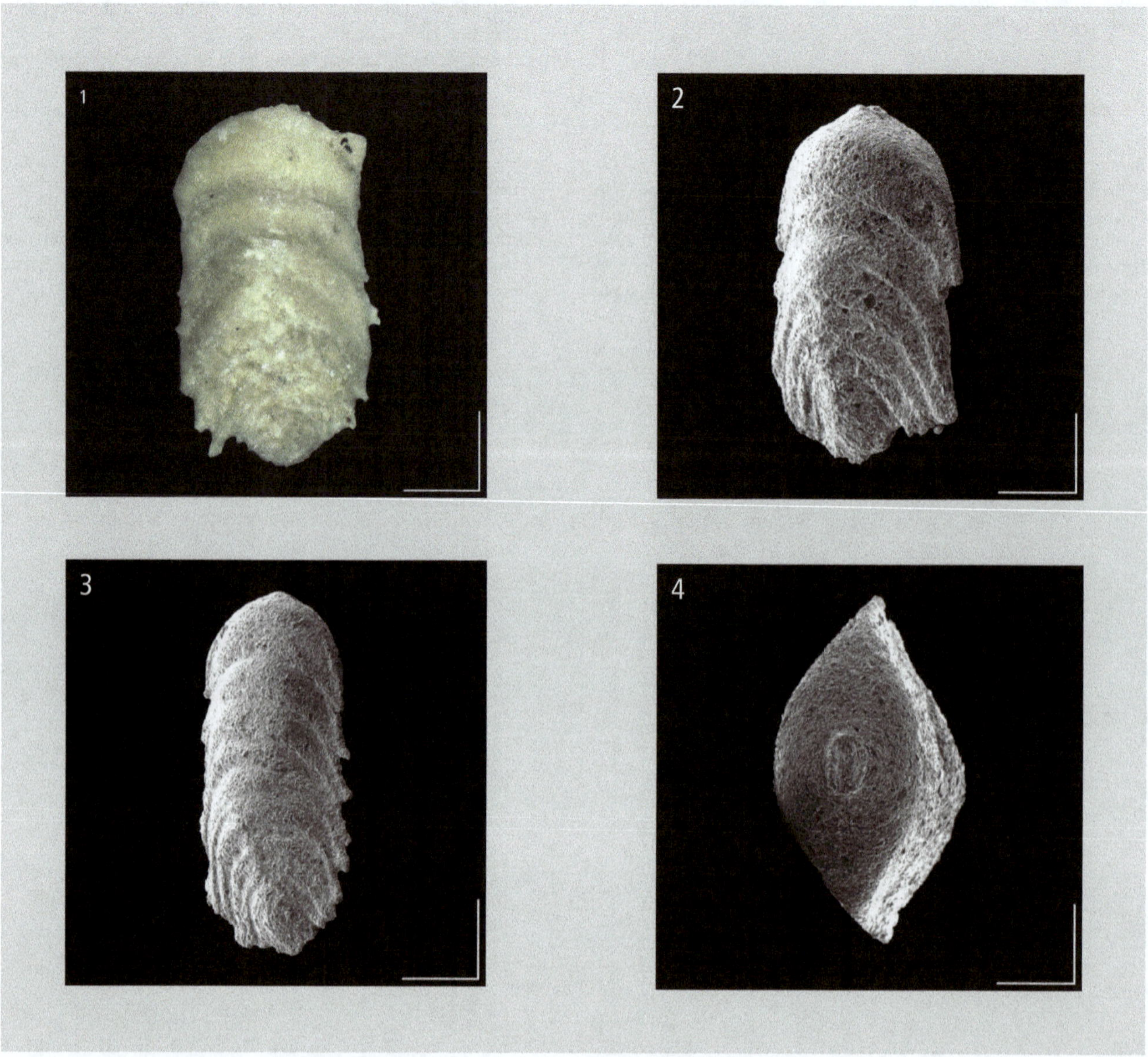


1. *Vulvulina spinosa.* Alaxan Clay, Rio Buena Vista, just south of crossing of Alazan to Moyutla Road. Vera Cruz, Mexico. Deposited in the collections of the United States National Museum, Washington, DC; CC 901. Holotype. Lateral view, NHMUK PM DI 0739, scale bar = 271 μm.

2. *Vulvulina spinosa.* Northeast continental slope of the South China Sea; ODP Leg 184, Hole 1148A, Core 49, Section 2, 70–72 cm. Lateral view, NHMUK PM DI 0740, scale bar = 258 μm.

3. *Vulvulina spinosa.* Northeast continental slope of the South China Sea; ODP Leg 184, Hole 1148A, Core 49, Section 2, 70–72 cm. Lateral view, NHMUK PM DI 0741, scale bar = 401 μm.

4. *Vulvulina spinosa.* Northeast continental slope of the South China Sea; ODP Leg 184, Hole 1148A, Core 49, Section 2, 70–72 cm. Apertural view, NHMUK PM DI 0742, scale bar = 184 μm.

## Original Designation

*Vulvulina spinosa* Cushman, 1927, p. 111, pl. 23, fig. 1.

## Type Specimen

Holotype (No. 901) deposited in the collections of the United States National Museum, Washington, DC.

## Type Level

Probably Lower Oligocene, Alazan Clay. Originally given as Upper Eocene by Cushman (1927).

## Type Locality

Rio Buena Vista, just south of crossing of Alazan to Moyutla Road, Vera Cruz, Mexico.

## Description

Test is elongate, flattened, lozenge-shaped or rhomboidal, with an acute periphery. Microspheric generation with initial planispire, later becomes biserial, then uniserial in well-developed specimens. The broad, low chambers are slightly inflated, increasing rapidly in size, and curving strongly downwards. Chambers possess a prominent projecting spine at the outer margins, and are separated by curved, depressed sutures. Chamber walls are coarsely agglutinated with calcareous cement, and roughly finished. The primary aperture is a broad, low interiomarginal arch in the early stage, later becoming a narrow, elongate, terminal slit.

## Synonyms and Citations

?*Vulvulina jarvisi* Cushman, 1932, p. 84, pl. 10, fig. 20.

*Vulvulina spinosa* Cushman. –Tjalsma and Lohmann, 1983, p. 38, pl. 10, figs. 4–5. –Miller and Katz, 1987, p. 140, pl. 1, fig. 2. –Boersma, 1990, pl. 2, fig. 2. –Thomas, 1990, p. 590. –Bolli et al., 1994, p. 325, figs. 87.3, 3a. –Kuhnt et al., 2002, p. 136, pl. 4. figs. 7–11.

## Remarks

Tjalsma and Lohmann (1983) considered *Vulvulina spinosa* Cushman and *Vulvulina jarvisi* Cushman to be conspecific.

## Biogeography

Worldwide. Reported from Trinidad, Venezuela, and Barbados (Bolli et al., 1994), Atlantic (Tjalsma and Lohmann, 1983), Weddell Sea, Antarctic (Thomas, 1990), North Atlantic (Miller and Katz, 1987), and West Pacific (Hermelin, 1989).

## Bathymetry

Restricted to shallower sites (bathyal) during the Early Eocene, expanding to deeper sites (abyssal) during the late Middle Eocene (Tjalsma and Lohmann, 1983). Lower bathyal (Thomas, 1990). Abyssal, greater than 3.5 kilometers (Miller and Katz, 1987). Lower bathyal and abyssal (Boersma, 1990).

## Chronostratigraphy

Early Eocene to Middle Miocene. Common to abundant in Oligocene to Middle Miocene, last occurrence in Serravallian with questionable occurrence in Tortonian (Miller and Katz, 1987).

# References


Boersma, A. 1990, Late Oligocene to late Pliocene benthic foraminifers from depth traverses in the central Indian Ocean. In Duncan, R.A., Backman, J., Peterson, L.C. et al., ed., Scientific Results of the Proceedings of the Ocean Drilling Program, vol. 115, pp. 315–379.

Bolli, H.M., Beckmann, J.P., and Saunders, J.B., 1994, Benthic Foraminiferal Biostratigraphy of the South Caribbean Region. Cambridge: Cambridge University Press, 408 pp.

Cushman, J.A., 1927, New and interesting foraminifera from Mexico and Texas. Contributions from the Cushman Laboratory for Foraminiferal Research, vol. 3, pp. 111–117.

Cushman, J.A., 1932, The genus Vulvulina and its species. Contributions from the Cushman Laboratory for Foraminiferal Research, vol. 8, pp. 75–85.

Hermelin, J.O.R., 1989, Pliocene benthic foraminifera from the Ontong-Java Plateau (Western Equatorial Pacific Ocean): faunal response to changing paleoenvironment. Special Publication of the Cushman Foundation for Foraminiferal Research, vol. 26, pp. 1–143.

Kuhnt, W., Holbourn, A.E., and Zhao, Q., 2002, The early history of the South China Sea: Evolution of Oligocene–Miocene deep water environments. Revue de Micropaléontologie, vol. 45, pp. 99–159.

Miller, K.G. and Katz, M.E., 1987, Oligocene to Miocene benthic foraminiferal and abyssal circulation changes in the North Atlantic. Micropaleontology, vol. 33, no. 2, pp. 97–149.

Thomas, E., 1990, Late Cretaceous through Neogene benthic foraminifers. In Barker, P.F., Kennett, J.P., et al., ed., Scientific Results of the Proceedings of the Ocean Drilling Program, vol. 113, pp. 571–594.

Tjalsma, R.C. and Lohmann, G.P., 1983, Paleocene–Eocene bathyal and abyssal benthic foraminifera from the Atlantic Ocean. Micropaleontology, Special Publication, no. 4, pp. 1–90.

# Glossary

For a complete glossary of terms used in the description of foraminifera please see Loeblich and Tappan (1987) and Hottinger (2006) – full reference details are given in the Introduction reference list.

## — A —

**Acuminate** – tapering, i.e., getting thinner or pointed; conical.

**Agglutinated** – foreign particles (mineral grains, sponge spicules, other foraminiferal tests, etc.) bound together with either organic or calcareous cement in the test wall.

**Alveolar** – wall with numerous small alveoli.

**Alveoli** – honeycomb-like structures in test wall, usually blind (found in fusulines, etc.). Wider than perforations, some are thought to allow light penetration into the test for symbiotic algae.

**Apertural plate** – a plate-like structure along the base of an interiomarginal aperture and restricting the latter.

**Aperture** – opening from the chamber of a test to the exterior.

**Arcuate** – having the form of a bow; curved; arc-shaped.

**Areal aperture** – aperture on face of final chamber.

## — B —

**Basal aperture** – aperture at the base of the apertural face adjacent to the previous whorl; interiomarginal.

**Biconvex** – in edge view, having both sides of test more or less equally convex.

**Bifid** – divided into two branches.

**Biloculine** – a test with only two chambers visible externally (e.g., *Pyrgo*).

**Biserial** – a test with chambers arranged in two alternating rows (e.g., *Bolivina*).

**Biumbilicate** – having a central depression (umbilicus) on either side of a planispiral test.

**Boss** – round knob in umbilicus.

**Buliminellid** – coiling as typified by *Buliminella*: comprising a high, close spiral formed of numerous very high, narrow chambers, with many chambers per whorl (but few whorls).

## — C —

**Canaliculate** – test wall perforated by fine tubular cavities that may be closed at one or both ends (e.g., *Marssonella*).

**Carinate** – possessing a keel or flange.

**Chamber** – in multilocular foraminifera, a test cavity with its surrounding wall formed by a single growth stage – usually connected to other chambers by pores, canals, etc.

**Chamberlet** – subdivision of main chamber.

**Concavo-convex** – in edge view, having one side of the test concave and the other convex.

**Conical** – test shaped like a cone, apex upwards.

**Costa(e)** – elevated ridge/rib(s) on test surface (a positive feature).

**Costate** – ornament of raised, straight, or gently curved ridges.

**Crenulate** – ornamentation consisting of finely scalloped or notched, wavy edged ridges.

**Cribrate** – perforated with round holes like a sieve.

**Crista** – crest.

**Cuneiform** – shaped like a wedge.

*Atlas of Benthic Foraminifera*, First Edition. Ann Holbourn, Andrew S. Henderson, and Norman MacLeod.

— D —

**Dendritic** – branched or tree-like test (e.g., *Rhizammina*).
**Denticulate** – having small, tooth-like projections.
**Diagenesis** – the chemical changes that accompany fossilization.
**Dimorphism** – two distinct forms – megalospheric and microspheric tests (formed from gamont and schizont generations, respectively).

— E —

**Entosolenian tube** – an internal tubular extension of the aperture (e.g., *Oolina*).
**Equatorial aperture** – symmetrical opening of an enrolled test, in the plane of coiling, commonly interiomarginal and against the periphery of the proceeding whorl but may be areal or peripheral.
**Evolute** – planispiral, uncoiled test with all chambers or whorls visible.
**Extraumbilical aperture** – sutural aperture midway between umbilicus and periphery, not connected with umbilicus.
**Extraumbilical–umbilical aperture** – aperture that extends from umbilicus forward toward the periphery of the test.

— F —

**Flabelliform** – fan-like test.
**Flange** – in edge view, plate-like peripheral extension of test (more pronounced than a keel).
**Fluted** – folded or corrugated.
**Folium** – acuminate umbilical extremity of the chamber that overlies a small cavity separate from the chamber lumen.
**Fusiform** – spindle-shaped planispiral test, tapering at poles.

— G —

**Glomospiral** – irregularly wound, coiled tubular chamber.
**Granular** – perforate lamellar test of minutely granular calcite, the equi-dimensional granules with c-axis preferentially orientated at an angle to the surface, neither parallel to nor perpendicular to the surface, resulting in a mottled appearance in crossed-polars of minute flecks of color.

— H —

**Hispid** – having very fine, hair-like spines.

— I —

**Imbricate flap** – overlapping flap.
**Imperforate** – wall of test without pores (e.g., miliolids and some agglutinating foraminifera).
**Interiomarginal aperture** – basal opening at margin of final chamber, along final suture.
**Involute** – planispiral test with strongly overlapping chambers, so that only the last whorl is seen.

— K —

**Keel** – in edge view, a peripheral non-perforate projection of the test (usually in ornate forms with limbate sutures).

— L —

**Lanceolate** – flat and narrow test, tapering to a point.
**Lateromarginal aperture** – primary aperture situated on lateral umbilical wall of coiled test, in subperipheral position.
**Lenticular** – test shaped like a double-convex lens in edge view (also called lenticuline, after *Lenticulina*).
**Limbate** – thickened, elevated border of chamber at the suture.
**Lip** – elevated border of aperture.
**Lobate/lobulate** – in inflated tests (in side view), chambers make lobe-like extensions of periphery, while sutures are indented.
**Lumen** – chamber cavity in the foraminiferal test.

— M —

**Megalospheric** – having large proloculus and representing the gamontic generation, but test usually smaller than schizont (microspheric) generation [in larger foraminifera often referred to as the A-form].
**Microgranular** – test composed of (secreted) minute calcite crystals, microscopically granulose (as found in the fusuline foraminifera).
**Microspheric** – having small proloculus and representing the schizont (or agamont) generation, but test much larger than megalospheric form [in larger foraminifera often referred to as the B-form].
**Milioline** – coiling with narrow "banana"-shaped chambers, two per whorl and added at different planes of coiling.

— P —

**Palmate** – flat test, like a hand with outspread fingers or a frond (e.g., *Frondicularia*).
**Papillate** – covered in papillae (small, nipple-like projections).
**Partitions** – subdivisions of chambers in various planes, typically in large conical foraminifera.
**Perforate** – wall of test pierced with numerous pores.
**Periphery** – extremity of test in edge view, sometimes keeled.
**Periumbilical extensions** – extensions around the umbilicus.
**Phialine** – apertural neck with pronounced rim, like the neck of a phial.

**Planispiral** – coiled in a single plane.

**Planoconvex** – in edge view, where one side of the trochospiral test is flat, the other convex.

**Ponticulus, pl. ponticuli** – bridge of lateral wall spanning an intraseptal interlocular space.

**Porcelaneous** –having a white, calcareous and (usually) imperforate wall resembling porcelain. Brown in transmitted light, becoming a characteristic dense black when diagenetically altered.

**Pore** – minute perforations through the test wall.

**Primary aperture** – main test opening – may be the only one, or can be accompanied by supplementary apertures.

**Produced** – not flush with chamber wall, but slightly protruding out.

**Proloculus** – the first chamber of the foraminiferal test.

**Pseudokeel** – a false keel (e.g., *Cibicidoides matanzasensis* is distinguished by its characteristic "pseudokeel" formed by circumperipheral extensions of the sutures).

**Pseudopodia** – projections of the cytoplasm which serve for locomotion, attachment, and for food capture.

**Punctate** – having small depressions or pits.

**Pustulate** – covered with small tubercles.

**Pyriform** – pear-shaped single-chambered test, like a Greek amphora (typified by *Lagena*).

## — Q —

**Quinqueloculine** – milioline coiling resulting in five visible chambers, with (in side view) four seen on one side and three on the other, as found in the intial chamber arrangement of *Sigmoilopsis schlumbergeri*.

## — R —

**Radiate** – aperture, as in *Lenticulina*, with slits arranged like the spokes of a wheel.

**Rectilinear** – growing in a straight line.

**Reniform** – kidney shaped.

**Reticulation** – network of ridges producing a honeycomb appearance.

## — S —

**Scroll-like** – form of internal apertural structure in *Patellina*, which looks like a rolled-up scroll.

**Septal flaps** – part of the inner lamella covering the preceding septal face.

**Septum(a)** – complete partition(s) between chambers.

**Sigmoidal** – s-shaped.

**Sinus** – indentation, embayment.

**Spine** – an elongate, solid structure on the surface of the test, comprising a single calcite crystal.

**Spinose** – test surface having fine elongate spines (usually preserved only as spine bases).

**Spiral side** – in trochospiral forms, the side of the test where all the whorls are visible. In theory, both sides of a trochospirally coiled test are spiral, but spiral is preferred to the term "dorsal", as the latter implies an orientation in life which is not always the case.

**Stellate** – star-shaped.

**Stolons** – tube-like projections interconnecting the chambers of some foraminifera (e.g., *Caudammina excelsa*).

**Streptospiral** – test coiled roughly like a ball of string.

**Striate** – ornamented by parallel grooves or lines (a negative feature, cf. costate).

**Sulcus** – peripheral infold of primary chamber wall, always imperforate.

**Supplementary apertures** – secondary openings in the test, additional to the primary aperture. In some foraminifera they can completely replace the primary aperture.

**Suture** – wall between two chambers or between two whorls (spiral suture).

## — T —

**Terminal aperture** – aperture positioned at the distal end of a chamber.

**Test** – shell of the foraminifera.

**Tooth** – inward projection of the inner portion of the chamber wall into the aperture; may be simple or complex, single or multiple.

**Toothplate (tooth plate)** – internal, apertural modification usually of the shape of a contorted plate, which extends from aperture to aperture.

**Tricarinate** – having three keels.

**Trilobate** – divided into three lobes.

**Triloculine** – milioline coiling resulting in three visible chambers, three on one side and two on the other, as found in *Triloculina*.

**Triserial** – usually a high trochospirally coiled test with three chambers per whorl, arranged in three columns.

**Trochospiral** – generally a low trochoid coil; chambers coiled spirally, evolute on one side (usually the spiral side) and involute on the other (usually the umbilical side).

## — U —

**Umbilical aperture** – primary aperture of a chamber leading into the umbilicus.

**Umbilical–extraumbilical aperture** – interiomarginal primary chamber aperture extending in enrolled species from umbilicus toward the periphery.

**Umbilical plug** – boss-like infilling of umbilicus (e.g., *Cibicidoides matanzasensis*), imperforate but often fissured.

**Umbilical side** – involute side in trochospiral foraminifera, with only chambers of the final whorl visible. Preferred to the term "ventral", which infers orientation in life.

**Umbilicus** – space at inner margins of chambers on umbilical side of test. May be filled with a plug or pillars, etc.

**Umbo** – central elevated area of lenticular or discoidal forms due to lamellar thickening on one or both sides of the test (e.g., *Lenticulina*).

**Umbonal boss** – elevated or knob-like secondary thickening of the umbonal region.

**Unilocular** – a single-chambered test (e.g., *Lagena*).

**Uniserial** – chambers arranged in a single row.

## — W —

**Whorl** – a single turn (of 360°) of a coiled test.

# References

Akpati, B.N., 1966, Some Lower Paleocene foraminifera from the Anita Formation: Santa Barbara County, California. Contributions from the Cushman Laboratory for Foraminiferal Research, vol. 17, no. 4, pp. 135–139.

Altenbach, A.V., Pflaumann, U., Schiebel, R., Thies, A., Timm, S., and Trauth, M., 1999, Scaling percentages and distributional patterns of benthic foraminifera with flux rates of organic carbon. Journal of Foraminiferal Research, vol. 29, pp. 173–185.

Asano, K., 1953, Miocene foraminifera from the Noto Peninsula, Ishikawa Prefecture. Short Papers, Institute of Geology and Paleontology, Tohoku University, vol. 5, pp. 1–21.

Bailey, J.W., 1851, Microscopical examination of soundings made by the United States Coast Survey, off the Atlantic Coast of the United States. Smithsonian Contributions to Knowledge, vol. 2, pp. 1–15.

Bailey, J.W., 1861, Notes on new species of microscopical organisms, chiefly from the Para River, South America. Boston Journal of Natural Histroy, vol. 7, pp. 329–352.

Bandy, O.L., 1944, Eocene foraminifera from Cape Blanco, Oregon. Journal of Paleontology, vol. 18, no. 4, pp. 366–377.

Bandy, O.L., 1949, Eocene and Oligocene foraminifera from Little Stave Creek, Clarke County: Alabama. Bulletins of American Paleontology, vol. 32, pp. 5–206.

Bandy, O.L., 1961, Distribution of foraminifera, radiolaria and diatoms in sediments from the Gulf of California. Micropaleontology, vol. 7, pp. 1–21.

Bandy, O.L. and Chierici, M.A., 1966, Depth–temperature evaluation of selected Californian and Mediterranean bathyal foraminifera. Marine Geology, vol. 4, pp. 259–271.

Banner, F.T., 1966, The morphology, classification and stratigraphic value of the Spirocyclinidae. Voprosy Mikropaleontologii, vol. 10, pp. 210–224 [in Russian].

Barker, R.W., 1960, Taxonomic notes on the species figured by H.B. Brady in his report on the foraminifera dredged by H.M.S. Challenger during the years 1873–1876. Society of Economic Paleontologists and Mineralogists Special Publication, vol. 9, pp. 1–238.

Bartenstein, H., 1952, Taxonomische Bemerkungen zu den Ammobaculites, Haplophragmium, Lituola und verwandten Gattungen. (For.) Senckenbergiana, vol. 33, pp. 313–342.

Bartenstein, H., 1954, Revision von Berthelin's Mémoire 1880 über die Alb-Foraminiferen von Montcley. Senckenbergiana Lethaea, vol. 35, no. 1/2, pp. 37–50.

Bartenstein, H., 1974, Upper Jurassic–Lower Cretaceous primitive arenaceous foraminifera from DSDP Sites 259 and 261, Eastern Indian Ocean. In Veevers, J.J., Heirtzler, J.R., et al., ed., Initial Repoprts of the Deep Sea Drilling Project, Washington (U.S. Government Printing Office), vol. 27, pp. 683–695.

Bartenstein, H. and Bolli, H.M., 1986, The foraminifera of the Lower Cretaceous of Trinidad, W.I. Part 5: Maridale Formation, upper part; Hedbergella rohri zone. Eclogae Geologicae Helvetiae, vol. 7, no. 3, pp. 945–999.

Bartenstein, H. and Brand, E., 1951, Mikropalaeontologische Untersuchungen zur Stratigraphie des nordwest-deutschen. Valendis Abhandlungen der Senckenbergischen Naturforschenden Gesellschaft, vol. 485, pp. 239–336.

Batsch, A.I.G.C., 1791, Sechs Kupfertafeln mit Conchylien des Seesandes, gezeichnet und gestochen. Jena: von A.J.G.K. Batsch, 6 pls.

Beckmann, J.P., 1953, Die Foraminiferen der Oceanic Formation (Eocaen–Oligocaen) von Barbados, Kl. Antillen. Eclogae Geologicae Helvetiae, vol. 46, pp. 301–412.

Beckmann, J.P., 1960, Distribution of benthonic foraminifera at the Cretaceous–Tertiary boundary of Trinidad (West Indies). Report of the International Geological Congress, 21st Session, Norden, Copenhagen, 1960, part 5, The Cretaceous–Tertiary Boundary, pp. 57–69.

Belford, D.J., 1960, Upper Cretaceous foraminifera from the Toolonga Calcilutite and Gingin Chalk, Western Australia. Bulletin of the Bureau of Mineral Resources, Geology and Geophysics, Australia, vol. 57, pp. 1–198.

Belford, D.J., 1966, Miocene and Pliocene smaller foraminifera from Papua and New Guinea. Australia Bureau of Mineral Resources Geology and Geophysics Bulletin, vol. 79, pp. 1–306.

*Atlas of Benthic Foraminifera*, First Edition. Ann Holbourn, Andrew S. Henderson, and Norman MacLeod.

Berggren, W.A. and Aubert, J., 1976, Eocene benthonic foraminiferal biostratigraphy and paleobathymetry of Orphan Knoll (Labrador Sea). Micropaleontology, vol. 22, no. 3, pp. 327–346.

Berggren, W.A. and Aubert, J., 1983, Paleogene benthonic foraminiferal biostratigraphy and bathymetry of the Central Coast Ranges of California. In Brabb, E.E., ed., Studies in Tertiary Stratigraphy of the California Coast Ranges. Professional Paper of the United States Geological Survey, vol. 1213, pp. 4–21.

Berggren, W.A. and Kaminski, M.A., 1990, Abyssal agglutinates: back to basics. In Hemleben, C., Kaminski, M.A., Kuhnt, W., and Scott, D.B., ed., Paleoecology, Biostratigraphy, Paleoceanography and Taxonomy of Agglutinated Foraminifera. NATO ASI Series, Series C: Mathematical and Physical Sciences, vol. 327. Dordrecht: Kluwer Academic, pp. 53–76.

Berggren, W.A., and Miller, K.G., 1989, Cenozoic bathyal and abyssal calcareous benthic foraminiferal zonations. Micropaleontology, vol. 35, pp. 308–320.

Bermúdez, P.J., 1937, Nuevas especies de foraminiferos del Eoceno de las cercanias de Guanajay, Provincia Pinar del Rio, Memorias de la Sociedad Cubana de Historia Natural "Felipe Poey", vol. 11, pp. 237–247.

Bermúdez, P.J., 1937, Nuevas especies de foraminiferos del Eoceno de Cuba. Memorias de la Sociedad Cubana de Historia Natural "Felipe Poey", vol. 11, pp. 137–150.

Bermúdez, P.J., 1949, Tertiary Smaller Foraminifera of the Dominican Republic. Special Publications of the Cushman Laboratory for Foraminiferal Research, no. 25, pp. 1–322.

Berthelin, G., 1880, Mémoire sur les foraminifères fossiles de l'étage albien de Montcley (Doubs). Mémoires de la Société géologique de France, ser. 31, no. 5, pp. 1–84.

Bock, W.D., 1971, A handbook of benthonic foraminifera of Florida Bay and adjacent waters. In Jones, J.I. and Bock, W.D., ed., A Symposium of Recent South Florida Foraminifera. Memoir of the Miami Geological Society, vol. 1, pp. 1–72.

Boersma, A., 1984, Handbook of Common Tertiary Uvigerina. Stony Point, NY: Microclimates Press, 207 pp.

Boersma, A., 1990, Late Oligocene to late Pliocene benthic foraminifers from depth traverses in the central Indian Ocean. In Duncan, R.A., Backman, J., Peterson, L.C. et al., ed., Scientific Results of the Proceedings of the Ocean Drilling Program, vol. 115, pp. 315–379.

Bolli, H.M., Beckmann, J.P., and Saunders, J.B., 1994, Benthic Foraminiferal Biostratigraphy of the South Caribbean Region. Cambridge: Cambridge University Press, 408 pp.

Boltovskoy, E., 1978, Late Cenozoic benthonic foraminifera of the Ninetyeast Ridge (Indian Ocean). Marine Geology, vol. 26, pp. 139–175.

Boltovskoy, E. and Watanabe, S., 1982, Foraminiferos bentónicos batiales del Cenozoico del Atlántico sudoriental. Physis (Buenos Aires), Secc. A, vol. 41, no. 100, pp. 29–41.

Bornemann, J.G., 1855, Die mikroskopische Fauna des Septarienthones von Hermsdorf bei Berlin. Zeitschrift der Deutschen Geologischen Gesellschaft, vol. 7, pp. 307–371.

Brady, H.B., 1877, Supplementary note on the foraminifera of the Chalk(?) of the New Britain Group. Geological Magazine (New Series), vol. 4, pp. 534–536.

Brady, H.B., 1878, On the Reticularian and Radiolarian Rhizopoda (Foraminifera and Polycystina) of the North Polar Expedition of 1875–76. Annals and Magazine of Natural History, ser. 5, no. 1, pp. 425–440.

Brady, H.B., 1879, Notes on some of the reticularian Rhizopoda of the Challenger Expedition, Part I. On new or little known Arenaceous types. Quarterly Journal of Microscopical Science (New Series), vol. 19, pp. 20–63.

Brady, H.B., 1881, Notes on some of the reticularian Rhizopoda of the Challenger Expedition, Part III. Quarterly Journal of Microscopical Science (New Series), vol. 21, pp. 31–71.

Brady, H.B., 1884, Report on the foraminifera dredged by H.M.S. Challenger during the years 1873–1876. In Brady, H.B., ed., Report on the Scientific Results of the Voyage of the H.M.S. Challenger during the years 1873–1876. Zoology, vol. 9, pp. 1–814.

Brady, H.B., Parker, W.K., and Jones, T.R., 1888, On some foraminifera from the Abrolhos Bank. Transactions of the Zoological Society of London, vol. 12 (1890), pp. 211–239.

Brönnimann, P. and Whittaker, J.E., 1984, A neotype for Trochammina inflata (Montagu) (Protozoa, Foraminifera) with notes on the wall structure. Bulletin of the British Museum (Natural History), (Zoology Series), vol. 46, pp. 311–315.

Brönnimann, P. and Whittaker, J.E., 1988, The Trochamminacea of the Discovery Reports. A review of the Trochamminacea (Protozoa: Foraminiferida) described from South Atlantic and Antarctic waters by Heron-Allen and Earland (1932) and Earland (1933; 1934; 1936). London: British Museum (Natural History), 152 pp.

Brönnimann, P. and Whittaker, J.E., 1988, The trochamminaceous test and the taxonomic criteria used in the classification of the Superfamily Trochamminacea. Abhandlungen der Geologischen Bundesanstalt, Wien, vol. 41, pp. 23–39.

Brotzen, F., 1936. Foraminiferen aus dem schwedischen untersten Senon von Eriksdal in Schonen. Årsbok Sveriges geologiska Undersökning, vol. 30, no. 3, pp. 1–206.

Brotzen, F., 1940, Flintraennans och Trindelraennans geologi (Oeresund). (Die Geologie der Flint- und Trindelrinne (Oeresund)). Årsbok Sveriges Geologiska Undersökning, vol. 34, no. 5, pp. 1–33.

Brotzen, F., 1945, De geologiska resultaten från borrningarna vid Hoellviken: Preliminar rapport – Del 1: Kritan. Årsbok Sveriges Geologiska Undersökning, vol. 38 (1944), no. 7, pp. 3–64.

Brotzen, F., 1948, The Swedish Paleocene and its Foraminiferal Fauna. Årsbok Sveriges Geologiska Undersökning, vol. 42, no. 2, pp. 1–140.

Bubik, M., 1995, Cretaceous to Paleogene agglutinated foraminifera from the Bile Karpaty unit. In Kaminski, M.A., Geroch, S., and Gasinski, M.A., ed., Proceedings of the Fourth International Workshop on Agglutinated Foraminifera, Grzybowski Foundation Special Publication, no. 3, pp. 71–116.

Bubik, M. and Kaminski, M.A., 2000, Remarks on the type locality and current status of the foraminiferal species Rzehakina epigona (Rzehak, 1895). In Hart, M.B., Kaminski, M.A., and Smart, C.W., ed., Proceedings of the Fifth International Workshop on Agglutinated Foraminifera. Grzybowski Special Publication, vol. 7, pp. 71–80.

Burke, S.C., 1981, Recent benthic foraminifera of the Ontong Java Plateau. Journal of Foraminiferal Research, vol. 11, pp. 1–19.

Carpenter, W.B., Parker, W.K., and Jones, J.P., 1862, Introduction to the Study of the Foraminifera. London: Ray Society, 319 pp.

Chapman, F., 1941, Report on the foraminiferal soundings and dredgings of the F.I.S. "Endeavour" along the continental shelf off the south-east coast of Australia. Transactions of the Royal Society of South Australia, vol. 65, no. 2, pp. 145–211.

Chapman, F. and Parr, W.J., 1935, Foraminifera and Ostracoda from soundings made by the trawler "Bonthorpe" in the Great Australian Bight. Journal of the Royal Society of Western Australia, vol. 21, pp. 1–7.

Chapman, F. and Parr, W.J., 1937, Foraminifera. In Johnston, T.H., ed., Australasian Antarctic Expedition 1911–1914, Scientific Reports, Series C (Zoology and Botany). Sydney: David Harold Paisley, vol. 1, pp. 1–190.

Charnock, M.A. and Jones, R.W., 1990, Agglutinated foraminifera from the Palaeogene of the North Sea. In Hemleben, C., Kaminski, M.A., Kuhnt, W., and Scott, D.B., ed., Paleoecology, Biostratigraphy, Paleoceanography and Taxonomy of Agglutinated Foraminifera, NATO ASI Series, C: Mathematical and Physical Sciences. Dordrecht: Kluwer Academic, pp. 139–244.

Cita, M.B., 1950, L'Eocene della sponda occidentale del Lago di Garda; Parte 1 – Piccoli foraminiferi. Rivista Italiana Paleontologia e Stratigrafia, vol. 56, no. 3, pp. 81–113.

Cole, W.S., 1928, A foraminiferal fauna from the Chapapote formation in Mexico. Bulletin of American Paleontology, vol. 14, pp. 200–232.

Colom, G., 1943, Nuevas especies y subespecies de foraminiferos fosiles de Mallorca. Boletín de la Real Sociedad Española de Historia Natural, vol. 41, pp. 317–335.

Colom, G., 1952, Foraminiferos de las costas de Galicia (campañas del "Xauen" en 1949 y 1950). Boletín del Instituto Español de Oceanografía, vol. 51, pp. 3–59.

Corliss, B.H., 1979, Recent deep-sea benthonic foraminiferal distributions in the southeast Indian Ocean: inferred bottom water routes and ecological implications. Marine Geology, vol. 31, pp. 115–138.

Corliss, B.H., 1979, Taxonomy of Recent deep-sea benthonic foraminifera from the Southeast Indian Ocean. Micropaleontology, vol. 25, pp. 1–19.

Corliss, B.H., 1985, Microhabitats of benthic foraminifera within deep-sea sediments. Nature, vol. 314, pp. 435–438.

Coryell, H.N. and Rivero, F.C., 1940, A Miocene microfauna of Haiti. Journal of Paleontology, vol. 14, pp. 324–344.

Crespin, I., 1963, Lower Cretaceous Arenaceous Foraminifera of Australia. Bulletin of the Bureau of Mineral Resources, Geology and Geophysics, Australia, vol. 66, pp. 1–110.

Crouch, R.W. and Poag, C.W., 1979, Amphistegina gibbosa d'Orbigny from the California borderlands. Journal of Foraminiferal Research, vol. 9, no. 2, pp. 85–105.

Cushman, J.A., 1910, A monograph of the Foraminifera of the North Pacific Ocean. Part 1. Astrorhizidae and Lituolidae. Bulletin of the United States National Museum, vol. 71, no. 1, pp. 1–134.

Cushman, J.A., 1911, A monograph of the Foraminifera of the North Pacific Ocean. Part II. Textulariidae. Bulletin of the United States National Museum, vol. 71, no. 2, pp. 1–108.

Cushman, J.A., 1913, A monograph of the Foraminifera of the North Pacific Ocean. Part 3. Lagenidae. Bulletin of the United States National Museum, vol. 71, no. 3, pp. 1–125.

Cushman, J.A., 1918, The smaller fossil foraminifera of the Panama Canal Biozone. Bulletin of the United States National Museum, vol. 103, pp. 45–87.

Cushman, J.A., 1920, The foraminifera of the Atlantic Ocean. Part 2. Lituolidae. Bulletin United States National Museum, vol. 104, no. 2, pp. 1–111.

Cushman, J.A., 1921, Foraminifera of the Philippine and adjacent seas. Bulletin of the United States National Museum, vol. 100, pp. 1–608.

Cushman, J.A., 1922, Shallow-water foraminifera of the Tortugas region. Publications of the Carnegie Institution of Washington, no. 311, Department of Marine Biology, vol. 17, 85 pp.

Cushman, J.A., 1922, The foraminifera of the Atlantic Ocean. Part 3. Textulariidae. Bulletin of the United States National Museum, vol. 104, pp. 1–149.

Cushman, J.A., 1923, The foraminifera of the Atlantic Ocean. Part 4. Lagenidae. Bulletin of the United States National Museum (US Nat. Mus. Bull.), vol. 104, no. 4, pp. 1–228.

Cushman, J.A., 1923, The foraminifera of the Vicksburg Group. Professional Paper of the United States Geological Survey, vol. 133, pp. 11–71.

Cushman, J.A., 1925, New foraminifera from the upper Eocene of Mexico. Contributions from the Cushman Laboratory for Foraminiferal Research, vol. 1, no. 1, pp. 4–9.

Cushman, J.A., 1925, Some new foraminifera from the Velasco Shale of Mexico. Contributions from the Cushman Laboratory for Foraminiferal Research, vol. 1, no. 1, pp. 18–23.

Cushman, J.A., 1926, Some foraminifera from the Mendez Shale of eastern Mexico. Contributions from the Cushman Laboratory for Foraminiferal Research, vol. 2, no. 1, pp. 16–26.

Cushman, J.A., 1926, Some fossil Bolivinas from Mexico. Contributions from the Cushman Laboratory for Foraminiferal Research, vol. 1, pp. 81–85.

Cushman, J.A., 1926, Some new foraminifera from the Upper Eocene of the southeastern Coastal Plain of the United States. Contributions from the Cushman Laboratory for Foraminiferal Research, vol. 2, no. 2, pp. 29–38.

Cushman, J.A., 1926, The foraminifera of the Velasco Shale of the Tampico Embayment. Bulletin of the American Association of Petroleum Geologists, vol. 10, no. 6, pp. 581–612.

Cushman, J.A., 1926. Foraminifera of the genera Siphogenerina and Pavonina. Proceedings of the United States National Museum, vol. 67, no. 25, pp. 1–24.

Cushman, J.A., 1927, American Upper Cretaceous species of Bolivina and related species. Contributions from the Cushman Laboratory for Foraminiferal Research, vol. 2, no. 4, pp. 85–91.

Cushman, J.A., 1927, An outline of a re-classification of the foraminifera. Contributions from the Cushman Laboratory for Foraminiferal Research, vol. 31, pp. 1–105.

Cushman, J.A., 1927, New and interesting foraminifera from Mexico and Texas. Contributions from the Cushman Laboratory for Foraminiferal Research, vol. 3, pp. 111–117.

Cushman, J.A., 1927, Recent foraminifera from off the West coast of America. Bulletin of the Scripps Institute of Oceanography, Technical Series, vol. 1, pp. 119–188.

Cushman, J.A., 1927, Some characteristic Mexican fossil foraminifera. Journal of Paleontology, vol. 1, pp. 147–172.

Cushman, J.A., 1930, The foraminifera of the Atlantic Ocean, Part 7. Nonionidae, Camerinidae, Peneroplidae and Alveolinellidae. Bulletin of the United States National Museum, vol. 104, no. 7, pp. 1–79.

Cushman, J.A., 1931, Notes on the Foraminifera described by Batsch in 1791. Contributions from the Cushman Laboratory for Foraminiferal Research, vol. 7, pp. 62–72.

Cushman, J.A., 1931, The foraminifera of the Atlantic Ocean; Part VIII – Rotaliidae, Amphisteginidae, Calcarinidae, Cymbaloporettidae, Globorotaliidae, Anomalinidae, Planorbulinidae, Rupertiidae and Homotremidae. Bulletin of the United States National Museum, vol. 104, no. 8, pp. 1–179.

Cushman, J.A., 1931, The Foraminifera of the Saratoga Chalk. Journal of Paleontology, vol. 5, pp. 297–315.

Cushman, J.A., 1932, The genus Vulvulina and its species. Contributions from the Cushman Laboratory for Foraminiferal Research, vol. 8, pp. 75–85.

Cushman, J.A., 1933, New American Cretaceous Foraminifera. Contributions from the Cushman Laboratory for Foraminiferal Research, vol. 9, pp. 49–64.

Cushman, J.A., 1933, Some new foraminiferal genera. Contributions from the Cushman Laboratory for Foraminiferal Research, vol. 9, pp. 32–38.

Cushman, J.A., 1933, Some new Recent foraminifera from the tropical Pacific. Contributions from the Cushman Laboratory for Foraminiferal Research, vol. 9, pp. 77–95.

Cushman, J.A., 1935, Notes on some American Cretaceous Flabellinas. Contributions from the Cushman Laboratory for Foraminiferal Research, vol. 11, no. 4, pp. 83–89.

Cushman, J.A., 1936, Cretaceous foraminifera of the family Chilostomellidae. Contributions from the Cushman Laboratory for Foraminiferal Research, vol. 12, no. 4, pp. 1–77.

Cushman, J.A., 1936, New genera and species of the families Verneuilinidae and Valvulinidae and of the subfamily Virgulininae. Contributions from the Cushman Laboratory for Foraminiferal Research, vol. 6, pp. 1–71.

Cushman, J.A., 1936, Some American Cretaceous species of Ellipsonodosaria and Chrysalogonium. Contributions from the Cushman Laboratory for Foraminiferal Research, vol. 12, no. 3, pp. 51–55.

Cushman, J.A., 1937, A few new species of American Cretaceous foraminifera. Contributions from the Cushman Laboratory for Foraminiferal Research, vol. 13, pp. 100–105.

Cushman, J.A., 1937, A monograph of the subfamily Virgulininae of the foraminiferal family Buliminidae. Special Publications of the Cushman Laboratory for Foraminiferal Research, vol. 9, pp. 1–228.

Cushman, J.A., 1938, Additional new species of American Cretaceous foraminifera. Contributions from the Cushman Laboratory for Foraminiferal Research, vol. 14, pp. 31–50.

Cushman, J.A. 1939, New American Cretaceous Foraminifera. Contributions from the Cushman Laboratory for Foraminiferal Research, vol. 15, no. 4, pp. 89–93.

Cushman, J.A., 1940, Midway foraminifera from Alabama. Contributions from the Cushman Laboratory for Foraminiferal Research, vol. 16, no. 3, pp. 51–73.

Cushman, J.A., 1943, Some new foraminifera from the Tertiary of the Island of St. Croix. Contributions from the Cushman Laboratory for Foraminiferal Research, vol. 19, pp. 90–93.

Cushman, J.A., 1946, Upper Cretaceous foraminifera of the Gulf coastal region of the United States and adjacent areas. Professional Paper of the United States Geological Survey, vol. 206, 241 pp.

Cushman, J.A. and Bermúdez, P.J., 1936, New genera and species of foraminifera from the Eocene of Cuba. Contributions from the Cushman Laboratory for Foraminiferal Research, vol. 12, no. 2, pp. 27–38.

Cushman, J.A. and Bermúdez, P.J., 1936, Additional new species of foraminifera and a new genus from the Eocene of Cuba. Contributions from the Cushman Laboratory for Foraminiferal Research, vol. 12, no. 3, pp. 55–63.

Cushman, J.A. and Bermúdez, P.J., 1937, Further new species of foraminifera from the Eocene of Cuba. Contributions from the Cushman Laboratory for Foraminiferal Research, vol. 13, pp. 1–29.

Cushman, J.A. and Campbell, A.S., 1934, A new Spiroplectoides from the Cretaceous of California. Contributions from the Cushman Laboratory for Foraminiferal Research, vol. 10, no. 3, p. 70.

Cushman, J.A. and Church, C.C., 1929, Some Upper Cretaceous foraminifera from near Coalinga, California. Proceedings of the California Academy of (Natural) Sciences, ser. 4, vol. 18, no. 16, pp. 497–530.

Cushman, J.A. and Dorsey, A.L., 1940, The genus Stensiöiona and its species. Contributions from the Cushman Laboratory for Foraminiferal Research, vol. 16, pp. 1–6.

Cushman, J.A. and Edwards, P.G., 1938, Notes on the Oligocene species of Uvigerina and Angulogerina. Contributions from the Cushman Laboratory for Foraminiferal Research, vol. 14, pp. 74–89.

Cushman, J.A. and Garret, J.B., 1938, Three new Rotaliform foraminifera from the Lower Oligocene and Upper Eocene of Alabama. Contributions from the Cushman Laboratory for Foraminiferal Research, vol. 14, no. 3, pp. 62–66.

Cushman, J.A. and Goudkoff, P.P., 1944, Some foraminifera from the upper Cretaceous of California. Contributions from the Cushman Laboratory for Foraminiferal Research, vol. 20, pp. 53–64.

Cushman, J.A. and Grant, U.S., 1927, Late Tertiary and Quaternary Elphidium of the West Coast of North America. Transactions of the San Diego Society of Natural History, vol. 5, pp. 69–82.

Cushman, J.A. and Hedberg, H.D., 1941, Upper Cretaceous foraminifera from Santander del Norte, Colombia. Contributions from the Cushman Laboratory for Foraminiferal Research, vol. 17, no. 4, pp. 79–102.

Cushman, J.A. and Jarvis, P.W., 1928, Cretaceous foraminifera from Trinidad. Contributions from the Cushman Laboratory for Foraminiferal Research, vol. 44, pp. 85–103.

Cushman, J.A. and Jarvis, P.W., 1929, New foraminifera from Trinidad. Contributions from the Cushman Laboratory for Foraminiferal Research, vol. 5, pp. 6–17.

Cushman, J.A. and Jarvis, P.W., 1930, Miocene foraminifera from Buff Bay, Jamaica. Journal of Paleontology, vol. 4, no. 4, pp. 353–368.

Cushman, J.A. and Jarvis, P.W., 1932, Upper Cretaceous foraminifera from Trinidad. Proceedings of the United States National Museum, vol. 80, no. 14, pp. 1–60.

Cushman, J.A. and Jarvis, P.W., 1934, Some interesting new uniserial foraminifera from Trinidad. Contributions from the Cushman Laboratory for Foraminiferal Research, vol. 10, pp. 71–75.

Cushman, J.A. and Kleinpell, R.M., 1934, New and unrecorded foraminifera from the California Miocene. Contributions from the Cushman Laboratory for Foraminiferal Research, vol. 10, pp. 1–23.

Cushman, J.A. and Moyer, D.A., 1930, Some Recent foraminifera from off San Pedro. Contributions from the Cushman Laboratory for Foraminiferal Research, vol. 13, pp. 36–40.

Cushman, J.A. and Parker, F.L., 1935, Some American Cretaceous Buliminas. Contributions from the Cushman Laboratory for Foraminiferal Research, vol. 11, no. 4, pp. 96–101.

Cushman, J.A. and Parker, F.L., 1936, Some American Eocene Buliminas. Contributions from the Cushman Laboratory for Foraminiferal Research, vol. 12, no. 2, pp. 39–45.

Cushman, J.A. and Parker, F.L., 1939, Bulimina macilenta Cushman and Parker a new name. Contributions from the Cushman Laboratory for Foraminiferal Research, vol. 15, no. 4, p. 93.

Cushman, J.A. and Parker, F.L., 1940, The species of the genus Bulimina having Recent types. Contributions from the Cushman Laboratory for Foraminiferal Research, vol. 16, no. 1, pp. 7–23.

Cushman, J.A. and Parker, F.L., 1947, Bulimina and related foraminiferal genera. Professional Paper of the United States Geological Survey, vol. 210-D, pp. 55–176.

Cushman, J.A. and Renz, H.H., 1941, New Oligocene–Miocene foraminifera from Venezuela. Contributions from the Cushman Laboratory for Foraminiferal Research, vol. 17, pp. 1–27.

Cushman, J.A. and Renz, H.H., 1946, The foraminiferal fauna of the Lizard Springs Formation of Trinidad, British West Indies. Special Publications of the Cushman Laboratory for Foraminiferal Research, vol. 18, pp. 1–48.

Cushman, J.A. and Renz, H.H., 1947, Further notes on the Cretaceous foraminifera of Trinidad. Contributions from the Cushman Laboratory for Foraminiferal Research, vol. 23, pp. 31–51.

Cushman, J.A. and Renz, H.H., 1948, Eocene foraminifera of the Navet and Hospital Hill formations of Trinidad, BWI. Special Publications of the Cushman Laboratory for Foraminiferal Research, vol. 24, pp. 1–42.

Cushman, J.A. and Siegfus, S.S., 1935, New species of foraminifera from the Kreyenhagen Shale of Fresno County, California. Contributions from the Cushman Laboratory for Foraminiferal Research, vol. 11, no. 4, pp. 90–95.

Cushman, J.A. and Siegfus, S.S., 1939, Some new and interesting foraminifera from the Kreyenhagen Shale of California. Contributions from the Cushman Laboratory for Foraminiferal Research, vol. 15, no. 2, pp. 1–26.

Cushman, J.A. and Stainforth, R.M., 1945, The foraminifera of the Cipero Marl Formation of Trinidad, British West Indies. Contributions from the Cushman Laboratory for Foraminiferal Research, vol. 14, pp. 3–75.

Cushman, J.A. and Stainforth, R.M., 1951, Tertiary foraminifera of coastal Ecuador – Part 1, Eocene. Journal of Paleontology, vol. 25, no. 2, pp. 129–164.

Cushman, J.A. and Todd, R., 1942, The genus Cancris and its species. Contributions from the Cushman Laboratory for Foraminiferal Research, vol. 18, no. 4, pp. 72–94.

Cushman, J.A. and Todd, R., 1942, The Recent and fossil species of Laticarinina. Contributions from the Cushman Laboratory for Foraminiferal Research, vol. 18, no. 1, pp. 14–20.

Cushman, J.A. and Todd, R., 1943, The genus Pullenia and its species. Contributions from the Cushman Laboratory for Foraminiferal Research, vol. 19, pp. 1–24.

Cushman, J.A. and Todd, R., 1945, Miocene foraminifera from Buff Bay, Jamaica. Special Publications of the Cushman Laboratory for Foraminiferal Research, vol. 15, pp. 1–73.

Cushman, J.A. and Todd, R., 1946, A foraminiferal fauna from the Paleocene of Arkansas. Contributions from the Cushman Laboratory for Foraminiferal Research, vol. 22, no. 2, pp. 45–65.

Cushman, J.A. and Todd, R., 1949, The genus Sphaeroidina and its species. Special Publications of the Cushman Laboratory for Foraminiferal Research, vol. 25, pp. 11–21.

Cushman, J.A. and Waters, J.A., 1927, Some arenaceous foraminifera from the Upper Cretaceous of Texas. Contributions from the Cushman Laboratory for Foraminiferal Research, vol. 2, pp. 81–85.

Cushman, J.A., Stewart, R.E., and Stewart, K.C., 1930, Tertiary Foraminifera from Humboldt County, California. Transactions of the San Diego Society for Natural History, vol. 6, pp. 41–94.

Czjzek, J., 1848, Beitrag zur Kenntniss der fossilen Foraminiferen des Wiener Beckens. Naturwissenschaftliche Abhadlungen, Wien, vol. 2, no. 1, pp. 137–150.

d'Orbigny, A., 1826, Tableau méthodique de la classe des Céphalopodes. Annales des Sciences Naturelles, vol. 7, pp. 245–314.

d'Orbigny, A.D., 1839, Foraminifères des Îles Canaries. In Barker-Webb, P. and Berthelot, S., ed., Historie Naturelle des Îles Canaries, vol. 2, part 2: Paris (Béthune), pp. 119–146.

d'Orbigny, A.D., 1839, Foraminifères. In Sagra, R. de la, ed., Histoire physique, politique et naturelle de l'île de Cuba. Paris: A. Bertrand, pp. 1–224, 12 pls.

d'Orbigny, A.D., 1839, Voyage dans l'Amérique méridionale; foraminifères, vol. 5, pt 5. Paris and Strasbourg: P. Bertrand, 86 pp.

d'Orbigny, A.D., 1846, Foraminifères fossiles du Bassin Tertiaire de Vienne (Autriche). Paris: Gide et Compe, xxxvii + 312 pp.

Dawson, J.W., Notice of Tertiary fossils from Labrador, Maine, etc., and remarks on the climate of Canada in the newer Pliocene or Pleistocene period. Canadian Naturalist, vol. 5, pp. 188–200.

de Klasz, I., 1953, Quadratobuliminella n. gen., eine neue Foraminiferengattung von der Wende Kreide-Tertiaer. Neues Jahrbuch für Geologie und Paläontologie, Monatshefte, vol. 10, pp. 434–436.

de Montfort, P.D., 1808, Conchyliologie systématique et classification méthodique des coquilles. Paris: F. Schoell, vol. 1, lxxxvii + 409 pp.

den Dulk, M., Reichart, G.J., van Heyst, S., Zachariasse, W.J., and van der Zwaan, 2000, Benthic foraminifera as proxies of organic matter flux and bottom water oxygenation? A case history from the northern Arabian Sea. Palaeogeography, Palaeoclimatology, Palaeoecology, vol. 161, pp. 337–359.

Desai, D. and Banner, F.T., 1987, The evolution of Early Cretaceous Dorothiinae (Foraminiferid). Journal of Micropalaeontology, vol. 6, no. 2, pp. 13–27.

Douglas, R.G., 1973, Benthonic foraminiferal biostratigraphy in the central North Pacific, Leg 17, Deep Sea Drilling Project. In Winterer, E.L., Ewing, J.I., et al., ed., Initial Reports of the Deep Sea Drilling Project, Washington (U.S. Government Printing Office), vol. 17, pp. 607–671.

Dylàzanka, M., 1923, Warstwy inoceramowe z lomu w Szymbarku kolo Gorlic. Rocznik Polskiego Towarzystwa Geologicznego w Krakowie, vol. 1, pp. 36–81.

Earland, A., 1933, Foraminifera. Part II. South Georgia Discovery Reports, vol. 7, pp. 27–138.

Earland, A., 1934, Foraminifera, Part III: the Falklands sector of the Antarctic (excluding South Georgia). Discovery Reports, vol. 10, pp. 1–208.

Egger, J.G., 1893, Foraminiferen aus Meeresgrundproben, gelothet von 1874 bis 1876 von S.M. Sch. Gazelle. Abhandlungen der Bayerischen Akademie der Wissenschaften, München, Math.-Phys. Cl., vol. 18, no. 2, pp. 193–458.

Ernst, S.R., Guasti, E., Dupuis, C., and Speijer, R.P., 2006, Environmental perturbation in the southern Tethys across the Paleocene/Eocene boundary (Dababiya, Egypt): Foraminiferal and clay mineral records. Marine Micropaleontology, vol. 60, pp. 89–111.

Escutia, C., Brinkhuis, H., Klaus, A., and the Expedition 318 Scientists, 2011, Proc. IODP, 318: Tokyo (Integrated Ocean Drilling Program Management International, Inc.). doi:10.2204/iodp.proc.318.104.2011.

Feyling-Hanssen, R.W. and Buzas, M.A., 1976, Emendation of Cassidulina and Islandiella helenae new species. Journal of Foraminiferal Research, vol. 6, pp. 154–158.

Fichtel, L. von and Moll, J.P.C. von, 1798, Testacea microscopica aliaque minuta ex generibus Argonauta et Nautilus ad naturam delineata et descripta. Microscopische und andere kleine Schalthiere aus den Geschlechtern Argonaute und Schiffer, nach der Natur gezeichnet und beschrieben. Vienna: Camesina, 124 pp.

Finlay, H.J., 1939, New Zealand foraminifera: key species in stratigraphy – no. 1. Transactions of the Royal Society of New Zealand, vol. 68, no. 4, pp. 504–543.

Finlay, H.J., 1940, New Zealand Foraminifera; key species in stratigraphy – No. 4. Transactions of the Royal Society of New Zealand, vol. 69, no. 4, pp. 448–472.

Finlay, H.J., 1947, New Zealand Foraminifera; Key species in stratigraphy – No. 5. New Zealand Journal of Science and Technology, vol. 28, no. 5 (sec. B), pp. 259–292.

Fisher, M.J., 1969, Benthonic foraminifera from the Maastrichtian chalk of Galicia Bank, west of Spain. Palaeontology, vol. 12, no. 2, pp. 189–200.

Flint, J.M., 1899, Recent foraminifera. A descriptive catalogue of specimens dredged by the US Fish Commission Steamer Albatross. Report of the United States National Museum for 1897, pt 1, pp. 249–349.

Fornasini, C., 1901, Le Bulimine e le Cassiduline fossili d'Italia. Società Geologica Italiana, Bollettino, Rome, vol. 20, pp. 159–214.

Fornasini, C., 1902, Contributo a la conoscenza de le Bulimine adriatiche. Reale Academia delle Scienze dell'Instituto di Bologna, Memorie delle Scienze Naturali, ser. 5, vol. 9, pp. 371–381.

Franke, A., 1927, Die Foraminiferen und Ostracoden des Palaozans von Rugaard in Jiitland und Sundkrogen bei Copenhagen. Danmarks geologiske Undersøgelse, række 2, no. 46, 49 pp.

Franke, A., 1936, Die Foraminiferen des deutschen Lias Preuss. Abhandlungen der Preussischen Geologischen Landesanstalt, Neue Folge, vol. 169, pp. 1–138.

Franzenau, A., 1884, Heterolepa, egy uj genus a foraminiferak rendjében. Természetrajzi Füzetek, Budapest, vol. 8, pp. 181–184 (214–217).

Franzenau, A., 1888, Adat Budapest altalajának ismeretéhez. Földtani Közlöny, Budapest, vol. 18, no. 3–4, pp. 87–106 (157–174).

Frizzell, D.L., 1943, Upper Cretaceous foraminifera from northwestern Peru. Journal of Paleontology, vol. 17, pp. 337.

Galloway, J.J. and Hemingway, C.E., 1941, The Tertiary Foraminifera of Porto Rico. New York Academy of Science, Scientific Survey of Porto Rico and the Virgin Islands, vol. 3, no. 4, pp. 275–491.

Galloway, J.J. and Morrey, M., 1929, A lower Tertiary foraminiferal fauna from Manta, Ecuador. Bulletins of American Paleontology, vol. 15, pp. 7–56.

Galloway, J.J. and Morrey, M., 1931, Late Cretaceous foraminifera from Tabasco, Mexico. Journal of Paleontology, vol. 5, no. 4, pp. 329–354.

Galloway, J.J. and Wissler, S.G., 1927, Pleistocene foraminifera from the Lomita Quarry, Palos Verdes Hills, California. Journal of Paleontology, vol. 1, pp. 35–87.

Geroch, S., 1959, Stratigraphic significance of arenaceous foraminifera in the Carpathian flysch. Paläontologische Zeitung, vol. 33, pp. 113–122.

Geroch, S., 1960, Microfaunal assemblages from the Cretaceous and Paleogene Silesian Unit in the Beskid Śląski Mts. (western Carpathians). Biuletyn Instytutu Geologicznego, vol. 153, pp. 7–138.

Geroch, S., 1961, Pseudoreophax a new genus of foraminifer from the Neocomian in the Flysch Carpathians. Rocznik Polskiego Towarzystwa Geologicznego, vol. 31, pp. 159–165.

Geroch, S., 1966, Lower Cretaceous small foraminifera of the Silesian series, Polish Carpathians. Rocznik Polskiego Towarzystwa Geologicznego, vol. 36, pp. 413–480.

Geroch, S. and Gradzinski, R., 1955, Stratigrafia serii podsląskiej Żywieckiego okna tektonicznego. Rocznik Polskiego Towarzystwa Geologicznego, vol. 24, pp. 3–62.

Geroch, S. and Kaminski, M.A., 1995, An emendation of some Cretaceous species of "Reophax" (Foraminiferida) from northwest Europe and Poland. In Kaminski, M.A., Geroch, S., and Gasinski, M.A., ed., Proceedings of the Fourth International Workshop on Agglutinated Foraminifera. Grzybowski Foundation Special Publication, no. 3, pp. 117–122.

Geroch, S. and Nowak, W., 1984, Proposal of zonation for the late Tithonian–late Eocene, based upon arenaceous foraminifera from the outer Carpathians, Poland. In Oertli, H., ed., Benthos '83, 2nd International Symposium on Benthic Foraminifera, Pau (France), April 11–15, 1983. Elf Aquitaine, ESSO REP and TOTAL CFP, Pau and Bordeaux, pp. 225–239.

Gibson, T.G., 1983, Key foraminifera from upper Oligocene to lower Pleistocene strata of the Central Atlantic Coastal Plain. Smithsonian Contributions to Paleobiology, vol. 53, pp. 355–453.

Glaessner, M.F., 1937, Studien über Foraminiferen aus der Kreide und dem Tertiär des Kaukasus; 1. Die Foraminiferen der ältesten Tertiärschichten des Nordwest-Kaukas. Problemy Paleontologii, Paleontologichesjaya Laboratoriya Moskovskogo Gosudarstvennogo Universiteta, vol. 2–3, pp. 349–408.

Goës, A., 1894, A synopsis of the Arctic and Scandinavian Recent marine Foraminifera hitherto discovered. Kongl. Svenska Vetenskaps Akademiens Handlingar, vol. 25, no. 9, pp. 3–127.

Gohrbandt, K., 1962, Die Kleinforaminiferenfauna des obereozaenen Anteils der Reingruber Serie bei Brunendorf (Bezirk Korneuburg, Niedersterreich). Mitteilungen der Geologischen Gesellschaft in Wien, vol. 54 (1961), pp. 55–145.

González-Donoso, J.M. and Linares, D., 1970, Datos sobre los foraminiferos del Tortonense de Alcalá la Real (Jaén). Revista Española de Micropaleontologia, vol. 2, pp. 235–242.

Gooday, A.J., 1986, The genus Rhabdammina in the northeast Atlantic: A new species, a redescription of R. major de Folin, 1887, and some speculations on species relationships. Journal of Foraminiferal research, vol. 16, pp. 150–160.

Gooday, A.J., 1988, A response by benthic foraminifera to the deposition of phytodetritus in the deep-sea. Nature, vol. 332, pp. 70–73.

Gooday, A.J., 1996, Epifaunal and shallow infaunal foraminiferal communities at three abyssal NE Atlantic sites subject to differing phytodetritus input regimes. Deep-Sea Research, vol. 43, p. 1395–1421.

Gradstein, F.M., 1983, Paleoecology and stratigraphy of Jurassic abyssal foraminifera in the Blake Bahama Basin, Deep Sea Drilling Project Site 534. In Sheridan, R.E., Gradstein, F.M., et al., ed., Initial Reports of the Deep Sea Drilling Project, Washington (U.S. Government Printing Office), vol. 76, pp. 537–559.

Gradstein, F.M. and Bäckström, S.A., 1996, Cenozoic biostratigraphy and paleobathymetry, northern North Sea and Haltenbanken. Norsk Geologisk Tidsskrift, vol. 76, pp. 3–32.

Gradstein, F.M. and Berggren, W.A., 1981, Flysch-type agglutinated foraminifera and the Maestrichtian to Paleogene history of the Labrador and North Seas. Marine Micropaleontology, vol. 6, pp. 448–472.

Gradstein, F.M. and Kaminski, M.A., 1997, New species of Paleogene deep-water agglutinated foraminifera from the North Sea and Norwegian Sea. Annales Societatis Geologorum Poloniae, vol. 67, pp. 217–229.

Gradstein, F.M., Agterberg, F.P., Ogg, J.G., Hardenbol, J., van Veen, P., Thierry, J., and Huang, Z., 1994, A Mesozoic time scale. Journal of Geophysical Research, vol. 99, B12, pp. 24051–24074.

Gradstein, F.M., Kaminski, M.A., Berggren, W.A., and D'Iorio, M.A., 1994, Cenozoic biostratigraphy of the Central North Sea and Labrador Shelf. Micropaleontology, vol. 40 (Supplement), pp. 1–152.

Grzybowski, J., 1896, Otwornice czerwonych iłów z Wadowic. Rozprawy Wydziału Matematyczno-Przyrodniczego, Akademia UmiejętnoŚci w Krakowie, serya 2, vol. 30, pp. 261–308.

Grzybowski, J., 1898, Otwornice pokładów naftonosnych okolicy Krosna. Rozprawy Wydziału Matematyczno-Przyrodniczego, Akademia Umiejętności w Krakowie, serya 2, vol. 33, pp. 257–305.

Grzybowski, J., 1901, Otwornice warstw inoceramowych okolicy Gorlic. Rozprawy Wydziału Matematyczno-Przyrodniczego, Akademia Umiejętności w Krakowie, serya 2, vol. 41, pp. 219–286.

Guérin-Meneville, F.E., 1843, Iconographie du Règne animal de G. Cuvier Mollusques. Paris: J.B. Baillière, 64 pp.

Gümbel, C.W., 1868, Beitrage zur Foraminiferenfauna der nordalpinen Eocängebilde. Abhandlungen der K. Bayerischen Akademie der Wissenschaften, cl. 2 (1870), vol. 10, no. 2, pp. 581–730.

Hadley, W.H., 1934, Some Tertiary foraminifera from the north coast of Cuba. Bulletin of American Paleontology, vol. 20, pp. 1–40.

Hagn, H., 1956, Geologische und palaontologische Untersuchungen im Tertiar des Monte Brione und seiner Umgebung. Palaeontographica, vol. 107(A), pp. 67–210.

Hantken, M., 1875, Die fauna der Clavulina szaboi-Schichten; Theil I – Foraminiferen. Könglich-Ungarische Geologische Anstalt, Mitteilugen Jahrbuch, vol. 4, pp. 1–93.

Hantken, M., 1883, Die Clavulina szabói-Schichten im Gebiete der Euganeen und der Meerealpen und die Cretacische Scaglia in den Euganeen. Mathematische und Naturwissenschaftliche Berichte aus Ungarn, Budapest, vol. 2 (1883–4), pp. 121–169.

Hart, M.B., Bailey, H.W., Crittenden, S., Fletcher, B.N., Price, R.J., and Swiecicki, A., 1989, Cretaceous. In Jenkins, D.G., and Murray, J.W., ed., Stratigraphical Atlas of Fossil Foraminifera, 2nd edn. Chichester: Ellis Horwood, pp. 273–371.

Haynes, J., 1954, Taxonomic position of some British Palaeocene Buliminidae. Contributions to the Cushman Foundation of Foraminiferal Research, vol. 5, no. 4, pp. 185–191.

Haynes, J.R., 1973, Cardigan Bay Recent Foraminifera. (Cruises of the R.V. Antur, 1962–1964). Bulletin of the British Museum (Natural History) Zoological Supplement, no. 4, pp. 1–245.

Hayward, B., 2002, Late Pliocene to Middle Pleistocene extinctions of deep-sea benthic foraminifera ("Stilostomella extinction") in the southwest Pacific. Journal of Foraminiferal Research, vol. 32, pp. 274–307.

Hayward, B. and Kawagata, S., 2005, Extinct foraminifera in Brady's Challenger Report. Journal of Micropalaeontology, vol. 24, pp. 171–175.

Hedberg, H.D., 1937, Foraminifera of the middle Tertiary Carapitana formation of northeastern Venezuela. Journal of Paleontology, vol. 11, pp. 661–697.

Hemleben, C. and Tröster, J., 1984, Campanian–Maestrichtian deep-water foraminifers from Hole 543A, Deep Sea Drilling Project. In Bij-Duval, B., Moore, J.C., et al., ed., Initial Reports of the Deep Sea Drilling Project, Washington (U.S. Government Printing Office), vol. 78A, pp. 509–532.

Hermelin, J.O.R., 1989, Pliocene benthic foraminifera from the Ontong-Java Plateau (Western Equatorial Pacific Ocean): faunal response to changing paleoenvironment. Special Publication of the Cushman Foundation for Foraminiferal Research, vol. 26, pp. 1–143.

Heron-Allen, E. and Earland, A., 1913, On some foraminifera from the North Sea, etc., dredged by the Fisheries Cruiser "Goldseeker" (International North Sea Investigations – Scotland). II. On the distribution of Saccammina sphaerica (M. Sars) and Psammosphaera fusca (Schulze) in the North Sea: particularly with reference to the suggested identity of the two species. Journal of the Royal Microscopical Society, vol. 1913, pp. 1–26.

Hillebrandt, A. von, 1962, Das Paleozaen und seine Foraminiferenfauna im Becken von Reichenhall und Salzburg. Abhandlungen, Bayerische Akademie der Wissenschaften, Mathematisch-Naturwissenschaftliche Klasse, N.F., no. 108, pp. 9–182.

Hofker, J., 1932, Notizen ueber die Foraminiferen des Golfes von Neapel; III – Die Foraminiferenfauna der Ammontatura. Pubblicazioni della Stazione Zoologica, Napoli, vol. 12, no. 1, pp. 61–145.

Hofker, J., 1951, The Foraminifera of the "Siboga" Expedition. Part III: Siboga-Expeditie, Monographie IVa. Leiden: E.J. Brill, 513 pp.

Hofker, J., 1972, Primitive Agglutinated Foraminifera. Leiden: E.J. Brill, 95 pp.

Höglund, H., 1947, Foraminifera in the Gullmar Fjord and the Skagerak. Zoologiska Bidrag Från Uppsala, vol. 26, pp. 1–328.

Holbourn, A.E.L. and Kaminski, M.A., 1997, Cretaceous deep-water benthic foraminifera of the Indian Ocean. Grzybowski Foundation Special Publication, no. 4, pp. 1–175.

Hornibrook, N. de, Brazier, R.C., and Strong, C.P., 1989, Manual of New Zealand Permian to Pleistocene foraminiferal biostratigraphy. New Zealand Geological Survey, Paleontological Bulletin 1989, pp. 1–175.

Howe, H.V., 1939, Louisiana Cook Mountain Eocene foraminifera. Bulletin of the Geological Survey of Louisiana, vol. 14, pp. 1–122.

Huss, F., 1966, Otwornice aglutynajace serii podslakiej jednostki roponosnej Weglowki (Polskie Karpaty Flisowe). Prace Geologiczne, Polska Akademia Nauk, no. 34, pp. 7–76.

Ingle, J.C. Jr, Keller, G. and Kolpack, R.L., 1980, Benthic foraminiferal biofacies, sediments and water masses of the southern Peru–Chile trench area, southeastern Pacific Ocean. Micropaleontology, vol. 26, pp. 113–150.

Inoue, Y., 1989, Northwest Pacific Foraminifera as paleoenvironmental indicators. Science Reports of the Institute of Geosciences, University of Tsukuba (Section B, Geological Sciences), vol. 10, pp. 57–162.

Israelsky, M.C., 1951, Foraminifera of the Lodo Formation, central California; General introduction and Part 1, arenaceous foraminifera. Professional Paper of the United States Geological Survey, no. 240-A, pp. 1–29.

Jendryka-Fuglewicz, B., 1975, Evolution of the Jurassic and Cretaceous smooth walled Lenticulina of Poland. Acta Palaeontologica Polonica, vol. 20, pp. 99–197.

Jennings, P.H., 1936, A microfauna from the Monmouth of basal Rancocas groups of New Jersey. Bulletin of American Paleontology, vol. 23, no. 78, pp. 3–76.

Jones, J.P. and Parker, W.K., 1860, On the Rhizopodal fauna of the Mediterranean compared with that of the Italian and some other Tertiary deposits. Quarterly Journal of the Geological Society of London, vol. 16, pp. 292–307.

Jones, R.W., 1984, On the designation of lectotypes for certain species and subspecies of unilocular hyaline foraminifera; Part 1 – Those housed in the British Museum (Natural History). Journal of Micropalaeontology, vol. 3, pp. 63–69.

Jones, R.W., 1994, The Challenger Foraminifera. Oxford, Oxford University Press, 149 pp.

Jones, R.W., Bender, H., Charnock, M.A., and Kaminski, M.A., 1993, Emendation of the foraminiferal genus Cribrostomoides Cushman, 1910, and its taxonomic implications. Journal of Micropalaeontology, vol. 12, pp. 181–193.

Jonkers, H.A., 1984, Pliocene benthonic foraminifera from homogeneous and laminated marls on Crete. Utrecht Micropaleontological Bulletins, vol. 31, pp. 1–179.

Jurkiewicz, H., 1960, Foraminifera in Czarnorzeki-Shales in the eastern part of the Silesian Unit (Carpathians). Rocznik Polskiego Towarzystwa Geologicznego, vol. 30, no. 3, pp. 333–343.

Kaminski, M.A., 1984, Shape variation in Spiroplectammina spectabilis (Grzybowski). Acta Palaeontologica Polonica, vol. 29, pp. 29–46.

Kaminski, M.A. and Cetean, C.G., 2011, Lectotypes of type species of Agglutinated Foraminiferal Genera in the Collections of the Natural History Museum, London. Part 2. Hippocrepinina and Hormosinina. In Kaminski, M.A. and Filipescu, S., ed., Proceedings of the Eighth International Workshop on Agglutinated Foraminifera, Grzybowski Foundation Special Publication, no. 16, pp. 61–69.

Kaminski, M.A. and Geroch, S., 1992, Trochamminoides grzybowskii, nom. nov., a new name for Trochammina elegens Grzybowski, 1898 (Foraminiferida). Journal of Micropalaeontology, vol. 11, p. 64.

Kaminski, M.A. and Geroch, S., 1993, A revision of foraminiferal species in the Grzybowski Collection. In Kaminski, M.A., Geroch, S., and Kaminski, D.G., ed., The Origins of Applied Micropalaeontology: The School of Jósef Grzybowski (with a taxonomic revision of the Grzyboswski Collection). Grzybowski Foundation Special Publication, no. 1, pp. 239–330.

Kaminski, M.A. and Gradstein, F.M., 2005, Atlas of Palaeogene Cosmopolitan Deep-Water Agglutinated Foraminifera. Grzybowski Foundation Special Publication, no. 10, 547 + vii pp.

Kaminski, M.A. and Kuhnt, W., 1991, Depth-related shape variation in Ammobaculites agglutinans (d'Orbigny). Annales Societatis Geologorum Poloniae, vol. 61, pp. 221–230.

Kaminski, M.A., Geroch, S., and Kaminski, D.G., 1993, The Origins of Applied Micropalaeontology: The School of Jósef Grzybowski (with a taxonomic revision of the Grzyboswski Collection). Grzybowski Foundation Special Publication, no. 1, pp. 1–336.

Kaminski, M.A., Gradstein, F.M., and Berggren, W.A., 1989, Paeogene benthic foraminifer biostratigraphy and paleoecology at site 647, southern Labrador Sea. In Srivastava, S.P., Arthur, M.A., Clement, B., et al., ed., Scientific Results of the Proceedings of the Ocean Drilling Program, vol. 105, pp. 705–730.

Kaminski, M.A., Gradstein, F.M., Berggren, W.A., Geroch, S., and Beckmann, J.P., 1988, Agglutinated foraminiferal assemblages from Trinidad: taxonomy, stratigraphy and paleobathymetry. In Gradstein, F.M. and Rögl, F., ed., Second International Workshop on Agglutinated Foraminifera, Vienna 1986, Abhandlungen der Geologischen Bundesanstalt, vol. 41, pp. 155–228.

Kaminski, M.A., Gradstein, F. and Geroch, S., 1992, Uppermost Jurassic to Lower Cretaceous deep-water benthic foraminiferal assemblages from Site 765 on the Argo Abyssal Plain. In Gradstein, F., Ludden, J.N., et al., ed., Scientific Results of the Proceedings of the Ocean Drilling Program, vol. 123, pp. 239–269.

Kaminski, M.A., Gradstein, F.M., Goll, R.M., and Greig, D., 1990, Biostratigraphy and Paleoecology of deep-water aggutinated foraminifera at ODP Site 643, Norwegian–Greenland Sea. In Hemleben, C., Kaminski, M.A., Kuhnt, W., and Scott, D.B., ed., Paleoecology, Biostratigraphy, Paleoceanography and Taxonomy of Agglutinated Foraminifera. NATO ASI Series, Series C: Mathematical and Physical Sciences, vol. 327. Dordrecht: Kluwer Academic, pp. 345–386.

Karrer, F., 1868, Die Miocene Foraminiferenfauna von Kostej im Banat. Sitzungsberichte der Kaiserlichen Akademie der Wissenschaften Wien, Mathematisch-Naturwissenschaftliche Klasse, vol. 58, no. 1, pp. 121–193.

Katz, M.E. and Miller, K.G., 1991, Early Paleogene benthic foraminiferal assemblages and stable isotopes in the Southern Ocean. In Ciesielski, P.F., Kristoffersen, Y., et al., ed., Scientific Results of the Proceedings of the Ocean Drilling Program, vol. 114, pp. 481–512.

Katz, M.E. and Miller, K.G., 1993, Latest Oligocene to Earliest Pliocene benthic foraminiferal biofacies of the northeastern Gulf of Mexico. Micropaleontology, vol. 39, no. 4, pp. 367–403.

Katz, M.E. and Miller, K.G., 1993. Neogene subsidence along the Northeast Australian margin: benthic foraminiferal evidence. In Davies, P.J., McKenzie, J.A., Julson, A., et al., ed., Scientific Results of the Proceedings of the Ocean Drilling Program, vol. 133, pp. 75–92.

Katz, M.E., Tjalsma, R.C., and Miller, K.G., 2003, Oligocene bathyal to abyssal benthic foraminifera of the Atlantic Ocean. Micropaleontology, vol. 49, supplement 2, pp. 1–45.

Keijzer, F.G., 1945, Outline of the geology of the eastern part of the province of Oriente, Cuba (E of 76 degrees WL), with notes on the geology of other parts of the island. Utrecht University, Geographische en Geologische Mededelingen, Physiographisch-Geologische Reeks, ser. 2, no. 6, pp. 1–238.

King, C., 1989, Cenozoic of the North Sea. In Jenkins, D.G. and Murray, J.W., ed., Stratigraphical Atlas of Fossil Foraminifera, 2nd edn. Chichester: Ellis Horwood, pp. 418–489.

King, C., Bailey, H.W., Burton, C.A., and King, A.D., 1989, Cretaceous of the North Sea. In Jenkins, D.G. and Murray, J.W., ed., Stratigraphical Atlas of Fossil Foraminifera, 2nd edn. Chichester: Ellis Horwood Ltd, pp. 312–417.

Knudsen, K.L., 1971, Late Quaternary Foraminifera from Vendsyssel, Denmark and Sandnes, Norway. Bulletin of the Geological Society of Denmark, vol. 21, pt 2–3, pp. 185–292.

Koch, R., 1926, Mitteltertiaere Foraminiferen aus Bulongan, Ost-Borneo. Eclogae Geologicae Helvetiae, vol. 19, no. 3, pp. 722–751.

Kohl, B., 1985. Early Pliocene benthic foraminifers from the Salina Basin, southeastern Mexico. Bulletin of American Paleontology, vol. 88, no. 322, 173 pp.

Krasheninnikov, V.A., 1973, Cretaceous benthonic foraminifera, Leg 20, Deep Sea Drilling Project. In Heezen, B.C,, MacGregor, I.D., et al., ed., Initial Reports of the Deep Sea Drilling Project, Washington (U.S. Government Printing Office), vol. 20, pp. 205–221.

Krasheninnikov, V.A., 1974, Upper Cretaceous benthonic agglutinated foraminifera, Leg 27 of the Deep Sea Drilling Project. In Veevers, J.J., Heirtzler, J.R., et al., ed., Initial Reports of the Deep Sea Drilling Project, Washington (U.S. Government Printing Office), vol. 27, pp. 531–662.

Krasheninnikov, V.A. and Pflaumann, U., 1978, Cretaceous agglutinated foraminifera of the Atlantic Ocean off West Africa (Leg 41, Deep Sea Drilling Project). In Lancelot, Y., Seibold, E., et al., ed., Initial Reports of the Deep Sea Drilling Project, Washington (U.S. Government Printing Office), vol. 41, pp. 565–580.

Kuhnt, W., 1990, Agglutinated foraminifera of western Mediterranean Upper Cretaceous pelagic limestones (Umbrian Apennines, Italy, and Betic Cordillera, Southern Spain). Micropaleontology, vol. 36, no. 4, pp. 297–330.

Kuhnt, W., 1995, Deep-water agglutinated foraminifera from the Lower Cretaceous (Neocomian) "complex à Aptychus" Formation (Corridor de Boyar, Betic Cordillera, southern Spain). Journal of Micropalaeontology, vol. 14, no. 1, pp. 37–52.

Kuhnt, W., and Kaminski, M.A., 1989, Upper Cretaceous deep-water agglutinated benthic foraminiferal assemblages from the western Mediterranean and adjacent areas. In Wiedmann, J., ed., Cretaceous of the western Tethys. Proceedings of the 3rd International Cretaceous Symposium, Tubingen, pp. 91–120.

Kuhnt, W. and Kaminski, M.A., 1990, Paleoecology of Late Cretaceous to Paleocene deep-water agglutinated foraminifera from the North Atlantic and western Tethys. In Hemleben, C., Kaminski, M.A., Kuhnt, W., and Scott, D.B., ed., Paleoecology, Biostratigraphy, Paleoceanography and Taxonomy of Agglutinated Foraminifera. NATO ASI Series, Series C: Mathematical and Physical Sciences. Dordrecht: Kluwer Academic, vol. 327, pp. 433–505.

Kuhnt, W. and Moullade, M., 1991, Quantitative analysis of upper Cretaceous abyssal agglutinated foraminiferal distribution in the North Atlantic – paleoceanographic implications. Revue de Micropaléontologie, vol. 34, no. 4, pp. 313–349.

Kuhnt, W., Collins, C., and Scott, D.B., 2000, Deep water agglutinated foraminiferal assemblages across the Gulf Stream: distribution patterns and taphonomy. In Hart, M., Kaminski, M., and Smart, C.W., ed., Proceedings of the 5th international workshop on Agglutinated Foraminifera. Grzybowski Foundation Special Publication, no. 7, pp. 261–298.

Kuhnt, W., Hess, S., and Jian, Z., 1999, Quantitative composition of benthic foraminiferal assemblages as a proxy indicator for organic carbon flux rates in the South China Sea. In Sarnthein, M. and Wang, P.X., ed., Response of West Pacific Marginal Seas to Global Climate Change. Marine Geology, vol. 156, pp. 123–157.

Kuhnt, W., Holbourn, A.E., and Zhao, Q., 2002, The early history of the South China Sea: Evolution of Oligocene–Miocene deep water environments. Revue de Micropaléontologie, vol. 45, pp. 99–159.

Kuznetsova, K.I., 1974, Distribution of benthonic Foraminifera in Upper Jurassic and Lower Cretaceous Deposits at Site 261, DSDP Leg 27, in the Eastern Indian Ocean. In Veevers, J.J., Heirtzler, J.R., et al., ed, Initial Reports of the Deep Sea Drilling Project, Washington (U.S. Government Printing Office), vol. 27, pp. 673–681.

Lalicker, C.G., 1935, New Cretaceous Textulariidae. Contributions from the Cushman Laboratory for Foraminiferal Research, vol. 11, pp. 39–51.

Lamarck, J.B., 1804, Suite des memoires sur les fossiles des environs de Paris. Annales Muséum National d'Histoire Naturelle, vol. 5, pp. 349–357.

Lamarck, J.B., 1816, Tableau encyclopédique et méthodique des trois règnes de la nature, partie 23 – Mollusques et polypes divers. Paris: Mme V. Agasse, pp. 1–16.

Lamarck, J.B., 1822, Histoire naturelle des animaux sans vertèbres, vol. 7. Paris: L'auteur, 711 pp.

Le Calvez, Y., 1974, Révision des foraminifères de la collection d'Orbigny. I. Foraminifères des Îles Canaries. Cahiers de Micropaléontologie, vol. 2, pp. 1–108.

LeRoy, L.W., 1939, Some small foraminifera, ostracoda and otoliths from the Neogene ("Miocene") of the Rokan-Tapanoeli area, Central Sumatra. Natuurkundig Tijdschrift voor Nederlandsch-Indië, vol. 99, no. 6, pp. 215–296.

LeRoy, L.W., 1941, A new species of Cibicides from the lower Pliocene (Repetto Formation) of Southern California. Journal of Paleontology, vol. 15, no. 6, pp. 622–623.

LeRoy, L.W., 1944, Miocene foraminifera from Sumatra and Java, Netherlands East Indies; Part 1. Miocene Foraminifera of Central Sumatra, Netherlands East Indies. Quarterly of the Colorado School of Mines, vol. 39, no. 3, pp. 8–69.

LeRoy, L.W., 1953, Biostratigraphy of the Maqfi section, Egypt. Memoirs of the Geological Society of America, vol. 34, pp. 1–73.

Lévy, A., Mathieu, R., Poignant, A., Rosset-Moulinier, M., and Rouvillois, A., 1975, Sur quelques foraminiféres actuels des plages de Dunkerque et des environs: néotypes et espéce nouvelle. Revue de Micropaléontologie, vol. 17, no. 4, pp. 171–181.

Lewis, K.B., 1979, Foraminifera on the continental shelf and slope off Southern Hawke's Bay. Memoirs of the New Zealand Oceanographic Institute, vol. 84, pp. 1–45.

Linnaeus, C., 1758, Systema Naturae, 10th edn. Lipsiae (Leipzig): G. Engelman, vol. 1, 824 pp.

Liszka, S. and Liszkowa, J., 1981, Revision of J. Grzybowski's paper (1896) "Foraminifera of the red clays from Wadowice". Rocznik Polskiego Towarzystwa Geologicznego, vol. 51, pp. 153–208.

Loeblich, A.R. Jr and Tappan, H., 1953, Studies of Arctic Foraminifera. Smithsonian Miscellaneous Collections, vol. 121, no. 7, pp. 1–150.

Loeblich, A.R. and Tappan, H., 1955, Revision of some Recent foraminiferal genera. Smithsonian Miscellaneous Collections, vol. 128, pp. 1–37.

Loeblich, A.R. Jr and Tappan, H., 1964, Sarcodina, chiefly "Thecamoebians" and Foraminiferida. In Moore, R.C., ed., Treatise on Invertebrate Paleontology, Part C, Protista 2 (2 vols). Lawrence, KS: Geological Society of America and University of Kansas Press, pp. 1–900.

Loeblich, A.R. and Tappan, H., 1986, Some new and revised genera and families of hyaline calcareous Foraminiferida (Protozoa). Transactions of the American Microscopical Society, vol. 105, pp. 239–265.

Loeblich, A.R. Jr and Tappan, H., 1987, Foraminiferal Genera and Their Classification. New York: Van Nostrand Reinhold, 970 pp.

Loeblich, A.R. Jr and Tappan, H., 1994, Foraminifera of the Sahul Shelf and Timor Sea. Special Publications of the Cushman Laboratory for Foraminiferal Research, vol. 31, pp. 1–661.

Lutze, G.F., 1980, Depth distribution of benthic foraminifera on the continental margin of NW-Africa. "Meteor" Forsch. Ergebnisse, Reihe C, vol. 32, pp. 31–80.

Lutze, G.F., 1986, Uvigerina species of the eastern North Atlantic. Utrecht Micropaleontological Bulletins, vol. 35, pp. 21–46.

MacFadyen, W.A., 1930, Miocene foraminifera from the Clysmic area of Egypt and Sinai. Cairo: Geological Survey of Egypt, 149 pp.

Mackensen, A. and Berggren, W.A. 1992. Paleogene benthic foraminifers from the southern Indian Ocean (Kerguelen Plateau): biostratigraphy and paleoecology. In Wise, S.W. Jr, Schlich, R., et al., ed., Scientific Results of the Proceedings of the Ocean Drilling Program, vol. 120, pp. 603–630.

Mackensen, A. and Bickert, T., 1999, Stable carbon isotopes in benthic foraminifera: proxies of deep and bottom water circulation and new production. In Fischer, G. and Wefer, G., ed., Use of Proxies in Paleoceanography: Examples from the South Atlantic. Berlin: Springer Verlag, pp. 229–254.

Mackensen, A., Grobe, H., Kuhn, G., and Fütterer, D.K., 1990, Benthic foraminiferal assemblages from the eastern Weddell Sea between 68° and 73° S: distribution, ecology and fossilization potential. Marine Micropaleontology, vol. 16, pp. 241–283.

Majzon, L., 1943, Adatok egyes Kárpátaljai Flisrétegekhez, tekintettel a Globotruncanákra: Beiträge zur Kenntniss einiger Flysch-Schichten des Karpaten-Vorlandes mit Rücksicht auf die Globotruncanen. A magyar Királyi Földtani Intézet, Évkönnyve, vol. 37, pp. 1–170.

Mallory, V.S., 1959, Lower Tertiary Biostratigraphy of the California Coast Ranges. Tulsa, OK: American Association of Petroleum Geologists, 416 pp.

Marsson, T., 1878, Die Foraminiferen der weissen Schreibkreide der Insel Ruegen. Mitteilungen des Naturwissenschaftlichen Vereins für Neu-Vorpommern und Rugen in Greifswald, vol. 10, pp. 115–196.

Martin, L., 1964, Upper Cretaceous and Lower Tertiary foraminifera from Fresno County, California. Jahrbuch der Geologischen Bundesanst, Austria, vol. 9, pp. 1–128.

Martin, L.T., 1943, Eocene foraminifera from the Type Lodo Formation, Fresno County, California. Stanford University Publications, Geological Sciences, vol. 3, no. 3, pp. 1–35.

Mathews, R.D., 1945, Rectuvigerina, a new genus of foraminifera from a restudy of Siphogenerina. Journal of Paleontology, vol. 19, no. 6, pp. 588–606.

Matsunaga, T., 1963, Benthonic smaller foraminifera from the oil fields of northern Japan. Science Reports of the Tôhoku University, Sendai, Japan, ser. 2 (Geology), vol. 35, no. 2, pp. 67–122.

McGowran, B., 1965, Two Paleocene foraminiferal faunas from the Wangerrip Group, Pebble Point Coastal Section, Western Victoria. Proceedings of the Royal Society of Victoria, vol. 79, no. 1, pp. 9–74.

McNeil, D.H. and Caldwell, W.G.E., 1981, Cretaceous rocks and their foraminifera in the Manitoba Escarpment. Special Papers of the Geological Association of Canada, vol. 21, pp. 1–439.

Mendes, I., Dias, J.A., Schönfeld, J., and Ferreira, O., 2012, Distribution of living foraminifera on the northern Gulf of Cadiz continental shelf. Journal of Foraminiferal Research, vol. 42, pp. 18–38.

Meyn, H. and Vespermann, J., 1994, Taxonomic revision of Lower Cretaceous Foraminifera described by Roemer (1839, 1841, 1842), Koch (1851), and Reuss (1863) from Lower Cretaceous Saxony. Senckenbergiana Lethaea, vol. 74, no. 1/2, pp. 49–272.

Michael, E., 1967, Die Mikrofauna der nordwestdeutschen Barreme. Teil I – Die Foraminiferen des NW-deutschen Barreme. Palaeontographica Supplement, vol. 12, pp. 1–176.

Milker, Y. and Schmiedl, G., 2012, A taxonomic guide to modern benthic shelf foraminifera of the western Mediterranean Sea. Palaeontologia Electronica, vol. 15, Issue 2;16A, 134 pp.

Miller, K.G., 1983, Eocene–Oligocene paleoceanography of the deep Bay of Biscay: benthic foraminiferal evidence. Marine Micropaleontology, vol. 7, pp. 403–440.

Miller, K.G. and Katz, M.E., 1987, Eocene benthic foraminiferal biofacies of the New Jersey transect. In Poag, C.W., Watts, A.B., et al., ed., Initial Reports of the Deep Sea Drilling Project, Washington (U.S. Government Printing Office), vol. 95, pp. 267–298.

Miller, K.G. and Katz, M.E., 1987, Oligocene to Miocene benthic foraminiferal and abyssal circulation changes in the north Atlantic. Micropaleontology, vol. 33, no. 2, pp. 97–149.

Miller, K.G., Curry, W.B. and Ostermann, D.R., 1985, Late Paleogene (Eocene to Oligocene) benthic foraminiferal oceanography of the

Goban Spur region, Deep Sea Drilling Project Leg 80. In de Graciansky, P.C., Poag, C.W., et al., ed., Initial Reports of the Deep Sea Drilling Project, Washington (U.S. Government Printing Office), vol. 80, pp. 505–538.

Miller, K.G., Gradstein, F.M., and Berggren, W.A., 1982, Late Cretaceous to early Tertiary agglutinated benthic foraminifera in the Labrador Sea. Micropaleontology, vol. 28, pp. 1–30.

Mjatliuk, E.V., 1939, Foraminifera of the Upper Jurassic and Lower Cretaceous of the region of the Middle Volga River and the Great Syrte (in Russian). Trudy Vsesoyuznogo Nauchno-Issledovatel'skogo Geologorazvedochnogo Instituta VNIGRI, vol. 120, pp. 1–76.

Mjatliuk, E.V., 1950, Stratigrafiya flishevykh osadkov Severnykh Karpat w svete dannykh fauny foraminifer. Trudy Vsesoyuznogo Nauchno-Issledovatel'skogo Geologorazvedochnogo Instituta VNIGRI (New Series), vol. 51, pp. 225–287.

Mjatliuk, E.V., 1970, Foraminifery flishevykh otlozhenii vostochnykh Karpat (Mel – Paleogen). Trudy Vsesoyuznogo Nauchno-Issledovatel'skogo Geologorazvedochnogo Instituta VNIGRI (New Series), vol. 282, pp. 1–360.

Montagne, D.G., 1941, Geologie und Palaeontologie der Umgebung von Sestanovac, Dalmatien. Utrecht University, Geographische en Geologische Mededelingen, Physiographisch-Geologische Reeks, ser. 2, no. 1, pp. 9–93.

Montagu, G., 1808, Supplement to Testacea Britannica. Exeter: S. Woolmer, v + 183 pp.

Morgiel, J. and Olszewska, B., 1981, Biostratigraphy of the Polish External Carpathians based on agglutinated foraminifera. Micropaleontology, vol. 27, no. 1, pp. 1–30.

Moullade, M., 1966, Étude stratigraphique et microplaéontologique du Crétacé Inférieur de la "Fosse Vocontienne". Documents des Laboratoires de Géologie de la Faculté des Sciences de Lyon, vol. 15. no. 1–2, pp. 1–369.

Moullade, M., Kuhnt, W., and Thurow, J., 1988, Agglutinated benthic foraminifers from Upper Cretaceous variegated clays of the North Atlantic Ocean (DSDP Leg 93 and ODP Leg 103). In Boillot, G., Winterer, E.L., et al., ed., Scientific Results of the Proceedings of the Ocean Drilling Program, vol. 103, pp. 349–377.

Murray, J.W., 1991, Ecology and Palaeoecology of Benthic Foraminifera. New York: John Wiley & Sons, 397 pp.

Nagy, J., Gradstein, F.M., Kaminski, M.A., and Holbourn, A.E.L., 1995, Foraminiferal morphogroups, paleoenvironments and new taxa from Jurassic and Cretaceous stata of Thakkhola, Nepal. In Kaminski, M.A., Geroch, S., and Gasinski, M.A., ed., Proceedings of the Fourth International Workshop on Agglutinated Foraminifera, Grzybowski Foundation Special Publication, no. 3, pp. 181–209.

Narayan, Y.R., Barnes, C.R., and Johns, M.J., 2005, Taxonomy and biostratigraphy of Cenozoic foraminifers from Shell Canada wells, Tofino Basin, offshore Vancouver Island. Micropaleontology, vol. 51, no. 2, pp. 101–167.

Neagu, T., 1962, Studiul foraminiferelor aglutinante din argilele cretacic-superioare de pe Valea Sadova (Cîmpulung Moldovenesc) si basinul superior al Vaii Buzaului. Studii Cercetari Geologie, Academia Republicii Populare Romine, Sectia de Geologie si Geografie si Istitutul de Geologie si Geografie, vol. 7, no. 1, pp. 45–81.

Neagu, T., 1975, Monographie de la faune des Foraminifères Éocrétacés du couloir de Dîmbovicioara, de Codlea et des Monts Persani (Couches de Carhaga). Memorii Institutul de Geologie si Geofizica, Bucaresti, vol. 35, pp. 1–141.

Nomura, R., 1991, Paleoceanography of upper Maestrichtian to Eocene benthic foraminiferal assemblages at Sites 752, 753 and 754, eastern Indian Ocean. In Weissel, J., Price, J., Taylor, E., Alt, J. et al., ed., Scientific Results of the Proceedings of the Ocean Drilling Program, vol. 121, pp. 3–29.

Nomura R., 1995, Paleogene to Neogene deep-sea paleoceanography in the eastern Indian Ocean: benthic foraminifera from ODP Sites 747, 757 and 758. Micropaleontology, vol. 41, pp. 251–290.

Nomura, R. and Takata, H., 2005, Data report: Paleocene/Eocene benthic foraminifers, ODP Leg 199, Sites 1215, 1220, and 1221, equatorial central Pacific Ocean. In Wilson, P.A., Lyle, M., and Firth, J.V., ed., Scientific Results of the Proceedings of the Ocean Drilling Program, vol. 199, pp. 1–34 [Online].

Nørvang, A., 1968, Interior characteristics of Bulimina (Foraminifera). Proceedings of the 23rd International Geological Congress, pp. 415–422.

Nuttall, W.L.F., 1928, Tertiary foraminifera from the Naparima region of Trinidad (British West Indies). Quarterly Journal of the Geological Society of London, vol. 84, pp. 57–117.

Nuttall, W.L.F., 1930, Eocene foraminifera from Mexico. Journal of Paleontology, vol. 4, pp. 271–293.

Nuttall, W.L.F., 1932, Lower Oligocene foraminifera from Mexico. Journal of Paleontology, vol. 6, pp. 3–35.

Nyong, E.E. and Olsson, R.K., 1984, A paleoslope model for Campanian to lower Maestrichtian foraminifera of the North American Basin and adjacent continental margin. Marine Micropaleontology, vol. 8, pp. 437–477.

Olsson, R.K., 1960, Foraminifera of latest Cretaceous and earliest Tertiary age in the New Jersey Coastal Plain. Journal of Paleontology, vol. 34, pp. 1–58.

Ortiz, S. and Thomas, E., 2006, Lower–middle Eocene benthic foraminifera from the Fortuna Section (Betic Cordillera, southeastern Spain). Micropaleontology, vol. 52, pp. 97–150.

Palmer, D.K., 1936, New genera and species of Cuban Oligocene Foraminifera. Memorias de la Sociedad Cubana de Historia Natural "Filipe Poey", vol. 10, no. 2, pp. 123–128.

Palmer, D.K., 1941, Foraminifera of the upper Oligocene Cojimar Formation of Cuba Part 4. Memorias de la Sociedad Cubana de Historia Natural "Filipe Poey", vol. 15, no. 2, pp. 181–200.

Palmer, D.K. and Bermúdez, P.J., 1936, An Oligocene foraminiferal fauna from Cuba; Part I. Memorias de la Sociedad Cubana de Historia Natural "Filipe Poey", vol. 10, no. 4, pp. 227–271.

Palmer, D.K. and Bermúdez, P.J., 1936, An Oligocene foraminiferal fauna from Cuba; Part II. Memorias de la Sociedad Cubana de Historia Natural "Filipe Poey", vol. 10, no. 5, pp. 273–317.

Papp, A. and Schmid, M.E., 1985, Die fossilen Foraminiferen des tertiaeren Beckens von Wien: Revision der Monographie Alcide d'Orbigny (1846). Abhandlungen der Geologischen Bundesanstalt, vol. 37, pp. 1–311.

Parker, F.L. and Bermúdez, P.J., 1937, Eocene species of the genera Bulimina and Buliminella from Cuba. Journal of Paleontology, vol. 11, pp. 513–516.

Parker, W.K. and Jones, T.R., 1857, Description of some foraminifera from the coast of Norway. Annals and Magazine of Natural History, ser. 2, vol. 19, pp. 273–303.

Parker, W.K. and Jones, T.R., 1865, On some foraminifera from the North Atlantic and Arctic Oceans, including Davis Straits and Baffin's Bay. Philosophical Transactions of the Royal Society, vol. 155, pp. 325–441.

Parker, W.K., Jones, T.R., and Brady, H.B., 1871, On the nomenclature of the foraminifera. Part XIV – The species enumerated by d'Orbigny in the "Annales des Sciences Naturelles" 1826. Annals and Magazine of Natural History, Series 4, vol. 7 (continued), – The species founded upon the figures in Soldani's "Testaceographia ac Zoophytographia", Annals and Magazine of Natural History, Series 4, vol. 8, pp. 145–179, 238–266.

Parr, W.J., 1932, Victorian and south Australian shallow-water foraminifera. Part I. Proceedings of the Royal Society of Victoria, vol. 44, pp. 1–14.

Parr, W.J., 1938, Upper Eocene foraminifera from deep borings in King's Park, Perth, Western Australia. Journal of the Royal Society of Western Australia, vol. 24, no. 8, pp. 69–101.

Parr, W.J., 1950, Foraminifera. Reports of the BANZ Antarctic Research Expedition 1929–1931, Series B (Zoology, Botany), vol. 5, no. 6, pp. 232–392.

Perconig, E., 1955, Due nuove specie di Uvigerina del Neogene della Planura Padana. Bollettino del Servizio Geologico d'Italia, vol. 77, p. 187.

Pflum, C.E. and Frerichs, W.E., 1976, Gulf of Mexico deep-water foraminifers. Special Publication of the Cushman Foundation, vol. 14, pp. 1–125.

Phleger, F.B. and Parker, F.L., 1951, Ecology of foraminifera, northwest Gulf of Mexico. Part II – Foraminifera species. Memoirs of the Geological Society of America, vol. 46, pp. 1–64.

Phleger, F.B. and Parker, F.L., 1952, New names for northwestern Gulf of Mexico foraminifera. Contributions from the Cushman Laboratory for Foraminiferal Research, vol. 3, no. 1, p. 14.

Phleger, F.B., Parker, F.L., and Pierson, J.F., 1953, North Atlantic foraminifera. Reports of the Swedish Deep-Sea Expedition, 1947–1948, vol. 7, pp. 3–122.

Plummer, H.J., 1927, Foraminifera of the Midway Formation in Texas. Bulletin of the University of Texas Bureau of Economic Geology and Technology, vol. 2644, pp. 1–206.

Poag, C.W., 1981, Ecologic Atlas of Benthic Foraminifera of the Gulf of Mexico. Stroudsburg, PA: Hutchinson Ross, 174 pp.

Proto Decima, F. and Bolli, H.M., 1978, Southeast Atlantic Leg 40 Paleogene benthic foraminifers. In Bolli, H.M., Ryan, W.B.F., et al., ed., Initial Reports of the Deep Sea Drilling Project, Washington (U.S. Government Printing Office), vol. 40, pp. 783–809.

Reicherter, K., Pletsch, T., Kuhnt, W., Manthey, J., Homeier, G., Wiedmann, J., and Thurow, J., 1994, Mid-Cretaceous paleogeography and paleoceanography of the Betic Seaway (Betic Cordillera, Spain). Palaeogeography, Palaeoclimatology, Palaeoecology, vol. 107, pp. 1–33.

Renz, H.H., 1948, Stratigraphy and fauna of the Agua Salada group, State of Falcón, Venezuela. Memoirs of the Geological Society of America, vol. 32, pp. 1–219.

Resig, J.M., 1976, Benthic foraminiferal stratigraphy, eastern margin, Nazca Plate. In Yeats, R.S., Hart, S.R., et al., ed. Initial Reports of the Deep Sea Drilling Project, Washington (U.S. Government Printing Office), vol. 34, pp. 743–759.

Resig., J., 1981, Biogeography of benthic foraminifera of the northern Nazca Plate and adjacent continental margin. Geological Society of America, Memoir 154, pp. 619–666.

Reuss, A.E., 1850, Neue Foraminiferen aus den Schichten des österreichischen Tertiärbeckens. Denkschriften der Kaiserlichen Akademie der Wissenschaften, Mathematisch-Naturwissenschaftliche Classe, vol. 1, pp. 365–390.

Reuss, A.E., 1851, Die Foraminiferen und Entomostraceen des Kreidemergels von Lemberg. Naturwissenschaftliche Abhandlungen, Wien, Österreichvol. 4, pp. 17–52.

Reuss, A.E., 1851, Ueber die fossilen Foraminiferen und Entomostraceen der Septarienthone der Umgegend von Berlin. Zeitschrift der Deutschen Geologischen Gesellschaft, Berlin, vol. 3, pp. 49–91.

Reuss, A.E., 1856, Beitraege zur Charakteristik der Tertiaerschichten des noerdlichen und mittleren Deutschlands. Sitzungsberichte der Kaiserlichen Akademie der Wissenschaften in Wien, Mathematisch-Naturwissenschaftliche Classe, vol. 18, no. 2, pp. 197–273.

Reuss, A.E., 1860, Die Foraminiferen der westphälischen Kreideformation. Sitzungsberichte der Kaiserlichen Akademie der Wissenschaften in Wien, Mathematisch-Naturwissenschaftliche Classe, vol. 40, pp. 147–238.

Reuss, A.E., 1863, Beiträge zur Kenntniss der tertiären Foraminiferen-Fauna (Zweite Folge). Sitzungsberichte der Kaiserlichen Akademie der Wissenschaften in Wien, Mathematisch-Naturwissenschaftliche Classe, vol. 48, no. 1, pp. 36–71.

Reuss, A.E., 1866, Die Foraminiferen, Anthozoen und Bryozoen des deutschen Septarienthones. Denkschriften der Kaiserlichen Akademie der Wissenschaften, Mathematisch-Naturwissenschaftliche Classe, vol. 25, no. 1, pp. 117–214.

Revets, S., 1991, The generic revision of the reussellids (Foraminiferida). Journal of Micropalaeontology, vol. 10, pp. 1–15.

Revets, S.A., 1993, The revision of the genus Buliminellita Cushman and Stainforth, 1947, and Eubuliminella gen. nov. Journal of Foraminiferal Research, vol. 23, no. 2, pp. 141–151.

Revets, S.A., 1996, The generic revision of five families of Rotaliine Foraminifera. Special Publications of the Cushman Foundation for Foraminiferal Research, vol. 34, pp. 1–108.

Rey, M., 1955, Description de quelques espèces nouvelles de foraminifères dans le Nummulitique nord-marocain. Bulletin de la Société géologique de France, series 6, vol. 4, no. 4–6, pp. 209–211.

Riegraf, W. and Luterbacher, H., 1989, Benthonische Foraminiferen aus de Unterkreide des Deep Sea Drilling Project (Leg 1-79). Geologische Rundschau, vol. 78, no. 3, pp. 1063–1120.

Riegraf, W. and Luterbacher, H., 1989, Oberjura-Foraminiferen aus dem Nord- und Südatlantik (Deep Sea Drilling Project 1–79). Geologische Rundschau, vol. 78, no. 3, pp. 999–1045.

Robertson, B.E., 1998, Systematics and paleoecology of the benthic foraminiferida from the Buff Bay section, Miocene of Jamaica. Micropaleontology, vol. 44, supplement 2, pp. 1–266.

Roemer, F.A., 1838, Die Cephalopoden des Nord-Deutschen tertiären Meersandes. Neues Jahrbuch für Mineralogie, Geognosie, Geologie und Petrefakten-Kunde, vol. 1838, pp. 381–394.

Roemer, F.A., 1839, Die Versteinerungen des norddeutschen Oolithen-Gebirges. Ein Nachtrag. Hannover: Hahn'schen Hofbuchhandlung, iv + 59 pp.

Rögl, F., 1976, Late Cretaceous to Pleistocene foraminifera from the Southeast Pacific Basin, DSDP Leg 35. In Hollister, C.D., Craddock, C. et al., ed., Initial Reports of the Deep Sea Drilling Project, Washington (U.S. Government Printing Office), vol. 35, pp. 1–930.

Rögl, F. and Hansen, J.H., 1984, Foraminifera described by Fichtel and Moll in 1798; A revision of Testacea Microscopica. Appendix Testacea Microscopica alique minuta ex Generibus Argonauta et Nautilus. Reprint of original plates. Neue Denkschriften des Naturhistorischen Museum in Wien, vol. 3, pp. 1–143.

Rzehak, A., 1886, Die Foraminiferenfauna der Neogenformation der Umgebung von Mähr: Ostrau. Verhandlungen des Naturforschenden Vereins in Brünn, vol. 24 (1885), pp. 77–126.

Rzehak, A., 1888, Die Foraminiferen der Nummuliten-schichten des Waschberges und Michelsberges bei Stockerau in Nieder-Österreich. Verhandlungen der Geologischen Bundesanstalt, vol. 1888, pp. 226–229.

Rzehak, A., 1895, Ueber einige merkwürdige Foraminiferen aus österreichischen Tertiär. Annalen Naturhistorisches Hofmuseum, Wien, vol. 10, pp. 213–230.

Said, R. and Kenawy, A., 1956, Upper Cretaceous and Lower Tertiary foraminifera from northern Sinai, Egypt. Micropaleontology, vol. 2, no. 2, pp. 105–173.

Saidova, Kh.M., 1961, Ekologiya foraminifer i paleogeografiya dal'nevostochnykh morey SSSR, I severo-zapadnoy chasti Tikhogo Okeans. Moscow: Instituta Okeanologii, Akademiya Nauk SSSR, 232 pp.

Saidova, Kh.M., 1975, Bentosnye foraminifery Tikhogo Okeana, 3 vols. Moscow: Institut Okeanologii P.P. Shirshova, Akademiya Nauk SSSR, 875 pp.

Samuel, O., 1977, Agglutinated foraminifers from Paleogene flysch formations in West Carpathians of Slovakia. Zapadne Karpaty, Serie Paleontologie, vol. 2–3, pp. 7–70.

Sars, G.O., 1872, Undersøgelser over Hardangerfjordens Fauna. Forhandlinger i Videnskasselskabet i Kristiania, vol. 1871, pp. 246–255.

Sars, M., 1869, Fortsatte bemaerkninger over det dyriske livs udbredning i havets dybder. Forhandlinger i Videnskasselskabet i Kristiania, vol. 1868, pp. 246–275.

Schlumberger, C., 1887, Note sur le genre Planispira. Bulletin de la Société Zoologique de France, vol. 12, pp. 105–118.

Schlumberger, C., 1891, Révision des Biloculines des grands fonds. Mémoires de la Société Zoologique de France, vol. 4, pp. 542–579.

Schnitker, D., 1979, Cenozoic deep water foraminifers, Bay of Biscay. In Montadert, L., Roberts, D.G., et al., ed. Initial Reports of the Deep Sea Drilling Project, Washington (U.S. Government Printing Office), vol. 48, pp. 377–413.

Schnitker, D. and Tjalsma, L.R.C., 1980, New genera and species of benthic foraminifera from Paleocene and Eocene deep-water deposits. Journal of Foraminiferal Research, vol. 10, no. 3, pp. 235–241.

Schröder, C.J., 1986, Deep-Water Arenaceous Foraminifera in the Northwest Atlantic Ocean. Canadian Technical Report of Hydrography and Ocean Sciences No. 71. Atlantic Geoscience Centre, Bedford Institute of Oceanography, 191 pp.

Schroeter, J.S., 1783, Einleitung in die Conchylienkenntniss nach Linné vol. 1. Halle: J.J. Gebauer, 860 pp.

Schulze, F.E., 1875, Zoologische Ergebnisse der Nordseefart, vom 21 Juli bis 9 September 1872: I, Rhizopoden. II. Jahresberichte Kommission zur Untersuchung der Deutschen Meer in Kiel für die Jahr 1872, 1873, pp. 99–114.

Schwager, C., 1866, Fossile Foraminiferen von Kar Nikobar, Reise der Österreichischen Fregatte Novara um Erde in den Jahren 1857, 1858, 1859 unten den Befehlen des Commodore B. Von Wüllerstorf-Urbair. Geologischer Theil, vol. 2, no. 1, Geologische Beobachtungen, no. 2, Paläontologische Mittheilungen, pp. 187–268.

Schwager, C., 1878, Nota su alcuni Foraminiferi nuovi del Tubo di Stretto presso Girgenti. Bolletino R. Comitato Geologico d'Italia, vol. 8, pp. 519–529.

Seguenza, G., 1862, Prime recerche intorno ai rizopodi fossili delle argille Pleistoceniche dei dintorni di Catania. Accademia Gioenia di Scienze Naturali di Catania, Atti, ser. 2, vol. 18, pp. 84–126.

Seguenza, G., 1880, Le formazioni terziarie nella provincia di Reggio (Calabria). Reale Accademia Nazionale dei Lincei, Classe di Scienze Fisiche, Matematiche e Naturali, Memorie, ser. 3, vol. 6, pp. 3–446.

Sen Gupta, B.K., 1989, Morphology and generic placement of the foraminifer "Anomalina" wuellerstorfi Schwager. Journal of Paleontology, vol. 63, no. 5, pp. 706–713.

Serova, M. Ya., 1987, Foraminifery i biostratigrafia Severnoy Patsifiki na rubiezhie Mela i Paleogena. Izdatelstvo "Nauka," Moscow, pp. 1–144.

Shchedrina, Z.G., 1969, On some changes in the systematics of the families Astrorhizidae and Reophacidae (Foraminifera). Voprosy Mikropaleontologii, vol. 11, pp. 157–170.

Silvestri, A., 1904, Richerche strutturali su alcune forme dei Trube di Bonfornello (Palermo). Memorie della Pontificia Accademia Romana dei Nuovi Lincei, vol. 22, pp. 235–276.

Sliter, W.V., 1968, Upper Cretaceous foraminifera from southern California and northwestern Baja California, Mexico. Paleontological Contributions of the University of Kansas, no. 49, Protozoa, art. 7, pp. 1–141.

Sliter, W.V., 1977, Cretaceous benthic foraminifers from the western South Atlantic Leg 39, Deep Sea Drilling Project. In Supko, P.R., Perch-Nielsen, K., et al., ed., Initial Reports of the Deep Sea Drilling Project, Washington (U.S. Government Printing Office), vol. 39, pp. 657–697.

Sliter, W.V., 1980, Mesozoic foraminifers and deep sea benthic environments from Deep Sea Drilling Project Sites 415 and 416, eastern North Atlantic. In Lancelot, Y., Winterer, E.L., et al., ed., Initial Reports of the Deep Sea Drilling Project, Washington (U.S. Government Printing Office), vol. 50, pp. 353–427.

Smart, C.W., King, S.C., Gooday, A.J., Murray, J.W., and Tomas, E. 1994, A benthic foraminiferal proxy of pulsed organic matter paleofluxes. Marine Micropaleontology, vol. 23, pp. 89–99.

Soldani, A., 1789, Testaceographiae ac Zoophytographiae parvae et microscopicae, tom. 1(1). Siena: F. Rossi.

Soldani, A., 1791, Testaceographiae ac Zoophytographiae parvae et microscopicae. Siena: F. Rossi, pp. 81–200.

Soldani, A., 1799, Testaceographiae ac Zoophytographiae parvae et microscopicae, tom. 1(4). Siena: F. Rossi.

Srinivasan, M.S. and Sharma, V., 1980, Schwager's Car Nicobar Foraminifera in the Reports of the Novara Expedition – A Revision. New Delhi: Today & Tomorrow's Printers and Publishers, 83 pp.

Stassen, P., Steurbaut, E., Morsi, A.-M.M., and Speijer, R.P., 2012, Biotic impact of Eocene thermal maximum 2 in a shelf setting (Dababiya, Egypt). Austrian Journal of Earth Sciences, vol. 105, pp. 154–160.

Subbotina, N.N., 1947, Microfauna of the oil fields of the Caucasus, Emba, and Central Asia. In Foraminifera of the Danian and Paleogene deposits of the North Caucasus. Leningrad: Vsesoyuznogo Neftyanogo Nauchno-Issledovatel'skogo, Geologo-razvedochnogo Instituta (VNIGRI), pp. 39–160.

Subbotina, N.N., 1953, Microfauna of the USSR, Volume 6. Upper Eocene Lagenidae and Buliminidae of the southern USSR. Trudy Vsesoyuznogo Neftyanogo Nauchno-Issledovatel'skogo, Geologo-razvedochnogo Instituta (VNIGRI) (new series), vol. 69, pp. 115–282.

Swiecicki, A., 1980, A foraminiferal biostratigraphy of the Campanian and Maastrichtian chalks of the United Kingdom. Unpublished PhD thesis, vol. 1, 358 pp.

Szarek, R., Kuhnt, W., Kawamura, H., and Nishi, H., 2009, Distribution of recent benthic foraminifera along the continental slope of the Sunda Shelf (South China Sea). Marine Micropaleontology, vol. 71, pp. 41–59.

Szczechura, J. and Pozaryska, K., 1974, Foraminiferida from the Paleocene of the Polish Carpathians (Babica Clays). Paleontologica Polonica, vol. 31, pp. 1–142.

Sztràkos, L., 1979, La stratigraphie, paléoécologie et les Foraminifères de l'Oligocène supérieur du Nord-Est de la Hongrie. Cahiers de Micropaléontologie, vol. 3, pp. 3–95.

Tappan, H., 1951, Northern Alaska index foraminifera. Contributions from the Cushman Foundation for Foraminiferal Research, vol. 2, pt 1, pp. 1–8.

Tappan, H., 1960, Cretaceous biostratigraphy of Northern Alaska. Bulletin of the American Association of Petroleum Geologists, vol. 44, no. 3, pp. 273–297.

ten Dam, A., 1944, Die Stratigraphische Gliederung des niederlaendlichen Paleozaens und Eozaens nach Foraminiferen (mit Ausnahme von Sued-Limburg). Mededeelingen van de Geologische Stichting, ser. C, vol. 5, no. 3, pp. 1–142.

Terquem, O., 1876, Essai sur le classement des animaux qui vivent sur la plage et dans les environs de Dunquerque. Paris: Deuxième fascicule, pp. 55–100.

Terquem, O., 1878, Les Foraminifères et les Entomostracés-Ostracodes du Pliocène Supérieur de l'île de Rhodes. Mémoires de la Société géologique de France, ser. 3, vol. 1, pp. 1–135.

Thomas, E., 1985, Late Eocene to Recent deep-sea benthic foraminifers from the central equatorial Pacific Ocean. In Mayer, L., Theyer, F., et al., ed., Initial Reports of the Deep Sea Drilling Project, Washington (U.S. Government Printing Office), vol. 85, pp. 655–694.

Thomas, E., 1990, Late Cretaceous through Neogene benthic foraminifers. In Barker, P.F., Kennett, J.P., et al., ed., Scientific Results of the Proceedings of the Ocean Drilling Program, vol. 113, pp. 571–594.

Thomas, F.C., 1988, Taxonomy and stratigraphy of selected Cenozoic benthic foraminifera, Canadian Atlantic margin. Micropaleontology, vol. 34, no. 1, pp. 67–82.

Tjalsma, R.C., 1983, Eocene to Miocene benthic foraminifera from DSDP site 516, Rio Grande Rise, South Atlantic. In Barker, P.F., Carlson, R.L., Johnson, D. A., et al., ed., Initial Reports of the Deep Sea Drilling Project, Washington (U.S. Government Printing Office), vol. 72, pp. 731–755.

Tjalsma, R.C. and Lohmann, G.P., 1983, Paleocene–Eocene bathyal and abyssal benthic foraminifera from the Atlantic Ocean. Micropaleontology, Special Publication, no. 4, pp. 1–90.

Todd, R. and Kniker, H.T., 1952, An Eocene foraminiferal fauna from the Agua Fresca shale of Magallanes Province, southernmost Chile. Special Publications of the Cushman Foundation for Foraminiferal Research, no. 1, pp. 1–28.

Trauth, F., 1918, Das Eozanvorkommen bei Radstadt im Pongou und seine Beziehungen zu den gleichalterigen Ablagerungen bei Kirchberg am Wechsel und Wimpassing am Leithagebirge. Denkschriften der Kaiserlichen Akademie der Wissenschaften Wien, Mathematisch-Naturwissenschaftliche Classe, vol. 95, pp. 171–278.

Trujillo, E.F., 1960, Upper Cretaceous foraminifera from near Redding, Shasta County, California. Journal of Paleontology, vol. 34, pp. 290–346.

van Bellen, R.C., 1941, Some Eocene foraminifera from the neighborhood of Ricice near Imotski, E. Dalmatia, Yugoslavia. Proceedings of the Koninklijke Nederlandse Akademie van Wetenschappen, vol. 44, no. 9, pp. 996–1005.

van Bellen, R.C., de Witt Puyt, J.F.C., Rutgers, A.C., and van Soest, J., 1941, Smaller foraminifera from the Lower Oligocene of Cuba. Proceedings of the Koninklijke Nederlandse Akademie van Wetenschappen, vol. 44, no. 9, pp. 1140–1146.

van den Broeck, E., 1876, Étude sur les foraminifères de la Barbade (Antilles) recueillis par L. Agassiz précédée de quelques considérations sur la classification et la nomenclature des Foraminifères. Annales de la Société Belge de Microscopie (Mémoires), vol. 1 (for 1876), pp. 55–152.

van Marle, L.J., 1991, Eastern Indonesian Late Cenozoic smaller benthic foraminifera. Verhandelingen der Koninklijke Nederlanse Akademie van Wetenschappen, Afdeeling Natuurkunde, Eerste Reeks, vol. 34, pp. 1–328.

van Morkhoven, F.P.C.M., 1981, Cosmopolitan Tertiary bathyal benthic foraminifera. Transactions of the Gulf Coast Association of Geological Societies, Supplement 31, 445 abstracts with accompanying range chart.

van Morkhoven, F.P.C.M., Berggren, W.A., Edwards, A.S., et al., 1986, Cenozoic cosmopolitan deep-water benthic foraminifera. Bulletin des Centres de Recherches Exploration-Production Elf-Aquitaine, Mémoire, vol. 11, pp. 1–421.

Vasicek, M., 1947, Poznámky k mikrobiostratigrafii magurského flyse na Morave. Vestník Státního Geologického Ústavu Ceskoslovenské Republiky, Praha, vol. 22, pp. 235–256.

Vasilenko, V.P., 1954, Fossil foraminifera of the USSR. Anomalinidae. Trudy Vsesoyuznogo Neftyanogo Nauchno-Issledovatel'skogo, Geologo-razvedochnogo Instituta (VNIGRI) (New Series), vol. 80, pp. 1–282.

Vasilenko, V.P., 1961, Foraminifera from the Upper Cretaceous of the Mangyshlak Peninsula. Trudy Vsesoyuznogo Neftyanogo Nauchno-Issledovatel'skogo, Geologo-razvedochnogo Instituta (VNIGRI) (New Series), vol. 171, pp. 1–487.

Vella, P., 1957, Studies in New Zealand foraminifera. New Zealand Geological Survey, Paleontological Bulletin, no. 28, pp. 1–64.

Voloshinova, N.A., 1958, On new systematics of the Nonionidae. Trudy Vsesoyuznogo Neftyanogo Nauchno-Issledovatel'skogo, Geologo-razvedochnogo Instituta (VNIGRI) (New Series), vol. 115, pp. 117–223.

Walker, G. and Jacob, E., 1798, Descriptions of Serpula (Lagena) sulcata and Nautilus lobatulus. In Adams, E., ed., Essays on the microscope, 2nd edn with considerable additions and improvements by F. Kanmacher. London: Dillon & Keating, pp. 634 and 642.

Weidich, K., 1990, Die kalkalpine Unterkreide und ihre Foraminiferenfauna. Zitteliana, vol. 17, pp. 1–312.

Weinelt, M., Kuhnt, W., Sarnthein, M., Altenbach, A., Costello O., Erlenkeuser, H., Matthiessen, J., Pflaumann, U., Simstich, J., Struck, U., Thies, A., Trauth, M., and Vogelsang, E., 2000, Paleoceanographic proxies in the Northern North Atlantic. In Schäfer, P., Schlüter, M., Schröder-Ritzrau, W., and Thiede, J., ed., The Northern North Atlantic: A Changing Environment. New York: Springer-Verlag, pp. 319–352.

White, M.P., 1928, Some index foraminifera of the Tampico Embayment area of Mexico (Part II). Journal of Paleontology, vol. 2, pp. 280–317.

White, M.P., 1929, Some index foraminifera of the Tampico Embayment area of Mexico (Part III). Journal of Paleontology, vol. 3, no. 1, pp. 30–58.

Whittaker, J.E., 1988, Benthic Foraminifera from Ecuador: Taxonomy and Distribution of Smaller Benthic Foraminifera from Coastal Ecuador (Late Oligocene–Late Pliocene). London: British Museum (Natural History), xi + 194 pp.

Wiesner, H., 1931, Die Foraminiferen der Deutsche Südpolar Expedition 1901–1903. Deutsche Südpolar Expedition, vol. 20, 165 pp.

Wightman, W.G. and Kuhnt, W., 1992, Biostratigraphy and Paleoecology of late Cretaceous abyssal agglutinated foraminifers from the western Pacific Ocean (Deep Sea Drilling Project Holes 196A and 198A and Ocean Drilling Program Holes 800A and 801A). In Larson, R.L., Lancelot, Y., et al., ed., Scientific Results of the Proceedings of the Ocean Drilling Program, vol. 129, pp. 247–264.

Williamson, W.C., 1848, On the Recent British species of the genus Lagena. Annals and Magazine of Natural History, ser. 2, vol. 1, pp. 1–20.

Williamson, W.C., 1858, On the Recent Foraminifera of Great Britain. London: Ray Society, 107 pp.

Wright, J., 1881, A list of the post-Tertiary foraminifera of the north-east of Ireland. Proceedings of the Belfast Naturalists' Field Club (New Series), vol. 1 (1873–80), pp. 149–163.

Zheng, S.Y., 1988, The Agglutinated and Porcelaneous Foraminifera of the East China Sea. Beijing: China Ocean Press, 337 pp.

# Taxonomic index

Assignations in **bold** type indicate those used in the Atlas.




*Atlas of Benthic Foraminifera*, First Edition. Ann Holbourn, Andrew S. Henderson, and Norman MacLeod.

# General index

Page numbers in *italic* type refer to tabulated information.



*Atlas of Benthic Foraminifera*, First Edition. Ann Holbourn, Andrew S. Henderson, and Norman MacLeod.

www.ingramcontent.com/pod-product-compliance
Lightning Source LLC
LaVergne TN
LVHW062152200926
848484LV00007B/1903
* 9 7 8 1 1 1 8 3 8 9 8 0 5 *